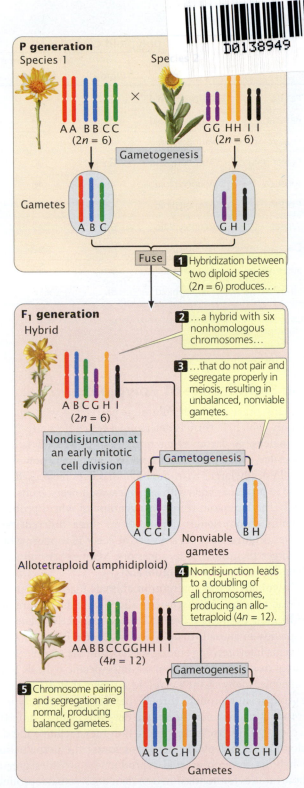

8.28 Most allopolyploids arise from hybridization between two species followed by chromosome doubling.

into plants that were viable and fertile. Analysis of their chromosomes revealed that the plants were allotetraploids, with $2n = 36$ chromosomes. To Karpechencko's great disappointment, however, the new plants possessed the roots of a cabbage and the leaves of a radish.

WORKED PROBLEM

Species I has $2n = 14$ and species II has $2n = 20$. Give all possible chromosome numbers that may be found in the following individuals.

a. An autotriploid of species I

b. An autotetraploid of species II

c. An allotriploid formed from species I and species II

d. An allotetraploid formed from species I and species II

Solution Strategy

What information is required in your answer to the problem?

All possible chromosome numbers for individuals with the type of polyploidy indicated.

What information is provided to solve the problem?

- Species I has $2n = 14$ and species II has $2n = 20$.
- The type of polyploidy the individual possesses.

Solution Steps

The haploid number of chromosomes (n) for species I is 7 and for species II is 10.

a. A triploid individual is $3n$. A common mistake is to assume that $3n$ means three times as many chromosomes as in a normal individual, but remember that normal individuals are $2n$. Because n for species I is 7 and all genomes of an autopolyploid are from the same species, $3n = 3 \times 7 = 21$.

b. An autotetraploid is $4n$ with all genomes from the same species. The n for species II is 10, so $4n = 4 \times 10 = 40$.

c. A triploid is $3n$. By definition, an allopolyploid must have genomes from two different species. An allotriploid could have $1n$ from species I and $2n$ from species II or $(1 \times 7) + (2 \times 10) = 27$. Alternatively, it might have $2n$ from species I and $1n$ from species II, or $(2 \times 7) + (1 \times 10) = 24$. Thus, the number of chromosomes in an allotriploid could be 24 or 27.

d. A tetraploid is $4n$. By definition, an allotetraploid must have genomes from at least two different species. An allotetraploid could have $3n$ from species I and $1n$ from species II or $(3 \times 7) + (1 \times 10) = 31$; or $2n$ from species I and $2n$ from species II or $(2 \times 7) + (2 \times 10) = 34$; or $1n$ from species I and $3n$ from species II or $(1 \times 7) + (3 \times 10) = 37$. Thus, the number of chromosomes could be 31, 34, or 37.

▶ For additional practice, try Problem 38 at the end of this chapter.

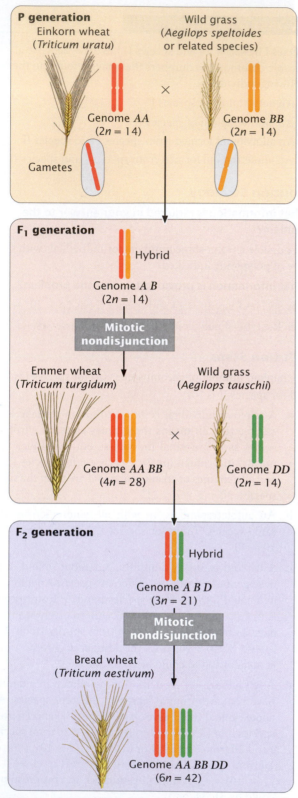

P generation

Einkorn wheat
(*Triticum uratu*)

Wild grass
(*Aegilops speltoides*
or related species)

×

Genome *AA*
(2n = 14)

Genome *BB*
(2n = 14)

Gametes

F₁ generation

Hybrid

Genome *A B*
(2n = 14)

Mitotic nondisjunction

Emmer wheat
(*Triticum turgidum*)

Wild grass
(*Aegilops tauschii*)

×

Genome *AA BB*
(4n = 28)

Genome *DD*
(2n = 14)

F₂ generation

Hybrid

Genome *A B D*
(3n = 21)

Mitotic nondisjunction

Bread wheat
(*Triticum aestivum*)

Genome *AA BB DD*
(6n = 42)

8.29 Modern bread wheat, *Triticum aestivum*, is a hexaploid with genes derived from three different species. Two diploid species, *T. uratu* (n = 14) and probably *Aegilops speltoides* or a related species (n = 14), originally crossed to produce a diploid hybrid (2n = 14) that underwent chromosome doubling to create *T. turgidum* (4n = 28). A cross between *T. turgidum* and *A. tauschii* (2n = 14) produced a triploid hybrid (3n = 21) that then underwent chromosome doubling to eventually produce *T. aestivum*, which is a hexaploid (6n = 42).

	Type of		Chromosome
Plant	Polyploidy	Ploidy	Number
Potato	Autopolyploid	4n	48
Banana	Autopolyploid	3n	33
Peanut	Autopolyploid	4n	40
Sweet potato	Autopolyploid	6n	90
Tobacco	Allopolyploid	4n	48
Cotton	Allopolyploid	4n	52
Wheat	Allopolyploid	6n	42
Oats	Allopolyploid	6n	42
Sugar cane	Allopolyploid	8n	80
Strawberry	Allopolyploid	8n	56

Table 8.2 Examples of polyploid crop plants

Source: After F. C. Elliot, *Plant Breeding and Cytogenetics* (New York: McGraw-Hill, 1958).

The Significance of Polyploidy

In many organisms, cell volume is correlated with nuclear volume, which, in turn, is determined by genome size. Thus, the increase in chromosome number in polyploidy is often associated with an increase in cell size, and many polyploids are physically larger than diploids. Breeders have used this effect to produce plants with larger leaves, flowers, fruits, and seeds. The hexaploid (6n = 42) genome of wheat probably contains chromosomes derived from three different wild species (**Figure 8.29**). As a result, the seeds of modern wheat are larger than those of its ancestors. Many other cultivated plants also are polyploid (**Table 8.2**).

Polyploidy is less common in animals than in plants for several reasons. As discussed, allopolyploids require hybridization between different species, which happens less frequently in animals than in plants. Animal behavior often prevents interbreeding among species, and the complexity of animal development causes most interspecific hybrids to be nonviable. Many of the polyploid animals that do arise are in groups that reproduce through parthenogenesis (a type of reproduction in which the animal develops from an unfertilized egg). Thus, asexual reproduction may facilitate the development of polyploids, perhaps because the perpetuation of hybrids through asexual reproduction provides greater opportunities for nondisjunction than does sexual reproduction. Only a few human polyploid babies have been reported, and most died within a few days of birth. Polyploidy—usually triploidy—is seen in about 10% of all spontaneously aborted human fetuses.

IMPORTANCE OF POLYPLOIDY IN EVOLUTION Polyploidy, particularly allopolyploidy, often gives rise to new species and has been particularly important in the evolution

Table 8.3   Different types of chromosome mutations

Chromosome Mutation	Definition
Chromosome rearrangement	Change in chromosome structure
Chromosome duplication	Duplication of a chromosome segment
Chromosome deletion	Deletion of a chromosome segment
Inversion	Chromosome segment inverted 180 degrees
Paracentric inversion	Inversion that does not include the centromere in the inverted region
Pericentric inversion	Inversion that includes the centromere in the inverted region
Translocation	Movement of a chromosome segment to a nonhomologous chromosome or to another region of the same chromosome
Nonreciprocal translocation	Movement of a chromosome segment to a nonhomologous chromosome or to another region of the same chromosome without reciprocal exchange
Reciprocal translocation	Exchange between segments of nonhomologous chromosomes or between regions of the same chromosome
Aneuploidy	Change in number of individual chromosomes
Nullisomy	Loss of both members of a homologous pair
Monosomy	Loss of one member of a homologous pair
Trisomy	Gain of one chromosome, resulting in three homologous chromosomes
Tetrasomy	Gain of two homologous chromosomes, resulting in four homologous chromosomes
Polyploidy	Addition of entire chromosome sets
Autopolyploidy	Polyploidy in which extra chromosome sets are derived from the same species
Allopolyploidy	Polyploidy in which extra chromosome sets are derived from two or more species

of flowering plants. Occasional genome doubling through polyploidy has been a major contributor to evolutionary success in several groups. For example, *Saccharomyces cerevisiae* (yeast) is a tetraploid, having undergone whole-genome duplication about 100 million years ago. The vertebrate genome has duplicated twice, once in the common ancestor of jawed vertebrates and again in the ancestor of fishes. Certain groups of vertebrates, such as some frogs and some fishes, have undergone additional polyploidy. Cereal plants have undergone several genome-duplication events. Different types of chromosome mutations are summarized in **Table 8.3**.

CONCEPTS

Polyploidy is the presence of extra chromosome sets: autopolyploids possess extra chromosome sets from the same species; allopolyploids possess extra chromosome sets from two or more species. Problems in chromosome pairing and segregation often lead to sterility in autopolyploids, but many allopolyploids are fertile.

✔ CONCEPT CHECK 7

Species A has $2n = 16$ chromosomes and species B has $2n = 14$. How many chromosomes would be found in an allotriploid of these two species?

a. 21 or 24   c. 22 or 23
b. 42 or 48   d. 45

CONCEPTS SUMMARY

■ Three basic types of chromosome mutations are: (1) chromosome rearrangements, which are changes in the structures of chromosomes; (2) aneuploidy, which is an increase or decrease in chromosome number; and (3) polyploidy, which is the presence of extra chromosome sets.

■ Chromosome rearrangements include duplications, deletions, inversions, and translocations.

■ In individuals heterozygous for a duplication, the duplicated region will form a loop when homologous

chromosomes pair in meiosis. Duplications often have pronounced effects on the phenotype owing to unbalanced gene dosage. Segmental duplications are common in the human genome.

■ In individuals heterozygous for a deletion, one of the chromosomes will loop out during pairing in meiosis. Deletions may cause recessive alleles to be expressed.

■ Pericentric inversions include the centromere; paracentric inversions do not. In individuals heterozygous for an inversion, the homologous chromosomes form inversion loops in meiosis, with reduced recombination taking place within the inverted region.

■ In translocation heterozygotes, the chromosomes form crosslike structures in meiosis, and the segregation of chromosomes produces unbalanced gametes.

■ Fragile sites are constrictions or gaps that appear at particular regions on the chromosomes of cells grown in culture and are prone to breakage under certain conditions.

■ Copy-number variations (CNVs) are differences in the number of copies of DNA sequences and include duplications and deletions. These variants are common in the human genome; some are associated with diseases and disorders.

■ Nullisomy is the loss of two homologous chromosomes; monosomy is the loss of one homologous chromosome; trisomy is the addition of

one homologous chromosome; tetrasomy is the addition of two homologous chromosomes.

■ Aneuploidy usually causes drastic phenotypic effects because it leads to unbalanced gene dosage.

■ Primary Down syndrome is caused by the presence of three full copies of chromosome 21, whereas familial Down syndrome is caused by the presence of two normal copies of chromosome 21 and a third copy that is attached to another chromosome through a translocation.

■ Uniparental disomy is the presence of two copies of a chromosome from one parent and no copy from the other. Mosaicism is caused by nondisjunction in an early mitotic division that leads to different chromosome constitutions in different cells of a single individual.

■ All the chromosomes in an autopolyploid derive from one species; chromosomes in an allopolyploid come from two or more species.

IMPORTANT TERMS

chromosome mutation (p. 210)
metacentric chromosome (p. 210)
submetacentric chromosome (p. 210)
acrocentric chromosome (p. 210)
telocentric chromosome (p. 210)
chromosome rearrangement (p. 212)
chromosome duplication (p. 212)
tandem duplication (p. 212)
displaced duplication (p. 212)
reverse duplication (p. 212)

segmental duplication (p. 214)
chromosome deletion (p. 214)
pseudodominance (p. 216)
haploinsufficient gene (p. 216)
chromosome inversion (p. 216)
paracentric inversion (p. 216)
pericentric inversion (p. 216)
position effect (p. 217)
dicentric chromatid (p. 218)
acentric chromatid (p. 218)
dicentric bridge (p. 218)
translocation (p. 219)
nonreciprocal translocation (p. 219)
reciprocal translocation (p. 219)

Robertsonian translocation (p. 219)
alternate segregation (p. 221)
adjacent-1 segregation (p. 221)
adjacent-2 segregation (p. 221)
fragile site (p. 221)
fragile-X syndrome (p. 221)
copy-number variation (CNV) (p. 222)
aneuploidy (p. 222)
polyploidy (p. 222)
nullisomy (p. 222)
monosomy (p. 222)
trisomy (p. 222)
tetrasomy (p. 222)
Down syndrome (trisomy 21) (p. 225)

primary Down syndrome (p. 225)
familial Down syndrome (p. 225)
translocation carrier (p. 226)
Edward syndrome (trisomy 18) (p. 227)
Patau syndrome (trisomy 13) (p. 227)
trisomy 8 (p. 227)
uniparental disomy (p. 227)
mosaicism (p. 228)
gynandromorph (p. 228)
autopolyploidy (p. 228)
allopolyploidy (p. 228)
unbalanced gametes (p. 229)
amphidiploidy (p. 229)

ANSWERS TO CONCEPT CHECKS

1. a

2. Pseudodominance is the expression of a recessive mutation. It is produced when the dominant wild-type allele in a heterozygous individual is absent due to a deletion on one chromosome.

3. c

4. b

5. 37

6. Dosage compensation prevents the expression of additional copies of X-linked genes in mammals, and there is little information in the Y chromosome so extra copies of the X and Y chromosomes do not have major effects on development. In contrast, there is no mechanism of dosage compensation for autosomes, and so extra copies of autosomal genes are expressed, upsetting development and causing the spontaneous abortion of aneuploid embryos.

7. c

WORKED PROBLEMS

Problem 1

A chromosome has the following segments, where · represents the centromere.

$$A B C D E \cdot F G$$

What types of chromosome mutations are required to change this chromosome into each of the following chromosomes? (In some cases, more than one chromosome mutation may be required.)

a. A B E · F G

b. A E D C B · F G

c. A B A B C D E · F G

d. A F · E D C B G

e. A B C D E E D C · F G

Solution Strategy

What information is required in your answer to the problem?

Types of chromosome mutations that would lead to the chromosome shown.

What information is provided to solve the problem?

- The original gene segments found on the chromosome.
- The altered gene segments that occur after the mutations.

For help with this problem, review:

Section 8.2.

Solution Steps

a. The mutated chromosome (A B E · F G) is missing segment C D; so this mutation is a deletion.

b. The mutated chromosome (A E D C B · F G) has one and only one copy of all the gene segments, but segment B C D E has been inverted 180 degrees. Because the centromere has not changed location and is not in the inverted region, this chromosome mutation is a paracentric inversion.

c. The mutated chromosome (A B A B C D E · F G) is longer than normal, and we see that segment A B has been duplicated. This mutation is a tandem duplication.

d. The mutated chromosome (A F · E D C B G) is normal length, but the gene order and the location of the centromere have changed; this mutation is therefore a pericentric inversion of region (B C D E · F).

e. The mutated chromosome (A B C D E E D C · F G) contains a duplication (C D E) that is also inverted; so this chromosome has undergone a duplication and a paracentric inversion.

Problem 2

Species I is diploid ($2n = 4$) with chromosomes AABB; related species II is diploid ($2n = 6$) with chromosomes MMNNOO. Give the chromosomes that would be found in individuals with the following chromosome mutations.

a. Autotriploidy in species I

b. Allotetraploidy including species I and II

c. Monosomy in species I

d. Trisomy in species II for chromosome M

e. Tetrasomy in species I for chromosome A

f. Allotriploidy including species I and II

g. Nullisomy in species II for chromosome N

Solution Strategy

What information is required in your answer to the problem?

The letter designations of chromosomes that will be found in individuals with each type of mutation.

What information is provided to solve the problem?

- Species I is diploid with $2n = 4$.

- Species I has chromosomes AABB.
- Species II is diploid with $2n = 6$).
- Species II has MMNNOO.

For help with this problem, review:

Sections 8.3 and 8.4.

Solution Steps

Hint: First determine the haploid genome complement for each species. For species I, $n = 2$ with chromosomes AB and, for species II, $n = 3$ with chromosomes MNO.

a. An autotriploid is $3n$, with all the chromosomes coming from a single species; so an autotriploid of species I would have chromosomes AAABBB ($3n = 6$).

b. An allotetraploid is $4n$, with the chromosomes coming from more than one species. An allotetraploid could consist of $2n$ from species I and $2n$ from species II, giving the allotetraploid ($4n = 2 + 2 + 3 + 3 = 10$) chromosomes AABBMMNNOO. An allotetraploid could also possess $3n$ from species I and $1n$ from species II ($4n = 2 + 2 + 2 + 3 = 9$; AAABBBMNO) or $1n$ from species I and $3n$ from species II ($4n = 2 + 3 + 3 + 3 = 11$; ABMMMNNNOOO).

c. A monosomic is missing a single chromosome; so a monosomic for species I would be $2n - 1 = 4 - 1 = 3$.

The monosomy might include either of the two chromosome pairs, with chromosomes ABB or AAB.

d. Trisomy requires an extra chromosome; so a trisomic of species II for chromosome M would be $2n + 1 = 6 + 1 = 7$ (MMMNNOO).

e. A tetrasomic has two extra homologous chromosomes; so a tetrasomic of species I for chromosome A would be $2n + 2 = 4 + 2 = 6$ (AAAABB).

f. An allotriploid is $3n$ with the chromosomes coming two from different species; so an allotriploid could be $3n = 2 + 2 + 3 = 7$ (AABBMNO) or $3n = 2 + 3 + 3 = 8$ (ABMMNNOO).

g. A nullisomic is missing both chromosomes of a homologous pair; so a nullisomic of species II for chromosome N would be $2n - 2 = 6 - 2 = 4$ (MMOO).

COMPREHENSION QUESTIONS

Section 8.1

1. List the different types of chromosome mutations and define each one.

Section 8.2

2. Why do extra copies of genes sometimes cause drastic phenotypic effects?

3. Draw a pair of chromosomes as they would appear during synapsis in prophase I of meiosis in an individual heterozygous for a chromosome duplication.

4. What is haploinsufficiency?

5. What is the difference between a paracentric and a pericentric inversion?

6. How can inversions in which no genetic information is lost or gained cause phenotypic effects?

7. Explain, with the aid of a drawing, how a dicentric bridge is produced when crossing over takes place in an individual heterozygous for a paracentric inversion.

8. Explain why recombination is suppressed in individuals heterozygous for paracentric and pericentric inversions.

9. How do translocations in which no genetic information is lost or gained produce phenotypic effects?

10. Sketch the chromosome pairing and the different segregation patterns that can arise in an individual heterozygous for a reciprocal translocation.

11. What is a Robertsonian translocation?

Section 8.3

12. List four major types of aneuploidy.

13. What is the difference between primary Down syndrome and familial Down syndrome? How does each type arise?

14. What is uniparental disomy and how does it arise?

15. What is mosaicism and how does it arise?

Section 8.4

16. What is the difference between autopolyploidy and allopolyploidy? How does each arise?

17. Explain why autopolyploids are usually sterile, whereas allopolyploids are often fertile.

> ▶ For more questions that test your comprehension of the key chapter concepts, go to **LEARNING**Curve for this chapter.

APPLICATION QUESTIONS AND PROBLEMS

Section 8.1

18. Examine the karyotypes shown in **Figure 8.1** and **Figure 8.2**. Are the individuals from whom these karyotype were made males or females?

*19. Which types of chromosome mutations

   a. increase the amount of genetic material in a particular chromosome?

   b. increase the amount of genetic material in all chromosomes?

   c. decrease the amount of genetic material in a particular chromosome?

   d. change the position of DNA sequences in a single chromosome without changing the amount of genetic material?

e. move DNA from one chromosome to a nonhomologous chromosome?

Section 8.2

*20. A chromosome has the following segments, where · represents the centromere:

A B · C D E F G

What types of chromosome mutations are required to change this chromosome into each of the following chromosomes? (In some cases, more than one chromosome mutation may be required.)

a. A B A B · C D E F G   **f.** A B · E D C F G

b. A B · C D E A B F G   **g.** C · B A D E F G

c. A B · C F E D G   **h.** A B · C F E D F E D G

d. A · C D E F G   **i.** A B · C D E F C D F E G

e. A B · C D E

21. A chromosome initially has the following segments:

A B · C D E F G

Draw the chromosome, identifying its segments, that would result from each of the following mutations.

a. Tandem duplication of DEF

b. Displaced duplication of DEF

c. Deletion of FG

d. Paracentric inversion that includes DEFG

e. Pericentric inversion of BCDE

22. The following diagram represents two nonhomologous chromosomes:

A B · C D E F G

R S · T U V W X

What type of chromosome mutation would produce each of the following chromosomes?

a. A B · C D   **c.** A B · T U V F G

   R S · T U V W X E F G   R S · C D E W X

b. A U V B · C D E F G   **d.** A B · C W G

   R S · T W X   R S · T U V D E F X

*23. The *Notch* mutation is a deletion on the X chromosome of *Drosophila melanogaster*. Female flies heterozygous for *Notch* have an indentation on the margins of their wings; *Notch* is lethal in the homozygous and hemizygous conditions. The *Notch* deletion covers the region of the X chromosome that contains the locus for white eyes, an X-linked recessive trait. Give the phenotypes and proportions of progeny produced in the following crosses.

a. A red-eyed, Notch female is mated with a white-eyed male.

b. A white-eyed, Notch female is mated with a red-eyed male.

c. A white-eyed, Notch female is mated with a white-eyed male.

24. The green-nose fly normally has six chromosomes: two metacentric and four acrocentric. A geneticist examines the chromosomes of an odd-looking green-nose fly and discovers that it has only five chromosomes; three of them are metacentric and two are acrocentric. Explain how this change in chromosome number might have taken place.

*25. A wild-type chromosome has the following segments:

A B C · D E F G H I

An individual is heterozygous for the following chromosome mutations. For each mutation, sketch how the wild-type and mutated chromosomes would pair in prophase I of meiosis, showing all chromosome strands.

a. A B C · D E F D E F G H I   **c.** A B C · D G F E H I

b. A B C · D H I   **d.** A B E D · C F G H I

26. For the chromosomes shown in **Figure 8.12**, draw the chromatids that would result from a two-strand double crossover: one crossover between C and D and the other crossover between D and E.

*27. As discussed in this chapter, crossing over within a pericentric inversion produces chromosomes that have extra copies of some genes and no copies of other genes. The fertilization of gametes containing such duplication or deficient chromosomes often results in children with syndromes characterized by developmental delay, intellectual disability, abnormal development of organ systems, and early death. Maarit Jaarola and colleagues examined individual sperm cells of a male who was heterozygous for a pericentric inversion on chromosome 8 and determined that crossing over took place within the pericentric inversion in 26% of the meiotic divisions (M. Jaarola, R. H. Martin, and T. Ashley. 1998. *American Journal of Human Genetics* 63:218–224).

Assume that you are a genetic counselor and that a couple seeks counseling from you. Both the man and the woman are phenotypically normal, but the woman is heterozygous for a pericentric inversion on chromosome 8. The man is karyotypically normal. What is the probability that this couple will produce a child with a debilitating syndrome as the result of crossing over within the pericentric inversion?

*28. An individual heterozygous for a reciprocal translocation possesses the following chromosomes:

A B · C D E F G

A B · C D V W X

R S · T U E F G

R S · T U V W X

a. Draw the pairing arrangement of these chromosomes in prophase I of meiosis.

b. Diagram the alternate, adjacent-1, and adjacent-2 segregation patterns in anaphase I of meiosis.

c. Give the products that result from alternate, adjacent-1, and adjacent-2 segregation.

Section 8.3

*29. Red–green color blindness is a human X-linked recessive disorder. A young man with a 47,XXY karyotype (Klinefelter syndrome) is color blind. His 46,XY brother also is color blind. Both parents have normal color vision. Where did the nondisjunction that gave rise to the young man with Klinefelter syndrome take place? Assume that no crossing over took place in prophase I of meiosis.

30. **DATA ANALYSIS** Junctional epidermolysis bullosa (JEB) is a severe skin disorder that results in blisters over the entire body. The disorder is caused by autosomal recessive mutations at any one of three loci that help to encode laminin 5, a major component in the dermal–epidermal basement membrane. Leena Pulkkinen and colleagues described a male newborn who was born with JEB and died at 2 months of age (L. Pulkkinen et al. 1997. *American Journal of Human Genetics* 61:611–619); the child had healthy unrelated parents. Chromosome analysis revealed that the infant had 46 normal-appearing chromosomes. Analysis of DNA showed that his mother was heterozygous for a JEB-causing allele at the *LAMB3* locus, which is on chromosome 1. The father had two normal alleles at this locus. DNA fingerprinting demonstrated that the male assumed to be the father had, in fact, conceived the child.

a. Assuming that no new mutations occurred in this family, explain the presence of an autosomal recessive disease in the child when the mother is heterozygous and the father is homozygous normal.

b. How might you go about proving your explanation? Assume that a number of genetic markers are available for each chromosome.

31. Some people with Turner syndrome are 45,X/46,XY mosaics. Explain how this mosaicism could arise.

*32. Bill and Betty have had two children with Down syndrome. Bill's brother has Down syndrome and his sister has two children with Down syndrome. On the basis of these observations, indicate which of the following statements are most likely correct and which are most likely incorrect. Explain your reasoning.

a. Bill has 47 chromosomes.

b. Betty has 47 chromosomes.

c. Bill and Betty's children each have 47 chromosomes.

d. Bill's sister has 45 chromosomes.

e. Bill has 46 chromosomes.

f. Betty has 45 chromosomes.

g. Bill's brother has 45 chromosomes.

33. In mammals, sex-chromosome aneuploids are more common than autosomal aneuploids but, in fish, sex-chromosome aneuploids and autosomal aneuploids are found with equal frequency. Offer a possible explanation for these differences in mammals and fish. (Hint: Think about why sex chromosome aneuploids are more common than autosomal aneuploids in mammals.)

*34. A young couple is planning to have children. Knowing that there have been a substantial number of stillbirths, miscarriages, and fertility problems on the husband's side of the family, they see a genetic counselor. A chromosome analysis reveals that, whereas the woman has a normal karyotype, the man possesses only 45 chromosomes and is a carrier of a Robertsonian translocation between chromosomes 22 and 13.

a. List all the different types of gametes that might be produced by the man.

b. What types of zygotes will develop when each of gametes produced by the man fuses with a normal gamete produced by the woman?

c. If trisomies and monosomies entailing chromosomes 13 and 22 are lethal, approximately what proportion of the surviving offspring are expected to be carriers of the translocation?

35. **DATA ANALYSIS** Using breeding techniques, Andrei Dyban and V. S. Baranov (*Cytogenetics of Mammalian Embryonic Development.* Oxford: Oxford University Press, Clarendon Press; New York: Oxford University Press, 1987) created mice that were trisomic for each of the different mouse chromosomes. They found that only mice with trisomy 19 developed. Mice trisomic for all other chromosomes died in the course of development. For some of these trisomics, they compared the length of development (number of days after conception before the embryo died) as a function of the size of the mouse chromosome that was present in three copies (see the adjoining graph). Summarize their findings as presented in this graph and provide a possible explanation for the results.

[E. Torres, B. R. Williams, and A. Amon. 2008. Genetics 179:737–746, Fig. 2B.]

Section 8.4

36. Species I has $2n = 16$ chromosomes. How many chromosomes will be found per cell in each of the following mutants in this species?

a. Monosomic   **e.** Double monosomic

b. Autotriploid   **f.** Nullisomic

c. Autotetraploid   **g.** Autopentaploid

d. Trisomic   **h.** Tetrasomic

37. Species I is diploid ($2n = 8$) with chromosomes AABBCCDD; related species II is diploid ($2n = 8$) with chromosomes MMNNOOPP. What types of chromosome mutations do individual organisms with the following sets of chromosomes have?

a. AAABBCCDD   **e.** AAABBCCDDD

b. MMNNOOOOPP   **f.** AABBDD

c. AABBCDD   **g.** AABBCCDDMMNNOOPP

d. AAABBBCCCDDD   **h.** AABBCCDDMNOP

*38. Species I has $2n = 8$ chromosomes and species II has $2n = 14$ chromosomes. What would the expected chromosome numbers be in individual organisms with the following chromosome mutations? Give all possible answers.

a. Allotriploidy including species I and II

b. Autotetraploidy in species II

c. Trisomy in species I

d. Monosomy in species II

e. Tetrasomy in species I

f. Allotetraploidy including species I and II

39. Suppose that Species I in **Figure 8.28** has $2n = 10$ and Species II in the figure has $2n = 12$. How many chromosomes would be present in the allotetraploid at the bottom of the figure?

40. Consider a diploid cell that has $2n = 4$ chromosomes—one pair of metacentric chromosomes and one pair of acrocentric chromosomes. Suppose that this cell undergoes nondisjunction giving rise to an autotriploid cell ($3n$). The triploid cell then undergoes meiosis. Draw the different types of gametes that may result from meiosis in the triploid cell, showing the chromosomes present in each type. To distinguish between the different metacentric and acrocentric chromosomes, use a different color to draw each metacentric chromosome; similarly, use a different color to draw each acrocentric chromosome. (Hint: See Figure 8.27).

41. Assume that the autotriploid cell in **Figure 8.27** has $3n = 30$ chromosomes. For each of the gametes produced by this cell, give the chromosome number of the resulting zygote if the gamete fused with a normal haploid gamete.

42. *Nicotiana glutinosa* ($2n = 24$) and *N. tabacum* ($2n = 48$) are two closely related plants that can be intercrossed, but the F_1 hybrid plants that result are usually sterile. In 1925, Roy Clausen and Thomas Goodspeed crossed *N. glutinosa* and *N. tabacum* and obtained one fertile F_1 plant (R. E. Clausen and T. H. Goodspeed. 1925 *Genetics* 10:278–284). They were able to self-pollinate the flowers of this plant to produce an F_2 generation. Surprisingly, the F_2 plants were fully fertile and produced viable seed. When Clausen and Goodspeed examined the chromosomes of the F_2 plants, they observed 36 pairs of chromosomes in metaphase I and 36 individual chromosomes in metaphase II. Explain the origin of the F_2 plants obtained by Clausen and Goodspeed and the numbers of chromosomes observed.

43. What would be the chromosome number of progeny resulting from the following crosses in wheat (see Figure 8.29)? What type of polyploid (allotriploid, allotetraploid, etc.) would result from each cross?

a. Einkorn wheat and emmer wheat

b. Bread wheat and emmer wheat

c. Einkorn wheat and bread wheat

44. Karl and Hally Sax crossed *Aegilops cylindrica* ($2n = 28$), a wild grass found in the Mediterranean region, with *Triticum vulgare* ($2n = 42$), a type of wheat (K. Sax and H. J. Sax. 1924. *Genetics* 9:454–464). The resulting F_1 plants from this cross had 35 chromosomes. Examination of metaphase I in the F_1 plants revealed the presence of 7 pairs of chromosomes (bivalents) and 21 unpaired chromosomes (univalents).

a. If the unpaired chromosomes segregate randomly, what possible chromosome numbers will appear in the gametes of the F_1 plants?

b. What does the appearance of the bivalents in the F_1 hybrids suggest about the origin of *Triticum vulgare* wheat?

Aegilops cylindrica, jointed goatgrass. [Sam Brinker, MNR-NHIC, 2008/Canadian Food Inspection Agency.]

Triticum vulgare, wheat. [Michael Hieber/123RF.com.]

CHALLENGE QUESTIONS

Section 8.3

45. Red–green color blindness is a human X-linked recessive disorder. Jill has normal color vision, but her father is color blind. Jill marries Tom, who also has normal color vision. Jill and Tom have a daughter who has Turner syndrome and is color blind.

  a. How did the daughter inherit color blindness?

  b. Did the daughter inherit her X chromosome from Jill or from Tom?

46. **DATA ANALYSIS** Progeny of triploid tomato plants often contain parts of an extra chromosome, in addition to the normal complement of 24 chromosomes (J. W. Lesley and M. M. Lesley. 1929. *Genetics* 14:321–336). Mutants with a part of an extra chromosome are referred to as secondaries. James and Margaret Lesley observed that secondaries arise from triploid ($3n$), trisomic ($3n + 1$), and double trisomic ($3n + 1 + 1$) parents, but never from diploids ($2n$). Give one or more possible reasons that secondaries arise from parents that have unpaired chromosomes but not from parents that are normal diploids.

47. **DATA ANALYSIS** Mules result from a cross between a horse ($2n = 64$) and a donkey ($2n = 62$), have 63 chromosomes, and are almost always sterile. However, in the summer of 1985, a female mule named Krause who was pastured with a male donkey gave birth to a newborn foal (O. A. Ryder et al. 1985. *Journal of Heredity* 76:379–381). Blood tests established that the male foal, appropriately named Blue Moon, was the offspring of Krause and that Krause was indeed a mule. Both Blue Moon and Krause were fathered by the same donkey (see the illustration). The foal, like his mother, had 63 chromosomes—half of them horse chromosomes and the other half donkey chromosomes. Analyses of genetic markers showed that, remarkably, Blue Moon seemed to have inherited a complete set of horse chromosomes from his mother, instead of the random mixture of horse and donkey chromosomes that would be expected with normal meiosis. Thus, Blue Moon and Krause were not only mother and son, but also brother and sister.

  a. With the use of a diagram, show how, if Blue Moon inherited only horse chromosomes from his mother, Blue Moon and Krause are both mother and son as well as brother and sister.

  b. Although rare, additional cases of fertile mules giving births to offspring have been reported. In these cases, when a female mule mates with a male horse, the offspring is horselike in appearance but, when a female mule mates with a male donkey, the offspring is mulelike in appearance. Is this observation consistent with the idea that the offspring of fertile female mules inherit only a set of horse chromosomes from their mule mothers? Explain your reasoning.

  c. Can you suggest a possible mechanism for how the offspring of fertile female mules might pass on a complete set of horse chromosomes to their offspring?

Section 8.4

48. Humans and many other complex organisms are diploid, possessing two sets of genes, one inherited from the mother and one from the father. However, a number of eukaryotic organisms spend most of their life cycles in a haploid state. Many of these eukaryotes, such as *Neurospora* and yeast, still undergo meiosis and sexual reproduction, but most of the cells that make up the organism are haploid.

  Considering that haploid organisms are fully capable of sexual reproduction and generating genetic variation, why are most complex eukaryotes diploid? In other words, what might be the evolutionary advantage of existing in a diploid state instead of a haploid state? And why might a few organisms, such as *Neurospora* and yeast, exist as haploids?

> Go to your **LaunchPad** to find additional learning resources and the Suggested Readings for this chapter.

9 | Bacterial and Viral Genetic Systems

Life in a Bacterial World

Humans like to think that that they rule the world but, compared with bacteria, we are clearly in a minor position. Bacteria first evolved some 3.5 billion years ago, 2 billion years before the first eukaryotes appeared (some evidence suggests bacteria evolved even earlier). Today, bacteria are found in every conceivable environment, including boiling springs, highly saline lakes, and beneath more than 2 miles of ice in Antarctica. They are found at the top of Mt. Everest and at the bottoms of the deepest oceans. They are also present on and in *us*—in alarming numbers! Within the average human gut, there are approximately 10 trillion bacteria, ten times the total number of cells in the entire human body. No one knows how many bacteria populate the world, but an analysis conducted by scientists in 1998 estimated that the total number of living bacteria on Earth exceeded 5 million trillion trillion (5×10^{30}).

Bacteria account for most of life's diversity and exist in almost every conceivable environment, including inhospitable habitats such as the highly saline Dead Sea. [PhotoStock-Israel/Alamy.]

Not only are bacteria numerically vast, they also constitute the majority of life's diversity. The total number of described species of bacteria is less than 10,000, compared with about 1.4 million plants, animals, fungi, and single-celled eukaryotes. But the number of described species of bacteria falls far short of the true microbial diversity.

Species of bacteria are typically described only after they have been cultivated and studied in the laboratory. Because only a few species are amenable to laboratory culture, for many years, it was impossible to identify and study most bacteria. Then, in the 1970s, molecular techniques for analyzing DNA became available and opened up a whole new vista on microbial diversity. These techniques revealed several important facts about bacteria. First, many of the relations among bacteria that microbiologists had worked out on the basis of physical and biochemical traits turned out to be incorrect. Bacteria once thought to be related were in fact genetically quite different. Second, molecular analysis showed that members of one group of microbes—now called the archaea—were as different from other bacteria (the eubacteria) as they are from eukaryotes. Third, molecular analysis revealed that the number of different types of bacteria is astounding.

In 2007, Luiz Roesch and his colleagues set out to determine exactly how many types of bacteria exist in a gram of soil. They obtained soil samples from four locations: Brazil, Florida, Illinois, and Canada. From the soil samples, they extracted and purified bacterial DNA. From this DNA, they determined the sequences of a gene present in all bacteria, the 16S rRNA gene. Each different species of bacteria has a unique 16S rRNA gene sequence, so they could determine how many species of bacteria existed in each soil sample by counting the number of different DNA sequences.

Roesch's results were amazing. The number of different eubacterial species in each gram of soil ranged from 26,140 for samples from Brazil to 53,533 for Canadian samples. Many unusual bacteria were detected that appeared dissimilar to all previously described groups of bacteria. Another interesting finding was that soil from agricultural fields harbored considerably fewer species than did soil from forests.

This study and others demonstrate that bacterial diversity far exceeds that of multicellular organisms and, undoubtedly, numerous groups of bacteria have yet to be discovered. Like it or not, we truly live in a bacterial world.

In this chapter, we examine some of the genetic properties of bacteria and viruses, and the mechanisms by which they exchange and recombine their genes. Since the 1940s, the genetic systems of bacteria and viruses have contributed to the discovery of many important concepts in genetics. The study of molecular genetics initially focused almost entirely on their genes; today, bacteria and viruses are still essential tools for probing the nature of genes in more-complex organisms, in part because they possess a number of characteristics that make them suitable for genetic studies (**Table 9.1**).

The genetic systems of bacteria and viruses are also studied because these organisms play important roles in human society. Bacteria are found naturally in the mouth, gut, and on the skin, where they are essential to human function and ecology. They have been harnessed to produce a number of economically important substances, and they are of immense medical significance, causing many human diseases. In this chapter, we focus on several unique aspects of bacterial and viral genetic systems. Important processes of gene transfer and recombination will be described, and we will see how these processes can be used to map bacterial and viral genes. ▶ TRY PROBLEM 17

Table 9.1   Advantages of using bacteria and viruses for genetic studies

1. Reproduction is rapid.

2. Many progeny are produced.

3. The haploid genome allows all mutations to be expressed directly.

4. Asexual reproduction simplifies the isolation of genetically pure strains.

5. Growth in the laboratory is easy and requires little space.

6. Genomes are small.

7. Techniques are available for isolating and manipulating their genes.

8. They have medical importance.

9. They can be genetically engineered to produce substances of commercial value.

9.1 Genetic Analysis of Bacteria Requires Special Methods

Heredity in bacteria is fundamentally similar to heredity in more-complex organisms, but the bacterial haploid genome and the small size of bacteria (which makes observation of their phenotypes difficult) require different approaches and methods.

Bacterial Diversity

Prokaryotes are unicellular organisms that lack nuclear membranes and membrane-bounded cell organelles. For many years, biologists considered all prokaryotes to be related, but DNA sequence information now provides convincing evidence that prokaryotes are divided into at least two distinct groups: the archaea and the eubacteria. The archaea are a group of diverse prokaryotes that are frequently found in extreme environments, such as hot springs and at the bottoms of oceans. The eubacteria are the remaining prokaryotes, including most of the familiar bacteria. Although superficially similar in their cell structure, eubacteria and archaea are distinct in their genetic makeup, and the differences between them are as great as those between eubacteria and eukaryotes. In fact, the archaea are *more* similar to eukaryotes than to eubacteria in a number of molecular features and genetic processes. In this book, the term bacteria is generally used to refer to eubacteria.

Bacteria are extremely diverse and come in a variety of shapes and sizes. Some are rod shaped, whereas others are spherical. Most are much smaller than eukaryotic cells, but at least one species isolated from the gut of fish is almost 1 mm long and can be seen with the naked eye. Some bacteria are photosynthetic. Others produce stalks and spores, superficially resembling fungi.

Bacteria have long been considered simple organisms that lack much of the cellular complexity of eukaryotes. However, recent evidence points to a number of similarities and parallels in bacterial and eukaryotic structure. For example, a bacterial protein termed FtsZ, which plays an integral part in cell division, is structurally similar to eukaryotic tubulin proteins, which are subunits of microtubules and help segregate chromosomes in mitosis and meiosis (Chapter 2). Like eukaryotes, bacteria have proteins that help condense DNA. Other bacterial proteins function much as cytoskeletal proteins do in eukaryotes, helping to give bacterial cells shape and structure.

(a) Sterile liquid medium → Inoculating loop → Inoculate medium with bacteria. → Bacteria grow and divide.

(b) Pipet · Lid · Glass rod · Dilute soluion of bacterial cells · Petri plate — A growth medium is suspended in gelatin-like agar. → Add a dilute solution of bacteria to petri plate. → Spread bacterial solution evenly with glass rod. → After incubation for 1 to 2 days, bacteria multiply, forming visible colonies.

9.1 Bacteria can be grown (a) in liquid medium or (b) on solid medium.

And, although bacteria don't go through mitosis and meiosis, replication of the bacterial chromosome precedes binary fission, and there are bacterial processes that ensure that one copy of the chromosome is allocated to each cell.

Techniques for the Study of Bacteria

The culture and study of bacteria require special techniques. Microbiologists have defined the nutritional needs of a number of bacteria and developed culture media for growing them in the laboratory. Culture media typically contain a carbon source, essential elements such as nitrogen and phosphorus, certain vitamins, and other required ions and nutrients. Wild-type, or **prototrophic**, bacteria can use these simple ingredients to synthesize all the compounds that they need for growth and reproduction. A medium that contains only the nutrients required by prototrophic bacteria is termed **minimal medium**.

Mutant strains called **auxotrophs** lack one or more enzymes necessary for synthesizing essential molecules and will grow only on medium supplemented with these essential molecules. For example, auxotrophic strains that are unable to synthesize the amino acid leucine will not grow on

minimal medium but *will* grow on medium to which leucine has been added. **Complete medium** contains all the substances, such as the amino acid leucine, required by bacteria for growth and reproduction.

Cultures of bacteria are often grown in test tubes that contain sterile liquid medium (**Figure 9.1a**). A few bacteria are added to a tube, and they grow and divide until all the nutrients are used up or—more commonly—until the concentration of their waste products becomes toxic. Bacteria are also grown on petri plates (**Figure 9.1b**). Growth medium suspended in agar is poured into the bottom half of the petri plate, providing a solid, gel-like base for bacterial growth. In a process called plating, a dilute solution of bacteria is spread over the surface of an agar-filled petri plate. As each bacterium grows and divides, it gives rise to a visible clump of genetically identical cells (a **colony**). Genetically pure strains of the bacteria can be isolated by collecting bacteria from a single colony and transferring them to a new test tube or petri plate. The chief advantage of this method is that it allows one to isolate and count bacteria, which individually are too small to see without a microscope.

Microbiologists often study phenotypes that affect the appearance of the colony (**Figure 9.2**) or can be detected by

9.2 Bacteria have a variety of phenotypes. (a) *Serratia marcescens* with color variation. (b) *Bacillus cereus*.
[Part a: Dr. Edward J. Bottone. Part b: Biophoto Associates/Photo Researchers.]

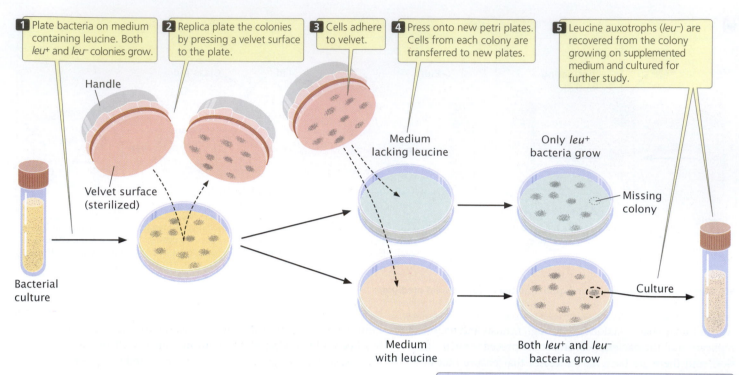

1 Plate bacteria on medium containing leucine. Both *leu⁺* and *leu⁻* colonies grow.

2 Replica plate the colonies by pressing a velvet surface to the plate.

3 Cells adhere to velvet.

4 Press onto new petri plates. Cells from each colony are transferred to new plates.

5 Leucine auxotrophs (*leu⁻*) are recovered from the colony growing on supplemented medium and cultured for further study.

Handle

Velvet surface (sterilized)

Bacterial culture

Medium lacking leucine

Only *leu⁺* bacteria grow

Missing colony

Medium with leucine

Both *leu⁺* and *leu⁻* bacteria grow

Culture

9.3 Mutant bacterial strains can be isolated on the basis of their nutritional requirements.

Conclusion: A colony that grows only on the supplemented medium has a mutation in a gene that encodes the synthesis of an essential nutrient.

simple chemical tests. Auxotrophs are commonly studied phenotypes. Suppose we want to detect auxotrophs that cannot synthesize leucine (*leu⁻* mutants). We first spread the bacteria on a petri plate containing medium that includes leucine; both prototrophs that have the *leu⁺* allele and auxotrophs that have *leu⁻* allele will grow on it (**Figure 9.3**). Next, using a technique called replica plating, we transfer a few cells from each of the colonies on the original plate to two new replica plates: one plate contains medium to which leucine has been added; the other plate contains a medium lacking leucine. A medium that lacks an essential nutrient, such as the medium lacking leucine, is called a selective medium. The *leu⁺* bacteria will grow on both media, but the *leu⁻* mutants will grow only on the selective medium supplemented by leucine, because they cannot synthesize their own leucine. Any colony that grows on medium that contains leucine but not on medium that lacks leucine consists of *leu⁻* bacteria. The auxotrophs that grow on the supplemented medium can then be cultured for further study.

The Bacterial Genome

Most bacterial genomes consist of a circular chromosome that contains a single DNA molecule several million base pairs (bp) in length (**Figure 9.4**). For example, the *E. coli*

genome has approximately 4.6 million base pairs of DNA. However, some bacteria contain multiple chromosomes. For example, *Vibrio cholerae*, which causes cholera, has two circular chromosomes, and *Rhizobium meliloti* has three chro-

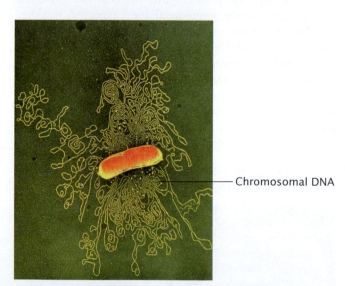

Chromosomal DNA

9.4 Most bacterial cells possess a single, circular chromosome, shown here emerging from a ruptured bacterial cell. [Dr. Gopal Murti/Science Source.]

1 Replication in a plasmid begins at the origin of replication, the *ori* site.

2 Strands separate and replication takes place in both directions,...

3 ...eventually producing two circular DNA molecules.

Origin of replication (*ori* site)

Newly synthesized DNA

Strand separation

Replication

Separation of daughter plasmids

Double-stranded DNA

Strands separate at *ori*

New strand

Old strand

9.5 A plasmid replicates independently of its bacterial chromosome. Replication begins at the origin of replication (*ori*) and continues around the circle. In this diagram, replication is taking place in both directions; in some plasmids, replication is in one direction only. [Photograph: Biology Pics/Science Source.]

mosomes. There are even a few bacteria that have linear chromosomes. Many bacterial chromosomes are organized efficiently. For example, more than 90% of the DNA in *E. coli* encodes proteins. In contrast, only about 1% of human DNA encodes proteins.

Plasmids

In addition to having a chromosome, many bacteria possess **plasmids**—small, usually circular DNA molecules. Some plasmids are present in many copies per cell, whereas others are present in only one or two copies. In general, plasmids carry genes that are not essential to bacterial function but that may play an important role in the life cycle and growth of their bacterial hosts. There are many different types of plasmids; *E. coli* alone is estimated to have more than 270 different naturally occurring plasmids. Some plasmids promote mating between bacteria; others contain genes that kill other bacteria. Plasmids are used extensively in genetic engineering (see Chapter 19), and some of them play a role in the spread of antibiotic resistance among bacteria.

Most plasmids are circular and several thousand base pairs in length, although plasmids consisting of several hundred thousand base pairs also have been found. Each plasmid possesses an origin of replication, a specific DNA sequence where DNA replication is initiated (see Chapter 2). The origin allows a plasmid to replicate independently of the bacterial chromosome (**Figure 9.5**). **Episomes** are plasmids that are capable of freely replicating and able to integrate into the bacterial chromosomes. The **F (fertility) factor** of *E. coli*

(**Figure 9.6**) is an episome that controls mating and gene exchange between *E. coli* cells, as will be discussed in the next section.

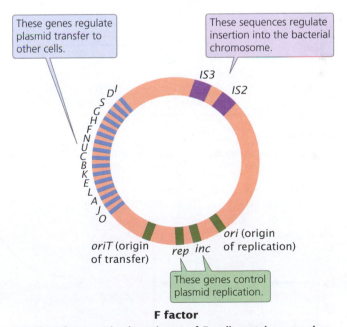

These genes regulate plasmid transfer to other cells.

These sequences regulate insertion into the bacterial chromosome.

IS3

IS2

ori (origin of replication)

oriT (origin of transfer)

rep inc

These genes control plasmid replication.

F factor

9.6 The F factor, a circular episome of *E. coli*, contains a number of genes that regulate transfer into the bacterial cell, replication, and insertion into the bacterial chromosome. Replication is initiated at *ori*. Insertion sequences (see Chapter 18) *IS3* and *IS2* control insertion into the bacterial chromosome and excision from it.

9.7 Conjugation, transformation, and transduction are three processes of gene transfer in bacteria. For the transferred DNA to be stably inherited, all three processes require the transferred DNA to undergo recombination with the bacterial chromosome.

9.2 Bacteria Exchange Genes Through Conjugation, Transformation, and Transduction

Bacteria exchange genetic material by three different mechanisms, all entailing some type of DNA transfer and recombination between the transferred DNA and the bacterial chromosome.

1. **Conjugation** takes place when genetic material passes directly from one bacterium to another (**Figure 9.7a**). In conjugation, two bacteria lie close together and a connection forms between them. A plasmid or a part of the bacterial chromosome passes from one cell (the donor) to the other (the recipient). Subsequent to conjugation, crossing over may take place between homologous sequences in the transferred DNA and the chromosome of the recipient cell. In conjugation, DNA is transferred only from donor to recipient, with no reciprocal exchange of genetic material.

2. **Transformation** takes place when a bacterium takes up DNA from the medium in which it is growing (**Figure 9.7b**). After transformation, recombination may take place between the introduced genes and those of the bacterial chromosome.

3. **Transduction** takes place when bacterial viruses (bacteriophages) carry DNA from one bacterium to another (**Figure 9.7c**). Inside the bacterium, the newly introduced DNA may undergo recombination with the bacterial chromosome.

Not all bacterial species exhibit all three types of genetic transfer. Conjugation takes place more frequently in some species than in others. Transformation takes place to a limited extent in many species of bacteria, but laboratory techniques increase the rate of DNA uptake. Most bacteriophages have a limited host range; so transduction is normally between bacteria of the same or closely related species only.

These processes of genetic exchange in bacteria differ from diploid eukaryotic sexual reproduction in two important ways. First, DNA exchange and reproduction are not coupled in bacteria; bacteria often undergo reproduction (cell division) without receiving any DNA from another cell. Second, donated genetic material that is not recombined into the host DNA is usually degraded, and so the recipient cell remains haploid. Each type of genetic transfer can be used to map genes, as will be discussed in the following sections.

Conjugation

In 1946, Joshua Lederberg and Edward Tatum demonstrated that bacteria can transfer and recombine genetic information, paving the way for the use of bacteria in genetic studies. In the course of their research, Lederberg and Tatum studied auxotrophic strains of *E. coli*. The Y10 strain required the amino acids threonine (and was genotypically thr^-) and leucine (leu^-) and the vitamin thiamine (thi^-) for growth but did not require the vitamin biotin (bio^+) or the amino acids phenylalanine (phe^+) and cysteine (cys^+); the genotype of this strain can be written as $thr^-\ leu^-\ thi^-\ bio^+\ phe^+\ cys^+$. The Y24 strain had the opposite set of alleles: it required biotin, phenylalanine, and cysteine in its medium, but it did not require threonine, leucine, or thiamine; its genotype was $thr^+\ leu^+\ thi^+\ bio^-\ phe^-\ cys^-$. In one experiment, Lederberg and Tatum mixed Y10 and Y24 bacteria together and plated them on minimal medium (**Figure 9.8**). Each strain was also plated separately on minimal medium.

Alone, neither Y10 nor Y24 grew on minimal medium: each strain required nutrients that were absent. Strain Y10 was unable to grow, because it required threonine, leucine, and thiamine, which were absent in the minimal medium; strain Y24 was unable to grow, because it required biotin, phenylalanine, and cysteine, which also were absent from the minimal medium. When Lederberg and Tatum mixed the two strains, however, a few colonies did grow on the minimal medium. These prototrophic bacteria must have had genotype $thr^+\ leu^+\ thi^+\ bio^+\ phe^+\ cys^+$. Where had they come from?

If mutations were responsible for the prototrophic colonies, then some colonies should also have grown on the plates containing Y10 or Y24 alone, but no bacteria grew on these plates. Multiple simultaneous mutations ($thr^- \rightarrow thr^+$, $leu^- \rightarrow leu^+$, and $thi^- \rightarrow thi^+$ in strain Y10 or $bio^- \rightarrow bio^+$, $phe^- \rightarrow phe^+$, and $cys^- \rightarrow cys^+$ in strain Y24) would have been required for either strain to become prototrophic by mutation, which was very improbable. Lederberg and Tatum concluded

Experiment

Question: Do bacteria exchange genetic information?

Methods   Y10   Y24

Bacterial chromosome

1 Auxotrophic bacterial strain Y10 cannot synthesize Thr, Leu, or Thi…

2 …and strain Y24 cannot synthesize biotin, Phe, or Cys,…

3 …and so neither auxotrophic strain can grow on minimal medium.

4 When strains Y10 and Y24 are mixed,…

Results

5 …some colonies grow…

6 …because genetic recombination has taken place and bacteria can synthesize all necessary nutrients.

Conclusion: Yes, genetic exchange and recombination took place between the two mutant strains.

9.8 Lederberg and Tatum's experiment demonstrated that bacteria undergo genetic exchange.

that some type of genetic transfer and recombination had taken place:

Auxotrophic strain

$$Y10 \quad\quad thr^- \; leu^- \; thi^- \; bio^+ \; phe^+ \; cys^+$$

$$Y24 \quad\quad thr^+ \; leu^+ \; thi^+ \; bio^- \; phe^- \; cys^-$$

$$thr^- \; leu^- \; thi^- \quad bio^+ \; phe^+ \; cys^+$$
$$thr^+ \; leu^+ \; thi^+ \quad bio^- \; phe^- \; cys^-$$

$$thr^- \; leu^- \; thi^- \; bio^- \; phe^- \; cys^-$$

Prototrophic strain $thr^+ \; leu^+ \; thi^+ \; bio^+ \; phe^+ \; cys^+$

What they did not know was *how* it had taken place.

To study this problem, Bernard Davis constructed a U-shaped tube (**Figure 9.9**) that was divided into two compartments by a filter with fine pores. This filter allowed liquid medium to pass from one side of the tube to the other, but the pores of the filter were too small to allow the passage of bacteria. Two auxotrophic strains of bacteria were placed on opposite sides of the filter, and suction was applied alternately to the ends of the U-tube, causing the medium to flow back and forth between the two compartments. Despite hours of incubation in the U-tube, bacteria plated out on minimal medium did not grow; there had been no genetic exchange between the strains. The exchange of bacterial genes clearly required direct contact, or conjugation, between the bacterial cells.

F⁺ AND F⁻ CELLS   In most bacteria, conjugation depends on a fertility (F) factor that is present in the donor cell and absent in the recipient cell. Cells that contain F are referred to as F⁺, and cells lacking F are F⁻.

The F factor contains an origin of replication and a number of genes required for conjugation (see Figure 9.6). For example, some of these genes encode sex **pili** (singular, pilus), slender extensions of the cell membrane. A cell containing F produces the sex pili, one of which makes contact with a receptor on an F⁻ cell (**Figure 9.10**) and pulls the two cells together. DNA is then transferred from the F⁺ cell to the F⁻ cell. Conjugation can take place only between a cell that possesses F and a cell that lacks F.

In most cases, the only genes transferred during conjugation between an F⁺ and F⁻ cell are those on the F factor (**Figure 9.11a** and **b**). Transfer is initiated when one of the DNA strands on the F factor is nicked at an origin (*oriT*). One end of the nicked DNA separates from the circle and passes into the recipient cell (**Figure 9.11c**). Replication takes place on the nicked strand, proceeding around the circular plasmid in the F⁺ cell and replacing the transferred strand (**Figure 9.11d**). Because the plasmid in the F⁺ cell is always nicked at the *oriT* site, this site always enters

Question: How did the genetic exchange seen in Lederberg and Tatum's experiment take place?

Methods   Auxotrophic strain A   Auxotrophic strain B

Airflow

Strain A   Strain B

Two auxotrophic strains were separated by a filter that allowed mixing of medium but not bacteria.

No prototrophic bacteria were produced

Results   Minimal medium   Minimal medium   Minimal medium   Minimal medium

No growth   No growth   No growth   No growth

Conclusion: Genetic exchange requires direct contact between bacterial cells.

9.9 Davis's U-tube experiment.

9.10 Sex pili connect F⁺ and F⁻ cells during bacterial conjugation. [Dr. L. Caro/Photo Researchers, Inc.]

the recipient cell first, followed by the rest of the plasmid. Thus, the transfer of genetic material has a defined direction. Inside the recipient cell, the single strand replicates, producing a circular, double-stranded copy of the F plasmid (**Figure 9.11e**). If the entire F factor is transferred to the recipient F⁻ cell, that cell becomes an F⁺ cell.

HFR CELLS   Conjugation transfers genetic material in the F plasmid from F⁺ to F⁻ cells but does not account for the transfer of chromosomal genes observed by Lederberg and Tatum. In Hfr (high-frequency) strains, the F factor is integrated into the bacterial chromosome (**Figure 9.12**). Hfr cells behave as F⁺ cells, forming sex pili and undergoing conjugation with F⁻ cells.

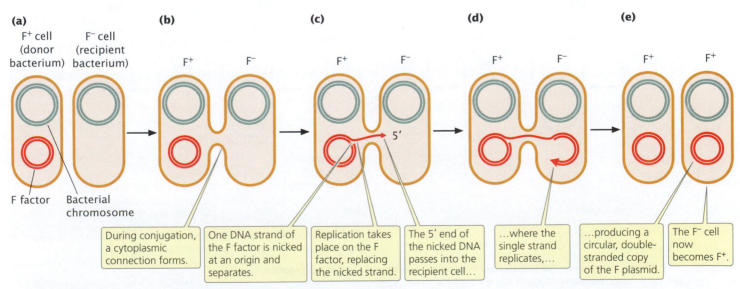

(a) F⁺ cell (donor bacterium)   F⁻ cell (recipient bacterium)   **(b)** F⁺ F⁻   **(c)** F⁺ F⁻   **(d)** F⁺ F⁻   **(e)** F⁺ F⁺

F factor   Bacterial chromosome

During conjugation, a cytoplasmic connection forms.

One DNA strand of the F factor is nicked at an origin and separates.

Replication takes place on the F factor, replacing the nicked strand.

The 5' end of the nicked DNA passes into the recipient cell…

…where the single strand replicates,…

…producing a circular, double-stranded copy of the F plasmid.

The F⁻ cell now becomes F⁺.

9.11 The F factor is transferred during conjugation between an F⁺ and F⁻ cell.

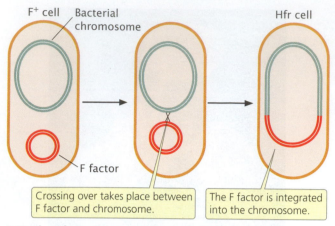

F⁺ cell   Bacterial chromosome

Hfr cell

F factor

Crossing over takes place between F factor and chromosome.

The F factor is integrated into the chromosome.

9.12 The F factor is integrated into the bacterial chromosome in an Hfr cell.

In conjugation between Hfr and F⁻ cells (**Figure 9.13a**), the integrated F factor is nicked, and the end of the nicked strand moves into the F⁻ cell (**Figure 9.13b**), just as it does in conjugation between F⁺ and F⁻ cells. Because, in an Hfr cell, the F factor has been integrated into the bacterial chromosome, the chromosome follows it into the recipient cell. How much of the bacterial chromosome is transferred depends on the length of time that the two cells remain in conjugation.

Inside the recipient cell, the donor DNA strand replicates (**Figure 9.13c**), and crossing over between it and the original chromosome of the F⁻ cell (**Figure 9.13d**) may take place. This gene transfer between Hfr and F⁻ cells is how the recombinant prototrophic cells observed by Lederberg and Tatum were produced. After crossing over has

taken place in the recipient cell, the donated chromosome is degraded and the recombinant recipient chromosome remains (**Figure 9.13e**), to be replicated and passed on to later generations by binary fission.

In a mating of Hfr × F⁻, the F⁻ cell almost never becomes F⁺ or Hfr because the F factor is nicked in the middle in the initiation of strand transfer, placing part of F at the beginning and part at the end of the strand to be transferred. To become F⁺ or Hfr, the recipient cell must receive the entire F factor, requiring the entire bacterial chromosome to be transferred. This event happens rarely, because most conjugating cells break apart before the entire chromosome has been transferred.

The F plasmid in F⁺ cells integrates into the bacterial chromosome, causing an F⁺ cell to become Hfr, at a frequency of only about 1 in 10,000. This low frequency accounts for the low rate of recombination observed by Lederberg and Tatum in their F⁺ cells. The F factor is excised from the bacterial chromosome at a similarly low rate, causing a few Hfr cells to become F⁺.

F′ CELLS   When an F factor does excise from the bacterial chromosome, a small amount of the bacterial chromosome may be removed with it, and these chromosomal genes will then be carried with the F plasmid (**Figure 9.14**). Cells containing an F plasmid with some bacterial genes are called F prime (F′). For example, if an F factor integrates into a chromosome adjacent to the *lac* genes (genes that enable a cell to metabolize the sugar lactose), the F factor may pick up *lac* genes when it excises, becoming F′ *lac*. F′ cells can conjugate with F⁻ cells because F′ cells possess the F plasmid with all the genetic information necessary for conjugation and gene

(a) Hfr cell   F⁻ cell

(b)

(c)

(d)

(e) Hfr cell   F⁻ cell

Bacterial chromosome

F factor

In conjugation, F is nicked and the 5′ end moves into the F⁻ cell.

The transferred strand replicates,…

Hfr chromosome (F factor plus bacterial genes)

…and crossing over takes place between the donated Hfr chromosome and the original chromosome of the F⁻ cell.

Crossing over may lead to the recombination of alleles (bright green in place of black segment).

The linear chromosome is degraded.

9.13 Bacterial genes may be transferred from an Hfr cell to an F⁻ cell in conjugation. In an Hfr cell, the F factor has been integrated into the bacterial chromosome.

9.14 An Hfr cell may be converted into an F′ cell when the F factor excises from the bacterial chromosome and carries bacterial genes with it. Conjugation produces a partial diploid.

transfer. Characteristics of different mating types of *E. coli* (cells with different types of F) are summarized in **Table 9.2**.

During conjugation between an F′ *lac* cell and an F⁻ cell, the F plasmid is transferred to the F⁻ cell, which means that any genes on the F plasmid, including those from the bacterial chromosome, may be transferred to F⁻ recipient cells. This process is called sexduction. It produces partial diploids, or merozygotes, which are cells with two copies of some genes, one the bacterial chromosome and one on the newly introduced F plasmid. The outcomes of conjugation between different mating types of *E. coli* are summarized in **Table 9.3**.

Table 9.2   Characteristics of *E. coli* cells with different types of F factor

Type	F Factor Characteristics	Role in Conjugation
F⁺	Present as separate circular DNA	Donor
F⁻	Absent	Recipient
Hfr	Present, integrated into bacterial chromosome	High-frequency donor
F′	Present as separate circular DNA, carrying some bacterial genes	Donor

Table 9.3   Results of conjugation between cells with different F factors

Conjugating	Cell Types Present after Conjugation
F⁺ × F⁻	Two F⁺ cells (F⁻ cell becomes F⁺)
Hfr × F⁻	One Hfr cell and one F⁻ (no change)*
F′ × F⁻	Two F′ cells (F⁻ cell becomes F′)

*Rarely, the F⁻ cell becomes F⁺ in an Hfr × F⁻ conjugation if the entire chromosome is transferred during conjugation.

CONCEPTS

Conjugation in *E. coli* is controlled by an episome called the F factor. Cells containing F (F⁺ cells) are donors during gene transfer; cells lacking F (F⁻ cells) are recipients. Hfr cells possess F integrated into the bacterial chromosome; they donate DNA to F⁻ cells at a high frequency. F′ cells contain a copy of F with some bacterial genes.

✓ CONCEPT CHECK 3

Conjugation between an F⁺ and an F⁻ cell usually results in

a. two F⁺ cells.

b. two F⁻ cells.

c. an F⁺ and an F⁻ cell.

d. an Hfr cell and an F⁺ cell.

MAPPING BACTERIAL GENES WITH INTERRUPTED CONJUGATION   The transfer of DNA that takes place during conjugation between Hfr and F⁻ cells allows bacterial genes to be mapped. In conjugation, the chromosome of the Hfr cell is transferred to the F⁻ cell. Transfer of the entire *E. coli* chromosome requires about 100 minutes; if conjugation is interrupted before 100 minutes have elapsed, only part of the chromosome will have passed into the F⁻ cell and have had an opportunity to recombine with the recipient chromosome.

Chromosome transfer always begins within the integrated F factor and proceeds in a continuous direction so genes are transferred according to their sequence on the chromosome. The time required for individual genes to be transferred indicates their relative positions on the chromosome. In most genetic maps, distances are expressed as percent recombination; however, in bacterial maps constructed with interrupted conjugation, the basic unit of distance is a minute. View **Animation 9.1** to see how genes are mapped using interrupted conjugation.

Worked Problem

To illustrate the method of mapping genes with interrupted conjugation, let's look at a cross analyzed by François Jacob and Elie Wollman, who developed this method of gene mapping (**Figure 9.15a**). They used donor Hfr cells that were sensitive to the antibiotic streptomycin (genotype str^s), resistant to sodium azide (azi^r) and infection by bacteriophage T1 (ton^r), prototrophic for threonine (thr^+) and leucine (leu^+), and able to break down lactose (lac^+) and galactose (gal^+). They used F⁻ recipient cells that were resistant to streptomycin (str^r), sensitive to sodium azide (azi^s) and to infection by bacteriophage T1 (ton^s), auxotrophic for threonine (thr^-) and leucine (leu^-), and unable to break down lactose (lac^-) and galactose (gal^-). Thus, the genotypes of the donor and recipient cells were:

Donor Hfr cells: $str^s\ leu^+\ thr^+\ azi^r\ ton^r\ lac^+\ gal^+$

Recipient F⁻ cells: $str^r\ leu^-\ thr^-\ azi^s\ ton^s\ lac^-\ gal^-$

The two strains were mixed in nutrient medium and allowed to conjugate. After a few minutes, the medium was diluted to prevent any new pairings. At regular intervals, a sample of cells was removed and agitated vigorously in a kitchen blender to halt all conjugation and DNA transfer. The cells from each sample were plated on a selective medium that contained streptomycin and lacked leucine and threonine. The original donor cells were streptomycin sensitive (str^s) and would not grow on this medium. The F⁻ recipient cells were auxotrophic for leucine and threonine, and they also failed to grow on this medium. Only cells that underwent conjugation and received at least the leu^+ and thr^+ genes from the Hfr donors could grow on this medium. All $str^r\ leu^+\ thr^+$ cells were then tested for the presence of other genes that might have been transferred from the donor Hfr strain.

Because Jacob and Wollman used streptomycin to kill all the donor cells, they were not able to examine the transfer of the str^s gene. All of the cells that grew on the selective medium were $leu^+\ thr^+$; so we know that these genes were transferred. The percentage of $str^r\ leu^+\ thr^+$ cells receiving specific alleles (azi^r, ton^r, leu^+, and gal^+) from the Hfr strain are plotted against the duration of conjugation (**Figure 9.15b**). What is the order in which the genes are transferred and the distances among them?

Solution Strategy

What information is required in your answer to the problem?

The order of the genes on the bacterial chromosome and the distances among them.

What information is provided to solve the problem?

- The donor cells were $str^s\ leu^+\ thr^+\ azi^r\ ton^r\ lac^+\ gal^+$ and the recipient cells were $str^r\ leu^-\ thr^-\ azi^s\ ton^s\ lac^-\ gal^-$.

- The percentage of recipient cells with different traits that appear at various times after the start of conjugation (Figure 9.15).

Experiment

Question: How can interrupted conjugation be used to map bacterial genes?

Methods

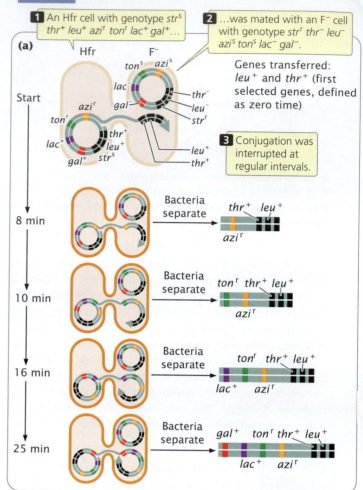

1 An Hfr cell with genotype str^s $thr^+\ leu^+\ azi^r\ ton^r\ lac^+\ gal^+$...

2 ...was mated with an F⁻ cell with genotype $str^r\ thr^-\ leu^-$ $azi^s\ ton^s\ lac^-\ gal^-$.

Genes transferred: leu^+ and thr^+ (first selected genes, defined as zero time)

3 Conjugation was interrupted at regular intervals.

Results

Conclusion: The transfer times indicate the order and relative distances between genes and can be used to construct a genetic map.

9.15 Jacob and Wollman used interrupted conjugation to map bacterial genes.

Solution Steps

The first donor gene to appear in all of the recipient cells (at about 9 minutes) was azi^r. Gene ton^r appeared next (after about 10 minutes), followed by *lac*⁺ (at about 18 minutes) and by *gal*⁺ (after 25 minutes). These transfer times indicate the order of gene transfer and the relative distances among the genes (see Figure 9.15b).

Notice that the frequency of gene transfer from donor to recipient cells decreased for the more distant genes. For example, about 90% of the recipients received the *azi*ʳ allele, but only about 30% received the *gal*⁺ allele. The lower percentage for *gal*⁺ is due to the fact that some conjugating cells spontaneously broke apart before they were disrupted by the blender. The probability of spontaneous disruption increases with time, so fewer cells had an opportunity to receive genes that were transferred later.

▶ For additional practice mapping bacterial genes with interrupted conjugation, try **Problem 23** at the end of the chapter.

DIRECTIONAL TRANSFER AND MAPPING Different Hfr strains of a given species of bacteria have the F factor integrated into the bacterial chromosome at different sites and in different orientations. Gene transfer always begins within F, and the orientation and position of F determine the direction and starting point of gene transfer. **Figure 9.16a** shows that, in strain Hfr1, F is integrated between *leu* and *azi*; the orientation of F at this site dictates that gene transfer will proceed in a counterclockwise direction around the circular chromosome. Genes from this strain will be transferred in the order of:

$$\longleftarrow leu–thr–thi–his–gal–lac–pro–azi$$

In strain Hfr5, F is integrated between the *thi* and the *his* genes (**Figure 9.16b**) and in the opposite orientation. Here gene transfer will proceed in a clockwise direction:

$$\longleftarrow thi–thr–leu–azi–pro–lac–gal–his$$

Although the starting point and direction of transfer may differ between two strains, the relative distance in time between any two pairs of genes is constant.

Notice that the order of gene transfer is not the same for different Hfr strains (**Figure 9.17a**). For example, *azi* is transferred just after *leu* in strain HfrH but long after *leu* in strain Hfr1. Aligning the sequences (**Figure 9.17b**) shows that the two genes on either side of *azi* are always the same: *leu* and *pro*. That they are the same makes sense when we recognize that the bacterial chromosome is circular and the

9.16 The orientation of the F factor in an Hfr strain determines the direction of gene transfer. Arrowheads indicate the origin and direction of transfer.

starting point of transfer varies from strain to strain. These data provided the first evidence that the bacterial chromosome is circular (**Figure 9.17c**). ▶ **TRY PROBLEM 22**

CONCEPTS

Conjugation can be used to map bacterial genes by mixing Hfr and F⁻ cells of different genotypes and interrupting conjugation at regular intervals. The amount of time required for individual genes to be transferred from the Hfr to the F⁻ cells indicates the relative positions of the genes on the bacterial chromosome.

✓ **CONCEPT CHECK 4**

Interrupted conjugation was used to map three genes in *E. coli*. The donor genes first appeared in the recipient cells at the following times: *gal*, 10 minutes; *his*, 8 minutes; *pro*, 15 minutes. Which gene is in the middle?

(a)

Hfr strain

H — thr leu azi pro lac gal his thi

1 — leu thr thi his gal lac pro azi

2 — pro azi leu thr thi his gal lac

3 — lac pro azi leu thr thi his gal

4 — thi his gal lac pro azi leu thr

5 — thi thr leu azi pro lac gal his

(b)

Hfr strain

5 — thi thr leu azi pro lac gal his

H — thr leu azi pro lac gal his thi

4 — thr leu azi pro lac gal his thi

1 — azi pro lac gal his thi thr leu

2 — lac gal his thi thr leu azi pro

3 — gal his thi thr leu azi pro lac

(c)

Strains 5, H, 4, 1, 2, 3 (circular chromosome maps with genes: thr, leu, azi, pro, lac, gal, his, thi)

Conclusion: The order of the genes on the chromosome is the same, but the position and orientation of the F factor differ among the strains.

9.17 The order of gene transfer in a series of different Hfr strains indicates that the *E. coli* chromosome is circular.

Natural Gene Transfer and Antibiotic Resistance

Antibiotics are substances that kill bacteria. Their development and widespread use has greatly reduced the threat of infectious disease and saved countless lives. But many pathogenic bacteria have developed resistance to antibiotics, particularly in environments where antibiotics are routinely used, such as hospitals, livestock operations, and fish farms. In these environments where antibiotics are continually present, the only bacteria to survive are those that possess antibiotic resistance. No longer in competition with other bacteria, resistant bacteria multiply quickly and spread. In this way, the presence of antibiotics selects for resistant bacteria and reduces the effectiveness of antibiotic treatment for infections.

Antibiotic resistance in bacteria frequently results from the action of genes located on *R plasmids*, small circular plasmids that can be transferred by conjugation. Drug-resistant R plasmids have evolved in the past 60 years (since the beginning of widespread use of antibiotics), and some of them convey resistance to several antibiotics simultaneously. Ironic but plausible sources of some of the resistance genes

found in R plasmids are the microbes that produce antibiotics in the first place. R plasmids can spread easily throughout the environment, passing between related and unrelated bacteria in a variety of common situations. For example, research shows that plasmids carrying genes for resistance to multiple antibiotics were transferred from a cow udder infected with *E. coli* to a human strain of *E. coli* on a hand towel: a farmer wiping his hands after milking an infected cow might unwittingly transfer antibiotic resistance from bovine- to human-inhabiting microbes. Conjugation taking place in minced meat on a cutting board allowed R plasmids to be passed from porcine (pig) to human *E. coli*. The transfer of R plasmids also takes place in sewage, soil, lake water, and marine sediments. The fact that R plasmids can easily spread throughout the environment and pass between related and unrelated bacteria underscores both the importance of limiting antibiotic use to the treatment of infections and the importance of hygiene in everyday life.

Transformation in Bacteria

A second way in which DNA can be transferred between bacteria is through transformation (see Figure 9.7b).

Recipient
DNA

Double-stranded
fragment of DNA

One strand of the DNA
fragment enters the cell;
the other is hydrolyzed.

The single-stranded fragment pairs
with the bacterial chromosome
and recombination takes place.

The remainder of the
single-stranded DNA
fragment is degraded.

Nontransformed

When the cell replicates and divides,
one of the resulting cells is transformed
and the other is not.

Transformed

9.18 Genes can be transferred between bacteria through transformation.

Transformation played an important role in the initial identification of DNA as the genetic material, which will be discussed in Chapter 10.

Transformation requires both the uptake of DNA from the surrounding medium and its incorporation into a bacterial chromosome or a plasmid. It may occur naturally when dead bacteria break up and release DNA fragments into the environment. In soil and marine environments, this may be an important route of genetic exchange for some bacteria.

MECHANISM OF TRANSFORMATION   Cells that take up DNA through their membranes are said to be **competent**. Some species of bacteria take up DNA more easily than others: competence is influenced by growth stage, the concentration of available DNA in the environment, and other environmental factors. The DNA that a competent cell takes up need not be bacterial: virtually any type of DNA (bacterial or otherwise) can be taken up by competent cells under the appropriate conditions.

As a DNA fragment enters the cell in the course of transformation (**Figure 9.18**), one of the strands is broken up, whereas the other strand moves across the membrane and may pair with a homologous region and become integrated into the bacterial chromosome. This integration requires two crossover events, after which the remaining single-stranded DNA is degraded by bacterial enzymes. In some bacteria, double-stranded DNA moves across the cell membrane and is integrated into the bacterial chromosome.

Bacterial geneticists have developed techniques for increasing the frequency of transformation in the laboratory in order to introduce particular DNA fragments into cells. They have also developed strains of bacteria that are more competent than wild-type cells. Treatment with calcium chloride, heat shock, or an electrical field makes bacterial membranes more porous and permeable to DNA, and the efficiency of transformation can also be increased by using high concentrations of DNA. These techniques enable researchers to transform bacteria such as *E. coli*, which are not naturally competent.

GENE MAPPING WITH TRANSFORMATION   Transformation, like conjugation, is used to map bacterial genes, especially in those species that do not undergo conjugation or transduction (see Figure 9.7a and c). Transformation mapping requires two strains of bacteria that differ in several genetic traits; for example, the recipient strain might be a^- b^- c^- (auxotrophic for three nutrients), and the donor cell prototrophic with alleles a^+ b^+ c^+ (**Figure 9.19**). DNA from the donor strain is isolated, purified, and fragmented. The recipient strain is treated to increase competency, and DNA from the donor strain is added to the medium. Fragments of the donor DNA enter the recipient cells and undergo recombination with homologous DNA sequences on the bacterial chromosome. Cells that receive genetic material through transformation are called **transformants**.

Genes can be mapped by observing the rate at which two or more genes are transferred to the host chromosome, or **cotransformed**, in transformation. When the donor DNA is fragmented before transformation, genes that are physically close on the chromosome are more likely to be present on the same DNA fragment and transferred together, as shown for genes a^+ and b^+ in Figure 9.19. Genes that are far apart are unlikely to be present on the same DNA fragment and will rarely be transferred together. Inside the cell, DNA becomes incorporated into the bacterial chromosome through recombination. If two genes are close together on the same fragment, any two crossovers are likely to take place on either side of the two genes, allowing both to become part of the recipient chromosome. If the two genes are far apart, there may be one crossover between them, allowing one gene but not the other to recombine with the bacterial chromosome. Thus, two genes are more likely to be transferred together when they are close together on the chromosome, and genes located far apart are rarely cotransformed. Therefore, the frequency of cotransformation can be used to map bacterial genes. If genes a and b as well as genes b and c are frequently cotransformed, but genes a and c are rarely cotransformed, then gene b must be between a and c—the gene order is a b c. ▶ **TRY PROBLEM 25**

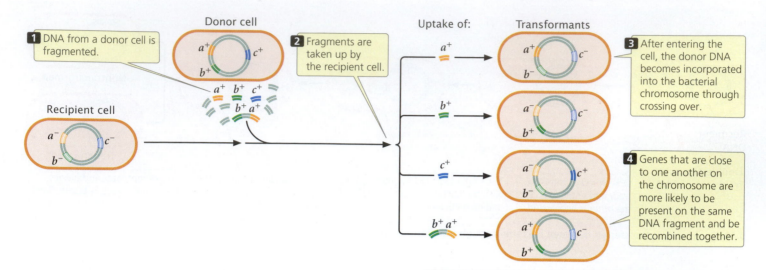

9.19 Transformation can be used to map bacterial genes.

CONCEPTS

Genes can be mapped in bacteria by taking advantage of transformation—the ability of cells to take up DNA from the environment and incorporate it into their chromosomes through crossing over. The relative rate at which pairs of genes are cotransformed indicates the distance between them: the higher the rate of cotransformation, the closer the genes are on the bacterial chromosome.

✓ **CONCEPT CHECK 5**

DNA from a bacterial strain with genotype *his⁻ leu⁻ thr⁻* is transformed with DNA from a strain that is *his⁺ leu⁺ thr⁺*. A few *leu⁺ thr⁺* cells and a few *his⁺ thr⁺* cells are found, but no *his⁺ leu⁺* cells are observed. Which genes are farthest apart?

Bacterial Genome Sequences

Genetic maps serve as the foundation for more-detailed information provided by DNA sequencing, such as gene content and organization (see Chapter 19 for a discussion of gene sequencing). Geneticists have now determined the complete nucleotide sequences of more than two thousand bacterial genomes (see Table 20.1), and many additional microbial sequencing projects are underway.

CHARACTERISTICS OF BACTERIAL GENOMES Most bacterial genomes contain from 1 million to 4 million base pairs of DNA, but a few are much smaller (e.g., 580,000 bp in *Mycoplasma genitalium*) and some are considerably larger (e.g., more than 7 million bp in *Mesorhizobium loti*). The small size of bacterial genomes relative to those found in multicellular eukaryotes, which often have billions of base pairs of DNA, is thought to be an adaptation for rapid cell

division, because the rate of cell division is limited by the time required to replicate the DNA. On the other hand, a lack of mobility in most bacteria requires metabolic and environmental flexibility, and so genome size and content are likely to be a balance between the opposing evolutionary forces of gene loss to maintain rapid reproduction and gene acquisition to ensure flexibility.

The function of a substantial proportion of genes in all bacteria has not been determined. Certain genes, particularly those with related functions, tend to reside next to one another, but these clusters are in very different locations in different species, suggesting that bacterial genomes are constantly being reshuffled. Comparisons of the gene sequences of disease-causing and benign bacteria are helping to identify genes implicated in disease and may suggest new targets for antibiotics and other antimicrobial agents.

Horizontal Gene Transfer

The availability of genome sequences has provided evidence that many bacteria have acquired genetic information from other species of bacteria—and sometimes even from eukaryotic organisms—in a process called **horizontal gene transfer**. In most eukaryotes, genes are passed only among members of the same species through the process of reproduction (called vertical transmission); that is, genes are passed down from one generation to the next. In horizontal transfer, genes can be passed between individual members of different species by nonreproductive mechanisms, such as conjugation, transformation, and transduction. Evidence suggests that horizontal gene transfer has taken place repeatedly among bacteria. For example, as much as 17% of *E. coli*'s genome has been acquired from other bacteria through horizontal gene transfer. Of medical significance, some pathogenic bacteria

(a)

(b)

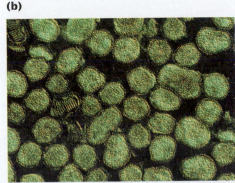

9.20 Viruses come in different structures and sizes.
(a) T4 bacteriophage (bright orange).
(b) Influenza A virus (green structures).
[Part a: Biozentrum, University of Basel/Photo Researchers. Part b: Eye of Science/Photo Researchers.]

have acquired the genes necessary for infection, whereas others have acquired genes that confer resistance to antibiotics.

Because of the widespread occurence of horizontal gene transfer, many bacterial chromosomes are a mixture of genes inherited through vertical transmission and genes acquired through horizontal transfer. This situation has caused some biologists to question whether the species concept is even appropriate for bacteria. A species is often defined as a group of organisms that are reproductively isolated from other groups, have a set of genes in common, and evolve together (see Chapter 26). Because of horizontal gene transfer, the genes of one bacterial species are not isolated from the genes of other species, making the traditional species concept difficult to apply. Horizontal gene transfer also muddies the determination of the ancestral relationships among bacteria. The reconstruction of ancestrial relationships is usually based on genetic similarities and differences: organisms that are genetically similar are assumed to have descended from a recent common ancestor, whereas organisms that are genetically distinct are assumed to be more distantly related. Through horizontal gene transfer, however, even distantly related bacteria may have genes in common and thus appear to have descended from a recent common ancestor. The nature of species and how to classify bacteria are currently controversial topics within the field of microbiology.

9.3 Viruses Are Simple, Replicating Systems Amenable to Genetic Analysis

All organisms—plants, animals, fungi, and bacteria—are infected by viruses. A **virus** is a simple replicating structure made up of nucleic acid surrounded by a protein coat (see Figure 2.4). Viruses come in a great variety of shapes and sizes (**Figure 9.20**). Some have DNA as their genetic material, whereas others have RNA; the nucleic acid may be double stranded or single stranded, linear or circular.

Viruses that infect bacteria (bacteriophages, or phages for short) have played a central role in genetic research since the late 1940s. They are ideal for many types of genetic research because they have small and easily manageable genomes, reproduce rapidly, and produce large numbers of progeny. Bacteriophages have two alternative life cycles: the *lytic* and the *lysogenic* cycles. In the lytic cycle, a phage attaches to a receptor on the bacterial cell wall and injects its DNA into the cell (**Figure 9.21**). Inside the host cell, the phage DNA is replicated, transcribed, and translated, producing more phage DNA and phage proteins. New phage particles are assembled from these components. The phages then produce an enzyme that breaks open the host cell, releasing the new phages. **Virulent phages** reproduce strictly through the lytic cycle and always kill their host cells.

Temperate phages can undergo either the lytic or the lysogenic cycle. The lysogenic cycle begins like the lytic cycle (see Figure 9.21) but, inside the cell, the phage DNA integrates into the bacterial chromosome, where it remains as an inactive **prophage**. The prophage is replicated along with the bacterial DNA and is passed on when the bacterium divides. Certain stimuli can cause the prophage to dissociate from the bacterial chromosome and enter into the lytic cycle, producing new phage particles and lysing the cell.

Techniques for the Study of Bacteriophages

Viruses reproduce only within host cells so bacteriophages must be cultured in bacterial cells. Phages can be grown in large liquid cultures of bacteria to generate large numbers of offspring, but to study the characteristics of individual offspring we must isolate them on plates. We mix phages and bacteria together and plate them on solid medium on a petri plate. A high concentration of bacteria is used so that the colonies grow into one another and produce a continuous layer of bacteria, or "lawn," on the agar. An individual phage infects a single bacterial cell and goes through its lytic cycle. Many new phages are released from the lysed cell and infect additional cells; the cycle is then repeated. Because the bacteria grow on solid medium the diffusion of the phages is restricted and only nearby cells are infected. After several rounds of phage

New phages are released to start the cycle again.

Lysis

The phage binds to the bacterium.

Phage

Phage DNA

1 Host DNA

The prophage may separate and the cell will enter the lytic cycle.

5

6

Assembly of new phages is complete. A phage-encoded enzyme causes the cell to lyse.

The phage DNA enters the host cell.

Lytic cycle

2

Lysogenic cycle

The prophage replicates. This replication can continue through many cell divisions.

4

The host DNA is digested.

3

3

Prophage

5

The host cell transcribes and translates the phage DNA, producing phage proteins.

4

The phage DNA integrates into the bacterial chromosome and becomes a prophage.

Replicated phage

Phage DNA replicates.

9.21 Bacteriophages have two alternative life cycles: lytic and lysogenic.

reproduction, a clear patch of lysed cells, a **plaque**, appears on the plate (**Figure 9.22**). Each plaque represents a single phage that multiplied and lysed many cells. Plating a known volume of a dilute solution of phages on a bacterial lawn and counting the number of plaques that appear can be used to determine the original concentration of phage in the solution.

9.22 Plaques are clear patches of lysed cells on a lawn of bacteria. [Courtesy of Maria P. MacWilliams and Tong Lee.]

CONCEPTS

Viral genomes may be DNA or RNA, circular or linear, and double or single stranded. Bacteriophages are used in many types of genetic research.

✓ CONCEPT CHECK 6

In which bacteriophage life cycle does the phage DNA become incorporated into the bacterial chromosome?

a. Lytic
b. Lysogenic
c. Both lytic and lysogenic
d. Neither lytic nor lysogenic

Transduction: Using Phages to Map Bacterial Genes

In the discussion of bacterial genetics, three mechanisms of gene transfer were identified: conjugation, transformation, and transduction (see Figure 9.7). Let's take a closer look at transduction, in which genes are transferred between bacteria by viruses. In **generalized transduction**, any gene may be transferred. In **specialized transduction**, only a few genes are transferred.

GENERALIZED TRANSDUCTION Joshua Lederberg and Norton Zinder discovered generalized transduction in 1952 while trying to produce recombination in the bacterium *Salmonella typhimurium* by conjugation. They mixed a strain of *S. typhimurium* that was *phe⁺ trp⁺ tyr⁺ met⁻ his⁻* with a strain that was *phe⁻ trp⁻ tyr⁻ met⁺ his⁺* (**Figure 9.23**) and plated them on minimal medium. A few prototrophic recombinants (*phe⁺ trp⁺ tyr⁺ met⁺ his⁺*) appeared, suggesting that conjugation had taken place. However, when they tested the two strains in a U-shaped tube similar to the one used by Davis, some *phe⁺ trp⁺ tyr⁺ met⁺ his⁺* prototrophs were obtained on one side of the tube (compare Figure 9.23 with Figure 9.9). This apparatus separated the two strains by a filter with pores too small for the passage of bacteria, so how were genes being transferred between bacteria in the absence of conjugation? The results of subsequent studies revealed that the agent of transfer was a bacteriophage.

In the lytic cycle of phage reproduction, the bacterial chromosome is broken into random fragments (**Figure 9.24**). For some types of bacteriophage, a piece of the bacterial chromosome instead of phage DNA occasionally gets packaged into a phage coat; these phage particles are called **transducing phages**. The transducing phage infects a new cell, releasing the bacterial DNA, and the introduced genes may then become integrated into the bacterial chromosome by a double crossover. Some transducing phages insert viral DNA, along with the bacterial gene, into the bacterial chromosome. In either case, bacterial genes can be moved from one bacterial strain to another, producing recombinant bacteria called **transductants**.

Not all phages are capable of transduction, a rare event that requires (1) that the phage degrade the bacterial chromosome, (2) that the process of packaging DNA into the phage protein not be specific for phage DNA, and (3) that the bacterial genes transferred by the virus recombine with the chromosome in the recipient cell. The overall rate of transduction ranges from only about 1 in 100,000 to 1 in 1,000,000.

Because of the limited size of a phage particle, only about 1% of the bacterial chromosome can be transduced. Only genes located close together on the bacterial chromosome will be transferred together, or **cotransduced**. Because the chance of a cell being transduced by two separate phages is exceedingly small, any cotransduced genes are usually located close together on the bacterial chromosome. Thus, rates of cotransduction, like rates of cotransformation, give an indication of the physical distances between genes on a bacterial chromosome.

To map genes by using transduction, two bacterial strains with different alleles at several loci are used. The donor strain

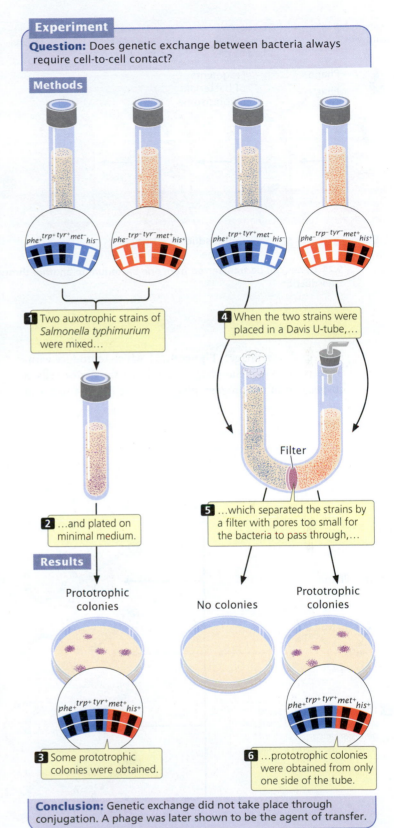

Experiment

Question: Does genetic exchange between bacteria always require cell-to-cell contact?

Methods

phe⁺ trp⁺ tyr⁺ met⁻ his⁻   *phe⁻ trp⁻ tyr⁻ met⁺ his⁺*   *phe⁺ trp⁺ tyr⁺ met⁻ his⁻*   *phe⁻ trp⁻ tyr⁻ met⁺ his⁺*

1 Two auxotrophic strains of *Salmonella typhimurium* were mixed…

4 When the two strains were placed in a Davis U-tube,…

Filter

2 …and plated on minimal medium.

5 …which separated the strains by a filter with pores too small for the bacteria to pass through,…

Results

Prototrophic colonies   No colonies   Prototrophic colonies

phe⁺ trp⁺ tyr⁺ met⁺ his⁺   *phe⁺ trp⁺ tyr⁺ met⁺ his⁺*

3 Some prototrophic colonies were obtained.

6 …prototrophic colonies were obtained from only one side of the tube.

Conclusion: Genetic exchange did not take place through conjugation. A phage was later shown to be the agent of transfer.

9.23 The Lederberg and Zinder experiment.

9.24 Genes can be transferred from one bacterium to another through generalized transduction.

is infected with phages (**Figure 9.25**), which reproduce within the cell. When the phages have lysed the donor cells, a suspension of the progeny phages is mixed with a recipient strain of bacteria, which is then plated on several different kinds of media to determine the phenotypes of the transducing progeny phages. ▶ TRY PROBLEM 32

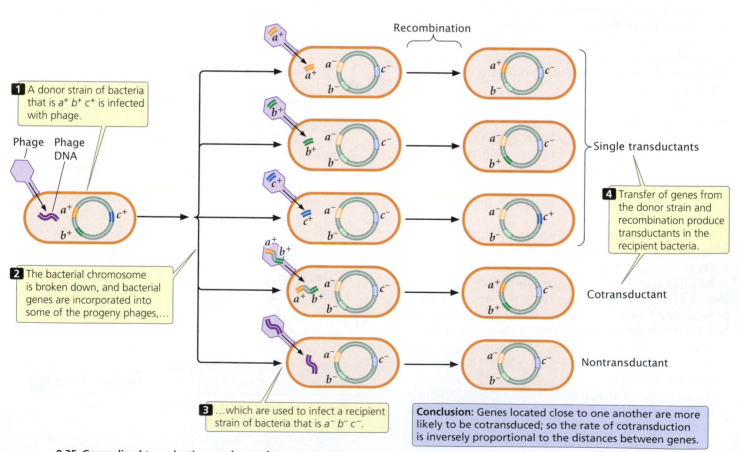

9.25 Generalized transduction can be used to map genes.

CONCEPTS

In transduction, bacterial genes become packaged into a viral coat, are transferred to another bacterium by the virus, and become incorporated into the bacterial chromosome by crossing over. Bacterial genes can be mapped with the use of generalized transduction.

✓ CONCEPT CHECK 7

In gene-mapping experiments using generalized transduction, bacterial genes that are cotransduced are
a. far apart on the bacterial chromosome.
b. on different bacterial chromosomes.
c. close together on the bacterial chromosome.
d. on a plasmid.

SPECIALIZED TRANSDUCTION Like generalized transduction, specialized transduction requires gene transfer from one bacterium to another through phages, but, here, only genes near particular sites on the bacterial chromosome are transferred. This process requires lysogenic bacteriophages. The prophage may imperfectly excise from the bacterial chromosome, carrying with it a small part of the bacterial DNA adjacent to the site of prophage integration. A phage carrying this DNA will then inject it into another bacterial cell in the next round of infection. This process resembles the situation in F' cells, in which the F plasmid carries genes from one bacterium into another (see Figure 9.14). Specialized transduction occurs in phages that utilize specific integration sites on the bacterial chromosome; many phages integrate randomly and only exhibit generalized transduction.

CONCEPTS

Specialized transduction transfers only those bacterial genes located near the site of prophage insertion.

CONNECTING CONCEPTS

Three Methods for Mapping Bacterial Genes

Three methods of mapping bacterial genes have now been outlined: (1) interrupted conjugation; (2) transformation; and (3) transduction. These methods have important similarities and differences.

Mapping with interrupted conjugation is based on the time required for genes to be transferred from one bacterium to another by means of cell-to-cell contact. The key to this technique is that the bacterial chromosome itself is transferred, and the order of genes and the time required for their transfer provide information about the positions of the genes on the chromosome. In contrast with other mapping methods, the distance between genes is measured not in recombination frequencies but in units of time required for genes to be transferred. Here, the basic unit of conjugation mapping is a minute.

In gene mapping with transformation, DNA from the donor strain is isolated, broken up, and mixed with the recipient strain.

Some fragments pass into the recipient cells, where the transformed DNA may recombine with the bacterial chromosome. The unit of transfer here is a random fragment of the chromosome. Loci that are close together on the donor chromosome tend to be on the same DNA fragment; so the rates of cotransformation provide information about the relative positions of genes on the chromosome.

Transduction mapping also relies on the transfer of genes between bacteria that differ in two or more traits, but, here, the vehicle of gene transfer is a bacteriophage. In a number of respects, transduction mapping is similar to transformation mapping. Small fragments of DNA are carried by the phage from donor to recipient bacteria, and the rates of cotransduction, like the rates of cotransformation, provide information about the relative distances between the genes.

All of the methods use a common strategy for mapping bacterial genes. The movement of genes from donor to recipient is detected by using strains that differ in two or more traits, and the transfer of one gene relative to the transfer of others is examined. Additionally, all three methods rely on recombination between the transferred DNA and the bacterial chromosome. In mapping with interrupted conjugation, the relative order and timing of gene transfer provide the information necessary to map the genes; in transformation and transduction, the rate of cotransfer provides this information.

In conclusion, the same basic strategies are used for mapping with interrupted conjugation, transformation, and transduction. The methods differ principally in their mechanisms of transfer: in conjugation mapping, DNA is transferred though contact between bacteria; in transformation, DNA is transferred as small naked fragments; and, in transduction, DNA is transferred by bacteriophages.

Gene Mapping in Phages

Mapping genes in the bacteriophages themselves depends on homologous recombination between phage chromosomes and therefore requires the application of the same principles as those applied to mapping genes in eukaryotic organisms (see Chapter 7). Crosses are made between viruses that differ in two or more genes, and recombinant progeny phages are identified and counted. The proportion of recombinant progeny is then used to estimate the distances between the genes and their linear order on the chromosome.

In 1949, Alfred Hershey and Raquel Rotman examined rates of recombination in the T2 bacteriophage, which has single-stranded DNA. They studied recombination between genes in two strains that differed in plaque appearance and host range (the bacterial strains that the phages could infect). One strain was able to infect and lyse type B *E. coli* cells but not B/2 cells (making this strain of phage wild type with normal host range, or h^+) and produced an abnormal plaque that was large with distinct borders (r^-). The other strain was able to infect and lyse *both* B *and* B/2 cells (mutant host range, h^-) and produced wild-type plaques that were small with fuzzy borders (r^+).

Hershey and Rotman crossed the $h^+ r^-$ and $h^- r^+$ strains of T2 by infecting type B *E. coli* cells with a mixture of the two strains. They used a high concentration of phages so that

Experiment

Question: How can we determine the position of a gene on a phage chromosome?

Method

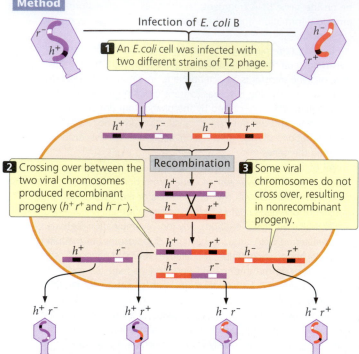

Infection of *E. coli* B

1 An *E.coli* cell was infected with two different strains of T2 phage.

h^+ r^- h^- r^+

2 Crossing over between the two viral chromosomes produced recombinant progeny ($h^+ r^+$ and $h^- r^-$).

Recombination

3 Some viral chromosomes do not cross over, resulting in nonrecombinant progeny.

h^+ r^-
h^- r^+

h^+ r^-
h^- r^-

h^+ r^+
h^- r^+

$h^+ r^-$ $h^+ r^+$ $h^- r^-$ $h^- r^+$

| Nonrecombinant phage produces cloudy, large plaques | Recombinant phage produces cloudy, small plaques | Recombinant phage produces clear, large plaques | Nonrecombinant phage produces clear, small plaques |

4 Progeny phages were then plated on a mixture of *E. coli* B and *E. coli* B/2 cells,...

Results

Genotype	Plaques	Designation
$h^- r^+$	42	Parental progeny 76%
$h^+ r^-$	34	
$h^+ r^+$	12	Recombinant 24%
$h^- r^-$	12	

5 ...which allowed all four genotypes of progeny to be identified.

6 The percentage of recombinant progeny allowed the h^- and r^- mutants to be mapped.

$$RF = \frac{\text{recombinant plaques}}{\text{total plaques}} = \frac{(h^+ r^+) + (h^- r^-)}{\text{total plaques}}$$

Conclusion: The recombination frequency indicates that the distance between *h* and *r* genes is 24%.

9.26 Hershey and Rotman developed a technique for mapping viral genes. [Courtesy Steven R. Spilatro.]

Table 9.4   Progeny phages produced from $h^- r^+ \times h^+ r^-$

Phenotype	Genotype
Clear and small	$h^- r^+$
Cloudy and large	$h^+ r^-$
Cloudy and small	$h^+ r^+$
Clear and large	$h^- r^-$

most cells could be simultaneously infected by both strains (**Figure 9.26**). Within the bacterial cells, homologous recombination occasionally took place between the chromosomes of the different bacteriophage strains, producing $h^+ r^+$ and $h^- r^-$ chromosomes, which were then packaged into new phage particles. When the cells lysed, the recombinant phages were released, along with the nonrecombinant $h^+ r^-$ phages and $h^- r^+$ phages.

Hershey and Rotman diluted and plated the progeny phages on a bacterial lawn that consisted of a *mixture* of B and B/2 cells. Phages carrying the h^+ allele (which conferred the ability to infect only B cells) produced a cloudy plaque because the B/2 cells did not lyse. Phages carrying the h^- allele produced a clear plaque because all the cells within the plaque were lysed. The r^+ phages produced small plaques, whereas the r^- phages produced large plaques. The genotypes of these progeny phages could therefore be determined by the appearance of the plaque (see Figure 9.26 and **Table 9.4**).

In this type of phage cross, the recombination frequency (*RF*) between the two genes can be calculated by using the following formula:

$$RF = \frac{\text{recombinant plaques}}{\text{total plaques}}$$

In Hershey and Rotman's cross, the recombinant plaques were $h^+ r^+$ and $h^- r^-$; so the recombination frequency was

$$RF = \frac{(h^+ r^+) + (h^- r^-)}{\text{total plaques}}$$

Recombination frequencies can be used to determine the distances between genes and their order on the phage chromosome, just as recombination frequencies are used to map genes in eukaryotes. ▶ TRY PROBLEMS 31 AND 35

CONCEPTS

To map phage genes, bacterial cells are infected with viruses that differ in two or more genes. Recombinant plaques are counted, and rates of recombination are used to determine the linear order of the genes on the chromosome and the distances between them.

Fine-Structure Analysis of Bacteriophage Genes

In the 1950s and 1960s, Seymour Benzer conducted a series of experiments to examine the structure of a gene. Because no molecular techniques were available at the time for directly examining nucleotide sequences, Benzer was forced to infer gene structure from analyses of mutations and their effects. The results of his studies showed that different mutational sites *within* a single gene could be mapped (referred to as **intragenic mapping**) by using techniques similar to those described for mapping bacterial genes by transduction. Different sites within a single gene are very close together; so recombination between them takes place at a very low frequency. Because large numbers of progeny are required to detect these recombination events, Benzer used the bacteriophage T4, which reproduces rapidly and produces large numbers of progeny.

BENZER'S MAPPING TECHNIQUES Wild-type T4 phages normally produce small plaques with rough edges when grown on a lawn of *E. coli*. Certain mutants, called *r* for rapid lysis, produce larger plaques with sharply defined edges. Benzer isolated phages with a number of different *r* mutations, concentrating on one particular subgroup called *rII* mutants.

Wild-type T4 phages produce typical plaques (**Figure 9.27**) on *E. coli* strains B and K. In contrast, the *rII* mutants produce *r* plaques on strain B and do not form plaques at all on strain K. Benzer recognized the *r* mutants by their distinctive plaques when grown on *E. coli* B. He then collected lysate from these plaques and used it to infect *E. coli* K. Phages that did not produce plaques on *E. coli* K were defined as the *rII* type.

Benzer collected thousands of *rII* mutations. He simultaneously infected bacterial cells with two different mutants and looked for recombinant progeny (**Figure 9.28**). Consider two *rII* mutations, a^- and b^- (their wild-type alleles are a^+ and b^+). Benzer infected *E. coli* B cells with two different strains of phages, one $a^-\ b^+$ and the other $a^+\ b^-$ (see Figure 9.28, step 3). Neither of these mutations is able to grow on *E. coli* K cells. While reproducing within the B cells, a few phages of

9.27 T4 phage *rII* mutants produce distinct plaques when grown on *E. coli* B cells. (Left) Plaque produced by wild-type phage. (Right) Plaque produced by *rII* mutant. [Dr. D. P. Snustad, College of Biological Sciences, University of Minnesota.]

the two strains recombined (see Figure 9.28, step 4). A single crossover produces two recombinant chromosomes; one with genotype $a^+\ b^+$ and the other with genotype $a^-\ b^-$:

The resulting recombinant chromosomes, along with the nonrecombinant (parental) chromosomes, were incorporated into progeny phages, which were then used to infect *E. coli* K cells. The resulting plaques were examined to determine the genotype of the infecting phage and map the *rII* mutants (see Figure 9.28, step 5).

Neither of the *rII* mutants grew on *E. coli* K (see Figure 9.28, step 2), but wild-type phages grew; so progeny phages that produced plaques on *E. coli* K must have the recombinant genotype $a^+\ b^+$. Each recombination event produces equal numbers of double mutants ($a^-\ b^-$) and wild-type chromosomes ($a^+\ b^+$); so the number of recombinant progeny should be twice the number of wild-type plaques that appeared on *E. coli* K. The recombination frequency between the two *rII* mutants would be:

$$RF = \frac{2 \times \text{number of plaques on } E.\ coli\ \text{K}}{\text{total number of plaques on } E.\ coli\ \text{B}}$$

Because phages produce large numbers of progeny, Benzer was able to detect a single recombinant among billions of progeny phages. Recombination frequencies are proportional to physical distances along the chromosome (see p. 174 in Chapter 7), revealing the positions of the different mutations within the *rII* region of the phage chromosome. In this way, Benzer eventually mapped more than 2400 *rII* mutations, many corresponding to single base pairs in the viral DNA. His work provided the first molecular view of a gene.

COMPLEMENTATION EXPERIMENTS Benzer's mapping experiments demonstrated that some *rII* mutations were very closely linked. This finding raised the question whether they were at the same locus or at different loci. To determine whether different *rII* mutations belonged to different functional loci, Benzer used the complementation (cis–trans) test (see p. 117 in Chapter 5).

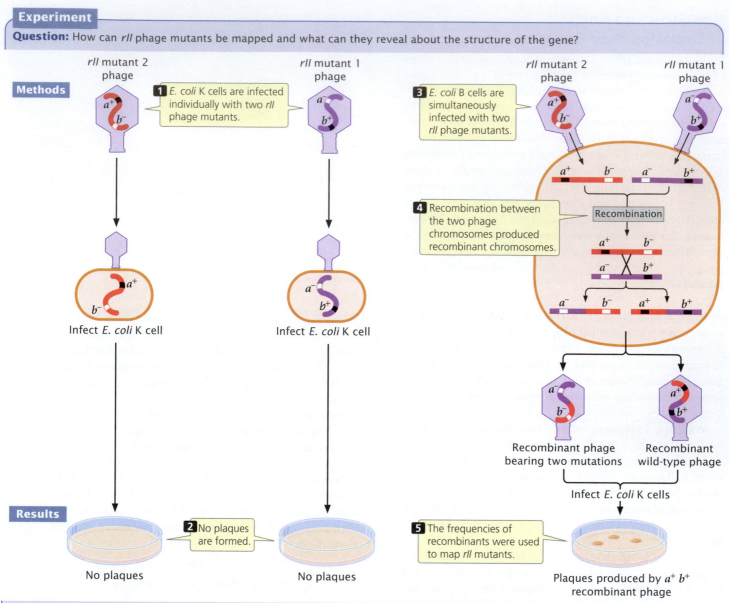

Experiment

Question: How can *rII* phage mutants be mapped and what can they reveal about the structure of the gene?

Methods

rII mutant 2 phage

rII mutant 1 phage

1 *E. coli* K cells are infected individually with two *rII* phage mutants.

a^+ b^-

a^- b^+

Infect *E. coli* K cell

a^+ b^-

Infect *E. coli* K cell

a^- b^+

3 *E. coli* B cells are simultaneously infected with two *rII* phage mutants.

rII mutant 2 phage

rII mutant 1 phage

a^+ b^-

a^- b^+

a^+ b^- a^- b^+

4 Recombination between the two phage chromosomes produced recombinant chromosomes.

Recombination

a^+ b^-

a^- b^+

a^- b^- a^+ b^+

a^- b^-

Recombinant phage bearing two mutations

a^+ b^+

Recombinant wild-type phage

Infect *E. coli* K cells

Results

2 No plaques are formed.

No plaques

No plaques

5 The frequencies of recombinants were used to map *rII* mutants.

Plaques produced by a^+ b^+ recombinant phage

Conclusion: Mapping more than 2400 *rII* mutants provided information about the internal structure of a gene at the base-pair level—the first view of the molecular structure of a gene.

9.28 Benzer developed a procedure for mapping *rII* mutants. Two different *rII* mutants (a^- b^+ and a^+ b^-) are isolated on *E. coli* B cells. Only the a^+ b^+ recombinant can grow on *E. coli* K, allowing these recombinants to be identified.

To carry out the complementation test in bacteriophage, Benzer infected cells of *E. coli* K with large numbers of two mutant strains of phage (**Figure 9.29**, step 1) so that cells would become doubly infected with both strains. Consider two *rII* mutations: r_{101}^- and r_{104}^-. Cells infected with both mutants

$$\frac{r_{101}^-\quad r_{104}^+}{r_{101}^+\quad r_{104}^-}$$

were effectively heterozygous for the phage genes, with the mutations in the trans configuration (see Figure 9.29, step 2).

In the complementation testing, the phenotypes of progeny phages were examined on the K strain, rather than the B strain as illustrated in Figure 9.28.

If the r_{101}^- and r_{104}^- mutations are at different functional loci that encode different proteins, then, in bacterial cells infected by both mutants, the wild-type sequences on the chromosome opposite each mutation will overcome the effects of the recessive mutations; the phages will produce normal plaques on *E. coli* K cells (Figure 9.29, steps 3, 4, and 5). If, on the other hand, the mutations are at the same locus, no functional protein is produced by either chromosome, and

Experiment

Question: How do we determine whether two different *rII* mutations are at the same locus?

Methods

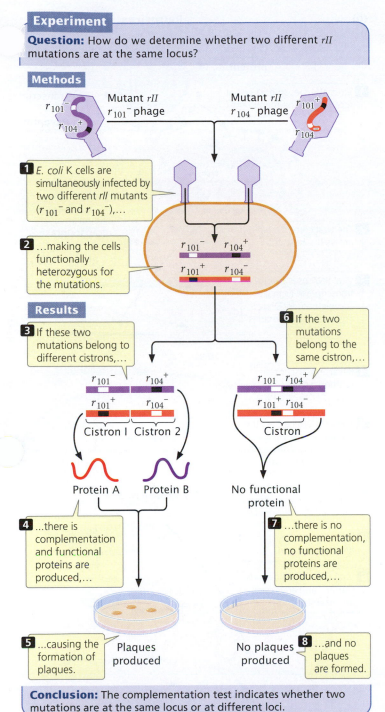

1 *E. coli* K cells are simultaneously infected by two different *rII* mutants (r_{101}^- and r_{104}^-),...

2 ...making the cells functionally heterozygous for the mutations.

Results

3 If these two mutations belong to different cistrons,...

6 If the two mutations belong to the same cistron,...

r_{101}^- r_{104}^+
r_{101}^+ r_{104}^-

Cistron I   Cistron 2

r_{101}^- r_{104}^+
r_{101}^+ r_{104}^-

Cistron

Protein A   Protein B   No functional protein

4 ...there is complementation and functional proteins are produced,...

7 ...there is no complementation, no functional proteins are produced,...

5 ...causing the formation of plaques.   Plaques produced   No plaques produced   **8** ...and no plaques are formed.

Conclusion: The complementation test indicates whether two mutations are at the same locus or at different loci.

9.29 Complementation tests are used to determine whether different mutations are at the same functional gene.

no plaques develop in the *E. coli* K cells (Figure 9.29, steps 6, 7, and 8). Thus, the absence of plaques indicates that the two mutations are at the same locus. Benzer coined the term *cistron* to designate a functional gene defined by the complementation test.

In the complementation test, the cis heterozygote is used as a control. Benzer simultaneously infected bacteria with

wild-type phages (r_{101}^+ r_{104}^+) and with phages carrying both mutations (r_{101}^- r_{104}^-). This test produced cells that were heterozygous and in cis configuration for the phage genes:

$$\frac{r_{101}^+ \quad r_{104}^+}{r_{101}^- \quad r_{104}^-}$$

Regardless of whether the r_{101}^- and r_{104}^- mutations are in the same functional unit, these cells contain a copy of the wild-type phage chromosome (r_{101}^+ r_{104}^+) and will produce normal plaques in *E. coli* K.

Benzer carried out complementation testing on many pairs of *rII* mutants. He found that the *rII* region consists of two loci, designated *rIIA* and *rIIB*. Mutations belonging to the *rIIA* and *rIIB* groups complemented each other, but mutations in the *rIIA* group did not complement others in *rIIA*; nor did mutations in the *rIIB* group complement others in *rIIB*.

CONCEPTS

In a series of experiments with the bacteriophage T4, Seymour Benzer showed that recombination could take place within a single gene and created the first molecular map of a gene. He used the complementation test to distinguish between functional genes (loci).

✓ **CONCEPT CHECK 8**

In complementation tests, Benzer simultaneously infected *E. coli* cells with two phages, each of which carried a different mutation. What conclusion did he make when the progeny phages produced normal plaques?

a. The mutations were at the same locus.
b. The mutations were at different loci.
c. The mutations were close together on the chromosome.
d. The genes were in the cis configuration.

At the time of Benzer's research, the relation between genes and DNA structure was unknown. A gene was defined as a functional unit of heredity that encoded a phenotype. Many geneticists believed that genes were indivisible and that recombination could not take place within them. Benzer demonstrated that intragenic recombination did indeed take place (although at a very low rate) and gave geneticists their first glimpse at the structure of an individual gene.

▶ **TRY PROBLEM 40**

RNA Viruses

Thus far, we have primarily considered viruses that infect bacteria. Viruses also infect plants and animals, and some are important pathogens in these organisms. What we learned about bacteriophages has important implications for viruses that infect these more-complex organisms.

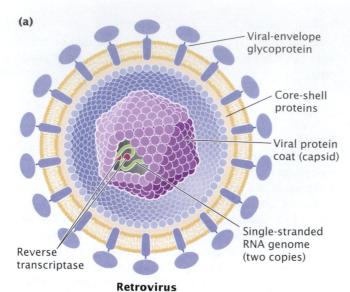

(a)

Viral-envelope glycoprotein

Core-shell proteins

Viral protein coat (capsid)

Single-stranded RNA genome (two copies)

Reverse transcriptase

Retrovirus

9.30 A retrovirus uses reverse transcription to incorporate its RNA into the host DNA. (a) Structure of a typical retrovirus. Two copies of the single-stranded RNA genome and the reverse transcriptase enzyme are shown enclosed within a protein capsid. The capsid is surrounded by a viral envelope that is studded with viral glycoproteins. (b) The retrovirus life cycle.

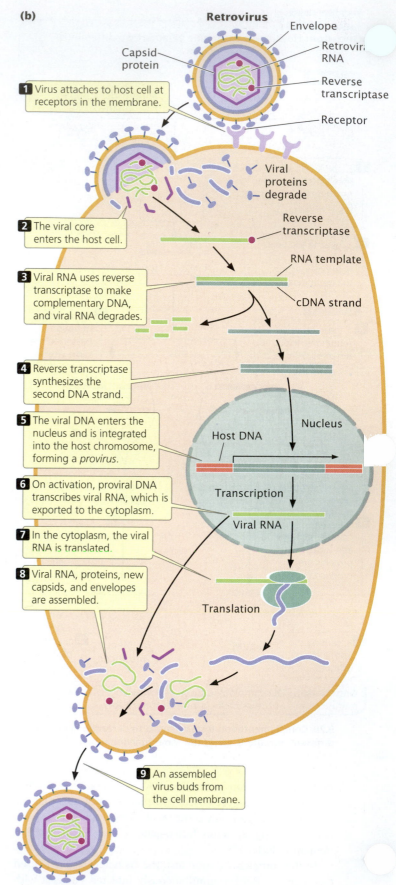

(b)

Retrovirus

Envelope

Capsid protein

Retroviral RNA

Reverse transcriptase

Receptor

1 Virus attaches to host cell at receptors in the membrane.

Viral proteins degrade

2 The viral core enters the host cell.

Reverse transcriptase

RNA template

3 Viral RNA uses reverse transcriptase to make complementary DNA, and viral RNA degrades.

cDNA strand

4 Reverse transcriptase synthesizes the second DNA strand.

Nucleus

5 The viral DNA enters the nucleus and is integrated into the host chromosome, forming a *provirus*.

Host DNA

6 On activation, proviral DNA transcribes viral RNA, which is exported to the cytoplasm.

Transcription

Viral RNA

7 In the cytoplasm, the viral RNA is translated.

8 Viral RNA, proteins, new capsids, and envelopes are assembled.

Translation

9 An assembled virus buds from the cell membrane.

Viral genomes may be encoded in either DNA or RNA. RNA is the genetic material of some human viruses, including those that cause colds, influenza, polio, and AIDS. Almost all viruses that infect plants have RNA genomes. The medical and economic importance of RNA viruses has encouraged their study.

RNA viruses capable of integrating into the genomes of their hosts, much as temperate phages insert themselves into bacterial chromosomes, are called **retroviruses** (**Figure 9.30a**). Because the retroviral genome is RNA, whereas that of the host is DNA, a retrovirus must produce **reverse transcriptase**, an enzyme that synthesizes complementary DNA (cDNA) from either an RNA or a DNA template. A retrovirus uses reverse transcriptase to copy its RNA genome into a single-stranded DNA molecule, and the reverse transcriptase enzyme—or sometimes the host DNA polymerase—copies this single-stranded DNA, creating a double-stranded DNA molecule. The DNA copy of the viral genome then integrates into the host chromosome. A viral genome incorporated into the host chromosome is called a **provirus**. The provirus is replicated by host enzymes when the host chromosome is duplicated (**Figure 9.30b**).

When conditions are appropriate, the provirus undergoes transcription to produce numerous copies of the original viral RNA genome. This RNA encodes viral proteins and serves as genomic RNA for new viral particles. As these viruses escape the cell, they collect patches of the cell membrane to use as their envelopes.

All known retroviral genomes have three genes in common: *gag*, *pol*, and *env*, each encoding a precursor protein that is cleaved into two or more functional proteins. The *gag* gene encodes proteins that make up the viral protein coat. The *pol* gene encodes reverse transcriptase and an enzyme called **integrase**, which inserts the viral DNA into the host chromosome. The *env* gene encodes the glycoproteins that appear on the surface of the virus.

Some retroviruses contain **oncogenes** (see Chapter 23) that may stimulate cell division and cause the formation of tumors. The first retrovirus to be isolated, the Rous sarcoma virus, was originally recognized by its ability to produce connective-tissue tumors (sarcomas) in chickens.

Human Immunodeficiency Virus and AIDS

An example of a retrovirus is human immunodeficiency virus (HIV), which causes acquired immune deficiency syndrome (AIDS). AIDS was first recognized in 1982, when a number of homosexual males in the United States began to exhibit symptoms of a new immune-system-deficiency disease. In that year, Robert Gallo proposed that AIDS was caused by a retrovirus. Between 1983 and 1984, as the AIDS epidemic became widespread, the HIV retrovirus was isolated from AIDS patients. AIDS is now known to be caused by two different immunodeficiency viruses, HIV-1 and HIV-2, which together have infected more than 60 million people worldwide. Of those infected, 30 million have died. Most cases of AIDS are caused by HIV-1, which now has a global distribution; HIV-2 is found primarily in western Africa.

HIV illustrates the importance of genetic recombination in viral evolution. Studies of the DNA sequences of HIV and other retroviruses reveal that HIV-1 is closely related to the simian immunodeficiency virus found in chimpanzees (SIV_{cpz}). Many wild chimpanzees in Africa are infected with SIV_{cpz}, although it doesn't cause AIDS-like symptoms in chimps. SIV_{cpz} is itself a hybrid that resulted from recombination between a retrovirus found in the red-capped mangabey (a monkey) and a retrovirus found in the greater spot-nosed monkey (**Figure 9.31**). Apparently, one or more chimpanzees became infected with both viruses; recombination between the viruses produced SIV_{cpz}, which was then transmitted to humans through contact with infected chimpanzees. In humans, SIV_{cpz} underwent significant evolution to become HIV-1, which then spread throughout the world to produce the AIDS epidemic. Several independent transfers of SIV_{cpz} to humans gave rise to different strains of HIV-1. HIV-2 evolved from a different retrovirus, SIV_{sm}, found in sooty mangabeys.

HIV is transmitted by sexual contact between humans and through any type of blood-to-blood contact, such as that caused by the sharing of dirty needles by drug addicts. HIV can also be transmitted between mother and child during pregnancy and after pregnancy in breast milk. Until screening tests could identify HIV-infected blood, transfusions and clotting factors used by hemophiliacs were sources of infection as well.

9.31 HIV-1 evolved from a similar virus (SIV_{cpz}) found in chimpanzees and was transmitted to humans. SIV_{cpz} arose from recombination taking place between retroviruses in red-capped mangabeys and greater spot-nosed monkeys.

HIV principally attacks a class of blood cells called helper T lymphocytes or, simply, helper T cells (**Figure 9.32**). HIV enters a helper T cell, undergoes reverse transcription, and integrates into the chromosome. The virus reproduces rapidly, destroying the T cell as new virus particles escape from the cell. Because helper T cells are central to immune function and are destroyed in the infection, AIDS patients have a diminished immune response; most AIDS patients die of secondary infections that develop because they have lost the ability to fight off pathogens.

The HIV genome is 9749 nucleotides long and carries *gag*, *pol*, *env*, and six other genes that regulate the life cycle of the virus. HIV's reverse transcriptase is very error prone, giving the virus a high mutation rate and allowing it to evolve

9.32 HIV principally attacks helper T lymphocytes. Electron micrograph showing a T cell (green) infected with HIV (orange). [Thomas Deerinck, NCMIR/Science Source.]

rapidly, even within a single host. This rapid evolution makes the development of an effective vaccine against HIV particularly difficult. Genetic variation within the human population also affects the virus. To date, more than 10 loci in humans that affect HIV infection and the progression of AIDS have been identified.

CONCEPTS

A retrovirus is an RNA virus that integrates into its host's chromosome by making a DNA copy of its RNA genome through the process of reverse transcription. Human immunodeficiency virus, the causative agent of AIDS, is a retrovirus. It evolved from related retroviruses found in other primates.

✔ **CONCEPT CHECK 9**

What enzyme is used by a retrovirus to make a DNA copy of its genome?

Influenza

Influenza demonstrates how rapid changes in a pathogen can arise through recombination of its genetic material. Influenza, commonly called flu, is a respiratory disease caused by influenza viruses. In the United States, from 5% to 20% of the entire population is infected with influenza annually and, though most cases are mild, an estimated 36,000 people die from influenza-related causes each year. At certain times, particularly when new strains of influenza virus enter the human population, there are worldwide epidemics (called pandemics); for example, in 1918, the Spanish flu

virus killed an estimated 20 million to 100 million people worldwide.

Influenza viruses are RNA viruses that infect birds and mammals. The three main types are influenza A, influenza B, and influenza C. Most cases of the common flu are caused by influenza A and B. Influenza A is divided into subtypes on the basis of two proteins, hemagglutinin (HA) and neuraminidase (NA), found on the surface of the virus. The HA and NA proteins affect the ability of the virus to enter host cells and the host organism's immune response to infection. There are 16 types of HA and 9 types of NA, which can exist in a virus in different combinations. For example, common strains of influenza A circulating in humans today are H1N1 and H3N2 (**Table 9.5**), along with several strains of influenza B. Most of the different subtypes of influenza A are found in birds.

Although influenza is an RNA virus, it is not a retrovirus: its genome is not copied into DNA and incorporated into the host chromosome as is that of a retrovirus. The influenza viral genome consists of seven or eight pieces of RNA that are enclosed in a viral envelope. Each piece of RNA encodes one or two of the virus's proteins. The virus enters a host cell by attaching to specific receptors on the cell membrane. After the viral particle has entered the cell, the viral RNA is released, copied, and translated into viral proteins. Viral RNA molecules and viral proteins are then assembled into new viral particles, which exit the cell and infect additional cells.

One of the dangers of the influenza virus is that it evolves rapidly, with new strains appearing frequently. Influenza evolves in two ways. First, each strain continually changes through mutations arising in the viral RNA. The enzyme that copies the RNA is especially prone to making mistakes, and so new mutations are continually introduced into the viral genome. This type of continual change is called **antigenic drift**. Occasionally, major changes in the viral genome take place through **antigenic shift**, in which genetic material from different strains is combined in a process called reassortment. Reassortment takes place when a host is simultaneously infected with two different strains. The RNAs of both strains are replicated within the cell, and RNA segments from two different strains are

Table 9.5	Strains of influenza virus responsible for major flu pandemics	
Year	**Influenza Pandemic**	**Strain**
1918	Spanish flu	H1N1
1957	Asian flu	H2N2
1968	Hong Kong flu	H3N2
2009	Swine flu	H1N1

Human   Avian   Swine

RNA segments

H1N1 swine
influenza virus

9.33 New strains of influenza virus are created by reassortment of genetic material from different strains. A new H1N1 virus (swine flu) that appeared in 2009 contained genetic material from avian, swine, and human viruses.

incorporated into the same viral particle, creating a new strain. For example, in 2002, reassortment occurred between the H1N1 and H3N2 subtypes, creating a new H1N2

strain that contained the hemagglutinin from H1N1 and the neuraminidase from H3N2. New strains produced by antigenic shift are responsible for most pandemics, because no one has immunity to the radically different virus that is produced.

Most different strains of influenza A infect birds. Humans are not easily infected with bird influenza. The appearance of new strains in humans is thought to often arise from viruses that reassort in pigs, which can be infected by viruses from both humans and birds. In 2009, a new strain of H1N1 influenza (called swine flu) emerged in Mexico and quickly spread throughout the world. This virus arose from a series of reassortment events that combined gene sequences from human, bird, and pig influenza viruses to produce the new H1N1 virus (**Figure 9.33**). Farming practices that raise pigs and birds in close proximity may facilitate reassortment among avian, swine, and human strains of influenza.

CONCEPTS

Influenza is caused by RNA influenza viruses. New strains of influenza appear through antigenic shift, in which new viral genomes are created through the reassortment of RNA molecules of different strains.

CONCEPTS SUMMARY

■ Bacteria and viruses are well suited to genetic studies: they are small, have a small haploid genome, undergo rapid reproduction, and produce large numbers of progeny through asexual reproduction.

■ The bacterial genome normally consists of a single, circular molecule of double-stranded DNA. Plasmids are small pieces of bacterial DNA that can replicate independently of the chromosome.

■ DNA can be transferred between bacteria by conjugation, transformation, or transduction.

■ Conjugation is the union of two bacterial cells and the transfer of genetic material between them. It is controlled by an episome called F. The rate at which individual genes are transferred during conjugation provides information about the order of the genes and the distances between them on the bacterial chromosome.

■ Bacteria take up DNA from the environment through the process of transformation. Frequencies of the cotransformation of genes provide information about the physical distances between chromosomal genes.

■ Complete DNA sequences of many bacterial species have been determined. This sequence information indicates that horizontal gene transfer—the movement of DNA between species—is common in bacteria.

■ Viruses are replicating structures with DNA or RNA genomes that may be double stranded or single stranded and linear or circular.

■ Bacterial genes become incorporated into phage coats and are transferred to other bacteria by phages through the process of transduction. Rates of cotransduction can be used to map bacterial genes.

■ Phage genes can be mapped by infecting bacterial cells with two different phage strains and counting the number of recombinant plaques produced by the progeny phages.

■ Benzer mapped a large number of mutations that occurred within the *rII* region of phage T4. The results of his complementation studies demonstrated that the *rII* region consists of two functional units that he called cistrons. He showed that intragenic recombination takes place.

■ A number of viruses have RNA genomes. Retroviruses encode reverse transcriptase, an enzyme used to make a DNA copy of the viral genome, which then integrates into the host genome as a provirus. HIV is a retrovirus that is the causative agent for AIDS.

■ Influenza is caused by RNA influenza viruses that evolve through small changes taking place by mutation (antigenic drift) and through major changes taking place by the reassortment of the genetic material from different strains.

IMPORTANT TERMS

ANSWERS TO CONCEPT CHECKS

1. d

2. b

3. a

4. *gal*

5. *his* and *leu*

6. b

7. c

8. b

9. Reverse transcriptase

WORKED PROBLEMS

Problem 1

DNA from a strain of bacteria with genotype $a^+ b^+ c^+ d^+ e^+$ was isolated and used to transform a strain of bacteria that was $a^- b^- c^- d^- e^-$. The transformed cells were tested for the presence of donated genes. The following genes were cotransformed:

   a^+ and d^+ b^+ and e^+ c^+ and d c^+ and e^+

What is the order of genes *a*, *b*, *c*, *d*, and *e* on the bacterial chromosome?

Solution Strategy

What information is required in your answer to the problem?

The order of genes *a*, *b*, *c*, *d*, and *e* on the bacterial chromosome.

What information is provided to solve the problem?

- The donor cells were $a^+ b^+ c^+ d^+ e^+$ and the recipient cells were $a^- b^- c^- d^- e^-$.

- The combinations of genes that were cotransformed.

For help with this problem, review:

Transformation in Bacteria in Section 9.2.

Solution Steps

> **Recall:** The rate at which genes are cotransformed is inversely proportional to the distance between them: genes that are close together are frequently cotransformed, whereas genes that are far apart are rarely cotransformed.

In this transformation experiment, gene c^+ is cotransformed with both genes e^+ and d^+, but genes e^+ and d^+ are not cotransformed; therefore the *c* locus must be between the *d* and *e* loci:

Gene e^+ is also cotransformed with gene b^+; so the *e* and *b* loci must be located close together. Locus *b* could be on either side of locus *e*. To determine whether locus *b* is on the same side of *e* as locus *c*, we look to see whether genes b^+ and c^+ are cotransformed. They are not; so locus *b* must be on the opposite side of *e* from *c*:

Gene a^+ is cotransformed with gene d^+; so they must be located close together. If locus *a* were located on the same side of *d* as locus *c*, then genes a^+ and c^+ would be cotransformed. Because these genes display no cotransformation, locus *a* must be on the opposite side of locus *d*:

Problem 2

Consider three genes in *E. coli*: thr^+ (the ability to synthesize threonine), ara^+ (the ability to metabolize arabinose), and leu^+ (the ability to synthesize leucine). All three of these genes are close together on the *E. coli* chromosome. Phages are grown in a thr^+ ara^+ leu^+ strain of bacteria (the donor strain). The phage lysate is collected and used to infect a strain of bacteria that is thr^- ara^- leu^-. The recipient bacteria are then tested on medium lacking leucine. Bacteria that grow and form colonies on this medium (leu^+ transductants) are then replica plated onto medium lacking threonine and onto medium lacking arabinose to see which are thr^+ and which are ara^+.

Another group of recipient bacteria are tested on medium lacking threonine. Bacteria that grow and form colonies on this medium (thr^+ transductants) are then replica plated onto medium lacking leucine and onto medium lacking arabinose to see which are ara^+ and which are leu^+. Results from these experiments are as follows:

Selected marker	Cells with cotransduced genes (%)
leu^+	3 thr^+
	76 ara^+
thr^+	3 leu^+
	0 ara^+

How are the loci arranged on the chromosome?

Solution Strategy

What information is required in your answer to the problem?

The order of genes *thr, leu*, and *ara* on the bacterial chromosome.

What information is provided to solve the problem?

- The genes are located close together on the *E. coli* chromosome.

- The donor strain is thr^+ ara^+ leu^+ and the recipient strain is thr^- ara^- leu^-.

- The percent of cells with cotransduced genes.

For help with this problem, review:

Transduction: Using Phages to Map Bacterial Genes in Section 9.3.

Solution Steps

Notice that, when we select for leu^+ (the top half of the table), most of the selected cells are also ara^+. This finding indicates that the *leu* and *ara* genes are located close together,

because they are usually cotransduced. In contrast, thr^+ is only rarely cotransduced with leu^+, indicating that *leu* and *thr* are much farther apart. On the basis of these observations, we know that *leu* and *ara* are closer together than are *leu* and *thr*, but we don't yet know the order of three genes—whether *thr* is on the same side of *ara* as *leu* or on the opposite side, as shown here:

Notice that, although the cotransduction frequency for *thr* and *leu* also is 3%, no thr^+ ara^+ cotransductants are observed. This finding indicates that *thr* is closer to *leu* than to *ara*, and therefore *thr* must be to the left of *leu*, as shown here:

> **Hint:** Genes located close together are more likely to be cotransduced than are genes located far apart.

> **Hint:** We can determine the position of *thr* with respect to the other two genes by looking at the cotransduction frequencies when *thr* is selected.

COMPREHENSION QUESTIONS

Section 9.1

1. Explain how auxotrophic bacteria are isolated.
2. What is the difference between complete medium and minimal medium? How are complete media and minimal media to which has been added one or more nutrients (supplemental media) used to isolate auxotrophic mutants of bacteria?

Section 9.2

3. Briefly explain the differences between F⁺, F⁻, Hfr, and F′ cells.

4. What types of matings are possible between F⁺, F⁻, Hfr, and F′ cells? What outcomes do these matings produce? What is the role of F factor in conjugation?

5. Explain how interrupted conjugation, transformation, and transduction can be used to map bacterial genes. How are these methods similar and how are they different?

6. What is horizontal gene transfer and how might it take place?

Section 9.2

7. List some of the characteristics that make bacteria and viruses ideal organisms for many types of genetic studies.

8. What types of genomes do viruses have?

9. Briefly describe the differences between the lytic cycle of virulent phages and the lysogenic cycle of temperate phages.

10. Briefly explain how genes in phages are mapped.

11. How does specialized transduction differ from generalized transduction?

12. Briefly explain the method used by Benzer to determine whether two different mutations occurred at the same locus.

13. Briefly describe the genetic structure of a typical retrovirus.

14. Explain how a retrovirus, which has an RNA genome, is able to integrate its genetic material into that of a host having a DNA genome.

15. What are the evolutionary origins of HIV-1 and HIV-2?

16. Most humans are not easily infected by avian influenza. How then do DNA sequences from avian influenza become incorporated into human influenza?

> ▶ For more questions that test your comprehension of the key chapter concepts, go to **LEARNING**Curve for this chapter.

APPLICATION QUESTIONS AND PROBLEMS

Introduction

*17. Suppose you want to compare the species of bacteria that exist in a polluted stream with the species that exist in an unpolluted stream. Traditionally, bacteria have been identified by growing them in the laboratory and comparing their physical and biochemical properties. You recognize that you will be unable to culture most of the bacteria that reside in the streams. How might you go about identifying the species in the two streams without culturing them in the laboratory?

Section 9.2

18. John Smith is a pig farmer. For the past 5 years, Smith has been adding vitamins and low doses of antibiotics to his pig food; he says that these supplements enhance the growth of the pigs. Within the past year, however, several of his pigs died from infections of common bacteria, which failed to respond to large doses of antibiotics. Can you explain the increased rate of mortality due to infection in Smith's pigs? What advice might you offer Smith to prevent this problem in the future?

19. Rarely, the conjugation of Hfr and F⁻ cells produces two Hfr cells. Explain how this event takes place.

20. In **Figure 9.8**, what do the red and blue parts of the DNA labeled by balloon 6 represent?

21. Austin Taylor and Edward Adelberg isolated some new strains of Hfr cells that they then used to map several genes in *E. coli* by using interrupted conjugation (A. L. Taylor and E. A. Adelberg. 1960.

Genetics 45:1233–1243). In one experiment, they mixed cells of Hfr strain AB-312, which were *xyl⁺ mtl⁺ mal⁺ met⁺* and sensitive to phage T6, with F⁻ strain AB-531, which was *xyl⁻ mtl⁻ mal⁻ met⁻* and resistant to phage T6. The cells were allowed to undergo conjugation. At regular intervals, the researchers removed a sample of cells and interrupted conjugation by killing the Hfr cells with phage T6. The F⁻ cells, which were resistant to phage T6, survived and were then tested for the presence of genes transferred from the Hfr strain. The results of this experiment are shown in the accompanying graph. On the basis of these data, give the order of the *xyl, mtl, mal,* and *met* genes on the bacterial chromosome and indicate the minimum distances between them.

*22. A series of Hfr strains that have genotype *m⁺ n⁺ o⁺ p⁺ q⁺ r⁺* are mixed with an F⁻ strain that has genotype

$m^- n^- o^- p^- q^- r^-$. Conjugation is interrupted at regular intervals and the order of the appearance of genes from the Hfr strain is determined in the recipient cells. The order of gene transfer for each Hfr strain is:

Hfr5	$m^+ q^+ p^+ n^+ r^+ o^+$
Hfr4	$n^+ r^+ o^+ m^+ q^+ p^+$
Hfr1	$o^+ m^+ q^+ p^+ n^+ r^+$
Hfr9	$q^+ m^+ o^+ r^+ n^+ p^+$

What is the order of genes on the circular bacterial chromosome? For each Hfr strain, give the location of the F factor in the chromosome and its polarity.

*23. Crosses of three different Hfr strains with separate samples of an F^- strain are carried out, and the following mapping data are provided from studies of interrupted conjugation:

Appearance of genes in F^- cells

Hfr1:	Genes	b^+	d^+	c^+	f^+	g^+
	Time	3	5	16	27	59
Hfr2:	Genes	e^+	f^+	c^+	d^+	b^+
	Time	6	24	35	46	48
Hfr3:	Genes	d^+	c^+	f^+	e^+	g^+
	Time	4	15	26	44	58

Construct a genetic map for these genes, indicating their order on the bacterial chromosome and the distances between them.

24. In **Figure 9.16**, which gene on the F factor will be transferred last in strain Hfr5?

*25. DNA from a strain of *Bacillus subtilis* with the genotype $trp^+ tyr^+$ was used to transform a recipient strain with the genotype $trp^- tyr^-$. The following numbers of transformed cells were recovered:

Genotype	Number of transformed cells
$trp^+ tyr^-$	154
$trp^- tyr^+$	312
$trp^+ tyr^+$	354

What do these results suggest about the linkage of the *trp* and *tyr* genes?

26. DNA from a strain of *Bacillus subtilis* with genotype $a^+ b^+ c^+ d^+ e^+$ is used to transform a strain with genotype $a^- b^- c^- d^- e^-$. Pairs of genes are checked for cotransformation and the following results are obtained:

Pair of genes	Cotransformation	Pair of genes	Cotransformation
a^+ and b^+	no	b^+ and d^+	no
a^+ and c^+	no	b^+ and e^+	yes
a^+ and d^+	yes	c^+ and d^+	no
a^+ and e^+	yes	c^+ and e^+	yes
b^+ and c^+	yes	d^+ and e^+	no

On the basis of these results, what is the order of the genes on the bacterial chromosome?

27. DNA from a bacterial strain that is $his^+ leu^+ lac^+$ is used to transform a strain that is $his^- leu^- lac^-$. The following percentages of cells were transformed:

Donor strain	Recipient strain	Genotype of transformed cells	Percentage
$his^+ leu^+ lac^+$	$his^- leu^- lac^-$	$his^+ leu^+ lac^+$	0.02
		$his^+ leu^+ lac^-$	0.00
		$his^+ leu^- lac^+$	2.00
		$his^+ leu^- lac^-$	4.00
		$his^- leu^+ lac^+$	0.10
		$his^- leu^- lac^+$	3.00
		$his^- leu^+ lac^-$	1.50

a. What conclusions can you make about the order of these three genes on the chromosome?

b. Which two genes are closest?

28. Rollin Hotchkiss and Julius Marmur studied transformation in the bacterium *Streptococcus pneumoniae* (R. D. Hotchkiss and J. Marmur. 1954. *Proceedings of the National Academy of Sciences* 40:55–60). They examined four mutations in this bacterium: penicillin resistance (*P*), streptomycin resistance (*S*), sulfanilamide resistance (*F*), and the ability to utilize mannitol (*M*). They extracted DNA from strains of bacteria with different combinations of different mutations and used this DNA to transform wild-type bacterial cells ($P^+ S^+ F^+ M^+$). The results from one of their transformation experiments are shown here.

Donor DNA	Recipient DNA	Transformants	Percentage of all cells
$M S F$	$M^+ S^+ F^+$	$M^+ S F^+$	4.0
		$M^+ S^+ F$	4.0
		$M S^+ F^+$	2.6
		$M S F^+$	0.41
		$M^+ S F$	0.22
		$M S^+ F$	0.0058
		$M S F$	0.0071

a. Hotchkiss and Marmur noted that the percentage of cotransformation was higher than would be expected on a random basis. For example, the results show that the 2.6% of the cells were transformed into *M* and 4% were transformed into *S*. If the *M* and *S* traits were inherited independently, the expected probability of cotransformation of *M* and *S* (*M S*) would be 0.026 × 0.04 = 0.001, or 0.1%. However, they observed 0.41% *M S* cotransformants, four times more than they expected. What accounts for the relatively high frequency of cotransformation of the traits they observed?

b. On the basis of the results, what conclusion can you make about the order of the *M, S,* and *F* genes on the bacterial chromosome?

c. Why is the rate of cotransformation for all three genes (*M S F*) almost the same as the rate of cotransformation for *M F* alone?

29. In the course of a study on the effects of the mechanical shearing of DNA, Eugene Nester, A. T. Ganesan, and Joshua Lederberg studied the transfer, by transformation, of sheared DNA from a wild-type strain of *Bacillus subtilis* (his_2^+ aro_3^+ try_2^+ aro_1^+ tyr_1^+ aro_2^+) to strains of bacteria carrying a series of mutations (E. W. Nester, A. T. Ganesan, and J. Lederberg. 1963. *Proceedings of the National Academy of Sciences* 49:61–68). They reported the following rates of cotransformation between his_2^+ and the other genes (expressed as cotransfer rate).

Genes	Rate of cotransfer
his_2^+ and aro_3^+	0.015
his_2^+ and try_2^+	0.10
his_2^+ and aro_1^+	0.12
his_2^+ and tyr_1^+	0.23
his_2^+ and aro_2^+	0.05

On the basis of these data, which gene is farthest from his_2^+? Which gene is closest?

30. C. Anagnostopoulos and I. P. Crawford isolated and studied a series of mutations that affected several steps in the biochemical pathway leading to tryptophan in the bacterium *Bacillus subtilis* (C. Anagnostopoulos and I. P. Crawford. 1961. *Proceedings of the National Academy of Sciences* 47:378–390). Seven of the strains that they used in their study are listed here, along with the mutation found in that strain.

Strain	Mutation
T3	T^-
168	I^-
168PT	I^-
TI	I^-
TII	I^-
T8	A^-
H25	H^-

To map the genes for tryptophan synthesis, they carried out a series of transformation experiments on strains having different mutations and determined the percentage of recombinants among the transformed bacteria. Their results were as follows:

Recipient	Donor	Percent recombinants
T3	168PT	12.7
T3	T11	11.8
T3	T8	43.5
T3	H25	28.6
168	H25	44.9
TII	H25	41.4
TI	H25	31.3
T8	H25	67.4
H25	T3	19.0
H25	TII	26.3
H25	TI	13.4
H25	T8	45.0

On the basis of these two-point recombination frequencies, determine the order of the genes and the distances between them. Where more than one cross was completed for a pair of genes, average the recombination rates from the different crosses. Draw a map of the genes on the chromosome.

Section 9.3

*31. Two mutations that affect plaque morphology in phages (a^- and b^-) have been isolated. Phages carrying both mutations (a^- b^-) are mixed with wild-type phages (a^+ b^+) and added to a culture of bacterial cells. Subsequent to infection and lysis, samples of the phage lysate are collected and cultured on bacterial cells. The following numbers of plaques are observed:

Plaque phenotype	Number
a^+ b^+	2043
a^+ b^-	320
a^- b^+	357
a^- b^-	2134

What is the frequency of recombination between the *a* and *b* genes?

*32. T. Miyake and M. Demerec examined proline-requiring mutations in the bacterium *Salmonella typhimurium* (T. Miyake and M. Demerec. 1960. *Genetics* 45:755–762). On the basis of complementation studies, they found four proline auxotrophs: *proA, proB, proC,* and *proD*. To determine whether *proA, proB, proC,* and *proD* loci were located close together on the bacterial chromosome, they conducted a transduction experiment. Bacterial strains that were *proC*$^+$ and had mutations at

[S. typhimurium. Kwangshin Kim/ Photo Researchers.]

proA, proB, or *proD,* were used as donors. The donors were infected with bacteriophages, and progeny phages were allowed to infect recipient bacteria with genotype *proC⁻ proA⁺ proB⁺ proD⁺*. The bacteria were then plated on a selective medium that allowed only *proC⁺* bacteria to grow. After this, the *proC⁺* transductants were plated on selective media to reveal their genotypes at the other three pro loci. The following results were obtained:

Donor genotype	Transductant genotype	Number
$proC^+ proA^- proB^+ proD^+$	$proC^+ proA^+ proB^+ proD^+$	2765
	$proC^+ proA^- proB^+ proD^+$	3
$proC^+ proA^+ proB^- proD^+$	$proC^+ proA^+ proB^+ proD^+$	1838
	$proC^+ proA^+ proB^- proD^+$	2
$proC^+ proA^+ proB^+ proD^-$	$proC^+ proA^+ proB^+ proD^+$	1166
	$proC^+ proA^+ proB^+ proD^-$	0

a. Why are there no *proC⁻* genotypes among the transductants?

b. Which genotypes represent single transductants and which represent cotransductants?

c. Is there evidence that *proA, proB,* and *proD* are located close to *proC*? Explain your answer.

*33. A geneticist isolates two mutations in a bacteriophage. One mutation causes clear plaques (*c*), and the other produces minute plaques (*m*). Previous mapping experiments have established that the genes responsible for these two mutations are 8 m.u. apart. The geneticist mixes phages with genotype $c^+ m^+$ and genotype $c^- m^-$ and uses the mixture to infect bacterial cells. She collects the progeny phages and cultures a sample of them on plated bacteria. A total of 1000 plaques are observed. What numbers of the different types of plaques ($c^+ m^+, c^- m^-, c^+ m^-, c^- m^+$) should she expect to see?

34. The geneticist carries out the same experiment described in Problem 33, but this time she mixes phages with genotypes $c^+ m^-$ and $c^- m^+$. What results are expected from *this* cross?

*35. A geneticist isolates two bacteriophage r mutants (r_{13} and r_2) that cause rapid lysis. He carries out the following crosses and counts the number of plaques listed here:

Genotype of parental phage	Progeny	Number of plaques
$h^+ r_{13}^- \times h^- r_{13}^+$	$h^+ r_{13}^+$	1
	$h^- r_{13}^+$	104
	$h^+ r_{13}^-$	110
	$h^- r_{13}^-$	2
Total		216

	$h^+ r_2^+$	6
$h^+ r_2^- \times h^- r_2^+$	$h^- r_2^+$	86
	$h^+ r_2^-$	81
	$h^- r_2^-$	7
Total		180

a. Calculate the recombination frequencies between r_2 and *h* and between r_{13} and *h*.

b. Draw all possible linkage maps for these three genes.

*36. *E. coli* cells are simultaneously infected with two strains of phage λ. One strain has a mutant host range, is temperature sensitive, and produces clear plaques (genotype is *h st c*); another strain carries the wild-type alleles (genotype is $h^+ st^+ c^+$). Progeny phages are collected from the lysed cells and are plated on bacteria. The numbers of different progeny phages are given here:

Progeny phage genotype	Number of plaques
$h^+ c^+ st^+$	321
$h \ c \ st$	338
$h^+ c \ st$	26
$h \ c^+ st^+$	30
$h^+ c \ st^+$	106
$h \ c^+ st$	110
$h^+ c^+ st$	5
$h \ c \ st^+$	6

a. Determine the order of the three genes on the phage chromosome.

b. Determine the map distances between the genes.

c. Determine the coefficient of coincidence and the interference (see p. 186 in Chapter 7).

37. A donor strain of bacteria with genes $a^+ b^+ c^+$ is infected with phages to map the donor chromosome with generalized transduction. The phage lysate from the bacterial cells is collected and used to infect a second strain of bacteria that are $a^- b^- c^-$. Bacteria with the a^+ gene are selected, and the percentage of cells with cotransduced b^+ and c^+ genes are recorded.

Donor	Recipient	Selected gene	Cells with cotransduced gene (%)
$a^+ b^+ c^+$	$a^- b^- c^-$	a^+	25 b^+
		a^+	3 c^+

Is gene *b* or gene *c* closer to gene *a*? Explain your reasoning.

38. A donor strain of bacteria with genotype *leu⁺ gal⁻ pro⁺* is infected with phages. The phage lysate from the bacterial cells is collected and used to infect a

second strain of bacteria that are *leu⁻ gal⁺ pro⁻*. The second strain is selected for *leu⁺*, and the following cotransduction data are obtained:

Donor	Recipient	Selected gene	Cells with cotransduced gene (%)
leu⁺ gal⁻ pro⁺	*leu⁻ gal⁺ pro⁻*	*leu⁺*	47 *pro⁺*
		leu⁺	26 *gal⁻*

Which genes are closest, *leu* and *gal* or *leu* and *pro*?

39. A geneticist isolates two new mutations, called *rII*$_x$ and *rII*$_y$, from the *rII* region of bacteriophage T4. *E. coli* B cells are simultaneously infected with phages carrying the *rII*$_x$ mutation and with phages carrying the *rII*$_y$ mutation. After the cells have lysed, samples of the phage lysate are collected. One sample is grown on *E. coli* K cells; a second sample is grown on *E. coli* B cells. There are 8322 plaques on *E. coli* B and 3 plaques on *E. coli* K. What is the recombination frequency between these two mutations?

*40. A geneticist is working with a new bacteriophage called phage Y3 that infects *E. coli*. He has isolated eight mutant phages that fail to produce plaques when grown on *E.*

coli strain K. To determine whether these mutations occur at the same functional gene, he simultaneously infects *E. coli* K cells with paired combinations of the mutants and looks to see whether plaques are formed. He obtains the following results. (A plus sign means that plaques were formed on *E. coli* K; a minus sign means that no plaques were formed on *E. coli* K.)

Mutant	1	2	3	4	5	6	7	8
1								
2	+							
3	+	+						
4	+	−	+					
5	−	+	+	+				
6	−	+	+	+	−			
7	+	−	+	−	+	+		
8	−	+	+	+	−	−	+	

a. To how many functional genes (cistrons) do these mutations belong?

b. Which mutations belong to the same functional gene?

41. For the H1N1 influenza virus shown at the bottom of **Figure 9.33**, viruses from which organism contributed the most RNA to the H1N1 virus?

CHALLENGE QUESTIONS

Section 9.2

42. As a summer project, a microbiology student independently isolates two mutations in *E. coli* that are auxotrophic for glycine (*gly⁻*). The student wants to know whether these two mutants are at the same functional unit. Outline a procedure that the student could use to determine whether these two *gly⁻* mutations occur within the same functional unit.

43. A group of genetics students mix two auxotrophic strains of bacteria: one is *leu⁺ trp⁺ his⁻ met⁻* and the other is *leu⁻ trp⁻ his⁺ met⁺*. After mixing the two strains, they plate the bacteria on minimal medium and observe a few prototrophic colonies

(*leu⁺ trp⁺ his⁺ met⁺*). They assume that some gene transfer has taken place between the two strains. How can they determine whether the transfer of genes is due to conjugation, transduction, or transformation?

Go to your ▶LaunchPad to find additional learning resources and the Suggested Readings for this chapter.

DNA: The Chemical Nature of the Gene

Greenland, one of Earth's most extreme environments, was originally settled by the Saqqaq people. The genome from a 4000-year-old male Saqqaq was sequenced in 2010: the remarkable stability of DNA makes analysis of genomes from ancient remains possible. [Alex Hibbert/age fotostock.]

Arctic Treks and Ancient DNA

Greenland is the world's largest island, consisting of over 830,000 square miles (2,200,000 square kilometers), but the vast majority of the land is permanently buried under hundreds of feet of ice. It is one of the Earth's most extreme environments. Temperatures along the coast rise a few degrees above freezing during summer days, but then drop to far below zero during much of the winter. With limited daylight (the Sun moves above the horizon for only a few hours on winter days), extreme cold, and winds reaching hurricane force, Greenland has a dangerously inhospitable environment.

Yet, in spite of the severe conditions, arctic peoples have continuously occupied Greenland for almost 5000 years. The earliest inhabitants were the Saqqaq people, who occupied small settlements on Greenland's coast from around 4800 to 2500 years ago. The Saqqaq lived in small tents and hunted marine mammals and seabirds. The origin of the Saqqaq people had long been a mystery. Did they descend from Native Americans, who migrated from Asia into the New World and later moved to Greenland? Or did they descend from the same group that gave rise to the Inuit people, who currently inhabit the New World Arctic? Or did they perhaps originate from yet another group that migrated independently from Asia to Greenland after the ancestors of both the Inuit and Native Americans entered the New World?

The mystery of the Saqqaq origin was solved in 2010, when geneticists determined the entire DNA sequence of a 4000 year-old Saqqaq male—nicknamed Inuk—whose remains were recovered from an archeological site on the western coast of Greenland. Scientists extracted DNA from four hair tufts found in the permafrost. Despite the great age of the sample, scientists were able to successfully determine Inuk's entire genome sequence, consisting of over 3 billion base pairs of DNA.

By comparing this DNA to sequences from known populations, scientists were able to demonstrate that the Saqqaq are most closely related to the Chukchis, a present-day group of indigenous people from Russia. This finding indicates that the Saqqaq originated from hunters who trekked from Siberia eastward across Alaska and Canada to Greenland, arriving in the New World independently of others who gave rise to Native Americans and the Inuit. Further analysis of Inuk's DNA revealed that he was dark-skinned, brown-eyed, blood-type A+, and probably was going bald.

DNA, with its double-stranded spiral, is among the most elegant of all biological molecules, but the double helix is not just a beautiful structure; it also gives DNA incredible stability and permanence, as evidenced by the sequencing of 4000-year-old DNA. In an even more remarkable feat, geneticists in 2009 sequenced the entire Neanderthal genome from DNA extracted from 38,000-year-old Neanderthal bones.

This chapter focuses on how DNA was identified as the source of genetic information and how it encodes the genetic instructions. We begin by considering the basic requirements of the genetic material and the history of the study of DNA—how its relation to genes was uncovered and its structure determined. The history of DNA illustrates several important points about the nature of scientific research. As with so many important scientific advances, the structure of DNA and its role as the genetic material were not discovered by any single person but were gradually revealed over a period of almost 100 years, thanks to the work of many investigators. Our understanding of the relation between DNA and genes was enormously enhanced in 1953, when James Watson and Francis Crick, analyzing data provided by Rosalind Franklin and Maurice Wilkins, proposed a three-dimensional structure for DNA that brilliantly illuminated its role in genetics.

After reviewing the discoveries that led to our current understanding of DNA, we will examine DNA structure. The structure of DNA is important in its own right, but the key genetic concept is the relation between the structure and the function of DNA—how its structure allows it to serve as the genetic material.

10.1 Genetic Material Possesses Several Key Characteristics

Life is characterized by tremendous diversity, but the coding instructions of all living organisms are written in the same genetic language—that of nucleic acids. Surprisingly, the idea that genes are made of nucleic acids was not widely accepted until after 1950. This skepticism was due in part to a lack of knowledge about the structure of deoxyribonucleic acid (DNA). Until the structure of DNA was understood, how DNA could store and transmit genetic information was unclear. Even before nucleic acids were identified as the genetic material, biologists recognized that, whatever the nature of the genetic material, it must possess four important characteristics.

1. **Genetic material must contain complex information.** First and foremost, the genetic material must be capable of storing large amounts of information—instructions for the traits and functions of an organism.

2. **Genetic material must replicate faithfully.** A second necessary feature is that genetic material must have the capacity to be copied accurately. Every organism begins life as a single cell. To produce a complex multicellular organism like yourself, this single cell must undergo billions of cell divisions. At each cell division, the genetic instructions must be transmitted to descendant cells with

great accuracy. When organisms reproduce and pass genes on to their progeny, the coding instructions must be copied with fidelity.

3. **Genetic material must encode the phenotype.** The genetic material (the genotype) must have the capacity to be expressed—to code for traits (the phenotype). The product of a gene is often a protein or an RNA molecule, so there must be a mechanism for genetic instructions in the DNA to be copied into RNAs and proteins.

4. **Genetic material must have the capacity to vary.** The genetic information must have the ability to vary, because different species and even individual members of a species differ in their genetic makeup.

CONCEPTS

The genetic material must be capable of carrying large amounts of information, replicating faithfully, expressing its coding instructions as phenotypes, and having the capacity to vary.

✔ CONCEPT CHECK 1

Why was the discovery of the structure of DNA so important for understanding genetics?

10.2 All Genetic Information Is Encoded in the Structure of DNA or RNA

Although our understanding of how DNA encodes genetic information is relatively recent, the study of DNA structure stretches back more than 100 years (**Figure 10.1**).

Early Studies of DNA

In 1868, Johann Friedrich Miescher graduated from medical school in Switzerland. Influenced by an uncle who believed that the key to understanding disease lay in the chemistry of tissues, Miescher traveled to Tubingen, Germany, to study under Ernst Felix Hoppe-Seyler, an early leader in the emerging field of biochemistry. Under Hoppe-Seyler's direction, Miescher turned his attention to the chemistry of pus, a substance of clear medical importance. Pus contains white blood cells with large nuclei; Miescher developed a method for isolating these nuclei. The minute amounts of nuclear material that he obtained were insufficient for a thorough chemical analysis, but he did establish that the nuclear material contained a novel substance that was slightly acidic and high in phosphorus. This material, which consisted of DNA

| 1833 Brown describes nucleus of the cell | 1869 Miescher discovers nuclein (DNA) in the nuclei of white blood cells | 1884 Histones isolated from nucleus | 1900 Mendel's work rediscovered | 1910 Levene proposes tetranucleotide theory | 1928 Griffith demonstrates transforming principle | 1947 Ashbury begins X-ray diffraction studies of DNA | 1952 Hershey and Chase demonstrate that DNA is genetic material in bacteriophage | 1953 Watson and Crick devise the secondary structure for DNA |

1830   1840   1850   1860   1870   1880   1890   1900   1910   1920   1930   1940   1950   1960

| 1839 Schleiden and Schwann propose cell theory | 1866 Mendel's work is first published | 1887 Recognition that nucleus is the physical basis of heredity | Late 1800's Kossel determines that DNA contains nitrogenous bases | 1944 Avery, MacLeod, and McCarty demonstrate that the transforming principle is DNA | 1948 Chargaff and colleagues discover regularity in base ratios of DNA | 1956 Fraenkel-Conrat and Singer show that some viruses use RNA as genetic material |

10.1 Many people have contributed to our understanding of the structure of DNA.

and protein, Miescher called *nuclein*. The substance was later renamed *nucleic acid* by one of his students.

By 1887, several researchers independently concluded that the physical basis of heredity lies in the nucleus. Chromatin was shown to consist of nucleic acid and proteins, but which of these substances was actually the genetic information was not clear. In the late 1800s, further work on the chemistry of DNA was carried out by Albrecht Kossel, who determined that DNA contains four nitrogenous bases: adenine, cytosine, guanine, and thymine (abbreviated A, C, G, and T).

In the early twentieth century, the Rockefeller Institute in New York City became a center for nucleic acid research. Phoebus Aaron Levene joined the Institute in 1905 and spent the next 40 years studying the chemistry of DNA. He discovered that DNA consists of a large number of linked, repeating units, called **nucleotides**; each nucleotide contains a sugar, a phosphate, and a base.

Nucleotide

Levene incorrectly proposed that DNA consists of a series of four-nucleotide units, each unit containing all four bases—adenine, guanine, cytosine, and thymine—in a fixed sequence. This concept, known as the tetranucleotide hypothesis, implied that the structure of DNA is not variable enough to be the genetic material. The tetranucleotide hypothesis contributed to the idea that protein is the genetic material because, with its 20 different amino acids, protein structure could be highly variable.

As additional studies of the chemistry of DNA were completed in the 1940s and 1950s, this notion of DNA as a simple, invariant molecule began to change. Erwin Chargaff and his colleagues carefully measured the amounts of the four bases in DNA from a variety of organisms and found that DNA from different organisms varies greatly in base composition. This finding disproved the tetranucleotide hypothesis.

They discovered that, within each species, there is some regularity in the ratios of the bases: the amount of adenine is always equal to the amount of thymine (A = T), and the amount of guanine is always equal to the amount of cytosine (G = C; **Table 10.1**). These findings became known as **Chargaff's rules**.

Table 10.1	**Base composition (percent*) of DNA from different sources and ratios of bases**						
						Ratio	
Source of DNA	**A**	**T**	**G**	**C**	**A/T**	**G/C**	**(A + G)/ (T + C)**
E. coli	26.0	23.9	24.9	25.2	1.09	0.99	1.04
Yeast	31.3	32.9	18.7	17.1	0.95	1.09	1.00
Sea urchin	32.8	32.1	17.7	18.4	1.02	0.96	1.00
Rat	28.6	28.4	21.4	21.5	1.01	1.00	1.00
Human	30.3	30.3	19.5	19.9	1.00	0.98	0.99

*Percent in moles of nitrogenous constituents per 100 g-atoms of phosphate in hydrolysate corrected for 100% recovery. From E. Chargaff and J. Davidson (eds). *The Nucleic Acids,* Vol 1. (New York: Academic Press, 1955)

CONCEPTS

Details of the structure of DNA were worked out by a number of scientists. At first, DNA was interpreted as being too regular in structure to carry genetic information but, by the 1940s, DNA from different organisms was shown to vary in its base composition.

✔ CONCEPT CHECK 2

Levene made which contribution to our understanding of DNA structure?
a. Determined that the nucleus contains DNA.
b. He determined that DNA contains four nitrogenous bases.
c. He determined that DNA consists of nucleotides.
d. He determined that the nucleotide bases of DNA are present in regular ratios.

DNA As the Source of Genetic Information

While chemists were working out the structure of DNA, biologists were attempting to identify the source of genetic information. Mendel identified the basic rules of heredity in 1866, but he had no idea about the physical nature of hereditary information. By the early 1900s, biologists had concluded that genes resided on chromosomes, which were known to contain both DNA and protein. Two sets of experiments, one conducted on bacteria and the other on viruses, provided pivotal evidence that DNA, rather than protein, was the genetic material.

THE DISCOVERY OF THE TRANSFORMING PRINCIPLE

The first clue that DNA was the carrier of hereditary information came with the demonstration that DNA was responsible for a phenomenon called *transformation*. This phenomenon was first observed in 1928 by Fred Griffith, an English physician whose special interest was the bacterium that causes pneumonia: *Streptococcus pneumoniae*. Griffith had succeeded in isolating several different strains of *S. pneumoniae* (type I, II, III, and so forth). In the virulent (disease-causing) forms of a strain, each bacterium is surrounded by a polysaccharide coat, which makes the bacterial colony appear smooth when grown on an agar plate; these forms are referred to as S, for smooth. Griffith found that these virulent forms occasionally mutated to nonvirulent forms, which lack a polysaccharide coat and produce a rough-appearing colony; these forms are referred to as R, for rough.

Griffith observed that small amounts of living type IIIS bacteria injected into mice caused the mice to develop pneumonia and die; when he examined the dead mice, he found large amounts of type IIIS bacteria in their blood (**Figure 10.2a**). When Griffith injected type IIR bacteria into mice, the mice lived, and no bacteria were recovered from their blood (**Figure 10.2b**). Griffith knew that boiling killed all the bacteria and destroyed their virulence; when he injected large amounts of heat-killed type IIIS bacteria into mice, the mice lived and no type IIIS bacteria were recovered from their blood (**Figure 10.2c**).

The results of these experiments were not unusual. However, Griffith got a surprise when he infected his mice with a small amount of living type IIR bacteria along with a large amount of heat-killed type IIIS bacteria. Because both the type IIR bacteria and the heat-killed type IIIS bacteria were nonvirulent, he expected these mice to live. Surprisingly, five days after the injections, the mice became infected with pneumonia and died (**Figure 10.2d**). When Griffith examined blood from the hearts of these mice, he observed live type IIIS bacteria. Furthermore, these bacteria retained their type IIIS characteristics through several generations; so the infectivity was heritable.

Griffith considered all of the possible interpretations of his results. First, it could have been the case that he had

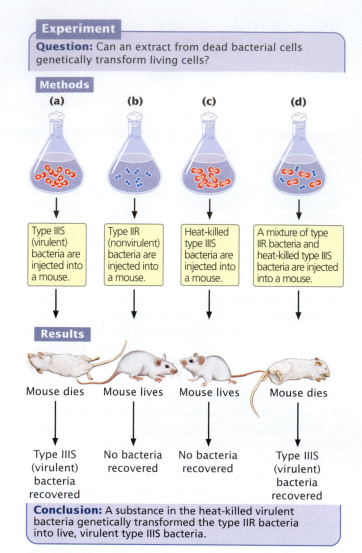

Experiment

Question: Can an extract from dead bacterial cells genetically transform living cells?

Methods

(a) Type IIIS (virulent) bacteria are injected into a mouse.

(b) Type IIR (nonvirulent) bacteria are injected into a mouse.

(c) Heat-killed type IIIS bacteria are injected into a mouse.

(d) A mixture of type IIR bacteria and heat-killed type IIIS bacteria are injected into a mouse.

Results

Mouse dies — Type IIIS (virulent) bacteria recovered

Mouse lives — No bacteria recovered

Mouse lives — No bacteria recovered

Mouse dies — Type IIIS (virulent) bacteria recovered

Conclusion: A substance in the heat-killed virulent bacteria genetically transformed the type IIR bacteria into live, virulent type IIIS bacteria.

10.2 Griffith's experiments demonstrated transformation in bacteria.

not sufficiently sterilized the type IIIS bacteria and thus a few live bacteria remained in the culture. Any live bacteria injected into the mice would have multiplied and caused pneumonia. Griffith knew that this possibility was unlikely, because he had used only heat-killed type IIIS bacteria in the control experiment, and they never produced pneumonia in the mice.

A second interpretation was that the live, type IIR bacteria had mutated to the virulent S form. Such a mutation would cause pneumonia in the mice, but it would produce type IIS bacteria, not the type IIIS that Griffith found in the dead mice. Because type II and type III bacteria differ in a number of traits, many mutations would be required for type II bacteria to mutate to type III bacteria, and the chance of all the mutations occurring simultaneously was impossibly low.

Griffith concluded that the type IIR bacteria had somehow been transformed, acquiring the genetic virulence of the dead type IIIS bacteria. This transformation had produced a permanent, genetic change in the bacteria. Although Griffith didn't understand the nature of this transformation, he theorized that some substance in the polysaccharide coat of the dead bacteria might be responsible. He called this substance the **transforming principle**. ▶ TRY PROBLEM 19

IDENTIFICATION OF THE TRANSFORMING PRINCIPLE

At the time of Griffith's report, Oswald Avery (see Figure 10.1) was a microbiologist at the Rockefeller Institute. At first Avery was skeptical but, after other microbiologists successfully repeated Griffith's experiments with other bacteria, Avery set out to identify the nature of the transforming substance.

After 10 years of research, Avery, Colin MacLeod, and Maclyn McCarty succeeded in isolating and partially purifying the transforming substance. They showed that it had a chemical composition closely matching that of DNA and quite different from that of proteins. Enzymes such as trypsin and chymotrypsin, known to break down proteins, had no effect on the transforming substance. Ribonuclease, an enzyme that destroys RNA, also had no effect. Enzymes capable of destroying DNA, however, eliminated the biological activity of the transforming substance (**Figure 10.3**). Avery, MacLeod, and McCarty showed that the transforming substance precipitated at about the same rate as purified DNA and that it absorbed ultraviolet light at the same wavelengths as DNA. These results, published in 1944, provided compelling evidence that the transforming principle—and therefore genetic information—resides in DNA. However, new theories in science are rarely accepted on the basis of a single experiment, and many biologists continued to prefer the hypothesis that the genetic material is protein.

CONCEPTS

The process of transformation indicates that some substance—the transforming principle—is capable of genetically altering bacteria. Avery, MacLeod, and McCarty demonstrated that the transforming principle is DNA, providing the first evidence that DNA is the genetic material.

✓ CONCEPT CHECK 3

If Avery, MacLeod, and McCarty had found that samples of heat-killed bacteria treated with RNase and DNase transformed bacteria, but samples treated with protease did not, what conclusion would they have made?

a. Protease carries out transformation.
b. RNA and DNA are the genetic materials.
c. Protein is the genetic material.
d. RNase and DNase are necessary for transformation.

Experiment

Question: What is the chemical nature of the transforming substance?

Methods

Type IIIS (virulent) bacteria

1 Use heat to kill virulent bacteria, homogenize, and filter.

Type IIIS bacterial filtrate

2 Treat samples with enzymes that destroy proteins, RNA, or DNA.

RNase (destroys RNA)   Protease (destroys proteins)   DNase (destroys DNA)

3 Add the treated samples to cultures of type IIR bacteria.

Type IIR bacteria   Type IIR bacteria   Type IIR bacteria

Results

Type IIIS and type IIR bacteria   Type IIIS and type IIR bacteria   Type IIR bacteria

4 Cultures treated with protease or RNase contain transformed type IIIS bacteria,...

5 ...but the culture treated with DNase does not.

Conclusion: Because only DNase destroyed the transforming substance, the transforming substance is DNA.

10.3 Avery, MacLeod, and McCarty's experiment revealed the nature of the transforming principle.

THE HERSHEY–CHASE EXPERIMENT   A second piece of evidence that indicated DNA was the genetic material resulted from a study of the T2 virus conducted by Alfred

Hershey and Martha Chase. The T2 virus is a *bacterio-phage* (phage) that infects the bacterium *Escherichia coli* (**Figure 10.4a**). As stated in Chapter 9, a phage reproduces by attaching to the outer wall of a bacterial cell and in-jecting its DNA into the cell, where it replicates and di-rects the cell to synthesize phage protein. The phage DNA becomes encapsulated within the proteins, producing progeny phages that lyse (break open) the cell and escape (**Figure 10.4b**).

At the time of the Hershey–Chase study (their paper was published in 1952), biologists did not understand exactly how phages reproduce. What they did know was that the T2 phage is approximately 50% protein and 50% DNA, that a phage infects a cell by first attaching to the cell wall, and that progeny phages are ultimately produced within the cell. Because the progeny carry the same traits as the infecting phage, genetic material from the infecting phage must be transmitted to the progeny, but how this genetic transmis-sion takes place was unknown.

Hershey and Chase designed a series of experiments to determine whether the phage *protein* or the phage *DNA* is transmitted in phage reproduction. To follow the fate of protein and DNA, they used radioactive forms, or **iso-topes**, of phosphorus and sulfur. A radioactive isotope can be used as a tracer to identify the location of a specific mol-ecule because any molecule containing the isotope will be radioactive and therefore easily detected. DNA contains phosphorus but not sulfur, so Hershey and Chase used ^{32}P to follow phage DNA during reproduction. Protein con-tains sulfur but not phosphorus, so they used ^{35}S to follow the protein.

Hershey and Chase grew one batch of *E. coli* in a medi-um containing ^{32}P and infected the bacteria with T2 phage so that all the new phages would have DNA labeled with ^{32}P (**Figure 10.5**). They grew a second batch of *E. coli* in a medium containing ^{35}S and infected these bacteria with T2 phage so that all these new phages would have protein labeled with ^{35}S. Hershey and Chase then infected separate batches of unlabeled *E. coli* with the ^{35}S- and ^{32}P-labeled phages. After allowing time for the phages to infect the cells, they placed the *E. coli* cells in a blender and sheared off the now-empty protein coats from the cell walls. They separated out the protein coats and cultured the infected bacterial cells.

When phages labeled with ^{35}S infected the bacteria, most of the radioactivity was detected in the protein coats and little was detected in the cells. Furthermore, when new phag-es emerged from the cell, they contained almost no ^{35}S (see Figure 10.5). This result indicated that the protein compo-nent of a phage does not enter the cell and is not transmitted to progeny phages.

In contrast, when Hershey and Chase infected bacteria with ^{32}P-labeled phages and removed the protein coats, the bacteria were radioactive. Most significantly, after the cells lysed and new progeny phages emerged, many of these phages

(a)
Phage genome is DNA.

All other parts of the bacteriophage are protein.

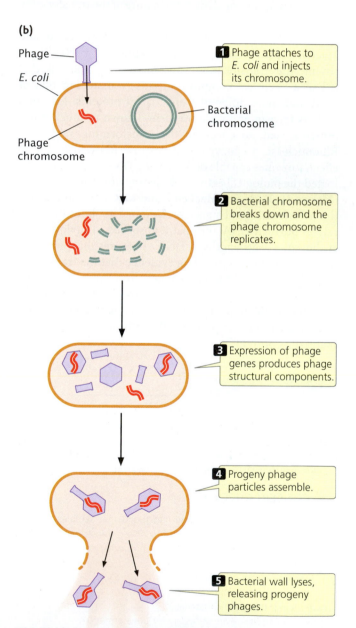

(b)

Phage

E. coli

Phage chromosome

Bacterial chromosome

1 Phage attaches to *E. coli* and injects its chromosome.

2 Bacterial chromosome breaks down and the phage chromosome replicates.

3 Expression of phage genes produces phage structural components.

4 Progeny phage particles assemble.

5 Bacterial wall lyses, releasing progeny phages.

10.4 T2 is a bacteriophage that infects *E. coli*. (a) T2 phage. (b) Its life cycle. [Micrograph: © Lee D. Simon/Photo Researchers.]

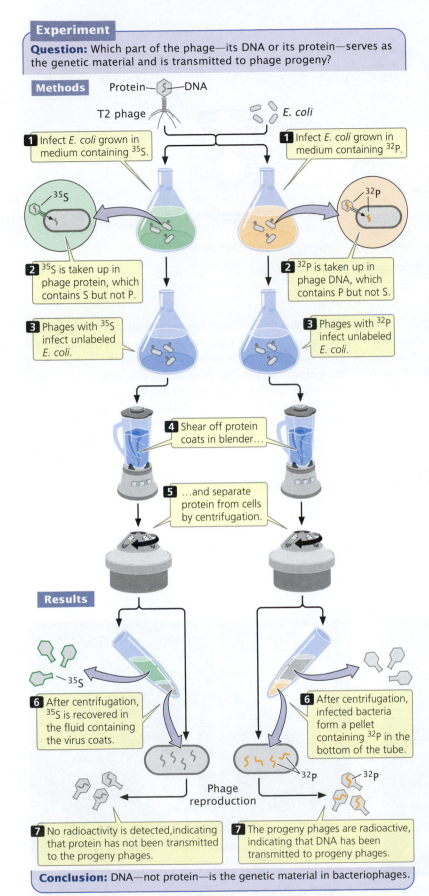

Experiment

Question: Which part of the phage—its DNA or its protein—serves as the genetic material and is transmitted to phage progeny?

Methods   Protein —— DNA

T2 phage   *E. coli*

1 Infect *E. coli* grown in medium containing ³⁵S.

1 Infect *E. coli* grown in medium containing ³²P.

³⁵S

³²P

2 ³⁵S is taken up in phage protein, which contains S but not P.

2 ³²P is taken up in phage DNA, which contains P but not S.

3 Phages with ³⁵S infect unlabeled *E. coli*.

3 Phages with ³²P infect unlabeled *E. coli*.

4 Shear off protein coats in blender…

5 …and separate protein from cells by centrifugation.

Results

6 After centrifugation, ³⁵S is recovered in the fluid containing the virus coats.

6 After centrifugation, infected bacteria form a pellet containing ³²P in the bottom of the tube.

³⁵S

³²P

Phage reproduction

³²P

7 No radioactivity is detected, indicating that protein has not been transmitted to the progeny phages.

7 The progeny phages are radioactive, indicating that DNA has been transmitted to progeny phages.

Conclusion: DNA—not protein—is the genetic material in bacteriophages.

10.5 Hershey and Chase demonstrated that DNA carries the genetic information in bacteriophages.

emitted radioactivity from ³²P, demonstrating that DNA from the infecting phages had been passed on to the progeny (see Figure 10.5). These results confirmed that DNA, not protein, is the genetic material of phages. ▶ **TRY PROBLEM 24**

CONCEPTS

Using radioactive isotopes, Hershey and Chase traced the movement of DNA and protein during phage infection. They demonstrated that DNA, not protein, enters the bacterial cell during phage reproduction and that only DNA is passed on to progeny phages.

✓ **CONCEPT CHECK 4**

Could Hershey and Chase have used a radioactive isotope of carbon instead of ³²P? Why or why not?

Watson and Crick's Discovery of the Three-Dimensional Structure of DNA

The experiments on the nature of the genetic material set the stage for one of the most important advances in the history of biology—the discovery of the three-dimensional structure of DNA by James Watson and Francis Crick in 1953.

Before Watson and Crick's breakthrough, much of the basic chemistry of DNA had already been determined by Miescher, Kossel, Levene, Chargaff, and others, who had established that DNA consists of nucleotides and that each nucleotide contains a sugar, a base, and a phosphate group. However, how the nucleotides fit together in the three-dimensional structure of the molecule was not at all clear.

In 1947, William Astbury began studying the three-dimensional structure of DNA by using a technique called **X-ray diffraction** (**Figure 10.6**), in which X-rays beamed at a molecule are reflected in specific patterns that reveal aspects of the structure of the molecule. However, his diffraction images did not provide enough resolution to reveal the structure. A research group at King's College in London, led by Maurice Wilkins, also used X-ray diffraction to study DNA. Working in Wilkins' laboratory, Rosalind Franklin obtained strikingly better images of the molecule. However, Wilkins and Franklin's progress in developing a complete structure of the molecule was impeded by the personal discord that existed between them.

Watson and Crick investigated the structure of DNA, not by collecting new data but by using all available information about the chemistry of

1 Crystals of a substance are bombarded with X-rays, which are diffracted (bounce off).

2 The spacing of the atoms within the crystal determines the diffraction pattern, which appears as spots on a photographic film.

3 The diffraction pattern provides information about the structure of the molecule.

X-ray source   Lead screen   Beam of X-rays   Detector (photographic plate)   Diffraction pattern

10.6 X-ray diffraction provides information about the structures of molecules. [Science Source.]

DNA to construct molecular models (**Figure 10.7a**). They used the excellent X-ray diffraction images taken by Rosalind Franklin (**Figure 10.7b**) and by applying the laws of structural chemistry they were able to limit the number of possible structures that DNA could assume. They tested various structures by building models made of wire and metal plates. With their models, they were able to see whether a structure was compatible with chemical principles and with the X-ray images.

The key to solving the structure came when Watson recognized that an adenine base could bond with a thymine base and that a guanine base could bond with a cytosine base; these pairings accounted for the base ratios that Chargaff had discovered earlier. The model developed by Watson and Crick showed that DNA consists of two strands of nucleotides that run in opposite directions (are antiparallel) and wind around each other to form a right-handed helix, with the sugars and phosphates on the outside and the bases in the interior. They recognized that the double-stranded structure of DNA with its specific base pairing provided an elegant means by which the DNA can be replicated. Watson and Crick published an electrifying description of their model in *Nature* in 1953. At the same time, Wilkins and Franklin each published their X-ray diffraction data, which demonstrated experimentally the hypothesis that DNA was helical in structure.

Many have called the solving of DNA's structure the most important biological discovery of the twentieth century. For their discovery, Watson and Crick, along with Maurice Wilkins, were awarded a Nobel Prize in 1962. Rosalind Franklin had died of cancer in 1958 and thus could not be

(a)

(b)

10.7 James Watson and Francis Crick (a) developed a three-dimensional model of the structure of DNA based in part on X-ray diffraction photographs taken by Rosalind Franklin (b). [Part a: A. Barrington Brown/Science Photo Library/Photo Researchers. Part b: Science Source/Photo Researchers.]

considered a candidate for the shared prize, but many scholars and historians believe that she should share equal credit for solving the structure of DNA.

After the discovery of DNA's structure, much research focused on how genetic information is encoded within the base sequence and how this information is copied and expressed. Even today, the details of DNA structure and function continue to be the subject of active research.

CONCEPTS

By collecting existing information about the chemistry of DNA and building molecular models, Watson and Crick were able to discover the three-dimensional structure of the DNA molecule.

✓ CONCEPT CHECK 5

What did Watson and Crick use to help solve the structure of DNA?
a. X-ray diffraction.
b. Laws of structural chemistry
c. Models of DNA
d. All the above

RNA As Genetic Material

In most organisms, DNA carries the genetic information. However, a few viruses use RNA, not DNA, as their genetic material. This was demonstrated in 1956 by Heinz Fraenkel-Conrat and Bea Singer, who worked with the tobacco mosaic virus (TMV), a virus that infects and causes disease in tobacco plants (**Figure 10.8**). The tobacco mosaic virus possesses a single molecule of RNA surrounded by a helically arranged cylinder of protein molecules. Fraenkel-Conrat found that, after separating the RNA and protein of TMV, he could remix the RNA and protein of different strains of TMV and obtain intact, infectious viral particles.

With Singer, Fraenkel-Conrat then created hybrid viruses by mixing RNA and protein from different strains of TMV. When these hybrid viruses infected tobacco leaves, new viral particles were produced. The new viral progeny were identical with the strain from which the RNA had been isolated and did not exhibit the characteristics of the strain that donated the protein. These results showed that RNA carries the genetic information in TMV.

Also in 1956, Alfred Gierer and Gerhard Schramm demonstrated that just RNA isolated from TMV is sufficient to infect tobacco plants and direct the production of new TMV particles. This finding confirmed that RNA carries genetic instructions. ▶ TRY PROBLEM 18

CONCEPTS

RNA serves as the genetic material in some viruses.

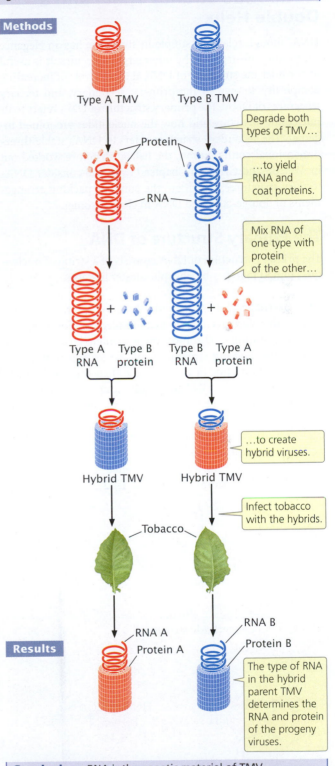

Experiment

Question: What substance—RNA or protein—carries the genetic material in tobacco mosaic virus (TMV)?

Methods

Type A TMV   Type B TMV

Degrade both types of TMV…

Protein

…to yield RNA and coat proteins.

RNA

Mix RNA of one type with protein of the other…

Type A RNA   Type B protein   Type B RNA   Type A protein

…to create hybrid viruses.

Hybrid TMV   Hybrid TMV

Infect tobacco with the hybrids.

Tobacco

Results

RNA A   RNA B
Protein A   Protein B

The type of RNA in the hybrid parent TMV determines the RNA and protein of the progeny viruses.

Conclusion: RNA is the genetic material of TMV.

10.8 Fraenkel-Conrat and Singer's experiment demonstrated that RNA in the tobacco mosaic virus carries the genetic information.

10.3 DNA Consists of Two Complementary and Antiparallel Nucleotide Strands That Form a Double Helix

DNA, though relatively simple in structure, has an elegance and beauty unsurpassed by other large molecules. It is useful to consider the structure of DNA at three levels of increasing complexity, known as the primary, secondary, and tertiary structures of DNA. The primary structure of DNA refers to its nucleotide structure and how the nucleotides are joined together. The secondary structure refers to DNA's stable three-dimensional configuration, the helical structure worked out by Watson and Crick. In Chapter 11, we will consider DNA's tertiary structures, which are the complex packing arrangements of double-stranded DNA in chromosomes.

The Primary Structure of DNA

The primary structure of DNA consists of a string of nucleotides joined together by phosphodiester linkages.

NUCLEOTIDES DNA is typically a very long molecule and is therefore termed a macromolecule. For example, within each human chromosome is a single DNA molecule that, if stretched out straight, would be several centimeters in length, thousands of times longer than the cell itself. In spite of its large size, DNA has a quite simple structure: it is a polymer—that is, a chain made up of many repeating units linked together. The repeating units of DNA are *nucleotides*, each comprised of three parts: (1) a sugar, (2) a phosphate, and (3) a nitrogen-containing base.

The sugars of nucleic acids—called pentose sugars—have five carbon atoms, numbered 1′, 2′, 3′, and so forth

Ribose   Deoxyribose

10.9 A nucleotide contains either a ribose sugar (in RNA) or a deoxyribose sugar (in DNA). The carbon atoms are assigned primed numbers.

(**Figure 10.9**). The sugars of DNA and RNA are slightly different in structure. RNA's sugar, called **ribose**, has a hydroxyl group (—OH) attached to the 2′-carbon atom, whereas DNA's sugar, or **deoxyribose**, has a hydrogen atom (—H) at this position and therefore contains one oxygen atom fewer overall. This difference gives rise to the names ribonucleic acid (RNA) and *deoxy*ribonucleic acid (DNA). This minor chemical difference is recognized by most of the cellular enzymes that interact with DNA or RNA, thus yielding specific functions for each nucleic acid. Furthermore, the additional oxygen atom in the RNA nucleotide makes it more reactive and less chemically stable than DNA. For this reason, DNA is better suited to serve as the long-term repository of genetic information.

The second component of a nucleotide is its **nitrogenous base**, which may be of two types—a **purine** or a **pyrimidine** (**Figure 10.10**). Each purine consists of a six-member ring attached to a five-member ring, whereas each pyrimidine consists of a six-member ring only. Both DNA and RNA contain two purines, **adenine** and **guanine** (A and G), which differ in the positions of their double bonds and in the

Purine
(basic structure)

Pyrimidine
(basic structure)

Adenine (A)   Guanine (G)   Cytosine (C)   Thymine (T)
(present in DNA)   Uracil (U)
(present in RNA)

10.10 A nucleotide contains either a purine or a pyrimidine base. The atoms of the rings in the bases are assigned unprimed numbers.

Phosphate

10.11 A nucleotide contains a phosphate group.

groups attached to the six-member ring. Three pyrimidines are common in nucleic acids: **cytosine** (C), **thymine** (T), and **uracil** (U). Cytosine is present in both DNA and RNA; however, thymine is restricted to DNA, and uracil is found only in RNA. The three pyrimidines differ in the groups or atoms attached to the carbon atoms of the ring and in the number of double bonds in the ring. In a nucleotide, the nitrogenous base always forms a covalent bond with the $1'$-carbon atom of the sugar (see Figure 10.9). A deoxyribose or a ribose sugar and a base together are referred to as a **nucleoside**.

The third component of a nucleotide is the **phosphate group**, which consists of a phosphorus atom bonded to four oxygen atoms (**Figure 10.11**). Phosphate groups are found in every nucleotide and frequently carry a negative charge, which makes DNA acidic. The phosphate group is always bonded to the $5'$-carbon atom of the sugar (see Figure 10.9) in a nucleotide.

The DNA nucleotides are properly known as **deoxyribonucleotides** or deoxyribonucleoside $5'$-monophosphates. Because there are four types of bases, there are four different kinds of DNA nucleotides (**Figure 10.12**). The equivalent RNA nucleotides are termed **ribonucleotides** or ribonucleoside $5'$-monophosphates. RNA molecules sometimes contain additional rare bases, which are modified forms of the four common bases. These modified bases will be discussed in more detail when we examine the function of RNA molecules in Chapter 14. The names for DNA bases, nucleotides, and nucleosides are shown in **Table 10.2**. ▶ TRY PROBLEM 26

CONCEPTS

The primary structure of DNA consists of a string of nucleotides. Each nucleotide consists of a five-carbon sugar, a phosphate, and a base. There are two types of DNA bases: purines (adenine and guanine) and pyrimidines (thymine and cytosine).

Deoxyadenosine 5′-monophosphate (dAMP)

Deoxyguanosine 5′-monophosphate (dGMP)

Deoxythymidine 5′-monophosphate (dTMP)

Deoxycytidine 5′-monophosphate (dCMP)

10.12 There are four types of DNA nucleotides.

✔ **CONCEPT CHECK 6**

How do the sugars of RNA and DNA differ?
a. RNA has a six-carbon sugar; DNA has a five-carbon sugar.
b. The sugar of RNA has a hydroxyl group that is not found in the sugar of DNA.
c. RNA contains uracil; DNA contains thymine.
d. DNA's sugar has a phosphorus atom; RNA's sugar does not.

Table 10.2	Names of DNA Bases, Nucleotides and Nucleosides			
	Adenine	**Guanine**	**Thymine**	**Cytosine**
Base symbol	A	G	T	C
Nucleotide	deoxyadenosine 5′ monophosphate	deoxyguanosine 5′ monophosphate	deoxythymidine 5′ monophosphate	deoxycytidine 5′ monophosphate
Nucleotide symbol	dAMP	dGMP	dTMP	dCMP
Nucleoside	deoxyadenosine	deoxyguanosine	deoxythymidine	deoxycytidine
Nucleoside symbol	dA	dG	dT	dC

POLYNUCLEOTIDE STRANDS DNA is made up of many nucleotides connected by covalent bonds, which join the 5′-phosphate group of one nucleotide to the 3′-hydroxyl of the next nucleotide (**Figure 10.13**); note that the structures shown in Figure 10.13 are flattened into two dimensions while the molecule itself is three dimensional, as shown in Figure 10.14). These bonds, called **phosphodiester linkages**, are strong covalent bonds; a series of nucleotides linked in this way constitutes a **polynucleotide strand**. The backbone of the polynucleotide strand is composed of alternating sugars and phosphates; the bases project away from the long axis of the strand. The negative charges of the phosphate groups are frequently neutralized by the association of positive charges on proteins, metals, or other molecules.

An important characteristic of the polynucleotide strand is its direction, or polarity. At one end of the strand, a free (meaning that it's unattached on one side) phosphate group is attached to the 5′-carbon atom of the sugar in the nucleotide.

This end of the strand is therefore referred to as the **5′ end**. The other end of the strand, referred to as the **3′ end**, has a free OH group attached to the 3′-carbon atom of the sugar.

RNA nucleotides also are connected by phosphodiester linkages to form similar 5′ to 3′ polynucleotide strands (see Figure 10.13).

> ### CONCEPTS
>
> The nucleotides of DNA are joined in polynucleotide strands by phosphodiester bonds that connect the 3′-carbon atom of one nucleotide to the 5′-phosphate group of the next. Each polynucleotide strand has polarity, with a 5′ end and a 3′ end.

Secondary Structures of DNA

The secondary structure of DNA refers to its three-dimensional configuration—its fundamental helical structure.

DNA polynucleotide strands

T–A pairs have two hydrogen bonds.

5′-to-3′ direction

5′-to-3′ direction

C–G pairs have three hydrogen bonds.

DNA has deoxyribose sugar (no oxygen here).

The strands run in opposite directions; they are antiparallel.

A phosphodiester linkage connects the 5′-phosphate group and the 3′-OH group of adjoining nucleotides.

RNA polynucleotide strand

In RNA, uracil (U) replaces thymine (T).

RNA has ribose sugar (an OH group here).

5′-to-3′ direction

10.13 DNA and RNA are composed of polynucleotide strands. DNA is usually composed of two polynucleotide strands, although single-stranded DNA is found in some viruses.

DNA's secondary structure can assume a variety of configurations, depending on its base sequence and the conditions in which it is placed.

THE DOUBLE HELIX   A fundamental characteristic of DNA's secondary structure is that it consists of two polynucleotide strands wound around each other—it's a *double* helix. The sugar–phosphate linkages are on the outside of the helix, and the bases are stacked in the interior of the molecule (see Figure 10.13). The two polynucleotide strands run in opposite directions—they are **antiparallel**, which means that the 5′ end of one strand is opposite the 3′ end of the other strand.

The strands are held together by two types of molecular forces. Hydrogen bonds link the bases on opposite strands (see Figure 10.13). These bonds are relatively weak compared with the covalent phosphodiester bonds that connect the sugar and phosphate groups of adjoining nucleotides on the same strand. As we will see, several important functions of DNA require the separation of its two nucleotide strands, and this separation can be readily accomplished because of the relative ease of breaking and reestablishing the hydrogen bonds.

The nature of the hydrogen bond imposes a limitation on the types of bases that can pair. Adenine normally pairs only with thymine through two hydrogen bonds, and cytosine normally pairs only with guanine through three hydrogen bonds (see Figure 10.13). Because three hydrogen bonds form between C and G and only two hydrogen bonds form between A and T, C–G pairing is stronger than A–T pairing. The specificity of the base pairing means that, wherever there is an A on one strand, there must be a T in the corresponding position on the other strand, and, wherever there is a G on one strand, a C must be on the other. The two polynucleotide strands of a DNA molecule are therefore not identical but are **complementary DNA strands**. The complementary nature of the two nucleotide strands provides for efficient and accurate DNA replication, as will be discussed in Chapter 12.

The second force that holds the two DNA strands together is the interaction between the stacked base pairs in the interior of the molecule. Stacking means that adjacent bases are aligned so that their rings are parallel and stack on top of one another. The stacking interactions stabilize the DNA molecule but do not require that any particular base follow another. Thus, the base sequence of the DNA molecule is free to vary, allowing DNA to carry genetic information.

▶ **TRY PROBLEMS 32 AND 37**

CONCEPTS

DNA consists of two polynucleotide strands. The sugar–phosphate groups of each polynucleotide strand are on the outside of the molecule, and the bases are in the interior. Hydrogen bonding joins the bases of the two strands: guanine pairs with cytosine, and adenine pairs with thymine. The two polynucleotide strands of a DNA molecule are complementary and antiparallel.

✔ **CONCEPT CHECK 7**

The antiparallel nature of DNA refers to
a. its charged phosphate groups.
b. the pairing of bases on one strand with bases on the other strand.
c. the formation of hydrogen bonds between bases from opposite strands.
d. the opposite direction of the two strands of nucleotides.

DIFFERENT SECONDARY STRUCTURES   As we have seen, DNA normally consists of two polynucleotide strands that are antiparallel and complementary (exceptions are single-stranded DNA molecules in a few viruses). The precise three-dimensional shape of the molecule can vary, however, depending on the conditions in which the DNA is placed and, in some cases, on the base sequence itself.

The three-dimensional structure of DNA described by Watson and Crick is termed the **B-DNA** structure (**Figure 10.14**). This structure exists when plenty of water surrounds the molecule and there is no unusual base sequence in the DNA—conditions that are likely to be present in cells. The B-DNA structure is the most stable configuration for a random sequence

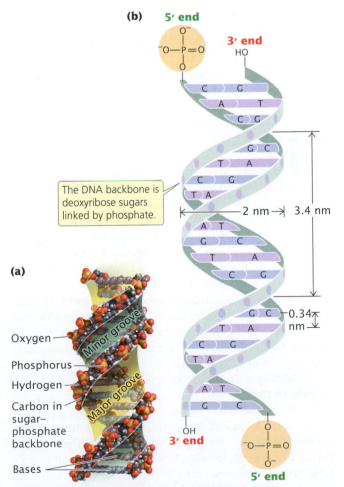

10.14 B-DNA consists of a right-handed helix with approximately 10 bases per turn. (a) Space-filling model of B-DNA showing major and minor grooves. (b) Diagrammatic representation.

Direction of helix

28 Å

(a) A form   (b) B form   (c) Z form

10.15 DNA can assume several different secondary structures.
[After J. M. Berg, J. L. Tymoczko, and L. Stryer, *Biochemistry*, 6th ed. (New York: W. H. Freeman and Company, 2002), pp. 785 and 787.]

of nucleotides under physiological conditions, and most evidence suggests that it is the predominant structure in the cell.

B-DNA is a right-handed helix, meaning that it has a clockwise spiral. There are approximately 10 base pairs (bp) per 360-degree rotation of the helix, so each base pair is twisted 36 degrees relative to the adjacent bases (see Figure 10.14b). The base pairs are 0.34 nanometer (nm) apart; so each complete rotation of the molecule encompasses 3.4 nm. The diameter of the helix is 2 nm, and the bases are perpendicular to the long axis of the DNA molecule. A space-filling model shows that B-DNA has a slim and elongated structure (see Figure 10.14a). The spiraling of the nucleotide strands creates major and minor grooves in the helix, features that are important for the binding of some proteins that regulate the expression of genetic information (see Chapter 16).

Another secondary structure that DNA can assume is the **A-DNA** structure, which exists if less water is present. Like B-DNA, A-DNA is a right-handed helix (**Figure 10.15a**), but it is shorter and wider than B-DNA (**Figure 10.15b**) and its bases are tilted away from the main axis of the molecule. A-DNA has been detected in some DNA-protein complexes and in spores of some bacteria.

A radically different secondary structure, called **Z-DNA** (**Figure 10.15c**), forms a left-handed helix. In this form, the sugar–phosphate backbone zigzags back and forth, giving rise to its name. A Z-DNA structure can result if the molecule contains particular base sequences, such as stretches of alternating C and G nucleotides. Researchers have found that Z-DNA-specific antibodies bind to regions of the DNA that are being transcribed into RNA, suggesting that Z-DNA may play some role in gene expression. Additional secondary structures

of DNA (C-DNA, D-DNA, etc.) can form under specialized laboratory conditions or in DNA with specific base sequences.

CONCEPTS

DNA can assume different secondary structures, depending on the conditions in which it is placed and on its base sequence. B-DNA is thought to be the most common configuration in the cell.

✓ **CONCEPT CHECK 8**

How does Z-DNA differ from B-DNA?

CONNECTING CONCEPTS

Genetic Implications of DNA Structure

Watson and Crick's great contribution was their elucidation of the genotype's chemical structure, making it possible for geneticists to begin to examine genes directly, instead of looking only at the phenotypic consequences of gene action. The determination of the structure of DNA led to the birth of molecular genetics—the study of the chemical and molecular nature of genetic information.

Watson and Crick's structure did more than just create the potential for molecular genetic studies; it was an immediate source of insight into key genetic processes. At the beginning of this chapter, four fundamental properties of the genetic material were identified. First, it must be capable of carrying large amounts of information. Watson and Crick's model suggested that genetic instructions are encoded in the base sequence, the only variable part of the molecule.

A second necessary property of genetic material is its ability to replicate faithfully. The complementary polynucleotide strands of DNA make this replication possible. Watson and Crick proposed that in replication, the two polynucleotide strands unzip, breaking the weak hydrogen bonds between the two strands, and each strand serves as a template on which a new strand is synthesized. The specificity of the base pairing means that only one possible sequence of bases—the complementary sequence—can be synthesized from each template. Newly replicated double-stranded DNA molecules will therefore be identical with the original double-stranded DNA molecule (see Chapter 12 on DNA replication).

A third essential property of genetic material is the ability to express its instructions into the phenotype. DNA expresses its genetic instructions by first transferring its information to an RNA molecule, in a process termed **transcription** (see Chapter 13). The term *transcription* is appropriate because, although the information is transferred from DNA to RNA, the information remains in the language of nucleic acids. In some cases, the RNA molecule then transfers the genetic information to a protein by specifying its amino acid sequence. This process is termed **translation** (see Chapter 15) because the information must be *translated* from the language of nucleotides into the language of amino acids. A fourth property of DNA is that it must be capable of varying. This variation consists of differences in the sequence of bases found among different individuals.

We can now identify three major pathways of information flow in the cell (**Figure 10.16a**): in **replication**, information passes from one DNA molecule to other DNA molecules; in transcription, information passes from DNA to RNA; and, in translation, information passes

(a) Major information pathways

DNA — DNA replication

Information is transferred from DNA to an RNA molecule. — Transcription

Information is transferred from one DNA molecule to another.

RNA

Information is transferred from RNA to a protein through a code that specifies the amino acid sequence. — Translation

PROTEIN

10.16 Pathways of information transfer within the cell.

(b) Special information pathways

DNA

Reverse transcription

In some viruses, information is transferred from RNA to DNA …

RNA — RNA replication

…or to another RNA molecule.

PROTEIN

from RNA to protein. This concept of information flow was formalized by Francis Crick in a concept that he called the **central dogma** of molecular biology. The central dogma states that genetic information passes from DNA to protein in a one-way information pathway. We now realize, however, that the central dogma is an oversimplification. In addition to the three general information pathways of replication, transcription, and translation, other transfers may take place in certain organisms or under special circumstances. Retroviruses (see Chapter 9) and some transposable elements (see Chapter 18) transfer information from RNA to DNA (in **reverse transcription**), and some RNA viruses transfer information from RNA to RNA (in **RNA replication; Figure 10.16b**).

10.4 Special Structures Can Form in DNA and RNA

Sequences *within* a single strand of nucleotides may be complementary to each other and can pair by forming hydrogen bonds, producing double-stranded regions (**Figure 10.17**). This internal base pairing imparts a secondary structure to a single-stranded molecule. One common type of secondary structure found in single strands of nucleotides is a **hairpin**, which forms when sequences of nucleotides on the same strand are inverted complements (see Figure 10.17a). A hairpin consists of a region of paired bases (the stem) and sometimes includes intervening unpaired bases (the loop). When the complementary sequences are contiguous, the hairpin has a stem but no loop (see Figure 10.17b). RNA molecules may contain numerous hairpins, allowing them to fold up into complex structures (see Figure 10.17c). Secondary structures play important roles in the functions of many RNA molecules, as we will see in Chapters 14 and 15.

DNA sequences can also sometimes form three-stranded (triplex) structures, called **H-DNA**, when some of the DNA

(a) Hairpin

TGCGATACTCATCGCA

Loop
Stem

(b) Stem

GGCAATGATTGCC

(c)

UUCA GUACGG
AAGU CAUGCC

10.17 Both DNA and RNA can form special secondary structures. (a) A hairpin, consisting of a region of paired bases (which form the stem) and a region of unpaired bases between the complementary sequences (which form a loop at the end of the stem). (b) A stem with no loop. (c) Secondary structure of RNA component of RNase P of *E. coli*. RNA molecules often have complex secondary structures.

10.18 H-DNA arises when three polynucleotide strands pair.

where H-DNA structures occur. Quadruplex structures involving four strands of DNA can also occur under certain conditions. ▶ TRY PROBLEM 39

CONCEPTS

In DNA and RNA, base pairing between nucleotides on the same strand produces special secondary structures such as hairpins. Triple-stranded DNA structures can arise when a single strand of DNA pairs with double-stranded DNA.

✓ **CONCEPT CHECK 9**

Hairpins are formed in DNA as a result of
a. sequences on the same strand that are inverted and complementary.
b. sequences on the opposite strand that are complements.
c. sequences on the same strand that are identical.
d. sequences on the opposite strand that are identical.

unwinds and a single polynucleotide strand from one part of the molecule pairs with double-stranded DNA from another part of the molecule (**Figure 10.18**). This is possible because under certain conditions one base can simultaneously pair with two other bases. H-DNA often occurs in long sequences containing only purine bases or only pyrimidine bases. Some triplex structures consist of one strand of purines paired with two strands of pyrimidines; other triplex structures consist of one strand of pyrimidine paired with two strands of purines. Sequences capable of adopting an H-DNA confirmation are common in mammalian genomes and evidence suggests that H-DNA occurs under natural conditions. Recent research has demonstrated that H-DNA breaks more readily than double-stranded DNA, which then leads to higher rates of mutation

The primary structure of DNA can be modified in various ways. One such modification is **DNA methylation**, a process in which methyl groups ($-CH_3$) are added (by specific enzymes) to certain positions on the nucleotide bases. Bacterial DNA is frequently methylated to distinguish it from foreign, unmethylated DNA that may be introduced by viruses; bacteria use proteins called restriction enzymes to cut up any unmethylated viral DNA (see Chapter 19). In eukaryotic cells, methylation is often related to gene expression. Sequences that are methylated typically show low levels of transcription while sequences lacking methylation are actively being transcribed (see Chapter 17). Methylation can also affect the three-dimensional structure of the DNA molecule.

Adenine and cytosine are commonly methylated in bacteria. In eukaryotic DNA, cytosine bases are sometimes methylated to form **5-methylcytosine** (**Figure 10.19**). The extent of cytosine methylation varies among eukaryotic organisms; in most animal cells, about 5% of the cytosine bases are methylated, but there is no methylation of cytosine in yeast and more than 50% of the cytosine bases in some plants are methylated. Why eukaryotic organisms differ so widely in their degree of methylation is not clear.

5-Methylcytosine

10.19 In eukaryotic DNA, cytosine bases are often methylated to form 5-methylcytosine.

CONCEPTS

Methyl groups may be added to certain bases in DNA, depending on the positions of the bases in the molecule. Both prokaryotic and eukaryotic DNA can be methylated. In eukaryotes, cytosine bases are most often methylated to form 5-methylcytosine and methylation is often related to gene expression.

CONCEPTS SUMMARY

- Genetic material must contain complex information, be replicated accurately, code for the phenotype, and have the capacity to vary.

- Evidence that DNA is the source of genetic information came from the finding by Avery, MacLeod, and McCarty that transformation depended on DNA and from the demonstration by Hershey and Chase that viral DNA is passed on to progeny phages. The results of experiments with TMV showed that RNA carries genetic information in some viruses.

- James Watson and Francis Crick, using data provided by Rosalind Franklin and Maurice Wilkins, proposed a model for the three-dimensional structure of DNA in 1953.

- A DNA nucleotide consists of a deoxyribose sugar, a phosphate group, and a nitrogenous base. RNA consists of a ribose sugar, a phosphate group, and a nitrogenous base.

- The bases of a DNA nucleotide are of two types: purines (adenine and guanine) and pyrimidines (cytosine and thymine). RNA contains the pyrimidine uracil instead of thymine.

- Nucleotides are joined together by phosphodiester linkages in a polynucleotide strand. Each polynucleotide strand has a free phosphate group at its 5′ end and a free hydroxyl group at its 3′ end.

- DNA consists of two nucleotide strands that wind around each other to form a double helix. The sugars and phosphates lie on the outside of the helix, and the bases are stacked in the interior. The two strands are joined together by hydrogen bonding between bases in each strand. The two strands are antiparallel and complementary.

- DNA molecules can form a number of different secondary structures, depending on the conditions in which the DNA is placed and on its base sequence.

- The structure of DNA has several important genetic implications. Genetic information resides in the base sequence of DNA, which ultimately specifies the amino acid sequence of proteins. Complementarity of the bases on DNA's two strands allows genetic information to be replicated.

- The central dogma of molecular biology proposes that information flows in a one-way direction, from DNA to RNA to protein. Exceptions to the central dogma are now known.

- Pairing between bases on the same nucleotide strand can lead to hairpins and other secondary structures.

- DNA may be modified by the addition of methyl groups to the nucleotide bases.

IMPORTANT TERMS

nucleotide (p. 279)
Chargaff's rules (p. 279)
transforming principle (p. 281)
isotope (p. 282)
X-ray diffraction (p. 283)
ribose (p. 286)
deoxyribose (p. 286)
nitrogenous base (p. 286)
purine (p. 286)
pyrimidine (p. 286)
adenine (A) (p. 286)
guanine (G) (p. 286)
cytosine (C) (p. 287)

thymine (T) (p. 287)
uracil (U) (p. 287)
nucleoside (p. 287)
phosphate group (p. 287)
deoxyribonucleotide (p. 287)
ribonucleotide (p. 287)
phosphodiester linkage (p. 288)
polynucleotide strand (p. 288)
5′ end (p. 288)
3′ end (p. 288)
antiparallel (p. 289)
complementary DNA strands (p. 289)
B-DNA (p. 289)

A-DNA (p. 290)
Z-DNA (p. 290)
transcription (p. 290)
translation (p. 290)
replication (p. 290)
central dogma (p. 291)
reverse transcription (p. 291)
RNA replication (p. 291)
hairpin (p. 291)
H-DNA (p. 291)
DNA methylation (p. 292)
5-methylcytosine (p. 292)

ANSWERS TO CONCEPT CHECKS

1. Without knowledge of the structure of DNA, an understanding of how genetic information was encoded or expressed was impossible.
2. c
3. c
4. No, because carbon is found in both protein and nucleic acid.
5. d

6. b
7. d
8. Z-DNA has a left-handed helix; B-DNA has a right-handed helix. The sugar–phosphate backbone of Z-DNA zigzags back and forth, whereas the sugar–phosphate backbone of B-DNA forms a smooth continuous ribbon.
9. a

WORKED PROBLEMS

Problem 1

The percentage of cytosine in a double-stranded DNA molecule is 40%. What is the percentage of thymine?

Solution Strategy

What information is required in your answer to the problem?

The percentage of thymine in the DNA molecule.

What information is provided to solve the problem?

- The DNA molecule is double stranded.
- The percentage of cytosine is 40%.

For help with this problem, review:

The Primary Structure of DNA and the Secondary Structures of DNA in Section 10.3.

Solution Steps

If C = 40%, then G also must be 40%. The total percentage of C + G is therefore 40% + 40% = 80%. All the remaining bases must be either A or T; so the total percentage of A + T = 100% − 80% = 20%; because the percentage of A equals the percentage of T, the percentage of T is 20%/2 = 10%.

> **Recall:** In double-stranded DNA, A pairs with T, whereas G pairs with C; so the percentage of A equals the percentage of T, and the percentage of G equals the percentage of C.

Problem 2

Which of the following relations will be true for the percentage of bases in double-stranded DNA?

a. $C + T = A + G$ **b.** $\dfrac{C}{A} = \dfrac{T}{G}$

Solution Strategy

What information is required in your answer to the problem?

Indicate whether $C + T = A + G$ and $\dfrac{C}{A} = \dfrac{T}{G}$ are true.

What information is provided to solve the problem?

- The DNA is double stranded.
- Ratios of different groups of bases.

For help with this problem, review:

The Primary Structure of DNA and the Secondary Structures of DNA in Section 10.3.

Solution Steps

An easy way to determine whether the relations are true is to arbitrarily assign percentages to the bases, remembering that, in double-stranded DNA, A = T and G = C. For example, if the percentages of A and T are each 30%, then the percentages of G and C are each 20%. We can substitute these values into the equations to see if the relations are true.

a. $20 + 30 = 30 + 20$ This relation is true.

b. $^{20}/_{30} \neq {}^{30}/_{20}$. This relation is not true.

COMPREHENSION QUESTIONS

Section 10.1

1. What four general characteristics must the genetic material possess?

Section 10.2

2. Briefly outline the history of our knowledge of the structure of DNA until the time of Watson and Crick. Which do you think were the principle contributions and developments?

3. What experiments demonstrated that DNA is the genetic material?

4. What is transformation? How did Avery and his colleagues demonstrate that the transforming principle is DNA?

5. How did Hershey and Chase show that DNA is passed to new phages in phage reproduction?

6. Why was the discovery of DNA structure so important?

Section 10.3

7. Draw and identify the three parts of a DNA nucleotide.

8. How does an RNA nucleotide differ from a DNA nucleotide?

9. How does a purine differ from a pyrimidine? What purines and pyrimidines are found in DNA and RNA?

10. Draw a short segment of a single polynucleotide strand, including at least three nucleotides. Indicate the polarity of the strand by identifying the 5′ end and the 3′ end.

11. Which bases are capable of forming hydrogen bonds with each other?

12. What different types of chemical bonds are found in DNA and where are they found?

13. What are some of the important genetic implications of the DNA structure?

14. What are the three major pathways of information flow within the cell?

Section 10.4

15. What are hairpins and how do they form?

16. What is DNA methylation?

> ▶ For more questions that test your comprehension of the key chapter concepts, go to **LEARNING**Curve for this chapter.

APPLICATION QUESTIONS AND PROBLEMS

Introduction

17. The introduction to this chapter, about the sequencing of 4000-year-old DNA, emphasizes DNA's extreme stability. What aspects of DNA's structure contribute to the stability of the molecule? Why is RNA less stable than DNA?

Section 10.2

*18. Match the researchers (a–j) with the discoveries listed.

a. Kossel
b. Fraenkel-Conrat
c. Watson and Crick
d. Levene
e. Miescher
f. Hershey and Chase
g. Avery, MacLeod, and McCarty
h. Griffith
i. Franklin and Wilkins
j. Chargaff

____ Took X-ray diffraction pictures used in constructing the structure of DNA.

____ Determined that DNA contains nitrogenous bases.

____ Identified DNA as the genetic material in bacteriophage.

____ Discovered regularity in the ratios of different bases in DNA.

____ Determined that DNA is responsible for transformation in bacteria.

____ Worked out the helical structure of DNA by building models.

____ Discovered that DNA consists of repeating nucleotides.

____ Determined that DNA is acidic and high in phosphorous.

____ Conducted experiments showing that RNA can serve as the genetic material in some viruses.

____ Demonstrated that heat-killed material from bacteria can genetically transform live bacteria.

*19. A student mixes some heat-killed type IIS *Streptococcus pneumoniae* bacteria with live type IIR bacteria and injects the mixture into a mouse. The mouse develops pneumonia and dies. The student recovers some type IIS bacteria from the dead mouse. If this is the only experiment conducted by the student, has the student demonstrated that transformation has taken place? What other explanations might explain the presence of the type IIS bacteria in the dead mouse?

20. Predict what would happen if Griffith had mixed some heat-killed type IIIS bacteria and some heat-killed type IIR bacteria and injected these into a mouse. Would the mouse have contracted pneumonia and died? Explain why or why not.

21. Explain how heat-killed type IIIS bacteria in Griffith's experiment genetically altered the live type IIR bacteria. (Hint: See the discussion of transformation in Chapter 9.)

22. What results would you expect if the Hershey and Chase experiment were conducted on tobacco mosaic virus?

23. Which of the processes of information transfer illustrated in **Figure 10.16** are required for the T2 phage reproduction illustrated in **Figure 10.4**?

*24. Imagine that you are a student in Alfred Hershey and Martha Chase's lab in the late 1940s. You are given five test tubes containing *E. coli* bacteria that were infected with T2 bacteriophage that have been labeled with either ^{32}P or ^{35}S. Unfortunately, you forgot to mark the tubes and are now uncertain about which were labeled with ^{32}P and which with ^{35}S. You place the contents of each tube in a blender and turn it on for a few seconds to shear off the protein coats. You then centrifuge the contents to separate the protein coats and the cells. You check for the presence of radioactivity and obtain the following results. Which tubes contained *E. coli* infected with ^{32}P-labeled phage? Explain your answer.

Tube number	Presence of radioactivity in
1	cells
2	protein coats
3	protein coats
4	cells
5	cells

25. **Figure 10.8** illustrates Fraenkel-Conrat and Singer's experiment on the genetic material of TMV. What results would you expect in this experiment if protein carried the genetic information of TMV instead of RNA?

Section 10.3

*26. DNA molecules of different sizes are often separated with the use of a technique called electrophoresis (see Chapter 19). With this technique, DNA molecules are placed in a gel, an electrical current is applied to the gel, and the DNA molecules migrate toward the positive (+) pole of the current. What aspect of its structure causes a DNA molecule to migrate toward the positive pole?

*27. Each nucleotide pair of a DNA double helix weighs about 1×10^{-21} g. The human body contains approximately 0.5 g of DNA. How many nucleotide pairs of DNA are in the human body? If you assume that all the DNA in human cells is in the B-DNA form, how far would the DNA reach if stretched end to end?

28. One nucleotide strand of DNA molecule has the base sequence illustrated below.

5′–ATTGCTACGG–3′

Give the base sequence and label the 5′ and 3′ ends of the complementary DNA nucleotide strand.

*29. Erwin Chargaff collected data on the proportions of nucleotide bases from the DNA of a variety of different organisms and tissues (E. Chargaff, in *The Nucleic Acids: Chemistry and Biology*, vol. 1, E. Chargaff and J. N. Davidson, Eds. New York: Academic Press, 1955). The following data are from the DNA of several organisms analyzed by Chargaff.

Erwin Chargaff. [Horst Tappe/Getty Images.]

Organism and tissue	Percent			
	A	G	C	T
Sheep thymus	29.3	21.4	21.0	28.3
Pig liver	29.4	20.5	20.5	29.7
Human thymus	30.9	19.9	19.8	29.4
Rat bone marrow	28.6	21.4	20.4	28.4
Hen erythrocytes	28.8	20.5	21.5	29.2
Yeast	31.7	18.3	17.4	32.6
E. coli	26.0	24.9	25.2	23.9
Human sperm	30.9	19.1	18.4	31.6
Salmon sperm	29.7	20.8	20.4	29.1
Herring sperm	27.8	22.1	20.7	27.5

a. For each organism, compute the ratio of (A + G)/(T + C) and the ratio of (A + T)/(C + G).

b. Are these ratios constant or do they vary among the organisms? Explain why.

c. Is the (A + G)/(T + C) ratio different for the sperm samples? Would you expect it to be? Why or why not?

30. Boris Magasanik collected data on the amounts of the bases of RNA isolated from a number of sources, expressed relative to a value of 10 for adenine (B. Magasanik, in *The Nucleic Acids: Chemistry and Biology*, vol. 1, E. Chargaff and J. N. Davidson, Eds. New York: Academic Press, 1955).

Organism and tissue	Percent			
	A	G	C	U
Rat liver nuclei	10	14.8	14.3	12.9
Rabbit liver nuclei	10	13.6	13.1	14.0
Cat brain	10	14.7	12.0	9.5
Carp muscle	10	21.0	19.0	11.0
Yeast	10	12.0	8.0	9.8

a. For each organism, compute the ratio of (A + G)/(U + C).

b. How do these ratios compare with the (A + G)/(T + C) ratio found in DNA (see Problem 29)? Explain.

31. Which of the following relations or ratios would be true for a double-stranded DNA molecule?

a. $A + T = G + C$

b. $A + T = T + C$

c. $A + C = G + T$

d. $\dfrac{A + T}{C + G} = 1.0$

e. $\dfrac{A + G}{C + T} = 1.0$

f. $\dfrac{A}{C} = \dfrac{G}{T}$

g. $\dfrac{A}{G} = \dfrac{C}{T}$

h. $\dfrac{A}{T} = \dfrac{G}{C}$

*32. If a double-stranded DNA molecule is 15% thymine, what are the percentages of all the other bases?

33. Suppose that each of the bases in DNA were capable of pairing with any other base. What effect would this capability have on DNA's capacity to serve as the source of genetic information?

34. Heinz Shuster collected the following data on the base composition of ribgrass virus (H. Shuster, in *The Nucleic Acids: Chemistry and Biology*, vol. 3, E. Chargaff and J. N. Davidson, Eds. New York: Academic Press, 1955). On the basis of this information, is the hereditary information of the ribgrass virus RNA or DNA? Is it likely to be single stranded or double stranded?

Ribgrass mosaic virus. [Jena Library of Biological Macromolecules.]

Organism	Percent				
	A	G	C	T	U
Ribgrass virus	29.3	25.8	18.0	0.0	27.0

*35. The relative amounts of each nucleotide base are tabulated here for four different viruses. For each virus listed in the following table, indicate whether its genetic material is DNA or RNA and whether it is single stranded or double stranded. Explain your reasoning.

Virus	T	C	U	G	A
I	0	12	9	12	9
II	23	16	0	16	23
III	34	42	0	18	39
IV	0	24	35	27	17

*36. A B-DNA molecule has 1 million nucleotide pairs. How many complete turns are there in this molecule?

*37. For entertainment on a Friday night, a genetics professor proposed that his children diagram a polynucleotide strand of DNA. Having learned about DNA in preschool, his 5-year-old daughter was able to draw a polynucleotide strand, but she made a few mistakes. The daughter's diagram (represented here) contained at least 10 mistakes.

a. Make a list of all the mistakes in the structure of this DNA polynucleotide strand.

b. Draw the correct structure for the polynucleotide strand.

38. Chapter 1 considered the theory of the inheritance of acquired characteristics and noted that this theory is no longer accepted. Is the central dogma consistent with the theory of the inheritance of acquired characteristics? Why or why not?

Section 10.4

*39. Write a sequence of bases in an RNA molecule that will produce a hairpin structure.

40. Write a sequence of nucleotides on a strand of DNA that will form a hairpin structure.

CHALLENGE QUESTIONS

Section 10.1

*41. Suppose that an automated, unmanned probe is sent into deep space to search for extraterrestrial life. After wandering for many light-years among the far reaches of the universe, this probe arrives on a distant planet and detects life. The chemical composition of life on this planet is completely different from that of life on Earth, and its genetic material is not composed of nucleic acids. What predictions can you make about the chemical properties of the genetic material on this planet?

Section 10.2

42. How might ^{32}P and ^{35}S be used to demonstrate that the transforming principle is DNA? Briefly outline an experiment that would show that DNA rather than protein is the transforming principle.

Section 10.3

43. Researchers have proposed that early life on Earth used RNA as its source of genetic information and that DNA eventually replaced RNA as the source of genetic information. What aspects of DNA structure might make it better suited than RNA to be the genetic material?

44. Scientists have reportedly isolated short fragments of DNA from fossilized dinosaur bones hundreds of millions of years old. The technique used to isolate this DNA is the polymerase chain reaction, which is capable of amplifying very small amounts of DNA a millionfold (see Chapter 19). Critics have claimed that the DNA isolated from dinosaur bones is not purely of ancient origin but instead has been contaminated by DNA from present-day organisms such as bacteria, mold, or humans. What precautions, analyses, and control experiments could be carried out to ensure that DNA recovered from fossils is truly of ancient origin?

Go to your **LaunchPad** to find additional learning resources and the Suggested Readings for this chapter.

Chromosome Structure and Organelle DNA

Child from a Romanian orphanage. Research demonstrated that children who lived in state-run orphanages had shorter telomeres than children from foster homes. [Jenny Matthews/Alamy.]

Telomeres and Childhood Adversity

Within each of our cells are 46 chromosomes, exquisitely complex structures of DNA and protein that carry the coding instructions for all of our traits. These chromosomes are passed down from our parents and constitute the basis of heredity, the passage of traits from one generation to the next. But chromosomes don't just carry a record of our genetic legacy. They also carry a record—in the lengths of their telomeres—of the stresses we encounter.

Telomeres are special protective structures found at the end of each of our chromosomes. Like the small plastic tips that keep the ends of a shoelace from unraveling, telomeres prevent chromosomes from being degraded at their ends. In spite of the protection of the telomeres, chromosomes of most cells shorten progressively with each cell division. Due to a quirk of DNA replication, most cells are unable to copy the very end of each linear chromosome (see Chapter 12 for a full discussion of the end-replication problem). Hence, with each round of replication, a chromosome becomes shorter, until it is so reduced that the cell stops dividing, becomes inactive, and eventually dies. For most cells, this shortening of telomeres limits the number of divisions possible. Exceptions occur in germ-line cells that produce future generations, certain stem cells, and—unfortunately—many cancer cells that have escaped normal constraints on cell division.

Because telomeres become shorter with each cell division, much research has focused on determining if telomere length is indicative of biological aging. Although the relationship between telomere length and aging is complex and not fully understood, considerable evidence suggests that telomeres do shorten with age, and that processes which lead to premature telomere shortening are associated with features of aging. In 2011, geneticists observed that hardships encountered early in life can play a part in shortening our telomeres.

To study the effects of early life experience on telomere length, geneticists studied 100 children living in state-run orphanages in Romania. At an early age, some of these children were placed in foster homes; others remained in the orphanages. Previous studies demonstrated that children in such orphanages receive less individual attention and care compared to children growing up with natural or foster parents, and institutional care is assumed to be more stressful than foster care.

When the children were 6 to 10 years old, the researchers collected samples of their DNA and measured the length of their telomeres. The results were striking: children who remained in the orphanages had significantly shorter telomeres than those that spent

E. coli bacterium

Bacterial chromosome

11.1 The DNA in *E. coli* is about 1000 times as long as the cell itself.

time in foster care. The researchers concluded that telomere length is affected by childhood adversity: children reared in stressful environments are more likely to have shorter telomeres than those raised in less stressful environments. Several other studies have found a similar association between telomere length in adults and early childhood stresses, such as abuse and chronic illness. How stress affects telomeres and results in their shortening is not known, but the research documents that chromosomes are more than just a repository of our genetic information—their structure is also affected by our environment.

In this chapter, we examine the molecular structure of chromosomes and the DNA found in cytoplasmic organelles. The first part of the chapter focuses on a storage problem: how to cram tremendous amounts of DNA into the limited confines of a cell. Even in those organisms having the smallest amounts of DNA, the length of genetic material far exceeds the length of the cell. Thus, cellular DNA must be highly folded and tightly packed, but this packing creates problems: it renders the DNA inaccessible, unable to be copied or read. Functional DNA must be capable of partly unfolding and expanding so that individual genes can undergo replication and transcription. The flexible, dynamic nature of DNA packing is a major theme of this chapter. We first consider supercoiling, an important tertiary structure of DNA found in both prokaryotic and eukaryotic cells. After a brief look at the bacterial chromosome, we examine the structure of eukaryotic chromosomes. We pay special attention to the working parts of a chromosome—specifically, centromeres and telomeres. We also consider the types of DNA sequences present in many eukaryotic chromosomes.

The second part of this chapter focuses on the organization of DNA sequences found in mitochondria and chloroplasts. Uniparental inheritance exhibited by genes found in these organelles was discussed in Chapter 5; here we examine molecular aspects of organelle DNA. We briefly consider the structures of mitochondria and chloroplasts, the inheritance of traits encoded by their genes, and the evolutionary origin of these organelles. We then examine the general characteristics of **mitochondrial DNA** (mtDNA), followed by a discussion of the organization and function of different types of mitochondrial genomes. Finally, we turn to **chloroplast DNA** (cpDNA) and examine its characteristics, organization, and function.

11.1 Large Amounts of DNA Are Packed into a Cell

The packaging of tremendous amounts of genetic information into the small space within a cell has been called the ultimate storage problem. Consider the chromosome of the bacterium *E. coli,* a single molecule of DNA with approximately 4.6 million base pairs. Stretched out straight, this DNA would be about 1000 times as long as the cell within which it resides (**Figure 11.1**). Human cells contain more than 6 billion base pairs of DNA, which would measure over 2 meters (over 6 feet) stretched end to end. Even DNA in the

smallest human chromosome would stretch 14,000 times the length of the nucleus. Clearly, DNA molecules must be tightly packed to fit into such small spaces.

The structure of DNA can be considered at three hierarchical levels: the primary structure of DNA is its nucleotide sequence; the secondary structure is the double-stranded helix; and the tertiary structure refers to higher-order folding that allows DNA to be packed into the confined space of a cell.

CONCEPTS

Chromosomal DNA exists in the form of very long molecules that are tightly packed to fit into the small confines of a cell.

Supercoiling

One type of DNA tertiary structure is **supercoiling**, which takes place when the DNA helix is subjected to strain by being overwound or underwound. The lowest energy state for B-DNA (see Chapter 10) is when it has approximately 10 bp per turn of its helix. In this **relaxed state**, a stretch of 100 bp of DNA would assume about 10 complete turns (**Figure 11.2a**). If energy is used to add or remove any turns, strain is placed on the molecule, causing the helix to supercoil, or twist, on itself (**Figure 11.2b** and **c**). Molecules that are overrotated exhibit **positive supercoiling** (see Figure 11.2b). Underrotated molecules exhibit **negative supercoiling** (see Figure 11.2c). Supercoiling is a partial solution to the cell's DNA packing problem because supercoiled DNA occupies less space than relaxed DNA.

Supercoiling takes place when the strain of overrotating or underrotating cannot be compensated by the turning of the ends of the double helix, which is the case if the DNA is circular—that is, there are no free ends. If the chains *can* turn freely, their ends will simply turn as extra rotations are added or removed, and the molecule will spontaneously revert to the relaxed state. Both bacterial and eukaryotic DNA usually fold into loops stabilized by proteins (which prevent free rotation of the ends, see Figure 11.3), and supercoiling takes place within the loops.

Supercoiling relies on **topoisomerases**, enzymes that add or remove rotations from the DNA helix by temporarily breaking the nucleotide strands, rotating the ends around each other, and then rejoining the broken ends. Thus topoisomerases can both induce and relieve supercoiling, although not all topoisomerase enzymes do both.

Most DNA found in cells is negatively supercoiled, which has two advantages over nonsupercoiled DNA. First, negative supercoiling makes the separation of the two strands of DNA easier during replication and transcription. Negatively supercoiled DNA is underrotated, so separation of the two strands during replication and transcription is more rapid and requires less energy. Second, supercoiled DNA can be packed into a smaller space than can relaxed DNA.

(a)

Relaxed circular DNA

A telephone cord is like relaxed circular DNA.

(b) Add two turns (overrotate)

(c) Remove two turns (underrotate)

Positive supercoil

Negative supercoil

Positive supercoiling occurs when DNA is overrotated; the helix twists on itself.

Negative supercoiling occurs when DNA is underrotated; the helix twists on itself in the opposite direction.

If you turn the receiver when you hang up, you induce a negative supercoil in the cord.

11.2 Supercoiled DNA is overwound or underwound, causing it to twist on itself. Electron micrographs are of relaxed DNA (top) and supercoiled DNA (bottom). [Dr. Gopal Murti/Phototake.]

CONCEPTS

Overrotation or underrotation of a DNA double helix places strain on the molecule, causing it to supercoil. Supercoiling is controlled by topoisomerase enzymes. Most cellular DNA is negatively supercoiled, which eases the separation of nucleotide strands during replication and transcription and allows DNA to be packed into small spaces.

✓ CONCEPT CHECK 1

A DNA molecule 300 bp long has 20 complete rotations. This DNA molecule is

a. positively supercoiled.

b. negatively supercoiled.

c. relaxed.

The Bacterial Chromosome

Most bacterial genomes consist of a single circular DNA molecule, although linear DNA molecules have been found in a few species. In circular bacterial chromosomes, the DNA does not exist in an open, relaxed circle; the 3 million to 4 million base pairs of DNA found in a typical bacterial genome would be much too large to fit into a bacterial cell (see Figure 11.1). Bacterial DNA is not attached to histone proteins as is eukaryotic DNA (discussed later in this chapter), but bacterial DNA is complexed to a number of proteins that help to compact it.

When a bacterial cell is viewed with the electron microscope, its DNA frequently appears as a distinct clump, the **nucleoid**, which is confined to a definite region of the cytoplasm. If a bacterial cell is broken open gently, its DNA spills out in a series of twisted loops (**Figure 11.3a**). The ends of the loops are most likely held in place by proteins (**Figure 11.3b**). Many bacteria contain additional DNA in the form of small circular molecules called plasmids, which replicate independently of the chromosome (see Chapter 9).

CONCEPTS

A typical bacterial chromosome consists of a large, circular molecule of DNA that is a series of twisted loops. Bacterial DNA appears as a distinct clump, the nucleoid, within the bacterial cell.

✓ CONCEPT CHECK 2

How does bacterial DNA differ from eukaryotic DNA?

(a)

11.3 Bacterial DNA is highly folded into a series of twisted loops. [Part a: Dr. Gopal Murti/Photo Researchers.]

(b) Twisted loops of DNA

Proteins

Eukaryotic Chromosomes

Individual eukaryotic chromosomes contain enormous amounts of DNA. Like the bacterial chromosome, each eukaryotic chromosome consists of a single, extremely long molecule of DNA. For this DNA to fit into the nucleus, tremendous packing and folding are required, the extent of which must change in the course of the cell cycle. The chromosomes are in an elongated, relatively uncondensed state during interphase of the cell cycle (see pp. 93–94 in Chapter 2), but the term *relatively* is important here. Although the DNA of interphase chromosomes is less tightly packed than the DNA of mitotic chromosomes, it is still highly condensed; it's just *less* condensed. In the course of the cell cycle, the level of DNA packing changes: chromosomes progress from a highly packed state to a state of extreme condensation, which is necessary for chromosome movement in mitosis and meiosis. DNA packing also changes locally in replication and transcription, when the two nucleotide strands must unwind so that particular base sequences are exposed. Thus, the packing of eukaryotic DNA (its tertiary chromosomal structure) is not static but changes regularly in response to cellular processes.

CHROMATIN   Eukaryotic DNA in the cell is closely associated with proteins. This combination of DNA and protein is called chromatin. The two basic types of chromatin are **euchromatin**, which undergoes the normal process of condensation and decondensation in the cell cycle, and **heterochromatin**, which remains in a highly condensed state throughout the cell cycle, even during interphase. Euchromatin constitutes the majority of the chromosomal material and is where most transcription takes place. All chromosomes have permanent heterochromatin (called constitutive heterochromatin) at the centromeres and telomeres; the Y chromosome also consists largely of constitutive heterochromatin. Heterochromatin may also occur during certain developmental stages; this is referred to as facultative heterochromatin. For example, facultative heterochromatin occurs along one entire X chromosome in female mammals

when this X becomes inactivated (see p. 25 in Chapter 4). In addition to remaining condensed throughout the cell cycle, heterochromatin is characterized by a general lack of transcription, the absence of crossing over, and replication late in the S phase. Differences between euchromatin and heterochromatin are summarized in **Table 11.1**.

The most abundant proteins in chromatin are the *histones*, which are small, positively charged proteins of five major types: H1, H2A, H2B, H3, and H4. All histones have a high percentage of arginine and lysine, positively charged amino acids that give the histones a net positive charge. The positive charges attract the negative charges on the phosphates of DNA; this attraction holds the DNA in contact with the histones. A heterogeneous assortment of **nonhistone chromosomal proteins** also are found in eukaryotic chromosomes. At times, variant histones, with somewhat different amino acid sequences, are incorporated into chromatin in place of one of the major histone proteins.

▶ **TRY PROBLEM 18**

Table 11.1	Characteristics of euchromatin and heterochromatin	
Characteristic	**Euchromatin**	**Heterochromatin**
Chromatin condensation	Less condensed	More condensed
Location	On chromosome arms	At centromeres, telomeres, and other specific places
Type of sequences	Unique sequences	Repeated sequences*
Presence of genes	Many genes	Few genes*
When replicated	Throughout S phase	Late S phase
Transcription	Often	Infrequent
Crossing over	Common	Uncommon

*Applies only to constitutive heterochromatin.

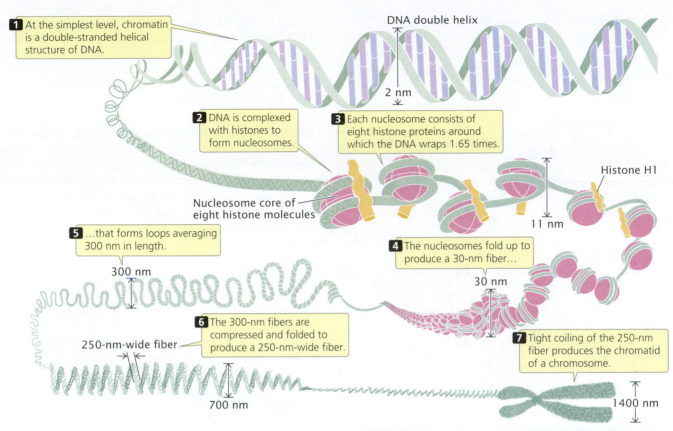

1 At the simplest level, chromatin is a double-stranded helical structure of DNA.

DNA double helix

2 nm

2 DNA is complexed with histones to form nucleosomes.

3 Each nucleosome consists of eight histone proteins around which the DNA wraps 1.65 times.

Histone H1

Nucleosome core of eight histone molecules

11 nm

5 ...that forms loops averaging 300 nm in length.

300 nm

4 The nucleosomes fold up to produce a 30-nm fiber...

30 nm

6 The 300-nm fibers are compressed and folded to produce a 250-nm-wide fiber.

250-nm-wide fiber

7 Tight coiling of the 250-nm fiber produces the chromatid of a chromosome.

700 nm

1400 nm

11.4 Chromatin has a highly complex structure with several levels of organization.

CONCEPTS

Chromatin, which consists of DNA complexed to proteins, is the material that makes up eukaryotic chromosomes. The most abundant of these proteins are the five types of positively charged histone proteins: H1, H2A, H2B, H3, and H4. Variant histones may at times be incorporated into chromatin in place of the normal histones.

✓ CONCEPT CHECK 3

Neutralizing their positive charges would have which effect on the histone proteins?

a. They would bind the DNA tighter.
b. They would bind less tightly to the DNA.
c. They would no longer be attracted to each other.
d. They would cause supercoiling of the DNA.

THE NUCLEOSOME Chromatin has a highly complex structure with several levels of organization (**Figure 11.4**). The simplest level is the double-helical structure of DNA discussed in Chapter 10. At a more complex level, the DNA molecule is associated with proteins and is highly folded to produce a chromosome.

When chromatin is isolated from the nucleus of a cell and viewed with an electron microscope, it frequently looks like beads on a string (**Figure 11.5a**). If a small amount of nuclease is added to this structure, the enzyme cleaves the "string" between the "beads," leaving individual beads attached to about 200 bp of DNA (**Figure 11.5b**). If more nuclease is added, the enzyme chews up all of the DNA between the beads and leaves a core of proteins attached to a fragment of DNA (**Figure 11.5c**). Such experiments demonstrated that chromatin is not a random association of proteins and DNA but has a fundamental repeating structure.

The repeating core of protein and DNA produced by digestion with nuclease enzymes is the simplest level of chromatin structure, the **nucleosome** (see Figure 11.4). The nucleosome is a core particle consisting of DNA wrapped about two times around an octamer of eight histone proteins (two copies each of H2A, H2B, H3, and H4), much like thread wound around a spool (**Figure 11.5d**). The DNA in direct contact with the histone octamer is between 145 and 147 bp in length.

Each of the histone proteins that make up the nucleosome core particle has a flexible "tail," containing from 11 to 37 amino acids, which extends out from the nucleosome. Positively charged amino acids in the tails of the histones interact with the negative charges of the phosphates on the DNA, keeping the DNA and histones tightly associated. The tails of one nucleosome may also interact with neighboring nucleosomes, which facilitates compaction of the nucleosomes

(a) Core histones of nucleosome | Linker DNA

"Beads-on-a-string" view of chromatin

Nuclease →

(b)

1 A small amount of nuclease cleaves the "string" between the beads,…

2 …releasing individual beads attached to about 200 bp of DNA.

Nuclease →

3 More nuclease destroys all of the unprotected DNA between the beads,…

(c)

11 nm

4 …leaving a core of proteins attached to 145–147 bp of DNA.

(d)

H2A
H2B
H3
H2B
H2A
H4

11.5 The nucleosome is the fundamental repeating unit of chromatin. Part (d) shows a space-filling model of the core particle, which consists of two copies each of H2A, H2B, H3, and H4, around which DNA (white) coils. [Part d: Reprinted by permission from Macmillan Publishers Ltd. From K. Luger et al., *Nature* 389:251. © 1997. Courtesy of T. H. Richmond.]

themselves. Chemical modifications of the histone tails bring about changes in chromatin structure (discussed in the next section) that are necessary for gene expression.

The fifth type of histone, H1, is not a part of the nucleosome core particle but plays an important role in nucleosome structure. H1 binds to 20 to 22 bp of DNA where the DNA joins and leaves the octamer (see Figure 11.4) and helps to lock the DNA into place, acting as a clamp around the nucleosome octamer.

Each nucleosome encompasses about 167 bp of DNA. Nucleosomes are located at regular intervals along the DNA

molecule and are separated from one another by **linker DNA**, which varies in size among cell types; in most cells, linker DNA comprises from about 30 to 40 bp. Nonhistone chromosomal proteins may be associated with this linker DNA, and a few also appear to bind directly to the core particle. ▶ TRY PROBLEMS 19 AND 22

HIGHER-ORDER CHROMATIN STRUCTURE When chromatin is in a condensed form, adjacent nucleosomes are not separated by space equal to the length of the linker DNA; rather, nucleosomes fold on themselves to form a dense, tightly packed structure (see Figure 11.4) that makes up a fiber with a diameter of about 30 nm. The precise molecular structure of the 30 nm fiber remains uncertain.

The next-higher level of chromatin structure is a series of loops of 30-nm fibers (Figure 11.4), each anchored at its base by proteins. On average, each loop encompasses some 20,000 to 100,000 bp of DNA and is about 300 nm in length, but the individual loops vary considerably. The 300-nm loops are packed and folded to produce a 250-nm-wide fiber. Tight helical coiling of the 250-nm fiber, in turn, produces the structure that appears in metaphase—individual chromatids approximately 700 nm in width.

CONCEPTS

The nucleosome consists of a core particle of eight histone proteins and DNA that wraps around the core. A single H1 histone associates with each core particle. Nucleosomes are separated by linker DNA. Nucleosomes fold to form a 30-nm chromatin fiber, which appears as a series of loops that pack to create a 250-nm-wide fiber. Helical coiling of the 250-nm fiber produces a chromatid.

✔ **CONCEPT CHECK 4**

How many copies of the H2B histone would be found in chromatin containing 50 nucleosomes?

a. 5   c. 50
b. 10   d. 100

Changes in Chromatin Structure

Although eukaryotic DNA must be tightly packed to fit into the cell nucleus, it must also periodically unwind to undergo transcription and replication.

POLYTENE CHROMOSOMES Giant chromosomes called **polytene chromosomes** are found in certain tissues of *Drosophila* and some other organisms (**Figure 11.6**). Polytene chromosomes have provided researchers with evidence of the changing nature of chromatin structure. These large, unusual chromosomes arise when repeated rounds of DNA replication take place without accompanying cell divisions, producing thousands of copies of DNA that lie side by side. When polytene chromosomes are stained with dyes, numerous bands are revealed. Under certain conditions, the bands may exhibit **chromosomal puffs**—localized swellings of the chromosome. Each puff is a region of the chromatin that has

(a)

A1-2 —
A4-5 —
A7 (
A8...10 [
B1(-2)—
B4(-5)—

Chromosome puff

Chromosome puff

C1 —

(b)

— A1-2

— C1

11.6 Chromosomal puffs are regions of relaxed chromatin where active transcription is taking place. Pictured here are (a) chromosomal puffs on giant polytene chromosomes isolated from the salivary glands of larval *Drosophila*, and (b) corresponding region without chromosomal puffs. [Courtesy of Dmitri V. Novikov.]

a relaxed structure and, consequently, a more open state. Research indicates that chromosomal puffs are regions of active transcription. This correlation between the occurrence of transcription and the relaxation of chromatin at a puff site indicates that chromatin structure undergoes dynamic change associated with gene activity. ▶ **TRY PROBLEM 20**

DNASE I SENSITIVITY A second piece of evidence indicating that chromatin structure changes with gene activity is sensitivity to DNase I, an enzyme that digests DNA. The ability of this enzyme to digest DNA depends on chromatin structure: when DNA is tightly bound to histone proteins, it is less sensitive to DNase I, whereas unbound DNA is more sensitive to DNase I. The results of experiments that examine the effect of DNase I on specific genes show that DNase sensitivity is correlated with gene activity. Studies of chicken globin genes give evidence for this correlation. Globin genes encode hemoglobin in the erythroblasts (precursors of red blood cells) of chickens (**Figure 11.7**). These types of experiments demonstrate that transcriptionally active genes are sensitive to DNase I, indicating that the chromatin structure is more exposed during transcription.

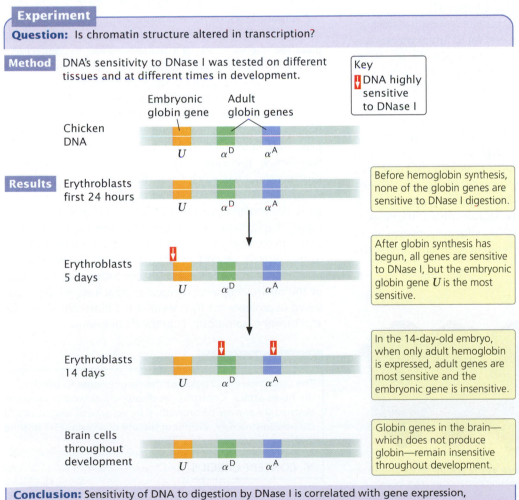

Experiment

Question: Is chromatin structure altered in transcription?

Method DNA's sensitivity to DNase I was tested on different tissues and at different times in development.

Key
↓ DNA highly sensitive to DNase I

Embryonic globin gene   Adult globin genes

Chicken DNA

U α^D α^A

Results Erythroblasts first 24 hours

U α^D α^A

Before hemoglobin synthesis, none of the globin genes are sensitive to DNase I digestion.

Erythroblasts 5 days

U α^D α^A

After globin synthesis has begun, all genes are sensitive to DNase I, but the embryonic globin gene U is the most sensitive.

Erythroblasts 14 days

U α^D α^A

In the 14-day-old embryo, when only adult hemoglobin is expressed, adult genes are most sensitive and the embryonic gene is insensitive.

Brain cells throughout development

U α^D α^A

Globin genes in the brain—which does not produce globin—remain insensitive throughout development.

11.7 DNase I sensitivity is correlated with the transcription of globin genes in erythroblasts of chick embryos. The U gene encodes embryonic hemoglobin; the α^D and α^A genes encode adult hemoglobin.

Conclusion: Sensitivity of DNA to digestion by DNase I is correlated with gene expression, suggesting that chromatin structure changes in the course of transcription.

What is the nature of the change in chromatin structure that produces chromosome puffs and DNase I sensitivity? In both cases, the chromatin relaxes; presumably, the histones loosen their grip on the DNA. One process that alters chromatin structure is acetylation. Enzymes called acetyltransferases attach acetyl groups to lysine amino acids on the histone tails. This modification reduces the positive charges that normally exist on lysine and destabilizes the nucleosome structure, and so the histones hold the DNA less tightly. Other chemical modifications of the histone proteins, such as methylation and phosphorylation, also alter chromatin structure, as do special chromatin-remodeling proteins that bind to the DNA.

EPIGENETIC CHANGES ASSOCIATED WITH CHROMATIN MODIFICATIONS We have now seen how chromatin structure can be altered by chemical modification of the histone proteins. A number of other changes can also affect chromatin structure, including the methylation of DNA (see Chapter 10), the use of variant histone proteins in the nucleosome, and the binding of proteins to DNA and chromatin. Although these changes do not alter the DNA sequence, they often have major effects on the expression of genes, which will be discussed in more detail in Chapter 17.

Some changes to chromatin structure are retained through cell division, and so they are passed on to future generations of cells and even occasionally to future generations of organisms. Stable alterations of chromatin structure that may be passed on to cells or individual organisms are frequently referred to as **epigenetic changes** or simply as epigenetics (see Chapter 5). For example, the *agouti* locus helps determine coat color in mice: parents that have identical DNA sequences but have different degrees of methylation on their DNA may give rise to offspring with different coat colors (**Figure 11.8**). Such epigenetic changes have been observed in a number of organisms and are responsible for a variety of phenotypic effects. Unlike mutations, epigenetic changes do not alter the DNA sequence, are capable of being reversed, and are often influenced by environmental factors.

11.8 Variation in DNA methylation at the *agouti* locus produces different coat colors in mice. [Cropley et al. ©2006 by The National Academy of Sciences of the USA.]

CONCEPTS

Epigenetic changes are alterations of chromatin or DNA structure that do not include changes in the base sequence but are stable and passed on to cells or organisms. Some epigenetic changes result from alterations of histone proteins.

11.2 Eukaryotic Chromosomes Possess Centromeres and Telomeres

Chromosomes segregate in mitosis and meiosis and remain stable over many cell divisions. These properties of chromosomes arise, in part, from special structural features of chromosomes, including centromeres and telomeres.

Centromere Structure

The centromere is a constricted region of the chromosome to which spindle fibers attach and is essential for proper chromosome movement in mitosis and meiosis (see Chapter 2). The essential role of the centromere in chromosome movement was recognized by early geneticists, who observed the consequences of chromosome breakage. When a chromosome break produces two fragments, one with a centromere and one without (**Figure 11.9**), the chromosome fragment containing the centromere attaches to spindle fibers and moves to the spindle pole. The fragment lacking a centromere fails to connect to a spindle fiber and is usually lost because it fails to move into the nucleus of a daughter cell during mitosis (see Figure 11.9).

Centromeres are the binding sites for the kinetochore, to which spindle fibers attach. In *Drosophila*, *Arabidopsis*, and humans, centromeres span hundreds of thousands of base pairs. Most of the centromere is made up of heterochromatin. Surprisingly, there are no specific sequences that are found in all centromeres, which raises the question of what exactly determines where the centromere is. Research suggests that most centromeres are not defined by DNA sequence but rather by epigenetic changes in chromatin structure. Nucleosomes in the centromeres of most eukaryotes have a variant histone protein called CenH3, which takes the place of the usual H3 histone. The CenH3 histone brings about a change in the nucleosome and chromatin structure, which is believed to promote the formation of the kinetochore and the attachment of spindle fibers to the chromosome.

CONCEPTS

The centromere is a region of the chromosome to which spindle fibers attach. Centromeres display considerable variation in structure and are distinguished by epigenetic alterations to chromatin structure, including the use of a variant H3 histone in the nucleosome.

✓ **CONCEPT CHECK 5**

What happens to a chromosome that loses its centromere?

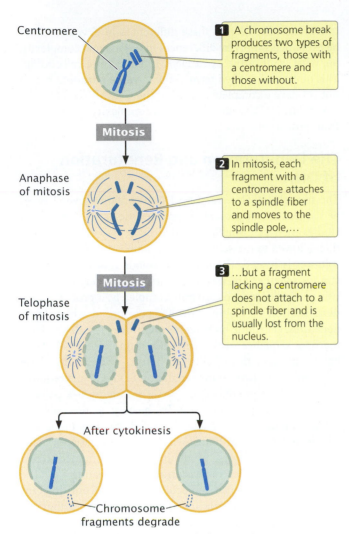

11.9 Chromosome fragments that lack centromeres are lost in mitosis.

Figure 11.9 annotations:
- Centromere
- Mitosis
- Anaphase of mitosis
- Mitosis
- Telophase of mitosis
- After cytokinesis
- Chromosome fragments degrade

1 A chromosome break produces two types of fragments, those with a centromere and those without.

2 In mitosis, each fragment with a centromere attaches to a spindle fiber and moves to the spindle pole,…

3 …but a fragment lacking a centromere does not attach to a spindle fiber and is usually lost from the nucleus.

Telomere Structure

Telomeres are the natural ends of a chromosome (see Figure 2.7 and the introduction to this chapter). Pioneering work by Hermann Muller (on fruit flies) and Barbara McClintock (on corn) showed that chromosome breaks produce unstable ends that have a tendency to stick together and enable the chromosome to be degraded. Because attachment and degradation do not happen to the ends of a chromosome that has telomeres, each telomere must serve as a cap that stabilizes the chromosome. Telomeres also provide a means of replicating the ends of the chromosome, which will be discussed in Chapter 12. In 2009, Elizabeth Blackburn, Carol Greider, and Jack Szostak were awarded the Nobel Prize in physiology or medicine for discovering the structure of telomeres and how they are replicated (discussed in Chapter 12).

Telomeres have now been isolated from protozoans, plants, humans, and other organisms; most are similar in structure (**Table 11.2**). These **telomeric sequences** usually consist of repeated units of a series of adenine or thymine nucleotides followed by several guanine nucleotides, taking the form

Table 11.2	DNA sequences typically found in telomeres of various organisms
Organism	**Sequence**
Tetrahymena (protozoan)	5′–T T GGGG–3′ 3′–AAC C C C–5′
Saccharomyces (yeast)	5′–T$_{1-6}$G TG$_{2-3}$–3′ 3′–A$_{1-6}$CAC$_{2-3}$–5′
Caenorhabditis (nematode)	5′–T TAGGC–3′ 3′–AATC C G–5′
Vertebrate	5′–T TAGGG–3′ 3′–AAT CC C–5′
Arabidopsis (plant)	5′–T T TAGGG–3′ 3′–AAAT CCC–5′

Source: V. A. Zakian, *Science* 270:1602, 1995.

5′-(A or T)$_m$G$_n$-3′, where m is from 1 to 4 and n is 2 or more. For example, the repeating unit in human telomeres is 5′-TTAGGG-3′, which may be repeated from hundreds to thousands of times. The sequence is always oriented with the string of Gs and Cs toward the end of the chromosome, as shown here:

toward centromere ← 5′–TTAGGGTTAGGGTTAGGG–3′
3′–AATCCCAATCCCAATCCC–5′ → end of chromosome

The G-rich strand often protrudes beyond the complementary C-rich strand at the end of the chromosome (**Figure 11.10a**) and is called the 3′ overhang. The 3′ overhang

(a)

DNA sequence at end of chromosome

5′ TTAGGGTTAGGGTTAGGGTTAGGGTTAGGG 3′
3′ AATCCCAATCCC 5′

(b)

t-loop

5′
5′
3′
TTAGGGTTAGGG 3′
AATCCCAATCCC

G-rich single-stranded overhang

11.10 DNA at the ends of eukaryotic chromosomes consists of telomeric sequences. (a) The G-rich strand at the telomere is longer than the C-rich strand. (b) In some cells, the G-rich strand folds over and pairs with a short stretch of DNA to form a t-loop.

in the telomeres of mammals is from 50 to 500 nucleotides long. Special proteins bind to the G-rich single-stranded sequence, protecting the telomere from degradation and preventing the ends of chromosomes from sticking together. A multiprotein complex called **shelterin** binds to telomeres and protects the ends of the DNA from being inadvertently repaired as a double-stranded break in the DNA. In some cells, the single-stranded overhang may fold over and pair with a short stretch of DNA to form a structure called a t-loop, which also functions in protecting the end of the telomere from degradation (**Figure 11.10b**).

> **CONCEPTS**
>
> A telomere is the stabilizing end of a chromosome. At the end of each telomere are many short telomeric sequences.

✔ **CONCEPT CHECK 6**

Which is a characteristic of DNA sequences at the telomeres?
a. One strand consists of guanine and adenine (or thymine) nucleotides.
b. They consist of repeated sequences.
c. One strand protrudes beyond the other, creating some single-stranded DNA at the end.
d. All of the above.

11.3 Eukaryotic DNA Contains Several Classes of Sequence Variation

Eukaryotic organisms differ dramatically in the amount of DNA per cell, a quantity termed an organism's **C value** (**Table 11.3**). Each cell of a fruit fly, for example, contains 35 times the amount of DNA found in a cell of the bacterium *E. coli*. In general, eukaryotic cells contain more DNA than prokaryotic cells do, but variability in the C values of different eukaryotes is huge. Human cells contain more than 10 times the amount of DNA found in *Drosophila* cells, whereas some salamander cells contain 20 times as much DNA as that

Table 11.3	Genome sizes of various organisms
Organism	**Approximate Genome Size (bp)**
λ (bacteriophage)	50,000
Escherichia coli (bacterium)	4,640,000
Saccharomyces cerevisiae (yeast)	12,000,000
Arabidopsis thaliana (plant)	125,000,000
Drosophila melanogaster (insect)	170,000,000
Homo sapiens (human)	3,200,000,000
Zea mays (corn)	4,500,000,000
Amphiuma (salamander)	765,000,000,000

in human cells. Clearly, these differences in C value cannot be explained simply by differences in organismal complexity. So, what is all of the extra DNA in eukaryotic cells doing? This question has been termed the **C-value paradox**. We do not yet have a complete answer to the C-value paradox, but eukaryotic DNA sequences reveal a complexity that is absent from prokaryotic DNA.

The Denaturation and Renaturation of DNA

The first clue that eukaryotic DNA contains several types of sequences not present in prokaryotic DNA came from studies in which double-stranded DNA was separated and then allowed to reassociate. When double-stranded DNA in solution is heated, the hydrogen bonds that hold the two strands together are weakened and, with enough heat, the two nucleotide strands separate completely, a process called **denaturation** or melting. The temperature at which DNA denatures, called the **melting temperature** (T_m), depends on the base sequence of the particular sample of DNA: G–C base pairs have three hydrogen bonds, whereas A–T base pairs only have two; so the separation of G–C pairs requires more heat (energy) than does the separation of A–T pairs.

The denaturation of DNA by heating is reversible; if single-stranded DNA is slowly cooled, single strands will collide and hydrogen bonds will form again between complementary base pairs, producing double-stranded DNA. This reaction is called **renaturation** or reannealing.

Two single-stranded molecules of DNA from different sources, such as different organisms, will anneal if they are complementary, a process termed **hybridization**. For hybridization to take place, the two strands do not have to be complementary at all their bases—just at enough bases to hold the two strands together. The extent of hybridization can be used to measure the similarity of nucleic acids from two different sources and for assessing evolutionary relationships. The rate at which hybridization takes place also provides information about the sequence complexity of DNA.

▶ **TRY PROBLEM 26**

Types of DNA Sequences in Eukaryotes

Eukaryotic DNA consists of at least three types of sequences: unique-sequence DNA, moderately repetitive DNA, and highly repetitive DNA. **Unique-sequence DNA** consists of sequences that are present only once or, at most, a few times in the genome. This DNA includes sequences that encode proteins, as well as a great deal of DNA whose function is unknown. Genes that are present in a single copy constitute from roughly 25% to 50% of the protein-encoding genes in most multicellular eukaryotes. Other genes within unique-sequence DNA are present in several similar, but not identical, copies and together are referred to as a **gene family**. Most gene families arose through duplication of an existing gene and include just a few member genes, but some, such as those that encode immunoglobulin proteins in vertebrates, contain hundreds of members. The genes that encode β-like globins are another example of a gene family. In humans, there are seven β-globin genes, clustered together on chromosome 11. The polypeptides encoded by these genes join with α-globin polypeptides to form hemoglobin molecules, which transport oxygen in the blood.

Other sequences exist in many copies and are called **repetitive DNA**. Some eukaryotic organisms have large amounts of repetitive DNA; for example, almost half of the human genome consists of repetitive DNA. A major class of repetitive DNA is called **moderately repetitive DNA**, which typically consists of sequences from 150 to 300 bp in length (although they may be longer) that are repeated many thousands of times. Some of these sequences perform important functions for the cell; for example, the genes for ribosomal RNAs (rRNAs) and transfer RNAs (tRNAs) make up a part of the moderately repetitive DNA. Moderately repetitive DNA itself is of two types of repeats. **Tandem repeat sequences** appear one after another and tend to be clustered at particular locations on the chromosomes. **Interspersed repeat sequences** are scattered throughout the genome. An example of an interspersed repeat is the *Alu* sequence, an approximately 300-bp sequence that is present more than a million times and comprises 11% of the human genome, although it has no obvious cellular function. Short repeats, such as the *Alu* sequences, are called **SINEs (short interspersed elements)**. Longer interspersed repeats consisting of several thousand base pairs are called **LINEs (long interspersed elements)**. One class of LINE, called LINE1, comprises about 17% of the human genome. Most interspersed repeats are the remnants of transposable elements, sequences that can multiply and move (see Chapter 18).

The other major class of repetitive DNA is **highly repetitive DNA**. These short sequences, often less than 10 bp in length, are present in hundreds of thousands to millions of copies that are repeated in tandem and clustered in certain regions of the chromosome, especially at centromeres and telomeres. Highly repetitive DNA is sometimes called satellite DNA, because its percentages of the four bases differ from those of other DNA sequences and, therefore, it separates as a satellite fraction when centrifuged at high speeds in a density gradient (see p. 327 in Chapter 12). Highly repetitive DNA is rarely transcribed into RNA. Although these sequences may contribute to centromere and telomere function, most highly repetitive DNA has no known function.

DNA renaturation reactions and, more recently, direct sequencing of eukaryotic genomes also tell us a lot about how genetic information is organized within chromosomes. We now know that the density of genes varies greatly among and within chromosomes. For example, human chromosome 19 has a high density of genes, with about 26 genes per million base pairs. Chromosome 13, on the other hand, has only about 6.5 genes per million base pairs. Gene density can also vary within different regions of the same chromosome: some

parts of the long arm of chromosome 13 have only 3 genes per million base pairs, whereas other parts have almost 30 genes per million base pairs. And the short arm of chromosome 13 contains almost no genes, consisting entirely of heterochromatin.

The functional role of DNA sequences that do not encode proteins, including repetitive DNA, has recently been addressed by the Encyclopedia of DNA Elements (ENCODE) project (see Chapter 20). The purpose of ENCODE was to identify all nucleotides within the human genome that have some function. The project concluded that much of the genome is transcribed and at least 80% of the sequences are functional. Many of the functional sequences appear to help control gene expression.

CONCEPTS

Eukaryotic DNA comprises three major classes: unique-sequence DNA, moderately repetitive DNA, and highly repetitive DNA. Unique-sequence DNA consists of sequences that exist in one or a few copies; moderately repetitive DNA consists of sequences that may be several hundred base pairs in length and is present in thousands to hundreds of thousands of copies. Highly repetitive DNA consists of very short sequences repeated in tandem and is present in hundreds of thousands to millions of copies. The density of genes varies greatly among and even within chromosomes.

✔ CONCEPT CHECK 7

Most of the genes that encode proteins are found in
a. unique-sequence DNA.
b. moderately repetitive DNA.
c. highly repetitive DNA.
d. all of the above.

11.4 Organelle DNA Has Unique Characteristics

As we have seen, eukaryotic chromosomes reside within the nucleus and have a complex structure consisting of DNA and associated histone proteins. However, some DNA found in eukaryotic cells occurs outside the nucleus, has a very different organization, and exhibits a different pattern of inheritance from nuclear DNA. This DNA occurs in mitochondria and chloroplasts, which are membrane-bounded organelles located in the cytoplasm of eukaryotic cells (**Figure 11.11**).

Mitochondrion and Chloroplast Structure

Mitochondria are present in almost all eukaryotic cells, whereas chloroplasts are found in plants and some protists. Both organelles generate ATP, the universal energy carrier of cells.

Mitochondria are tubular structures that are from 0.5 to 1.0 micrometer (μm) in diameter, about the size of a typical bacterium, whereas chloroplasts are typically from about 4 to 6 μm in diameter. Both are surrounded by two membranes enclosing a region (called the matrix in mitochondria and the

Mitochondrion

Chloroplast

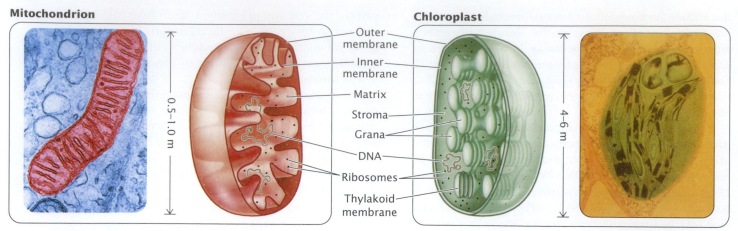

11.11 Comparison of the structures of mitochondria and chloroplasts. [Left: Don W. Fawcett/ Science Source /Photo Researchers, Inc. Right: Biophoto Associates/Photo Researchers.]

stroma in chloroplasts) that contains enzymes, ribosomes, RNA, and DNA. In mitochondria, the inner membrane is highly folded; embedded within it are the enzymes that catalyze electron transport and oxidative phosphorylation. Chloroplasts have a thylakoid membrane, which is highly folded and stacked to form aggregates called grana. This membrane bears the pigments and enzymes required for photophosphorylation. New mitochondria and chloroplasts arise by the division of existing organelles—divisions that take place throughout the cell cycle and are independent of mitosis and meiosis.

Mitochondria and chloroplasts possess DNA that encodes some polypeptides used by the organelle, as well as RNA found in the ribosome (ribosomal RNA or rRNA) and some transfer RNAs (tRNAs) needed for the translation of these proteins. The genes for most of the 900 or so structural proteins and enzymes found in mitochondria are actually encoded by nuclear DNA; the mitochondrial genome typically encodes only a few proteins and a few rRNA and tRNA molecules needed for mitochondrial protein synthesis.

The Endosymbiotic Theory

Chloroplasts and mitochondria are similar to bacteria in many ways. This resemblance is not superficial; indeed there is compelling evidence that these organelles evolved from eubacteria (see p. 19 in Chapter 2). The **endosymbiotic theory** (**Figure 11.12**) proposes that mitochondria and chloroplasts were once free-living bacteria that became internal

1 Approximately 1 billion to 1.5 billion years ago, an anaerobic eukaryotic cell engulfed an aerobic eubacterial cell through endocytosis.

2 The aerobic endosymbiont evolved into mitochondria.

3 Likewise, endocytosis of a photosynthesizing eubacterium…

4 …led to the evolution of modern eukaryotic cells with mitochondria and chloroplasts.

11.12 The endosymbiotic theory proposes that mitochondria and chloroplasts in eukaryotic cells arose from eubacteria.

inhabitants (endosymbionts) of early eukaryotic cells. It is assumed that over evolutionary time, many of the endosymbiont's original genes were subsequently lost (because nuclear genes existed that provided the same function) or were transferred to the nucleus.

A great deal of evidence supports the idea that mitochondria and chloroplasts originated as eubacterial cells. Many modern single-celled eukaryotes (protists) are hosts to endosymbiotic bacteria. Mitochondria and chloroplasts are similar in size to present-day eubacteria and possess their own DNA, which has many characteristics in common with eubacterial DNA. Mitochondria and chloroplasts possess ribosomes, some of which are similar in size and structure to eubacterial ribosomes. In addition, antibiotics that inhibit protein synthesis in eubacteria but do not affect protein synthesis in eukaryotic cells also inhibit protein synthesis in these organelles.

The strongest evidence for the endosymbiotic theory comes from the study of DNA sequences, which demonstrate that sequences in mtDNA and cpDNA are more closely related to sequences in the genes of eubacteria than they are to those found in the eukaryotic nucleus. All of this evidence indicates that mitochondria and chloroplasts are more closely related to eubacterial cells than they are to the eukaryotic cells in which they are now found.

11.13 Individual cells may contain many mitochondria, each with several copies of the mitochondrial genome. Shown is a cell of *Euglena gracilis*, a protist, stained so that the nucleus appears red, mitochondria green, and mtDNA yellow. [From Y. Huyashi and K. Veda, *Journal of Cell Sciences* 93:565, 1989.]

CONCEPTS

Mitochondria and chloroplasts are membrane-bounded organelles of eukaryotic cells that generally possess their own DNA. The well-supported endosymbiotic theory proposes that these organelles began as free-living eubacteria that developed stable endosymbiotic relations with early eukaryotic cells.

✓ **CONCEPT CHECK 8**

What evidence supports the endosymbiotic theory?

Uniparental Inheritance of Organelle-Encoded Traits

Mitochondria and chloroplasts are present in the cytoplasm, as already stated, and are usually inherited from a single parent. Thus, traits encoded by mitochondrial DNA (mtDNA) and chloroplast DNA (cpDNA) exhibit uniparental inheritance (see Chapter 5). In animals, mtDNA is inherited almost exclusively from the female parent, although occasional male transmission of mtDNA has been documented. Maternal inheritance of animal mtDNA may be partly a function of gamete size—sperm are much smaller than eggs and hold fewer mitochondria. However, recent research has found that in some eukaryotes, paternal mitochondria are selectively eliminated by autophagy, a process in which mitochondria are digested by the cell. Paternal mitochondria are targeted for destruction, whereas maternal mitochondria are not; the mechanism that produces this difference is not known. Paternal

inheritance of organelles is common in gymnosperms (conifers) and in a few angiosperms (flowering plants). Some plants even exhibit biparental inheritance of mtDNA and cpDNA.

REPLICATIVE SEGREGATION Individual cells may contain from dozens to hundreds of organelles, each with numerous copies of the organelle genome, so each cell typically possesses from hundreds to thousands of copies of mitochondrial and chloroplast genomes (**Figure 11.13**). A mutation arising within one organelle DNA molecule generates a mixture of organelles within the cell, some with a mutant DNA sequence and others with a wild-type DNA sequence. The occurrence of two distinct varieties of DNA within the cytoplasm of a single cell is termed **heteroplasmy**. When a heteroplasmic cell divides, the organelles segregate randomly into the two progeny cells in a process called **replicative segregation** (**Figure 11.14**), and chance determines the proportion of mutant organelles in each cell. Although most progeny cells will inherit a mixture of mutant and normal organelles, just by chance some cells may receive organelles with only mutant or only wild-type sequences; this situation, in which all organelles are genetically identical, is known as **homoplasmy**. Fusion of mitochondria also takes place frequently.

When replicative segregation takes place in somatic cells, it may create phenotypic variation within a single organism; different cells of the organism may possess different proportions of mutant and wild-type sequences, resulting in different degrees of phenotypic expression in different tissues. When replicative segregation takes place in the germ cells of a heteroplasmic cytoplasmic donor there may be different phenotypes among the offspring.

The disease known as myoclonic epilepsy and ragged red fiber syndrome (MERRF) is caused by a mutation in an mtDNA gene. A 20-year-old person who carried this mutation in 85% of his mtDNAs displayed a normal phenotype,

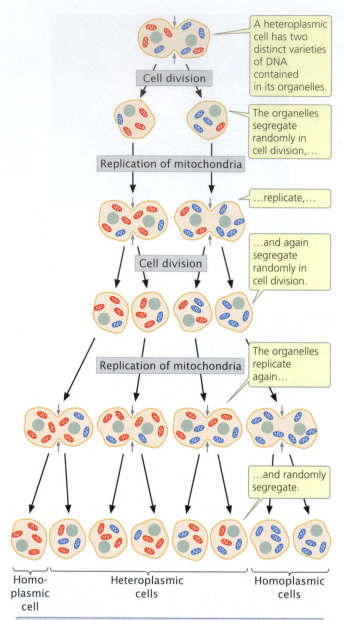

A heteroplasmic cell has two distinct varieties of DNA contained in its organelles.

Cell division

The organelles segregate randomly in cell division,…

Replication of mitochondria

…replicate,…

…and again segregate randomly in cell division.

Cell division

Replication of mitochondria

The organelles replicate again…

…and randomly segregate.

Homoplasmic cell   Heteroplasmic cells   Homoplasmic cells

Conclusion: Most of the resulting cells are heteroplasmic, but, just by chance, some cells may receive only one type of organelle (e.g., they may receive all normal or all mutant).

11.14 Organelles in a heteroplasmic cell divide randomly into the progeny cells. This diagram illustrates replicative segregation in mitosis; the same process also takes place in meiosis.

whereas a cousin who had the mutation in 96% of his mtDNAs was severely affected. In diseases caused by mutations in mtDNA, the severity of the disease is frequently related to the proportion of mutant mtDNA sequences inherited at birth. ▶ TRY PROBLEM 28

TRAITS ENCODED BY mtDNA A number of traits affected by organelle DNA have been studied. One of the first to be examined in detail was the phenotype produced by

petite mutations in yeast (**Figure 11.15**). In the late 1940s, Boris Ephrussi and his colleagues noticed that, when grown on solid medium, some colonies of yeast were much smaller than normal. Examination of these *petite* colonies revealed that the growth rates of the cells within the colonies were greatly reduced. The results of biochemical studies demonstrated that *petite* mutants were unable to carry out aerobic respiration; they obtained all of their energy from anaerobic metabolism (glycolysis and fermentation), which is much less efficient than aerobic respiration and results in the smaller colony size.

Some *petite* mutations are defects in nuclear DNA, but most *petite* mutations occur in mitochondrial DNA. Mitochondrial *petite* mutants often have large deletions in mtDNA or, in some cases, are missing mtDNA entirely. Much of the mtDNA encodes enzymes that catalyze aerobic respiration; therefore the *petite* mutants are unable to carry out aerobic respiration and cannot produce normal quantities of ATP, which inhibits their growth.

Another known mtDNA mutation occurs in *Neurospora* (see pp. 412-415 in Chapter 15). Isolated by Mary Mitchell in 1952, *poky* mutants grow slowly, display cytoplasmic inheritance, and have abnormal amounts of cytochromes. Cytochromes are protein components of the electron-transport chain of the mitochondria and play an integral role in the production of ATP. Most organisms have three primary types of cytochromes: cytochrome *a*, cytochrome *b*, and cytochrome *c*. *Poky* mutants have cytochrome *c* but no cytochrome *a* or *b*. Like *petite* mutants, *poky* mutants are defective in ATP synthesis and therefore grow more slowly than do normal, wild-type cells. ▶ TRY PROBLEM 32

In recent years, a number of genetic diseases that result from mutations in mtDNA have been identified in humans. In addition to MERRF syndrome mentioned earlier, Leber hereditary optic neuropathy (LHON) results from mutations in the mtDNA genes that encode electron-transport proteins.

Normal colony

petite colony

11.15 The *petite* mutants have large deletions in their mtDNA and are unable to carry out oxidative phosphorylation. Colonies of normal yeast cells and colonies of *petite* mutants. [From Xin Jie Chen and G. Desmond Clark-Walker, *Genetics* 144: 1445-1454, Fig1, 1996. © Genetics Society of America. Courtesy of Xin Jie Chen, Department of Biochemistry and Molecular Biology, SUNY Upstate Medical University.]

LHON typically leads to sudden loss of vision in middle age. Another disease caused by mtDNA mutations is neurogenic muscle weakness, ataxia, and retinitis pigmentosa (NARP), which is characterized by seizures, dementia, and developmental delay. Other mitochondrial diseases include Kearns–Sayre syndrome (KSS) and chronic external opthalmoplegia (CEOP), both of which result in paralysis of the eye muscles, droopy eyelids, and, in severe cases, vision loss, deafness, and dementia. All of these diseases exhibit cytoplasmic inheritance and variable expression (see Chapter 5).

A trait in plants that is produced by mutations in mitochondrial genes is cytoplasmic male sterility, a mutant phenotype found in more than 140 different plant species and inherited only from the maternal parent. These mutations inhibit pollen development but do not affect female fertility.

A number of cpDNA mutants also have been discovered. One of the first to be recognized was leaf variegation in the four o'clock plant *Mirabilis jalapa*, which was studied by Carl Correns in 1909 (see pp. 122–123 in Chapter 5). In the green alga *Chlamydomonas*, streptomycin-resistant mutations occur in cpDNA, and a number of mutants exhibiting altered pigmentation and growth in higher plants have been traced to defects in cpDNA.

CONCEPTS

In most organisms, genes encoded by mtDNA and cpDNA are inherited from a single parent. A gamete may contain more than one distinct type of mtDNA or cpDNA; in these cases, random segregation of the organelle DNA may produce phenotypic variation within a single organism or it may produce different degrees of phenotypic expression among progeny.

✓ CONCEPT CHECK 9

In a few organisms, traits encoded by mtDNA can be inherited from either parent. This observation indicates that in these organisms
a. mitochondria do not exhibit replicative segregation.
b. heteroplasmy is present.
c. both sperm and eggs contribute cytoplasm to the zygote.
d. there are multiple copies of mtDNA in each cell.

WORKED PROBLEM

To illustrate the inheritance of a trait encoded by organelle DNA, consider the following problem. A physician examines a young man who has a progressive muscle disorder and visual abnormalities. A number of the patient's relatives have the same condition, as shown in the adjoining pedigree. The degree of expression of the trait is highly variable among members of the family: some are only slightly affected, whereas others developed severe symptoms at an early age. The physician concludes that this disorder is due to a mutation in the mitochondrial genome. Do you agree with the physician's conclusion? Why or why not? Could the disorder be due to a mutation in a nuclear gene? Explain your reasoning.

Solution Strategy

What information is required in your answer to the problem?

An explanation of whether this disorder could be due to a mutation in the mitochondrial genome and why, as well as whether the disorder could be due to a mutation in a nuclear gene and why.

What information is provided to solve the problem?

- The young man has a progressive muscle disorder and visual abnormalities.
- A pedigree illustrating the young man's family.
- The trait is highly variable among members of the family.

Solution Steps

The conclusion that the disorder is caused by a mutation in the mitochondrial genome is supported by the pedigree and the observation of variable expression in affected members of the same family. The disorder is passed only from affected mothers to both male and female offspring; when fathers are affected, none of their children have the trait (as seen in the children of II-2 and III-6). This outcome is expected of traits determined by mutations in mtDNA, because mitochondria are in the cytoplasm and usually inherited only from a single (in humans, the maternal) parent. The trait cannot be X-linked recessive, because a cross between a female with the trait ($X^a X^a$) and a male without the trait ($X^+ Y$) would *not* produce daughters with the trait ($X^a X^a$), which we see in III-10, IV-3, and IV-4. It cannot be X-linked dominant because II-2 and III-6 would have to pass it to their daughters, who are unaffected (unless the trait exhibited incomplete penetrance).

The facts that some offspring of affected mothers do not show the trait (III-9 and IV-5) and that expression varies from one person to another suggest that affected persons are heteroplasmic, with both mutant and wild-type mitochondria. Random segregation of mitochondria in meiosis may produce gametes having different proportions of mutant and wild-type sequences, resulting in different degrees of phenotypic expression among the offspring.

Most likely, symptoms of the disorder develop when some minimum proportion of the mitochondria are mutant. Just by chance, some of the gametes produced by an affected mother contain few mutant mitochondria and result in offspring that lack the disorder.

Another possible explanation for the disorder is that it results from an autosomal dominant gene. When an affected (heterozygous) person mates with an unaffected (homozygous) person, about half of the offspring are expected to have the trait, but, just by chance, some affected parents will have no affected offspring. Individuals II-2 and III-6 in the pedigree could just have happened to be male and their sex could be unrelated to the mode of transmission. The variable expression could be explained by variable expressivity (see Chapter 5).

▶ For more experience with the inheritance of organelle encoded traits, try working Problem 29 at the end of the chapter.

Table 11.4	Sizes of mitochondrial genomes in selected organisms
Organism	**Size of mtDNA (bp)**
Pichia canadensis (fungus)	27,694
Podospora anserina (fungus)	100,314
Saccharomyces cerevisiae (fungus)	85,779*
Drosophila melanogaster (fruit fly)	19,517
Lumbricus terrestris (earthworm)	14,998
Xenopus laevis (frog)	17,553
Mus musculus (house mouse)	16,295
Homo sapiens (human)	16,569
Chlamydomonas reinhardtii (green alga)	15,758
Plasmodium falciparum (protist)	5,966
Paramecium aurelia (protist)	40,469
Arabidopsis thaliana (plant)	166,924
Cucumis melo (plant)	2,400,000

*Size varies among strains.

The Mitochondrial Genome

In most animals and fungi, the entire mitochondrial genome exists on a single, highly coiled, circular DNA molecule, although there may be many copies of this genome in each cell. The circular structure of the mitochondria is similar in structure to a eubacterial chromosome. Plant mitochondrial genomes often exist as a complex collection of multiple circular DNA molecules. In some species, the mitochondrial genome consists of a single, linear DNA molecule.

Each mitochondrion contains multiple copies of the mitochondrial genome, and a cell may contain many mitochondria. A typical rat liver cell, for example, has from 5 to 10 mtDNA molecules in each of about 1000 mitochondria; so each cell possesses from 5000 to 10,000 copies of the mitochondrial genome. Mitochondrial DNA constitutes about 1% of the total cellular DNA in a rat liver cell. Like eubacterial chromosomes, mtDNA lacks the histone proteins normally associated with eukaryotic nuclear DNA, although it is complexed with other proteins that have some histone-like properties. The guanine–cytosine (GC) content of mtDNA is often sufficiently different from that of nuclear DNA in that mtDNA can be separated from nuclear DNA by density-gradient centrifugation.

Mitochondrial genomes are small compared with nuclear genomes and vary greatly in size among different organisms (**Table 11.4**). The sizes of mitochondrial genomes of most species range from 15,000 bp to 65,000 bp, but those of a few species are much smaller (e.g., the genome of *Plasmodium falciparum*, the parasite that causes

malaria, is only 6,000 bp) while those of some plants are several million base pairs. Although the amount of DNA in mitochondrial genomes varies widely, there is no correlation between genome size and number of genes. The number of genes is more constant than genome size; most species have only from 40 to 50 genes. These genes encode five basic functions: respiration and oxidative phosphorylation, translation, transcription, RNA processing, and the import of proteins into the cell. Most of the variation in size of mitochondrial genomes is due to differences in noncoding DNA sequences. As mentioned earlier, genes for most of the proteins and enzymes found in mitochondria are actually encoded by nuclear DNA.

HUMAN mtDNA Human mtDNA is a circular molecule encompassing 16,569 bp that encode two rRNAs, 22 tRNAs, and 13 proteins. The two nucleotide strands of the molecule differ in their base composition: the heavy (H) strand has more guanine nucleotides, whereas the light (L) strand has more cytosine nucleotides. The H strand is the template for both rRNAs, 14 of the 22 tRNAs, and 12 of the 13 proteins, whereas the L strand serves as template for only 8 of the tRNAs and 1 protein. The D loop (**Figure 11.16**) is a region of the mtDNA that contains sites where replication and transcription of the mtDNA is initiated. Human mtDNA is highly economical in its organization: there are few noncoding nucleotides between the genes and almost all the messenger RNA codes for proteins. Human mtDNA also contains very little repetitive DNA. The one region

(a)

(b)

11.16 The human mitochondrial genome, consisting of 16,569 bp, is highly economical in its organization. (a) The outer circle represents the heavy (H) strand, and the inner circle represents the light (L) strand. The origins of replication for the H and L strands are *ori* H and *ori* L, respectively. ND identifies genes that encode subunits of NADH dehydrogenase. (b) Electron micrograph of isolated mtDNA. [Part b: CNRI/Photo Researchers.]

CONCEPTS

The mitochondrial genome consists of circular DNA with no associated histone proteins, although it is complexed with other proteins that have some histone-like properties. The sizes and structures of mtDNA differ greatly among organisms. Human mtDNA exhibits extreme economy, but mtDNAs found in yeast and flowering plants contain many noncoding nucleotides and repetitive sequences. Mitochondrial DNA in most flowering plants is large and typically has one or more large direct repeats that can recombine to generate smaller or larger molecules.

of the human mtDNA that does contain some noncoding nucleotides is the D loop.

YEAST mtDNA   The organization of yeast mtDNA is quite different from that of human mtDNA. Although the yeast mitochondrial genome with 78,000 bp is nearly five times as large, it encodes only six additional genes, for a total of 2 rRNAs, 25 tRNAs, and 16 polypeptides (**Figure 11.17**). Most of the extra DNA in the yeast mitochondrial genome consists of noncoding sequences found within and between genes.

FLOWERING-PLANT mtDNA   Flowering plants (angiosperms) have the largest and most-complex mitochondrial genomes known; their mitochondrial genomes range in size from 186,000 bp in white mustard to 2,400,000 bp in muskmelon. Even closely related plant species may differ greatly in the sizes of their mtDNA.

Part of the extensive size variation in the mtDNA of flowering plants can be explained by the presence of long sequences that are direct repeats. Crossing over between these repeats can generate multiple circular chromosomes of different sizes. The mitochondrial genome in turnips, for example, consists of a "master circle" consisting of 218,000 bp that has direct repeats. Homologous recombination between the repeats can generate two smaller circles of 135,000 bp and 83,000 bp (**Figure 11.18**). Other species contain several direct repeats, providing possibilities for complex crossing-over events that may increase or decrease the number and sizes of the circles.

11.17 The yeast mitochondrial genome, consisting of 78,000 bp, contains much noncoding DNA.

11.18 Size variation in plant mtDNA can be generated through recombination between direct repeats. In turnips, the mitochondrial genome consists of a "master circle" of 218,000 bp; crossing over between the direct repeats produces two smaller circles of 135,000 bp and 83,000 nucleotide pairs.

The Evolution of Mitochondrial DNA

As already mentioned, comparisons of mitochondrial DNA sequences with DNA sequences in bacteria strongly support a common eubacterial origin for all mtDNA. Nevertheless, patterns of evolution seen in mtDNA vary greatly among different groups of organisms.

The sequences of vertebrate mtDNA exhibit an accelerated rate of evolution: the sequences in mammalian mtDNA, for example, typically change from 5 to 10 times as fast as those in mammalian nuclear DNA. The accelerated rate of evolution seen in vertebrate mtDNA is due to its high mutation rate, which allows DNA sequences to change more quickly. In spite of the high rate of sequence evolution, the number of genes present and the organization of vertebrate mitochondrial genomes are relatively constant. In contrast, sequences of plant mtDNA evolve slowly at a rate only $\frac{1}{10}$ that of the nuclear genome, but their gene content and organization change rapidly. The reason for these basic differences in rates of evolution is not yet known.

Mitochondrial DNA has been studied extensively to reconstruct patterns of evolution in humans and many other organisms. Some of the advantages of mtDNA for studying evolution include: (1) the small size and abundance of mtDNA in the cell; (2) the rapid evolution of mtDNA sequences in some organisms, facilitating study of closely related groups; and (3) the maternal inheritance of mtDNA and lack of recombination, which makes it possible to trace female lines of descent. Samples of human mtDNA have been analyzed from thousands of people belonging to hundreds of different ethnic groups throughout the world. These mtDNA samples are helping to unravel many aspects of human evolution and history. For example, initial studies on mtDNA sequences led to the proposal that small groups of humans migrated out of Africa about 85,000 years ago and populated the rest of the world. This is called the Out of Africa hypothesis or the African Replacement hypothesis, and has now gained wide acceptance. The Out of Africa hypothesis is supported by additional studies of DNA sequences from the Y chromosome and nuclear genes. The use of mtDNA in evolutionary studies will be described in more detail in Chapter 26.

At conception, a mammalian zygote inherits approximately 100,000 copies of mtDNA inherited from the egg. Because of the large number of mtDNA molecules in each cell and the high rate of mutation in mtDNA, most cells would be expected to contain a mixture of wild-type and mutant mtDNA molecules (heteroplasmy). However, heteroplasmy is rarely present: in most organisms, the copies of mtDNA are genetically identical (homoplasmy). To account for the uniformity of mtDNA within individual organisms, geneticists hypothesize that, in early development or gamete formation, mtDNA goes through some type of bottleneck, during which the mtDNAs within a cell are reduced to just a few copies, which then replicate and give rise to all subsequent copies of mtDNA. Through this process, genetic variation in mtDNA within a cell is eliminated and most copies of mtDNA are identical. Recent studies have provided evidence that a bottleneck does exist, but there is contradictory evidence concerning where in development it arises.

CONCEPTS

All mtDNA appears to have evolved from a common eubacterial ancestor, but the patterns of evolution seen in different mitochondrial genomes vary greatly. Vertebrate mtDNA exhibits rapid change in sequence but little change in gene content and organization, whereas the mtDNA of plants exhibits little change in sequence but much variation in gene content and organization. Mitochondrial DNA sequences are frequently used to study patterns of evolution.

Damage to Mitochondrial DNA Is Associated with Aging

The symptoms of many human genetic diseases caused by defects in mtDNA first appear in middle age or later and increase in severity as people age. One hypothesis to explain this is related to the decline in oxidative phosphorylation with aging.

Oxidative phosphorylation is the process that generates ATP, the primary carrier of energy in the cell. This process takes place on the inner membrane of the mitochondrion and requires a number of different proteins, some encoded by mtDNA and others encoded by nuclear genes. Oxidative phosphorylation normally declines with age and, if it falls below a critical threshold, tissues do not make enough ATP to sustain vital functions and disease symptoms appear. Most people start life with an excess capacity for oxidative phosphorylation; this capacity decreases with age, but most people reach old age or die before the critical threshold is passed. Persons born with mitochondrial diseases carry mutations in their mtDNA that lower their oxidative phosphorylation capacity. At birth, their capacity may be sufficient to support their ATP needs but, as their oxidative phosphorylation capacity declines with age, they cross the critical threshold and begin to experience disease symptoms.

Why does oxidative phosphorylation capacity decline with age? A possible explanation is that damage to mtDNA accumulates with age: deletions and base substitutions in mtDNA increase with age. For example, a common 5000-bp deletion in mtDNA is absent in normal heart muscle cells before the age of 40, but, afterward, this deletion is present with increasing frequency. The same deletion is found at a low frequency in normal brain tissue before age 75 but is found in 11% to 12% of mtDNAs in the basal ganglia by age 80. People with mtDNA genetic diseases may age prematurely because they begin life with damaged mtDNA.

The mechanism of age-related increases in mtDNA damage is not yet known. Oxygen radicals—highly reactive compounds that are natural by-products of oxidative phosphorylation—are known to damage DNA (see p. 508 in Chapter 18). Because mtDNA is physically close to the enzymes taking part in oxidative phosphorylation, mtDNA may be more prone to oxidative damage than is nuclear DNA. When mtDNA has been damaged, the cell's capacity to produce ATP drops.

The Chloroplast Genome

Geneticists have long recognized that many traits associated with chloroplasts exhibit cytoplasmic inheritance, indicating that these traits are not encoded by nuclear genes. In 1963, chloroplasts were shown to have their own DNA (**Figure 11.19**).

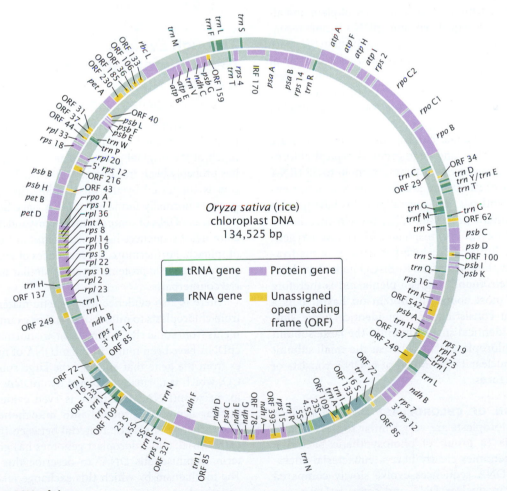

11.19 Chloroplast DNA of rice.

Table 11.5	Sizes of chloroplast genomes in selected organisms
Organism	**Size of cpDNA (bp)**
Euglena gracilis (protist)	143,172
Porphyra purpurea (red alga)	191,028
Chlorella vulgaris (green alga)	150,613
Marchantia polymorpha (liverwort)	121,024
Nicotiana tabacum (tobacco)	155,939
Zea mays (corn)	140,387
Pinus thunbergii (black pine)	119,707

Among different plants, the chloroplast genome ranges in size from 80,000 to 600,000 bp, but most chloroplast genomes range from 120,000 to 160,000 bp (**Table 11.5**). Chloroplast DNA is usually a single, double-stranded DNA molecule that is circular, highly coiled, and lacks associated histone proteins. As in mtDNA, multiple copies of the chloroplast genome are found in each chloroplast, and there are multiple organelles per cell; so there are several hundred to several thousand copies of cpDNA in a typical plant cell.

The chloroplast genomes from a number of plant and algal species have been sequenced, and cpDNA is now recognized to be basically eubacterial in its organization: the order of some groups of genes is the same as that observed in *E. coli*, and many chloroplast genes are organized into clusters similar to those found in bacteria. Many of the sequences in cpDNA are quite similar to those found in equivalent eubacterial genes.

Among vascular plants, chloroplast chromosomes are similar in gene content and gene order. A typical chloroplast genome encodes 4 rRNA genes, from 30 to 35 tRNA genes, a number of ribosomal proteins, many proteins engaged in photosynthesis, and several proteins having roles in nonphotosynthetic processes. A key protein encoded by cpDNA is ribulose-1,5-bisphosphate carboxylase-oxygenase (abbreviated RuBisCO), which participates in the fixation of carbon in photosynthesis. RuBisCO makes up about 50% of the protein found in green plants and is therefore considered the most abundant protein on Earth. It is a complex protein consisting of eight identical large subunits and eight identical small subunits. The large subunit is encoded by chloroplast DNA, whereas the small subunit is encoded by nuclear DNA. Much of cpDNA consists of noncoding sequences.

THE EVOLUTION OF CHLOROPLAST DNA The DNA sequences of chloroplasts are very similar to those found in cyanobacteria (a group of photosynthetic bacteria), so chloroplast genomes clearly have a eubacterial ancestry. Overall, cpDNA sequences evolve slowly compared with sequences in nuclear DNA and some mtDNA. For most chloroplast genomes, size and gene organization are similar, although there are some notable exceptions. Because they evolve slowly and, like mtDNA, are inherited from only one parent, cpDNA is often useful for determining the evolutionary relationships among different plant species.

CONCEPTS

Most chloroplast genomes consist of a single circular DNA molecule not complexed with histone proteins. Although there is considerable size variation among species, the cpDNAs found in most vascular plants are about 150,000 bp. Chloroplast DNA sequences are most similar to DNA sequences in cyanobacteria, which supports the endosymbiotic theory.

✔ CONCEPT CHECK 10

In its organization, chloroplast DNA is most similar to
a. eubacteria.
b. archaea.
c. nuclear DNA of plants.
d. nuclear DNA of primitive eukaryotes.

Through Evolutionary Time, Genetic Information Has Moved Between Nuclear, Mitochondrial, and Chloroplast Genomes

Many proteins found in modern mitochondria and chloroplasts are encoded by nuclear genes, which suggests that much of the original genetic material in the endosymbiont has probably been transferred to the nucleus. This assumption is supported by the observation that some DNA sequences normally found in mtDNA have been detected in the nuclear DNA of some strains of yeast and maize. Likewise, chloroplast sequences have been found in the nuclear DNA of spinach. Furthermore, the sequences of nuclear genes that encode organelle proteins are most similar to their eubacterial counterparts.

There is also evidence that genetic material has moved from chloroplasts to mitochondria. For example, DNA fragments from some rRNA genes that are normally encoded by cpDNA have been found in the mtDNA of maize. Sequences from the gene that encodes the large subunit of RuBisCO, which is normally encoded by cpDNA, are duplicated in maize mtDNA. And there is even evidence that some nuclear genes have moved into mitochondrial genomes. The exchange of genetic material between the nuclear, mitochondrial, and chloroplast genomes has given rise to the term "promiscuous DNA" to describe this phenomenon. The mechanism by which this exchange takes place is not entirely clear.

CONCEPTS SUMMARY

- Chromosomes contain very long DNA molecules that are tightly packed.

- Supercoiling results from strain produced when rotations are added to a relaxed DNA molecule or removed from it. Overrotation produces positive supercoiling; underrotation produces negative supercoiling. Supercoiling is controlled by topoisomerase enzymes.

- A bacterial chromosome consists of a single, circular DNA molecule that is bound to proteins and exists as a series of large loops. It usually appears in the cell as a distinct clump known as the nucleoid.

- Each eukaryotic chromosome contains a single, long linear DNA molecule that is bound to histone and nonhistone chromosomal proteins. Euchromatin undergoes the normal cycle of decondensation and condensation in the cell cycle. Heterochromatin remains highly condensed throughout the cell cycle.

- The nucleosome is a core of eight histone proteins and the DNA that wraps around the core. Nucleosomes are folded into a 30-nm fiber that forms a series of 300-nm-long loops; these loops are anchored at their bases by proteins. The 300-nm loops are condensed to form a fiber that is itself tightly coiled to produce a chromatid.

- Chromosome regions that are undergoing active transcription are sensitive to digestion by DNase I, indicating that DNA unfolds during transcription.

- Epigenetic changes are stable alterations of gene expression that do not require changes in DNA sequences. Epigenetic changes can take place through alterations of chromatin structure.

- Centromeres are chromosomal regions where spindle fibers attach; chromosomes without centromeres are usually lost in the course of cell division. Most centromeres are defined by epigenetic changes to chromatin structure. Telomeres stabilize the ends of chromosomes.

- Eukaryotic DNA exhibits three classes of sequences. Unique-sequence DNA exists in very few copies. Moderately repetitive DNA consists of moderately long sequences that are repeated from hundreds to thousands of times. Highly repetitive DNA consists of very short sequences that are repeated in tandem from many thousands to millions of times.

- Mitochondria and chloroplasts are eukaryotic organelles that possess their own DNA. The endosymbiotic theory proposes that mitochondria and chloroplasts originated as free-living prokaryotic (specifically eubacterial) organisms that entered into a beneficial association with eukaryotic cells.

- Traits encoded by mtDNA and cpDNA are usually inherited from a single parent, most often the mother. Random segregation of organelles in cell division may produce phenotypic variation among cells within an individual organism and among the offspring of a single female.

- The mitochondrial genome usually consists of a single circular DNA molecule that lacks histone proteins. Mitochondrial DNA varies in size among different groups of organisms. Human mtDNA is highly economical, with few noncoding nucleotides. Fungal and plant mtDNAs contain much noncoding DNA between genes.

- Comparisons of mtDNA sequences suggest that mitochondria evolved from a eubacterial ancestor. Vertebrate mtDNA exhibits rapid change in sequence but little change in gene content and organization. Plant mtDNA exhibits little change in sequence but much variation in gene content and organization.

- Mitochondrial DNA sequences are widely used to study evolution.

- Chloroplast genomes consist of a single circular DNA molecule that lacks histone proteins and varies little in size. Each plant cell contains multiple copies of cpDNA. Chloroplast DNA sequences are most similar to those in cyanobacteria and tend to evolve slowly.

- Through evolutionary time, many mitochondrial and chloroplast genes have moved to nuclear chromosomes. In some plants, there is evidence that copies of chloroplast genes have moved to the mitochondrial genome.

IMPORTANT TERMS

mitochondrial DNA (mtDNA) (p. 300)
chloroplast DNA (cpDNA) (p. 300)
supercoiling (p. 300)
relaxed state of DNA (p. 300)
positive supercoiling (p. 300)
negative supercoiling (p. 300)
topoisomerase (p. 300)
nucleoid (p. 301)

euchromatin (p. 302)
heterochromatin (p. 302)
nonhistone chromosomal protein (p. 302)
nucleosome (p. 303)
linker DNA (p. 304)
polytene chromosome (p. 304)
chromosomal puff (p. 304)
epigenetic change (p. 306)

telomeric sequence (p. 307)
shelterin (p. 308)
C value (p. 308)
C-value paradox (p. 308)
denaturation (melting) (p. 308)
melting temperature (T_m) (p. 308)
renaturation (reannealing) (p. 308)

hybridization (p. 308)
unique-sequence DNA (p. 308)
gene family (p. 308)
repetitive DNA (p. 309)
moderately repetitive DNA (p. 309)
tandem repeat sequence (p. 309)

interspersed repeat sequence
  (p. 309)
short interspersed element
  (SINE) (p. 309)

long interspersed element
  (LINE) (p. 309)
highly repetitive DNA
  (p. 309)

endosymbiotic theory
  (p. 310)
heteroplasmy
  (p. 311)

replicative segregation
  (p. 311)
homoplasmy (p. 311)
D loop (p. 314)

ANSWERS TO CONCEPT CHECKS

1. b

2. Bacterial DNA is not complexed to histone proteins and is circular.

3. b

4. d

5. A chromosome that loses its centromere will not segregate into the nucleus in mitosis and is usually lost.

6. d

7. a

8. Many modern protists are hosts to endosymbiotic bacteria. Mitochondria and chloroplasts are similar in size to eubacteria and have their own DNA, as well as ribosomes that are similar in size and shape to eubacterial ribosomes. Antibiotics that inhibit protein synthesis in eubacteria also inhibit protein synthesis in mitochondria and chloroplasts. Gene sequences in mtDNA and cpDNA are most similar to eubacterial DNA sequences.

9. c

10. a

WORKED PROBLEMS

Problem 1

A diploid plant cell contains 2 billion base pairs of DNA.

a. How many nucleosomes are present in the cell?

b. Give the number of molecules of each type of histone protein associated with the genomic DNA.

Solution Strategy

What information is required in your answer to the problem?
The number of nucleosomes per cell and the numbers of each type of histone protein associated with the DNA.

What information is provided to solve the problem?
The cell contains 2 billion base pairs of DNA.

For help with this problem, review:
The Nucleosome in Section 11.1.

Solution Steps

> **Recall:** The repeating unit of the chromosome is a nucleosome, which consists of DNA complexed to histone proteins.

Each nucleosome encompasses about 200 bp of DNA: from 145 to 147 bp of DNA wrapped around the histone core, from 20 to 22 bp of DNA associated with the H1 protein, and another 30 to 40 bp of linker DNA.

a. To determine how many nucleosomes are present in the cell, we simply divide the total number of base pairs of DNA (2×10^9 bp) by the number of base pairs per nucleosome:

$$\frac{2 \times 10^9 \text{ nucleotides}}{2 \times 10^2 \text{ nucleotides per nucleosome}}$$

$$= 1 \times 10^7 \text{ nucleosomes}$$

Thus, there are approximately 10 million nucleosomes in the cell.

b. Each nucleosome includes two molecules each of H2A, H2B, H3, and H4 histones. Therefore, there are 2×10^7 molecules each of H2A, H2B, H3, and H4 histones. Each nucleosome has associated with it one copy of the H1 histone; so there are 1×10^7 molecules of H1.

Problem 2

Suppose that a new organelle is discovered in an obscure group of protists. This organelle contains a small DNA genome, and some scientists are arguing that, like chloroplasts and mitochondria, this organelle originated as a free-living eubacterium that entered into an endosymbiotic relation with the protist. Outline a research plan to determine if the new organelle evolved from a free-living eubacterium. What kinds of data would you collect and what predictions would you make if the theory were correct?

Solution Strategy

What information is required in your answer to the problem?

A research plan with the types of data you would collect and predictions.

What information is provided to solve the problem?

- A new organelle is discovered.
- The organelle contains a small DNA genome.
- The organelle might have evolved from an endosymbiotic relationship.

For help with this problem, review:

Sections on Endosymbiotic Theory, The Mitochondrial Genome, and The Chloroplast Genome in Section 11.4.

Solution Steps

We should examine the structure, organization, and sequences of the organelle genome. If the organelle shows only characteristics of eukaryotic DNA, then it most likely has a eukaryotic origin but, if it displays some characteristics of eubacterial DNA, then this supports the theory of a eubacterial origin.

We could start by examining the overall characteristics of the organelle DNA. If it has a eubacterial origin, we might expect that the organelle genome will consist of a circular molecule and will lack histone proteins. We could compare the DNA sequences found in the organelle genome with homologous sequences from eubacteria and eukaryotic genomes. If the theory of an endosymbiotic origin is correct, then the organelle sequences should be most similar to homologous sequences found in eubacteria.

> **Recall:** The endosymbiotic theory proposes that organelles evolved from eubacteria.

COMPREHENSION QUESTIONS

Section 11.1

1. How does supercoiling arise? What is the difference between positive and negative supercoiling?

2. What functions does supercoiling serve for the cell?

3. Describe the composition and structure of the nucleosome.

4. Describe in steps how the double helix of DNA, which is 2 nm in width, gives rise to a chromosome that is 700 nm in width.

5. What are polytene chromosomes and chromosomal puffs?

6. What are epigenetic changes and how are they brought about?

Section 11.2

7. Describe the function of the centromere. How are centromeres different from other regions of the chromosome?

8. Describe the function and molecular structure of a telomere.

9. What is the difference between euchromatin and heterochromatin?

Section 11.3

10. What is the C value of an organism?

11. Describe the different classes of DNA sequence variation that exist in eukaryotes.

Section 11.4

12. Explain why many traits encoded by mtDNA and cpDNA exhibit considerable variation in their expression, even among members of the same family.

13. What is the endosymbiotic theory? How does it help to explain some of the characteristics of mitochondria and chloroplasts?

14. What evidence supports the endosymbiotic theory?

15. Briefly describe the organization of genes on the chloroplast genome.

16. What is meant by the term "promiscuous DNA"?

> For more questions that test your comprehension of the key chapter concepts, go to *LEARNINGCurve* for this chapter.

APPLICATION QUESTIONS AND PROBLEMS

Introduction

17. The introduction to this chapter discussed a study of telomere length in Romanian children. The study demonstrated that children raised in orphanages had shorter telomeres than children raised in foster homes. What effect, if any, do you think having shorter telomeres in childhood might have on adult life?

Section 11.1

*18. Compare and contrast prokaryotic and eukaryotic chromosomes. How are they alike and how do they differ?

*19. (a) In a typical eukaryotic cell, would you expect to find more molecules of the H1 histone or more molecules of the H2A histone? Explain your reasoning. (b) Would you expect to find more molecules of H2A or more molecules of H3? Explain your reasoning.

*20. Based on the DNA sensitivity to DNase I illustrated in **Figure 11.7**, which type of chicken hemoglobin (embryonic or adult) is likely produced in highest quantity at the following tissues and developmental stages?

   a. Erythroblasts during the first 24 hours.

   b. Erythroblasts at day 5

   c. Erythroblasts at day 14

   d. Brain cells throughout development

21. Suppose that a researcher briefly added radioactively labeled uridine to the *Drosophila* larva whose polytene chromosome is shown in **Figure 11.6**. Label on the figure where you would expect to see accumulation of radioactive uridine.

*22. A diploid human cell contains approximately 6.4 billion base pairs of DNA.

   a. How many nucleosomes are present in such a cell? (Assume that the linker DNA encompasses 40 bp.)

   b. How many histone proteins are complexed to this DNA?

*23. Would you expect to see more or less acetylation in regions of DNA that are sensitive to digestion by DNase I? Why?

24. Gunter Korge examined several proteins that are secreted from the salivary glands of *Drosophila melanogaster* during larval development (G. Korge. 1975. *Proceedings of the National Academy of Sciences of the United States of America* 72:4550–4554). One protein, called protein fraction 4, was encoded by a gene found by deletion mapping to be located on the X chromosome at position 3C. Korge observed that, about 5 hours before the first synthesis of protein fraction 4, an expanded and puffed-out region formed on the X chromosome at position 3C. This chromosome puff disappeared before the end of the third larval instar stage, when the synthesis of protein fraction 4 ceased. He observed that there was no puff at position 3C in a special strain of flies that lacked secretion of protein fraction 4. Explain these results. What is the chromosome puff at region 3 and why does its appearance and disappearance roughly coincide with the secretion of protein fraction 4?

25. Suppose a chemist develops a new drug that neutralizes the positive charges on the tails of histone proteins. What would be the most likely effect of this new drug on chromatin structure? Would this drug have any effect on gene expression? Explain your answers.

Section 11.3

*26. Which of the following two molecules of DNA has the lower melting temperature? Why?

AGTTACTAAAGCAATACATC
TCAATGATTTCGTTATGTAG

AGGCGGGTAGGCACCCTTA
TCCGCCCATCCGTGGGAAT

27. In a DNA hybridization study, DNA was isolated from a particular species, labeled with ^{32}P, and sheared into small fragments (S. K. Dutta et al. 1967. *Genetics* 57:719–727). Hybridizations between these labeled fragments and denatured DNA from different species were then compared. The following table gives the percentages of labeled wheat DNA that hybridized to DNA molecules of wheat, corn, radish, and cabbage.

Species	Percentage of bound wheat DNA hybridized relative to wheat
Wheat	100
Cabbage	23
Corn	63
Radish	30

What do these results indicate about the evolutionary differences among these organisms?

Section 11.4

*28. A wheat plant that is light green in color is found growing in a field. Biochemical analysis reveals that chloroplasts in this plant produce only 50% of the chlorophyll normally found in wheat chloroplasts. Propose a set of crosses to determine whether the light-green phenotype is caused by a mutation in a nuclear gene or in a chloroplast gene.

*29. A rare neurological disease is found in the family illustrated in the following pedigree. What is the most likely mode of inheritance for this disorder? Explain your reasoning.

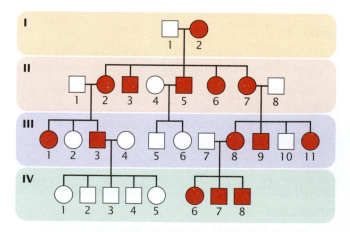

30. Assume that the disorder shown in the pedigree in the Worked Problem on p. 313 is a rare disease that results from a defect in mitochrondrial DNA. If individual III-8 has a daughter, what is the probability that the daughter will inherit the muscle disorder from her affected parent?

31. Fredrick Wilson and his colleagues studied members of a large family who had low levels of magnesium in their blood (see the pedigree below). They argued that this disorder of magnesium (and associated high blood pressure and high cholesterol) is caused by a mutation in mtDNA (F. H. Wilson et al. 2004. *Science* 306:1190–1194).

a. What evidence suggests that a gene in the mtDNA is causing this disorder?

b. Could this disorder be caused by an autosomal dominant gene? Why or why not?

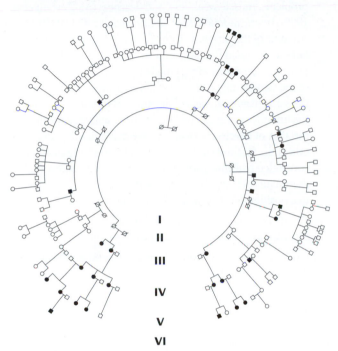

[After F. H. Wilson et al. 2004. *Science* 306:1190–1194.]

*32. In a particular strain of *Neurospora*, a *poky* mutation exhibits biparental inheritance, whereas *poky* mutations in other strains are inherited only from the maternal parent. Explain these results.

*33. A scientist collects cells at various points in the cell cycle and isolates DNA from them. Using density-gradient centrifugation, she separates the nuclear and mtDNA. She then measures the amount of mtDNA and nuclear DNA present at different points in the cell cycle. On the following graph, draw a line to represent the relative amounts of nuclear DNA that you expect her to find per cell throughout the cell cycle. Then, draw a dotted line on the same graph to indicate the relative amount of mtDNA that you would expect to see at different points throughout the cell cycle.

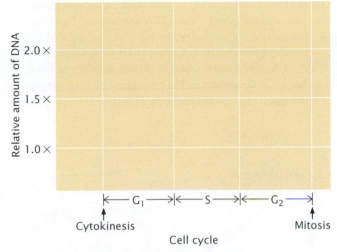

34. In 1979, bones found outside Ekaterinburg, Russia, were shown to be those of Tsar Nicholas and his family, who were executed in 1918 by a Bolshevik firing squad in the Russian Revolution (see the introduction to Chapter 14). To prove that the skeletons were those of the royal family, mtDNA was extracted from the bone samples, amplified by PCR, and compared with mtDNA from living relatives of the tsar's family.

a. Why was DNA from the mitochondria analyzed instead of nuclear DNA? What are some of the advantages of using mtDNA for this type of study?

b. Mitochrondrial DNA from which living relatives would provide useful information for verifying that the skeletons were those of the royal family?

35. Antibiotics such as chloramphenicol, tetracycline, and erythromycin inhibit protein synthesis in eubacteria but have no effect on protein synthesis encoded by nuclear genes. Cycloheximide inhibits protein synthesis encoded by nuclear genes but has no effect on eubacterial protein synthesis. How might these compounds be used to determine which proteins are encoded by mitochondrial and chloroplast genomes?

CHALLENGE QUESTIONS

Section 11.1

36. An explorer discovers a strange new species of plant and sends some of the plant tissue to a geneticist to study. The geneticist isolates chromatin from the plant and examines it with an electron microscope. She observes what appear to be beads on a string. She then adds a small amount of nuclease, which cleaves the string into individual beads that each contain 280 bp of DNA. After digestion with more nuclease, a 120-bp fragment of DNA remains attached to a core of histone proteins. Analysis of the histone core reveals histones in the following proportions:

H1	12.5%
H2A	25%
H2B	25%
H3	0%
H4	25%
H7 (a new histone)	12.5%

On the basis of these observations, what conclusions could the geneticist make about the probable structure of the nucleosome in the chromatin of this plant?

Section 11.3

37. In DNA-hybridization experiments on six species of plants in the genus *Vicia*, DNA was isolated from each of the six species, denatured by heating, and sheared into small fragments (W. Y. Chooi. 1971. *Genetics* 68:213–230). In one experiment, DNA from each species and from *E. coli* was allowed to renature. The graph shows the results of this renaturation experiment.

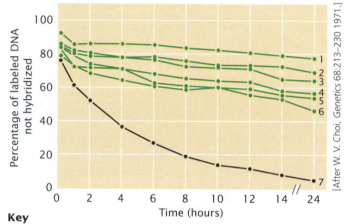

[After W. V. Choi, *Genetics* 68:213–230 1971.]

Key

1 = *V. melanops*, 2 = *V. sativa*, 3 = *V. benghalensis*,
4 = *V. atropurpurea*, 5 = *V. faba*, 6 = *V. narbonensis*, 7 = *E. coli*

a. Can you explain why the *E. coli* DNA renatures at a much faster rate than does DNA from all of the *Vicia* species?

Vodder vetch (*Vicia sativa*). [Bob Gibbons/Alamy.]

b. Notice that, for the *Vicia* species, the rate of renaturation is much faster in the first hour and then slows down. What might cause this initial rapid renaturation and the subsequent slowdown?

Section 11.4

38. Steven Frank and Laurence Hurst argued that a cytoplasmically inherited mutation in humans that has severe effects in males but no effect in females will not be eliminated from a population by natural selection, because only females pass on mtDNA (S. A. Frank and L. D. Hurst. 1996. *Nature* 383:224). Using this argument, explain why males with Leber hereditary optic neuropathy are more severely affected than females.

39. In a study of a myopathy, several families exhibited vision problems, muscle weakness, and deafness (M. Zeviani et al. 1990. *American Journal of Human Genetics* 47:904–914). Analysis of the mtDNA from affected persons in these families revealed that large numbers of their mtDNA possessed deletions of varying length. Different members of the same family and even different mitochondria from the same person possessed deletions of different sizes; so the underlying defect appeared to be a tendency for the mtDNA of affected persons to have deletions. A pedigree of one of the families studied is shown here. The researchers concluded that this disorder is inherited as an autosomal dominant trait and they mapped the disease-causing gene to a position on chromosome 10 in the nucleus.

a. What characteristics of the pedigree rule out inheritance of a trait encoded by a gene in the mtDNA?

b. Explain how a mutation in a nuclear gene might lead to deletions in mtDNA.

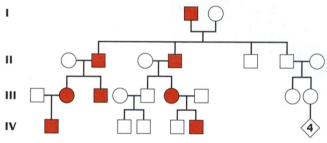

[After M. Zeviani et al. 1990. *American Journal of Human Genetics* 47:904–914.]

40. Mitochondrial DNA sequences have been detected in the nuclear genomes of many organisms, and cpDNA sequences are sometimes found in the mitochondrial genome. Propose a mechanism for how such "promiscuous DNA" might move between nuclear, mitochondrial, and chloroplast genomes.

▶ Go to your ⌑LaunchPad to find additional learning resources and the Suggested Readings for this chapter.

12

DNA Replication and Recombination

The happy tree, *Camptotheca acuminata*, contains camptothecin, a substance used to treat cancer. Camptothecin inhibits cancer by blocking an important component of the replication machinery. [Johnny Pan/Getty Images.]

Topoisomerase, Replication, and Cancer

In 1966, Monroe Wall and Mansukh Wani found a potential cure for cancer in the bark of the happy tree (*Camptotheca acuminata*), a rare plant native to China. Wall and Wani were in the process of screening a large number of natural substances for anticancer activity, hoping to find chemicals that might prove effective in cancer treatment. They discovered that an extract from the happy tree was effective in treating leukemia in mice. Through chemical analysis, they were able to isolate the active compound, which was dubbed camptothecin.

In the 1970s, physicians administered camptothecin to patients with incurable cancers. Although the drug showed some anticancer activity, it had toxic side effects. Eventually, chemists synthesized several analogs of camptothecin that were less toxic and more effective in cancer treatment. Two of these analogs, topotecan and irinotecan, are used today for the treatment of ovarian cancer, small-cell lung cancer, and colon cancer.

For many years, the mechanism by which camptothecin compounds inhibited cancer was unknown. In 1985, almost 20 years after its discovery, scientists at Johns Hopkins University and Smith Kline and French Laboratories (now GlaxoSmithKline) showed that camptothecin worked by inhibiting an important component of the DNA-synthesizing machinery in humans, an enzyme called topoisomerase I.

Cancer chemotherapy is a delicate task because the target cells are the patient's own and the drugs must kill the cancer cells without killing the patient. One of the hallmarks of cancer is proliferation: cancer-cell division is unregulated and many cancer cells divide at a rapid rate, giving rise to tumors with the ability to grow and spread. As we learned in Chapter 2, before a cell can divide it must successfully replicate its DNA so that each daughter cell receives an exact copy of the genetic material. Checkpoints in the cell cycle ensure that cell division does not proceed if DNA replication is inhibited or faulty, and many cancer treatments focus on interfering with the process of DNA replication.

DNA replication is a complex process that requires a large number of components, the actions of which must be intricately coordinated to ensure that DNA is accurately copied. An essential component of replication is topoisomerase. As the DNA unwinds in the course of replication, strain builds up ahead of the separation and the two strands writhe around each other, much as a rope knots up as you pull apart two of its strands. This writhing of the DNA is called supercoiling (see Chapter 11). If the supercoils are not removed, they eventually stop strand separation and replication comes to a halt.

Topoisomerase enzymes remove the supercoils by clamping tightly to the DNA and breaking one or both of its strands. The strands then revolve around each other, removing the supercoiling and strain. After the DNA has relaxed, the topoisomerase that removed the supercoils reseals the the broken ends of the DNA.

Camptothecin works by interfering with topoisomerase I. The drug inserts itself into the gap created by the break in the DNA strand, blocking the topoisomerase from resealing the broken ends. Researchers originally assumed that camptothecin trapped the topoisomerase and blocked the action of other enzymes needed for synthesizing DNA. However, recent research indicates that camptothecin poisons the topoisomerase so that it is unable to remove supercoils ahead of replication. Accumulating supercoils halt the replication machinery and prevent the proliferation of cancer cells. Like many other cancer drugs, camptothecin also inhibits the replication of normal, noncancerous cells, which is why chemotherapy makes many patients sick.

This chapter focuses on DNA replication, the process by which a cell doubles its DNA before division. We begin with the basic mechanism of replication that emerged from the Watson-and-Crick structure of DNA. We then examine several different modes of replication, the requirements of replication, and the universal direction of DNA synthesis. We also examine the enzymes and proteins that participate in the process. Finally, we consider the molecular details of recombination, which is closely related to replication and is essential for the segregation of homologous chromosomes, the production of genetic variation, and DNA repair.

12.1 Genetic Information Must Be Accurately Copied Every Time a Cell Divides

In a schoolyard game, a verbal message, such as "John's brown dog ran away from home," is whispered to a child, who runs to a second child and repeats the message. The message is relayed from child to child around the schoolyard until it returns to the original sender. Inevitably, the last child returns with an amazingly transformed message, such as "Joe Brown has a pig living under his porch." The larger the number of children playing the game, the more garbled the message becomes. This game illustrates an important principle: errors arise whenever information is copied; the more times it is copied, the greater the potential number of errors.

A complex, multicellular organism faces a problem analogous to that of the children in the schoolyard game: how to faithfully transmit genetic instructions each time that its cells divide. The solution to this problem is central to replication. A single-celled human zygote contains 6.4 billion base pairs of DNA; even a low rate of error during copying, such as once per million base pairs, would result in 6400 mistakes made every time a cell divided—errors that would be compounded at each of the millions of cell divisions that take place in human development.

Not only must the copying of DNA be astoundingly accurate, it must also take place at breakneck speed. The single circular chromosome of *E. coli* contains about 4.6 million base pairs. At a rate of more than 1000 nucleotides per minute, replication of the entire chromosome would require almost three days. Yet, as already stated, these bacteria are capable of dividing every 20 minutes. *Escherichia coli* actually replicates its DNA at a rate of 1000 nucleotides per second, with less than one error in a billion nucleotides. How is this extraordinarily accurate and rapid process accomplished?

12.2 All DNA Replication Takes Place in a Semiconservative Manner

From the three-dimensional structure of DNA proposed by Watson and Crick in 1953 (see Figure 10.7), several important genetic implications were immediately apparent. The complementary nature of the two nucleotide strands in a DNA molecule suggested that, during replication, each strand can serve as a template for the synthesis of a new strand. The specificity of base pairing (adenine with thymine; guanine with cytosine) implied that only one sequence of bases can be specified by each template, and so the two DNA molecules built on the pair of templates will be identical with the original. This process is called **semiconservative replication** because each of the original nucleotide strands remains intact (conserved), despite no longer being combined in the same molecule; the original DNA molecule is half (semi) conserved during replication.

Initially, three models were proposed for DNA replication. In conservative replication (**Figure 12.1a**), the entire double-stranded DNA molecule serves as a template for a whole new molecule of DNA, and the original DNA molecule is *fully* conserved during replication. In dispersive replication (**Figure 12.1b**), both nucleotide strands break down (disperse) into fragments, which serve as templates for the synthesis of new DNA fragments, and then somehow reassemble into two complete DNA molecules. In this model, each resulting DNA molecule is interspersed with fragments of old and new DNA; none of the original molecule is conserved. Semiconservative replication (**Figure 12.1c**) is intermediate between these two models; the two nucleotide strands unwind and each serves as a template for a new DNA molecule.

(a) Conservative replication   **(b) Dispersive replication**   **(c) Semiconservative replication**

Original DNA

First replication

Second replication

12.1 Three proposed models of replication are conservative replication, dispersive replication, and semiconservative replication.

These three models allow different predictions to be made about the distribution of original DNA and newly synthesized DNA after replication. With conservative replication, after one round of replication, 50% of the molecules would consist entirely of the original DNA and 50% would consist entirely of new DNA. After a second round of replication, 25% of the molecules would consist entirely of the original DNA and 75% would consist entirely of new DNA. With each additional round of replication, the proportion of molecules with new DNA would increase, although the number of molecules with the original DNA would remain constant. Dispersive replication would always produce hybrid molecules, containing some original and some new DNA, but the proportion of new DNA within the molecules would increase with each replication event. In contrast, with semiconservative replication, one round of replication would produce two hybrid molecules, each consisting of half original DNA and half new DNA. After a second round of replication, half the molecules would be hybrid, and the other half would consist of new DNA only. Additional rounds of replication would produce more and more molecules consisting entirely of new DNA, and a few hybrid molecules would persist.

Meselson and Stahl's Experiment

To determine which of the three models of replication applied to *E. coli* cells, Matthew Meselson and Franklin Stahl needed a way to distinguish old and new DNA. They did so by using two isotopes of nitrogen, ^{14}N (the common form) and ^{15}N (a rare, heavy form). Meselson and Stahl grew a culture of *E. coli* in a medium that contained ^{15}N as the sole nitrogen source; after many generations, all the *E. coli* cells had ^{15}N incorporated into all of the purine and pyrimidine bases of their DNA (see Figure 10.10). Meselson and Stahl took a sample of these bacteria, switched the rest of the bacteria to a medium that contained only ^{14}N, and then took additional samples of bacteria over the next few cellular generations. In each sample, the bacterial DNA that was synthesized before

the change in medium contained ^{15}N and was relatively heavy, whereas any DNA synthesized after the switch contained ^{14}N and was relatively light.

Meselson and Stahl distinguished between the heavy ^{15}N-laden DNA and the light ^{14}N-containing DNA with the use of **equilibrium density gradient centrifugation** (**Figure 12.2**).

A centrifuge tube is filled with a heavy salt solution and DNA fragments.

It is then spun in a centrifuge at high speeds for several days.

DNA with ^{14}N

DNA with ^{15}N

A density gradient develops within the tube. Heavy DNA (with ^{15}N) will move toward the bottom; light DNA (with ^{14}N) will remain closer to the top.

12.2 Meselson and Stahl used equilibrium density gradient centrifugation to distinguish between heavy, ^{15}N-laden DNA and lighter, ^{14}N-laden DNA.

In this technique, a centrifuge tube is filled with a heavy salt solution and a substance of unknown density—in this case, DNA fragments. The tube is then spun in a centrifuge at high speeds. After several days of spinning, a gradient of density develops within the tube, with high density at the bottom and low density at the top. The density of the DNA fragments matches that of the salt: light molecules rise and heavy molecules sink.

Meselson and Stahl found that DNA from bacteria grown only on medium containing ^{15}N produced a single band at the position expected of DNA containing only ^{15}N (**Figure 12.3a**). DNA from bacteria transferred to the medium with ^{14}N and allowed one round of replication also produced a single band but at a position intermediate between that expected of DNA containing only ^{15}N and that expected of DNA containing only ^{14}N (**Figure 12.3b**). This result is inconsistent with the conservative replication model, which predicts one heavy band (the original DNA molecules) and one light band (the new DNA molecules). A single band of intermediate density is predicted by both the semiconservative and the dispersive models.

To distinguish between these two models, Meselson and Stahl grew the bacteria in medium containing ^{14}N for a second generation. After a second round of replication in medium with ^{14}N, two bands of equal intensity appeared, one in the intermediate position and the other at the position expected of DNA containing only ^{14}N (**Figure 12.3c**). All samples taken after additional rounds of replication produced the same two bands, and the band representing light DNA became progressively stronger (**Figure 12.3d**). Meselson and Stahl's results were exactly as expected for semiconservative replication and are incompatible with those predicated for both conservative and dispersive replication. ▶ TRY PROBLEM 22

Experiment

Question: Which model of DNA replication—conservative, dispersive or semiconservative—applies to *E. coli*?

(a)   (b)   (c)   (d)

^{15}N medium

Transfer to ^{14}N; one round of replication

Second round of replication in ^{14}N medium

Additional rounds of replication in ^{14}N medium

Spin   Spin   Spin   Spin

Light (^{14}N)

Heavy (^{15}N)

DNA from bacteria that had been grown on medium containing ^{15}N appeared as a single band.

After one round of replication, the DNA appeared as a single band at intermediate weight.

After a second round of replication, DNA appeared as two bands, one light and the other intermediate in weight.

Samples taken after additional rounds of replication appeared as two bands, as in part c.

Original DNA

Parental strand   New strand

Conclusion: DNA replication in *E.coli* is semiconservative.

12.3 Meselson and Stahl demonstrated that DNA replication is semiconservative.

CONCEPTS

Replication is semiconservative: each DNA strand serves as a template for the synthesis of a new DNA molecule. Meselson and Stahl convincingly demonstrated that replication in *E. coli* is semiconservative.

✓ CONCEPT CHECK 1

How many bands of DNA would be expected in Meselson and Stahl's experiment after two rounds of *conservative* replication?

Modes of Replication

After Meselson and Stahl's work, investigators confirmed that other organisms also use semiconservative replication. No evidence was found for conservative or dispersive replication. There are, however, several different ways in which semiconservative replication can take place, differing principally in the nature of the template DNA—that is, whether it is linear or circular.

Individual units of replication are called **replicons**, each of which contains a **replication origin**. Replication starts at the origin and continues until the entire replicon has been replicated. Bacterial chromosomes have a single replication origin, whereas eukaryotic chromosomes contain many.

THETA REPLICATION   A common type of replication that takes place in circular DNA, such as that found in *E. coli* and other bacteria, is called **theta replication** (**Figure 12.4a**) because it generates a structure that resembles the Greek letter theta (θ). In this and all subsequent figures in this chapter, the original (template) strand of DNA is shown in gray and the newly synthesized strand of DNA is shown in red.

In theta replication, double-stranded DNA begins to unwind at the replication origin, producing single-stranded nucleotide strands that then serve as templates on which new DNA can be synthesized. The unwinding of the double helix generates a loop, termed a **replication bubble**. Unwinding may be at one or both ends of the bubble, making it progressively larger. DNA replication on both of the template strands is simultaneous with unwinding. The point of unwinding, where the two single nucleotide strands separate from the double-stranded DNA helix, is called a **replication fork**.

If there are two replication forks, one at each end of the replication bubble, the forks proceed outward in both directions in a process called **bidirectional replication**, simultaneously

(a)

Origin of replication

1 Double-stranded DNA unwinds at the replication origin,…

Replication fork

Newly synthesized DNA

Replication bubble

2 …producing single-stranded templates for the synthesis of new DNA. A replication bubble forms, usually with a replication fork at each end.

3 The forks proceed around the circle.

4 Eventually two circular DNA molecules are produced.

Conclusion: The products of theta replication are two circular DNA molecules.

(b)

Replication fork

Origin of replication

Replication bubble

12.4 Theta replication is a type of replication common in *E. coli* and other organisms possessing circular DNA. [Part b: Bernhard Hirt, L'Institut Suisse de Recherche Expérimentale sur le Cancer.]

unwinding and replicating the DNA until they eventually meet. If a single replication fork is present, it proceeds around the entire circle. Both bidirectional and unidirectional replication produce two complete circular DNA molecules, each consisting of one old and one new nucleotide strand.

John Cairns provided the first visible evidence of theta replication in 1963 by growing bacteria in the presence of radioactive nucleotides. After replication, each DNA molecule consisted of one "hot" (radioactive) strand and one "cold" (nonradioactive) strand. Cairns isolated DNA from the bacteria after replication, placed it on an electron-microscope grid, and then covered it with a photographic emulsion. Radioactivity present in the sample exposed the emulsion and produced a picture of the molecule (called an autoradiograph), similar to the way in which light exposes a photographic film. Because the newly synthesized DNA contained radioactive nucleotides, Cairns was able to produce an electron micrograph of the replication process, similar to those shown in **Figure 12.4b**.

ROLLING-CIRCLE REPLICATION   Another form of replication, called **rolling-circle replication** (**Figure 12.5**), takes place in some viruses and in the F factor (a small circle of extrachromosomal DNA that controls mating, discussed in Chapter 9) of *E. coli*. This form of replication is initiated by a break in one of the nucleotide strands that creates a 3′-OH group and a 5′-phosphate group. New nucleotides are added to the 3′ end of the broken strand, with the inner (unbroken) strand used as a template. As new nucleotides are added to the 3′ end, the 5′ end of the broken strand is displaced from the template, rolling out like thread being pulled off a spool. The 3′ end grows around the circle, giving rise to the name rolling-circle.

The replication fork may continue around the circle a number of times, producing several linked copies of the same sequence. With each revolution around the circle, the growing 3′ end displaces the nucleotide strand synthesized in the preceding revolution. Eventually, the linear DNA molecule is cleaved from the circle, resulting in a double-stranded circular DNA molecule and a single-stranded linear DNA molecule. The linear molecule circularizes either before or after serving as a template for the synthesis of a complementary strand.

LINEAR EUKARYOTIC REPLICATION   Circular DNA molecules that undergo theta or rolling-circle replication have a single origin of replication. Because of the limited size of these DNA molecules, replication starting from one origin can traverse the entire chromosome in a reasonable amount of time. The large linear chromosomes in eukaryotic cells, however, contain far too much DNA to be replicated speedily from a single origin. Eukaryotic replication proceeds at a rate ranging from 500 to 5000 nucleotides per minute at each replication fork (considerably slower than bacterial replication). Even at 5000 nucleotides per minute at each fork, DNA synthesis starting from a single origin would require 7 days to replicate a typical human chromosome consisting of 100 million base pairs of DNA. The replication of eukaryotic chromosomes actually takes place in a matter of minutes or hours, not days. This rate is possible because replication initiates at thousands of origins.

Typical eukaryotic replicons are from 20,000 to 300,000 base pairs in length (**Table 12.1**). At each replication origin, the DNA unwinds and produces a replication bubble. Replication takes place on both strands at each end of the bubble, with the two replication forks spreading outward. Eventually, the replication forks of adjacent replicons run into each other, and the replicons fuse to form long stretches of newly synthesized DNA (**Figure 12.6**). Replication and fusion of all the replicons leads to two identical DNA

Replication is initiated by a break in one of the nucleotide strands.

DNA synthesis begins at the 3′ end of the broken strand; the inner strand is used as a template. The 5′ end of the broken strand is displaced.

Cleavage releases a single-stranded linear DNA and a double-stranded circular DNA.

The linear DNA may circularize and serve as a template for synthesis of a complementary strand.

Conclusion: The products of rolling-circle replication are multiple circular DNA molecules.

Cleavage

The cycle may be repeated

12.5 Rolling-circle replication takes place in some viruses and in the F factor of *E. coli*.

Table 12.1 Number and length of replicons

Organism	Number of Replication Origins	Average Length of Replicon (bp)
Escherichia coli (bacterium)	1	4,600,000
Saccharomyces cerevisiae (yeast)	500	40,000
Drosophila melanogaster (fruit fly)	3,500	40,000
Xenopus laevis (frog)	15,000	200,000
Mus musculus (mouse)	25,000	150,000

Source: Data from B. L. Lewin, *Genes V* (Oxford: Oxford University Press, 1994), p. 536.

molecules. Important features of theta replication, rolling-circle replication, and linear eukaryotic replication are summarized in **Table 12.2**. ▶ **TRY PROBLEM 23**

CONCEPTS

Theta replication, rolling-circle replication, and linear replication differ with respect to the initiation and progression of replication, but all produce new DNA molecules by semiconservative replication.

✓ CONCEPT CHECK 2

Which type of replication requires a break in the nucleotide strand to get started?

a. Theta replication
b. Rolling-circle replication
c. Linear eukaryotic replication
d. All of the above

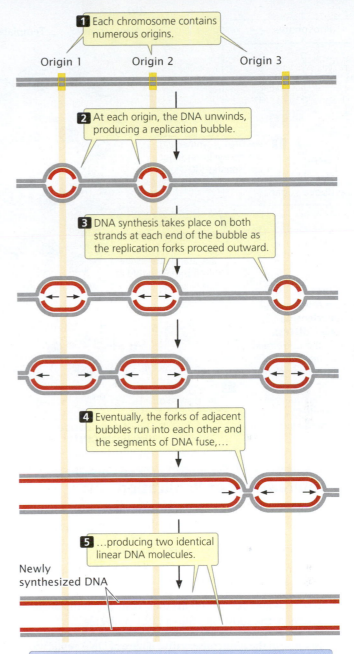

1. Each chromosome contains numerous origins.

Origin 1   Origin 2   Origin 3

2. At each origin, the DNA unwinds, producing a replication bubble.

3. DNA synthesis takes place on both strands at each end of the bubble as the replication forks proceed outward.

4. Eventually, the forks of adjacent bubbles run into each other and the segments of DNA fuse,…

5. …producing two identical linear DNA molecules.

Newly synthesized DNA

Conclusion: The products of eukaryotic DNA replication are two linear DNA molecules.

12.6 Linear DNA replication takes place in eukaryotic chromosomes.

Table 12.2 Characteristics of theta, rolling-circle, and linear eukaryotic replication

Replication Model	DNA Template	Breakage of Nucleotide Strand	Number of Replicons	Unidirectional or Bidirectional	Products
Theta	Circular	No	1	Unidirectional or bidirectional	Two circular molecules
Rolling circle	Circular	Yes	1	Unidirectional	One circular molecule and one linear molecule that may circularize
Linear eukaryotic	Linear	No	Many	Bidirectional	Two linear molecules

(a)

Phosphates

base

H_2C

3' OH H

Deoxyribose sugar

1 New DNA is synthesized from deoxyribonucleoside triphosphates (dNTPs).

(b)

New strand · Template strand
5' **3'** OH

5' **3'** OH

T ··· A

C ··· G

A ··· T

C ··· G

2 In replication, the 3'-OH group of the last nucleotide on the strand attacks the 5'-phosphate group of the incoming dNTP.

OH
3'

5'

C

3 Two phosphates are cleaved off.

3' OH

3'

Deoxyribonucleoside triphosphate (dNTP)

G

G

C

5'

4 A phosphodiester bond forms between the two nucleotides,...

T ··· A

C ··· G

A ··· T

C ··· G

C ··· G

OH **3'**

C

5'

12.7 New DNA is synthesized from deoxyribonucleoside triphosphates (dNTPs). The newly synthesized strand is complementary and antiparallel to the template strand; the two strands are held together by hydrogen bonds (represented by red dotted lines) between the bases.

Requirements of Replication

Although the process of replication includes many components, they can be combined into three major groups:

1. A template consisting of single-stranded DNA

2. Raw materials (substrates) to be assembled into a new nucleotide strand

3. Enzymes and other proteins that "read" the template and assemble the substrates into a DNA molecule

Because of the semiconservative nature of DNA replication, a double-stranded DNA molecule must unwind to expose the bases that act as a template for the assembly of new polynucleotide strands, which will be complementary and antiparallel to the template strands. The raw materials from which new DNA molecules are synthesized are deoxyribonucleoside triphosphates (dNTPs), each consisting of a deoxyribose sugar and a base (a nucleoside) attached to three phosphate groups (**Figure 12.7a**). In DNA synthesis, nucleotides are added to the 3'-OH group of the growing nucleotide strand (**Figure 12.7b**). The 3'-OH group of the last nucleotide on the strand attacks the 5'-phosphate group of the incoming dNTP. Two phosphate groups are cleaved from the incoming dNTP, and a phosphodiester bond is created between the two nucleotides.

DNA synthesis does not happen spontaneously. Rather, it requires a host of enzymes and proteins that function in a coordinated manner. We will examine this complex array of proteins and enzymes as we consider the replication process in more detail.

Direction of Replication

In DNA synthesis, new nucleotides are joined one at a time to the 3' end of the newly synthesized strand. **DNA polymerases**, the enzymes that synthesize DNA, can add nucleotides only to the 3' end of the growing strand (not the 5' end), and so new DNA strands always elongate in the same 5'-to-3' direction (5'→3'). Because the two single-stranded DNA templates are antiparallel and strand elongation is always 5'→3', if synthesis on one template proceeds from, say, right to left, then synthesis on the other template must proceed in the opposite direction, from left to right (**Figure 12.8**). As DNA unwinds during replication, the antiparallel nature of the two DNA strands means that one template is exposed in the 5'→3' direction and the other template is exposed in the 3'→5' direction. So how can synthesis take place simultaneously on both strands at the fork?

CONTINUOUS AND DISCONTINUOUS REPLICATION As the DNA unwinds, the template strand that is exposed in the 3'→5' direction (the lower strand in Figures 12.8 and 12.9) allows the new strand to be synthesized continuously, in the 5'→3' direction. This new strand, which undergoes **continuous replication**, is called the **leading strand**.

12.8 DNA synthesis takes place in opposite directions on the two DNA template strands. DNA replication at a single replication fork begins when a double-stranded DNA molecule unwinds to provide two single-strand templates.

The other template strand is exposed in the 5′→3′ direction (the upper strand in Figures 12.8 and 12.9). After a short length of the DNA has been unwound, synthesis must proceed 5′→3′; that is, in the direction *opposite* that of unwinding (**Figure 12.9**). Because only a short length of DNA needs to be unwound before synthesis on this strand gets started, the replication machinery soon runs out of template. By that time, more DNA has unwound, providing new template at the 5′ end of the new strand. DNA synthesis must start anew at the replication fork and proceed in the direction opposite that of the movement of the fork until it runs into the previously replicated segment of DNA. This process is repeated again and again, so synthesis of this strand is in short, discontinuous bursts. The newly made strand that undergoes **discontinuous replication** is called the **lagging strand**.

OKAZAKI FRAGMENTS   The short lengths of DNA produced by discontinuous replication of the lagging strand are called **Okazaki fragments**, after Reiji Okazaki, who discovered them. In bacterial cells, each Okazaki fragment ranges from about 1000 to 2000 nucleotides in length; in eukaryotic cells, they are about 100 to 200 nucleotides long. Okazaki fragments on the lagging strand are linked together to create a continuous new DNA molecule. To see how replication occurs continuously on one strand and discontinuously on the other, view **Animation 12.1**.

Ⓐ

CONCEPTS

All DNA synthesis is 5′→3′, meaning that new nucleotides are always added to the 3′ end of the growing nucleotide strand. At each replication fork, synthesis of the leading strand proceeds continuously and that of the lagging strand proceeds discontinuously.

✔ **CONCEPT CHECK 3**

Discontinuous replication is a result of which property of DNA?

a. Complementary bases   c. Antiparallel nucleotide strands
b. Charged phosphate group   d. Five-carbon sugar

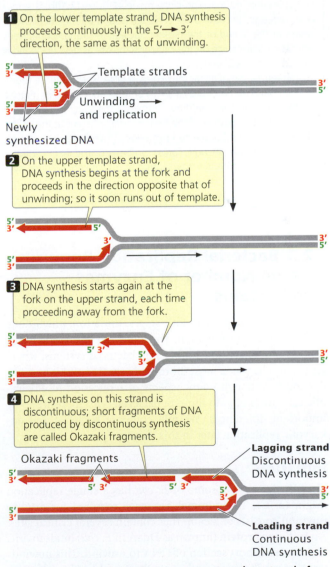

12.9 DNA synthesis is continuous on one template strand of DNA and discontinuous on the other.

The Direction of Synthesis in Different Models of Replication

Let's relate the direction of DNA synthesis to the modes of replication examined earlier. In the theta model (**Figure 12.10a**), the DNA unwinds at one particular location, the origin, and a replication bubble is formed. If the bubble has two forks, one at each end, synthesis takes place simultaneously at both forks (bidirectional replication). At each fork, synthesis on one of the template strands proceeds in the same direction as that of unwinding; this newly replicated strand is the leading strand with continuous replication. On the other template strand, synthesis proceeds in the direction opposite that of unwinding; this newly synthesized strand is the lagging strand with discontinuous replication. Focus on just one of the template strands within the bubble. Notice that synthesis on this template strand is continuous at one fork but discontinuous at the other. This difference arises because DNA synthesis is always in the same direction (5′→3′), but the two forks are moving in opposite directions.

Replication in the rolling-circle model (**Figure 12.10b**) is somewhat different, because there is no replication bubble. Replication begins at the 3′ end of the broken nucleotide strand. Continuous replication takes place on the circular template as new nucleotides are added to this 3′ end.

The replication of linear molecules of DNA, such as those found in eukaryotic cells, produces a series of replication bubbles (**Figure 12.10c**). DNA synthesis in these bubbles is the same as that in the single replication bubble of the theta model; it begins at the center of each replication bubble and proceeds at two forks, one at each end of the bubble. At both forks, synthesis of the leading strand proceeds in the same direction as that of unwinding, whereas synthesis of the lagging strand proceeds in the direction opposite that of unwinding. ▶ TRY PROBLEM 25a–c

12.3 Bacterial Replication Requires a Large Number of Enzymes and Proteins

Replication takes place in four stages: initiation, unwinding, elongation, and termination. The following discussion of the process of replication will focus on bacterial systems, where replication has been most thoroughly studied and is best understood. Although many aspects of replication in eukaryotic cells are similar to those in bacterial cells, there are some important differences. We will compare bacterial and eukaryotic replication later in this chapter.

Initiation

The circular chromosome of *E. coli* has a single replication origin (*oriC*). The minimal sequence required for *oriC* to function consists of 245 bp that contain several critical sites. An **initiator protein** (known as DnaA in *E. coli*) binds to *oriC* and causes a short section of DNA to unwind. This unwinding allows helicase and other single-strand-binding proteins to attach to the polynucleotide strand (**Figure 12.11**).

(a) Theta model

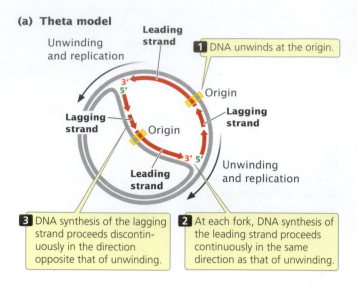

1. DNA unwinds at the origin.

3. DNA synthesis of the lagging strand proceeds discontinuously in the direction opposite that of unwinding.

2. At each fork, DNA synthesis of the leading strand proceeds continuously in the same direction as that of unwinding.

(b) Rolling-circle model

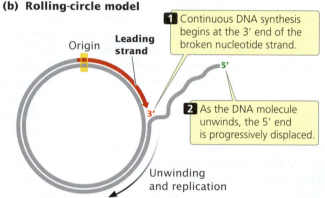

1. Continuous DNA synthesis begins at the 3′ end of the broken nucleotide strand.

2. As the DNA molecule unwinds, the 5′ end is progressively displaced.

(c) Linear eukaryotic replication

1. At each fork, the leading strand is synthesized continuously in the same direction as that of unwinding.

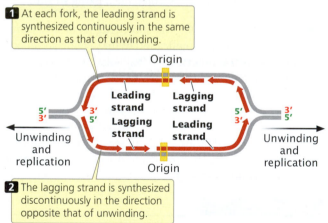

2. The lagging strand is synthesized discontinuously in the direction opposite that of unwinding.

12.10 The process of replication differs in theta replication, rolling-circle replication, and linear replication.

Unwinding

Because DNA synthesis requires a single-stranded template and because double-stranded DNA must be unwound before DNA synthesis can take place, the cell relies on several proteins and enzymes to accomplish the unwinding.

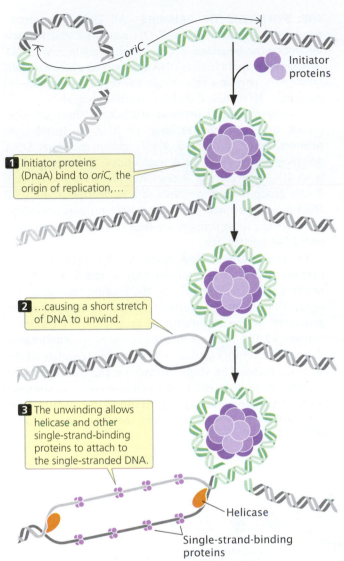

12.11 *E. coli* DNA replication begins when initiator proteins bind to *oriC*, the origin of replication.

① Initiator proteins (DnaA) bind to *oriC*, the origin of replication,…

② …causing a short stretch of DNA to unwind.

③ The unwinding allows helicase and other single-strand-binding proteins to attach to the single-stranded DNA.

Initiator proteins

Helicase

Single-strand-binding proteins

DNA HELICASE   A **DNA helicase** breaks the hydrogen bonds that exist between the bases of the two nucleotide strands of a DNA molecule. Helicase cannot *initiate* the unwinding of double-stranded DNA; the initiator protein first separates DNA strands at the origin, providing a short stretch of single-stranded DNA to which a helicase binds. Helicase binds to the lagging-strand template at each replication fork and moves in the 5′→3′ direction along this strand, thus also moving the replication fork (**Figure 12.12**).

SINGLE-STRAND-BINDING PROTEINS   After DNA has been unwound by helicase, **single-strand-binding proteins** (SSBs) attach tightly to the exposed single-stranded DNA (see Figure 12.12). These proteins protect the single-stranded nucleotide chains and prevent the formation of secondary structures such as hairpins (see Figure 10.17) that interfere with replication. Unlike many DNA-binding proteins, SSBs are indifferent to base sequence: they will bind to any single-stranded DNA. Single-strand-binding proteins form tetramers (groups of four); each tetramer covers from 35 to 65 nucleotides.

DNA GYRASE   Another protein essential for the unwinding process is the enzyme **DNA gyrase**, a topoisomerase. As discussed in Chapter 11 and the introduction to this chapter, topoisomerases control the supercoiling of DNA. They come in two major types: type I topoisomerases alter supercoiling by making single-strand breaks in DNA, while type II topoisomerases create double-stranded breaks. DNA gyrase is a type II topoisomerase. In replication, it reduces the torsional strain (torque) that builds up ahead of the replication fork as a result of unwinding (see Figure 12.12). It reduces torque by making a double-stranded break in one segment of the DNA helix, passing another segment of the helix through the break, and then resealing the broken ends of the DNA. This action, which requires ATP, removes a twist in the DNA and reduces the supercoiling.

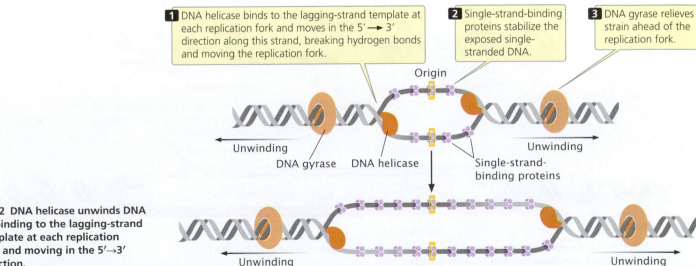

① DNA helicase binds to the lagging-strand template at each replication fork and moves in the 5′ → 3′ direction along this strand, breaking hydrogen bonds and moving the replication fork.

② Single-strand-binding proteins stabilize the exposed single-stranded DNA.

③ DNA gyrase relieves strain ahead of the replication fork.

Origin

Unwinding

Unwinding

DNA gyrase

DNA helicase

Single-strand-binding proteins

Unwinding

Unwinding

12.12 DNA helicase unwinds DNA by binding to the lagging-strand template at each replication fork and moving in the 5′→3′ direction.

A group of antibiotics called 4-quinolones kill bacteria by binding to DNA gyrase and inhibiting its action. The inhibition of DNA gyrase results in the cessation of DNA synthesis and bacterial growth. An example of a 4-quinolone is nalidixic acid, which was first introduced in the 1960s and is commonly used to treat urinary infections. Many bacteria have acquired resistance to quinolones through mutations in the gene for DNA gyrase.

CONCEPTS

Replication is initiated at a replication origin, where an initiator protein binds and causes a short stretch of DNA to unwind. DNA helicase breaks hydrogen bonds at a replication fork, and single-strand-binding proteins stabilize the separated strands. DNA gyrase reduces the torsional strain that develops as the two strands of double-helical DNA unwind.

✓ CONCEPT CHECK 4

Place the following components in the order in which they are first used in the course of replication: helicase, single-strand-binding protein, DNA gyrase, initiator protein.

Elongation

In the elongation phase of replication, single-stranded DNA is used as a template for the synthesis of DNA. This process requires a series of enzymes.

THE SYNTHESIS OF PRIMERS All DNA polymerases require a nucleotide with a 3′-OH group to which a new nucleotide can be added. Because of this requirement, DNA polymerases cannot initiate DNA synthesis on a bare template; rather, they require a primer—an existing 3′-OH group—to get started. How, then, does DNA synthesis begin?

An enzyme called **primase** synthesizes short stretches (about 10–12 nucleotides long) of RNA nucleotides, or **primers**, which provide a 3′-OH group to which DNA polymerase can attach DNA nucleotides. (Because primase is an RNA polymerase, it does not require a pre-existing 3′-OH group to start synthesis of a nucleotide strand.) All DNA molecules initially have short RNA primers embedded within them; these primers are later removed and replaced with DNA nucleotides.

On the leading strand, where DNA synthesis is continuous, a primer is required only at the 5′ end of the newly synthesized strand. On the lagging strand, where replication is discontinuous, a new primer must be generated at the beginning of each Okazaki fragment (**Figure 12.13**). Primase forms a complex with helicase at the replication fork and moves along the template of the lagging strand. The single primer on the leading strand is probably synthesized by the primase–helicase complex on the template of the lagging strand of the *other* replication fork, at the opposite end of the replication bubble.

▶ TRY PROBLEM 30

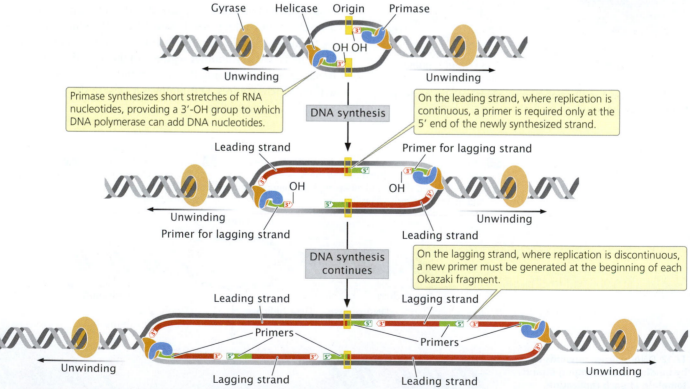

Gyrase   Helicase   Origin   Primase

Primase synthesizes short stretches of RNA nucleotides, providing a 3′-OH group to which DNA polymerase can add DNA nucleotides.

DNA synthesis

On the leading strand, where replication is continuous, a primer is required only at the 5′ end of the newly synthesized strand.

Leading strand   Primer for lagging strand

Primer for lagging strand   Leading strand

DNA synthesis continues

On the lagging strand, where replication is discontinuous, a new primer must be generated at the beginning of each Okazaki fragment.

Leading strand   Lagging strand

Primers   Primers

Lagging strand   Leading strand

12.13 Primase synthesizes short stretches of RNA nucleotides, providing a 3′-OH group to which DNA polymerase can add DNA nucleotides.

DNA SYNTHESIS BY DNA POLYMERASES   After DNA has unwound and a primer has been added, DNA polymerases elongate the new polynucleotide strand by catalyzing DNA polymerization. The best-studied polymerases are those of *E. coli*, which has at least five different DNA polymerases. Two of them, DNA polymerase I and DNA polymerase III, carry out DNA synthesis in replication (**Table 12.3**); the other three have specialized functions in DNA repair.

DNA polymerase III is a large multiprotein complex that acts as the main workhorse of replication. DNA polymerase III synthesizes nucleotide strands by adding new nucleotides to the 3' end of a growing DNA molecule. This enzyme has two enzymatic activities (see Table 12.3). Its 5'→3' polymerase activity allows it to add new nucleotides in the 5'→3' direction. Its 3'→5' exonuclease activity allows it to remove nucleotides in the 3'→5' direction, enabling it to correct errors. If a nucleotide having an incorrect base is inserted into the growing DNA molecule, DNA polymerase III uses its 3'→5' exonuclease activity to back up and remove the incorrect nucleotide. It then resumes its 5'→3' polymerase activity. These two functions together allow DNA polymerase III to efficiently and accurately synthesize new DNA molecules. DNA polymerase III has high processivity, which means that it is capable of adding many nucleotides to the growing DNA strand without releasing the template: it normally holds on to the template and continues synthesizing DNA until the template has been completely replicated. The high processivity of DNA polymerase III is ensured by one of the polypeptides that constitutes the enzyme. This polypeptide, termed the β subunit, serves as a clamp for the polymerase enzyme: it encircles the DNA and keeps the DNA polymerase attached to the template strand during replication. DNA polymerase III adds DNA nucleotides to the primer, synthesizing the DNA of both the leading and the lagging strands.

The first *E. coli* polymerase to be discovered, **DNA polymerase I**, also has 5'→3' polymerase and 3'→5' exonuclease activities (see Table 12.3), permitting the enzyme to synthesize DNA and to correct errors. Unlike DNA polymerase III, however, DNA polymerase I also possesses 5'→3' exonuclease activity, which is used to remove the primers laid down by primase and to replace them with DNA nucleotides by synthesizing in a 5'→3' direction. DNA polymerase I has lower processivity than DNA polymerase III. The removal and replacement of primers appear to constitute the main function of DNA polymerase I. After DNA polymerase III has initiated synthesis at the primer and moved downstream, DNA polymerase I removes the RNA nucleotides of the primer, replacing them with DNA nucleotides. DNA polymerases II, IV, and V function in DNA repair.

Despite their differences, all of *E. coli*'s DNA polymerases

1. synthesize any sequence specified by the template strand;
2. synthesize in the 5'→3' direction by adding nucleotides to a 3'-OH group;
3. use dNTPs to synthesize new DNA;
4. require a 3'OH-group to initiate synthesis;
5. catalyze the formation of a phosphodiester bond by joining the 5'-phosphate group of the incoming nucleotide to the 3'-OH group of the preceding nucleotide on the growing strand, cleaving off two phosphates in the process;
6. produce newly synthesized strands that are complementary and antiparallel to the template strands; and
7. are associated with a number of other proteins.

▶ TRY PROBLEM 27

Table 12.3   Characteristics of DNA Polymerases in *E. coli*

DNA Polymerase	5'→3' Polymerization	3'→5' Exonuclease	5'→3' Exonuclease	Function
I	Yes	Yes	Yes	Removes and replaces primers
II	Yes	Yes	No	DNA repair; restarts replication after damaged DNA halts synthesis
III	Yes	Yes	No	Elongates DNA
IV	Yes	No	No	DNA repair
V	Yes	No	No	DNA repair; translesion DNA synthesis

(a)

Template strand

DNA nucleotides have been added to the primer by DNA polymerase III.

RNA primer added by primase

(b)

DNA polymerase I

DNA polymerase I replaces the RNA nucleotides of the primer with DNA nucleotides.

RNA nucleotide

DNA dNTP

(c)

Nick

After the last nucleotide of the RNA primer has been replaced, a nick remains in the sugar–phosphate backbone of the strand.

(d)

DNA ligase

DNA ligase seals this nick with a phosphodiester bond between the 5′-P group of the initial nucleotide added by DNA polymerase III and the 3′-OH group of the final nucleotide added by DNA polymerase I.

12.14 DNA ligase seals the break left by DNA polymerase I in the sugar–phosphate backbone.

DNA LIGASE After DNA polymerase III attaches a DNA nucleotide to the 3′-OH group on the last nucleotide of the RNA primer, each new DNA nucleotide then provides the 3′-OH group needed for the next DNA nucleotide to be added. This process continues as long as a template is available (**Figure 12.14a**). DNA polymerase I follows DNA polymerase III and, using its 5′→3′ exonuclease activity, removes the RNA primer. It then uses its 5′→3′ polymerase activity to replace the RNA nucleotides with DNA nucleotides. DNA

polymerase I attaches the first nucleotide to the OH group at the 3′ end of the preceding Okazaki fragment and then continues, in the 5′→3′ direction along the nucleotide strand, removing and replacing, one at a time, the RNA nucleotides of the primer (**Figure 12.14b**).

After polymerase I has replaced the last nucleotide of the RNA primer with a DNA nucleotide, a break remains in the sugar–phosphate backbone of the new DNA strand. The 3′-OH group of the last nucleotide to have been added by DNA polymerase I is not attached to the 5′-phosphate group of the first nucleotide added by DNA polymerase III (**Figure 12.14c**). This break is sealed by the enzyme **DNA ligase**, which catalyzes the formation of a phosphodiester bond without adding another nucleotide to the strand (**Figure 12.14d**). Some of the major enzymes and proteins required for prokaryotic DNA replication are summarized in **Table 12.4**.

Table 12.4	Components required for replication in bacterial cells
Component	**Function**
Initiator protein	Binds to origin and separates strands of DNA to initiate replication
DNA helicase	Unwinds DNA at replication fork
Single-strand-binding proteins	Attach to single-stranded DNA and prevent secondary structures from forming
DNA gyrase	Moves ahead of the replication fork, making and resealing breaks in the double-helical DNA to release the torque that builds up as a result of unwinding at the replication fork
DNA primase	Synthesizes a short RNA primer to provide a 3′-OH group for the attachment of DNA nucleotides
DNA polymerase III	Elongates a new nucleotide strand from the 3′-OH group provided by the primer
DNA polymerase I	Removes RNA primers and replaces them with DNA
DNA ligase	Joins Okazaki fragments by sealing breaks in the sugar–phosphate backbone of newly synthesized DNA

ELONGATION AT THE REPLICATION FORK   Now that the major enzymatic components of elongation—DNA polymerases, helicase, primase, and ligase—have been introduced, let's consider how these components interact at the replication fork. Because the synthesis of both strands takes place simultaneously, two units of DNA polymerase III must be present at the replication fork, one for each strand. In one model of the replication process, the two units of DNA polymerase III are connected (**Figure 12.15**); the lagging-strand template loops around so that it is in position for 5′→3′ replication. In this way, the DNA polymerase III complex is able to carry out 5′→3′ replication simultaneously on both templates, even though they run in opposite directions. After about 1000 bp of new DNA has been synthesized, DNA polymerase III releases the lagging-strand template, and a new loop forms (see Figure 12.15). Primase synthesizes a new primer on the lagging strand and DNA polymerase III then synthesizes a new Okazaki fragment. See how replication takes place on both strands simultaneously by viewing **Animation 12.2**.

In summary, each active replication fork requires five basic components:

1. helicase to unwind the DNA,

2. single-strand-binding proteins to protect the single nucleotide strands and prevent secondary structures,

3. the topoisomerase gyrase to remove strain ahead of the replication fork,

4. primase to synthesize primers with a 3′-OH group at the beginning of each DNA fragment, and

5. DNA polymerase to synthesize the leading and lagging nucleotide strands.

You can see how the different components of the replication process work together by viewing **Animations 12.3** and **12.4**.

Termination

In some DNA molecules, replication is terminated whenever two replication forks meet. In others, specific termination sequences (called *Ter* sites) block further replication. A termination protein, called Tus in *E. coli*, binds to these sequences, creating a Tus-*Ter* complex that blocks the movement of helicase, thus stalling the replication fork and preventing further DNA replication. Each Tus-*Ter* complex blocks a replication fork moving in one direction but not the other.

The Fidelity of DNA Replication

Overall, the error rate in replication is less than one mistake per billion nucleotides. How is this incredible accuracy achieved?

DNA polymerases are very particular in pairing nucleotides with their complements on the template strand. Errors in nucleotide selection by DNA polymerase arise only about once per 100,000 nucleotides. Most of the errors that do arise

in nucleotide selection are corrected in a second process called **proofreading**. When a DNA polymerase inserts an incorrect nucleotide into the growing strand, the 3′-OH

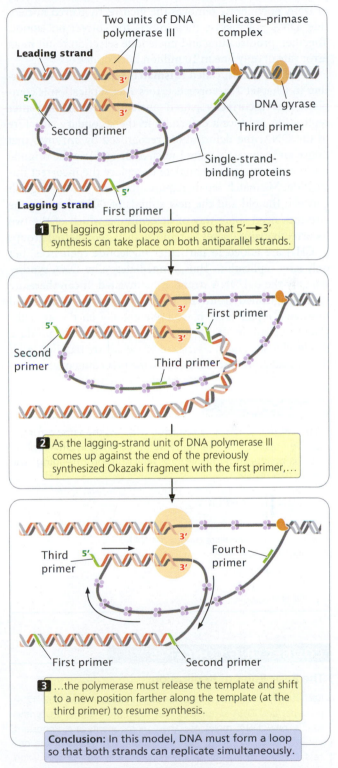

1 The lagging strand loops around so that 5′ → 3′ synthesis can take place on both antiparallel strands.

2 As the lagging-strand unit of DNA polymerase III comes up against the end of the previously synthesized Okazaki fragment with the first primer,…

3 …the polymerase must release the template and shift to a new position farther along the template (at the third primer) to resume synthesis.

Conclusion: In this model, DNA must form a loop so that both strands can replicate simultaneously.

12.15 In one model of DNA replication in *E. coli*, the two units of DNA polymerase III are connected. The lagging-strand template forms a loop so that replication can take place on the two antiparallel DNA strands. Components of the replication machinery at the replication fork are shown at the top.

group of the mispaired nucleotide is not correctly positioned in the active site of the DNA polymerase for accepting the next nucleotide. The incorrect positioning stalls the polymerization reaction, and the $3' \rightarrow 5'$ exonuclease activity of DNA polymerase removes the incorrectly paired nucleotide. DNA polymerase then inserts the correct nucleotide. Together, proofreading and nucleotide selection result in an error rate of only one in 10 million nucleotides.

A third process, called **mismatch repair** (discussed further in Chapter 18), corrects errors after replication is complete. Any incorrectly paired nucleotides remaining after replication produce a deformity in the secondary structure of the DNA; the deformity is recognized by enzymes that excise an incorrectly paired nucleotide and use the original nucleotide strand as a template to replace the incorrect nucleotide. Mismatch repair requires the ability to distinguish between the old and the new strands of DNA, because the enzymes need some way of determining which of the two incorrectly paired bases to remove. In *E. coli*, methyl groups ($-CH_3$) are added to particular nucleotide sequences, but only *after* replication. Thus, immediately after DNA synthesis, only the old DNA strand is methylated. It can therefore be distinguished from the newly synthesized strand, and mismatch repair takes place preferentially on the unmethylated nucleotide strand. No single process could produce this level of accuracy; a series of processes are required, each process catching errors missed by the preceding ones.

CONCEPTS

Replication is extremely accurate, with less than one error per billion nucleotides. The high level of accuracy in DNA replication is produced by nucleotide selection, proofreading, and mismatch repair.

✔ CONCEPT CHECK 7

Which mechanism requires the ability to distinguish between newly synthesized and template strands of DNA?

a. Nucleotide selection
b. DNA proofreading
c. Mismatch repair
d. All of the above

CONNECTING CONCEPTS

The Basic Rules of Replication

Bacterial replication requires a number of enzymes (see Table 12.4), proteins, and DNA sequences that function together to synthesize a new DNA molecule. These components are important, but we must not become so immersed in the details of the process that we lose sight of the general principles of replication.

1. Replication is always semiconservative.
2. Replication begins at sequences called origins.
3. DNA synthesis is initiated by short segments of RNA called primers.

4. The elongation of DNA strands is always in the $5' \rightarrow 3'$ direction.
5. New DNA is synthesized from dNTPs; in the polymerization of DNA, two phosphate groups are cleaved from a dNTP and the resulting nucleotide is added to the 3'-OH group of the growing nucleotide strand.
6. Replication is continuous on the leading strand and discontinuous on the lagging strand.
7. New nucleotide strands are complementary and antiparallel to their template strands.
8. Replication takes place at very high rates and is astonishingly accurate, thanks to precise nucleotide selection, proofreading, and mismatch repair.

12.4 Eukaryotic DNA Replication Is Similar to Bacterial Replication but Differs in Several Aspects

Although eukaryotic replication resembles bacterial replication in many respects, replication in eukaryotic cells presents several additional challenges. First, the much greater size of eukaryotic genomes requires that replication be initiated at multiple origins. Second, eukaryotic chromosomes are linear, whereas prokaryotic chromosomes are circular. Third, the DNA template is associated with histone proteins in the form of nucleosomes, and nucleosome assembly must immediately follow DNA replication.

Eukaryotic Origins

Researchers first isolated eukaryotic origins of replication from yeast cells by demonstrating that certain DNA sequences confer the ability to replicate when transferred from a yeast chromosome to small circular pieces of DNA (plasmids). These **autonomously replicating sequences** (ARSs) enabled any DNA to which they were attached to replicate. They were subsequently shown to be the origins of replication in yeast chromosomes. The origins of replication of different eukaryotic organisms vary greatly in sequence, although they usually contain a number of A-T base pairs. A multiprotein complex, the origin-recognition complex (ORC), binds to origins and unwinds the DNA in this region.

CONCEPTS

Eukaryotic DNA contains many origins of replication. At each origin, a multiprotein origin-recognition complex binds to initiate the unwinding of the DNA.

✔ CONCEPT CHECK 8

In comparison with prokaryotes, what are some differences in the genome structure of eukaryotic cells that affect how replication takes place?

The Licensing of DNA Replication

Eukaryotic cells utilize thousands of origins, and so the entire genome can be replicated in a timely manner. The use of multiple origins, however, creates a special problem in the timing of replication: the entire genome must be precisely replicated once and only once in each cell cycle so that no genes are left unreplicated and no genes are replicated more than once. How does a cell ensure that replication is initiated at thousands of origins only once per cell cycle?

The precise replication of DNA is accomplished by the separation of the initiation of replication into two distinct steps. In the first step, the origins are licensed—approved for replication. This step takes place early in the cell cycle when a **replication licensing factor** attaches to an origin. In the second step, the replication machinery initiates replication at each *licensed* origin. The key is that the replication machinery functions only at licensed origins. As the replication forks move away from the origin, the licensing factor is removed, leaving the origin in an unlicensed state, where replication cannot be initiated again until the license is renewed. To ensure that replication takes place only once per cell cycle, the licensing factor is active only after the cell has completed mitosis and before the replication is initiated.

One eukaryotic licensing factor is a complex called MCM (for minichromosome maintenance), which contains a DNA helicase that unwinds a short stretch of DNA in the initiation of replication. MCM must bind to the DNA for replication to initiate at an origin. After replication has begun at an origin, a protein called Geminin prevents MCM from binding to DNA and reinitiating replication at that origin. At the end of mitosis, Geminin is degraded, allowing MCM to bind once again to DNA and relicense the origin. MCM also functions as the DNA helicase during the replication process.

Unwinding

Several different helicases that separate double-stranded DNA have been isolated from eukaryotic cells, as have single-strand-binding proteins and topoisomerases (which have a function equivalent to the DNA gyrase in bacterial cells). These enzymes and proteins are assumed to function in unwinding eukaryotic DNA in much the same way as their bacterial counterparts do.

Eukaryotic DNA Polymerases

Some significant differences in the processes of bacterial and eukaryotic replication are in the number and functions of DNA polymerases. Eukaryotic cells contain a number of different DNA polymerases that function in replication, recombination, and DNA repair.

Three DNA polymerases carry out most of nuclear DNA synthesis during replication: DNA polymerase α, DNA polymerase δ, and DNA polymerase ε (**Table 12.5**).

Table 12.5 DNA polymerases in eukaryotic cells

DNA Polymerase	5′→3′ Polymerase Activity	3′→5′ Exonuclease Activity	Cellular Function
α (alpha)	Yes	No	Initiation of nuclear DNA synthesis and DNA repair; has primase activity
δ (delta)	Yes	Yes	Lagging-strand synthesis of nuclear DNA, DNA repair, and translesion DNA synthesis
ε (epsilon)	Yes	Yes	Leading-strand synthesis
γ (gamma)	Yes	Yes	Replication and repair of mitochondrial DNA
ξ (zeta)	Yes	No	Translesion DNA synthesis
η (eta)	Yes	No	Translesion DNA synthesis
θ (theta)	Yes	No	DNA repair
ι (iota)	Yes	No	Translesion DNA synthesis
κ (kappa)	Yes	No	Translesion DNA synthesis
λ (lambda)	Yes	No	DNA repair
μ (mu)	Yes	No	DNA repair
σ (sigma)	Yes	No	Nuclear DNA replication (possibly), DNA repair, and sister-chromatid cohesion
ϕ (phi)	Yes	No	Translesion DNA synthesis
Rev1	Yes	No	DNA repair

Note: The polymerases listed at the top of the table are those that carry out DNA replication.

DNA polymerase α contains primase activity and initiates nuclear DNA synthesis by synthesizing an RNA primer, followed by a short string of DNA nucleotides. After DNA polymerase α has laid down from 30 to 40 nucleotides, **DNA polymerase δ** completes replication on the lagging strand. Similar in structure and function to DNA polymerase δ, **DNA polymerase ε** replicates the leading strand. Other DNA polymerases take part in repair and recombination or catalyze the replication of organelle DNA.

Some DNA polymerases, such as DNA polymerase δ and DNA polymerase ε, are capable of replicating DNA at high speed and with high fidelity (few mistakes) because they have active sites that snugly and exclusively accommodate the four normal DNA nucleotides, adenosine, guanosine, cytidine, and thymidine monophosphates. As a result of this specificity, distorted DNA templates and abnormal bases are not readily accommodated within the active site of the enzyme. When these errors are encountered in the DNA template, the high-fidelity DNA polymerases stall and are unable to bypass the lesion.

Other DNA polymerases have lower fidelity but are able to bypass distortions in the DNA template. These specialized **translesion DNA polymerases** generally have a more open active site and are able to accommodate and copy templates with abnormal bases, distorted structures, and bulky lesions. Thus, these specialized enzymes can bypass such errors but, because their active sites are more open and accommodating, they tend to make more errors. In replication, high-speed, high-fidelity enzymes are generally used until they encounter a replication block. At that point, one or more of the translesion polymerases takes over, bypasses the lesion, and continues replicating a short section of DNA. Then, the translesion polymerases detach from the replication fork and high-fidelity enzymes resume replication with high speed and accuracy. DNA-repair enzymes often repair errors produced by the translesion polymerases, although some of these errors may escape detection and lead to mutations.

CONCEPTS

There are a large number of different DNA polymerases in eukaryotic cells. DNA polymerases α, δ, and ε carry out replication on the leading and lagging strands. Other DNA polymerases carry out DNA repair. Specialized translesion polymerases are used to bypass distortions of the DNA template that normally stall the main DNA polymerases.

✔ **CONCEPT CHECK 9**

Some of the eukaryotic DNA polymerases have a tendency to make errors in replication. Why would a cell use an error-prone DNA polymerase instead of one that is more accurate?

Nucleosome Assembly

Eukaryotic DNA is complexed to histone proteins in nucleosome structures that contribute to the stability and packing of the DNA molecule (see Figure 11.4). In replication, chromatin structure is disrupted by the replication fork, but nucleosomes are quickly reassembled on the two new DNA molecules. Electron micrographs of eukaryotic DNA, such as that in **Figure 12.16**, show recently replicated DNA already covered with nucleosomes, indicating that nucleosomes are reassembled quickly.

The creation of new nucleosomes requires three steps: (1) the disruption of the original nucleosomes on the parental DNA molecule ahead of the replication fork; (2) the redistribution of preexisting histones on the new DNA molecules; and (3) the addition of newly synthesized histones to complete the formation of new nucleosomes. Before replication, a single DNA molecule is associated with histone proteins. After replication and nucleosome assembly, two DNA molecules are associated with histone proteins. Do the original histones of a nucleosome remain together, attached to one of the new DNA molecules, or do they disassemble and mix with new histones on both DNA molecules?

0.1 μm

12.16 Nucleosomes are quickly reassembled onto newly synthesized DNA. This electron micrograph of eukaryotic DNA in the process of replication clearly shows that newly replicated DNA is already covered with nucleosomes (dark circles). [Victoria Foe.]

Techniques similar to those employed by Meselson and Stahl to determine the mode of DNA replication were used to address this question. Cells were cultivated for several generations in a medium containing amino acids labeled with a heavy isotope. The histone proteins incorporated these heavy amino acids and were dense (**Figure 12.17**). The cells were then transferred to a culture medium that contained amino acids labeled with a light isotope. Histones assembled after the transfer possessed the new, light amino acids and were less dense.

After replication, the histone octamers were isolated and centrifuged in a density gradient. Results showed that, after replication, the octamers were in a continuous band between high density (representing old octamers) and low density (representing new octamers). This finding indicates that newly assembled octamers consist of a mixture of old and new histones. Further evidence indicates that reconstituted nucleosomes appear on the new DNA molecules quickly after the new DNA emerges from the replication machinery.

The reassembly of nucleosomes during replication is facilitated by proteins called histone chaperones, which are associated with the helicase enzyme that unwinds the DNA. The histone chaperones accept old histones from the original DNA molecule and deposit them, along with newly synthesized histones, on the two new DNA molecules. Current evidence suggests that the original nucleosome is broken down into two H2A-H2B dimers (each dimer consisting of one H2A and one H2B) and a single H3-H4 tetramer (each tetramer consisting of two H3 histones and two H4 histones). The old H3-H4 tetramer is then transferred randomly to one of the new DNA molecules and serves as a foundation onto which either new or old copies of H2A-H2B dimers are added. Newly synthesized H3-H4 tetramers and H2A-H2b dimers also are added to each new DNA molecule to complete the formation of new nucleosomes. The assembly of the new nucleosomes is facilitated by a protein called chromatin-assembly factor 1 (CAF-1). ▶ **TRY PROBLEM 33**

CONCEPTS

After DNA replication, new nucleosomes quickly reassemble on the molecules of DNA. Nucleosomes break down in the course of replication and reassemble from a mixture of old and new histones. The reassembly of nucleosomes during replication is facilitated by histone chaperones and chromatin-assembly factors.

The Location of Replication Within the Nucleus

The DNA polymerases that carry out replication are frequently depicted as moving down the DNA template, much as a locomotive travels along a train track. Recent evidence

Experiment

Question: What happens to histones in eukaryotic DNA replication?

Methods

1 Grow cells for several generations in medium that contains amino acids labeled with a heavy isotope.

2 Transfer the cells to a medium that contains amino acids labeled with a light isotope.

Change medium   Replication

Isolate octamers   Isolate octamers

3 Isolate histone octamers before and after replication…

Spin   Spin

4 …and subject them to density-gradient centrifugation.

Results

5 Newly synthesized octamers are less dense and thus will be higher in the tube.

6 Old octamers are dense and will move toward the bottom of the tube.

Single band; old octamers with heavy amino acids

Broad band; octamers with mixture of old and new histones (heavy and light amino acids)

Conclusion: After DNA replication, the new reassembled octamers are a random mixture of old and new histones.

12.17 Experimental procedure for studying how nucleosomes dissociate and reassociate in the course of replication.

suggests that this view is incorrect. A more accurate view is that the polymerase is fixed in location and template DNA is threaded through it, with newly synthesized DNA molecules emerging from the other end.

Techniques of fluorescence microscopy, which are able to reveal active sites of DNA synthesis, show that most replication in the nucleus of a eukaryotic cell takes place at a limited number of fixed sites, often referred to as replication factories. Time-lapse micrographs reveal that newly duplicated DNA is extruded from these particular sites. Similar results have been obtained for bacterial cells.

DNA Synthesis and the Cell Cycle

In rapidly dividing bacteria, DNA replication is continuous. In eukaryotic cells, however, replication is coordinated with the cell cycle. Passage through the cell cycle, including the onset of replication, is controlled by cell-cycle checkpoints. The important G_1/S checkpoint (see Chapter 2) holds the cell cycle in G_1 until the DNA is ready to be replicated. After the G_1/S checkpoint is passed, the cell enters S phase and the

DNA is replicated. The replication licensing system then ensures that the DNA is not replicated again until after the cell has passed through mitosis.

Replication at the Ends of Chromosomes

A fundamental difference between eukaryotic and bacterial replication arises because eukaryotic chromosomes are linear and thus have ends. As already stated, the 3′-OH group needed for replication by DNA polymerases is provided at the initiation of replication by RNA primers that are synthesized by primase. This solution is temporary because, eventually, the primers must be removed and replaced by DNA nucleotides. In a circular DNA molecule, elongation around the circle eventually provides a 3′-OH group immediately in front of the primer (**Figure 12.18a**). After the primer has been removed, the replacement DNA nucleotides can be added to this 3′-OH group.

THE END-REPLICATION PROBLEM In linear chromosomes with multiple origins, the elongation of DNA in adjacent replicons also provides a 3′-OH group preceding each primer (**Figure 12.18b**). At the very end of a linear chromosome, however, there is no adjacent stretch of replicated DNA to provide this crucial 3′-OH group. When the primer at the end of the chromosome has been removed, it cannot be replaced by DNA nucleotides, which produces a gap at the end of the chromosome, suggesting that the chromosome should become progressively shorter with each round of replication. Chromosome shortening would mean that, when an organism reproduced, it would pass on shorter chromosomes than it had inherited. Chromosomes would become shorter with each new generation and would eventually destabilize. This situation has been termed the end-replication problem. Chromosome shortening does in fact take place in many somatic cells but, in single-celled organisms, germ cells, and early embryonic cells, chromosomes do not shorten and self-destruct. So how are the ends of linear chromosomes replicated?

TELOMERES AND TELOMERASE The ends of chromosomes—the telomeres—possess several unique features, one of which is the presence of many copies of a short repeated sequence. In the protozoan *Tetrahymena* (where these repeated sequences were first discovered) this telomeric repeat is TTGGGG (see Table 11.2), with this G-rich strand typically protruding beyond the C-rich strand (**Figure 12.19a**; also see the section on Telomere Structure in Chapter 11):

toward ← 5′–TTGGGGTTGGGG–3′ → end of
centromere 3′–AACCCC–5′ chromosome

The single-stranded protruding end of the telomere, known as the **G overhang** can be extended by **telomerase**, an enzyme with both a protein and an RNA component (also known as a ribonucleoprotein). The RNA part of the enzyme contains from 15 to 22 nucleotides that are complementary to the sequence on the G-rich strand. This sequence pairs with the overhanging

(a) Circular DNA

Replication around the circle provides a 3′-OH group in front of the primer; nucleotides can be added to the 3′-OH group when the primer is replaced.

(b) Linear DNA

Telomeres

1 In linear DNA with multiple origins of replication, elongation of DNA in adjacent replicons provides a 3′-OH group for replacement of each primer.

Lagging strand

Origin Primer

Leading strand

Unwinding

2 Primers at the ends of chromosomes cannot be replaced, because there is no adjacent 3′-OH to which DNA nucleotides can be attached.

Primer

3′OH

3 When the primer at the end of a chromosome is removed,…

Primer

4 …there is no 3′-OH group to which DNA nucleotides can be attached, producing a gap.

Gap left by removal of primer

Conclusion: In the absence of special mechanisms, DNA replication would leave gaps due to the removal of primers at the ends of chromosomes.

12.18 DNA synthesis at the ends of circular and linear chromosomes must differ.

3′ end of the DNA (**Figure 12.19b**) and provides a template for the synthesis of additional DNA copies of the repeats. DNA nucleotides are added to the 3′ end of the strand one at a time (**Figure 12.19c**) and, after several nucleotides have been added, the RNA template moves down the DNA and more nucleotides are added to the 3′ end (**Figure 12.19d**). Usually, from 14 to 16 nucleotides are added to the 3′ end of the G-rich strand.

In this way, the telomerase can extend the 3′ end of the chromosome without the use of a complementary DNA template (**Figure 12.19e**). How the complementary C-rich

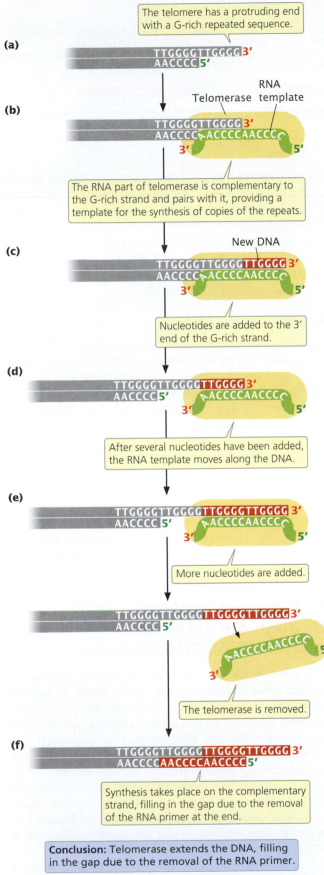

(a) The telomere has a protruding end with a G-rich repeated sequence.

TTGGGGTTGGGG 3'
AACCCC 5'

(b)

Telomerase   RNA template

TTGGGGTTGGGG 3'
AACCCC ACCCCAACCCC
3'   5'

The RNA part of telomerase is complementary to the G-rich strand and pairs with it, providing a template for the synthesis of copies of the repeats.

(c)

New DNA

TTGGGGTTGGGGTTGGGG 3'
AACCCC ACCCCAACCCC
3'   5'

Nucleotides are added to the 3' end of the G-rich strand.

(d)

TTGGGGTTGGGGTTGGGG 3'
AACCCC 5'
ACCCCAACCCC
3'   5'

After several nucleotides have been added, the RNA template moves along the DNA.

(e)

TTGGGGTTGGGGTTGGGGTTGGGG 3'
AACCCC 5'
ACCCCAACCCC
3'   5'

More nucleotides are added.

TTGGGGTTGGGGTTGGGGTTGGGG 3'
AACCCC 5'

ACCCCAACCCC
3'   5'

The telomerase is removed.

(f)

TTGGGGTTGGGGTTGGGGTTGGGG 3'
AACCCCAACCCCAACCCC 5'

Synthesis takes place on the complementary strand, filling in the gap due to the removal of the RNA primer at the end.

Conclusion: Telomerase extends the DNA, filling in the gap due to the removal of the RNA primer.

12.19 The enzyme telomerase is responsible for the replication of chromosome ends.

strand is synthesized (**Figure 12.19f**) is not clear. It may be synthesized by conventional replication, with DNA polymerase α synthesizing an RNA primer on the 5' end of the extended (G-rich) template. The removal of this primer once again leaves a gap at the 5' end of the chromosome, but this gap does not matter, because the end of the chromosome is extended at each replication by telomerase; so, the chromosome does not become shorter overall.

Telomerase is present in single-celled organisms, germ cells, early embryonic cells, and certain proliferative somatic cells (such as bone-marrow cells and cells lining the intestine), all of which must undergo continuous cell division. Most somatic cells have little or no telomerase activity, and chromosomes in these cells progressively shorten with each cell division. These cells are capable of only a limited number of divisions; when the telomeres have shortened beyond a critical point a chromosome becomes unstable, has a tendency to undergo rearrangements, and is degraded. These events lead to cell death.

CONCEPTS

The ends of eukaryotic chromosomes are replicated by an RNA–protein enzyme called telomerase. This enzyme adds extra nucleotides to the G-rich DNA strand of the telomere.

✔ CONCEPT CHECK 10

What would be the result if an organism's telomerase were mutated and nonfunctional?
a. No DNA replication would take place.
b. The DNA polymerase enzyme would stall at the telomere.
c. Chromosomes would shorten with each new generation.
d. RNA primers could not be removed.

TELOMERASE, AGING, AND DISEASE The shortening of telomeres may contribute to the process of aging. The telomeres of genetically engineered mice that lack a functional telomerase gene (and therefore do not express telomerase in somatic or germ cells) undergo progressive shortening in successive generations. After several generations, these mice show some signs of premature aging, such as graying, hair loss, and delayed wound healing. Through genetic engineering, it is also possible to create somatic cells that express telomerase. In these cells, telomeres do not shorten, cell aging is inhibited, and the cells will divide indefinitely.

Some of the strongest evidence that telomere length is related to aging comes from studies of telomeres in birds. In 2012, scientists in the United Kingdom measured telomere length in red blood cells taken from 99 zebra finches at various times during their lives. The scientists found a strong correlation between telomere length and longevity: birds with longer telomeres lived longer than birds with short telomeres. The strongest predictor of life span was when telomere length was measured early in life, at 25 days, which is roughly equivalent to human adolescence. Although these

1 Homologous chromosomes align and single-strand breaks occur in the same position on both DNA molecules.

2 A free end of each broken strand migrates to the other DNA molecule.

3 Each invading strand joins to the broken end of the other DNA molecule, creating a Holliday junction, and begins to displace the original complementary strand.

Holliday junction

observations suggest that telomere length is associated with aging in some animals, the precise role of telomeres in *human* aging remains uncertain.

Some diseases are associated with abnormalities of telomere replication. People who have Werner syndrome, an autosomal recessive disease, show signs of premature aging that begins in adolescence or early adulthood, including wrinkled skin, graying of the hair, baldness, cataracts, and muscle atrophy. They often develop cancer, osteoporosis, heart and artery disease, and other ailments typically associated with aging. The causative gene, *WRN*, has been mapped to human chromosome 8 and normally encodes a RecQ helicase enzyme. This enzyme is necessary for the efficient replication of telomeres. In people who have Werner syndrome, this helicase is defective and, consequently, the telomeres shorten prematurely.

Another disease associated with abnormal maintenance of telomeres is dyskeratosis congenita, which leads to progressive bone-marrow failure, in which the bone marrow fails to produce enough new blood cells. People with an X-linked form of the disease have a mutation in a gene that encodes dyskerin, a protein that normally helps process the RNA component of telomerase. People who have the disease typically inherit short telomeres from a parent who carries the mutation and who is unable to maintain telomere length in his or her germ cells owing to defective dyskerin. In families that carry this mutation, telomere length typically shortens with each successive generation, leading to anticipation, a progressive increase in the severity of the disease over generations (see Chapter 5).

Telomerase also appears to play a role in cancer. Cancer tumor cells have the capacity to divide indefinitely, and the telomerase enzyme is expressed in 90% of all cancers. Some recent evidence indicates that telomerase may stimulate cell proliferation independently of its effect on telomere length, and so the mechanism by which telomerase contributes to cancer is not clear. As will be discussed in Chapter 23, cancer is a complex, multistep process that usually requires mutations in at least several genes. Telomerase activation alone does not lead to cancerous growth in most cells, but it does appear to be required, along with other mutations, for cancer to develop. Some experimental cancer drugs work by inhibiting the action of telomerase.

One of the difficulties in studying the effect of telomere shortening on the aging process is that the expression of telomerase in somatic cells also promotes cancer, which may shorten a person's life span. To circumvent this problem,

Antonia Tomas-Loba and her colleagues created genetically engineered mice that expresssed telomerase and carried genes that made them resistant to cancer. These mice had longer telomeres, lived longer, and exhibited fewer age-related changes, such as skin alterations, a decrease in neuromuscular coordination, and degenerative diseases. These results support the idea that telomere shortening contributes to aging. ▶ **TRY PROBLEM 35**

Replication in Archaea

The process of replication in archaea has a number of features in common with replication in eukaryotic cells; many of the proteins taking part are more similar to those in eukaryotic cells than to those in eubacteria. Like eubacteria, some archaea have a single replication origin, but the archaean *Sulfolobus solfataricus* has two origins of replication, similar to the multiple origins seen in eukaryotic genomes. The replication origins of archaea do not contain the typical sequences recognized by bacterial initiator proteins; instead, they have sequences that are similar to those found in eukaryotic origins. The initiator proteins of archaea also are more similar to those of eukaryotes than to those of eubacteria. These similarities in replication between archaeal and eukaryotic cells reinforce the conclusion that the archaea are more closely related to eukaryotic cells than to the prokaryotic eubacteria.

12.5 Recombination Takes Place Through the Breakage, Alignment, and Repair of DNA Strands

Recombination is the exchange of genetic information between DNA molecules; when the exchange is between homologous DNA molecules, it is called **homologous recombination**. This process takes place in crossing over, in which homologous regions of chromosomes are exchanged (see Figure 7.5) and alleles are shuffled into new combinations. Recombination is an extremely important genetic process because it increases genetic variation. Rates of recombination provide important information about linkage relations among genes, which is used to create genetic maps (see Figures 7.13 and 7.14). Recombination is also essential for some types of DNA repair (as will be discussed in Chapter 18).

Homologous recombination is a remarkable process: a nucleotide strand of one chromosome aligns precisely with a nucleotide strand of the homologous chromosome, breaks

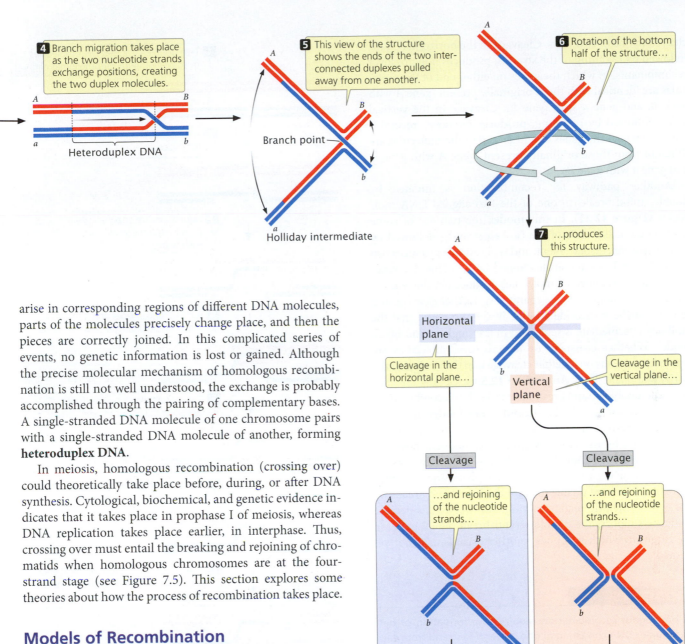

4 Branch migration takes place as the two nucleotide strands exchange positions, creating the two duplex molecules.

Heteroduplex DNA

5 This view of the structure shows the ends of the two interconnected duplexes pulled away from one another.

Branch point

Holliday intermediate

6 Rotation of the bottom half of the structure...

7 ...produces this structure.

Horizontal plane

Cleavage in the horizontal plane...

Vertical plane

Cleavage in the vertical plane...

Cleavage

...and rejoining of the nucleotide strands...

Cleavage

...and rejoining of the nucleotide strands...

Noncrossover recombinants

Crossover recombinants

...produces noncrossover recombinants consisting of two heteroduplex molecules.

...produces crossover recombinants consisting of two heteroduplex molecules.

Conclusion: The Holliday model predicts noncrossover or crossover recombinant DNA, depending on whether cleavage is in the horizontal or the vertical plane.

12.20 The Holliday model of homologous recombination. In this model, recombination takes place through a single-strand break in each DNA duplex, strand displacement, branch migration, and resolution of a single Holliday junction.

arise in corresponding regions of different DNA molecules, parts of the molecules precisely change place, and then the pieces are correctly joined. In this complicated series of events, no genetic information is lost or gained. Although the precise molecular mechanism of homologous recombination is still not well understood, the exchange is probably accomplished through the pairing of complementary bases. A single-stranded DNA molecule of one chromosome pairs with a single-stranded DNA molecule of another, forming **heteroduplex DNA**.

In meiosis, homologous recombination (crossing over) could theoretically take place before, during, or after DNA synthesis. Cytological, biochemical, and genetic evidence indicates that it takes place in prophase I of meiosis, whereas DNA replication takes place earlier, in interphase. Thus, crossing over must entail the breaking and rejoining of chromatids when homologous chromosomes are at the four-strand stage (see Figure 7.5). This section explores some theories about how the process of recombination takes place.

Models of Recombination

Homologous recombination can take place through several different pathways. One pathway is initiated by a single-strand break in each of two DNA molecules and includes the formation of a special structure called the **Holliday junction** (**Figure 12.20**). In this model, double-stranded DNA molecules from two homologous chromosomes align precisely. A single-strand break in one of the DNA molecules provides a free end that invades and joins the free end of the other DNA molecule. Strand invasion and joining take place on both DNA molecules, creating two heteroduplex DNAs, each consisting of one original strand plus one new strand from the other DNA molecule. The point at which nucleotide strands pass from one DNA molecule to the other is the Holliday junction. The junction moves along the molecules in a process called branch migration. The exchange of nucleotide strands and branch migration produce a structure termed the Holliday intermediate, which can be cleaved in one of two ways. Cleavage in the horizontal plane, followed by rejoining of the strands, produces noncrossover recombinants, in which the genes on either end of the molecules are identical with those originally present (gene *A* with gene *B*, and gene *a* with gene *b*). Cleavage in the vertical plane, followed by rejoining, produces crossover recombinants, in which the genes on either end of the molecules are different from those originally present (gene *A* with gene *b*, and gene *a* with gene *B*).

Another pathway for recombination is initiated by double-strand breaks in one of the two aligned DNA molecules (**Figure 12.21**). In this model, the removal of some nucleotides at the ends of the broken strands—followed by strand invasion, displacement, and replication—produces two heteroduplex DNA molecules joined by two Holliday junctions. The interconnected molecules produced in the double-strand-break model can be separated by further cleavage and reunion of the nucleotide strands in the same way that the Holliday intermediate is separated in the single-strand-break model. Whether crossover or noncrossover molecules are produced depends on whether cleavage is in the vertical or the horizontal plane. View **Animation 12.5** to see how the Holliday and double-strand break models lead to recombination.

Evidence for the double-strand-break model originally came from results of genetic crosses in yeast that could not be explained by the Holliday model. Subsequent observations showed that double-strand breaks appear in yeast in prophase I, when crossing over takes place, and that mutant strains that are unable to form double-strand breaks do not exhibit meiotic recombination. Although considerable evidence supports the double-strand-break model in yeast, the extent to which it applies to other organisms is not yet known.

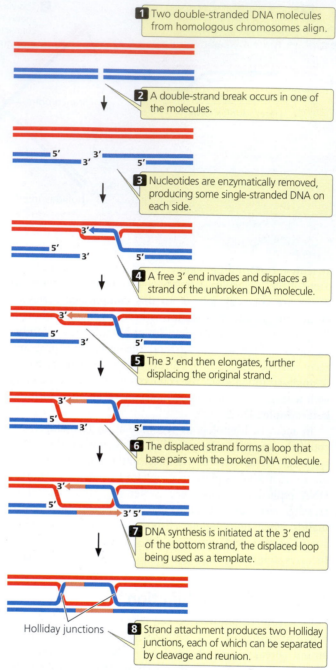

1 Two double-stranded DNA molecules from homologous chromosomes align.

2 A double-strand break occurs in one of the molecules.

3 Nucleotides are enzymatically removed, producing some single-stranded DNA on each side.

4 A free 3′ end invades and displaces a strand of the unbroken DNA molecule.

5 The 3′ end then elongates, further displacing the original strand.

6 The displaced strand forms a loop that base pairs with the broken DNA molecule.

7 DNA synthesis is initiated at the 3′ end of the bottom strand, the displaced loop being used as a template.

Holliday junctions

8 Strand attachment produces two Holliday junctions, each of which can be separated by cleavage and reunion.

12.21 The double-strand-break model of recombination. In this model, recombination takes place through a double-strand break in one DNA duplex, strand displacement, DNA synthesis, and the resolution of two Holliday junctions.

CONCEPTS

Homologous recombination requires the formation of heteroduplex DNA consisting of one nucleotide strand from each of two homologous chromosomes. In the Holliday model, homologous recombination is accomplished through a single-strand break in the DNA, strand displacement, and branch migration. In the double-strand-break model, recombination is accomplished through double-strand breaks, strand displacement, and branch migration.

✓ **CONCEPT CHECK 11**

Why is recombination important?

Enzymes Required for Recombination

Recombination between DNA molecules requires the unwinding of DNA helices, the cleavage of nucleotide strands, strand invasion, and branch migration, followed by further strand cleavage and union to remove Holliday junctions. Much of what we know about these processes arises from studies of gene exchange in *E. coli*. Although bacteria do not undergo meiosis, they do have a type of sexual reproduction (conjugation), in which one bacterium donates its chromosome to another (discussed more fully in Chapter 9). Subsequent to conjugation, the recipient bacterium has two chromosomes, which may undergo homologous recombination. Geneticists have isolated mutant strains of *E. coli* that are deficient in recombination; the study of these strains has resulted in the identification of genes and proteins that take part in bacterial recombination, revealing several different pathways by which it can take place.

Three genes that play pivotal roles in *E. coli* recombination are *recB*, *recC*, and *recD*, which encode three polypeptides that together form the RecBCD protein. This protein unwinds double-stranded DNA and is capable of cleaving nucleotide strands. The *recA* gene encodes the RecA protein; this protein allows a single strand to invade a DNA helix and the subsequent displacement of one of the original strands. In eukaryotes, the formation and branch migration of Holliday structures is facilitated by the enzyme Rad51.

In *E. coli*, *ruvA* and *ruvB* genes encode proteins that catalyze branch migration, and the *ruvC* gene produces a protein, called resolvase, that cleaves Holliday structures. Cleavage and resolution of Holliday structures in eukaryotes is carried out by an analogous enzyme called GEN1. Single-strand-binding proteins, DNA ligase, DNA polymerases, and DNA gyrase also play roles in various types of recombination, in addition to their functions in DNA replication.

CONCEPTS

A number of proteins have roles in recombination, including RecA, RecBCD, RuvA, RuvB, resolvase, single-strand-binding proteins, ligase, DNA polymerases, and gyrase.

✓ CONCEPT CHECK 12

What is the function of resolvase in recombination?

a. Unwinds double-stranded DNA.

b. Allows a single DNA strand to invade a DNA helix.

c. Displaces one of the original DNA strands during branch migration.

d. Cleaves the Holliday structure.

Gene Conversion

As we have seen, homologous recombination is the mechanism that produces crossing over. It is also responsible for a related phenomenon known as **gene conversion**, a process of nonreciprocal genetic exchange that can produce abnormal ratios of gametes following meiosis. For example, an individual organism with genotype *Aa* is expected to produce $\frac{1}{2}$ *A* gametes and $\frac{1}{2}$ *a* gametes. Sometimes, however, meiosis in an *Aa* individual produces $\frac{3}{4}$ *A* and $\frac{1}{4}$ *a* or $\frac{1}{4}$ *A* and $\frac{3}{4}$ *a*. Gene conversion arises from heteroduplex formation that takes place in recombination. During heteroduplex formation, a single-stranded DNA molecule of one chromosome pairs with a single-stranded DNA molecule of another chromosome. If the two strands in a heteroduplex come from chromosomes with different alleles, there will be a mismatch of bases in the heteroduplex DNA (**Figure 12.22**). Such mismatches are often repaired by the cell. Repair mechanisms frequently excise nucleotides on one of the strands and replace them with new DNA by using the complementary

strand as a template. One copy of an allele may be converted into the other allele, leading to a gene-conversion event (see Figure 12.22), depending on which strand serves as a template.

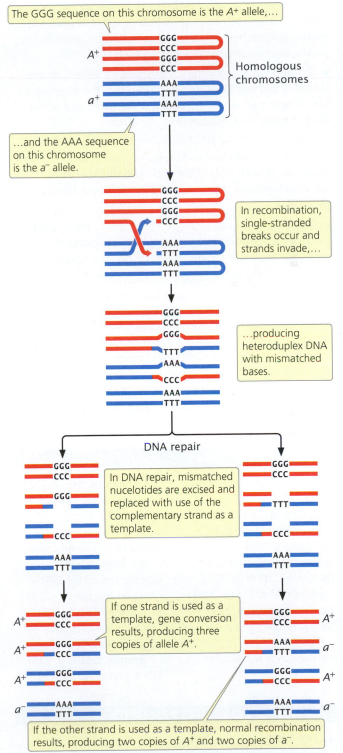

The GGG sequence on this chromosome is the *A*+ allele,...

...and the AAA sequence on this chromosome is the *a*− allele.

In recombination, single-stranded breaks occur and strands invade,...

...producing heteroduplex DNA with mismatched bases.

DNA repair

In DNA repair, mismatched nucleotides are excised and replaced with use of the complementary strand as a template.

If one strand is used as a template, gene conversion results, producing three copies of allele *A*+.

If the other strand is used as a template, normal recombination results, producing two copies of *A*+ and two copies of *a*−.

12.22 Gene conversion takes place through the repair of mismatched bases in heteroduplex DNA.

CONCEPTS SUMMARY

■ Replication is semiconservative: DNA's two nucleotide strands separate, and each serves as a template on which a new strand is synthesized.

■ In theta replication of DNA, the two nucleotide strands of a circular DNA molecule unwind, creating a replication bubble; within each replication bubble, DNA is normally synthesized on both strands and at both replication forks, producing two circular DNA molecules.

■ Rolling-circle replication is initiated by a break in one strand of circular DNA, which produces a 3′-OH group to which new nucleotides are added while the 5′ end of the broken strand is displaced from the circle.

■ Linear eukaryotic DNA contains many origins of replication. Unwinding and replication take place on both templates at both ends of the replication bubble until adjacent replicons meet, resulting in two linear DNA molecules.

■ All DNA synthesis is in the 5′→3′ direction. Because the two nucleotide strands of DNA are antiparallel, replication takes place continuously on one strand (the leading strand) and discontinuously on the other (the lagging strand).

■ Replication begins when an initiator protein binds to a replication origin and unwinds a short stretch of DNA to which DNA helicase attaches. DNA helicase unwinds the DNA at the replication fork, single-strand-binding proteins bind to single nucleotide strands to prevent secondary structures, and DNA gyrase (a topoisomerase) removes the strain ahead of the replication fork that is generated by unwinding.

■ During replication, primase synthesizes short primers of RNA nucleotides, providing a 3′-OH group to which DNA polymerase can add DNA nucleotides.

■ DNA polymerase adds new nucleotides to the 3′ end of a growing polynucleotide strand. Bacteria have two DNA polymerases that have primary roles in replication: DNA polymerase III, which synthesizes new DNA on the leading and lagging strands, and DNA polymerase I, which removes and replaces primers.

■ DNA ligase seals the breaks that remain in the sugar–phosphate backbones when the RNA primers are replaced by DNA nucleotides.

■ Several mechanisms ensure the high rate of accuracy in replication, including precise nucleotide selection, proofreading, and mismatch repair.

■ Precise replication at multiple origins in eukaryotes is ensured by a licensing factor that must attach to an origin before replication can begin.

■ Eukaryotic nucleosomes are quickly assembled on new molecules of DNA; newly assembled nucleosomes consist of a random mixture of old and new histone proteins.

■ The ends of linear eukaryotic DNA molecules are replicated by the enzyme telomerase.

■ Homologous recombination takes place through breaks in nucleotide strands, alignment of homologous DNA segments, and rejoining of the strands. Homologous recombination requires a number of enzymes and proteins.

■ Gene conversion is nonreciprocal genetic exchange and produces abnormal ratios of gametes.

IMPORTANT TERMS

semiconservative replication (p. 326)
equilibrium density gradient centrifugation (p. 327)
replicon (p. 329)
replication origin (p. 329)
theta replication (p. 329)
replication bubble (p. 329)
replication fork (p. 329)
bidirectional replication (p. 329)
rolling-circle replication (p. 330)

DNA polymerase (p. 332)
continuous replication (p. 332)
leading strand (p. 332)
discontinuous replication (p. 333)
lagging strand (p. 333)
Okazaki fragment (p. 333)
initiator protein (p. 334)
DNA helicase (p. 335)
single-strand-binding protein (SSB) (p. 335)
DNA gyrase (p. 335)

primase (p. 336)
primer (p. 336)
DNA polymerase III (p. 337)
DNA polymerase I (p. 337)
DNA ligase (p. 338)
proofreading (p. 339)
mismatch repair (p. 340)
autonomously replicating sequence (ARS) (p. 340)
replication licensing factor (p. 341)
DNA polymerase α (p. 342)

DNA polymerase δ (p. 342)
DNA polymerase ε (p. 342)
translesion DNA polymerase (p. 342)
G overhang (p. 344)
telomerase (p. 344)
homologous recombination (p. 346)
heteroduplex DNA (p. 347)
Holliday junction (p. 347)
gene conversion (p. 349)

ANSWERS TO CONCEPT CHECKS

1. Two bands

2. b

3. c

4. Initiator protein, helicase, single-strand-binding protein, DNA gyrase.

5. c

6. b

7. c

8. The size of eukaryotic genomes, the linear structure of eukaryotic chromosomes, and the association of DNA with histone proteins.

9. Because error-prone DNA polymerases can bypass lesions in the DNA helix that stall accurate, high-speed DNA polymerases.

10. c

11. Recombination is important for genetic variation and for some types of DNA repair.

12. d

WORKED PROBLEMS

Problem 1

The following diagram represents the template strands of a replication bubble in a DNA molecule. Draw in the newly synthesized strands and identify the leading and lagging strands.

Solution Strategy

What information is required in your answer to the problem?

The diagram above with the newly synthesized strands drawn in and the leading and lagging strands labeled.

What information is provided to solve the problem?

A diagram of the template DNA with 5′ and 3′ ends labeled.

For help with this problem, review:

Direction of Replication in Section 12.2 and Figure 12.10c.

Next, determine the direction of replication for each new strand, which must be 5′→3′. You might draw arrows on the new strands to indicate the direction of replication. After you have established the direction of replication for each strand, look at each fork and determine whether the direction of replication for a strand is the same as the direction of unwinding. The strand on which replication is in the same direction as that of unwinding is the leading strand. The strand on which replication is in the direction opposite that of unwinding is the lagging strand.

> **Recall:** DNA synthesis is always 5′ to 3′.

Solution Steps

> **Recall:** The two strands of DNA are antiparallel, so the newly synthesized strand should have the opposite polarity (direction) as the template strand.

To determine the leading and lagging strands, first note which end of each template strand is 5′ and which end is 3′. With a pencil, draw in the strands being synthesized on these templates, and identify their 5′ and 3′ ends.

> **Hint** Each replication fork should have one leading and one lagging strand.

Problem 2

Consider the experiment conducted by Meselson and Stahl in which they used ^{14}N and ^{15}N in cultures of *E. coli* and equilibrium density gradient centrifugation. Draw pictures to represent the bands produced by bacterial DNA in the density-gradient tube before the switch to medium containing ^{14}N and after one, two, and three rounds of replication after the switch to the medium containing ^{14}N. Use a separate set of drawings to show the bands that would appear if replication were (a) semiconservative; (b) conservative; (c) dispersive.

Solution Strategy

What information is required in your answer to the problem?

Drawings that represent the bands produced by bacterial DNA in density-gradient tubes before the switch to medium containing ^{14}N and after one, two, and three rounds of replication following the switch to the medium containing ^{14}N; thus, you should have drawings of four tubes for each model of replication. You will need a separate set of drawings for semiconservative, conservative, and dispersive replication.

What information is provided to solve the problem?

- The bacterial DNA was originally labeled with ^{15}N and then the bacteria were switched to a medium with ^{14}N (see discussion of experiment on pp. 327–328).

- Original DNA will have ^{15}N. Newly synthesized DNA will have ^{14}N.

- Density gradient centrifugation was performed before switching to ^{14}N and after one, two, and three rounds of replication following the switch.

For help with this problem, review:

Meselson and Stahl's Experiment in Section 12.2.

Solution Steps

DNA labeled with ^{15}N will be denser than DNA labeled with ^{14}N; therefore ^{15}N-labeled DNA will sink lower in the density-gradient tube. Before the switch to medium containing ^{14}N, all DNA in the bacteria will contain ^{15}N and will produce a single band in the lower end of the tube.

a. With semiconservative replication, the two strands separate, and each serves as a template on which a new strand is synthesized. After one round of replication, the original template strand of each molecule will contain ^{15}N and the new strand of each molecule will contain ^{14}N; so a single band will appear in the density gradient halfway between the positions expected of

DNA containing only ^{15}N and of DNA containing only ^{14}N. In the next round of replication, the two strands again separate and serve as templates for new strands. Each of the new strands contains only ^{14}N, thus some DNA molecules will contain one strand with the original ^{15}N and one strand with new ^{14}N, whereas the other molecules will contain two strands with ^{14}N. This labeling will produce two bands, one at the intermediate position and one at a higher position in the tube. Additional rounds of replication should produce increasing amounts of DNA that contains only ^{14}N; so the higher band will get darker.

Hint: Review the distribution of new and old DNA in semiconservative, conservative, and dispersive replication in Figure 12.1.

| Before the switch to ^{14}N | After one round of replication | After two rounds of replication | After three rounds of replication |

b. With conservative replication, the entire molecule serves as a template. After one round of replication, some molecules will consist entirely of ^{15}N, and others will consist entirely of ^{14}N; so two bands should be present. Subsequent rounds of replication will increase the fraction of DNA consisting entirely of new ^{14}N; thus the upper band will get darker. However, the original DNA with ^{15}N will remain, and so two bands will be present.

| Before the switch to ^{14}N | After one round of replication | After two rounds of replication | After three rounds of replication |

c. In dispersive replication, both nucleotide strands break down into fragments that serve as templates for the synthesis of new DNA. The fragments then reassemble into DNA molecules. After one round of replication, all DNA should contain approximately half ^{15}N and half ^{14}N, producing a single band that is halfway between the positions expected of DNA labeled with ^{15}N and of DNA labeled with ^{14}N. With further rounds of replication, the proportion of ^{14}N in each molecule increases; so a single hybrid band remains, but its position in the density gradient will move upward. The band is also expected to get darker as the total amount of DNA increases.

| Before the switch to ^{14}N | After one round of replication | After two rounds of replication | After three rounds of replication |

COMPREHENSION QUESTIONS

Section 12.2

1. What is semiconservative replication?

2. How did Meselson and Stahl demonstrate that replication in *E. coli* takes place in a semiconservative manner?

3. Draw a molecule of DNA undergoing theta replication. On your drawing, identify (a) origin, (b) polarity (5′ and 3′ ends) of all template strands and newly synthesized strands, (c) leading and lagging strands, (d) Okazaki fragments, and (e) location of primers.

4. Draw a molecule of DNA undergoing rolling-circle replication. On your drawing, identify (a) origin, (b) polarity (5′ and 3′ ends) of all template and newly synthesized strands, (c) leading and lagging strands, (d) Okazaki fragments, and (e) location of primers.

5. Draw a molecule of DNA undergoing eukaryotic linear replication. On your drawing, identify (a) origin, (b) polarity (5′ and 3′ ends) of all template and newly synthesized strands, (c) leading and lagging strands, (d) Okazaki fragments, and (e) location of primers.

6. What are three major requirements of replication?

7. What substrates are used in the DNA-synthesis reaction?

Section 12.3

8. List the different proteins and enzymes taking part in bacterial replication. Give the function of each in the replication process.

9. Why is DNA gyrase necessary for replication?

10. What similarities and differences exist in the enzymatic activities of DNA polymerases I and III? What is the function of each DNA polymerase in bacterial cells?

11. Why is primase required for replication?

12. What three mechanisms ensure the accuracy of replication in bacteria?

Section 12.4

13. How does replication licensing ensure that DNA is replicated only once at each origin per eukaryotic cell cycle?

14. In what ways is eukaryotic replication similar to bacterial replication, and in what ways is it different?

15. What is the end-of-chromosome problem for linear replication? Why, in the absence of telomerase, do the ends of chromosomes get progressively shorter each time the DNA is replicated?

16. Outline in words and pictures how telomeres at the ends of eukaryotic chromosomes are replicated.

Section 12.5

17. Explain how the type of cleavage of the Holliday intermediate leads to noncrossover recombinants and crossover recombinants.

18. What are some of the enzymes taking part in recombination in *E. coli* and what roles do they play?

19. What is gene conversion? How does it arise?

▶ For more questions that test your comprehension of the key chapter concepts, go to ***LEARNING***Curve for this chapter.

APPLICATION QUESTIONS AND PROBLEMS

Section 12.2

20. Suppose a future scientist explores a distant planet and discovers a novel form of double-stranded nucleic acid. When this nucleic acid is exposed to DNA polymerases from *E. coli*, replication takes place continuously on both strands. What conclusion can you make about the structure of this novel nucleic acid?

21. Phosphorus is required to synthesize the deoxyribonucleoside triphosphates used in DNA replication. A geneticist grows some *E. coli* in a medium containing nonradioactive phosphorous for many generations. A sample of the bacteria is then transferred to a medium that contains a radioactive isotope of phosphorus (^{32}P). Samples of the bacteria are removed immediately after the transfer and after one and two rounds of replication. Assume that newly synthesized DNA contains ^{32}P and the original DNA contains nonradioactive phosphorous. What will be the distribution of radioactivity in the DNA of the bacteria in each sample? Will radioactivity be detected in neither, one, or both strands of the DNA?

*22. A line of mouse cells is grown for many generations in a medium with ^{15}N. Cells in G_1 are then switched to a new medium that contains ^{14}N. Draw a pair of homologous chromosomes from these cells at the following stages, showing the two strands of DNA molecules found in the chromosomes. Use different colors to represent strands with ^{14}N and ^{15}N.

   a. Cells in G_1, before switching to medium with ^{14}N

   b. Cells in G_2, after switching to medium with ^{14}N

   c. Cells in anaphase of mitosis, after switching to medium with ^{14}N

   d. Cells in metaphase I of meiosis, after switching to medium with ^{14}N

   e. Cells in anaphase II of meiosis, after switching to medium with ^{14}N

*23. A circular molecule of DNA contains 1 million base pairs. If the rate of DNA synthesis at a replication fork is 100,000 nucleotides per minute, how much time will theta replication require to completely replicate the molecule, assuming that theta replication is bidirectional? How long will replication of this circular chromosome take by rolling-circle replication? Ignore replication of the displaced strand in rolling-circle replication.

24. A bacterium synthesizes DNA at each replication fork at a rate of 1000 nucleotides per second. If this bacterium completely replicates its circular chromosome by theta replication in 30 minutes, how many base pairs of DNA will its chromosome contain?

Section 12.3

*25. The following diagram represents a DNA molecule that is undergoing replication. Draw in the strands of newly synthesized DNA and identify (a) the polarity of newly synthesized strands, (b) the leading and lagging strands, (c) Okazaki fragments, and (d) RNA primers.

26. In **Figure 12.8**, which is the leading strand and which is the lagging strand?

*27. What would be the effect on DNA replication of mutations that destroyed each of the following activities in DNA polymerase I?

   a. $3' \rightarrow 5'$ exonuclease activity

   b. $5' \rightarrow 3'$ exonuclease activity

   c. $5' \rightarrow 3'$ polymerase activity

28. Which of the DNA polymerases shown in **Table 12.3** has the ability to proofread?

29. How would DNA replication be affected in a bacterial cell that is lacking DNA gyrase?

*30. If the gene for primase were mutated so that no functional primase was produced, what would be the effect on theta replication? On rolling-circle replication?

31. DNA polymerases are not able to prime replication, yet primase and other RNA polymerases can. Some geneticists have speculated that the inability of DNA polymerase to prime replication is due to its proofreading function. This hypothesis argues that proofreading is essential for the faithful transmission of genetic information and that, because DNA polymerases have evolved the ability to proofread, they cannot prime DNA synthesis. Explain why proofreading and priming functions in the same enzyme might be incompatible.

Section 12.4

32. Marina Melixetian and her colleagues suppressed the expression of Geminin protein in human cells by treating the cells with small interfering RNAs (siRNAs) complementary to Geminin messenger RNA (M. Melixetian et al. 2004. *Journal of Cell Biology* 165:473–482). (Small interfering RNAs form a complex with proteins and pair with complementary sequences on mRNAs; the complex then cleaves the mRNA, so there is no translation of the mRNA; pp. 403–404 in Chapter 14). Forty-eight hours after treatment with siRNA, the Geminin-depleted cells were enlarged and contained a single giant nucleus. Analysis of DNA content showed that many of these Geminin-depleted cells were 4*n* or greater. Explain these results.

*33. What results would be expected in the experiment outlined in **Figure 12.17** if, during replication, all the original histone proteins remained on one strand of the DNA and new histones attached to the other strand?

34. A number of scientists who study cancer treatment have become interested in telomerase. Why? How might cancer-drug therapies that target telomerase work?

*35. The enzyme telomerase is part protein and part RNA. What would be the most likely effect of a large deletion in the gene that encodes the RNA part of telomerase? How would the function of telomerase be affected?

36. Dyskeratosis congenita (DKC) is a rare genetic disorder characterized by abnormal fingernails and skin pigmentation, the formation of white patches on the tongue and cheek, and progressive failure of the bone marrow. An autosomal dominant form of DKC results from mutations in the gene that encodes the RNA component of telomerase. Tom Vulliamy and his colleagues examined 15 families with autosomal dominant DKC (T. Vulliamy et al. 2004. *Nature Genetics* 36:447–449). They observed that the median age of onset of DKC in parents was 37 years, whereas the median age of onset in the children of affected parents was 14.5 years. Thus, DKC in these families arose at progressively younger ages in successive generations, a phenomenon known as anticipation (see p. 126 in Chapter 5). The researchers measured telomere length of members of these families; the measurements are given in the adjoining table. Telomere length normally shortens with age, and so telomere length was adjusted for age. Note that the age-adjusted telomere length of all members of these families is negative, indicating that their telomeres are shorter than normal. For age-adjusted telomere length, the more negative the number, the shorter the telomere.

Parent telomere length	Child telomere length
−4.7	−6.1
	−6.6
	−6.0
−3.9	−0.6
−1.4	−2.2
−5.2	−5.4
−2.2	−3.6
−4.4	−2.0
−4.3	−6.8
−5.0	−3.8
−5.3	−6.4
−0.6	−2.5
−1.3	−5.1
	−3.9
−4.2	−5.9

a. How does the telomere length of the parents compare with the telomere length of the children? (Hint: Calculate the average telomere length of all parents and the average telomere length of all children.)

b. Explain why the telomeres of people with DKC are shorter than normal.

c. Explain why DKC arises at an earlier age in subsequent generations.

37. An individual is heterozygous at two loci (*Ee Ff*) and the genes are in repulsion (see p. 172 in Chapter 7). Assume that single-strand breaks and branch migration occur at the positions shown below. Using different colors to represent the two homologous chromosomes, draw out the noncrossover recombinant and crossover recombinant DNA molecules that will result from homologous recombination (Hint: see **Figure 12.20**).

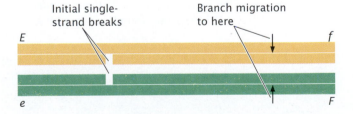

CHALLENGE QUESTIONS

Section 12.3

38. A conditional mutation expresses its mutant phenotype only under certain conditions (the restrictive conditions) and expresses the normal phenotype under other conditions (the permissive conditions). One type of conditional mutation is a temperature-sensitive mutation, which expresses the mutant phenotype only at certain temperatures.

    Strains of E. coli have been isolated that contain temperature-sensitive mutations in the genes encoding different components of the replication machinery. In each of these strains, the protein produced by the mutated gene is nonfunctional under the restrictive conditions. These strains are grown under permissive conditions and then abruptly switched to the restrictive condition. After one round of replication under the restrictive condition, the DNA from each strain is isolated and analyzed. What characteristics would you expect to see in the DNA isolated from each strain with a temperature-sensitive mutation in its gene that encodes in the following proteins?

    a. DNA ligase

    b. DNA polymerase I

    c. DNA polymerase III

    d. Primase

    e. Initiator protein

Section 12.4

39. DNA topoisomerases play important roles in DNA replication and in supercoiling (see Chapter 11). These enzymes are also the targets for certain anticancer drugs. Eric Nelson and his colleagues studied m-AMSA, one of the anticancer compounds that acts on topisomerase enzymes (E. M. Nelson, K. M. Tewey, and L. F. Liu. 1984. *Proceedings of the National Academy of Sciences* 81:1361–1365). They found that m-AMSA stabilizes an intermediate produced in the course of the topoisomerase's action. The intermediate consisted of the topoisomerase bound to the broken ends of the DNA. Breaks in DNA that are produced by anticancer compounds such as m-AMSA inhibit the replication of the cellular DNA and thus stop cancer cells from proliferating. Propose a mechanism for how m-AMSA and other anticancer agents that target topoisomerase enzymes taking part in replication might lead to DNA breaks and chromosome rearrangments.

*40. The regulation of replication is essential to genomic stability, and, normally, the DNA is replicated just once every eukaryotic cell cycle (in the S phase). Normal cells produce protein A, which increases in concentration in the S phase. In cells that have a mutated copy of the gene for protein A, the protein is not functional and replication takes place continuously throughout the cell cycle, with the result that cells may have 50 times the normal amount of DNA. Protein B is normally present in G_1 but disappears from the cell nucleus in the S phase. In cells with a mutated copy of the gene for protein A, the levels of protein B fail to disappear in the S phase and, instead, remain high throughout the cell cycle. When the gene for protein B is mutated, no replication takes place.
Propose a mechanism for how protein A and protein B might normally regulate replication so that each cell gets the proper amount of DNA. Explain how mutation of these genes produces the effects just described.

Go to your LaunchPad to find additional learning resources and the Suggested Readings for this chapter.

13

Transcription

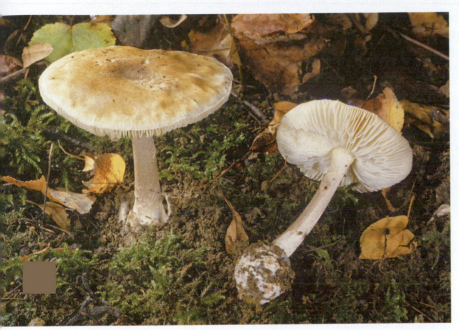

The death cap mushroom, *Amanita phalloides*, causes death by inhibiting the process of transcription. [© MAP/Jean-Yves Grospas/Age FotoStock America, Inc.]

Death Cap Poisoning

On November 8, 2009, 31-year-old Tomasa was hiking the Lodi Lake nature trail east of San Francisco with her husband and cousin when they came across some large white mushrooms that looked very much like the edible mushrooms that they enjoyed in their native Mexico. They picked the mushrooms and took them home, cooking and consuming them for dinner. Within hours, Tomasa and her family were sick and went to the hospital. They were later transferred to the critical care unit at California Pacific Medical Center in San Francisco, where Tomasa died of liver failure 3 weeks later. Her husband eventually recovered after a lengthy hospitalization; her cousin required a liver transplant to survive.

The mushrooms consumed by Tomasa and her family were *Amanita phalloides*, commonly known as the death cap. A single death cap contains enough toxin to kill an adult human. The death rate among those who consume death caps is 22%; among children under the age of 10, it's more than 50%. Death cap mushrooms appear to be spreading in California, leading to a recent surge in the number of mushroom poisonings.

Death cap poisoning is insidious. Gastrointestinal symptoms—abdominal pain, cramping, vomiting, diarrhea—begin within 6 to 12 hours of consuming the mushrooms, but these symptoms usually subside within a few hours and the patient seems to recover. Because of this initial remission, the poisoning is often not taken seriously until it's too late to pump the stomach and remove the toxin from the body. After a day or two, serious symptoms begin. Cells in the liver die, often causing permanent liver damage and death within a few days. There is no effective treatment, other than a liver transplant to replace the damaged organ.

How do death caps kill? Their deadly toxin, contained within the fruiting bodies that produce reproductive spores, is the protein α-amanitin, which consists of a short peptide of eight amino acids that forms a circular loop. α-Amanitin is a potent inhibitor of RNA polymerase II, the enzyme that transcribes protein-encoding genes in eukaryotes. RNA polymerase II binds to genes and synthesizes RNA molecules that are complementary to the DNA template. In the process of transcription, the RNA polymerase moves down the DNA template, adding one nucleotide at a time to the growing RNA chain. α-Amanitin binds to RNA polymerase and jams the moving parts of the enzyme, interfering with its ability to move along the DNA template. In the presence of α-amanitin, RNA synthesis slows from its normal rate of several thousand nucleotides per minute to just a few nucleotides per minute. The results are catastrophic. Without transcription, protein synthesis—required for cellular function—ceases and cells die. The liver, where the toxin accumulates, is irreparably damaged and stops functioning. In severe cases, the patient dies.

Death cap poisoning illustrates the extreme importance of transcription and the central role that RNA polymerase plays in the process. This chapter is about the process of transcription—the first step in the central dogma, the pathway of information transfer from DNA (genotype) to protein (phenotype). Transcription is a complex process that requires precursors to RNA nucleotides, a DNA template, and a number of protein components. As we examine the stages of transcription, try to keep all the details in perspective and focus on understanding how they relate to the overall purpose of transcription—the selective synthesis of an RNA molecule.

This chapter begins with a brief review of RNA structure and a discussion of the different classes of RNA. We then consider the major components required for transcription. Finally, we explore the process of transcription. At several points in the text, we'll pause to consider some general principles that emerge.

13.1 RNA, Consisting of a Single Strand of Ribonucleotides, Participates in a Variety of Cellular Functions

Before we begin our study of transcription, we will consider the past and present importance of RNA, review the structure of RNA, and examine some of the different types of RNA molecules.

An Early RNA World

Life requires two basic functions. First, living organisms must be able to store and faithfully transmit genetic information during reproduction. Second, they must have the ability to catalyze the chemical transformations that drive life processes. A long-held belief was that the functions of information storage and chemical transformation are handled by two entirely different types of molecules: genetic information is stored in nucleic acids, whereas chemical transformations are catalyzed by protein enzymes. This biochemical dichotomy created a dilemma. Which came first: proteins or nucleic acids? If nucleic acids carry the coding instructions for proteins, how could proteins be generated without them? Because nucleic acids are unable to copy themselves, how could they be generated without proteins? If DNA and proteins each require the other, how could life begin?

This apparent paradox was answered in 1981 when Thomas Cech and his colleagues discovered that RNA can serve as a biological catalyst. They found that RNA from the protozoan *Tetrahymena thermophila* can excise 400 nucleotides from

its RNA in the absence of any protein. Other examples of catalytic RNAs have now been discovered in different types of cells. Called **ribozymes**, these catalytic RNA molecules can cut out parts of their own sequences, connect some RNA molecules together, replicate others, and even catalyze the formation of peptide bonds between amino acids. The discovery of ribozymes complements other evidence suggesting that the original genetic material was RNA.

Self-replicating ribozymes probably first arose between 3.5 billion and 4 billion years ago and may have begun the evolution of life on Earth. Early life was probably an RNA world, with RNA molecules serving both as carriers of genetic information and as catalysts that drove the chemical reactions needed to sustain and perpetuate life. These catalytic RNAs may have acquired the ability to synthesize protein-based enzymes, which are more-efficient catalysts. With enzymes taking over more and more of the catalytic functions, RNA probably became relegated to the role of information storage and transfer. DNA, with its chemical stability and faithful replication, eventually replaced RNA as the primary carrier of genetic information. Nevertheless, RNA is either produced by or plays a vital role in many biological processes, including transcription, replication, RNA processing, and translation. Research in the past 15 years has also determined that newly discovered small RNA molecules play a fundamental role in many basic biological processes, demonstrating that life today is still very much an RNA world. These small RNA molecules will be discussed in more detail in Chapter 14.

CONCEPTS

Early life probably centered on RNA, which served as the original genetic material and as biological catalysts.

The Structure of RNA

RNA, like DNA, is a polymer consisting of nucleotides joined together by phosphodiester bonds (see Chapter 10 for a discussion of RNA structure). However, there are several important differences in the structures of DNA and RNA. Whereas DNA nucleotides contain deoxyribose sugars, RNA nucleotides have ribose sugars (**Figure 13.1a**). With a free hydroxyl group on the 2′-carbon atom of the ribose sugar, RNA is degraded rapidly under alkaline conditions. The deoxyribose sugar of DNA lacks this free hydroxyl group; so DNA is a more-stable molecule. Another important difference is that thymine, one of the two pyrimidines found in DNA, is replaced by uracil in RNA.

A final difference in the structures of DNA and RNA is that RNA usually consists of a single polynucleotide

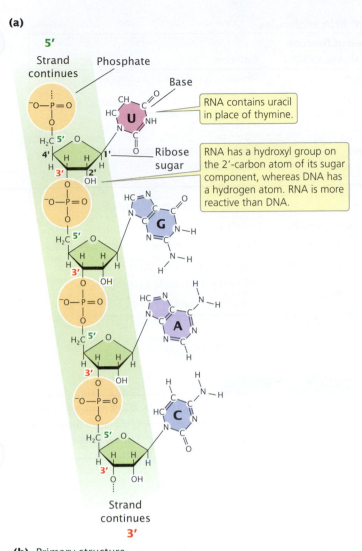

(a)

- Strand continues 5′
- Phosphate
- Base

> RNA contains uracil in place of thymine.

- Ribose sugar

> RNA has a hydroxyl group on the 2′-carbon atom of its sugar component, whereas DNA has a hydrogen atom. RNA is more reactive than DNA.

Strand continues 3′

(b) Primary structure

5′ AUGCGGCUACGUAACGAGCUUAGCGCGUAUACCGAAAGGGUAGAAC 3′

> An RNA molecule folds to form secondary structures...

Folding

> ...owing to hydrogen bonding between complementary bases on the same strand.

Secondary structure

13.1 RNA has a primary and a secondary structure.

and form secondary structures (see Figure 13.1b). These RNA secondary structures are often called hairpin-loop or stem-loop structures. When two regions within a single RNA molecule pair up, the strands in those regions must be antiparallel, with pairing between cytosine and guanine and between adenine and uracil (although occasionally guanine pairs with uracil).

The formation of secondary structures plays an important role in RNA function. Secondary structure is determined by the base sequence of the nucleotide strand, so different RNA molecules can assume different structures. Because their structure determines their function, RNA molecules have the potential for tremendous variation in function. With its two complementary strands forming a helix, DNA is much more restricted in the range of secondary structures that it can assume and so has fewer functional roles in the cell. Similarities and differences in DNA and RNA structures are summarized in **Table 13.1**.

▶ **TRY PROBLEM 14**

Classes of RNA

RNA molecules perform a variety of functions in the cell. **Ribosomal RNA** (rRNA) and ribosomal protein subunits make up the ribosome, the site of protein assembly. We'll take a more-detailed look at the ribosome in Chapter 14. **Messenger RNA** (mRNA) carries the coding instructions for polypeptide chains from DNA to a ribosome. After attaching to the ribosome, an mRNA molecule specifies the sequence of the amino acids in a polypeptide chain and provides a template for joining amino acids. Large precursor molecules, which are termed **pre-messenger RNAs** (pre-mRNAs), are the immediate products of transcription in eukaryotic cells. Pre-mRNAs are modified extensively before becoming mRNA and exiting the nucleus for translation into protein. Bacterial cells do not possess pre-mRNA; in these cells, transcription takes place concurrently with translation.

Table 13.1	The structures of DNA and RNA compared	
Characteristic	**DNA**	**RNA**
Composed of nucleotides	Yes	Yes
Type of sugar	Deoxyribose	Ribose
Presence of 2′-OH group	No	Yes
Bases	A, G, C, T	A, G, C, U
Nucleotides joined by phosphodiester bonds	Yes	Yes
Double or single stranded	Usually double	Usually single
Secondary structure	Double helix	Many types
Stability	Stable	Easily degraded

strand (**Figure 13.1b**), whereas DNA normally consists of two polynucleotide strands joined by hydrogen bonding between complementary bases (although some viruses contain double-stranded RNA genomes, as discussed in Chapter 9). Although RNA is usually single stranded, short complementary regions within a nucleotide strand can pair

Table 13.2 Location and functions of different classes of RNA molecules

Class of RNA	Cell Type	Location of Function in Eukaryotic Cells*	Function
Ribosomal RNA (rRNA)	Bacterial and eukaryotic	Cytoplasm	Structural and functional components of the ribosome
Messenger RNA (mRNA)	Bacterial and eukaryotic	Nucleus and cytoplasm	Carries genetic code for proteins
Transfer RNA (tRNA)	Bacterial and eukaryotic	Cytoplasm	Helps incorporate amino acids into polypeptide chain
Small nuclear RNA (snRNA)	Eukaryotic	Nucleus	Processing of pre-mRNA
Small nucleolar RNA (snoRNA)	Eukaryotic	Nucleus	Processing and assembly of rRNA
MicroRNA (miRNA)	Eukaryotic	Nucleus and cytoplasm	Inhibits translation of mRNA
Small interfering RNA (siRNA)	Eukaryotic	Nucleus and cytoplasm	Triggers degradation of other RNA molecules
Piwi-interacting RNA (piRNA)	Eukaryotic	Nucleus and cytoplasm	Suppresses the transcription of transposable elements in reproductive cells
CRISPR RNA (crRNA)	Prokaryotic	—	Assists destruction of foreign DNA

*All eukaryotic RNAs are synthesized in the nucleus.

Transfer RNA (tRNA) serves as the link between the coding sequence of nucleotides in the mRNA and the amino acid sequence of a polypeptide chain. Each tRNA attaches to one particular type of amino acid and helps to incorporate that amino acid into a polypeptide chain (discussed in Chapter 15).

Additional classes of RNA molecules are found in the nuclei of eukaryotic cells. **Small nuclear RNAs** (snRNAs) combine with small protein subunits to form **small nuclear ribonucleoproteins** (snRNPs, affectionately known as "snurps"). Some snRNAs participate in the processing of RNA, converting pre-mRNA into mRNA. **Small nucleolar RNAs** (snoRNAs) take part in the processing of rRNA.

A class of very small and abundant RNA molecules, termed **microRNAs** (miRNAs) and **small interfering RNAs** (siRNAs), are found in eukaryotic cells and carry out RNA interference (RNAi), a process in which these small RNA molecules help trigger the degradation of mRNA or inhibit its translation into protein. More will be said about RNA interference in Chapter 14. Recent research has uncovered another class of small RNA molecules called **Piwi-interacting RNAs** (piRNAs; named after Piwi proteins, with which they interact). Found in mammalian testes, these RNA molecules are similar to miRNAs and siRNAs; they are thought have a role in suppressing the expression of transposable elements (see Chapter 18) in reproductive cells. Recently, an RNA interference-like system has been discovered in prokaryotes, in which small **CRISPR RNAs** (crRNAs) assist in the destruction of foreign DNA molecules. Some of the different classes of RNA molecules are summarized in **Table 13.2**.

CONCEPTS

RNA differs from DNA in that RNA possesses a hydroxyl group on the 2′-carbon atom of its sugar, contains uracil instead of thymine, and is usually single stranded. Several classes of RNA exist within bacterial and eukaryotic cells.

✓ **CONCEPT CHECK 1**

Which class of RNA is correctly paired with its function?
a.   Small nuclear RNA (snRNA): processes rRNA
b.   Transfer RNA (tRNA): attaches to an amino acid
c.   MicroRNA (miRNA): carries information for the amino acid sequence of a protein
d.   Ribosomal RNA (rRNA): carries out RNA interference

13.2 Transcription Is the Synthesis of an RNA Molecule from a DNA Template

All cellular RNAs are synthesized from DNA templates through the process of transcription (**Figure 13.2**). Transcription is in many ways similar to the process of replication, but a fundamental difference relates to the length of the template used. In replication, all the nucleotides in the DNA template are copied, but, in transcription, only small parts of the DNA molecule—usually a single gene or, at most, a few genes—are transcribed into RNA. Because not all gene products are needed at the same time or in the same cell, the constant transcription of all of a cell's genes would be highly inefficient. Furthermore, much of the DNA does not encode

1 Some RNAs are transcribed in both prokaryotic and eukaryotic cells;...

- Messenger RNA (mRNA)
- Ribosomal RNA (rRNA)
- Transfer RNA (tRNA)

2 ...some are produced only in eukaryotes,...

- Pre-messenger RNA (pre-mRNA)
- Small nuclear RNA (snRNA)
- Small nucleolar RNA (snoRNA)
- MicroRNA (miRNA)
- Small interfering RNA (siRNA)
- Piwi-interacting RNA (piRNA)

3 and others are produced only in prokaryotes.

- CRISPR RNA (crRNA)

4 Some viruses copy RNA directly from RNA.

13.2 All cellular types of RNA are transcribed from DNA.

a functional product, and transcription of such sequences would be pointless. Transcription is, in fact, a highly selective process: individual genes are transcribed only as their products are needed. However, this selectivity imposes a fundamental problem on the cell—how to recognize individual genes and transcribe them at the proper time and place.

Like replication, transcription requires three major components:

1. a DNA template;

2. the raw materials (ribonucleotide triphosphates) needed to build a new RNA molecule; and

3. the transcription apparatus, consisting of the proteins necessary for catalyzing the synthesis of RNA.

The Template

In 1970, Oscar Miller, Jr., Barbara Hamkalo, and Charles Thomas used electron microscopy to examine cellular contents and demonstrate that RNA is transcribed from a DNA template. They saw Christmas-tree-like structures within the cell: thin central fibers (the trunk of the tree), to which were attached strings (the branches) with granules (**Figure 13.3a**). The addition of deoxyribonuclease (an enzyme that degrades DNA) caused the central fibers to disappear, indicating that the "tree trunks" were DNA molecules. Ribonuclease (an enzyme that degrades RNA) removed the granular strings, indicating that the branches were RNA. Their conclusion was that each "Christmas tree" represented a gene undergoing transcription (**Figure 13.3b**). The transcription of each gene begins at the top of the tree; there, little of the DNA has been transcribed and the RNA branches are short. As the transcription apparatus moves down the tree, transcribing more of the template, the RNA molecules lengthen, producing the long branches at the bottom.

(a)

(b)

DNA molecule

RNA molecules transcribed from DNA

13.3 Under the electron microscope, DNA molecules undergoing transcription exhibit Christmas-tree-like structures. (a) Electron micrograph of structures similar in appearance to Christmas trees. (b) The trunk of each "Christmas tree" (a transcription unit) represents a DNA molecule; the tree branches are RNA molecules that have been transcribed from the DNA. As the transcription apparatus moves down the DNA, transcribing more of the template, the RNA molecules become longer and longer. [Part a: Dr. Thomas Broker/Phototake.]

13.4 RNA molecules are synthesized that are complementary and antiparallel to one of the two nucleotide strands of DNA, the template strand.

THE TRANSCRIBED STRAND The template for RNA synthesis, as for DNA synthesis, is a single strand of the DNA double helix. Unlike replication, however, the transcription of a gene takes place on only one of the two nucleotide strands of DNA (**Figure 13.4**). The nucleotide strand used for transcription is termed the **template strand**. The other strand, called the **nontemplate strand**, is not ordinarily transcribed. Thus, within a gene, only one of the nucleotide strands is normally transcribed into RNA (there are some exceptions to this rule).

During transcription, an RNA molecule that is complementary and antiparallel to the DNA template strand is synthesized (see Figure 13.4). The RNA transcript has the same polarity and base sequence as that of the nontemplate strand, with the exception that RNA contains U rather than T. In most organisms, each gene is transcribed from a single strand, but different genes may be transcribed from different strands, as shown in **Figure 13.5**. ▶TRY PROBLEM 15

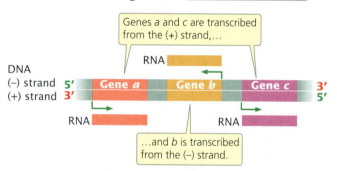

13.5 RNA is transcribed from one DNA strand. In most organisms, each gene is transcribed from a single DNA strand, but different genes may be transcribed from either DNA strand.

CONCEPTS

Within a single gene, only one of the two DNA strands, the template strand, is usually transcribed into RNA.

✓ **CONCEPT CHECK 2**

What is the difference between the template strand and the nontemplate strand?

THE TRANSCRIPTION UNIT A **transcription unit** is a stretch of DNA that encodes an RNA molecule and the sequences necessary for its transcription. How does the complex of enzymes and proteins that performs transcription—the transcription apparatus—recognize a transcription unit? How does it know which DNA strand to read and where to start and stop? This information is encoded by the DNA sequence.

Included within a transcription unit are three critical regions: a promoter, an RNA-coding sequence, and a terminator (**Figure 13.6**). The **promoter** is a DNA sequence that the transcription apparatus recognizes and binds. It indicates which of the two DNA strands is to be read as the template and the direction of transcription. The promoter also determines the transcription start site, the first nucleotide that will be transcribed into RNA. In many transcription units, the promoter is located next to the transcription start site but is not, itself, transcribed.

The second critical region of the transcription unit is the **RNA-coding region**, a sequence of DNA nucleotides that is copied into an RNA molecule. The third component of the

13.6 A transcription unit includes a promoter, an RNA-coding region, and a terminator.

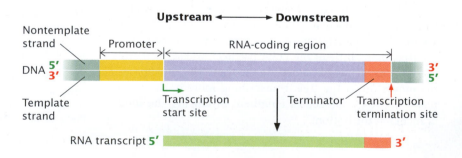

transcription unit is the **terminator**, a sequence of nucleotides that signals where transcription is to end. Terminators are usually part of the RNA-coding sequence; transcription stops only after the terminator has been copied into RNA.

Molecular biologists often use the terms *upstream* and *downstream* to refer to the direction of transcription and the location of nucleotide sequences surrounding the RNA-coding sequence. The transcription apparatus is said to move downstream during transcription: it binds to the promoter (which is usually upstream of the start site) and moves toward the terminator (which is downstream of the start site).

When DNA sequences are written out, often the sequence of only one of the two strands is listed. Molecular biologists typically write the sequence of the nontemplate strand, because it will be the same as the sequence of the RNA transcribed from the template (with the exception that U in RNA replaces T in DNA). By convention, the sequence on the nontemplate strand is written with the 5′ end on the left and the 3′ end on the right. The first nucleotide transcribed (the transcription start site) is numbered +1; nucleotides downstream of the start site are assigned positive numbers, and nucleotides upstream of the start site are assigned negative numbers. So, nucleotide +34 would be 34 nucleotides downstream of the start site, whereas nucleotide −75 would be 75 nucleotides upstream of the start site. There is no nucleotide numbered 0.

CONCEPTS

A transcription unit is a piece of DNA that encodes an RNA molecule and the sequences necessary for its proper transcription. Each transcription unit includes a promoter, an RNA-coding region, and a terminator.

✔ CONCEPT CHECK 3

Which of the following phrases does *not* describe a function of the promoter?
a. Serves as sequence to which transcription apparatus binds
b. Determines the first nucleotide that is transcribed into RNA
c. Determines which DNA strand is template
d. Signals where transcription ends

The Substrate for Transcription

RNA is synthesized from **ribonucleoside triphosphates** (rNTPs; **Figure 13.7**). In synthesis, nucleotides are added one at a time to the 3′-OH group of the growing RNA molecule. Two phosphate groups are cleaved from the incoming ribonucleoside triphosphate; the remaining phosphate group participates in a phosphodiester bond that connects the nucleotide to the growing RNA molecule. The overall chemical reaction for the addition of each nucleotide is

$$RNA_n + rNTP \rightarrow RNA_{n+1} + PP_i$$

where PP_i represents pyrophosphate. Nucleotides are always added to the 3′ end of the RNA molecule, and the direction

13.7 Ribonucleoside triphosphates are substrates used in RNA synthesis.

of transcription is therefore 5′→3′ (**Figure 13.8**), the same as the direction of DNA synthesis during replication. The synthesis of RNA is complementary and antiparallel to one of the DNA strands (the template strand). Unlike DNA synthesis, RNA synthesis does not require a primer.

CONCEPTS

RNA is synthesized from ribonucleoside triphosphates. Transcription is 5′→3′: each new nucleotide is joined to the 3′-OH group of the last nucleotide added to the growing RNA molecule.

The Transcription Apparatus

Recall that DNA replication requires a number of different enzymes and proteins. Although transcription might initially appear to be quite different because a single enzyme—**RNA polymerase**—carries out all the required steps of transcription, on closer inspection, the processes are actually similar. The action of RNA polymerase is enhanced by a number of accessory proteins that join and leave the polymerase at different stages of the process. Each accessory

1 Initiation of RNA synthesis does not require a primer.

2 New nucleotides are added to the 3′ end of the RNA molecule.

3 DNA unwinds at the front of the transcription bubble...

4 ...and then rewinds.

13.8 In transcription, nucleotides are always added to the 3′ end of the RNA molecule.

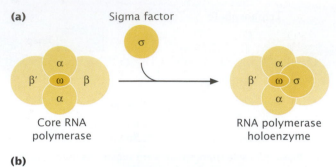

(a)

Sigma factor

Core RNA polymerase

RNA polymerase holoenzyme

(b)

13.9 In bacterial RNA polymerase, the core enzyme consists of five subunits: two copies of alpha (α), a single copy of beta (β), a single copy of beta prime (β'), and a single copy of omega (ω). The core enzyme catalyzes the elongation of the RNA molecule by the addition of RNA nucleotides. (a) The sigma factor (σ) joins the core to form the holoenzyme, which is capable of binding to a promoter and initiating transcription. (b) The molecular model shows RNA polymerase (blue), binding DNA (purple), and synthesizing mRNA (red). [Part b: Laguna Design/Science Photo Library.]

protein is responsible for providing or regulating a special function. Thus, transcription, like replication, requires an array of proteins.

BACTERIAL RNA POLYMERASE   Bacterial cells typically possess only one type of RNA polymerase, which catalyzes the synthesis of all classes of bacterial RNA: mRNA, tRNA, and rRNA. Bacterial RNA polymerase is a large, multimeric enzyme (meaning that it consists of several polypeptide chains).

At the heart of most bacterial RNA polymerases are five subunits (individual polypeptide chains) that make up the **core enzyme**: two copies of a subunit called alpha (α) and single copies of subunits beta (β), beta prime (β'), and omega (ω) (**Figure 13.9**). The ω subunit is not essential for transcription, but it helps stabilize the enzyme. The core enzyme catalyzes the elongation of the RNA molecule by the addition of RNA nucleotides. Other functional subunits join

and leave the core enzyme at particular stages of the transcription process. The **sigma (σ) factor** controls the binding of RNA polymerase to the promoter. Without sigma, RNA polymerase will initiate transcription at a random point along the DNA. After sigma has associated with the core enzyme (forming a **holoenzyme**), RNA polymerase binds stably only to the promoter region and initiates transcription at the proper start site. Sigma is required only for promoter binding and initiation; when a few RNA nucleotides have been joined together, sigma usually detaches from the core enzyme. Many bacteria have multiple types of sigma factors; each type of sigma initiates the binding of RNA polymerase to a particular set of promoters.

Rifamycins are a group of antibiotics that kill bacterial cells by inhibiting RNA polymerase. These antibiotics are widely used to treat tuberculosis, a disease that kills almost 2 million people worldwide each year. The structures of bacterial and eukaryotic RNA polymerases are sufficiently different that rifamycins inhibit bacterial RNA polymerases without interfering with eukaryotic RNA polymerases. Recent research has demonstrated that several rifamycins inibit RNA polymerase by binding to the part of the RNA polymerase that clamps on to DNA and jamming it, thus preventing the RNA polymerase from interacting with the promoter on the DNA.

EUKARYOTIC RNA POLYMERASES   Most eukaryotic cells possess three distinct types of RNA polymerase, each of which is responsible for transcribing a different class of RNA: **RNA polymerase I** transcribes rRNA; **RNA polymerase II** transcribes pre-mRNAs, snoRNAs, some miRNAs, and some snRNAs; and **RNA polymerase III** transcribes other small RNA molecules—specifically tRNAs, small rRNA, some miRNAs, and some snRNAs (**Table 13.3**). RNA polymerases I, II, and III are found in all eukaryotes. Two additional RNA polymerases, **RNA polymerase IV** and **RNA polymerase V**, have been found in plants. RNA polymerases IV and V transcribe RNAs that play a role in DNA methylation and chromatin structure.

Table 13.3	Eukaryotic RNA polymerases	
Type	**Present in**	**Transcribes**
RNA polymerase I	All eukaryotes	Large rRNAs
RNA polymerase II	All eukaryotes	Pre-mRNA, some snRNAs, snoRNAs, some miRNAs
RNA polymerase III	All eukaryotes	tRNAs, small rRNAs, some snRNAs, some miRNAs
RNA polymerase IV	Plants	Some siRNAs
RNA polymerase V	Plants	RNA molecules taking part in heterochromatin formation

All eukaryotic polymerases are large, multimeric enzymes, typically consisting of more than a dozen subunits. Some subunits are common to all RNA polymerases, whereas others are limited to one of the polymerases. As in bacterial cells, a number of accessory proteins bind to the core enzyme and affect its function.

> **CONCEPTS**
>
> Bacterial cells possess a single type of RNA polymerase, consisting of a core enzyme and other subunits that participate in various stages of transcription. Eukaryotic cells possess several distinct types of RNA polymerase that transcribe different kinds of RNA molecules.
>
> ✔ **CONCEPT CHECK 4**
>
> What is the function of the sigma factor?

13.10 **A consensus sequence consists of the most commonly encountered bases at each position in a group of related sequences.**

13.3 Bacterial Transcription Consists of Initiation, Elongation, and Termination

Now that we've considered some of the major components of transcription, we're ready to take a detailed look at the process. Transcription can be conveniently divided into three stages:

1. initiation, in which the transcription apparatus assembles on the promoter and begins the synthesis of RNA;

2. elongation, in which DNA is threaded through RNA polymerase, the polymerase unwinding the DNA and adding new nucleotides, one at a time, to the 3′ end of the growing RNA strand; and

3. termination, the recognition of the end of the transcription unit and the separation of the RNA molecule from the DNA template.

We will first examine each of these steps in bacterial cells, where the process is best understood; then we will consider eukaryotic and archaeal transcription.

Initiation

Initiation comprises all the steps necessary to begin RNA synthesis, including (1) promoter recognition, (2) formation of the transcription bubble, (3) creation of the first bonds between rNTPs, and (4) escape of the transcription apparatus from the promoter.

Transcription initiation requires that the transcription apparatus recognize and bind to the promoter. At this step, the selectivity of transcription is enforced; the binding of RNA polymerase to the promoter determines which parts of the DNA template are to be transcribed and how often. Different genes are transcribed with different frequencies, and promoter binding is primarily responsible for determining the frequency of transcription for a particular gene. Promoters

also have different affinities for RNA polymerase. Even within a single promoter, the affinity can vary with the passage of time, depending on the promoter's interaction with RNA polymerase and a number of other factors.

BACTERIAL PROMOTERS Essential information for the transcription unit—where it will start transcribing, which strand is to be read, and in what direction the RNA polymerase will move—is imbedded in the nucleotide sequence of the promoter. Promoters are DNA sequences that are recognized by the transcription apparatus and are required for transcription to take place. In bacterial cells, promoters are usually adjacent to an RNA-coding sequence.

An examination of many promoters in *E. coli* and other bacteria reveals a general feature: although most of the nucleotides within the promoters vary in sequence, short stretches of nucleotides are common to many. Furthermore, the spacing and location of these nucleotides relative to the transcription start site are similar in most promoters. These short stretches of common nucleotides are called **consensus sequences**; "consensus sequence" refers to sequences that possess considerable similarity, or consensus (**Figure 13.10**) The presence of consensus in a set of nucleotides usually implies that the sequence is associated with an important function. ▶ **TRY PROBLEM 21**

The most commonly encountered consensus sequence, found in almost all bacterial promoters, is centered about 10 bp upsteam of the start site. Called the **–10 consensus sequence** or, sometimes, the Pribnow box, its consensus sequence is

$$5'-TATAAT-3'$$
$$3'-ATATTA-5'$$

and is often written simply as TATAAT (**Figure 13.11**). Remember that TATAAT is just the *consensus* sequence—representing the most commonly encountered nucleotides at

13.11 In bacterial promoters, consensus sequences are found upstream of the start site, approximately at positions −10 and −35.

each of these positions. In most prokaryotic promoters, the actual sequence is not TATAAT.

Another consensus sequence common to most bacterial promoters is TTGACA, which lies approximately 35 nucleotides upstream of the start site and is termed the **−35 consensus sequence** (see Figure 13.11). The nucleotides on either side of the −10 and −35 consensus sequences and those between them vary greatly from promoter to promoter, suggesting that these nucleotides are not very important in promoter recognition.

The function of these consensus sequences in bacterial promoters has been studied by inducing mutations at various positions within the consensus sequences and observing the effect of the changes on transcription. The results of these studies reveal that most base substitutions within the −10 and −35 consensus sequences reduce the rate of transcription; these substitutions are termed *down mutations* because they slow down the rate of transcription. Occasionally, a particular change in a consensus sequence increases the rate of transcription; such a change is called an *up mutation.*

As mentioned earlier, the sigma factor associates with the core enzyme (**Figure 13.12a**) to form a holoenzyme, which binds to the −35 and −10 consensus sequences in the DNA promoter (**Figure 13.12b**). Although it binds only the nucleotides of consensus sequences, the enzyme extends from −50 to +20 when bound to the promoter. The holoenzyme initially binds weakly to the promoter but then undergoes a change in structure that allows it to bind more tightly and unwind the double-stranded DNA (**Figure 13.12c**). Unwinding begins within the −10 consensus sequence and extends downstream for about 14 nucleotides, including the start site (from nucleotides −12 to +2).

Some bacterial promoters contain a third consensus sequence that also takes part in the initiation of transcription. Called the **upstream element**, this sequence contains a number of A–T pairs and is found at about −40 to −60. A number of proteins may bind to sequences in and near the promoter; some stimulate the rate of transcription and others repress it. We will consider these proteins, which regulate gene expression, in Chapter 16. ▶**TRY PROBLEM 24**

CONCEPTS

A promoter is a DNA sequence that is adjacent to a gene and required for transcription. Promoters contain short consensus sequences that are important in the initiation of transcription.

✔ **CONCEPT CHECK 5**

What binds to the −10 consensus sequence found in most bacterial promoters?
a. The holoenzyme (core enzyme + sigma)
b. The sigma factor alone
c. The core enzyme alone
d. mRNA

INITIAL RNA SYNTHESIS After the holoenzyme has attached to the promoter, RNA polymerase is positioned over the start site for transcription (at position +1) and has unwound the DNA to produce a single-stranded template. The orientation and spacing of consensus sequences on a DNA strand determine which strand will be the template for transcription and thereby determine the direction of transcription.

The position of the start site is determined not by the sequences located there but by the location of the consensus sequences, which positions RNA polymerase so that the enzyme's active site is aligned for the initiation of transcription at +1. If the consensus sequences are artificially moved upstream or downstream, the location of the starting point of transcription correspondingly changes.

To begin the synthesis of an RNA molecule, RNA polymerase pairs the base on a ribonucleoside triphosphate with its complementary base at the start site on the DNA template strand (**Figure 13.12d**). No primer is required to initiate the synthesis of the 5′ end of the RNA molecule. Two of the three phosphate groups are cleaved from the ribonucleoside triphosphate as the nucleotide is added to the 3′ end of the growing RNA molecule. However, because the 5′ end of the first ribonucleoside triphosphate does not take part in the formation of a phosphodiester bond, all three of its phosphate groups remain. An RNA molecule therefore possesses, at least initially, three phosphate groups at its 5′ end (**Figure 13.12e**).

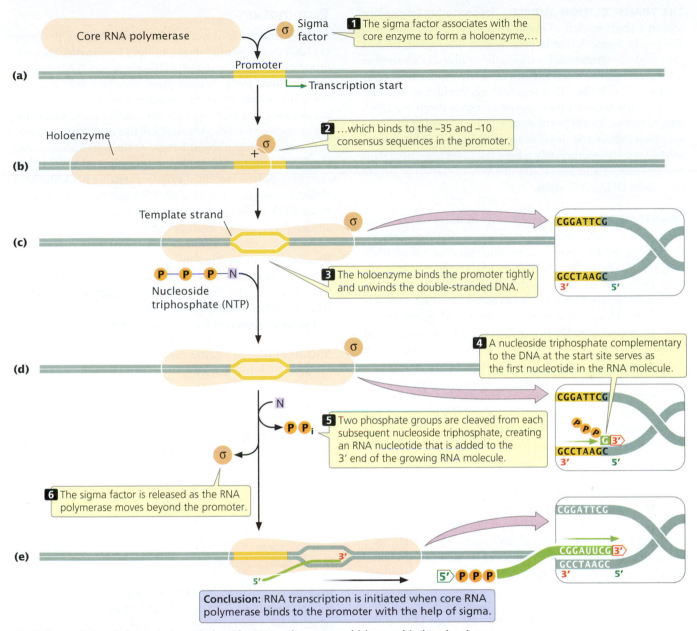

Core RNA polymerase

Sigma factor σ

1 The sigma factor associates with the core enzyme to form a holoenzyme,...

Promoter

Transcription start

(a)

Holoenzyme

σ

2 ...which binds to the −35 and −10 consensus sequences in the promoter.

(b) +

Template strand

σ

(c)

CGGATTCG

GCCTAAGC
3'   5'

P P P N

Nucleoside triphosphate (NTP)

3 The holoenzyme binds the promoter tightly and unwinds the double-stranded DNA.

σ

(d)

4 A nucleoside triphosphate complementary to the DNA at the start site serves as the first nucleotide in the RNA molecule.

CGGATTCG

P P P G 3'
GCCTAAGC
3'   5'

N

P Pi

5 Two phosphate groups are cleaved from each subsequent nucleoside triphosphate, creating an RNA nucleotide that is added to the 3' end of the growing RNA molecule.

σ

6 The sigma factor is released as the RNA polymerase moves beyond the promoter.

(e)

3'

5'

5' P P P

CGGATTCG

CGGAUUCG 3'
GCCTAAGC
3'   5'

Conclusion: RNA transcription is initiated when core RNA polymerase binds to the promoter with the help of sigma.

13.12 Transcription in bacteria is carried out by RNA polymerase, which must bind to the sigma factor to initiate transcription.

Often, in the course of initiation, RNA polymerase repeatedly generates and releases short transcripts, from 2 to 6 nucleotides in length, while still bound to the promoter. This process, termed **abortive initiation**, occurs in both prokaryotes and eukaryotes. After several abortive attempts, the polymerase synthesizes an RNA molecule from 9 to 12 nucleotides in length, which allows the RNA polymerase to transition to the elongation stage.

Elongation

At the end of initiation, RNA polymerase undergoes a change in conformation (shape) and thereafter is no longer able to bind to the consensus sequences in the promoter. This change

allows the polymerase to escape from the promoter and begin transcribing downstream. The sigma subunit is usually released after initiation, although some populations of RNA polymerase may retain sigma throughout elongation.

As it moves downstream along the template, RNA polymerase progressively unwinds the DNA at the leading (downstream) edge of the transcription bubble, joining nucleotides to the RNA molecule according to the sequence on the template, and rewinds the DNA at the trailing (upstream) edge of the bubble. In bacterial cells at 37°C, about 40 nucleotides are added per second. This rate of RNA synthesis is much lower than that of DNA synthesis, which is 1000 to 2000 nucleotides per second in bacterial cells.

THE TRANSCRIPTION BUBBLE Transcription takes place within a short stretch of about 18 nucleotides of unwound DNA—the transcription bubble. Within this region, RNA is continuously synthesized, with single-stranded DNA used as a template. About 8 nucleotides of newly synthesized RNA are paired with the DNA-template nucleotides at any one time. As the transcription apparatus moves down the DNA template, it generates positive supercoiling ahead of the transcription bubble and negative supercoiling behind it. Topoisomerase enzymes probably relieve the stress associated with the unwinding and rewinding of DNA in transcription, as they do in DNA replication.

TRANSCRIPTIONAL PAUSING A number of features of RNA or DNA, such as secondary structures, specific sequences, or the presence of nucleosomes cause RNA polymerase to pause the elongation stage of transcription. Pauses often are caused by backtracking—when the RNA polymerase slides backward along the DNA template strand. Backtracking disengages the 3′ OH group of the RNA molecule from the active site of RNA polymerase and temporarily halts further RNA synthesis. Cells use several mechanisms to minimize backtracking, including proteins that cleave the backtracked RNA in the active site, generating a new 3′OH to which new nucleotides can then be added. In bacterial cells, translation of mRNA by ribosomes closely follows transcription (see Chapter 15) and the presence of ribosomes moving along the mRNA in a 5′→3′ direction also prevents backtracking of the RNA polymerase at the 3′ end of the mRNA.

Transitory pauses in transcription are important in the coordination of transcription and translation in bacteria, as well as in the coordination of RNA processing in eukaryotes. Pausing also affects the rates of RNA synthesis. Sometimes a pause may be stabilized by sequences in the DNA that ultimately lead to the termination of transcription (see the next section on termination).

ACCURACY OF TRANSCRIPTION Although RNA polymerase is quite accurate in incorporating nucleotides into the growing RNA chain, errors do occasionally arise. Research has demonstrated that RNA polymerase is capable of a type of proofreading in the course of transcription. When RNA polymerase incorporates a nucleotide that does not match the DNA template, it backs up and cleaves the last two nucleotides (including the misincorporated nucleotide) from the growing RNA chain. RNA polymerase then proceeds forward, transcribing the DNA template again.

CONCEPTS

Transcription is initiated at the start site, which, in bacterial cells, is set by the binding of RNA polymerase to the consensus sequences of the promoter. No primer is required. Transcription takes place within the transcription bubble. DNA is unwound ahead of the bubble and rewound behind it. There are frequent pauses in the process of transcription.

Termination

RNA polymerase adds nucleotides to the 3′ end of the growing RNA molecule until it transcribes a terminator. Most terminators are found upstream of the site at which termination actually takes place. Transcription therefore does not suddenly stop when polymerase reaches a terminator, as does a car stopping at a stop sign. Rather, transcription stops after the terminator has been transcribed, like a car that stops only after running over a speed bump. At the terminator, several overlapping events are needed to bring an end to transcription: RNA polymerase must stop synthesizing RNA, the RNA molecule must be released from RNA polymerase, the newly made RNA molecule must dissociate fully from the DNA, and RNA polymerase must detach from the DNA template.

Bacterial cells possess two major types of terminators. **Rho-dependent terminators** are able to cause the termination of transcription only in the presence of an ancillary protein called the **rho factor**. **Rho-independent terminators** (also known as intrinsic terminators) are able to cause the end of transcription in the absence of rho.

RHO-DEPENDENT TERMINATORS Rho-dependent terminators have two features. The first is the terminator itself, which consists of DNA sequences that cause the RNA polymerase to pause. The second feature is a DNA sequence that encodes a stretch of RNA upstream of the terminator that is usually rich in cytosine nucleotides and devoid of any secondary structures. This sequence is called the rho utilization (rut) site; it serves as a binding site for the rho protein. Once rho binds to the RNA, it moves toward its 3′ end, following the RNA polymerase (**Figure 13.13**). When RNA polymerase encounters the terminator, it pauses, allowing rho to catch up. The rho protein has helicase activity, which it uses to unwind the RNA–DNA hybrid in the transcription bubble, bringing transcription to an end.

RHO-INDEPENDENT TERMINATORS Rho-independent terminators, which make up about 50% of all terminators in prokaryotes, have two common features. First, they contain inverted repeats, which are sequences of nucleotides on one strand that are inverted and complementary. When inverted repeats have been transcribed into RNA, a hairpin secondary structure forms (**Figure 13.14**). Second, in rho-independent terminators, a string of seven to nine adenine nucleotides follows the second inverted repeat in the template DNA. Their transcription produces a string of uracil nucleotides after the hairpin in the transcribed RNA.

The string of uracils in the RNA molecule causes the RNA polymerase to pause, allowing time for the hairpin structure to form. Evidence suggests that the formation of the hairpin destabilizes the DNA–RNA pairing, causing the RNA molecule to separate from its DNA template. Separation may be facilitated by the adenine–uracil base pairings, which are relatively weak compared with other types of base pairings. When the RNA transcript has separated from the template, RNA synthesis can no longer continue (see Figure 13.13).

▶ **TRY PROBLEM 29**

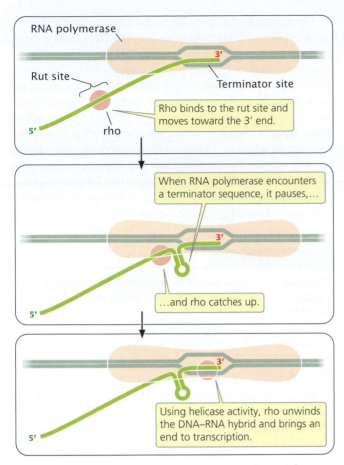

Rho binds to the rut site and moves toward the 3′ end.

When RNA polymerase encounters a terminator sequence, it pauses,...

...and rho catches up.

Using helicase activity, rho unwinds the DNA–RNA hybrid and brings an end to transcription.

13.13 The termination of transcription in some bacterial genes requires the presence of the rho protein.

POLYCISTRONIC MRNA   In bacteria, a group of genes is often transcribed into a single RNA molecule, which is termed a **polycistronic RNA**. Thus, polycistronic RNA is produced when a single terminator is present at the end of a group of several genes that are transcribed together, instead of each gene having its own terminator. Polycistronic mRNA does occur in some eukaryotes such as *Caenorhabditis elegans*, but it is uncommon. You can view the process of transcription, including initiation, elongation, and termination in **Animation 13.1**. The animation shows how the different parts of the transcriptional unit interact to bring about the complete synthesis of an RNA molecule.

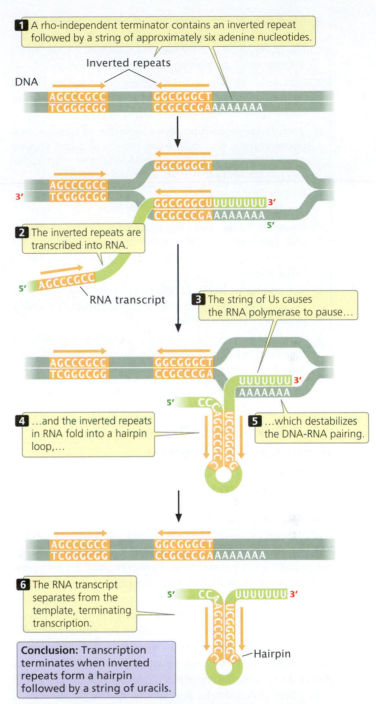

1 A rho-independent terminator contains an inverted repeat followed by a string of approximately six adenine nucleotides.

Inverted repeats

DNA

2 The inverted repeats are transcribed into RNA.

RNA transcript

3 The string of Us causes the RNA polymerase to pause...

4 ...and the inverted repeats in RNA fold into a hairpin loop,...

5 ...which destabilizes the DNA-RNA pairing.

6 The RNA transcript separates from the template, terminating transcription.

Hairpin

Conclusion: Transcription terminates when inverted repeats form a hairpin followed by a string of uracils.

13.14 Rho-independent termination in bacteria is a multistep process.

CONCEPTS

Transcription ends after RNA polymerase transcribes a terminator. Bacterial cells possess two types of terminator: a rho-independent terminator, which RNA polymerase can recognize by itself; and a rho-dependent terminator, which RNA polymerase can recognize only with the help of the rho protein.

✓ CONCEPT CHECK 6

What characteristics are most commonly found in rho-independent terminators?

CONNECTING CONCEPTS

The Basic Rules of Transcription

Before we examine the process of eukaryotic transcription, let's summarize some of the general principles of bacterial transcription.

1. Transcription is a selective process; only certain parts of the DNA are transcribed at any one time.

2. RNA is transcribed from single-stranded DNA. Within a gene, only one of the two DNA strands—the template strand—is usually copied into RNA.

3. Ribonucleoside triphosphates are used as the substrates in RNA synthesis. Two phosphate groups are cleaved from a ribonucleoside triphosphate, and the resulting nucleotide is joined to the 3′-OH group of the growing RNA strand.

4. RNA molecules are antiparallel and complementary to the DNA template strand. Transcription is always in the 5′→3 direction, meaning that the RNA molecule grows at the 3′ end.

5. Transcription depends on RNA polymerase—a complex, multimeric enzyme. RNA polymerase consists of a core enzyme, which is capable of synthesizing RNA, and other subunits that may join transiently to perform additional functions.

6. A sigma factor enables the core enzyme of RNA polymerase to bind to a promoter and initiate transcription.

7. Promoters contain short sequences crucial in the binding of RNA polymerase to DNA; these consensus sequences are interspersed with nucleotides that play no known role in transcription.

8. RNA polymerase binds to DNA at a promoter, begins transcribing at the start site of the gene, and ends transcription after a terminator has been transcribed.

9. Topoisomerase enzymes remove supercoiling that develops ahead and behind the transcription bubble as the DNA is unwound and rewound during transcription.

13.4 Eukaryotic Transcription Is Similar to Bacterial Transcription but Has Some Important Differences

Transcription in eukaryotes is similar to bacterial transcription in that it includes initiation, elongation, and termination, and the basic principles of transcription already outlined apply to eukaryotic transcription. However, there are some important differences. Eukaryotic cells possess three different RNA polymerases, each of which transcribes a different class of RNA and recognizes a different type of promoter. Thus, a generic promoter cannot be described for eukaryotic cells; rather, a promoter's description depends on whether the promoter is recognized by RNA polymerase I, II, or III. Another difference is in the nature of promoter recognition and initiation. Many proteins take part in the binding of eukaryotic RNA polymerases to DNA templates, and the different types of promoters require different proteins.

Transcription and Nucleosome Structure

Transcription requires sequences on DNA to be accessible to RNA polymerase and other proteins. However, in eukaryotic cells, DNA is complexed with histone proteins in highly compressed chromatin (see Figure 11.4). How can the proteins necessary for transcription gain access to eukaryotic DNA when it is complexed with histones?

The answer to this question is that chromatin structure is modified before transcription so that the DNA is in a more open configuration and is more accessible to the transcription machinery. Several types of proteins have roles in chromatin modification. Acetyltransferases add acetyl groups to amino acids at the ends of the histone proteins, which destabilizes nucleosome structure and makes the DNA more accessible. Other types of histone modification also can affect chromatin packing. In addition, proteins called chromatin-remodeling proteins may bind to the chromatin and displace nucleosomes from promoters and other regions important for transcription. We will take a closer look at the role of changes in chromatin structure associated with gene expression in Chapter 17.

CONCEPTS

The initiation of transcription requires modification of chromatin structure so that DNA is accessible to the transcriptional machinery.

Promoters

A significant difference between bacterial and eukaryotic transcription is the existence of three different eukaryotic RNA polymerases, which recognize different types of promoters. In bacterial cells, the holoenzyme (RNA polymerase plus the sigma factor) recognizes and binds directly to sequences in the promoter. In eukaryotic cells, promoter recognition is carried out by accessory proteins that bind to the promoter and then recruit a specific RNA polymerase (I, II, or III) to the promoter.

One class of accessory proteins comprises **general transcription factors**, which, along with RNA polymerase, form the **basal transcription apparatus**—a group of proteins that assemble near the start site and are sufficient to initiate minimal levels of transcription. Another class of accessory proteins consists of **transcriptional activator proteins**, which bind to specific DNA sequences and bring about higher levels of transcription by stimulating the assembly of the basal transcription apparatus at the start site.

We will focus our attention on promoters recognized by RNA polymerase II, which transcribes the genes that encode proteins. A promoter for a gene transcribed by RNA polymerase II typically consists of two primary parts: the core promoter and the regulatory promoter.

CORE PROMOTER The **core promoter** is located immediately upstream of the gene (**Figure 13.15**) and is the site to which the basal transcription apparatus binds. The core promoter typically includes one or more consensus sequences. One of the most common of these sequences is the **TATA box**,

13.15 The promoters of genes transcribed by RNA polymerase II consist of a core promoter and a regulatory promoter that contain consensus sequences. Not all the consensus sequences shown are found in all promoters.

which has the consensus sequence TATAAA and is located from −25 to −30 bp upstream of the start site. Additional consensus sequences that may be found in the core promoters of genes transcribed by RNA polymerase II are shown in Figure 13.15. These consensus sequences are recognized by transcription factors that bind to them and serve as a platform for the assembly of the basal transcription apparatus.

REGULATORY PROMOTER The **regulatory promoter** is located immediately upstream of the core promoter. A variety of different consensus sequences can be found in the regulatory promoters, and they can be mixed and matched in different combinations. Transcriptional activator proteins bind to these sequences and either directly or indirectly make contact with the basal transcription apparatus and affect the rate at which transcription is initiated. Transcriptional activator proteins also regulate transcription by binding to more-distant sequences called **enhancers**. The DNA between an enhancer and the promoter loops out, and so transcriptional activator proteins bound to the enhancer can interact with the basal transcription machinery at the core promoter. Enhancers will be discussed in more detail in Chapter 17.

POLYMERASE I AND III PROMOTERS RNA polymerase I and RNA polymerase III each recognize promoters that are distinct from those recognized by RNA polymerase II. For example, promoters for small rRNA and tRNA genes, transcribed by RNA polymerase III, contain **internal promoters** that are downstream of the start site and are transcribed into the RNA.

CONCEPTS

General transcription factors and RNA polymerase assemble into the basal transcription apparatus, which binds to DNA near the start site and is necessary for transcription to take place at minimal levels. Additional proteins called transcriptional activators bind to other consensus sequences in promoters and enhancers and affect the rate of transcription.

✓ CONCEPT CHECK 7

What is the difference between the core promoter and the regulatory promoter?

a. Only the core promoter has consensus sequences.
b. The regulatory promoter is farther upstream of the gene.
c. Transcription factors bind to the core promoter; transcriptional activator proteins bind to the regulatory promoter.
d. Both b and c.

Initiation

Transcription in eukaryotes is initiated through the assembly of the transcriptional machinery on the promoter. This machinery consists of RNA polymerase II and a series of transcription factors that form a giant complex consisting of 50 or more polypeptides. Assembly of the transcription machinery begins when regulatory proteins bind DNA near the promoter and modify the chromatin structure so that transcription can take place. These proteins and other regulatory proteins then recruit the basal transcriptional apparatus to the core promoter.

The basal transcription apparatus consists of RNA polymerase, a series of general transcription factors, and a complex of proteins known as the mediator (**Figure 13.16**). The general transcription factors include TFIIA, TFIIB, TFIID, TFIIE, TFIIF, and TFIIH, in which TFII stands for transcription factor for RNA polymerase II and the final letter designates the individual factor.

RNA polymerase II and the general transcription factors assemble at the core promoter, forming a pre-initiation complex that is analogous to the closed complex seen in bacterial initiation. Recall that, in bacteria, the sigma factor recognizes and binds to the promoter sequence. In eukaryotes, the function of sigma is replaced by that of the general transcription factors. A first step in initiation is the binding of TFIID to the TATA box on the DNA template. TFIID consists of at least nine polypeptides. One of them is the **TATA-binding protein** (TBP), which recognizes and binds to the TATA consensus sequence. The TATA-binding protein binds to the minor groove and straddles the DNA as a molecular saddle

Regulatory promoter

Core promoter

DNA

TATA box

1 TFIID binds to TATA box
in the core promoter.

TFIID

TBP

Transcription
start

TFIIE

TFIIF

TFIIA   TFIIB

TFIIH

Mediator

RNA polymerase II

2 Then transcription factors and RNA
polymerase II bind to the core promoter.

3 Transcription activator proteins
bind to sequences in enhancers.

Enhancer

Transcriptional
activator protein

4 DNA loops out, allowing the proteins
bound to the enhancer to interact with
the basal transcription apparatus.

Coactivator

Transcriptional
activator protein

Basal transcription
apparatus

5 Transcriptional activator proteins bind to sequences in the regulatory promoter
and interact with the basal transcription apparatus through the mediator.

13.16 Transcription is initiated at RNA polymerase II promoters. Transcription is initiated when the TFIID
transcription factor binds to the TATA box, followed by the binding of a preassembled holoenzyme containing
general transcription factors, RNA polymerase II, and the mediator. TBP stands for TATA-binding protein.

(**Figure 13.17**), bending the DNA and partly unwinding it.
Other transcription factors bind to additional consensus
sequences in the core promoter and to RNA polymerase and
position the polymerase over the transcription start site.

After the RNA polymerase and transcription factors have
assembled on the core promoter, conformational changes take
place in both the DNA and the polymerase. These changes
cause from 11 to 15 bp of DNA surrounding the transcription
start site to separate, producing the single-stranded DNA that
will serve as a template for transcription.

The single-stranded DNA template is positioned within
the active site of RNA polymerase, creating a structure called
the open complex. After the open complex has formed, the
synthesis of RNA begins as phosphate groups are cleaved
off nucleoside triphosphates and nucleotides are joined
together to form an RNA molecule. As in bacterial transcription, RNA polymerase may generate and release several short RNA molecules in abortive transcription before

the polymerase initiates the synthesis of a full-length RNA
molecule. ▶ TRY PROBLEM 35

TBP

DNA

**13.17 The TATA-binding protein (TBP) binds to the minor groove
of DNA, straddling the double helix of DNA like a saddle.**

CONCEPTS

Transcription is initiated when the basal transcription apparatus, consisting of RNA polymerase and transcription factors, assembles on the core promoter and becomes an open complex.

✓ **CONCEPT CHECK 8**

What is the role of TFIID in transcription initiation?

Elongation

After about 30 bp of RNA have been synthesized, the RNA polymerase leaves the promoter and enters the elongation stage of transcription. Many of the transcription factors are left behind at the promoter and can serve to quickly reinitiate transcription with another RNA polymerase enzyme.

The molecular structure of eukaryotic RNA polymerase II and how it functions during elongation have been revealed through the work of Roger Kornberg and his colleagues, for which Kornberg was awarded a Nobel Prize in chemistry in 2006. The RNA polymerase maintains a transcription bubble during elongation, in which about eight nucleotides of RNA remain base paired with the DNA template strand. The DNA double helix enters a cleft in the polymerase and is gripped by jawlike extensions of the enzyme (**Figure 13.18**). The two strands of the DNA are unwound and RNA nucleotides that are complementary to the template strand are added to the growing 3′ end of the RNA molecule. As it funnels through the polymerase, the DNA–RNA hybrid hits a wall of amino acids and bends at almost a right angle; this bend positions the end of the DNA–RNA hybrid at the active site of the polymerase, and new nucleotides are added to the 3′ end of the growing RNA molecule. The newly synthesized RNA is separated from the DNA and runs through another groove before exiting from the polymerase.

Termination

The three eukaryotic RNA polymerases use different mechanisms for termination. RNA polymerase I requires a termination factor like the rho factor utilized in the termination of some bacterial genes. Unlike rho, which binds to the newly transcribed RNA molecule, the termination factor for RNA polymerase I binds to a DNA sequence downstream of the termination site.

RNA polymerase III ends transcription after transcribing a terminator sequence that produces a string of uracil nucleotides in the RNA molecule, like that produced by the rho-independent terminators of bacteria. Unlike rho-independent terminators in bacterial cells, however, RNA polymerase III does not require that a hairpin structure precede the string of Us.

The termination of transcription by RNA polymerase II is not at specific sequences. Instead, RNA polymerase II often continues to synthesize RNA hundreds or even thousands of nucleotides past the coding sequence necessary to produce the mRNA. As we will see in Chapter 14, the end of pre-mRNA is cleaved at a specific site, designated by a consensus sequence, while transcription is still taking place at the 3′ end of the molecule. Cleavage cuts the pre-mRNA into two pieces: the mRNA that will eventually encode the protein and another piece of RNA that has its 5′ end trailing out of the RNA polymerase (**Figure 13.19**). An enzyme (called Rat1 in yeast) attaches to the 5′ end of this RNA and moves toward the 3′ end where RNA polymerase continues the transcription of RNA. Rat1 is a 5′→3′ exonuclease—an enzyme capable of degrading RNA in the 5′→3′ direction. Like a guided torpedo, Rat1 homes in on the polymerase, chewing up the RNA as it moves. When Rat1 reaches the transcriptional machinery, transcription terminates. Note that this mechanism is similar to that of rho-dependent termination in bacteria (see Figure 13.13), except that rho does not degrade the RNA molecule.

RNA polymerase II

13.18 The structure of RNA polymerase II is a source of insight into its function. The DNA double helix enters the polymerase through a groove and unwinds. The DNA–RNA duplex is bent at a right angle, which positions the 3′ end of the RNA at the active site of the enzyme. New nucleotides are added to the 3′ end of the RNA.

CONCEPTS

The different eukaryotic RNA polymerases use different mechanisms of termination. Transcription at genes transcribed by RNA polymerase II is terminated when an exonuclease enzyme attaches to the cleaved 5′ end of the RNA, moves down the RNA, and reaches the polymerase enzyme.

✓ **CONCEPT CHECK 9**

How are the processes of RNA polymerase II termination and rho-dependent termination in bacteria similar and how are they different?

1 RNA polymerase II transcribes well past the coding sequences of most genes.

RNA polymerase

DNA

3'

Cleavage site

5'
Pre-mRNA

DNA

3'

2 Cleavage is near the 3' end of the RNA...

5'

3'

5'
Pre-mRNA

3 ...while RNA polymerase continues transcribing.

DNA

4 The Rat1 endonuclease attaches to the 5' end of the trailing RNA...

Rat1

Pre-mRNA
5' ——————— 3'

DNA

5 ...and moves toward the RNA polymerase, degrading the RNA as it goes.

Pre-mRNA
5' ——————— 3'

DNA

6 When Rat1 reaches the polymerase, transcription is terminated.

Pre-mRNA
5' ——————— 3'

DNA

13.19 Termination of transcription by RNA polymerase II requires the Rat1 exonuclease. Cleavage of the pre-mRNA produces a 5' end to which Rat1 attaches. Rat1 degrades the RNA molecule in the 5'→3' direction. When Rat1 reaches the polymerase, transcription is halted.

13.5 Transcription in Archaea Is More Similar to Transcription in Eukaryotes Than to Transcription in Eubacteria

Some 2 billion to 3 billion years ago, life diverged into three lines of evolutionary descent: the eubacteria, the archaea, and the eukaryotes (see Chapter 2). Although eubacteria and archaea are superficially similar—both are unicellular and lack a nucleus—the results of studies of their DNA sequences and other biochemical properties indicate that they are as distantly related to each other as they are to eukaryotes. The evolutionary distinction between archaea, eubacteria, and eukaryotes is clear. However, did eukaryotes first diverge from an ancestral prokaryote, with the later separation of prokaryotes into eubacteria and archaea, or did the archaea and the eubacteria split first, with the eukaryotes later evolving from one of these groups?

Studies of transcription in eubacteria, archaea, and eukaryotes have yielded important findings about the evolutionary relationships of these organisms. Archaea, like eubacteria, have a single RNA polymerase, but this enzyme is most similar to the RNA polymerases of eukaryotes. As discussed earlier, bacterial RNA polymerase consists of 5 subunits, whereas eukaryotic RNA polymerases are much more complex; for example, RNA polymerase II is composed of 12 subunits. Archaeal RNA polymerase is similarly complex, with 11 or more subunits. Furthermore, the amino acid sequence of RNA polymerase in archaea is similar to the amino acid sequence of eukaryotic RNA polymerase II.

Archaeal promoters contain a consensus sequence similar to the TATA box found in eukaryotic promoters. The archaeal TATA box is found approximately 27 bp upstream of the transcription start site and, as in eukaryotes, helps to determine the location of the transcription start site. Archaea possess a TATA-binding protein (TBP), which is a critical transcription factor found in all three of the eukaryotic polymerases but not in eubacterial RNA polymerase. TBP binds the TATA box in archaea with the help of another transcription factor, TFIIB, which also is found in eukaryotes but not in eubacteria. However, some other regulators of transcription found in archaea are more similar to those found in bacteria, emphasizing that transcription in archaea is not entirely eukaryotic in nature. As prokaryotes, archaea lack a nuclear membrane, but many species do produce histone proteins, which help compact the DNA and form nucleosome-related structures.

Thus, transcription, one of the most basic of life processes, has strong similarities in eukaryotes and archaea, suggesting that these two groups are more closely related to each other than either is to the eubacteria. This conclusion is supported by other data, including those obtained from a comparison of gene sequences.

CONCEPTS

The process of transcription in archaea has many similarities to transcription in eukaryotes.

CONCEPTS SUMMARY

- Early life used RNA as both the carrier of genetic information and as biological catalysts.

- RNA is a polymer, consisting of nucleotides joined together by phosphodiester bonds. Each RNA nucleotide consists of a ribose sugar, a phosphate, and a base. RNA contains the base uracil and is usually single stranded, which allows it to form secondary structures.

- Cells possess a number of different classes of RNA. Ribosomal RNA is a component of the ribosome, messenger RNA carries coding instructions for proteins, and transfer RNA helps incorporate the amino acids into a polypeptide chain.

- The template for RNA synthesis is single-stranded DNA. In transcription, RNA synthesis is complementary and antiparallel to the DNA template strand. A transcription unit consists of a promoter, an RNA-coding region, and a terminator.

- The substrates for RNA synthesis are ribonucleoside triphosphates.

- RNA polymerase in bacterial cells consists of a core enzyme, which catalyzes the addition of nucleotides to an RNA molecule, and other subunits. The sigma factor controls the binding of the core enzyme to the promoter.

- Eukaryotic cells contain several different RNA polymerases.

- The process of transcription consists of three stages: initiation, elongation, and termination.

- Transcription begins at the start site, which is determined by consensus sequences. A short stretch of DNA is unwound near the start site, RNA is synthesized from a single strand of DNA as a template, and the DNA is rewound at the lagging end of the transcription bubble. RNA polymerases are capable of proofreading.

- RNA synthesis ceases after a terminator sequence has been transcribed. Bacterial cells have two types of terminators: rho-independent terminators and rho-dependent terminators.

- The initiation of transcription in eukaryotes requires the modification of chromatin structure. Different types of RNA polymerases in eukaryotes recognize different types of promoters.

- For genes transcribed by RNA polymerase II, general transcription factors bind to the core promoter and are part of the basal transcription apparatus. Transcriptional activator proteins bind to sequences in regulatory promoters and enhancers and interact with the basal transcription apparatus at the core promoter.

- The three RNA polymerases found in all eukaryotic cells use different mechanisms of termination.

- Transcription in archaea has many similarities to transcription in eukaryotes.

IMPORTANT TERMS

ribozyme (p. 358)
ribosomal RNA (rRNA) (p. 359)
messenger RNA (mRNA) (p. 359)
pre-messenger RNA (pre-mRNA) (p. 359)
transfer RNA (tRNA) (p. 360)
small nuclear RNA (snRNA) (p. 360)
small nuclear ribonucleo-protein (snRNP) (p. 360)
small nucleolar RNA (snoRNA) (p. 360)

microRNA (miRNA) (p. 360)
small interfering RNA (siRNA) (p. 360)
Piwi-interacting RNA (piRNA) (p. 360)
CRISPR RNA (crRNA) (p. 360)
template strand (p. 362)
nontemplate strand (p. 362)
transcription unit (p. 362)
promoter (p. 362)

RNA-coding region (p. 362)
terminator (p. 363)
ribonucleoside triphosphate (rNTP) (p. 363)
RNA polymerase (p. 363)
core enzyme (p. 364)
sigma (σ) factor (p. 364)
holoenzyme (p. 364)
RNA polymerase I (p. 364)
RNA polymerase II (p. 364)

RNA polymerase III (p. 364)
RNA polymerase IV (p. 364)
RNA polymerase V (p. 364)
consensus sequence (p. 365)
−10 consensus sequence (Pribnow box) (p. 365)
−35 consensus sequence (p. 366)
upstream element (p. 366)

abortive initiation (p. 367)
rho-dependent terminator
(p. 368)
rho factor (p. 368)
rho-independent terminator
(p. 368)

polycistronic mRNA
(p. 369)
general transcription factor
(p. 370)
basal transcription apparatus
(p. 370)

transcriptional activator
protein (p. 370)
core promoter (p. 370)
TATA box (p. 370)
regulatory promoter
(p. 371)

enhancer (p. 371)
internal promoter
(p. 371)
TATA-binding protein (TBP)
(p. 371)

ANSWERS TO CONCEPT CHECKS

1. b

2. The template strand is the DNA strand that is copied into an RNA molecule, whereas the nontemplate strand is not copied.

3. d

4. The sigma factor recognizes the promoter and controls the binding of RNA polymerase to the promoter.

5. a

6. Inverted repeats followed by a string of adenine nucleotides.

7. d

8. TFIID binds to the TATA box and helps to center the RNA polymerase over the start site of transcription.

9. Both processes use a protein that binds to the RNA molecule and moves down the RNA toward the RNA polymerase. They differ in that rho does not degrade the RNA, whereas Rat1 does so.

WORKED PROBLEMS

Problem 1

The diagram at right represents a sequence of nucleotides surrounding an RNA-coding sequence.

```
5′–CATGTT...TTGATGT– ┌──────────┐ –GACGA...TTTATA...GGCGCGC–3′
                     │ RNA-     │
                     │ coding   │
3′–GTACAA...AACTACA– │ sequence │ –CTGCT...AAATAT..CCGCGCG–5′
                     └──────────┘
```

a. Is the RNA-coding sequence likely to be from a bacterial cell or from a eukaryotic cell? How can you tell?

b. Which DNA strand will serve as the template strand during the transcription of the RNA-coding sequence?

Solution Strategy

What information is required in your answer to the problem?

a. Whether the sequence is likely to be from a bacteria or eukaryotic cell and why.

b. Which strand is the template strand.

What information is provided to solve the problem?

- The nucleotide sequences of both strands of DNA.
- The 5′ and 3′ ends of the strands.

For help with this problem, review:

The Template in Section 13.2, Bacterial Promoters in Section 13.3 and Promoters in Section 13.4.

Solution Steps

a. Bacterial and eukaryotic cells use the same DNA bases (A, T, G, and C); so the bases themselves provide no clue to the origin of the sequence. The RNA-coding

sequence must have a promoter, and bacterial and eukaryotic cells do differ in the consensus sequences found in their promoters; so we should examine the sequences for the presence of familiar consensus sequences. On the bottom strand to the right of the RNA-coding sequence, we find AAATAT, which, written in the conventional manner (5′ on the left), is 5′—TATAAA—3′. This sequence is the TATA box found in most eukaryotic promoters. However, the sequence is also quite similar to the −10 consensus sequence (5′—TATAAT—3′) found in bacterial promoters.

Farther to the right on the bottom strand, we also see 5′—GCGCGCC—3′, which is the TFIIB recognition element (BRE, see Figure 13.15) in eukaryotic RNA polymerase II promoters. No similar consensus sequence is found in bacterial promoters; so we can be fairly certain that this sequence is a eukaryotic promoter and an RNA-coding sequence.

Hint: Review the consensus sequences found in bacterial and eukaryotic promoters in Figures 13.11 and 13.15.

b. The TATA box and BRE of RNA polymerase II promoters are upstream of the RNA-coding sequences; so RNA polymerase must bind to these sequences and then proceed downstream, transcribing the RNA-coding sequence. Thus RNA polymerase must proceed from right (upstream) to left (downstream). The RNA molecule is always synthesized in the $5' \rightarrow 3'$ direction and is antiparallel to the DNA template strand; so the template strand must be read $3' \rightarrow 5'$. If the enzyme proceeds from right to left and reads the template in the $3' \rightarrow 5'$ direction, the upper strand must be the template, as shown in the diagram below.

> **Recall:** During transcription, the template strand is read $3' \rightarrow 5'$.

Template strand	RNA-coding sequence	RNA polymerase
5'–CATGTT…TTGATGT–		–GACGA…TTTATA…GGCGCGC–3'
3'–GTACAA…AACTACA–		–CTGCT…AAATAT…CCGCGCG–5'

←———— **Direction of transcription**

Problem 2

Suppose that a consensus sequence in the regulatory promoter of a eukaryotic gene that encodes enzyme A were deleted. Which of the following effects would result from this deletion? Explain your reasoning.

a. Enzyme A would have a different amino acid sequence.

b. The mRNA for enzyme A would be abnormally short.

c. Enzyme A would be missing some amino acids.

d. The mRNA for enzyme A would be transcribed but not translated.

e. The amount of mRNA transcribed would be affected.

Solution Strategy

What information is required in your answer to the problem?

Selection of the result (*a*, *b*, *c*. *d*. or *e*) that would occur when a consensus sequence in the regulatory promoter were deleted.

What information is provided to solve the problem?

The deleted consensus sequence is in the regulatory promoter.

For help with this problem, review:

Section 13.4.

Solution Steps

The correct answer is part *e*. The regulatory promoter contains binding sites for transcriptional activator proteins. These sequences are not part of the RNA-coding sequence for enzyme A; so the mutation would have no effect on the length or the amino acid sequence of the enzyme, eliminating answers *a*, *b*, and *c*. Transcriptional activator proteins bind to the regulatory promoter and affect the amount of transcription that takes place through interactions with the basal transcription apparatus at the core promoter.

COMPREHENSION QUESTIONS

Section 13.1

1. Draw an RNA nucleotide and a DNA nucleotide, highlighting the differences. How is the structure of RNA similar to that of DNA? How is it different?

2. What are the major classes of cellular RNA?

3. Why is DNA more stable than RNA?

Section 13.2

4. What parts of DNA make up a transcription unit? Draw a typical bacterial transcription unit and identify its parts.

5. What is the substrate for RNA synthesis? How is this substrate modified and joined together to produce an RNA molecule?

6. Describe the structure of the holoenzyme of bacterial RNA polymerase.

7. Give the names of the RNA polymerases found in eukaryotic cells and the types of RNA that they transcribe.

Section 13.3

8. What are the three basic stages of transcription? Describe what happens at each stage.

9. Draw a typical bacterial promoter and identify any common consensus sequences.

10. What are the two basic types of terminators found in bacterial cells? Describe the structure of each type.

Section 13.4

11. Compare the roles of general transcription factors and transcriptional activator proteins.

12. How are transcription and replication similar and how are they different?

13. How is transcription different in bacteria and eukaryotes? How is it similar?

> For more questions that test your comprehension of the key chapter concepts, go to *LEARNINGCurve* for this chapter.

APPLICATION QUESTIONS AND PROBLEMS

Section 13.1

*14. An RNA molecule has the following percentages of bases: A = 23%, U = 42%, C = 21%, and G = 14%.

a. Is this RNA single stranded or double stranded? How can you tell?

b. What would be the percentages of bases in the template strand of the DNA that contains the gene for this RNA?

Section 13.2

*15. The following diagram represents DNA that is part of the RNA-coding sequence of a transcription unit. The bottom strand is the template strand. Give the sequence found on the RNA molecule transcribed from this DNA and identify the 5′ and 3′ ends of the RNA.

    5′−ATAGGCGATGCCA−3′
    3′−TATCCGCTACGGT−5′ ← Template strand

16. For the RNA molecule shown in **Figure 13.1a**, write out the sequence of bases on the template and nontemplate strands of DNA from which this RNA is transcribed. Label the 5′ and 3′ ends of each strand.

17. The following sequence of nucleotides is found in a single-stranded DNA template:

    ATTGCCAGATCATCCCAATAGAT

Assume that RNA polymerase proceeds along this template from left to right.

a. Which end of the DNA template is 5′ and which end is 3′?

b. Give the sequence and identify the 5′ and 3′ ends of the RNA copied from this template.

18. RNA polymerases carry out transcription at a much lower rate than that at which DNA polymerases carry out replication. Why is speed more important in replication than in transcription?

19. Assume that a mutation occurs in the gene that codes for each of the following RNA polymerases. Match the mutation with possible effects by placing the correct letter in the blank below. There may be more than one effect for each mutated polymerase.

| A Mutation in the Gene | |
That Codes for	Effects
RNA Polymerase I	_____
RNA Polymerase II	_____
RNA Polymerase III	_____

Possible Effects

a. tRNA not synthesized

b. some ribosomal RNA not synthesized

c. ribosomal RNA not processed

d. pre-mRNA not processed

e. some mRNA molecules not degraded

f. pre-mRNA not synthesized

Section 13.3

20. Provide the consensus sequence for the *first three* actual sequences shown in **Figure 13.10**.

*21. Write the consensus sequence for the following set of nucleotide sequences.

    AGGAGTT
    AGCTATT
    TGCAATA
    ACGAAAA
    TCCTAAT
    TGCAATT

22. List at least five properties that DNA polymerases and RNA polymerases have in common. List at least three differences.

23. Most RNA molecules have *three* phosphate groups at the 5′ end, but DNA molecules never do. Explain this difference.

*24. Write a hypothetical sequence of bases that might be found in the first 20 nucleotides of a promoter of a bacterial gene. Include both strands of DNA and identify the 5′ and 3′ ends of both strands. Be sure to include the start site for transcription and any consensus sequences found in the promoter.

*25. What would be the most likely effect of a mutation at the following locations in an *E. coli* gene?

a. −8   c. −20

b. −35   d. Start site

26. A strain of bacteria possesses a temperature-sensitive mutation in the gene that encodes the sigma factor. The mutant bacteria produce a sigma factor that is unable to bind to RNA polymerase at elevated temperatures. What effect will this mutation have on the process of transcription when the bacteria are raised at elevated temperatures?

27. On **Figure 13.5**, indicate the location of the promoters and terminators for genes *a*, *b*, and *c*.

28. The following diagram represents a transcription unit on a DNA molecule.

Transcription start site

Template strand

a. Assume that this DNA molecule is from a bacterial cell. Draw the approximate location of the promoter and terminator for this transcription unit.

b. Assume that this DNA molecule is from a eukaryotic cell. Draw the approximate location of an RNA polymerase II promoter.

*29. The following DNA nucleotides are found near the end of a bacterial transcription unit.

3′−AGCATACAGCAGACCGTTGGTCTGAAAAAAGCATACA−5′

a. Mark the point at which transcription will terminate.

b. Is this terminator rho independent or rho dependent?

c. Draw a diagram of the RNA that will be transcribed from this DNA, including its nucleotide sequence and any secondary structures that form.

30. A strain of bacteria possesses a temperature-sensitive mutation in the gene that encodes the rho subunit. At high temperatures, rho is not functional. When these bacteria are raised at elevated temperatures, which of the following effects would you expect to see? Explain your reasoning for accepting or rejecting each of these five options.

a. Transcription does not take place.

b. All RNA molecules are shorter than normal.

c. All RNA molecules are longer than normal.

d. Some RNA molecules are longer than normal.

e. RNA is copied from both DNA strands.

31. The following diagram represents one of the Christmas-tree-like structures as shown in **Figure 13.3**. On the diagram, identify parts *a* through *i*.

a. DNA molecule

b. 5′ and 3′ ends of the template strand of DNA

c. At least one RNA molecule

d. 5′ and 3′ ends of at least one RNA molecule

e. Direction of movement of the transcription apparatus on the DNA molecule

f. Approximate location of the promoter

g. Possible location of a terminator

h. Upstream and downstream directions

i. Molecules of RNA polymerase (use dots to represent these molecules)

32. Suppose that the string of A nucleotides following the inverted repeat in a rho-independent terminator was deleted but that the inverted repeat was left intact. How will this deletion affect termination? What will happen when RNA polymerase reached this region?

Section 13.4

33. The following diagram represents a transcription unit in a hypothetical DNA molecule.

5′...TTGACA...TATAAT...3′
3′...AACTGT...ATATTA...5′

a. On the basis of the information given, is this DNA from a bacterium or from a eukaryotic organism?

b. If this DNA molecule is transcribed, which strand will be the template strand and which will be the nontemplate strand?

c. Where, approximately, will the start site of transcription be?

34. Computer programmers, working with molecular geneticists, have developed computer programs that can identify genes within long stretches of DNA sequences. Imagine that you are working with a computer programmer on such a project. On the basis of what

you know about the process of transcription, what sequences should be used to identify the beginning and end of a gene with the use of this computer program?

*35. Through genetic engineering, a geneticist mutates the gene that encodes TBP in cultured human cells. This mutation destroys the ability of TBP to bind to the TATA box. Predict the effect of this mutation on cells that possess it.

36. Elaborate repair mechanisms are associated with replication to prevent permanent mutations in DNA, yet no similar repair is associated with transcription. Can you think of a reason for this difference in replication and transcription? (Hint: Think about the relative effects of a permanent mutation in a DNA molecule compared with one in an RNA molecule.)

CHALLENGE QUESTIONS

Section 13.3

37. Many genes in both bacteria and eukaryotes contain numerous sequences that potentially cause pauses or premature terminations of transcription. Nevertheless, the transcription of these genes within a cell normally produces multiple RNA molecules thousands of nucleotides long without pausing or terminating prematurely. However, when a single round of transcription takes place on such templates in a test tube, RNA synthesis is frequently interrupted by pauses and premature terminations, which reduce the rate at which transcription takes place and frequently shortens the length of the mRNA molecules produced. Most pauses and premature terminations occur when RNA polymerase temporarily backtracks (i.e., backs up) for one or two nucleotides along the DNA. Experimental findings have demonstrated that most transcriptional delays and premature terminations disappear if several RNA polymerases are simultaneously transcribing the DNA molecule. Propose an explanation for faster transcription and longer mRNA when the template DNA is being transcribed by multiple RNA polymerases.

Section 13.4

38. Enhancers are sequences that affect the initiation of the transcription of genes that are hundreds or thousands of nucleotides away. Transcriptional activator proteins that bind to enhancers usually interact directly with transcription factors at promoters by causing the intervening DNA to loop out. An enhancer of bacteriophage T4 does not function by looping of the DNA (D. R. Herendeen et al. 1992. *Science* 256: 1298–1303). Propose some additional mechanisms (other than DNA looping) by which this enhancer might affect transcription at a gene thousands of nucleotides away.

*39. The locations of the TATA box in two species of yeast, *Saccharomyces pombe* and *Saccharomyces cerevisiae*, differ dramatically. The TATA box of *S. pombe* is about 30 nucleotides upstream of the start site, similar to the location in most other eukaryotic cells. However, the TATA box of *S. cerevisiae* is 40 to 120 nucleotides upstream of the start site.

To better understand what sets the start site in these organisms, researchers at Stanford University conducted a series of experiments to determine which components of the transcription apparatus of these two species could be interchanged (Y. Li et al. 1994. *Science* 263:805–807). In these experiments, different transcription factors and RNA polymerases were switched in *S. pombe* and *S. cerevisiae*, and the effects of the switch on the level of RNA synthesis and on the start point of transcription were observed. The results from one set of experiments are shown in the table below. Components cTFIIB, cTFIIE, cTFIIF, cTFIIH are transcription factors from *S. cerevisisae*. Components pTFIIB, pTFIIE, pTFIIF, pTFIIH are transcription factors from *S. pombe*. Components cPol II and pPol II are RNA polymerase II from *S. cerevisiae* and *S. pombe*, respectively. The table indicates whether the component was present (+) or missing (−) in the experiment. In the accompanying gel, the presence of a band indicates that RNA was produced and the position of the band indicates whether it was the length predicated when transcription begins 30 bp downstream from the TATA box or from 40 to 120 bp downstream from the TATA box.

Components	Experiment						
	1	2	3	4	5	6	7
cTFIIE	+	−	+	+	+	+	−
cTFIIH + cTFIIF	+	−	+	+	+	−	+
cTFIIB + cPol II	+	−	−	−	−	−	−
pPol II	−	+	+	+	−	+	+
pTFIIB	−	+	+	−	+	+	+
pTFIIE + pTFIIH + pTFIIF	−	+	−	−	−	−	−

Transcription at *S. pombe* start site (30 bp downstream from TATA box) →

Transcription at *S. cerevisiae* start site (40–120 bp downstream from TATA box) →

Gel

a. What conclusion can you draw from these data about what components determine the start site for transcription?

b. What conclusions can you draw about the interactions of the different components of the transcription apparatus?

S. pombe. [Steve Gschmeissner/Photo Researchers.]

c. Propose a mechanism for why the start site for transcription in *S. pombe* is about 30 bp downstream from the TATA box, whereas the start site for transcription in *S. cerevisiae* is 40 to 120 bp downstream from the TATA box.

S. cerevisiae. [Steve Gschmeissner/Science Source.]

40. Glenn Croston and his colleagues studied the relation between chromatin structure and transcription activity. In one set of experiments, they measured the level of in vitro transcription of a *Drosophila* gene by RNA polymerase II with the use of DNA and various combinations of histone proteins (G. E. Croston et al. 1991. *Science* 251:643–649).

First, they measured the level of transcription for naked DNA, with no associated histone proteins. Then, they measured the level of transcription after nucleosome octamers (without H1) were added to the DNA. The addition of the octamers caused the level of transcription to drop by 50%. When both the nucleosome octamers and the H1 proteins were added to the DNA, transcription was greatly repressed, dropping to less than 1% of that obtained with naked DNA, as shown in the table below.

GAL4-VP16 is a protein that binds to the DNA of certain eukaryotic genes. When GAL4-VP16 is added to DNA, the level of RNA polymerase II transcription is greatly elevated.

Treatment	Relative amount of transcription
Naked DNA	100
DNA + octamers	50
DNA + octamers + H1	<1
DNA + GAL4-VP16	1000
DNA + octamers + GAL4-VP16	1000
DNA + octamers + H1 + GAL4-VP16	1000

Even in the presence of the H1 protein, GAL4-VP16 stimulates high levels of transcription.

Propose a mechanism for how the H1 protein represses transcription and how GAL4-VP16 overcomes this repression. Explain how your proposed mechanism would produce the results obtained in these experiments.

> ▶ Go to your ▶LaunchPad to find additional learning resources and the Suggested Readings for this chapter.

14

RNA Molecules and RNA Processing

Family of Tsar Nicholas Romanov II of Russia. The boy in front is the tsar's son Alexei, who suffered from hemophilia. [Mondadori via Getty Images.]

A Royal Disease

On August 12, 1904, Tsar Nicholas Romanov II of Russia wrote in his diary: "A great never-to-be forgotten day when the mercy of God has visited us so clearly." That day Alexei, Nicholas's first son and heir to the Russian throne, had been born.

At birth, Alexei was a large and vigorous baby with yellow curls and blue eyes, but, at 6 weeks of age, he began spontaneously hemorrhaging from the navel. The bleeding persisted for several days and caused great alarm. As he grew and began to walk, Alexei often stumbled and fell, as all children do. Even his small scrapes bled profusely, and minor bruises led to significant internal bleeding. It soon became clear that Alexei had hemophilia.

Marked by slow clotting and excessive bleeding, hemophilia is caused by a mutation in one of several genes that encode proteins involved in the process of blood clotting. In those with hemophilia, minor injuries can result in life-threatening blood loss, and spontaneous bleeding into joints erodes the bone with crippling consequences.

Alexei suffered from classic hemophilia, which is an X-linked genetic disorder. He inherited the hemophilia gene from his mother, Alexandra, who was a carrier. The gene appears to have originated with Queen Victoria of England (see Figure 6.8). In all, ten of Queen Victoria's male descendants suffered from hemophilia. Six female descendants, including her granddaughter Alexandra (Alexei's mother), were carriers.

During his childhood, Alexei experienced a number of severe bleeding episodes. The royal physicians were often helpless during these crises—they had no treatment that would stop the bleeding. At this time the Russian Revolution broke out. Bolsheviks captured the tsar and his family and held them captive in the city of Yekaterinburg. On the night of July 16, 1918, a firing squad executed the royal family and their attendants, including Alexei and his four sisters. For many years, the bodies of the Tsar's family were lost, but their skeletons were eventually recovered from two graves outside Yekaterinburg. Comparisons of mitochondrial DNA and nuclear DNA from the bones and from descendants of relatives of the family verified that the bones were indeed those of the royal family. Although this analysis determined the identity of the remains, the molecular nature of the royal hemophilia long remained unknown.

In 2009, geneticists analyzed DNA from the bones of the Tsar's family to determine the genetic nature of the royal hemophilia. In people with X-linked hemophilia, a mutation commonly occurs in one of two genes on the X chromosome, either the gene for blood coagulation factor VIII or the gene for factor IX.

In eukaryotic cells, genes are often interrupted by noncoding sequences. Those parts of a gene that encode the amino acids of a protein are called exons, while the noncoding sequences are called introns. Initially both exons and introns are transcribed into pre-mRNA but, in a process called RNA splicing, the introns are later cut out and the exons pasted together. Examination of the DNA sequences for the two clotting-factor genes in Alexandra revealed no mutations in the nucleotides that code for amino acids, but a mutation did occur in one of the introns of her gene for factor IX. This mutation altered the splicing of the exons, creating a new stop signal in the coding sequence of the mRNA and producing a truncated, defective clotting factor IX. Both normal and mutant alleles were detected in DNA from Alexandra, as is expected of a heterozygous carrier. DNA from Alexei carried only the mutant allele, which caused his hemophilia.

As illustrated by Alexei's hemophilia, RNA processing is critically important for the proper synthesis of proteins. In eukaryotic cells, RNA molecules are often extensively modified after transcription: for genes that encode proteins, a special nucleotide called the cap is added to the 5′ end, a tail of adenine nucleotides is added to the 3′ end, and introns are cut out of the middle. In both prokaryotes and eukaryotes, rRNAs and tRNAs are also modified after transcription. In Chapter 13, we focused on transcription—the process by which RNA molecules are synthesized. In this chapter, we examine the function and processing of RNA.

We begin by taking a careful look at the nature of the gene. We then examine messenger RNA (mRNA), its structure, and how it is modified in eukaryotes after transcription. Then we turn to transfer RNA (tRNA), the adapter molecule that forms the interface between amino acids and mRNA in protein synthesis. We examine ribosomal RNA (rRNA), the structure and organization of rRNA genes, and how rRNAs are processed. Finally, we consider a newly discovered group of very small RNAs that play important roles in numerous biological functions.

As we explore the world of RNA and its role in gene function, we will see evidence of two important characteristics of this nucleic acid. First, RNA is extremely versatile, both structurally and biochemically. It can assume a number of different secondary structures, which provide the basis for its functional diversity. Second, RNA processing and function frequently include interactions between two or more RNA molecules.

14.1 Many Genes Have Complex Structures

What is a gene? As noted in Chapter 3, the definition of a gene often changes as we explore different aspects of heredity. A gene was defined in Chapter 3 as an inherited factor that determines a characteristic. This definition may have seemed vague because it says only what a gene does rather than what a gene is. Nevertheless, this definition was appropriate at the time because our focus was on how genes influence the inheritance of traits. We did not have to consider the physical nature of the gene in learning the rules of inheritance.

Knowing something about the chemical structure of DNA and the process of transcription now enables us to be more precise about what a gene is. Chapter 10 described how genetic information is encoded in the base sequence of DNA: a gene consists of a set of DNA nucleotides. But how many nucleotides constitute a gene, and how is the information in these nucleotides organized? In 1902, Archibald Garrod suggested, correctly, that genes encode proteins. Proteins are made of amino acids, so a gene contains the nucleotides that specify the amino acids of a protein. Therefore, for many years the working definition of a gene was a set of nucleotides that specifies the amino acid sequence of a protein. As geneticists learned more about the structure of genes, however, it became clear that this concept of a gene was an oversimplification.

Gene Organization

Early work on gene structure was carried out largely through the examination of mutations in bacteria and viruses. This research led Francis Crick in 1958 to propose that genes and proteins are **colinear**—that there is a direct correspondence between the nucleotide sequence of DNA and the amino acid sequence of a protein (**Figure 14.1**). The concept of colinearity suggests that the number of nucleotides in a gene should be proportional to the number of amino acids in the protein encoded by that gene. In a general sense, this concept is true for genes found in bacterial cells and many viruses, although these genes are slightly longer than would be expected if colinearity were strictly applied, because the mRNAs encoded by the genes contain sequences at their ends that do not specify amino acids. At first, eukaryotic genes and proteins also were generally assumed to be colinear, but there were hints that eukaryotic gene structure is fundamentally different. Eukaryotic cells contain far more DNA than is required to encode proteins (see Chapter 11). Furthermore, many large RNA molecules observed in the nucleus were absent from the cytoplasm, suggesting that nuclear RNAs undergo some type of change before they are exported to the cytoplasm.

Most geneticists were nevertheless surprised by the announcement in the 1970s that not all genes are continuous. Researchers observed four coding sequences in a gene

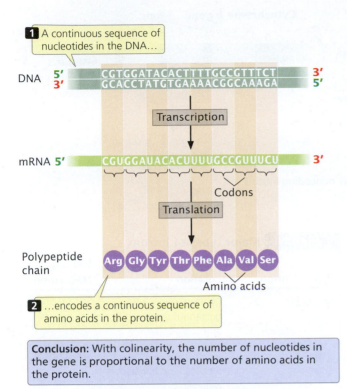

1 A continuous sequence of nucleotides in the DNA...

DNA 5′ CGTGGATACACTTTTGCCGTTTCT 3′
3′ GCACCTATGTGAAAACGGCAAAGA 5′

Transcription

mRNA 5′ CGUGGAUACACUUUUGCCGUUUCU 3′

Codons

Translation

Polypeptide
chain   Arg Gly Tyr Thr Phe Ala Val Ser

Amino acids

2 ...encodes a continuous sequence of amino acids in the protein.

Conclusion: With colinearity, the number of nucleotides in the gene is proportional to the number of amino acids in the protein.

14.1 The concept of colinearity suggests that a continuous sequence of nucleotides in DNA encodes a continuous sequence of amino acids in a protein.

from a eukaryotic virus that were interrupted by nucleotides that did not specify amino acids. This discovery was made when the viral DNA was hybridized with the mRNA transcribed from it and the hybridized structure was examined using an electron microscope (**Figure 14.2**). The DNA was clearly much longer than the mRNA because regions of DNA looped out from the hybridized molecules. These regions contained nucleotides in the DNA that were absent from the coding nucleotides in the mRNA. Many other examples of interrupted genes were subsequently discovered; it quickly became apparent that most eukaryotic genes consist of stretches of coding and noncoding nucleotides.

CONCEPTS

When a continuous sequence of nucleotides in DNA encodes a continuous sequence of amino acids in a protein, the two are said to be colinear. In eukaryotes, not all genes are colinear with the proteins that they encode.

✓ CONCEPT CHECK 1

What evidence indicated that eukaryotic genes are not colinear with their proteins?

Introns

Many eukaryotic genes contain coding regions called **exons** and noncoding regions called intervening sequences or **introns**. For example, the gene encoding the protein

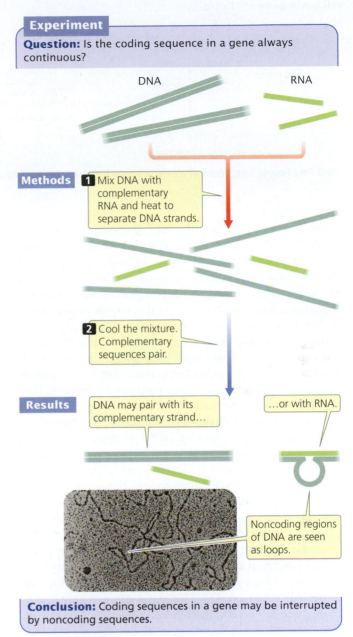

Experiment

Question: Is the coding sequence in a gene always continuous?

DNA   RNA

Methods **1** Mix DNA with complementary RNA and heat to separate DNA strands.

2 Cool the mixture. Complementary sequences pair.

Results DNA may pair with its complementary strand...   ...or with RNA.

Noncoding regions of DNA are seen as loops.

Conclusion: Coding sequences in a gene may be interrupted by noncoding sequences.

14.2 The noncolinearity of eukaryotic genes was discovered by hybridizing DNA and mRNA. [From Susan M. Berget, Claire Moore, and Phillip A. Sharp, *Proc. Nati. Acad. Sci. USA* 74(8), pp.3171–3175, Fig. 4g August 1977 Biochemistry.]

ovalbumin has eight exons and seven introns; the gene for cytochrome *b* has five exons and four introns (**Figure 14.3**). The average human gene contains from eight to nine introns. All the introns and the exons are initially transcribed into RNA but, after transcription, the introns are removed by splicing and the exons are joined to yield the mature RNA.

Introns are common in eukaryotic genes but are rare in bacterial genes. For a number of years after their discovery, introns were thought to be entirely absent from prokaryotic genomes, but they have now been observed in archaea, bacteriophages, and even some eubacteria. Introns are present in mitochondrial and chloroplast genes as well as the nuclear

14.3 The coding sequences of many eukaryotic genes are disrupted by noncoding introns.

genes of eukaryotes. In eukaryotic genomes, the size and number of introns appear to be directly related to increasing organismal complexity: yeast genes contain only a few short introns; *Drosophila* introns are longer and more numerous; and most vertebrate genes are interrupted by long introns. All classes of eukaryotic genes—those that encode rRNA, tRNA, and proteins—may contain introns. The number and size of introns vary widely: some eukaryotic genes have no introns, whereas others may have more than 60; intron length varies from fewer than 200 nucleotides to more than 50,000. Introns tend to be longer than exons, and most eukaryotic genes contain more noncoding nucleotides than coding nucleotides. Finally, most introns do not encode proteins: an intron of one gene is not usually an exon for a different gene.

Geneticists have long debated the evolutionary origin of introns. One idea, called the *intron late hypothesis*, proposes that introns were absent from ancient organisms but were later acquired by eukaryotes. Another idea, termed the *intron early hypothesis*, suggests that early ancestors to bacteria, archaea, and eukaryotes possessed introns that were later lost by prokaryotes and simple eukaryotes. Evidence suggests that introns have been lost *and* gained through evolutionary time. Many researchers now assume that the earliest eukaryotes possessed introns, because divergent eukaryotes have introns in the same positions in their genes, suggesting that these introns were present in the ancestors to all eukaryotes.

There are four major types of introns, differentiated by how the intron is removed (**Table 14.1**). **Group I introns**, found in some genes of eubacteria, bacteriophages, and eukaryotes, are self-splicing: they can catalyze their own removal. **Group II introns** are present in some genes of mitochondria, chloroplasts, archaea, and a few eubacteria; they also are self-splicing, but their mechanism of splicing differs from that of the group I introns. **Nuclear pre-mRNA introns** are the best studied; they include introns located in the protein-encoding genes of the eukaryotic nucleus. The splicing mechanism by which these introns are removed is similar to that of the group II introns, but nuclear introns are not self-splicing; their removal requires snRNAs (discussed later in this chapter) and a number of proteins. **Transfer RNA introns**, found in tRNA genes of eubacteria, archaea, and eukaryotes, utilize yet another splicing mechanism that relies on enzymes to cut

Table 14.1	Major types of introns	
Type of Intron	**Location**	**Splicing Mechanism**
Group I	genes of eubacteria, bacteriophages, and eukaryotes	Self-splicing
Group II	genes of eubacteria, archaea, and eukaryotic organelles	Self-splicing
Nuclear pre-mRNA	Protein-encoding genes in the nucleus of eukaryotes	Spliceosomal
tRNA	tRNA genes of eubacteria, archaea, and eukaryotes	Enzymatic

Note: There are also several types of minor introns, including group III introns, twintrons, and archaeal introns.

and reseal the RNA. In addition to these major groups, there are several other types of introns.

We'll take a detailed look at the chemistry and mechanics of RNA splicing later in this chapter. For now, we should keep in mind two general characteristics of the splicing process: (1) the splicing of all pre-mRNA introns takes place in the nucleus; and (2) the order of exons in DNA is usually maintained in the spliced RNA: the coding sequences of a gene may be split up, but they are not usually jumbled up. ▶ **TRY PROBLEM 19**

CONCEPTS

Many eukaryotic genes contain exons and introns. Both are transcribed into RNA, but introns are later removed by RNA processing. The number and size of introns vary from gene to gene; they are common in many eukaryotic genes but uncommon in bacterial genes.

✓ CONCEPT CHECK 2

What are the four major types of introns?

The Concept of the Gene Revisited

How does the presence of introns affect our concept of a gene? To define a gene as a sequence of nucleotides that encodes amino acids in a protein no longer seems appropriate because this definition excludes introns, which do not specify amino acids. This definition also excludes nucleotides that encode the 5′ and 3′ ends of an mRNA molecule, which are required for translation but do not encode amino acids. Defining a gene in these terms also excludes sequences that encode rRNA, tRNA, and other RNAs that do not encode proteins. Given our current understanding of DNA structure and function, we need a more precise definition of gene.

Many geneticists have broadened the concept of a gene to include all sequences in DNA that are transcribed into a single RNA molecule. Defined this way, a gene includes all exons, introns, and those sequences at the beginning and end of the RNA that are not translated into a protein. This definition also includes DNA sequences that encode rRNAs, tRNAs, and other types of nonmessenger RNA. Some geneticists have expanded the definition of a gene even further, to include the entire transcription unit—the promoter, the RNA coding sequence, and the terminator. However, new evidence now calls into question even this definition. Recent research suggests that much of the genome is transcribed into RNA, although it is unclear what, if anything, much of this RNA does. What is certain is that the process of transcription is more complex than formerly thought, and defining a gene as a sequence that is transcribed into an RNA molecule is not as straightforward as formerly thought. The more we learn about the nature of genetic information, the more elusive the definition of a gene seems to become.

CONCEPTS

The discovery of introns forced a reevaluation of the definition of the gene. Today, a gene is often defined as a DNA sequence that encodes an RNA molecule or the entire DNA sequence required to transcribe and encode an RNA molecule.

14.2 Messenger RNAs, Which Encode the Amino Acid Sequences of Proteins, Are Modified after Transcription in Eukaryotes

As soon as DNA was identified as the source of genetic information, it became clear that DNA cannot directly encode proteins. In eukaryotic cells, DNA resides in the nucleus, yet protein synthesis takes place in the cytoplasm. Geneticists recognized that an additional molecule must take part in transferring genetic information.

The results of studies of bacteriophage infection conducted in the late 1950s and early 1960s pointed to RNA as a likely candidate for this transport function. Bacteriophages inject their DNA into bacterial cells, where the DNA is replicated, and large amounts of phage protein are produced on the bacterial ribosomes. As early as 1953, Alfred Hershey discovered a type of RNA that was synthesized rapidly after bacteriophage infection. Findings from later studies showed that this short-lived RNA had a nucleotide composition similar to that of the phage DNA but quite different from that of the bacterial RNA. These observations were consistent with the idea that RNA was copied from DNA and that this RNA then directed the synthesis of proteins.

At the time, ribosomes were known to be *somehow* implicated in protein synthesis, and much of the RNA in a cell was known to be in the form of ribosomes. Ribosomes were believed to be the agents by which genetic information was moved to the cytoplasm for the production of protein. Using equilibrium density gradient centrifugation (see Figure 12.2), Sydney Brenner, François Jacob, and Matthew Meselson demonstrated in 1961 that it is not the case. They showed that new ribosomes are *not* produced during the burst of protein synthesis that accompanies phage infection (**Figure 14.4**). Ribosomes did not carry the genetic information needed to produce new phage proteins.

In a related experiment, François Gros and his colleagues infected *E. coli* cells with bacteriophages while adding radioactively labeled uracil, which would become incorporated into newly produced phage RNA, to the medium. Gros and his coworkers found that the newly produced phage RNA was short-lived, lasting only a few minutes, and was associated with ribosomes but was distinct from them. They concluded that newly synthesized, short-lived RNA carries the genetic information for protein structure to the ribosome. The term *messenger RNA* was coined for this carrier.

The Structure of Messenger RNA

Messenger RNA functions as the template for protein synthesis; it carries genetic information from DNA to a ribosome and helps to assemble amino acids in their correct order. In bacteria, mRNA is transcribed directly from DNA but, in eukaryotes, a pre-mRNA (also called the primary transcript) is first transcribed from DNA and then processed to yield the mature mRNA. We will reserve the term mRNA for RNA molecules that have been completely processed and are ready to undergo translation.

In the mRNA, each amino acid in a protein is specified by a set of three nucleotides called a **codon**. Both prokaryotic and eukaryotic mRNAs contain three primary regions (**Figure 14.5**). The **5′ untranslated region** (5′ UTR; sometimes called the leader) is a sequence of nucleotides at the 5′ end of the mRNA that does not encode any of the amino acids of a protein. In bacterial mRNA, this region contains a consensus sequence (UAAGGAGGU) called the **Shine–Dalgarno** sequence, which serves as the ribosome-binding site during translation (see Chapter 15); it is found approximately seven nucleotides upstream of the first codon

Experiment

Question: Do ribosomes carry genetic information?

Methods

1 *E. coli* were grown in medium containing heavy isotopes through several generations so that the heavy isotopes would become incorporated into all *E. coli* ribosomes.

Medium with ^{15}N and ^{13}C
E. coli culture

2 The cells were moved into medium containing light isotopes (^{14}N and ^{12}C)...

3 ... and infected with bacteriophage.

Medium with ^{14}N and ^{12}C
E. coli culture

4 If new ribosomes were produced after phage infection, they would contain ^{14}N and ^{12}C and would be relatively light.

5 After phage proteins were produced, ribosomes were separated by equilibrium density gradient centrifugation.

Spin

Results

Increasing density

6 Only old ribosomes containing heavy isotopes (^{15}N and ^{13}C) were found.

Conclusion: Ribosomes are not produced in phage reproduction and therefore do not carry genetic information.

14.4 Brenner, Jacob, and Meselson demonstrated that ribosomes do not carry genetic information.

translated into an amino acid (called the start codon). During translation, the Shine-Dalgarno sequence is complementary to and pairs with sequences found in one of the RNA molecules that make up the ribosome. Eukaryotic mRNA has no equivalent consensus sequence in its 5′ untranslated

Shine–Dalgarno sequence in prokaryotes only

mRNA

Start codon   **Stop codon**

5′   3′

5′ untranslated region   Protein-coding region   3′ untranslated region

14.5 Three primary regions of mature mRNA are the 5′ untranslated region, the protein-coding region, and the 3′ untranslated region.

region. In eukaryotic cells, ribosomes bind to a modified 5′ end of mRNA, as discussed later in this chapter.

The next section of mRNA is the **protein-coding region**, which comprises the codons that specify the amino acid sequence of the protein. The protein-coding region begins with a start codon and ends with a stop codon. The last region of mRNA is the **3′ untranslated region** (3′ UTR; sometimes called a trailer), a sequence of nucleotides at the 3′ end of the mRNA and not translated into protein. The 3′ UTR affects the stability of mRNA and the translation of the mRNA protein-coding sequence. View **Animation 14.1** to see how mutations in different regions of a gene affect the flow of information from genotype to phenotype.

(A)

CONCEPTS

Messenger RNA molecules contain three main regions: a 5′ untranslated region, a protein-coding region, and a 3′ untranslated region. The 5′ and 3′ untranslated regions do not encode any amino acids of a protein, but contain information that is important in translation, RNA stability, and regulation of gene expression.

Pre-mRNA Processing

In bacterial cells, transcription and translation take place simultaneously; while the 3′ end of an mRNA is undergoing transcription, ribosomes attach to the Shine–Dalgarno sequence near the 5′ end and begin translation. Because transcription and translation are coupled, bacterial mRNA has little opportunity to be modified before protein synthesis. In contrast, transcription and translation are separated in both time and space in eukaryotic cells. Transcription takes place in the nucleus, whereas translation takes place in the cytoplasm; this separation provides an opportunity for eukaryotic RNA to be modified before it is translated. Indeed, eukaryotic mRNA is extensively altered after transcription. Changes are made to the 5′ end, the 3′ end, and the protein-coding section of the RNA molecule (**Table 14.2**).

The Addition of the 5′ Cap

One type of modification of eukaryotic pre-mRNA is the addition of a structure called a **5′ cap**. The cap consists of an

Table 14.2	Posttranscriptional modifications to eukaryotic pre-mRNA
Modification	**Function**
Addition of 5′ cap	Facilitates binding of ribosome to 5′ end of mRNA, increases mRNA stability, enhances RNA splicing
3′ cleavage and addition of poly(A) tail	Increases stability of mRNA, facilitates binding of ribosome to mRNA
RNA splicing	Removes noncoding introns from pre-mRNA, facilitates export of mRNA to cytoplasm, allows for multiple proteins to be produced through alternative splicing
RNA editing	Alters nucleotide sequence of mRNA

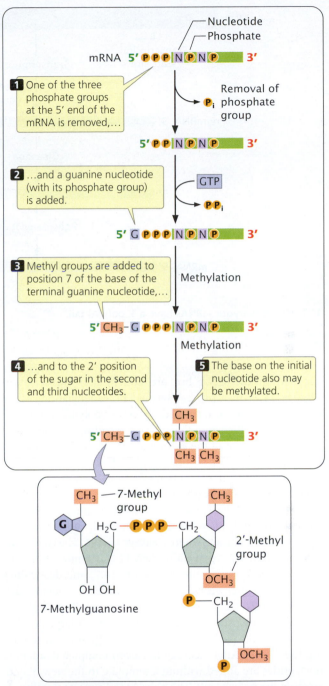

14.6 Most eukaryotic mRNAs have a 5′ cap. The cap consists of a nucleotide with 7-methylguanosine attached to the pre-mRNA by a unique 5′–5′ bond (shown in detail in the bottom box).

extra nucleotide at the 5′ end of the mRNA and methyl groups (CH_3) on the base in the newly added nucleotide and on the 2′-OH group of the sugar of one or more nucleotides at the 5′ end (**Figure 14.6**). The addition of the cap takes place rapidly after the initiation of transcription and, as will be discussed in more depth in Chapter 15, functions in the initiation of translation. Cap-binding proteins recognize the cap and attach to it; a ribosome then binds to these proteins and moves downstream along the mRNA until the start codon is reached and translation begins. The presence of a 5′ cap also increases the stability of mRNA and influences the removal of introns.

As noted in the discussion of transcription in Chapter 13, three phosphate groups are present at the 5′ end of all RNA molecules because phosphate groups are not cleaved from the first ribonucleoside triphosphate in the transcription reaction. The 5′ end of pre-mRNA can be represented as 5′–pppNpNpN . . . , in which the letter "N" represents a ribonucleotide and "p" represents a phosphate. Shortly after the initiation of transcription, one of these phosphate groups is removed and a guanine nucleotide is added (see Figure 14.6). This guanine nucleotide is attached to the pre-mRNA by a unique 5′–5′ bond, which is quite different from the usual 5′–3′ phosphodiester bond that joins all the other nucleotides in RNA. One or more methyl groups are then added to the 5′ end; the first of these methyl groups is added to position 7 of the base of the terminal guanine nucleotide, making the base 7-methylguanine. Next, a methyl group may be added to the 2′ position of the sugar in the second and third nucleotides (see Figure 14.6). Rarely, additional methyl groups may be attached to the bases of the second and third nucleotides of the pre-mRNA.

Several different enzymes take part in the addition of the 5′ cap. The initial step is carried out by an enzyme that associates with RNA polymerase II. Because neither RNA polymerase I nor RNA polymerase III have this associated enzyme, RNA molecules transcribed by these polymerases (rRNAs, tRNAs, and some snRNAs) are not capped.

The Addition of the Poly(A) Tail

A second type of modification to eukaryotic mRNA is the addition of 50 to 250 or more adenine nucleotides at the 3′ end, forming a **poly(A) tail**. These nucleotides are not

DNA

Transcription start site

Transcription

Consensus sequence | 11–30 nucleotides

Pre-mRNA is cleaved, at a position from 11 to 30 nucleotides downstream of the consensus sequence, in the 3' untranslated region.

Pre-mRNA 5' ... AAUAAA ... 3'

Cleavage site | U-rich sequence

Cleavage

5' ... AAUAAA ... 3'

The addition of adenine nucleotides (polyadenylation) takes place at the 3' end of the pre-mRNA, generating the poly(A) tail.

Polyadenylation

Poly (A) tail

mRNA 5' ... AAUAAA ... AAAAAAAAAAAAAAAAAAAAAA 3'

Conclusion: In pre-mRNA processing, a poly(A) tail is added through cleavage and polyadenylation.

14.7 Most eukaryotic mRNAs have a 3' poly(A) tail.

encoded in the DNA but are added after transcription (**Figure 14.7**) in a process termed polyadenylation. Many eukaryotic genes transcribed by RNA polymerase II are transcribed well beyond the end of the coding sequence (see Chapter 13); most of the extra material at the 3' end is then cleaved and the poly(A) tail is added. For some pre-mRNA molecules, more than 1000 nucleotides may be removed from the 3' end before polyadenylation.

Processing of the 3' end of pre-mRNA requires sequences, termed the polyadenylation signal, both upstream and downstream of the site where cleavage occurs. The consensus sequence AAUAAA is usually from 11 to 30 nucleotides upstream of the cleavage site (see Figure 14.7) and determines the point at which cleavage will take place. A sequence rich in uracil nucleotides (or in guanine and uracil nucleotides) is typically downstream of the cleavage site. A large number of proteins take part in finding the cleavage site and removing the 3' end. After cleavage has been completed, adenine nucleotides are added without a template to the new 3' end, creating the poly(A) tail. The poly(A) tail confers stability on many mRNAs, increasing the time during which the mRNA remains intact and available for translation before it is degraded by cellular enzymes. The stability conferred by the poly(A) tail depends on the proteins that attach to the tail and on its length. The poly(A) tail also facilitates attachment of the ribosome to the mRNA and plays a role in export of the mRNA into the cytoplasm.

Poly(U) tails are added to the 3' ends of some mRNAs, microRNAs, and small nuclear RNAs. Although the function of poly(U) tails is still under investigation, evidence suggests that poly(U) tails on some mRNAs may facilitate their degradation. ▶ TRY PROBLEM 29

CONCEPTS

Eukaryotic pre-mRNAs are processed at their 5' and 3' ends. A cap, consisting of a modified nucleotide and several methyl groups, is added to the 5' end. The cap facilitates the binding of a ribosome, increases the stability of the mRNA, and may affect the removal of introns. Processing at the 3' end includes cleavage downstream of an AAUAAA consensus sequence and the addition of a poly(A) tail.

✓ CONCEPT CHECK 3

Why are pre-mRNAs capped, but tRNAs and rRNAs aren't?

RNA Splicing

The other major type of modification of eukaryotic pre-mRNA is the removal of introns by **RNA splicing**. This modification takes place in the nucleus, before the RNA moves to the cytoplasm.

CONSENSUS SEQUENCES AND THE SPLICEOSOME Splicing requires the presence of three sequences in the intron. One end of the intron is referred to as the **5' splice site**, and the other end is the **3' splice site** (**Figure 14.8**); these splice sites possess short consensus sequences. Most introns in pre-mRNAs begin with GU and end with AG, indicating that these sequences play a crucial role in splicing. Indeed, changing a single nucleotide at either of these sites prevents splicing.

The third sequence important for splicing is at the **branch point**, which is an adenine nucleotide that lies from 18 to 40 nucleotides upstream of the 3' splice site (see Figure 14.8).

14.8 Splicing of pre-mRNA requires consensus sequences. Critical consensus sequences are present at the 5′ splice site and the 3′ splice site. A weak consensus sequence (not shown) exists at the branch point.

The sequence surrounding the branch point does not have a strong consensus. The deletion or mutation of the adenine nucleotide at the branch point prevents splicing.

Splicing takes place within a large structure called the **spliceosome**, which is one of the largest and most complex of all molecular structures. The spliceosome consists of five RNA molecules and almost 300 proteins. The RNA components are small nuclear RNAs (snRNAs, see Chapter 13) ranging in length from 107 to 210 nucleotides; these snRNAs associate with proteins to form small nuclear ribonucleoprotein particles (snRNPs). Each snRNP contains a single snRNA molecule and multiple proteins. The spliceosome is composed of five snRNPs (U1, U2, U4, U5, and U6), and some proteins not associated with an snRNA.

CONCEPTS

Introns in nuclear genes contain three consensus sequences critical to splicing: a 5′ splice site, a 3′ splice site, and a branch point. The splicing of pre-mRNA takes place within a large complex called the spliceosome, which consists of snRNAs and proteins.

✓ CONCEPT CHECK 4

If a splice site were mutated so that splicing did not take place, what would be the effect on the mRNA?

a. It would be shorter than normal.

b. It would be longer than normal.

c. It would be the same length but would encode a different protein.

THE PROCESS OF SPLICING   Before splicing takes place, an intron lies between an upstream exon (exon 1) and a downstream exon (exon 2), as shown in **Figure 14.9**. Pre-mRNA is spliced in two distinct steps. In the first step of splicing, the pre-mRNA is cut at the 5′ splice site. This cut frees exon 1 from the intron, and the 5′ end of the intron attaches to the branch point; the intron folds back on itself, forming a structure called a **lariat**. In this reaction, the guanine nucleotide in the consensus sequence at the 5′ splice site bonds with the adenine nucleotide at the branch point through a transesterification reaction. In this reaction, both 5′ cleavage and lariat formation occur in a single step. The result is that the 5′ phosphate group of the guanine nucleotide is now attached to the 2′-OH group of the adenine nucleotide at the branch point (see Figure 14.9).

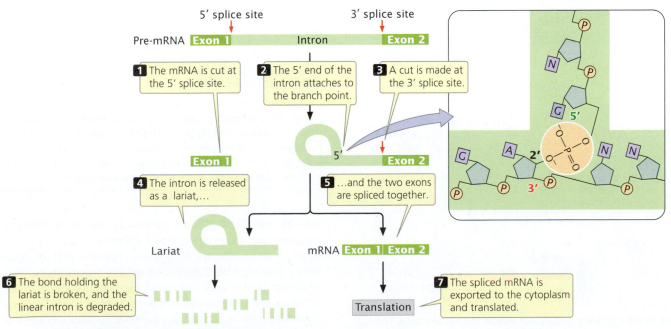

14.9 The splicing of nuclear introns requires a two-step process.

In the second step of RNA splicing, a cut is made at the 3′ splice site and, simultaneously, the 3′ end of exon 1 becomes covalently attached (spliced) to the 5′ end of exon 2. The intron is released as a lariat. Eventually, a lariat debranching enzyme breaks the bond at the branch point, producing a linear intron that is rapidly degraded by nuclear enzymes. The mature mRNA consisting of the exons spliced together is exported to the cytoplasm, where it is translated.

These splicing reactions take place within the spliceosome, which assembles on the pre-mRNA in a step-by-step fashion and carries out the splicing reactions (**Figure 14.10**). A crucial feature of the process is a series of interactions between the mRNA and the snRNAs and between different snRNAs. These interactions depend on complementary base pairing between the different RNA molecules and bring the essential components of the pre-mRNA transcript and the spliceosome close together, which make splicing possible. Key catalytic steps in the splicing process are carried out by the snRNAs that constitute the spliceosome.

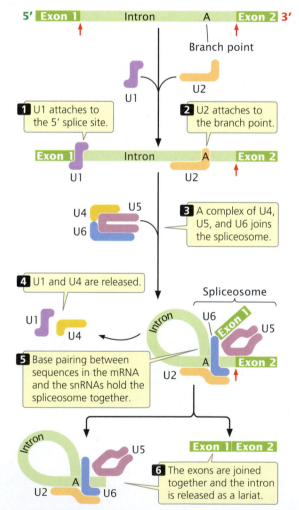

14.10 RNA splicing takes place within the spliceosome. The spliceosome assembles sequentially.

First, snRNP U1 attaches to the 5′ splice site, and then U2 attaches to the branch point. A complex consisting of U4, U5, and U6 (which form a single snRNP) joins the spliceosome. This addition causes a conformational change in the spliceosome, the intron loops over, and the 5′ splice site is brought close to the branch point. Particles U1 and U4 dissociate from the spliceosome, with the subsequent formation of base pairs between U6 and U2 and between U6 and the 5′ splice site. The 5′ splice site, 3′ splice site, and branch point are in close proximity, held together by the spliceosome. The two transesterification reactions take place, joining the two exons together and releasing the intron as a lariat.

Most mRNAs are produced from a single pre-mRNA molecule from which the exons are spliced together. However, in a few organisms (principally nematodes and trypanosomes), mRNAs may be produced by splicing together sequences from two or more different RNA molecules; this process is called **trans-splicing.**

Many human genetic diseases arise from mutations that affect pre-mRNA splicing; indeed, about 15% of single-base substitutions that result in human genetic diseases alter pre-mRNA splicing. Some of these mutations interfere with recognition of the normal 5′ and 3′ splice sites. Others create new splice sites, as was the case with the mutation that caused the royal hemophilia discussed in the introduction to this chapter. ▶ TRY PROBLEM 24

RNA splicing, which takes place in the nucleus, must be done before the RNA can move into the cytoplasm. Incompletely spliced RNAs remain in the nucleus until splicing is complete or until the pre-mRNA is degraded. Immediately after splicing, a group of proteins called the exon-junction complex (EJC) is deposited approximately 20 nucleotides upstream of each exon–exon junction on the mRNA. The EJC promotes the export of the mRNA from the nucleus into the cytoplasm.

CONCEPTS

Intron splicing of nuclear genes is a two-step process: (1) the 5′ end of the intron is cleaved and attached to the branch point to form a lariat and (2) the 3′ end of the intron is cleaved and the ends of the two exons are spliced together. In the process, the exons are joined and the intervening intron is removed. These reactions take place within the spliceosome.

MINOR SPLICING Some introns in the pre-mRNAs of multicellular eukaryotes utilize a different process of intron removal known as minor splicing. Introns that undergo minor splicing have different consensus sequences at the 5′ splice site and branch point and use a minor spliceosome, which contains a somewhat different set of snRNAs. Some 700-800 genes in the human genome contain introns that undergo minor splicing.

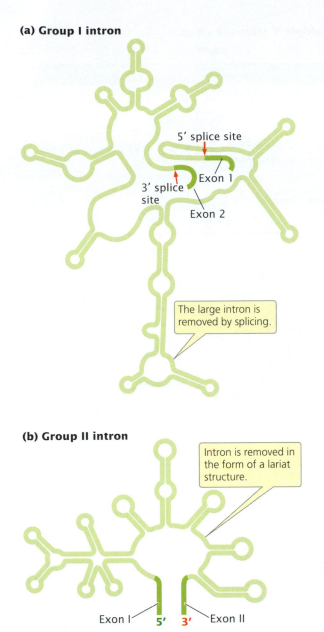

(a) Group I intron

5′ splice site

Exon 1

3′ splice site

Exon 2

The large intron is removed by splicing.

(b) Group II intron

Intron is removed in the form of a lariat structure.

Exon I   **5′   3′**   Exon II

14.11 Group I and group II introns fold into characteristic secondary structures.

SELF-SPLICING INTRONS Some introns are self-splicing—they possess the ability to remove themselves from an RNA molecule. These self-splicing introns fall into two major categories. Group I introns are found in a variety of genes, including some rRNA genes in protists, some mitochondrial genes in fungi, and even some bacterial and bacteriophage genes. Although the lengths of group I introns vary, all of them fold into a common secondary structure with nine looped stems (**Figure 14.11a**), which are necessary for splicing.

Group II introns, present in genes of eubacteria, archaea, and eukaryotic organelles, also have the ability to self-splice. All group II introns also fold into secondary structures (**Figure 14.11b**). The splicing of group II introns is

accomplished by a mechanism that has some similarities to the spliceosomal-mediated splicing of nuclear genes, and splicing generates a lariat structure. Because of these similarities, group II introns and nuclear pre-mRNA introns have been suggested to be evolutionarily related; perhaps the nuclear introns evolved from self-splicing group II introns and later adopted the proteins and snRNAs of the spliceosome to carry out the splicing reaction.

CONCEPTS

Some introns are removed by the minor splicing system. Other introns are self-splicing and consist of two types: group I introns and group II introns. These introns have complex secondary structures that enable them to catalyze their excision from RNA molecules without the aid of enzymes or other proteins.

Alternative Processing Pathways

A finding that complicates the view of a gene as a sequence of nucleotides that specifies the amino acid sequence of a protein (see section on The Concept of the Gene Revisited) is the existence of **alternative processing pathways**. In these pathways, a single pre-mRNA is processed in different ways to produce alternative types of mRNA, resulting in the production of different proteins from the same DNA sequence.

One type of alternative processing is **alternative splicing**, in which the same pre-mRNA can be spliced in more than one way to yield multiple mRNAs that are translated into different amino acid sequences and thus different proteins (**Figure 14.12a**). Another type of alternative processing requires the use of **multiple 3′ cleavage sites** (**Figure 14.12b**) where two or more potential sites for cleavage and polyadenylation are present in the pre-mRNA. In the example in Figure 14.12b, cleavage at the first site produces a relatively short mRNA compared with the mRNA produced through cleavage at the second site. The use of an alternative cleavage site may or may not produce a different protein, depending on whether the position of the site is before or after the termination codon.

Both alternative splicing and multiple 3′ cleavage sites can exist in the same pre-mRNA transcript. An example is seen in the mammalian gene that encodes calcitonin; this gene contains six exons and five introns (**Figure 14.13a**). The entire gene is transcribed into pre-mRNA (**Figure 14.13b**). There are two possible 3′ cleavage sites. In cells of the thyroid gland, 3′cleavage and polyadenylation take place after the fourth exon to produce a mature mRNA consisting of exons 1, 2, 3, and 4 (**Figure 14.13c**). This mRNA is translated into the hormone calcitonin, which is produced by the thyroid gland and regulates levels of calcium. In brain cells, the *identical* pre-mRNA is transcribed from DNA, but cleavage and polyadenylation take place after the sixth exon, yielding an initial transcript that includes all six exons. During splicing, exon 4 is removed, and

(a) Alternative splicing

(b) Multiple 3′ cleavage sites

Conclusion: Both alternative splicing and multiple 3′ cleavage sites produce different mRNAs from a single pre-mRNA.

14.12 Eukaryotic cells have alternative pathways for processing pre-mRNA. (a) With alternative splicing, pre-mRNA can be spliced in different ways to produce different mRNAs. (b) With multiple 3′ cleavage sites, there are two or more potential sites for cleavage and polyadenylation; use of the different sites produces mRNAs of different lengths.

so only exons 1, 2, 3, 5, and 6 are present in the mature mRNA (**Figure 14.13d**). When translated, this mRNA produces a protein called calcitonin-gene-related peptide (CGRP), which has an amino acid sequence quite different from that of calcitonin. CGRP causes dilation of blood vessels and can function in transmission of pain. Some research suggests that CGRP is involved in the development of migraine headaches. Alternative splicing may produce different combinations of exons in the mRNA, but the order of the exons is not usually changed.

Alternative processing of pre-mRNAs is common in multicellular eukaryotes. For example, researchers estimate that more than 90% of all human genes undergo alternative splicing. Often the form of splicing differs between human tissues: human brain and liver tissues have more alternatively spliced RNA compared with other tissues. Sometimes splicing even varies from one person to another. Different processing pathways contribute to gene regulation, as will be discussed in Chapter 17.

Alternative splicing may play a role in organism complexity. The complete sequencing of the genomes of numerous organisms (see Chapter 20) has led to the conclusion that an organism's number of genes is not correlated with the organism's complexity. For example, fruit flies have only about 14,000 genes, whereas anatomically simpler nematode worms have 19,000 genes. The plant *Arabidopsis thaliana* has about 20,000 genes, almost as many as humans have. If anatomically simple organisms have as many genes as complex organisms have, how is developmental complexity encoded in the genome? A possible answer is alternative processing, which can produce multiple proteins from a single gene and is an important source of protein diversity in vertebrates.

Recent research demonstrates that even closely related species often differ in how their pre-mRNAs are spliced, and alternative splicing may have played an important role in speciation (see Chapter 26).

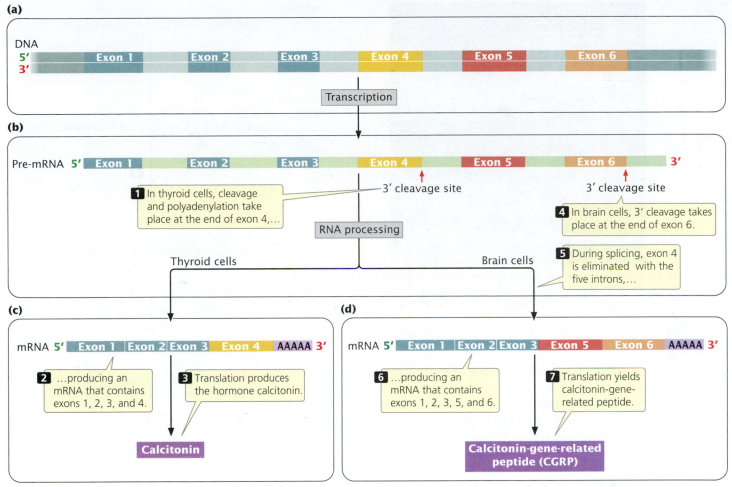

14.13 Pre-mRNA encoded by the gene for calcitonin undergoes alternative processing.

Alternative splicing enables exons to be spliced together in different combinations to yield mRNAs that encode different proteins. Alternative 3′ cleavage sites allow pre-mRNA to be cleaved at different sites.

✓ **CONCEPT CHECK 5**

Alternative 3′ cleavage sites result in
a. multiple genes of different lengths.
b. multiple pre-mRNAs of different lengths.
c. multiple mRNAs of different lengths.
d. all of the above.

RNA Editing

The assumption that all information about the amino acid sequence of a protein resides in DNA is violated by a process called **RNA editing**. In RNA editing, the coding sequence of an mRNA molecule is altered after transcription, so the protein has an amino acid sequence that differs from that encoded by the gene.

RNA editing was first detected in 1986 when the coding sequences of mRNAs were compared with the coding sequences of the DNA from which they had been transcribed. In some nuclear genes in mammalian cells and in some mitochondrial genes in plant cells, there had been substitutions in some of the nucleotides of the mRNA. More extensive RNA editing has been found in the mRNA for some mitochondrial genes in trypanosome parasites (which cause African sleeping sickness; **Figure 14.14**). In some mRNAs of these organisms, more than 60% of the sequence is determined by RNA editing. Different types of RNA editing have now been observed in mRNAs, tRNAs, and rRNAs from a wide range of organisms; the types include the insertion and the deletion of nucleotides and the conversion of one base into another.

If the modified sequence in an edited RNA molecule doesn't come from a DNA template, then how is it specified? A variety of mechanisms can bring about changes in RNA sequences. In some cases, molecules called **guide RNAs**

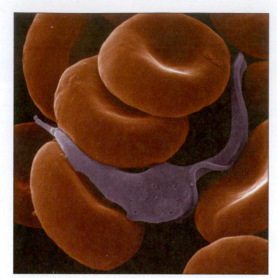

14.14 *Trypanosoma brucei* **causes African sleeping sickness.** Messenger RNA produced from mitochondrial genes of this parasite (in purple) undergoes extensive RNA editing. [Davic Spears/Last Refuge Ltd./Phototake.]

(gRNAs) play a crucial role. A gRNA contains sequences that are partly complementary to segments of the pre-edited RNA, and the two molecules undergo base pairing in these sequences (**Figure 14.15**). After the mRNA is anchored to the gRNA, the mRNA undergoes cleavage and nucleotides are added, deleted, or altered according to the template provided by gRNA. In other cases, enzymes bring about base conversion. In humans, for example, a gene is transcribed into mRNA that encodes a lipid-transporting polypeptide called apolipoprotein-B100, which has 4563 amino acids and is synthesized in liver cells. A truncated form of the protein called apolipoprotein-B48—with only 2153 amino acids—is synthesized in intestinal cells through editing of the apolipoprotein-B100 mRNA. In this editing, an enzyme deaminates a cytosine base, converting it into uracil. This conversion changes a codon that specifies the amino acid glutamine into a stop codon that prematurely terminates translation, resulting in the shortened protein. ▶ TRY PROBLEM 34

CONCEPTS

Individual nucleotides in the interior of pre-mRNA may be changed, added, or deleted by RNA editing. The amino acid sequence produced by the edited mRNA is not the same as that encoded by DNA.

✓ CONCEPT CHECK 6

What specifies the modified sequence of nucleotides found in an edited RNA molecule?

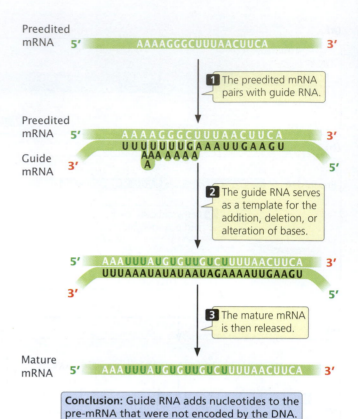

14.15 RNA editing is carried out by guide RNAs. The guide mRNA has sequences that are partly complementary to those of the pre-edited mRNA and pairs with it. After pairing, the mRNA undergoes cleavage and new nucleotides are added, with sequences in the gRNA serving as a template. The ends of the mRNA are then joined together.

CONNECTING CONCEPTS

Eukaryotic Gene Structure and Pre-mRNA Processing

Chapters 13 and 14 introduced a number of different components of genes and RNA molecules, including promoters, 5′ untranslated regions, coding sequences, introns, 3′ untranslated regions, poly(A) tails, and caps. Let's see how some of these components are combined to create a typical eukaryotic gene and how a mature mRNA is produced from them.

The promoter, which typically lies upstream of the transcription start site, is necessary for transcription to take place but is itself not usually transcribed when protein-encoding genes are transcribed by RNA polymerase II (**Figure 14.16a**). Farther upstream or downstream of the start site, there may be enhancers—DNA sequences that also regulate transcription.

In transcription, all the nucleotides between the transcription start site and the termination site are transcribed into pre-mRNA, including exons, introns, and a long 3′ end that is later cleaved from the transcript (**Figure 14.16b**). Notice that the 5′ end of the first exon contains the sequence that encodes the 5′ untranslated region and that the 3′ end of the last exon contains the sequence that encodes the 3′ untranslated region.

(a)

← Enhancer is typically upstream, but could be downstream or in an intron

Promoter

RNA coding

DNA
5′
3′

Exon 1   Intron   Exon 2   Intron   Exon 3

→ Transcription start

1 Introns, exons, and a long 3′ end are all transcribed into pre-mRNA.

End of transcription

(b)
Pre-mRNA
5′   Consensus sequence   AAUAAA   3′

5′ untranslated region   3′ untranslated region

(c)
Pre-mRNA
5′   AAUAAA   3′

2 A 5′ cap is added.   **3** Cleavage at the 3′ end is approximately 10 nucleotides downstream of the consensus sequence.   3′ cleavage site

(d)
Pre-mRNA
5′   AAUAAA   3′

4 Polyadenylation at the cleavage site produces the poly(A) tail.   3′ cleavage site

(e)
Pre-mRNA
5′   AAUAAA AAAAA   3′

3′ cleavage site

(f)
5 Finally, the introns are removed,…   **6** …producing the mature mRNA.   RNA splicing

Introns

mRNA
5′   Poly(A) tail   AAAAA   3′

5′ untranslated region   Protein-coding region   3′ untranslated region

14.16 Mature eukaryotic mRNA is produced when pre-mRNA is transcribed and undergoes several types of processing.

The pre-mRNA is then processed to yield a mature mRNA. The first step in this processing is the addition of a cap to the 5′ end of the pre-mRNA (**Figure 14.16c**). Next, the 3′ end is cleaved at a site downstream of the AAUAAA consensus sequence in the last exon (**Figure 14.16d**). Immediately after cleavage, a poly(A) tail is added to the 3′ end (**Figure 14.16e**). Finally, the introns are removed to yield the mature mRNA (**Figure 14.16f**). The mRNA now contains 5′ and 3′ untranslated regions, which are not translated into amino acids, and the nucleotides that carry the protein-coding sequences. You can explore the consequences of failed RNA processing by viewing and interacting with **Animation 14.2**.

The nucleotide sequence of a small gene (the human interleukin 2 gene), with these components identified, is presented in **Figure 14.17**.

14.3 Transfer RNAs, Which Attach to Amino Acids, Are Modified after Transcription in Bacterial and Eukaryotic Cells

In 1956, Francis Crick proposed the idea of a molecule that transports amino acids to the ribosome and interacts with codons in mRNA, placing amino acids in their proper order in protein synthesis. By 1963, the existence of such an adapter molecule, called transfer RNA, had been confirmed. Transfer RNA (tRNA) serves as a link between the genetic code in mRNA and the amino acids that make up a protein. Each tRNA attaches to a particular amino acid and carries it

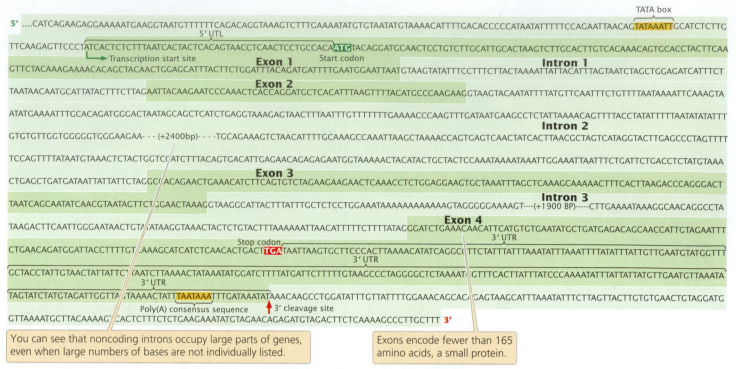

TATA box

5' ....CATCAGAAGAGGAAAAATGAAGGTAATGTTTTTTCAGACAGGTAAAGTCTTTGAAAATATGTGTAATATGTAAAACATTTTGACACCCCCATAATATTTTTCCAGAATTAACAGTATAAATTGCATCTCTTG

5' UTL

TTCAAGAGTTCCCTATCACTCTCTTTAATCACTACTCACAGTAACCTCAACTCCTGCCACAATGTACAGGATGCAACTCCTGTCTTGCATTGCACTAAGTCTTGCACTTGTCACAAACAGTGCACCTACTTCAA

→ Transcription start site   Start codon

Exon 1

GTTCTACAAAGAAAACACAGCTACAACTGGAGCATTTACTTCTGGATTTACAGATGATTTTGAATGGAATTAATGTAAGTATATTTCCTTTCTTACTAAAATTATTACATTTAGTAATCTAGCTGGAGATCATTTCT

Exon 2   Intron 1

TAATAACAATGCATTATACTTTCTTAGAATTACAAGAATCCCAAACTCACCAGGATGCTCACATTTAAGTTTTACATGCCCAAGAAGGTAAGTACAATATTTTATGTTCAATTTCTGTTTTAATAAAATTCAAAGTA

ATATGAAAATTTGCACAGATGGGACTAATAGCAGCTCATCTGAGGTAAAGAGTAACTTTAATTTGTTTTTTTGAAAACCCAAGTTTGATAATGAAGCCTCTATTAAAACAGTTTTACCTATATTTTTAATATATATTT

Intron 2

GTGTGTTGGTGGGGGTGGGAAGAA- - -(+2400bp)- - -TGCAGAAAGTCTAACATTTTGCAAAGCCAAATTAAGCTAAAACCAGTGAGTCAACTATCACTTAACGCTAGTCATAGGTACTTGAGCCCTAGTTTT

TCCAGTTTTATAATGTAAACTCTACTGGTCCATCTTTACAGTGACATTGAGAACAGAGAGAATGGTAAAAACTACATACTGCTACTCCAAATAAAATAAATTGGAAATTAATTTCTGATTCTGACCTCTATGTAAA

Exon 3

CTGAGCTGATGATAATTATTATTCTAGGCCACAGAACTGAAACATCTTCAGTGTCTAGAAGAAGAACTCAAACCTCTGGAGGAAGTGCTAAATTTAGCTCAAAGCAAAAACTTTCACTTAAGACCCAGGGACT

TAATCAGCAATATCAACGTAATAGTTCTGGAACTAAAGGTAAGGCATTACTTTATTTGCTCTCCTGGAAATAAAAAAAAAAAAAAGTAGGGGGAAAAGT----(+1900 BP)-----CTTGAAAATAAAGGCAACAGGCCTA

Intron 3

TAAGACTTCAATTGGGAATAACTGTATAAAGGTAAACTACTCTGTACTTTAAAAAAATTAACATTTTTCTTTTATAGGGATCTGAAACAACATTCATGTGTGAATATGCTGATGAGACAGCAACCATTGTAGAATTT

Exon 4   3' UTR

CTGAACAGATGGATTACCTTTTGTAAAGCATCATCTCAACACTGACTTGATAATTAAGTGCTTCCCACTTAAAAACATATCAGGCCTTCTATTTATTTAAAATATTTAAATTTTATATTTATTGTTGAATGTATGGTTT

Stop codon   TGA   3' UTR

GCTACCTATTGTAACTATTATTCTTAATCTTAAAACTATAAATATGGATCTTTTATGATTCTTTTTGTAAGCCCTAGGGGCTCTAAAATGTTTCACTTATTTATCCCAAAATATTTATTATTATGTTGAATGTTAAATA

3' UTR

TAGTATCTATGTAGATTGGTTAGTAAAACTATTTAATAAATTTGATAAATATAAACAAGCCTGGATATTTGTTATTTTGGAAACAGCACGAGTAAGCATTTAAATATTTCTTAGTTACTTGTGTGAACTGTAGGATG

Poly(A) consensus sequence   ↑ 3' cleavage site

GTTAAAATGCTTACAAAAGTCACTCTTTCTCTGAAGAAATATGTAGAACAGAGATGTAGACTTCTCAAAAGCCCTTGCTTT 3'

You can see that noncoding introns occupy large parts of genes, even when large numbers of bases are not individually listed.

Exons encode fewer than 165 amino acids, a small protein.

14.17 This representation of the nucleotide sequence of the gene for human interleukin 2 includes the TATA box, transcription start site, start and stop codons, introns, exons, poly(A) consensus sequence, and 3' cleavage site.

to the ribosome, where the tRNA adds its amino acid to the growing polypeptide chain at the position specified by the genetic instructions in the mRNA. We'll take a closer look at the mechanism of this process in Chapter 15.

Each tRNA is capable of attaching to only one type of amino acid. The complex of tRNA plus its amino acid can be written in abbreviated form by adding a three-letter superscript representing the amino acid to the term tRNA. For example, a tRNA that attaches to the amino acid alanine is written as tRNA^{Ala}. Because 20 different amino acids are found in proteins, there must be a minimum of 20 different types of tRNA. In fact, most organisms possess at least 30 to 40 different types of tRNA, each encoded by a different gene (or, in some cases, multiple copies of a gene) in DNA.

The Structure of Transfer RNA

A unique feature of tRNA is the occurrence of rare **modified bases**. All RNAs have the four standard bases (adenine, cytosine, guanine, and uracil) specified by DNA, but tRNAs have additional bases, including ribothymine, pseudouridine (which is also occasionally present in snRNAs and rRNA), and dozens of others. The structures of two of these modified bases are shown in **Figure 14.18**.

If there are only four bases in DNA and all RNA molecules are transcribed from DNA, how do tRNAs acquire these additional bases? Modified bases arise from chemical changes made to the four standard bases after transcription. These changes are carried out by special **tRNA-modifying enzymes.** For example, the addition of a methyl group to uracil creates the modified base ribothymidine.

14.18 Two of the modified bases found in tRNAs. All the modified bases in tRNAs are produced by the chemical alteration of the four standard RNA bases.

The structures of all tRNAs are similar, a feature critical to tRNA function. Most tRNAs contain between 74 and 95 nucleotides, some of which are complementary to each other and form intramolecular hydrogen bonds. As a result, each tRNA has a **cloverleaf structure** (**Figure 14.19**). The cloverleaf has four major arms. If we start at the top and proceed clockwise around the tRNA shown at the right in Figure 14.19, the four major arms are the acceptor arm, the TΨC arm, the anticodon arm, and the DHU arm. Three of the arms (the TΨC, anticodon, and DHU arms) consist of a

This computer-generated space-filling molecular model shows the three-dimensional structure of a tRNA.

This ribbon model emphasizes the internal regions of base pairing.

This flattened cloverleaf model shows pairing between complementary nucleotides.

Hydrogen bonds between paired bases

This icon for tRNA will be used in subsequent chapters.

Amino acid attachment site (always CCA)

Acceptor arm

DHU arm

TΨC arm

Rare base (●)

Anticodon arm

Extra arm (size varies)

The anticodon comprises three bases and interacts with a codon in mRNA.

14.19 All tRNAs possess a common secondary structure, the cloverleaf structure. The base sequence in the flattened model is for tRNA^Ala.

stem and a loop. The stem is formed by the pairing of complementary nucleotides, and the loop lies at the terminus of the stem, where there is no nucleotide pairing.

Instead of having a loop, the acceptor arm includes the 5′ and 3′ ends of the tRNA molecule. All tRNAs have the same sequence (CCA) at the 3′ end, where the amino acid attaches to the tRNA; so, clearly, this sequence is not responsible for specifying which amino acid will attach to the tRNA.

The TΨC arm is named for the bases of three nucleotides in the loop of this arm: thymine (T), pseudouridine (Ψ), and cytosine (C). The anticodon arm lies at the bottom of the tRNA. Three nucleotides at the end of this arm make up the **anticodon**, which pairs with the corresponding codon on mRNA to ensure that the amino acids link in the correct order. The DHU arm is so named because it often contains the modified base dihydrouridine.

Although each tRNA molecule folds into a cloverleaf owing to the complementary pairing of bases, the cloverleaf is not the three-dimensional (tertiary) structure of tRNAs found in the cell. The results of X-ray crystallographic studies have shown that the cloverleaf folds on itself to form an L-shaped structure, as illustrated by the space-filling and ribbon models in Figure 14.19. Notice that the acceptor stem is at one end of the tertiary structure and the anticodon is at the other end.

Transfer RNA Gene Structure and Processing

The genes that produce tRNAs may be in clusters or scattered about the genome. In *E. coli*, the genes for some tRNAs are present in a single copy, whereas the genes for other tRNAs are present in several copies; eukaryotic cells usually have many copies of each tRNA gene. All tRNA molecules in both bacterial and eukaryotic cells undergo processing after transcription.

In *E. coli*, several tRNAs are usually transcribed together as one large precursor tRNA, which is then cut up into pieces, each containing a single tRNA. Additional nucleotides may then be removed one at a time from the 5′ and 3′ ends of the tRNA in a process known as trimming. Base-modifying enzymes may then change some of the standard bases into modified bases (**Figure 14.20**). In some prokaryotes, the CCA sequences found at the 3′ ends of tRNAs are encoded in the tRNA gene and are transcribed into the tRNA; in other prokaryotes and in eukaryotes, these sequences are added by a special enzyme that adds the nucleotides without the use of any template.

There is no generic processing pathway for all tRNAs: different tRNAs are processed in different ways. Eukaryotic tRNAs are processed in a manner similar to that for bacterial tRNAs: most are transcribed as larger precursors that are then cleaved, trimmed, and modified to produce mature tRNAs.

Some eukaryotic and archaeal tRNA genes possess introns of variable length that must be removed in processing. For example, about 40 of the 400 tRNA genes in yeast contain a single intron that is always found adjacent to the 3′ side of the anticodon. The tRNA introns are shorter than those found in pre-mRNA and do not have the consensus sequences found at the intron–exon junctions of pre-mRNAs. The splicing process for tRNA genes is quite different from the spliceosome-mediated reactions that remove introns from protein-encoding genes.

CONCEPTS

All tRNAs are similar in size and have a common secondary structure known as the cloverleaf. Transfer RNAs contain modified bases and are extensively processed after transcription in both bacterial and eukaryotic cells.

14.20 Transfer RNAs are processed in both bacterial and eukaryotic cells. Different tRNAs are modified in different ways. One example is shown here.

✓ **CONCEPT CHECK 7**

How are rare bases incorporated into tRNAs?
a. Encoded by guide RNAs
b. By chemical changes to one of the standard bases
c. Encoded by rare bases in DNA
d. Encoded by sequences in introns

14.4 Ribosomal RNA, a Component of the Ribosome, Is Also Processed after Transcription

Within ribosomes, the genetic instructions contained in mRNA are translated into the amino acid sequences of polypeptides. Thus, ribosomes play an integral part in the transfer of genetic information from genotype to phenotype. We will examine the role of ribosomes in the process of translation in Chapter 15. Here, we consider ribosome structure and examine how ribosomes are processed before becoming functional.

The Structure of the Ribosome

The ribosome is one of the most abundant molecular complexes in the cell: a single bacterial cell may contain as many as 20,000 ribosomes, and eukaryotic cells possess even more. Ribosomes typically contain about 80% of the total cellular RNA. They are complex structures, each consisting of more than 50 different proteins and RNA molecules (**Table 14.3**). A functional ribosome consists of two subunits, a **large ribosomal subunit** and a **small ribosomal subunit**, each of which consists of one or more pieces of RNA and a number of proteins. The sizes of the ribosomes and their RNA components are given in Svedberg (S) units (a measure of how rapidly an object sediments in a centrifugal field). It is important to note that S units are not additive; combining a 10S structure and a 20S structure does not necessarily produce a 30S structure, because the sedimentation rate is affected by the three-dimensional structure as well as the mass. The three-dimensional structure of the bacterial ribosome has been elucidated in great detail through X-ray crystallography. More will be said about the ribosome's structure in Chapter 15.

Table 14.3	**Composition of ribosomes in bacterial and eukaryotic cells**			
Cell Type	Ribosome Size	Subunit	rRNA Component	Proteins
Bacterial	70S	Large (50S)	23S (2900 nucleotides), 5S (120 nucleotides)	31
		Small (30S)	16S (1500 nucleotides)	21
Eukaryotic	80S	Large (60S)	28S (4700 nucleotides), 5.8S (160 nucleotides), 5S (120 nucleotides)	49
		Small (40S)	18S (1900 nucleotides)	33

Note: The letter "S" stands for "Svedberg unit."

Ribosomal RNA Gene Structure and Processing

The genes for rRNA, like those for tRNA, can be present in multiple copies, and the numbers vary among species (**Table 14.4**); all copies of the rRNA gene in a species are identical or nearly identical. In bacteria, rRNA genes are dispersed, but in eukaryotic cells they are clustered, with the genes arrayed in tandem, one after another.

Eukaryotic cells possess two types of rRNA genes: a large gene that encodes 18S rRNA, 28S rRNA, and 5.8S rRNA, and a small gene that encodes the 5S rRNA. All three bacterial rRNAs (23S rRNA, 16S rRNA, and 5S rRNA) are encoded by a single type of gene.

Ribosomal RNA is processed in both bacterial and eukaryotic cells. In *E. coli*, each RNA gene is transcribed into a 30S rRNA precursor (**Figure 14.21a**). This 30S precursor is methylated in several places, and then cleaved and trimmed to produce 16S rRNA, 23S rRNA, and 5S rRNA, along with one or more tRNAs. A series of enzymes bring about cleavage, methylation, and trimming.

Eukaryotic rRNAs undergo similar processing (**Figure 14.21b**). Small nucleolar RNAs (snoRNAs) help to cleave and modify eukaryotic rRNAs and assemble the processed rRNAs into mature ribosomes. Like the snRNAs taking part in pre-mRNA splicing, snoRNAs associate with proteins to form ribonucleoprotein particles (snoRNPs). The snoRNAs have extensive complementarity to the rRNA sequences in which modification takes place. Interestingly, some snoRNAs are encoded by sequences in the introns

Table 14.4	Number of rRNA genes in different organisms
Species	**Number of Copies of rRNA Genes per Genome**
Escherichia coli	7
Yeast	100–200
Human	280
Frog	450

of other protein-encoding genes. The processing of rRNA and ribosome assembly in eukaryotes take place in the nucleolus.

CONCEPTS

A ribosome is a complex organelle consisting of several rRNA molecules and many proteins. Each functional ribosome consists of a large and a small subunit. Ribosomal RNAs in both bacterial and eukaryotic cells are modified after transcription. In eukaryotes, rRNA processing is carried out by small nucleolar RNAs.

✓ CONCEPT CHECK 8

What types of changes take place in rRNA processing?
a. Methylation of bases
b. Cleavage of a larger precursor
c. Nucleotides are trimmed from the ends of rRNAs.
d. All of the above

(a) Prokaryotic rRNAs
Precursor rRNA transcript (30S)

Methylation

Methyl groups

Intermediates
16S   tRNA   23S   5S

Mature RNAs
16S rRNA   tRNA   23S rRNA   5S rRNA

(b) Eukaryotic rRNAs
Precursor rRNA transcript (45S)

Methylation

18S rRNA   5.8S rRNA   28S rRNA

1 Methyl groups are added to specific bases and to the 2'-carbon atom of some ribose sugars.

2 The RNA is cleaved into several intermediates…

3 …and trimmed.

4 Mature rRNA molecules are the result.

14.21 Ribosomal RNA is processed after transcription. Prokaryotic rRNA (a) and eukaryotic rRNA (b) are produced from precursor RNA transcripts that are methylated, cleaved, and processed to produce mature rRNAs. Eukaryotic 5S rRNA is transcribed separately from a different gene.

14.5 Small RNA Molecules Participate in a Variety of Functions

Much evidence suggests that the first genetic material was RNA and early life was dominated by RNA molecules (see Chapter 13). This time period, when RNA dominated life's essential processes, has been termed the "early RNA world." The common perception is that this RNA world died out billions of years ago, when many of RNA's functions were replaced by more-stable DNA molecules and more-efficient protein catalysts. However, within the past 10 years, numerous small RNA molecules (most of them 20–30 nucleotides long) have been discovered that greatly influence many basic biological processes, including the formation of chromatin structure, transcription, and translation. These small RNA molecules play important roles in gene expression, development, cancer, and defense against foreign DNA. They are also being harnessed by researchers to study gene function and treat genetic diseases. The discovery of small RNA molecules has greatly influenced our understanding of how genes are regulated and the importance of DNA sequences that do not encode proteins. These new findings demonstrate that we still live very much in an RNA world.

In this section, we will examine RNA interference (which led to the discovery of small RNAs), different types of small RNAs, and how miRNAs are processed. In Chapter 17, we will look further at the role of small RNAs in controlling gene expression; in Chapter 19, we will see how small RNAs are being used as important tools in biotechnology.

RNA Interference

In 1998, Andrew Fire, Craig Mello, and their colleagues observed a strange phenomenon. They were inhibiting the expression of genes in the nematode *Caenorhabditis elegans* by inserting single-stranded RNA molecules that were complementary to a gene's DNA sequence. Called antisense RNA, such molecules are known to inhibit gene expression by binding to the mRNA sequences and inhibiting translation. Fire, Mello, and colleagues found that even more potent gene silencing was triggered when double-stranded RNA was injected into the animals. This finding was puzzling, because no mechanism by which double-stranded RNA could inhibit translation was known. Several other, previously described types of gene silencing also were found to be triggered by double-stranded RNA. These initial studies led to the discovery of small RNA molecules that are important in gene silencing.

Subsequent research revealed an astonishing array of small RNA molecules with important cellular functions in eukaryotes, which now include at least three major classes: small interfering RNAs (siRNAs), microRNAs (miRNAs), and Piwi-interacting RNAs (piRNAs), depending on their origin and mode of function. These small RNAs are found in many eukaryotes and are responsible for a variety of different functions, including the regulation of gene expression, defense against viruses, suppression of transposons, and

modification of chromatin structure. An analogous group of small RNAs with silencing functions—called CRISPR RNAs (crRNAs)—have been detected in prokaryotes. For their discovery of RNA interference, Fire and Mello were awarded the Nobel Prize in physiology or medicine in 2006.

RNA interference (RNAi) is a powerful and precise mechanism used by eukaryotic cells to limit the invasion of foreign genes (from viruses and transposons) and to censor the expression of their own genes. RNA interference is triggered by double-stranded RNA molecules, which may arise in several ways (**Figure 14.22**): by the transcription of

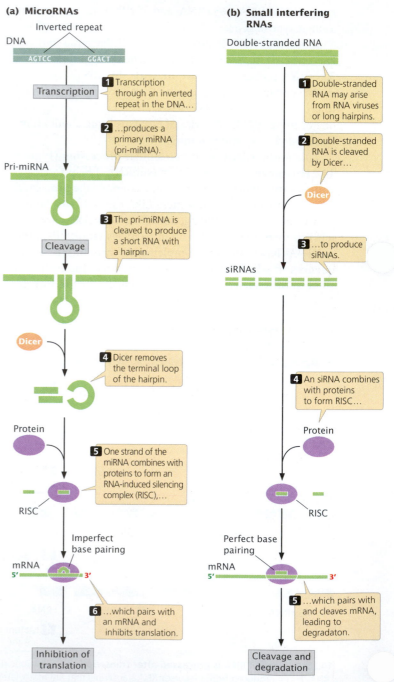

14.22 Small interfering RNAs and microRNAs are produced from double-stranded RNAs.

inverted repeats into an RNA molecule that then base pairs with itself to form double-stranded RNA; by the simultaneous transcription of two different RNA molecules that are complementary to one another and that pair, forming double-stranded RNA; or by infection by viruses that make double-stranded RNA. These double-stranded RNA molecules are chopped up by an enzyme appropriately called Dicer, resulting in tiny RNA molecules that are unwound to produce siRNAs and miRNAs (see Figure 14.22).

Some geneticists speculate that RNA interference evolved as a defense mechanism against RNA viruses and transposable elements that move through RNA intermediates (see Chapter 18); indeed, some have called RNAi the immune system of the genome. However, RNA interference is also responsible for regulating a number of key genetic and developmental processes, including changes in chromatin structure, translation, cell fate and proliferation, and cell death. Geneticists also use the RNAi machinery as an effective tool for blocking the expression of specific genes (see Chapter 19).

Small Interfering and Micro RNAs

Two abundant classes of RNA molecules that function in RNA interference in eukaryotes are small interfering RNAs and microRNAs. Although these two types of RNA differ in how they originate (**Table 14.5**; see Figure 14.22), they have a number of features in common and their functions overlap considerably. Both are about 22 nucleotides long. Small interfering RNAs arise from the cleavage of mRNAs, RNA transposons, and RNA viruses. Some miRNAs are cleaved from RNA molecules transcribed from sequences that encode miRNA only, but others are encoded in the introns and exons of mRNAs. Each miRNA is cleaved from a single-stranded RNA precursor that forms small hairpins, whereas multiple siRNAs are produced from the cleavage of an RNA duplex consisting of two different RNA molecules.

Usually, siRNAs have exact complementarity with their target mRNA or DNA sequences, whereas miRNAs often have limited complementarity with their target mRNAs. Small interfering RNAs suppress gene expression by degrading mRNA or inhibiting transcription, while miRNAs often suppress gene expression by inhibiting translation. Finally, miRNAs usually silence genes that are distinct from those from which the miRNAs were transcribed, whereas siRNAs typically silence the genes from which the siRNAs were transcribed. Note, however, that these differences between siRNAs and miRNAs are not hard and fast, and scientists are increasingly finding small RNAs that exhibit characteristics of both. For example, some miRNAs (such as those in plants) have exact complementarity with mRNA sequences and cleave these sequences, characteristics that are usually associated with siRNAs.

Both siRNA and miRNA molecules combine with proteins to form an **RNA-induced silencing complex** (RISC; see Figure 14.22). Key to the functioning of RISCs is a protein called Argonaute. The RISC pairs with an mRNA molecule that possesses a sequence complementary to its siRNA or miRNA component and either cleaves the mRNA, leading to degradation of the mRNA, or represses translation of the mRNA. Some siRNAs also serve as guides for the methylation of complementary sequences in DNA and others alter chromatin structure, both of which affect transcription. To see how small interfering RNAs and microRNAs affect gene expression, see **Animation 14.3**.

MicroRNAs have been found in all eukaryotic organisms examined to date, as well as viruses: they control the expression of genes taking part in many biological processes, including growth, development, and metabolism. Humans have more than 450 distinct miRNAs; scientists estimate that more than one-third of all human genes are regulated by miRNAs. Most miRNA genes are found in regions of noncoding DNA or within the introns of other genes.

PROCESSING AND FUNCTION OF miRNAs AND siRNAs The genes that encode miRNAs are transcribed into longer precursors, called primary miRNA (pri-miRNA), that range from several hundred to several thousand nucleotides in length (Figure 14.22a). The pri-miRNA is then cleaved into one or more smaller RNA molecules with a hairpin. Dicer binds to this hairpin structure and removes the terminal loop. One of the miRNA strands is incorporated into the RISC; the other strand is released and degraded.

Table 14.5	Differences between siRNAs and miRNAs	
Feature	**siRNA**	**miRNA**
Origin	mRNA, transposon, or virus	RNA transcribed from distinct gene
Cleavage of	RNA duplex or single-stranded RNA that forms long hairpins	Single-stranded RNA that forms short hairpins of double-stranded RNA
Size	21–25 nucleotides	21–25 nucleotides
Action	Degradation of mRNA, inhibition of transcription, chromatin modification	Degradation of mRNA, inhibition of translation, chromatin modification
Target	Genes from which they were transcribed	Genes other than those from which they were transcribed

The RISC attaches to a complementary sequence on the mRNA, usually in the 3′ untranslated region of the mRNA. The region of close complementarity, called the seed region, is quite short, usually only about seven nucleotides long. Because the seed sequence is so short, each miRNA can potentially pair with sequences on hundreds of different mRNAs. Furthermore, a single mRNA molecule may possess multiple miRNA-binding sites. The inhibition of translation may require binding by several RISC complexes to the same mRNA molecule. **Animation 14.2** illustrates the process by which microRNAs are produced.

Small interfering RNAs are processed in a similar way (Figure 14.22b). Double-stranded RNA from viruses or long hairpins is cleaved by Dicer to produce siRNAs, which combine with proteins to form a RISC. The RISC then pairs with sequences on target mRNA and cleaves the mRNA, after which the mRNA is degraded.

> ### CONCEPTS
>
> Small interfering RNAs and microRNAs are tiny RNAs produced when larger, double-stranded RNA molecules are cleaved by the enzyme Dicer. Small interfering RNAs and microRNAs participate in a variety of processes, including mRNA degradation, the inhibition of translation, the methylation of DNA, and chromatin remodeling.
>
> ### ✓ CONCEPT CHECK 9
>
> How do siRNAs and miRNAs target specific mRNAs for degradation or for the repression of translation?

Piwi-Interacting RNAs

Piwi-interacting RNAs (piRNAs) were discovered in 2001. They are somewhat longer than siRNAs and miRNAs, consisting of 24 to 30 nucleotides, and are derived from long, single-stranded RNA transcripts, in contrast with siRNAs and miRNAs, which are processed from double-stranded RNA. Also unlike siRNAs and miRNAs, Dicer is not involved in the production of piRNAs.

Piwi-interacting RNAs combine with Piwi proteins, which are related to Argonaute, and suppress the expression and movement of transposons in the germ cells of animals. Although the mechanism of transposon silencing by piRNAs is not fully understood, we know that it includes the degradation of mRNA transcribed from transposons, changes in chromatin structure that inhibit the transcription of transposons, and inhibition of the translation of proteins encoded by transposons. What, if any, function piRNAs have outside of germ cells is not known.

CRISPR RNA

After the discovery of small RNAs in eukaryotes, similar small RNAs called CRISPR RNAs (crRNAs) were discovered in prokaryotes: crRNAs are encoded by DNA sequences found in bacterial and archaeal genomes termed Clustered Regularly Interspaced Short Palindromic Repeats (CRISPR). Palindromic sequences are sequences that read the same forwards and backwards on two complementary DNA strands. CRISPR consists of a series of such palindromic sequences, separated by unique sequences that are homologous to DNA from bacteriophage or plasmid genomes. For example, in the bacterium *Pseudomonas auruginosa*, CRISPR palindromic repeats are 28 bp in length, separated by 32 bp spacers.

CRISPR RNAs play a role in defense against the invasion of specific foreign DNAs, such as DNA originating from bacteriophage and plasmids (see Chapter 9). Because they target specific DNA molecules, the CRISPR system has been compared to the immune system of vertebrates.

Through a mechanism that is still incompletely understood, when a bacteriophage or plasmid invades a prokaryotic cell, small portions of the invader genome are inserted as spacers between the palindromic repeats in CRISPR (**Figure 14.23a**). The spacer DNA then serves as a memory of this invader's DNA. The CRISPR region is transcribed

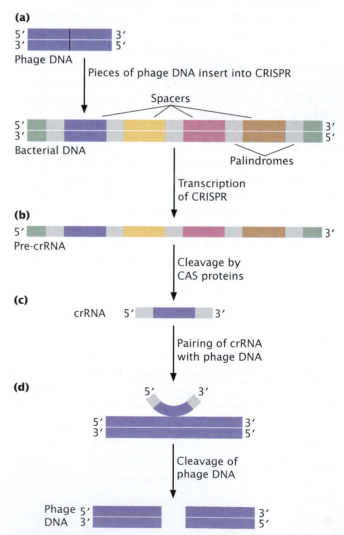

14.23 CRISPR RNAs function in defense against invasion of foreign DNAs, such as DNA from bacteriophage and plasmids.

into a single long CRISPR precursor RNA (**Figure 14.23b**), which is then cleaved by CRISPR-associated proteins (Cas proteins) into crRNAs, each of which consists of a spacer sequence (homologous to the invader's DNA) flanked by a part of the palindromic sequence (**Figure 14.23c**). When additional foreign DNA from the same source enters the cell, crRNAs pair with it and help bring about cleavage of the invader DNA (**Figure 14.23d**). In this way, crRNAs serve as an adaptive, RNA defense system against foreign invaders.

> **CONCEPTS**
>
> Piwi-interacting RNAs are found in the germ cells of animals and inhibit transposons. CRISPR RNAs are found in prokaryotes, where they function in defense against foreign DNA.

14.6 Long Noncoding RNAs Regulate Gene Expression

For many years, our knowledge of RNA was limited to those molecules that play a central role in the synthesis of proteins: mRNAs, tRNAs, and rRNAs. Later, small nuclear RNAs that participate in the post-transcriptional processing of RNA (snRNAs and snoRNAs) were added to the list. Starting in the late 1990s, geneticists began to recognize that numerous small RNAs (siRNAs, miRNAs, piRNAs, and crRNAs) were also abundant and fundamentally important to cell function. More recently, it has become apparent that most of eukaryotic genomes are transcribed—although only about 1% of the human genome directly codes for proteins, over 80% is transcribed, producing many long RNA molecules that do not code for proteins. Called long noncoding RNAs (lncRNAs), these RNAs are typically over 100 nucleotides in length and lack an open reading frame (a sequence with a start and a stop codon, which is translated by ribosomes). Thousands of lncRNAs have been discovered in the last five years. The DNA sequences that encode them, along with other DNA of unknown function, have been called "the dark matter of the genome."

Although the function of many lncRNAs is still unclear, there is increasing evidence that at least some play a role in controlling gene expression. Some lncRNAs interact with proteins that regulate transcription. For example, a lncRNA called lincRNA-p21 interacts with a protein called p53, a transcription factor that activates numerous genes, including genes involved in control of the cell cycle and cancer. By repressing p53, lincRNA-21 affects the transcription of hundreds of genes. Other lncRNAs modify chromatin structure, which also regulates transcription (see Chapter 17). Some lncRNA have sites that are recognized by miRNAs, and the lncRNAs serve as decoys for miRNA attachment. Thus, the lncRNAs and mRNAs compete for a limited number of miRNAs and regulate one another's translation and degradation. Still other lncRNAs are complementary to mRNA sequences and function by base pairing with the mRNA and preventing translation or splicing.

One of the best-studied lncRNAs is Xist RNA, which plays a central role in dosage compensation in mammalian cells (see Chapter 4). To balance expression of X-linked genes in males (with one X chromosome) and females (with two X chromosomes), one of the X chromosomes in each mammalian female cell is inactivated. Which X chromosome is inactivated is random and set early in development; once inactivated, this chromosome remains inactive through multiple rounds of cell division. Xist RNA is transcribed only from the X chromosome destined to become inactive; Xist RNA coats it and recruits proteins that methylate histones in the chromatin. Methylation of the chromatin then leads to the inhibition of transcription of genes on the inactive X chromosome. At least two additional lncRNAs act to regulate the expression of Xist RNA.

Evidence suggests that other lncRNAs also bring about genomic imprinting (see Chapters 4 and 21). Imprinting occurs when a gene is expressed differently depending on whether it is inherited from a male or female parent. Many clusters of imprinted genes contain sequences that encode lncRNAs and evidence suggests that some imprinted genes are controlled by lncRNAs.

> **CONCEPTS**
>
> Long noncoding RNAs are long RNA molecules that do not encode proteins. Evidence increasingly suggests that many of these molecules function in the control of gene expression.

CONCEPTS SUMMARY

- A gene is often defined as a sequence of DNA nucleotides that is transcribed into a single RNA molecule.

- Introns—noncoding sequences that interrupt the coding sequences (exons) of genes—are common in eukaryotic cells but rare in bacterial cells.

- An mRNA molecule has three primary parts: a 5′ untranslated region, a protein-coding sequence, and a 3′ untranslated region.

- Bacterial mRNA is translated immediately after transcription and undergoes little processing. The pre-mRNA of a eukaryotic protein-encoding gene is extensively processed: a modified nucleotide and methyl group, collectively termed the cap, are added to the 5′ end of pre-mRNA; the 3′ end is cleaved and a poly(A) tail is added; and introns are removed. Introns are removed within a structure called the spliceosome, which is composed of several small nuclear RNAs and proteins.

■ Some introns found in rRNA genes and mitochondrial genes are self-splicing.

■ Some pre-mRNAs undergo alternative splicing, in which different combinations of exons are spliced together or different 3′ cleavage sites are used.

■ Messenger RNAs may be altered by the addition, deletion, or modification of nucleotides in the coding sequence, a process called RNA editing.

■ Transfer RNAs, which attach to amino acids, are short molecules that assume a common secondary structure and contain modified bases.

■ Ribosomes, the sites of protein synthesis, are composed of several ribosomal RNA molecules and numerous proteins.

■ Small interfering RNAs, microRNAs, Piwi-interacting RNAs, and CRISPR RNAs play important roles in gene silencing and in a number of other biological processes.

■ Long noncoding RNAs are RNA molecules that do not encode proteins. Evidence increasingly suggests that many of these molecules function in the control of gene expression.

IMPORTANT TERMS

colinearity (p. 384)
exon (p. 385)
intron (p. 385)
group I intron (p. 386)
group II intron (p. 386)
nuclear pre-mRNA intron (p. 386)
transfer RNA intron (p. 386)
codon (p. 387)
5′ untranslated region (5′ UTR) (p. 387)

Shine–Dalgarno sequence (p. 387)
protein-coding region (p. 388)
3′ untranslated region (3′ UTR) (p. 388)
5′ cap (p. 388)
poly(A) tail (p. 389)
RNA splicing (p. 390)
5′ splice site (p. 390)
3′ splice site (p. 390)
branch point (p. 390)
spliceosome (p. 391)

lariat (p. 391)
trans-splicing (p. 392)
alternative processing pathway (p. 393)
alternative splicing (p. 393)
multiple 3′ cleavage sites (p. 393)
RNA editing (p. 395)
guide RNA (p. 395)
modified base (p. 398)
tRNA-modifying enzyme (p. 398)

cloverleaf structure (p. 398)
anticodon (p. 399)
large ribosomal subunit (p. 400)
small ribosomal subunit (p. 400)
RNA interference (RNAi) (p. 402)
RNA-induced silencing complex (RISC) (p. 403)

ANSWERS TO CONCEPT CHECKS

1. When DNA was hybridized to the mRNA transcribed from it, regions of DNA that did not correspond to RNA looped out.

2. Group I introns, group II introns, nuclear pre-mRNA introns, and transfer RNA introns.

3. A protein that adds the 5′ cap is associated with RNA polymerase II, which transcribes pre-mRNAs but is absent from RNA polymerase I and III, which transcribe rRNA and tRNAs.

4. b

5. c

6. Guide RNA

7. b

8. d

9. An siRNA or miRNA combines with proteins to form RISC, which then pairs with mRNA through complementary pairing between bases on the siRNA or miRNA and bases on the mRNA.

WORKED PROBLEMS

Problem 1

DNA from a eukaryotic gene was isolated, denatured, and hybridized to the mRNA transcribed from the gene; the hybridized structure was then observed with an electron microscope. The adjoining structure was observed.

a. How many introns and exons are there in this gene? Explain your answer.

b. Identify the exons and introns in this hybridized structure.

Solution Strategy

What information is required in your answer to the problem?

- The number of introns and exons and how you arrived at your answer.
- The location of the introns and exons labeled on the figure.

What information is provided to solve the problem?

- The DNA and mRNA are from a eukaryote.
- The DNA was denatured and hybridized to the mRNA.
- A picture of the hybridized structure.

For help with this problem, review:
Introns in Section 14.1 and Figure 14.2.

Solution Steps

a. Each of the loops represents a region in which sequences in the DNA do not have corresponding sequences in the RNA; these regions are introns. There are five loops in the hybridized structure; so there must be five introns in the DNA and six exons.

Recall: Introns are noncoding sequences found within eukaryotic genes.

b.

Hint: The number of introns will be one less than the number of exons.

Problem 2

Draw a typical bacterial mRNA and the gene from which it was transcribed. Identify the 5′ and 3′ ends of the RNA and DNA molecules, as well as the following regions or sequences:

a. Promoter
b. 5′ untranslated region
c. 3′ untranslated region
d. Protein-coding sequence

e. Transcription start site
f. Terminator
g. Shine–Dalgarno sequence
h. Start and stop codons

Hint: Review the structure of a transcription unit in Figure 13.6 and the structure of mRNA in Figure 14.5.

Solution Strategy

What information is required in your answer to the problem?

A drawing of the mRNA and gene from which it is transcribed. The 5′ and 3′ ends of the mRNA and DNA molecules. Locations of the listed structures on the drawing.

What information is provided to solve the problem?

- The gene is from a bacterium.
- Different parts of the DNA and RNA that are to be labeled.

For help with this problem, review:
The Template in Section 13.2 and The Structure of Messenger RNA in Section 14.2.

Solution Steps

COMPREHENSION QUESTIONS

Section 14.1

1. What is the concept of colinearity? In what way is this concept fulfilled in bacterial and eukaryotic cells?

2. What are some characteristics of introns?

3. What are the four basic types of introns? In which organisms are they found?

Section 14.2

4. What are the three principal elements in mRNA sequences in bacterial cells?

5. What is the function of the Shine–Dalgarno consensus sequence?

6. (a) What is the 5′ cap? (b) How is the 5′ cap added to eukaryotic pre-mRNA? (c) What is the function of the 5′ cap?

7. How is the poly(A) tail added to pre-mRNA? What is the purpose of the poly(A) tail?

8. What makes up the spliceosome? What is the function of the spliceosome?

9. Explain the process of pre-mRNA splicing in nuclear genes.

10. Describe two types of alternative processing pathways. How do they lead to the production of multiple proteins from a single gene?

11. What is RNA editing? Explain the role of guide RNAs in RNA editing.

12. Summarize the different types of processing that can take place in pre-mRNA.

Section 14.3

13. What are some of the modifications in tRNA that take place through processing?

Section 14.4

14. Describe the basic structure of ribosomes in bacterial and eukaryotic cells.

15. Explain how rRNA is processed.

Section 14.5

16. What is the origin of small interfering RNAs, microRNAs, and Piwi-interacting RNAs? What do these RNA molecules do in the cell?

17. What are some similarities and differences between siRNAs and miRNAs?

18. How are miRNAs processed?

> ▶ For more questions that test your comprehension of the key chapter concepts, go to **LEARNING**Curve for this chapter.

APPLICATION QUESTIONS AND PROBLEMS

Section 14.1

*19. Duchenne muscular dystrophy is caused by a mutation in a gene that comprises 2.5 million nucleotides and specifies a protein called dystrophin. However, less than 1% of the gene actually encodes the amino acids in the dystrophin protein. On the basis of what you now know about gene structure and RNA processing in eukaryotic cells, provide a possible explanation for the large size of the dystrophin gene.

20. What would happen in the experiment illustrated in **Figure 14.2** if the DNA and RNA that are mixed together came from very different organisms, for example a worm and a pig?

21. For the ovalbumin gene shown in **Figure 14.3**, where would the 5′ untranslated region and 3′ untranslated regions be located in the DNA and in the RNA?

Section 14.2

22. How do the mRNAs of bacterial cells and the pre-mRNAs of eukaryotic cells differ? How do the mature mRNAs of bacterial and eukaryotic cells differ?

23. Are the 5′ untranslated regions (5′ UTR) of eukaryotic mRNAs encoded by sequences in the promoter, exon, or intron of the gene? Explain your answer.

*24. Draw a typical eukaryotic gene and the pre-mRNA and mRNA derived from it. Assume that the gene contains three exons. Identify the following items and, for each item, give a brief description of its function:

a. 5′ untranslated region   f. Introns

b. Promoter   g. Exons

c. AAUAAA consensus sequence   h. Poly(A) tail

d. Transcription start site   i. 5′ cap

e. 3′ untranslated region

25. How would the deletion of the Shine–Dalgarno sequence affect a bacterial mRNA?

26. What would be the most likely effect of moving the AAUAAA consensus sequence shown in **Figure 14.7** ten nucleotides upstream?

27. How would the deletion of the following sequences or features most likely affect a eukaryotic pre-mRNA?

a. AAUAAA consensus sequence

b. 5′ cap

c. Poly(A) tail

28. Suppose that a mutation occurs in the middle of a large intron of a gene encoding a protein. What will the most likely effect of the mutation be on the amino acid sequence of that protein? Explain your answer.

*29. A geneticist induces a mutation in a line of cells growing in the laboratory. The mutation occurs in one of the genes that encodes proteins that participate in the cleavage and polyadenylation of eukaryotic mRNA. What will the immediate effect of this mutation be on RNA molecules in the cultured cells?

30. A geneticist mutates the gene for proteins that bind to the poly(A) tail in a line of cells growing in the laboratory. What will the immediate effect of this mutation be in the cultured cells?

31. A geneticist isolates a gene that contains eight exons. He then isolates the mature mRNA produced by this gene. After making the DNA single stranded, he mixes the single-stranded DNA and RNA. Some of the single-stranded DNA hybridizes (pairs) with the complementary mRNA. Draw a picture of what the DNA–RNA hybrids will look like under the electron microscope.

32. A geneticist discovers that two different proteins are encoded by the same gene. One protein has 56 amino acids, and the other has 82 amino acids. Provide a possible explanation for how the same gene can encode both of these proteins.

33. What conclusion can you make about the relative sizes of the two proteins produced by alternative splicing in **Figure 14.12**?

*34. Explain how each of the following processes complicates the concept of colinearity.

a. Trans-splicing

b. Alternative splicing

c. RNA editing

Section 14.5

35. RNA interference may be triggered when inverted repeats are transcribed into an RNA molecule that then folds to form double-stranded RNA. Write out a sequence of inverted repeats within an RNA molecule. Using a diagram, show how the RNA with the inverted repeats can fold to form double-stranded RNA.

36. In the early 1990s, Carolyn Napoli and her colleagues were working on petunias, attempting to genetically engineer a variety with dark purple petals by introducing numerous copies of a gene that codes for purple petals (C. Napoli, C. Lemieux, and R. Jorgensen. 1990. *Plant Cell* 2:279–289). Their thinking was that extra copies of the gene would cause more purple pigment to be produced and would result in a petunia with an even darker hue of purple. However, much to their surprise, many of the plants carrying extra copies of the purple gene were completely white or had only patches of color. Molecular analysis revealed that the level of the mRNA produced by the purple gene was reduced 50-fold in the engineered plants compared with levels of mRNA in wild-type plants. Somehow, the introduction of extra copies of the purple gene silenced both the introduced copies and the plant's own purple genes. Provide a possible explanation for how the introduction of numerous copies of the purple gene silenced all copies of the purple gene.

White petunia. [roger ashford/Alamy.]

CHALLENGE QUESTIONS

Section 14.2

37. Alternative splicing takes place in more than 90% of the human genes that encode proteins. Researchers have found that how a pre-mRNA is spliced is affected by the pre-mRNA's promoter sequence (D. Auboeuf et al. 2002. *Science* 298:416–419). In addition, factors that affect the rate of elongation of the RNA polymerase during transcription affect the type of splicing that takes place. These findings suggest that the process of transcription affects splicing. Propose one or more mechanisms that would explain how transcription might affect alternative splicing.

38. Duchenne muscular dystrophy (DMD) is an X-linked recessive genetic disease caused by mutations in the gene that encodes dystrophin, a large protein that plays an important role in the development of normal muscle fibers. The dystrophin gene is immense, spanning 2.5 million base pairs, and includes 79 exons and 78 introns. Many of the mutations that cause DMD produce premature stop codons, which bring protein synthesis to a halt, resulting in a greatly shortened and nonfunctional form of dystrophin. Some geneticists have proposed treating DMD patients by introducing small RNA molecules that cause the spliceosome to skip the exon containing the stop codon.

The introduction of the small RNAs will produce a protein that is somewhat shortened (because an exon is skipped and some amino acids are missing) but may still result in a protein that has some function (A. Goyenvalle et al. 2004. *Science* 306:1796–1799). The small RNAs used for exon skipping are complementary to bases in the pre-mRNA. If you were designing small RNAs to bring about exon skipping for the treatment of DMD, what sequences should the small RNAs contain?

39. In eukaryotic cells, a poly(A) tail is normally added to pre-mRNA molecules but not to rRNA or tRNA. With the use of recombinant DNA techniques, a protein-encoding gene (which is normally transcribed by RNA polymerase II) can be connected to a promoter for RNA polymerase I. This hybrid gene is subsequently transcribed by RNA polymerase I and the appropriate pre-mRNA is produced, but this pre-mRNA is not cleaved at the 3′ end and a poly(A) tail is not added.

    Propose a mechanism to explain how the type of promoter found at the 5′ end of a gene can affect whether a poly(A) tail is added to the 3′ end.

40. SR proteins are essential to proper spliceosome assembly and are known to take part in the regulation of alternative splicing. Surprisingly, the role of SR proteins in splice-site selection and alternative splicing is affected by the promoter used for the transcription of the pre-mRNA. For example, through genetic engineering, RNA polymerase II promoters that have somewhat different sequences can be created. When pre-mRNAs with exactly the same sequences are transcribed from two different RNA polymerase II promoters that differ slightly in sequence, which promoter is used can affect how the pre-mRNA is spliced.

    Propose a mechanism for how the DNA sequence of an RNA polymerase II promoter could affect alternative splicing of the pre-mRNA.

▶ Go to your ▶LaunchPad to find additional learning resources and the Suggested Readings for this chapter.

15

The Genetic Code and Translation

The Hutterites are a religious branch of Anabaptists who live on communal farms in the prairie states and provinces of North America. A small number of founders, coupled with a tendency to intermarry, have caused a high frequency of the mutation for Bowen–Conradi syndrome among Hutterites. Bowen–Conradi syndrome results from defective ribosome biosynthesis, affecting the process of translation. Shown here are healthy Hutterite children. [Kevin Fleming/Corbis.]

Hutterites, Ribosomes, and Bowen–Conradi Syndrome

The essential nature of the ribosome—the cell's protein factory—is poignantly illustrated by children with Bowen–Conradi syndrome. Born with a prominent nose, small head, and an unusual curvature of the small finger, these children fail to thrive and gain weight, usually dying within the first year of life.

Almost all children with Bowen–Conradi syndrome are Hutterites, a branch of Anabaptists who originated in the 1500s in the Tyrolean Alps of Austria. After years of persecution, the Hutterites immigrated to South Dakota in the 1870s and subsequently spread to neighboring prairie states and the Canadian provinces. Today, the Hutterites in North America number about 40,000 persons. They live on communal farms, are strict pacifists, and rarely marry outside of the Hutterite community.

Bowen–Conradi syndrome is inherited as an autosomal recessive disorder, and the association of Bowen–Conradi syndrome with the Hutterite community is a function of the group's unique genetic history. The gene pool of present-day Hutterites in North America can be traced to fewer than 100 persons who immigrated to South Dakota in the late 1800s. The increased incidence of Bowen–Conradi syndrome in Hutterites today is due to the founder effect—the presence of an allele that causes Bowen-Conradi in one or more of the original founders—and its spread as Hutterites intermarried within their community. Because of the founder effect and inbreeding (see Chapter 25), many Hutterites today are as closely related as first cousins. This close genetic relationship among the Hutterites increases the probability that a child will inherit two copies of the recessive allele and have Bowen–Conradi syndrome; indeed, almost 1 in 10 Hutterites is a heterozygous carrier of the allele that causes the disease.

Although Bowen–Conradi syndrome was first described in 1976, the genetic and biochemical basis of the disease long remained a mystery. After a 7-year quest to find the causative gene, researchers at the University of Manitoba determined in 2009 that Bowen–Conradi syndrome results from the mutation of a single base pair in the *EMG1* gene, located on chromosome 12.

The discovery of the gene for Bowen–Conradi syndrome gave immediate insight into the biochemical nature of the disease. Although little is known about the function of the *EMG1* gene in humans, earlier studies in yeast revealed that it encodes a protein that aids in assembling the ribosome. As discussed in Chapter 14, the ribosome is composed of small and large subunits. The small subunit in humans consists of a single piece of ribosomal RNA, known as 18S rRNA, and a large number of proteins. The protein encoded by the *EMG1* gene plays an essential role in processing 18S rRNA and helps assemble it into the small subunit of the ribosome. Because of a mutation in the *EMG1* gene, babies with Bowen–Conradi syndrome produce ribosomes that function poorly and the process by which all proteins are synthesized is affected.

Bowen–Conradi syndrome illustrates the extreme importance of translation, the process of protein synthesis, which is the focus of this chapter. We begin by examining the molecular relation between genotype and phenotype. Next, we study the genetic code—the instructions that specify the amino acid sequence of a protein—and then examine the mechanism of protein synthesis. Our primary focus will be on protein synthesis in bacterial cells, but we will examine some of the differences in eukaryotic cells. At the end of the chapter, we look at some additional aspects of protein synthesis.

15.1 Many Genes Encode Proteins

The first person to suggest the existence of a relation between genotype and proteins was English physician Archibald Garrod. In 1908, Garrod correctly proposed that genes encode enzymes, but, unfortunately, his theory made little impression on his contemporaries. Not until the 1940s, when George Beadle and Edward Tatum examined the genetic basis of biochemical pathways in *Neurospora*, did the relation between genes and proteins become widely accepted. Beadle and Tatum's work helped define the relation between genotype and phenotype by leading to the one gene, one enzyme hypothesis, the idea that each gene encodes a separate enzyme.

The One Gene, One Enzyme Hypothesis

Beadle and Tatum used the bread mold *Neurospora* to study the biochemical results of mutations. *Neurospora* is easy to cultivate in the laboratory, and the main vegetative part of the fungus is haploid, which allows the effects of recessive mutations to be easily observed (**Figure 15.1**).

Wild-type *Neurospora* grows on minimal medium, which contains only inorganic salts, nitrogen, a carbon source such as sucrose, and the vitamin biotin. The fungus can synthesize all the biological molecules that it needs from these basic compounds. However, mutations may arise that disrupt fungal growth by destroying the fungus's ability to synthesize one or more essential biological molecules. These nutritionally deficient mutants, termed auxotrophs (see Chapter 9), will not grow on minimal medium, but they can grow on medium that contains the substance that they are no longer able to synthesize.

Beadle and Tatum first irradiated spores of *Neurospora* to induce mutations (**Figure 15.2**). Then they placed individual spores into different culture tubes containing complete medium (medium having all the biological substances needed for growth). These spores grew into fungi and produced spores by mitosis. Next, they transferred spores from each culture to tubes containing minimal medium. Fungi containing auxotrophic mutations grew on complete medium but would not grow on minimal medium, which allowed Beadle and Tatum to identify cultures that contained mutations.

After they had determined that a particular culture had an auxotrophic mutation, Beadle and Tatum set out to determine the specific *effect* of the mutation. They transferred spores of each mutant strain from complete medium to a series of tubes (see Figure 15.2), each of which possessed minimal medium plus one of a variety of essential biological molecules, such as an amino acid. If the spores in a tube grew, Beadle and Tatum were able to identify the added substance as the biological molecule whose synthesis had been affected by the mutation. For example, an auxotrophic mutant that would grow only on minimal medium to which arginine had been added must have possessed a mutation that disrupts the synthesis of arginine.

Adrian Srb and Norman H. Horowitz patiently applied this procedure to genetically dissect the multistep

15.1 Beadle and Tatum used the fungus *Neurospora*, which has a complex life cycle, to work out the relation of genes to proteins. [Namboori B. Raju, Stanford University.]

15.2 Beadle and Tatum developed a method for isolating auxotrophic mutants in *Neurospora*.

Experiment

Question: How can auxotrophic mutations be isolated and identified?

Methods

1 A culture of *Neurospora* was irradiated to induce mutations.

X-rays

Neurospora

Medium

2 Individual spores were transferred to tubes containing complete medium.

Complete medium

3 Spores from each culture were transferred to tubes containing minimal medium.

4 Fungi with auxotrophic mutations would not grow on minimal medium,…

Minimal medium

5 …whereas fungi that grew on minimal medium did not have auxotrophic mutations.

6 Spores from mutant cultures were transferred to tubes, each with minimal medium plus one amino acid.

Results

7 The mutant *Neurospora* grew only when supplemented with arginine, indicating that the mutant was defective in the synthesis of arginine.

Glycine Alanine Leucine Isoleucine Valine Arginine Phenylalanine Tyrosine Methionine Proline

Tryptophan Lysine Histidine Glutamic acid Aspartic acid Glutamine Asparagine Serine Threonine Cysteine

Conclusion: The mutation could be identified by the spores' ability to grow in minimal medium supplemented by the substance that the spores could not synthesize.

biochemical pathway of arginine synthesis (**Figure 15.3**). They first isolated a series of auxotrophic mutants whose growth required arginine. Then they tested these mutants for their ability to grow on minimal medium supplemented with three compounds: ornithine, citrulline, and arginine. From the results, they were able to place the mutants into three groups on the basis of which of the substances allowed growth (**Table 15.1**).

Based on these results, Srb and Horowitz proposed that the biochemical pathway leading to the amino acid arginine has at least three steps:

$$\text{precursor} \xrightarrow{\text{Step}\ 1} \text{ornithine} \xrightarrow{\text{Step}\ 2} \text{citrulline} \xrightarrow{\text{Step}\ 3} \text{arginine}$$

They concluded that the mutations in group I affect step 1 of this pathway, mutations in group II affect step 2, and mutations in group III affect step 3. But how did they know that the order of the compounds in the biochemical pathway was correct?

Notice that if step 1 is blocked by a mutation, then the addition of either ornithine or citrulline allows growth, because these compounds can still be converted into arginine (see Figure 15.3). Similarly, if step 2 is blocked, the addition of citrulline allows growth, but the addition of ornithine has no effect. If step 3 is blocked, the spores will grow only if arginine is added to the medium. The underlying principle is that an auxotrophic mutant cannot synthesize any compound that comes after the step blocked by a mutation.

Using this reasoning with the information in Table 15.1, we can see that the addition of arginine to the medium allows all three groups of mutants to grow. Therefore, biochemical steps affected by all the mutants precede the step that results in arginine. The addition of citrulline allows group I and group II mutants to grow but not group III mutants; therefore, group III mutations must affect a biochemical step that takes place after the production of citrulline but before the production of arginine.

Table 15.1	Growth of arginine auxotrophic mutants on minimal medium with various supplements		
Mutant Strain Number	**Ornithine**	**Citrulline**	**Arginine**
Group I	+	+	+
Group II	−	+	+
Group III	−	−	+

Note: A plus sign (+) indicates growth; a minus sign (−) indicates no growth.

15.3 Method used to determine the relation between genes and enzymes in *Neurospora*. This biochemical pathway leads to the synthesis of arginine in *Neurospora*. Steps in the pathway are catalyzed by enzymes affected by mutations.

Group
III
mutations
citrulline ⟶ arginine

The addition of ornithine allows the growth of group I mutants but not group II or group III mutants; thus, mutations in groups II and III affect steps that come after the production of ornithine. We've already established that group II mutations affect a step before the production of citrulline; so group II mutations must block the conversion of ornithine into citrulline.

Group   Group
II   III
mutations   mutations
ornithine ⟶ citrulline ⟶ arginine

Because group I mutations affect some step before the production of ornithine, we can conclude that they must affect the conversion of some precursor into ornithine. We can now outline the biochemical pathway yielding ornithine, citrulline, and arginine.

Group   Group   Group
I   II   III
mutations   mutations   mutations
precursor ⟶ ornithine ⟶ citrulline ⟶ arginine

Importantly, this procedure does not necessarily detect all steps in a pathway; rather, it only detects the steps that produce the compounds tested.

Using mutations and this type of reasoning, Beadle, Tatum, and others were able to identify genes that control several biosynthetic pathways in *Neurospora*. They established that each step in a pathway is controlled by a different enzyme, as shown in Figure 15.3 for the arginine pathway. They also conducted genetic crosses and mapping experiments (see Chapter 7) and were able to demonstrate that mutations affecting any one step in a pathway always occurred at the same chromosomal location. Beadle and Tatum reasoned that mutations affecting a particular biochemical step occurred at a single locus that encoded a particular enzyme. This idea became known as the **one gene, one enzyme hypothesis**: genes function by encoding enzymes, and each gene encodes a separate enzyme. Although the genes Beadle and Tatum examined encoded enzymes, many genes encode proteins that are not enzymes, so more generally their idea was that each gene encodes a protein. When research findings showed that some proteins are composed of more than one polypeptide chain and that different polypeptide chains are encoded by separate genes, this model was modified to become the **one gene, one polypeptide hypothesis.** ▶TRY PROBLEM 16

CONCEPTS

Beadle and Tatum's studies of biochemical pathways in the fungus *Neurospora* helped define the relation between genotype and phenotype by establishing the one gene, one enzyme hypothesis, the idea that each gene encodes a separate enzyme. This was later modified to the one gene, one polypeptide hypothesis.

✔ **CONCEPT CHECK 1**

Auxotrophic mutation 103 grows on minimal medium supplemented with A, B, or C; mutation 106 grows on medium supplemented with A and C but not B; and mutation 102 grows only on medium supplemented with C. What is the order of A, B, and C in a biochemical pathway?

The Structure and Function of Proteins

Proteins are central to all living processes (**Figure 15.4**). Many proteins are enzymes, the biological catalysts that drive the chemical reactions of the cell; others are structural components, providing scaffolding and support for membranes, filaments, bone, and hair. Some proteins help transport substances; others have a regulatory, communication, or defense function.

AMINO ACIDS   All proteins are composed of **amino acids**, linked end to end. Twenty common amino acids are found in proteins; these amino acids are shown in **Figure 15.5** with both their three-letter and one-letter abbreviations (other amino acids sometimes found in proteins are modified forms of the common amino acids). The 20 common amino acids are similar in structure: each consists of a central carbon atom bonded to an amino group, a hydrogen atom, a carboxyl group, and an R (radical) group that differs for each amino acid. The amino acids in proteins are joined together by **peptide bonds** (**Figure 15.6**) to form **polypeptide** chains, and a protein consists of one or more polypeptide chains. Like nucleic acids, polypeptides have polarity, with one end having a free amino group (NH_3^+) and the other end possessing a free carboxyl group (COO^-). Some proteins consist of only a few amino acids, whereas others may have thousands.

PROTEIN STRUCTURE   Like that of nucleic acids, the molecular structure of proteins has several levels of organization. The *primary structure* of a protein is its sequence of amino acids (**Figure 15.7a**). Through interactions between neighboring amino acids, a polypeptide chain folds and twists into a *secondary structure* (**Figure 15.7b**); two common secondary structures found in proteins are the beta (β) pleated sheet and the alpha (α) helix. Secondary structures interact and fold further to form a *tertiary structure* (**Figure 15.7c**), which is the overall, three-dimensional shape of the protein. The secondary and tertiary structures of a protein are largely determined by the primary

(a)

(b)

(c)

15.4 Proteins serve a number of biological functions. (a) The light produced by fireflies is the result of a light-producing reaction between luciferin and ATP catalyzed by the enzyme luciferase. (b) The protein fibroin is the major structural component of spider webs. (c) Castor beans contain a highly toxic protein called ricin. [Part a: Darwin Dale/Science Source. Part b: Rosemary Calvert/Imagestate. Part c: Paroli Galperti/ © Cuboimages/Photoshot.]

3 Hydrogen

1 Amino group

2 Carboxyl group

^+H_3N——C_α——COO^-

4 Radical group (side chain)

15.5 The common amino acids have similar structures. Each amino acid consists of a central carbon atom (C_α) attached to: (1) an amino group (NH_3^+); (2) a carboxyl group (COO^-); (3) a hydrogen atom (H); and (4) a radical group, designated R. In the structures of the 20 common amino acids, the parts in black are common to all amino acids and the parts in red are the R groups.

Nonpolar, aliphatic R groups

Glycine (Gly, G)

Alanine (Ala, A)

Valine (Val, V)

Leucine (Leu, L)

Isoleucine (Ile, I)

Methionine (Met, M)

Polar, uncharged R groups

Serine (Ser, S)

Threonine (Thr, T)

Cysteine (Cys, C)

Proline (Pro, P)

Asparagine (Asn, N)

Glutamine (Gln, Q)

Aromatic R groups

Phenylalanine (Phe, F)

Tyrosine (Try, Y)

Tryptophan (Trp, W)

Positively charged R groups

Lysine (Lys, K)

Arginine (Arg, R)

Histidine (His, H)

Negatively charged R groups

Aspartate (Asp, D)

Glutamate (Glu, E)

$$^+H_3N—CH—C—O^- \quad ^+H—N—CH—COO^-$$

15.6 Amino acids are joined together by peptide bonds. In a peptide bond (red), the carboxyl group of one amino acid is covalently attached to the amino group of another amino acid.

structure—the amino acid sequence—of the protein. Finally, some proteins consist of two or more polypeptide chains that associate to produce a *quaternary structure* (**Figure 15.7d**).

CONCEPTS

The products of many genes are proteins whose actions produce the traits specified by these genes. Proteins are polymers consisting of amino acids linked by peptide bonds. The amino acid sequence of a protein is its primary structure. This structure folds to create the secondary and tertiary structures; two or more polypeptide chains may associate to create a quaternary structure.

✔ CONCEPT CHECK 2

What primarily determines the secondary and tertiary structures of a protein?

15.2 The Genetic Code Determines How the Nucleotide Sequence Specifies the Amino Acid Sequence of a Protein

In 1953, James Watson and Francis Crick solved the structure of DNA and identified the base sequence as the carrier of genetic information (see Chapter 10). However, the way in which the base sequence of DNA specifies the amino acid sequences of proteins (the genetic code) remained elusive for another 10 years.

One of the first questions about the genetic code to be addressed was how many nucleotides are necessary to specify a single amino acid. The set of nucleotides that encode a single amino acid—the basic unit of the genetic code—is a *codon* (see Chapter 14). Many early investigators recognized that codons must contain a minimum of three nucleotides. Each nucleotide position in mRNA can be occupied by one of four bases: A, G, C, or U. If a codon consisted of a single nucleotide, only four different codons (A, G, C, and U) would be possible, which is not enough to encode the 20 different amino acids commonly found in proteins. If codons were made up of two nucleotides each (i.e., GU, AC, etc.), there would be $4 \times 4 = 16$ possible codons—still not enough to encode all 20 amino acids. With three nucleotides per codon, there are $4 \times 4 \times 4 = 64$ possible codons, which is more than enough to specify 20 different amino acids. Therefore, a *triplet code* requiring three nucleotides per codon is the most efficient way to encode all 20 amino acids. Using mutations in bacteriophage, Francis Crick and his colleagues confirmed in 1961 that the genetic code is indeed a triplet code.

▶ **TRY PROBLEMS 20 AND 21**

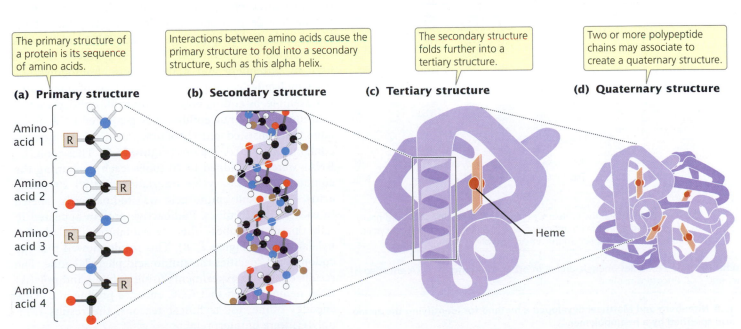

The primary structure of a protein is its sequence of amino acids.

Interactions between amino acids cause the primary structure to fold into a secondary structure, such as this alpha helix.

The secondary structure folds further into a tertiary structure.

Two or more polypeptide chains may associate to create a quaternary structure.

(a) Primary structure   **(b) Secondary structure**   **(c) Tertiary structure**   **(d) Quaternary structure**

Amino acid 1

Amino acid 2

Amino acid 3

Amino acid 4

Heme

15.7 Proteins have several levels of structural organization.

Question: What amino acids are specified by codons composed of only one type of base?

Methods

Uracil nucleotides → Polynucleotide phosphorylase → Poly(U) homopolymer

1 A homopolymer—in this case, poly(U) mRNA—was added to a test tube containing a cell-free translation system, 1 radioactively labeled amino acid, and 19 unlabeled amino acids.

2 The tube was incubated at 37°C.

3 Translation took place.

Precipitate protein

4 The protein was filtered, and the filter was checked for radioactivity.

Free amino acids

Protein

Suction

5 The procedure was repeated in 20 tubes, with each tube containing a different labeled amino acid.

Results

Pro Lys Arg His Tyr Ser Thr Asn Gln Cys

Phe Asp Glu Trp Gly Ala Val Ile Leu Met

6 The tube in which the protein was radioactively labeled contained newly synthesized protein with the amino acid specified by the homopolymer. In this case, UUU specified the amino acid phenylalanine.

Conclusion: UUU encodes phenylalanine; in other experiments, AAA encoded lysine, and CCC encoded proline.

15.8 Nirenberg and Matthaei developed a method for identifying the amino acid specified by a homopolymer.

Breaking the Genetic Code

When it had been firmly established that the genetic code consists of codons that are three nucleotides in length, the next step was to determine which groups of three nucleotides specify which amino acids. Logically, the easiest way to break the code would have been to determine the base sequence of a piece of RNA, add it to a test tube containing all the components necessary for translation, and allow it to direct the synthesis of a protein. The amino acid sequence of the newly synthesized protein could then be determined, and its sequence could be compared with that of the RNA. Unfortunately, there was no way at that time to determine the nucleotide sequence of a piece of RNA, so indirect methods were necessary to break the code.

THE USE OF HOMOPOLYMERS The first clues to the genetic code came in 1961, from the work of Marshall Nirenberg and Johann Heinrich Matthaei. These investigators created synthetic RNAs by using an enzyme called polynucleotide phosphorylase. Unlike RNA polymerase, polynucleotide phosphorylase does not require a template; it randomly links together any RNA nucleotides that happen to be available. The first synthetic mRNAs used by Nirenberg and Matthaei were homopolymers, RNA molecules consisting of a single type of nucleotide. For example, by adding polynucleotide phosphorylase to a solution of uracil nucleotides, they generated RNA molecules that consisted entirely of uracil nucleotides and thus contained only UUU codons (**Figure 15.8**). These poly(U) RNAs were then added to 20 tubes, each containing the components necessary for translation and all 20 amino acids. A different amino acid was radioactively labeled in each of the 20 tubes. Radioactive protein appeared in only one of the tubes—the one containing labeled phenylalanine (see Figure 15.8). This result showed that the codon UUU specifies the amino acid phenylalanine. The results of similar experiments using poly(C) and poly(A) RNA demonstrated that CCC encodes proline and AAA encodes lysine; for technical reasons, the results from poly(G) were uninterpretable.

THE USE OF RANDOM COPOLYMERS   To gain information about additional codons, Nirenberg and his colleagues created synthetic RNAs containing two or three different bases. Because polynucleotide phosphorylase incorporates nucleotides randomly, these RNAs contained random mixtures of the bases and are thus called random copolymers. For example, when adenine and cytosine nucleotides are mixed with polynucleotide phosphorylase, the RNA molecules produced have eight different codons: AAA, AAC, ACC, ACA, CAA, CCA, CAC, and CCC. These poly(AC) RNAs produced proteins containing six different amino acids: asparagine, glutamine, histidine, lysine, proline, and threonine.

The proportions of the different amino acids in the proteins depended on the ratio of the two nucleotides used in creating the synthetic mRNA, and the theoretical probability of finding a particular codon could be calculated from the ratios of the bases. If a 4 : 1 ratio of C to A were used in making the RNA, then the probability of C being at any given position in a codon is $\frac{4}{5}$ and the probability of A being in it is $\frac{1}{5}$. With random incorporation of bases, the probability of any one of the codons with two Cs and one A (CCA, CAC, or ACC) should be $\frac{4}{5} \times \frac{4}{5} \times \frac{1}{5} = \frac{16}{125} = 0.13$, or 13%, and the probability of any codon with two As and one C (AAC, ACA, or CAA) should be $\frac{1}{5} \times \frac{1}{5} \times \frac{4}{5} = \frac{4}{125} = 0.032$, or about 3%. Therefore, an amino acid encoded by two Cs and one A should be more common than an amino acid encoded by two As and one C. By comparing the percentages of amino acids in proteins produced by random copolymers with the theoretical frequencies expected for the codons, Nirenberg and his colleagues could derive information about the *base composition* of the codons. These experiments revealed nothing, however, about the codon base *sequence*; histidine was clearly encoded by a codon with two Cs and one A, but whether that codon was ACC, CAC, or CCA was unknown. There were other problems with this method: the theoretical calculations depended on the random incorporation of bases, which did not always occur, and, because the genetic code is redundant, sometimes several different codons specify the same amino acid.

THE USE OF RIBOSOME-BOUND tRNAs   To overcome the limitations of random copolymers, Nirenberg and Philip Leder developed another technique in 1964 that used ribosome-bound tRNAs. They found that a very short sequence of mRNA—even one consisting of a single codon—would bind to a ribosome. The codon on the short mRNA would then base pair with the matching anticodon on a transfer RNA that carried the amino acid specified by the codon (**Figure 15.9**). Short mRNAs that were bound to ribosomes were mixed with tRNAs and amino acids, and this mixture was passed through a nitrocellulose filter. The tRNAs that were paired with the ribosome-bound mRNA stuck to the filter, whereas unbound tRNAs passed through it. The advantage of this system was that it could be used with very

short synthetic mRNA molecules that could be synthesized with a known sequence. Nirenberg and Leder synthesized more than 50 short mRNAs with known codons and added them individually to a mixture of ribosomes and tRNAs. They then isolated the tRNAs that were bound to the mRNA and ribosomes and determined which amino acids were

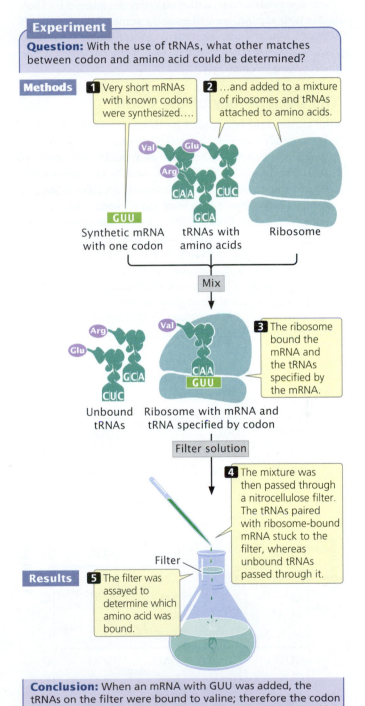

Experiment

Question: With the use of tRNAs, what other matches between codon and amino acid could be determined?

Methods

1 Very short mRNAs with known codons were synthesized….

2 …and added to a mixture of ribosomes and tRNAs attached to amino acids.

Synthetic mRNA with one codon   tRNAs with amino acids   Ribosome

Mix

3 The ribosome bound the mRNA and the tRNAs specified by the mRNA.

Unbound tRNAs   Ribosome with mRNA and tRNA specified by codon

Filter solution

4 The mixture was then passed through a nitrocellulose filter. The tRNAs paired with ribosome-bound mRNA stuck to the filter, whereas unbound tRNAs passed through it.

Filter

Results

5 The filter was assayed to determine which amino acid was bound.

Conclusion: When an mRNA with GUU was added, the tRNAs on the filter were bound to valine; therefore the codon GUU specifies valine. Many other codons were determined by using this method.

15.9 Nirenberg and Leder used ribosome-bound tRNAs to provide additional information about the genetic code.

present on the bound tRNAs. For example, synthetic RNA with the codon GUU retained a tRNA to which valine was attached, whereas RNAs with the codons UGU and UUG did not. Using this method, Nirenberg and his colleagues were able to determine the amino acids encoded by more than 50 codons.

Other experiments provided additional information about the genetic code, and it was fully deciphered by 1968. In the next section, we will examine some of the features of the code, which is so important to modern biology that Francis Crick compared its place to that of the periodic table of the elements in chemistry.

The Degeneracy of the Code

One amino acid is encoded by three consecutive nucleotides in mRNA, and each nucleotide can have one of four possible bases (A, G, C, and U) at each nucleotide position, thus permitting $4^3 = 64$ possible codons (**Figure 15.10**). Three of these codons are stop codons, specifying the end of translation. Thus, 61 codons, called **sense codons**, encode amino acids. Because there are 61 sense codons and only 20 different amino acids commonly found in proteins, the code contains more information than is needed to specify the amino acids and is said to be a **degenerate code**. This expression does not mean that the genetic code is depraved; *degenerate* is a term that Francis Crick borrowed from quantum physics, where it describes multiple physical states that have equivalent meaning.

15.11 Wobble may exist in the pairing of a codon and anticodon. The mRNA and tRNA pair in an antiparallel fashion. Pairing at the first and second codon positions is in accord with the Watson-and-Crick pairing rules (A with U, G with C); however, pairing rules are relaxed at the third position of the codon, and G on the anticodon can pair with either U or C on the codon in this example.

The degeneracy of the genetic code means that amino acids may be specified by more than one codon. Only tryptophan and methionine are encoded by a single codon (see Figure 15.10). Other amino acids are specified by two codons, and some, such as leucine, are specified by six different codons. Codons that specify the same amino acid are said to be **synonymous**, just as synonymous words are different words that have the same meaning.

As we learned in Chapter 14, tRNAs serve as adapter molecules, binding particular amino acids and delivering them to a ribosome, where the amino acids are then assembled into polypeptide chains. Each type of tRNA attaches to a single type of amino acid. The cells of most organisms possess from about 30 to 50 different tRNAs, and yet there are only 20 different amino acids in proteins. Thus, some amino acids are carried by more than one tRNA. Different tRNAs that accept the same amino acid but have different anticodons are called **isoaccepting tRNAs**.

Even though some amino acids have multiple (isoaccepting) tRNAs, there are still more codons than anticodons, because one anticodon can pair with different codons through flexibility in base pairing at the third position of the codon. Examination of Figure 15.10 reveals that many synonymous codons differ only in the third position. For example, alanine is encoded by the codons GCU, GCC, GCA, and GCG, all of which begin with GC. When the codon on the mRNA and the anticodon of the tRNA join (**Figure 15.11**), the first (5′) base of the codon pairs with the third (3′) base of the anticodon, strictly according to Watson-and-Crick rules: A with U; C with G. Next, the middle bases of codon and anticodon pair, also strictly following the Watson-and-Crick rules. After these pairs have hydrogen bonded, the third bases pair weakly and there may be flexibility, or **wobble**, in their pairing.

In 1966, Francis Crick developed the wobble hypothesis, which proposed that some nonstandard pairings of bases could take place at the third position of a codon. For example, a G in the anticodon may pair with either a C or a U in the third position of the codon (see p. 399 in Chapter 14 and

Second base

		U	C	A	G	
First base	U	UUU UUC Phe / UUA UUG Leu	UCU UCC UCA UCG Ser	UAU UAC Tyr / UAA Stop UAG Stop	UGU UGC Cys / UGA Stop / UGG Trp	U C A G
	C	CUU CUC CUA CUG Leu	CCU CCC CCA CCG Pro	CAU CAC His / CAA CAG Gln	CGU CGC CGA CGG Arg	U C A G
	A	AUU AUC Ile / AUA / AUG Met	ACU ACC ACA ACG Thr	AAU AAC Asn / AAA AAG Lys	AGU AGC Ser / AGA AGG Arg	U C A G
	G	GUU GUC Val / GUA GUG	GCU GCC GCA GCG Ala	GAU GAC Asp / GAA GAG Glu	GGU GGC GGA GGG Gly	U C A G

Third base

15.10 The genetic code consists of 64 codons. The amino acids specified by each codon are given in their three-letter abbreviation. The codons are written 5′ → 3′, as they appear in the mRNA. AUG is an initiation codon; UAA, UAG, and UGA are termination (stop) codons.

Table 15.2	The wobble rules, indicating which bases in the third position (3′ end) of the mRNA codon can pair with bases at the first position (5′ end) of the anticodon of the tRNA	
First Position of Anticodon	**Third Position of Codon**	**Pairing**
C	G	Anticodon 3′–X—Y—C–5′ | | | 5′–Y—X—G–3′ Codon
G	U or C	Anticodon 3′–X—Y—G–5′ | | | 5′–Y—X—U–3′ C Codon
A	U	Anticodon 3′–X—Y—A–5′ | | | 5′–Y—X—U–3′ Codon
U	A or G	Anticodon 3′–X—Y—U–5′ | | | 5′–Y—X—A–3′ G Codon

I (inosine)*	A, U, or C	Anticodon
		3′–X—Y—I–5′
		| | |
		5′–Y—X—A–3′
		      U
		      C
		Codon

*Inosine is one of the modified bases found in tRNAs.

Table 15.2: note that inosine in this table is one of the modified base found in tRNAs). The important thing to remember about wobble is that it allows some tRNAs to pair with more than one codon on an mRNA; thus from 30 to 50 tRNAs can pair with 61 sense codons. Some codons are synonymous through wobble. ▶ TRY PROBLEM 26

CONCEPTS

The genetic code consists of 61 sense codons that specify the 20 common amino acids; the code is degenerate, meaning that some amino acids are encoded by more than one codon. Isoaccepting tRNAs are different tRNAs with different anticodons that specify the same amino acid. Also, wobble at the third position of the codon allows different codons to specify the same amino acid.

✓ **CONCEPT CHECK 4**

Through wobble, a single _____ can pair with more than one _____.
a. codon, anticodon.
b. group of three nucleotides in DNA, codon in mRNA.
c. tRNA, amino acid.
d. anticodon, codon.

The Reading Frame and Initiation Codons

Findings from early studies of the genetic code indicated that the code is generally **nonoverlapping**. An overlapping code is one in which a single nucleotide may be included in more than one codon, as follows:

Nucleotide sequence   A U A C G A G U C

Nonoverlapping code   A U A C G A G U C
                       Ile   Arg   Val

Overlapping code   A U A C G A G U
                      Ile
                      U A C
                      Tyr
                        A C G
                        Thr

However, usually each nucleotide is part of a single codon. A few overlapping genes are found in viruses, but codons within the same gene do not overlap, and the genetic code is generally considered to be nonoverlapping.

For any sequence of nucleotides, there are three potential sets of codons—three ways in which the sequence can be read in groups of three. Each different way of reading the sequence is called a **reading frame**, and any sequence of nucleotides has three potential reading frames. The three reading frames have completely different sets of codons and will therefore specify proteins with entirely different amino acid sequences. Thus, it is essential for the translational machinery to use the correct reading frame. How is the correct reading frame established? The reading frame is set by the **initiation codon**, which is the first codon of the mRNA to specify an amino acid. After the initiation codon, the other codons are read as successive groups of three nucleotides. No bases are skipped between the codons; so there are no punctuation marks to separate the codons.

The initiation codon is usually AUG, although GUG and UUG are used on rare occasions. The initiation codon is not just a sequence that marks the beginning of translation; it specifies an amino acid. In bacterial cells, the first AUG encodes a modified type of methionine, *N*-formylmethionine; all proteins in bacteria initially begin with this amino acid, but the formyl group (or, in some cases, the entire amino acid) may be removed after the protein has been synthesized. When the codon AUG is at an internal position in a gene, it encodes unformylated methionine. In archaeal and eukaryotic cells,

AUG specifies unformylated methionine both at the initiation position and at internal positions. In both bacteria and eukaryotes there are different tRNAs for the initiator methionine (designated tRNA$_i^{fMet}$ in bacteria and tRNA$_i^{Met}$ in eukaryotes) and internal methionine (designated tRNA^{Met}).

Termination Codons

Three codons—UAA, UAG, and UGA—do not encode amino acids. These codons signal the end of the protein in both bacterial and eukaryotic cells and are called **stop codons, termination codons,** or **nonsense codons**. No tRNA molecules have anticodons that pair with termination codons.

The Universality of the Code

For many years the genetic code was assumed to be **universal**, meaning that each codon specifies the same amino acid in all organisms. We now know that the genetic code is almost, but not completely, universal; a few exceptions have been found. Most of these exceptions are termination codons, but there are a few cases in which one sense codon substitutes for another. Most exceptions are found in mitochondrial genes; a few nonuniversal codons have also been detected in the nuclear genes of protozoans and in bacterial DNA (**Table 15.3**). ▶ TRY PROBLEM 22

CONCEPTS

Each sequence of nucleotides possesses three potential reading frames. The correct reading frame is set by the initiation codon. The end of a protein-encoding sequence is marked by a termination codon. With a few exceptions, all organisms use the same genetic code.

✓ CONCEPT CHECK 5

Do the initiation and termination codons specify an amino acid? If so, which ones?

Table 15.3 Some exceptions to the universal genetic code

Genome	Codon	Universal Code	Altered Code
Bacterial DNA			
Mycoplasma capricolum	UGA	Stop	Trp
Mitochondrial DNA			
Human	UGA	Stop	Trp
Human	AUA	Ile	Met
Human	AGA, AGG	Arg	Stop
Yeast	UGA	Stop	Trp
Trypanosomes	UGA	Stop	Trp
Plants	CGG	Arg	Trp
Nuclear DNA			
Tetrahymena	UAA	Stop	Gln
Paramecium	UAG	Stop	Gln

CONNECTING CONCEPTS

Characteristics of the Genetic Code

We have now considered a number of characteristics of the genetic code. Let's take a moment to review these characteristics.

1. The genetic code consists of a sequence of nucleotides in DNA or RNA. There are four letters in the code, corresponding to the four bases—A, G, C, and U (T in DNA).
2. The genetic code is a triplet code. Each amino acid is encoded by a sequence of three consecutive nucleotides, called a codon.
3. The genetic code is degenerate; of 64 codons, 61 codons encode only 20 amino acids in proteins (3 codons are termination codons). Some codons are synonymous, specifying the same amino acid.
4. Isoaccepting tRNAs are tRNAs with different anticodons that accept the same amino acid; wobble allows the anticodon on one type of tRNA to pair with more than one type of codon on mRNA.
5. The code is generally nonoverlapping; each nucleotide in an mRNA sequence belongs to a single reading frame.
6. The reading frame is set by an initiation codon, which is usually AUG.
7. When a reading frame has been set, codons are read as successive groups of three nucleotides.
8. Any one of three termination codons (UAA, UAG, and UGA) can signal the end of a protein; no amino acids are encoded by the termination codons.
9. The code is almost universal.

15.3 Amino Acids Are Assembled into a Protein Through Translation

Now that we are familiar with the genetic code, we can begin to study how amino acids are assembled into proteins. Because more is known about translation in bacteria, we will focus primarily on bacterial translation. In most respects, eukaryotic translation is similar, although some significant differences will be noted.

Remember that only mRNAs are translated into proteins. Translation takes place on ribosomes; indeed, ribosomes can be thought of as moving protein-synthesizing machines. Through a variety of techniques, a detailed view of the structure of the ribosome has been produced in recent years, which has greatly improved our understanding of translation. A ribosome attaches near the 5′ end of an mRNA strand and moves toward the 3′ end, translating the codons as it goes (**Figure 15.12**). Synthesis begins at the amino end of the protein, and the protein is elongated by the addition of new amino acids to the carboxyl end. Protein synthesis includes a series of RNA–RNA interactions: interactions between the mRNA and the rRNA that hold the mRNA in the ribosome, between the codon on the mRNA and the anticodon on the tRNA, and between the tRNA and the rRNAs of the ribosome.

15.12 The translation of an mRNA molecule takes place on a ribosome. The letter N represents the amino end of the protein; C represents the carboxyl end.

Protein synthesis can be conveniently divided into four stages: (1) tRNA charging, in which tRNAs bind to amino acids; (2) initiation, in which the components necessary for translation are assembled at the ribosome; (3) elongation, in which amino acids are joined, one at a time, to the growing polypeptide chain; and (4) termination, in which protein synthesis halts at the termination codon and the translation components are released from the ribosome.

The Binding of Amino Acids to Transfer RNAs

The first stage of translation is the binding of tRNA molecules to their appropriate amino acids, called tRNA charging. Each tRNA is specific for a particular amino acid. All tRNAs have the sequence CCA at the 3′ end, and the carboxyl group (COO⁻) of the amino acid is attached to the adenine nucleotide at the 3′ end of the tRNA (**Figure 15.13**). If each tRNA is specific for a particular amino acid but all amino acids are attached to the same nucleotide (A) at the 3′ end of a tRNA, how does a tRNA link up with its appropriate amino acid?

The key to specificity between an amino acid and its tRNA is a set of enzymes called **aminoacyl-tRNA synthetases**. A cell has 20 different aminoacyl-tRNA synthetases, one for each of the 20 amino acids. Each synthetase recognizes a particular amino acid, as well as all the tRNAs that accept that amino acid. Recognition of the appropriate amino acid by a synthetase is based on the different sizes, charges, and R groups of the amino acids. The recognition of tRNAs by a synthetase depends on the differing nucleotide sequences of the tRNAs. Researchers have identified which nucleotides are important in recognition by altering different nucleotides in a particular tRNA and determining whether the altered tRNA is still recognized by its synthetase (**Figure 15.14**).

The attachment of a tRNA to its appropriate amino acid, termed **tRNA charging**, requires energy, which is supplied by adenosine triphosphate (ATP):

$$\text{amino acid} + \text{tRNA} + \text{ATP} \rightarrow \text{aminoacyl-tRNA} + \text{AMP} + \text{PP}_i$$

This reaction takes place in two steps (**Figure 15.15**). To identify the resulting aminoacylated tRNA, we write the three-letter abbreviation for the amino acid in front of the tRNA; for example, the amino acid alanine (Ala) attaches to its tRNA (tRNA^{Ala}), giving rise to its aminoacyl-tRNA (Ala-tRNA^{Ala}).

Errors in tRNA charging are rare: they occur in only about 1 in 10,000 to 1 in 100,000 reactions. This fidelity is due in part to the presence of editing (proofreading) activity in many of the synthetases. Editing activity detects and removes incorrectly paired amino acids from the tRNAs. Some antifungal chemical agents work by trapping tRNAs in the editing site of the enzyme, preventing their release and thus inhibiting the process of translation in the fungi.

CONCEPTS

Amino acids are attached to specific tRNAs by aminoacyl-tRNA synthetases in a two-step reaction that requires ATP.

✓ **CONCEPT CHECK 6**

Amino acids bind to which part of the tRNA?
a. anticodon   c. 3′ End
b. DHU arm.   d. 5′ end

15.13 An amino acid attaches to the 3′ end of a tRNA. The carboxyl group (COO⁻) of the amino acid attaches to the hydroxyl group of the 2′- or 3′-carbon atom of the final nucleotide at the 3′ end of the tRNA, in which the base is always adenine.

1 Positions in blue are the same in all tRNAs and cannot be used to differentiate among tRNAs.

2 Positions in red are important in the recognition of tRNAs by one synthetase.

3 Positions in yellow are used by more than one synthetase.

Acceptor stem

DHU arm

TψC arm

Extra arm

Anticodon arm

Anticodon

15.14 Certain positions on tRNA molecules are recognized by the appropriate aminoacyl-tRNA synthetase.

The Initiation of Translation

The second stage in the process of protein synthesis is initiation. At this stage, all the components necessary for protein synthesis assemble: (1) mRNA; (2) the small and large subunits of the ribosome; (3) a set of three proteins called initiation factors; (4) initiator tRNA with *N*-formylmethionine attached (fMet-tRNA^{fMet}); and (5) guanosine triphosphate (GTP). Initiation comprises three major steps. First, mRNA binds to the small subunit of the ribosome. Second, initiator tRNA binds to the mRNA through base pairing between the codon and the anticodon. Third, the large ribosome joins the initiation complex. Let's look at each of these steps more closely.

INITIATION IN BACTERIA The functional ribosome of bacteria exists as two subunits, the small 30S subunit and the large 50S subunit (**Figure 15.16a**). An mRNA molecule can bind to the small ribosome subunit only when the subunits

are separate. **Initiation factor 3** (IF-3) binds to the small subunit of the ribosome and prevents the large subunit from binding during initiation (**Figure 15.16b**). Another factor, **initiation factor 1** (IF-1), enhances the disassociation of the large and small ribosomal subunits.

Where on the mRNA does the ribosome bind during initiation of translation? Key sequences on the mRNA required for ribosome binding have been identified in experiments designed to allow the ribosome to bind to mRNA but not proceed with protein synthesis; the ribosome is thereby stalled at the initiation site where binding occurs. A ribonuclease is added, which degrades all the mRNA except the region covered by the ribosome. The intact mRNA can be separated from the ribosome and studied. The sequence covered by the ribosome during initiation is from 30 to 40 nucleotides long and includes the AUG initiation codon. Within the ribosome-binding site is the Shine–Dalgarno consensus sequence (**Figure 15.17**; see also Chapter 14), which is complementary to a sequence of nucleotides at the 3′ end of 16S rRNA (part of the small subunit of the ribosome). During initiation, the nucleotides in the Shine–Dalgarno sequence pair with their complementary nucleotides in the 16S rRNA, allowing the small subunit of the ribosome to attach to the mRNA and positioning the ribosome directly over the initiation codon. These ribosome-binding sequences are within the 5′ untranslated region of the mRNA.

The initiator tRNA, fMet-tRNA^{fMet}, attaches to the initiation codon (see Figure 15.16c). This requires **initiation factor 2** (IF-2), which forms a complex with GTP.

At this point, the initiation complex consists of (1) the small subunit of the ribosome; (2) the mRNA; (3) the initiator tRNA with its amino acid (fMet-tRNA^{fMet}); (4) one molecule of GTP; and (5) several initiation factors. These components are collectively known as the **30S initiation complex** (see Figure 15.16c). In the final step of initiation, IF-3 dissociates from the small subunit, allowing the large subunit of the ribosome to join the initiation complex. The molecule of GTP (provided by IF-2) is hydrolyzed to guanosine diphosphate (GDP), and the initiation factors dissociate (see Figure 15.16d). When the large subunit has joined the initiation complex, the complex is called the **70S initiation complex**.

1 In the first step, the amino acid reacts with ATP,...

2 ...producing aminocyl-AMP and PP$_i$.

3 In the second step, the amino acid is transferred to the tRNA,...

Amino acid

ATP

R group

$^+\text{H}_3\text{N}-\text{C}-\text{C}$
O
O$^-$
H

P P$_i$

R group

$^+\text{H}_3\text{N}-\text{C}-\text{C}$
O
AMP
H

tRNA

4 ...and AMP is released.

AMP

R group

$^+\text{H}_3\text{N}-\text{C}-\text{C}$
O
O
H

Aminoacyl tRNA

Conclusion: At the end of tRNA charging, an amino acid is linked to its appropriate tRNA.

15.15 An amino acid becomes attached to the appropriate tRNA in a two-step reaction.

15.17 The Shine–Dalgarno consensus sequence in mRNA is required for the attachment of the small subunit of the ribosome.

INITIATION IN EUKARYOTES Similar events take place in the initiation of translation in eukaryotic cells, but there are some important differences. In bacterial cells, sequences in 16S rRNA of the small subunit of the ribosome bind to the Shine–Dalgarno sequence in mRNA. No analogous consensus sequence exists in eukaryotic mRNA. Instead, the cap at the 5′ end of eukaryotic mRNA plays a critical role in the initiation of translation. In a series of steps, the small subunit of the eukaryotic ribosome, initiation factors, and the initiator tRNA with its amino acid (Met-tRNA$_i^{Met}$) form an initiation complex that recognizes the cap and binds there. The initiation complex then moves along (scans) the mRNA until it locates the first AUG codon. The identification of the start codon is facilitated by the presence of a consensus sequence (called the Kozak sequence) that surrounds the start codon:

Another important difference is that eukaryotic initiation requires at least seven initiation factors. Some factors keep the ribosomal subunits separated, just as IF-3 does in bacterial cells. Others recognize the 5′ cap on mRNA and allow the small subunit of the ribosome to bind there. Still others possess RNA helicase activity, which is used to unwind secondary structures that may exist in the 5′ untranslated region of mRNA, allowing the small subunit to move down the mRNA until the initiation codon is reached. Other initiation factors help bring Met-tRNA$_i^{Met}$ to the initiation complex.

In eukaryotes, the cap is initially bound by several proteins, one of which is the **cap-binding complex (CBC)**. The CBC aids in exporting the mRNA from the nucleus and then promotes the "pioneer" or initial round of translation in the cytoplasm. This first round of translation plays an important role in checking for errors in the mRNA (see Messenger RNA Surveillance in the next section). After the pioneer round of translation, the CBC is replaced by eukaryotic initiation factor 4E (eIF-4E), which promotes continued translation of the mRNA.

The poly(A) tail at the 3′ end of eukaryotic mRNA also plays a role in the initiation of translation. During initiation, proteins that attach to the poly(A) tail interact with proteins that bind to the 5′ cap, enhancing the binding of the small subunit of the ribosome to the 5′ end of the mRNA. This interaction indicates that the 3′ end of mRNA bends over and associates with the 5′ cap during the initiation of translation, forming a circular

15.16 The initiation of translation requires several initiation factors and GTP.

15.18 The poly(A) tail of eukaryotic mRNA plays a role in the initiation of translation.

structure known as the closed loop (**Figure 15.18**). A few eukaryotic mRNAs contain internal ribosome entry sites, where ribosomes can bind directly without first attaching to the 5′ cap.

CONCEPTS

In the initiation of translation in bacterial cells, the small ribosomal subunit attaches to mRNA, and initiator tRNA attaches to the initiation codon. This process requires several initiation factors (IF-1, IF-2, and IF-3) and GTP. In the final step, the large ribosomal subunit joins the initiation complex.

✓ CONCEPT CHECK 7

During the initiation of translation, the small ribosome binds to which consensus sequence in bacteria?

Elongation

The next stage in protein synthesis is elongation, in which amino acids are joined to create a polypeptide chain. Elongation requires (1) the 70S complex just described; (2) tRNAs charged with their amino acids; (3) several elongation factors; and (4) GTP.

A ribosome has three sites that can be occupied by tRNAs; the **aminoacyl**, or **A**, **site**, the **peptidyl**, or **P**, **site**, and the **exit**, or **E**, **site** (**Figure 15.19a**). The initiator tRNA immediately occupies the P site (the only site to which the fMet-tRNA^fMet is able to bind), but all other tRNAs first enter the A site. At the end of initiation, the ribosome is attached to the mRNA, and fMet-tRNA^fMet is positioned over the AUG start codon in the P site; the adjacent A site is unoccupied (see Figure 15.19a).

Elongation takes place in three steps. In the first step (**Figure 15.19b**), a charged tRNA binds to the A site. This binding takes place when **elongation factor Tu** (EF-Tu) joins with GTP and then with a charged tRNA to form a three-part complex. This complex enters the A site of the ribosome, where the anticodon on the tRNA pairs with the codon on the mRNA. After the charged tRNA is in the A site, GTP is cleaved to GDP, and the EF-Tu–GDP complex is released (**Figure 15.19c**). **Elongation factor Ts** (EF-Ts) regenerates EF-Tu–GDP to EF-Tu–GTP. In eukaryotic cells, a similar set of reactions delivers the charged tRNA to the A site.

The second step of elongation is the formation of a peptide bond between the amino acids that are attached to tRNAs in the P and A sites (**Figure 15.19d**). The formation of this peptide bond releases the amino acid in the P site from its tRNA. Peptide bond formation occurs within the peptidyl transferase center, which is part of the large subunit of the ribosome. Evidence indicates that the catalytic activity is a property of ribosomal RNA in the large subunit of the ribosome (the 23S rRNA in bacteria, the 28S RNA in eukaryotes); this rRNA acts as a ribozyme (see p. 358 in Chapter 13).

The third step in elongation is **translocation** (**Figure 15.19e**), the movement of the ribosome down the mRNA in

15.19 The elongation of translation comprises three steps.

the 5′→3′ direction. This step positions the ribosome over the next codon and requires **elongation factor G (EF-G)** and the hydrolysis of GTP to GDP. Because the tRNAs in the P and A sites are still attached to the mRNA through codon–anticodon pairing, they do not move with the ribosome as it translocates. Consequently, the ribosome shifts so that the tRNA that previously occupied the P site now occupies the E site, from which it moves into the cytoplasm where it can be recharged with another amino acid. Translocation also causes the tRNA that occupied the A site (which is attached to the growing polypeptide chain) to be in the P site, leaving the A site open. Thus, the progress of each tRNA through the ribosome in the course of elongation can be summarized as follows: cytoplasm → A site → P site → E site → cytoplasm. As stated earlier, the initiator tRNA is an exception: it attaches directly to the P site and never occupies the A site.

After translocation, the A site of the ribosome is empty and ready to receive the tRNA specified by the next codon. The elongation cycle (see Figure 15.19b through e) repeats itself: a charged tRNA and its amino acid occupy the A site, a peptide bond is formed between the amino acids in the A and P sites, and the ribosome translocates to the next codon. Throughout the cycle, the polypeptide chain remains attached to the tRNA in the P site. Another protein called translational elongation factor P (EF-P), enhances the translation of proteins that contain consecutive copies of the amino acid proline. If EF-P is absent, ribosomes often stall during the translation of these polyproline-containing proteins.

Messenger RNAs, although single stranded, often contain secondary structures formed by pairing of complementary bases on different parts of the mRNA (see Figure 13.1b). As the ribosome moves along the mRNA, these secondary structures are unwound by helicase activity located in the small subunit of the ribosome.

Recently, researchers have developed methods for following a single ribosome as it translates individual codons of an mRNA molecule. These studies revealed that translation does not take place in a smooth continuous fashion. Each translocation step typically requires less than a tenth of a second, but sometimes there are distinct pauses, often lasting a few seconds, between each translocation event when the ribosome moves from one codon to another. Thus, translation takes place in a series of quick translocations interrupted by brief pauses. In addition to the short pauses between translocation events, translation may be interrupted by longer pauses—lasting from 1 to 2 minutes—that may play a role in regulating the process of translation.

Elongation in eukaryotic cells takes place in a similar manner. Eukaryotes possess at least three elongation factors, one of which also acts in initiation and termination. Another of the elongation factors used in eukaryotes, called eukaryotic elongation factor 2 (eEF-2), is the target of a toxin produced by bacteria that causes diphtheria, a disease that until recently was a leading killer of children. The diphtheria toxin inhibits eEF-2, preventing the translocation of the ribosome along the mRNA, and protein synthesis ceases.

CONCEPTS

Elongation consists of three steps: (1) a charged tRNA enters the A site, (2) a peptide bond is created between amino acids in the A and P sites, and (3) the ribosome translocates to the next codon. Elongation requires several elongation factors and GTP.

✔ CONCEPT CHECK 8

In elongation, the creation of peptide bonds between amino acids is catalyzed by

a. rRNA.   c. protein in the large subunit.
b. protein in the small subunit.   d. tRNA.

Termination

Protein synthesis terminates when the ribosome translocates to a termination codon. Because there are no tRNAs with anticodons complementary to the termination codons, no tRNA enters the A site of the ribosome when a termination codon is encountered (**Figure 15.20a**). Instead, proteins called **release factors** bind to the ribosome (**Figure 15.20b**). *Escherichia coli*

6 A peptide bond forms between the amino acids in the P and A sites, and the tRNA in the P site releases its amino acid.

7 The ribosome moves down the mRNA to the next codon (translocation) which requires EF-G and GTP.

8 The tRNA that was in the P site is now in the E site from which it moves into the cytoplasm.

9 The tRNA that occupied the A site is now in the P site. The A site is now open and ready to receive another tRNA.

(d)   Dipeptide

Step 2

UAC GGG
AUG CCC ACG
5′   E P A   3′

EF-G

GTP   GDP + Pᵢ

Step 3

(e)

fMet Pro
UAC
5′   GGG
AUG CCC ACG   3′
E P A

Conclusion: At the end of each cycle of elongation, the amino acid that was in the A site is added to the polypeptide chain and the A site is free to accept another tRNA.

(a)

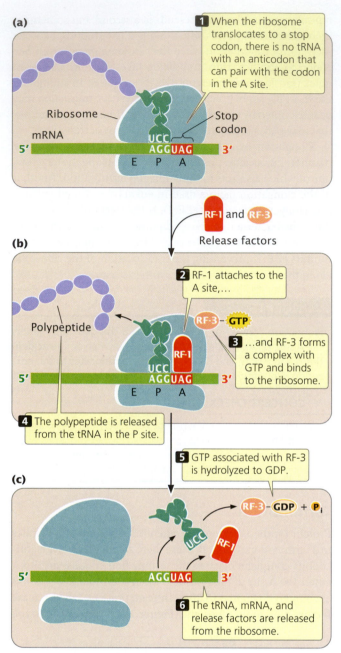

1 When the ribosome translocates to a stop codon, there is no tRNA with an anticodon that can pair with the codon in the A site.

Ribosome

Stop codon

mRNA

5' UCC
AGGUAG 3'
E P A

RF-1 and RF-3
Release factors

(b)

2 RF-1 attaches to the A site,…

RF-3 - GTP

Polypeptide

3 …and RF-3 forms a complex with GTP and binds to the ribosome.

RF-1

UCC
AGGUAG 3'
E P A

4 The polypeptide is released from the tRNA in the P site.

5 GTP associated with RF-3 is hydrolyzed to GDP.

(c)

RF-3 - GDP + Pᵢ

UCC RF-1

5' AGGUAG 3'

6 The tRNA, mRNA, and release factors are released from the ribosome.

15.20 Translation ends when a stop codon is encountered. Because UAG is the termination codon in this illustration, the release factor is RF-1.

15.21 Translation requires tRNA charging, initiation, elongation, and termination. In this process, amino acids are linked together in the order specified by mRNA to create a polypeptide chain. A number of initiation, elongation, and release factors take part in the process, and energy is supplied by ATP and GTP.

tRNA charging

AA — Amino acid
— tRNA
Ribosomal subunits
UAC — Anticodon
Large

Start codon

5' AUGCCCACGACUGCGAGCGUUCCGCUAAGGUAG 3'
mRNA

Small

Stop codon

Initiation

AA₁

UAC
5' AUGCCCACGACUGCGAGCGUUCCGCUAAGGUAG 3'
E P A

Elongation

AA₁ AA₂ AA₃ AA₄ AA₅ AA₆ AA₇

Charged tRNA

AA₈

UCG CAA

5' AUGCCCACGACUGCGAGCGUUCCGCUAAGGUAG 3'
E P A

Termination

AA₁ AA₂ AA₃ AA₄ AA₅ AA₆ AA₇ AA₈ AA₉ AA₁₀

Release factor

UCG

5' AUGCCCACGACUGCGAGCGUUCCGCUAAGGUAG 3'
E P A

Peptide release

Completed polypeptide

5' AUGCCCACGACUGCGAGCGUUCCGCUAAGGUAG 3'

Conclusion: Through the process of translation, amino acids are linked in the order specified by the mRNA.

has three release factors—RF-1, RF-2, and RF-3. Release factor 1 binds to the termination codons UAA and UAG, and RF-2 binds to UGA and UAA. The binding of release factor RF-1 or RF-2 to the A site of the ribosome promotes the cleavage of the tRNA in the P site from the polypeptide chain and the release of the polypeptide. Release factor 3 binds to the ribosome and forms a complex with GTP. This complex brings about a conformational change in the ribosome, releasing RF-1 or RF-2 from the A site and causing the tRNA in the P site to move to the E site; in the process, GTP is hydrolyzed to GDP. Additional factors help bring about the release of the tRNA from the P site, the release of the mRNA from the ribosome, and the dissociation of the ribosome (**Figure 15.20c**). It is important to note that the termination codon is not located at the 3′ end of the mRNA; rather, the termination codon is followed by a number of nucleotides that constitute the 3′ untranslated region (UTR) of the mRNA. The 3′ UTR often contains sequences that affect the stability of the mRNA and influence whether translation takes place (see Chapter 17).

Recent research shows that some bacterial ribosomes engage in a type of proofreading, similar to the way that DNA polymerases proofread during replication. After translocation, the ribosome checks the interaction between the mRNA and the tRNA in the P site. If the wrong tRNA was added, the alignment between the mRNA and tRNA will be incorrect, which triggers premature termination of translation. Evidence suggests that an important function of RF-3 is to help bring about termination of translation when the wrong tRNA has been used. Explore the process of bacterial translation by examining the consequences of various mutations in the coding region of a gene in **Animation 15.1**.

Translation in eukaryotic cells terminates in a similar way, except that there are two release factors: eRF-1, which recognizes all three termination codons, and eRF-2, which binds GTP and stimulates the release of the polypeptide from the ribosome. No translational proofreading, such as that stimulated by RF-3 in bacteria, has been observed eukaryotic cells. ▶ **TRY PROBLEMS 30 AND 34**

CONCEPTS

Termination takes place when the ribosome reaches a termination codon. Release factors bind to the termination codon, causing the release of the polypeptide from the last tRNA, of the tRNA from the ribosome, and of the mRNA from the ribosome.

The overall process of protein synthesis, including tRNA charging, initiation, elongation, and termination, is summarized in **Figure 15.21**. The components taking part in this process are listed in **Table 15.4**.

Table 15.4	**Components required for protein synthesis in bacterial cells**	
Stage	**Component**	**Function**
tRNA charging	Amino acids	Building blocks of proteins
	tRNAs	Deliver amino acids to ribosomes
	Aminoacyl-tRNA synthetases	Attaches amino acids to tRNAs
	ATP	Provides energy for binding amino acid to tRNA
Initiation	mRNA	Carries coding instructions
	fMet-tRNA^{fMet}	Provides first amino acid in peptide
	30S ribosomal subunit	Attaches to mRNA
	50S ribosomal subunit	Stabilizes tRNAs and amino acids
	Initiation factor 1	Enhances dissociation of large and small subunits of ribosome
	Initiation factor 2	Binds GTP; delivers fMet-tRNA^{fMet} to initiation codon
	Initiation factor 3	Binds to 30S subunit and prevents association with 50S subunit
Elongation	70S initiation complex	Functional ribosome with A, P, and E sites where protein synthesis takes place
	Charged tRNAs	Bring amino acids to ribosome and help assemble them in order specified by mRNA
	Elongation factor Tu	Binds GTP and charged tRNA; delivers charged tRNA to A site
	Elongation factor Ts	Generates active elongation factor Tu
	Elongation factor G	Stimulates movement of ribosome to next codon
	GTP	Provides energy
	Peptidyl transferase center	Creates peptide bond between amino acids in A site and P site
Termination	Release factors 1, 2, and 3	Bind to ribosome when stop codon is reached and terminate translation

A Comparison of Bacterial and Eukaryotic Translation

We have now considered the process of translation in bacterial cells and noted some distinctive differences that exist in eukaryotic cells. Let's reflect on some of the important similarities and differences of protein synthesis in bacterial and eukaryotic cells.

First, we should emphasize that the genetic code of bacterial and eukaryotic cells is virtually identical; the only difference is in the amino acid specified by the initiation codon. In bacterial cells, the initiation AUG encodes a modified type of methionine, N-formylmethionine, whereas, in eukaryotic cells, initiation AUG encodes unformylated methionine. One consequence of the fact that bacteria and eukaryotes use the same code is that eukaryotic genes can be translated in bacterial systems, and vice versa; this feature makes genetic engineering possible, as we will see in Chapter 19.

Another difference is that transcription and translation take place simultaneously in bacterial cells, but the nuclear envelope separates these processes in eukaryotic cells. The physical separation of transcription and translation has important implications for the control of gene expression, which we will consider in Chapter 17, and it allows for extensive modification of eukaryotic mRNAs, as discussed in Chapter 14.

Yet another difference is that mRNA in bacterial cells is short-lived, typically lasting only a few minutes, but mRNA in eukaryotic cells can last hours or days. The 5′ cap and 3′ poly(A) tail found on eukaryotic mRNAs add to their stability (see Chapter 14).

In both bacterial and eukaryotic cells, aminoacyl-tRNA synthetases attach amino acids to their appropriate tRNAs and the chemical process is the same. There are significant differences in the sizes and compositions of bacterial and eukaryotic ribosomal subunits. For example, the large subunit of the eukaryotic ribosome contains three rRNAs, whereas the bacterial ribosome contains only two. These differences allow antibiotics and other substances to inhibit bacterial translation while having no effect on the translation of eukaryotic nuclear genes, as will be discussed later in this chapter.

Other fundamental differences lie in the process of initiation. In bacterial cells, the small subunit of the ribosome attaches directly to the region surrounding the start codon through hydrogen bonding between the Shine–Dalgarno consensus sequence in the 5′ untranslated region of the mRNA and a sequence at the 3′ end of the 16S rRNA. In contrast, the small subunit of a eukaryotic ribosome first binds to proteins attached to the 5′ cap on mRNA and then migrates down the mRNA, scanning the sequence until it encounters the first AUG initiation codon. Additionally, a larger number of initiation factors take part in eukaryotic initiation than in bacterial initiation.

Elongation and termination are similar in bacterial and eukaryotic cells, although different elongation and termination factors are used. In both types of organisms, mRNAs are translated multiple times and are simultaneously attached to several ribosomes, forming polyribosomes, as discussed in Section 15.4.

Much less is known about the process of translation in archaea, but they appear to possess a mixture of eubacterial and eukaryotic features. Because archaea lack nuclear membranes, transcription and translation take place simultaneously, just as they do in eubacterial cells. Archaea utilize unformylated methionine as the initiator amino acid, a characteristic of eukaryotic translation. Some of the initiation and release factors in archaea are similar to those found in eubacteria, whereas others are similar to those found in eukaryotes. Finally, some of the antibiotics that inhibit translation in eubacteria have no effect on translation in archaea, providing further evidence of the fundamental differences between eubacteria and archaea.

15.4 Additional Properties of RNA and Ribosomes Affect Protein Synthesis

Now that we have considered in some detail the process of translation, we will examine some additional aspects of protein synthesis and the protein-synthesis machinery.

The Three-Dimensional Structure of the Ribosome

The central role of the ribosome in protein synthesis was recognized in the 1950s, and numerous aspects of its structure have been studied since then. Nevertheless, many aspects remained a mystery until detailed, three-dimensional reconstructions were completed. In 2009, the Nobel Prize in chemistry was awarded to Venkatraman Ramakrishnan, Thomas Steitz, and Ada Yonath for their research on the molecular structure of the ribosome.

Figure 15.22a shows a model of the *E. coli* ribosome at low resolution. A high-resolution image of the bacterial ribosome as determined by X-ray crystallography is represented in the model depicted in **Figure 15.22b**. The mRNA is bound to the small subunit of the ribosome, and the tRNAs are located in the A, P, and E sites (**Figure 15.22c**) that bridge the small and large subunits (see Figure 15.22a). Initiation factors 1 and 3 bind to sites on the outside of the small subunit of the ribosome. EF-Tu, EF-G, and other factors complexed with GTP interact with a *factor-binding center*. High-resolution crystallographic images provide information indicating that a *decoding center* resides in the small subunit of the ribosome (the decoding center cannot be seen in Figure 15.22b). This center senses the fit between the codon on the mRNA and the anticodon on the incoming charged tRNA. Only tRNAs with the correct anticodon are bound tightly by the ribosome. The structural analyses also indicate that the large subunit of the ribosome contains the *peptidyl transferase center*, where peptide-bond formation takes place. There are no ribosomal proteins in the peptidyl transferase center; peptide-bond formation is carried out by the RNA molecule. These analyses also reveal that a tunnel connects the site of peptide-bond formation with the back of the ribosome; the growing polypeptide chain passes through this tunnel to the outside of the ribosome. The tunnel can accommodate about 35 amino acids of the growing polypeptide chain.

15.22 Structure of the ribosome. (a) Low-resolution model of the ribosome, showing the A, P, and E sites where tRNAs, the mRNA, and the growing polypeptide chain reside. (b) High-resolution model of the ribosome. The tRNA in the E site is reddish brown, the tRNA in the P site is red, and the tRNA in the A site is green. The 16S rRNA is shown in yellow, the 23S rRNA is in cyan, and the 5S rRNA is green. Small ribosomal subunit proteins are shown in orange, whereas large ribosomal subunit proteins are shown in purple. (c) Positions of tRNAs in E, P, and A sites of the ribosome shown in part *b*. [Part b: MRC Lab of Molecular Biology, Wellcome Images.]

Polyribosomes

In both prokaryotic and eukaryotic cells, mRNA molecules are translated simultaneously by multiple ribosomes (**Figure 15.23**). The resulting structure—an mRNA with several ribosomes attached—is called a **polyribosome** (or often just a polysome). Each ribosome successively attaches to the ribosome-binding site at the 5′ end of the mRNA and moves toward the 3′ end; the polypeptide associated with each ribosome becomes progressively longer as the ribosome moves along the mRNA. In prokaryotic cells, transcription and translation are simultaneous; multiple ribosomes may be attached to the 5′ end of the mRNA while transcription is still taking place at the 3′ end, as shown in Figure 15.23. In eukaryotes, transcription and translation are separated in time and space, with transcription taking place in the nucleus and translation taking place in the cytoplasm.

CONCEPTS

In both prokaryotic and eukaryotic cells, multiple ribosomes may be attached to a single mRNA, generating a structure called a polyribosome.

✔ CONCEPT CHECK 9

In a polyribosome, the polypeptides associated with which ribosomes will be the longest?

a. Those at the 5′ end of mRNA.
b. Those at the 3′ end of mRNA.
c. Those in the middle of mRNA.
d. All polypeptides will be the same length.

Messenger RNA Surveillance

The accurate transfer of genetic information from one generation to the next and from genotype to phenotype is critical for the proper development and functioning of an organism. Consequently, cells have evolved a number of quality-control mechanisms to ensure the accuracy of information transfer.

Protein synthesis is no exception: several mechanisms, collectively termed **mRNA surveillance**, exist to detect and deal with errors in mRNAs that may create problems in the course of translation. These mechanisms keep the cell from wasting

15.23 An mRNA molecule may be translated simultaneously by several ribosomes. (a) Four ribosomes are translating an mRNA molecule; the ribosomes move from the 5′ end to the 3′ end of the mRNA. (b) In this electron micrograph, the long horizontal filament is DNA, the dark-staining spheres are polyribosomes, and the thin filaments connecting the ribosomes are mRNAs. Transcription of the DNA is proceeding from right to left; the mRNAs on the right are shorter than those left. Each mRNA is being translated by multiple ribosomes. [Part b: O. L. Miller, Jr., and B.A. Hamaklo, and C.A. Thomas, Jr. Science 169(1970):392. Reprinted with permission from AAAS.]

resources translating aberrant mRNAs and prevent the production of truncated proteins, which may be toxic to the cell.

NONSENSE-MEDIATED mRNA DECAY   A common mutation is one in which a codon that specifies an amino acid is altered to become a termination codon (called a nonsense mutation, see Chapter 18). A nonsense mutation does not affect transcription, but translation ends prematurely when the termination codon is encountered. The resulting protein is truncated and often nonfunctional. A common way that nonsense mutations arise in eukaryotic cells is when one or more of the exons is skipped or improperly spliced. Improper splicing leads to the deletion or addition of nucleotides in the mRNA, which alters the reading frame and often introduces premature termination codons.

To prevent the synthesis of aberrant proteins resulting from nonsense mutations, eukaryotic cells have evolved a mechanism called **nonsense-mediated mRNA decay** (NMD), which results in the rapid elimination of mRNA containing premature termination codons. The mechanism responsible for nonsense-mediated mRNA decay is still poorly understood. In mammals, it appears to entail proteins that bind to the exon–exon junctions. These exon-junction proteins may interact with enzymes that degrade the mRNA. One possibility is that the first ribosome to translate the mRNA (in the pioneer round of translation) removes the exon-junction proteins, thus protecting the mRNA from degradation. However, when the ribosome encounters a premature termination codon, the ribosome does not traverse the entire mRNA, and some of the exon-junction proteins are not removed, resulting in nonsense-mediated mRNA decay. ▶ TRY PROBLEM 38

STALLED RIBOSOMES AND NONSTOP mRNAs   A problem that occasionally arises in translation is when a ribosome stalls on the mRNA before translation is terminated. This situation can arise when a mutation in the DNA changes a termination codon into a codon that specifies an amino acid. It can also arise when transcription terminates prematurely, producing a truncated mRNA lacking a termination codon. In these cases, the ribosome reaches the end of the mRNA without encountering a termination codon and stalls, still attached to the mRNA. Attachment to the mRNA prevents the ribosome from being recycled for use on other mRNAs and, if such occurrences are frequent, the result is a shortage of ribosomes that diminishes overall levels of protein synthesis.

Bacteria have evolved a kind of molecular tow truck called **transfer–messenger RNA** (tmRNA) that removes stalled ribosomes. This RNA molecule has properties of both tRNA and mRNA; its tRNA component is normally charged with the amino acid alanine. When a ribosome becomes stalled on an mRNA, EF-Tu delivers the tmRNA to the ribosome's A site, where tmRNA acts as surrogate tRNA (**Figure 15.24**). A peptide bond is created between the amino acid in the P site of the ribosome and alanine (attached to tmRNA) now in the A site, transferring the polypeptide chain to the tmRNA and releasing the tRNA in the P site.

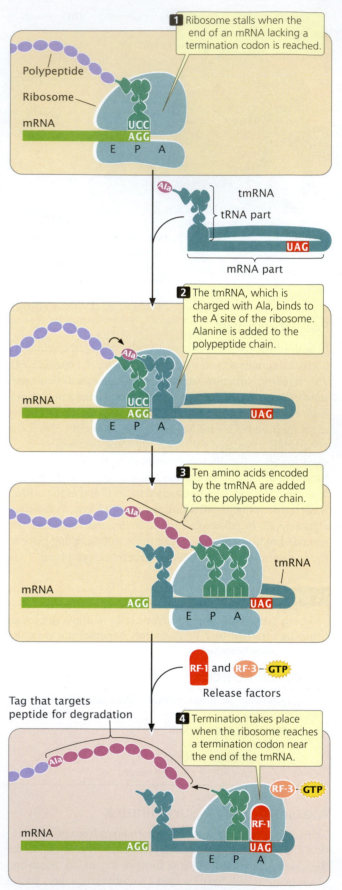

15.24 The tmRNA in bacteria allows stalled ribosomes to resume translation.

The ribosome then resumes translation, switching from the original, aberrant mRNA to the mRNA part of tmRNA. Translation adds 10 amino acids encoded by the tmRNA, and then a termination codon is reached at the 3′ end of the tmRNA, which terminates translation and releases the ribosome. The added amino acids are a special tag that targets the incomplete polypeptide chain for degradation. Some evidence suggests that the tmRNA also targets the aberrant mRNA for degradation. How stalled ribosomes are recognized by the tmRNA is not clear, but this method is efficient at recycling stalled ribosomes and eliminating abnormal proteins that result from truncated transcription.

Eukaryotes have evolved a different mechanism to deal with mRNAs that are missing termination codons. Instead of restarting the stalled ribosome and degrading the abnormal protein that results, eukaryotic cells use a mechanism called **nonstop mRNA decay**, which results in the rapid degradation of abnormal mRNA. In this mechanism, the codon-free A site of the stalled ribosome is recognized by a special protein that binds to the A site and recruits other proteins, which then degrade the mRNA from its 3′ end.

NO-GO DECAY Another mRNA surveillance system found in eukaryotes is **no-go decay** (NGD), which helps remove stalled ribosomes resulting from secondary structures in the mRNA, chemical damage to the mRNA, premature stop codons, and ribosomal defects. A series of proteins bring about termination, recycling of the ribosomes, and degradation of the mRNA.

CONCEPTS

Cells possess mRNA surveillance mechanisms to detect and eliminate mRNA molecules containing errors that create problems in the course of translation.

Folding and Posttranslational Modifications of Proteins

The functions of many proteins critically depend on the proper folding of the polypeptide chain; some proteins spontaneously fold into their correct shapes, but, for others, correct folding may initially require the participation of other molecules called **molecular chaperones**. Some molecular chaperones are associated with the ribosome and fold newly synthesized polypeptide chains as they emerge from the ribosome tunnel, in which case protein folding takes place during ongoing translation.

Many proteins must be modified after translation to become active. Proteins in both prokaryotic and eukaryotic cells often undergo alterations following translation, which are termed posttranslational modifications. A number of different types of modifications are possible. Some proteins are synthesized as larger precursor proteins and must be cleaved and trimmed by enzymes before the proteins can become functional. As mentioned earlier, the formyl group or the entire methionine residue may be removed from the amino end of a protein. Some proteins require the attachment of

carbohydrates for activation. Amino acids within a protein may be modified: phosphates, carboxyl groups, and methyl groups are added to some amino acids. In eukaryotic cells, the amino end of a protein is often acetylated after translation.

A common posttranslational modification in eukaryotes is the attachment of a protein called ubiquitin, which targets the protein for degradation. Another modification of some proteins is the removal of 15 to 30 amino acids, called the **signal sequence**, at the amino end of the protein. The signal sequence helps direct a protein to a specific location within the cell, after which the sequence is removed by special enzymes.

CONCEPTS

Many proteins undergo posttranslational modifications after their synthesis.

Translation and Antibiotics

Antibiotics are drugs that kill microorganisms. To make an effective antibiotic, not just any poison will do: the trick is to kill the bacteria without harming the patient.

Translation is frequently the target of antibiotics because it is essential to all living organisms and differs significantly between bacterial and eukaryotic cells. A number of antibiotics bind selectively to bacterial ribosomes and inhibit various steps in translation, but they do not affect eukaryotic ribosomes. Tetracyclines, for instance, are a class of antibiotics that bind to the A site of a bacterial ribosome and block the entry of charged tRNAs, yet they have no effect on eukaryotic ribosomes. Neomycin binds to the ribosome near the A site and induces translational errors, probably by causing mistakes in the binding of charged tRNAs to the A site. Chloramphenicol binds to the large subunit of the ribosome and blocks peptide-bond formation. Streptomycin binds to the small subunit of the ribosome and inhibits initiation, and erythromycin blocks translocation. Although chloramphenicol and streptomycin are potent inhibitors of translation in bacteria, they do not inhibit translation in archaea.

The three-dimensional structure of puromycin resembles the 3′ end of a charged tRNA, permitting puromycin to enter the A site of a ribosome efficiently and inhibit the entry of tRNAs. A peptide bond can form between the puromycin molecule in the A site and an amino acid on the tRNA in the P site of the ribosome, but puromycin cannot bind to the P site and translocation does not take place, blocking further elongation of the protein. Because tRNA structure is similar in all organisms, puromycin inhibits translation in both bacterial and eukaryotic cells; consequently, puromycin kills eukaryotic cells along with bacteria and is sometimes used in cancer therapy to destroy tumor cells.

Many antibiotics act by blocking specific steps in translation, and different antibiotics block different steps in protein synthesis, such as initiation or elongation. Because of this specificity, antibiotics are frequently used to study the process of protein synthesis.

CONCEPTS SUMMARY

■ George Beadle and Edward Tatum developed the one gene, one enzyme hypothesis, which proposed that each gene specifies one enzyme; this hypothesis was later modified to become the one gene, one polypeptide hypothesis.

■ Proteins are composed of twenty different amino acids. The amino acids in a protein are linked together by peptide bonds. Chains of amino acids fold and associate to produce the secondary, tertiary, and quaternary structures of proteins.

■ Solving the genetic code required several different approaches including the use of synthetic mRNAs with random sequences and short mRNAs that bind charged tRNAs.

■ The genetic code is a triplet code: three nucleotides specify a single amino acid. It is also degenerate (meaning that more than one codon may specify an amino acid), nonoverlapping, and universal (almost).

■ Different tRNAs (isoaccepting tRNAs) may accept the same amino acid. Different codons may pair with the same anticodon through wobble, which can exist at the third position of the codon and allows some nonstandard pairing of bases in this position.

■ The reading frame is set by the initiation codon. The end of the protein-coding section of an mRNA is marked by one of three termination codons.

■ Protein synthesis comprises four steps: (1) the binding of amino acids to the appropriate tRNAs, (2) initiation, (3) elongation, and (4) termination.

■ The binding of an amino acid to a tRNA requires the presence of a specific aminoacyl-tRNA synthetase and ATP. The amino acid is attached by its carboxyl end to the 3′ end of the tRNA.

■ In bacterial translation initiation, the small subunit of the ribosome attaches to the mRNA and is positioned over the initiation codon. It is joined by the first tRNA and its associated amino acid (N-formylmethionine in bacterial cells) and, later, by the large subunit of the ribosome. Initiation requires several initiation factors and GTP.

■ In elongation, a charged tRNA enters the A site of a ribosome, a peptide bond is formed between amino acids in the A and P sites, and the ribosome moves (translocates) along the mRNA to the next codon. Elongation requires several elongation factors and GTP.

■ Translation is terminated when the ribosome encounters one of the three termination codons. Release factors and GTP are required to bring about termination.

■ Each mRNA may be simultaneously translated by several ribosomes, producing a structure called a polyribosome.

■ Cells possess RNA surveillance mechanisms that eliminate mRNAs with errors that may create problems in translation.

■ Antibiotics frequently work by interfering with translation, because many aspects of translation differ in bacteria and eukaryotes.

■ Many proteins undergo posttranslational modification.

IMPORTANT TERMS

one gene, one enzyme hypothesis (p. 415)
one gene, one polypeptide hypothesis (p. 415)
amino acid (p. 415)
peptide bond (p. 415)
polypeptide (p. 415)
sense codon (p. 420)
degenerate genetic code (p. 420)
synonymous codons (p. 420)
isoaccepting tRNAs (p. 420)
wobble (p. 420)

nonoverlapping genetic code (p. 421)
reading frame (p. 421)
initiation codon (p. 421)
stop (termination or nonsense) codon (p. 422)
universal genetic code (p. 422)
aminoacyl-tRNA synthetase (p. 423)
tRNA charging (p. 423)
initiation factor (IF-1, IF-2, IF-3) (p. 424)
30S initiation complex (p. 424)

70S initiation complex (p. 424)
cap-binding complex (CBC) (p. 425)
aminoacyl (A) site (p. 426)
peptidyl (P) site (p. 426)
exit (E) site (p. 426)
elongation factor Tu (EF-Tu) (p. 426)
elongation factor Ts (EF-Ts) (p. 426)
translocation (p. 426)
elongation factor G (EF-G) (p. 427)

release factor (RF-1, RF-2, RF-3) (p. 427)
polyribosome (p. 427)
mRNA surveillance (p. 431)
nonsense-mediated mRNA decay (NMD) (p. 432)
transfer–messenger RNA (tmRNA) (p. 432)
nonstop mRNA decay (p. 433)
no-go decay (p. 433)
molecular chaperone (p. 433)
signal sequence (p. 433)

ANSWERS TO CONCEPT CHECKS

1. B → A → C

2. The amino acid sequence (primary structure) of the protein

3. b

4. d

5. The initiation codon in bacteria encodes *N*-formylmethionine; in eukaryotes, it encodes methionine. Termination codons do not specify amino acids.

6. c

7. The Shine–Dalgarno sequence

8. a

9. b

WORKED PROBLEMS

Problem 1

A series of auxotrophic mutants were isolated in *Neurospora*. Examination of fungi containing these mutations revealed that they grew on minimal medium to which various compounds (A, B, C, D) were added; growth responses to each of the four compounds are presented in the following table. Give the order of compounds A, B, C, and D in a biochemical pathway. Outline a biochemical pathway that includes these four compounds and indicate which step in the pathway is affected by each of the mutations.

Mutation number	Compound			
	A	**B**	**C**	**D**
134	+	+	−	+
276	+	+	+	+
987	−	−	−	+
773	+	+	+	+
772	−	−	−	+
146	+	+	−	+
333	+	+	−	+
123	−	+	−	+

> **Hint:** Group the mutations by which compounds allow growth.

Solution Strategy

What information is required in your answer to the problem?

The order of compounds in a biochemical pathway; for each mutation, which step in the pathway is affected by the mutation.

What information is provided to solve the problem?

For each mutant, whether it grew on minimal medium to which compounds A, B, C, and D were added.

For help with this problem, review:

The One Gene One Enzyme Hypothesis in Section 15.1.

Solution Steps

Mutation		Compound			
Group	Number	A	B	C	D
I	276	+	+	+	+
	773	+	+	+	+
II	134	+	+	−	+
	146	+	+	−	+
	333	+	+	−	+
III	123	−	+	−	+
IV	987	−	−	−	+
	772	−	−	−	+

Hint: If a compound is added after the block, it will allow the mutant to grow; if a compound is added before the block, it will have no effect.

Mutants in group I will grow if compound A, B, C, or D is added to the medium; so these mutations must affect a step before the production of all four compounds:

$$\longrightarrow \text{compounds A, B, C, D}$$

Group I mutations

Group II mutants will grow if compound A, B, or D is added but not if compound C is added. Thus, compound C comes before A, B, and D; and group II mutations affect the conversion of compound C into one of the other compounds:

$$\longrightarrow \text{compound C} \longrightarrow \text{compounds A, B, D}$$

Group I mutations   Group II mutations

Group III mutants allow growth if compound B or D is added but not if compound A or C is added. Thus, group III mutations affect steps that follow the production of A and C; we have already determined that compound C

precedes A in the pathway, and so A must be the next compound in the pathway:

$$\longrightarrow \text{compound C} \longrightarrow$$

Group I mutations   Group II mutations

$$\text{compound A} \longrightarrow \text{compounds B, D}$$

Group III mutations

Finally, mutants in group IV will grow if compound D is added but not if compound A, B, or C is added. Thus, compound D is the fourth compound in the pathway, and mutations in group IV block the conversion of B into D:

$$\longrightarrow \text{compound C} \longrightarrow \text{compound A} \longrightarrow$$

Group I mutations   Group II mutations   Group III mutations

$$\text{compound B} \longrightarrow \text{compound D}$$

Group IV mutations

Problem 2

A template strand in bacterial DNA has the following base sequence:

$$5'–AGGTTTAACGTGCAT–3'$$

What amino acids are encoded by this sequence?

Solution Strategy

What information is required in your answer to the problem?
The list of amino acids encoded by the given sequence.

What information is provided to solve the problem?
- The DNA sequence of the template strand.
- The 5′ and 3′ ends of the template sequence.
- The amino acids encoded by different codons (Figure 15.10).

For help with this problem, review:
The Degeneracy of the Code in Section 15.2.

Solution Steps

Recall: The mRNA is antiparallel and complementary to the DNA template strand.

To answer this question, we must first work out the mRNA sequence that will be transcribed from this DNA sequence:

DNA template strand: 5′–AGGTTTAACGTGCAT–3′
mRNA copied from DNA: 3′–UCCAAAUUGCACGUA–5′

An mRNA is translated 5′→3′; so it will be helpful if we turn the RNA molecule around with the 5′ end on the left:

mRNA copied from DNA: 5′–AUGCACGUUAAACCU–3′

The codons consist of groups of three nucleotides that are read successively after the first AUG codon; using Figure 15.10, we can determine that the amino acids are

5′–AUG—CAC—GUU—AAA—CCU–3′

fMet——His——Val——Lys——Pro

Problem 3

The following triplets constitute anticodons found on a series of tRNAs. Name the amino acid carried by each of these tRNAs.

a. 5′–UUU–3′

b. 5′–GAC–3′

c. 5′–UUG–3′

d. 5′–CAG–3′

Solution Strategy

What information is required in your answer to the problem?

The amino acid carried by each tRNA.

What information is provided to solve the problem?

■ The sequence on the anticodon of each tRNA.

■ The amino acids encoded by different codons (Figure 15.10).

For help with this problem, review:

The Degeneracy of the Code in Section 15.2.

Solution Steps

...odons are antiparallel and complementary to the anticodons.

Hint: Consult the wobble rules in Table 15.2.

To solve this problem, we first determine the codons with which these anticodons pair and then look up the amino acid specified by the codon in Figure 15.10. For part a, the anticodon is 5′–UUU–3′. According to the wobble rules, U in the first position of the anticodon can pair with either A or G in the third position of the codon, so there are two codons that can pair with this anticodon:

Anticodon: 5′–UUU–3′
Codon: 3′–AAA–5′
Codon: 3′–GAA–5′

Listing these codons in the conventional manner, with the 5′ end on the right, we have:

Codon: 5′–AAA–3′
Codon: 5′–AAG–3′

According to Figure 15.10, both codons specify the amino acid lysine (Lys). Recall that the wobble in the third position allows more than one codon to specify the same amino acid; so any wobble that exists should produce the same amino acid as the standard base pairings would, and we do not need to figure the wobble to answer this question. The answers for parts *b*, *c*, and *d* are:

b. Anticodon: 5′–GAC–3′
        Anticodon: 3′–CAG–5′
        Codon: 5′–GUC–3′ encodes Val

c. Anticodon: 5′–UUG–3′
        Anticodon: 3′–GUU–5′
        Codon: 5′–CAA–3′ encodes Gln

d. Anticodon: 5′–CAG–3′
        Anticodon: 3′–GUC–5′
        Codon: 5′–CUG–3′ encodes Leu

COMPREHENSION QUESTIONS

Section 15.1

1. What is the one gene, one enzyme hypothesis? Why was this hypothesis an important advance in our understanding of genetics?

Section 15.2

2. What different methods were used to help break the genetic code? What did each method reveal and what were the advantages and disadvantages of each one?

3. What are isoaccepting tRNAs?

4. What is the significance of the fact that many synonymous codons differ only in the third nucleotide position?

5. Define the following terms as they apply to the genetic code:

   a. Reading frame   f. Sense codon
   b. Overlapping code   g. Nonsense codon
   c. Nonoverlapping code   h. Universal code
   d. Initiation codon   i. Nonuniversal codons
   e. Termination codon

6. How is the reading frame of a nucleotide sequence set?

Section 15.3

7. How are tRNAs linked to their corresponding amino acids?

8. What role do the initiation factors play in protein synthesis?

9. How does the process of initiation differ in bacterial and eukaryotic cells?

10. Give the elongation factors used in bacterial translation and explain the role played by each factor in translation.

11. What events bring about the termination of translation?

12. Compare and contrast the process of protein synthesis in bacterial and eukaryotic cells, giving similarities and differences in the process of translation in these two types of cells.

Section 15.4

13. How do prokaryotic cells overcome the problem of a stalled ribosome on an mRNA that has no termination codon? How do eukaryotic cells solve this problem?

14. What are some types of posttranslational modification of proteins?

15. Explain how some antibiotics work by affecting the process of protein synthesis.

> ▶ For more questions that test your comprehension of the key chapter concepts, go to *LEARNINGCurve* for this chapter.

APPLICATION QUESTIONS AND PROBLEMS

Section 15.1

*16. Sydney Brenner isolated *Salmonella typhimurium* mutants that were implicated in the biosynthesis of tryptophan and would not grow on minimal medium. When these mutants were tested on minimal medium to which one of four compounds (indole glycerol phosphate, indole, anthranilic acid, and tryptophan) had been added, the growth responses shown in the following table were obtained.

Mutant	Minimal medium	Anthranilic acid	Indole glycerol phosphate	Indole	Tryptophan
trp-1	−	−	−	−	+
trp-2	−	−	+	+	+
trp-3	−	−	−	+	+
trp-4	−	−	+	+	+
trp-6	−	−	−	−	+
trp-7	−	−	−	−	+
trp-8	−	+	+	+	+
trp-9	−	−	−	−	+
trp-10	−	−	−	−	+
trp-11	−	−	−	−	+

Give the order of indole glycerol phosphate, indole, anthranilic acid, and tryptophan in a biochemical pathway leading to the synthesis of tryptophan. Indicate which step in the pathway is affected by each of the mutations.

17. Compounds I, II, and III are in the following biochemical pathway:

precursor $\xrightarrow[\text{enzyme A}]{}$ compound I $\xrightarrow[\text{enzyme B}]{}$

compound II $\xrightarrow[\text{enzyme C}]{}$ compound III

Mutation *a* inactivates enzyme A, mutation *b* inactivates enzyme B, and mutation *c* inactivates enzyme C. Mutants, each having one of these defects, were tested on minimal medium to which compound I, II, or III was added. Fill in the results expected of these tests by placing a plus sign (+) for growth or a minus sign (−) for no growth in the table below.

Strain with mutation	Minimal medium to which is added		
	Compound I	Compound II	Compound III
a			
b			
c			

Section 15.2

18. A geneticist conducts the experiment outlined in **Figure 15.8**, but this time she combines guanine nucleotides (instead of uracil) with polynucleotide phosphorylase. Radioactively labeled protein should appear in which tube?

19. For the experiment outlined in **Figure 15.8**, could Nirenberg and Mattaei have substituted RNA polymerase instead of polynucleotide phosphorylase without otherwise modifying the experiment? Why or why not?

*20. Assume that the number of different types of bases in RNA is four. What would be the minimum codon size (number of nucleotides) required to specify all amino acids if the number of different types of amino acids in proteins were: (a) 2, (b) 8, (c) 17, (d) 45, (e) 75?

*21. How many codons would be possible in a triplet code if only three bases (A, C, and U) were used?

*22. Referring to the genetic code presented in **Figure 15.10**, give the amino acids specified by the following bacterial mRNA sequences.

  a. 5′–AUGUUUAAAUUUAAAUUUUGA–3′

  b. 5′–AGGGAAAUCAGAUGUAUAUAUAUAUGA–3′

  c. 5′–UUUGGAUUGAGUGAAACGAUG
  GAUGAAAGAUUUCUCGCUUGA–3′

  d. 5′–GUACUAAGGAGGUUGUAUGGG
  UUAGGGGACAUCAUUUUGA–3′

23. A nontemplate strand on bacterial DNA has the following base sequence. What amino acid sequence will be encoded by this sequence?

  5′–ATGATACTAAGGCCC–3′

24. The following amino acid sequence is found in a tripeptide: Met-Trp-His. Give all possible nucleotide sequences on the mRNA, on the template strand of DNA, and on the nontemplate strand of DNA that can encode this tripeptide.

25. How many different mRNA sequences can encode a polypeptide chain with the amino acid sequence Met-Leu-Arg? (Be sure to include the stop codon.)

*26. A series of tRNAs have the following anticodons. Consider the wobble rules listed in **Table 15.2** and give all possible codons with which each tRNA can pair.

  a. 5′–GGC–3′

  b. 5′–AAG–3′

  c. 5′–IAA–3′

  d. 5′–UGG–3′

  e. 5′–CAG–3′

27. A researcher creates random copolymers of three nucleotides by mixing polynucletide phosphorylase with guanine and adenine nucleotides in a ratio of 5 guanine nucleotides to 1 adenine. Give the different copolymers produced and their theoretical proportions.

28. Assume that the nucleotide at the 5′ end of the first tRNA's anticodon (the tRNA on the left) in **Figure 15.11** were mutated from G to U. Give all codons with which the new, mutated anticodon could pair.

29. Which of the following amino acid changes could result from a mutation that changed a single base? For each change that could result from the alteration of a single base, determine which position of the codon (first, second, or third nucleotide) in the mRNA must be altered for the change to result.

  a. Leu → Gln

  b. Phe → Ser

  c. Phe → Ile

  d. Pro → Ala

  e. Asn → Lys

  f. Ile → Asn

Section 15.3

*30. Arrange the following components of translation in the approximate order in which they would appear or be used in prokaryotic protein synthesis:

  70S initiation complex

  30S initiation complex

  release factor 1

  elongation factor G

  initiation factor 3

  elongation factor Tu

  fMet-tRNA^{fMet}

31. Examine **Figure 15.14** of a tRNA. What do you think would be the potential effect of a mutation in the part of the tRNA gene that encodes: (a) the acceptor stem; (b) the anticodon; (c) one of red-colored nucleotides?

32. The following diagram illustrates a step in the process of translation. Sketch the diagram and identify the following elements on it.

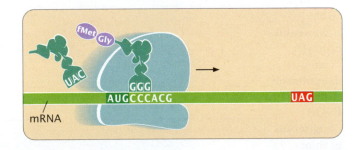

  a. 5′ and 3′ ends of the mRNA

  b. A, P, and E sites

  c. Start codon

  d. Stop codon

  e. Amino and carboxyl ends of the newly synthesized polypeptide chain

  f. Approximate location of the next peptide bond that will be formed

  g. Place on the ribosome where release factor 1 will bind

*33. Refer to the diagram in Problem 32 to answer the following questions.

   a. What will be the anticodon of the next tRNA added to the A site of the ribosome?

   b. What will be the next amino acid added to the growing polypeptide chain?

*34. A synthetic mRNA added to a cell-free protein-synthesizing system produces a peptide with the following amino acid sequence: Met-Pro-Ile-Ser-Ala. What would be the effect on translation if the following components were omitted from the cell-free protein-synthesizing system? What, if any, type of protein would be produced? Explain your reasoning.

   a. Initiation factor 3

   b. Initiation factor 2

   c. Elongation factor Tu

   d. Elongation factor G

   e. Release factors RF-1, RF-2, and RF-3

   f. ATP

   g. GTP

35. For each of the following sequences, place a check mark in the appropriate space to indicate the process *most immediately* affected by deleting the sequence. Choose only one process for each sequence (i.e., one check mark per sequence).

Process most immediately affected by deletion

Sequence deleted	Replication	Transcription	RNA processing	Translation
a. *ori* site	___	___	___	___
b. 3′ splice-site consensus	___	___	___	___
c. poly(A) tail	___	___	___	___
d. terminator	___	___	___	___
e. start codon	___	___	___	___
f. −10 consensus	___	___	___	___
g. Shine–Dalgarno	___	___	___	___

36. MicroRNAs are small RNA molecules that bind to the 3′ end of mRNAs and suppress translation (see Chapter 14). How miRNAs suppress translation is still being investigated. Some eukaryotic mRNAs have internal ribosome-binding sites downstream of the 5′ cap, where ribosomes normally bind. In one investigation, miRNAs did not suppress the translation of ribosomes that attach to internal ribosome-binding sites (R. S. Pillai et al. 2005. *Science* 309:1573–1576). What does this finding suggest about how miRNAs suppress translation?

37. Give the amino acid sequence of the protein encoded by the mRNA in **Figure 15.21**.

Section 15.4

*38. Mutations that introduce stop codons cause a number of genetic diseases. For example, from 2% to 5% of the people who have cystic fibrosis possess a mutation that causes a premature stop codon in the gene encoding the cystic fibrosis transmembrane conductance regulator (CFTR). This premature stop codon produces a truncated form of CFTR that is nonfunctional and results in the symptoms of cystic fibrosis. One possible way to treat people with genetic diseases caused by these types of mutations is to trick the ribosome into reading through the stop codon, inserting an amino acid into its place. Although the protein produced may have one altered amino acid, it is more likely to be at least partly functional than is the truncated protein produced when the ribosome stalls at the stop codon. Indeed, geneticists have conducted clinical trials on people with cystic fibrosis with the use of a drug called PTC124, which interferes with the ribosome's ability to correctly read stop codons (C. Ainsworth. 2005. *Nature* 438:726–728). On the basis of what you know about the mechanism of nonsense-mediated mRNA decay (NMD), would you expect NMD to be a problem with this type of treatment? Why or why not?

Child with cystic fibrosis. [Lisa Eastman/Alamy.]

CHALLENGE QUESTION

Section 15.2

39. The redundancy of the genetic code means that some amino acids are specified by more than one codon. For example, the amino acid leucine is encoded by six different codons. Within a genome, synonymous codons are not present in equal numbers; some synonymous codons appear much more frequently than others, and the preferred codons differ among different species. For example, in one species, the codon UUA might be used most often to encode leucine, whereas, in another species, the codon CUU might be used most often. Speculate on a reason for this bias in codon usage and why the preferred codons are not the same in all organisms.

Section 15.3

40. In what ways are spliceosomes and ribosomes similar? In what ways are they different? Can you suggest some possible reasons for their similarities.

*41. Several experiments were conducted to obtain information about how the eukaryotic ribosome recognizes the AUG start codon. In one experiment, the gene that encodes methionine initiator tRNA ($tRNA_i^{Met}$) was located and changed. The nucleotides that specify the anticodon on $tRNA_i^{Met}$ were mutated so that the anticodon in the tRNA was 5′-CCA-3′ instead of 5′-CAU-3′. When this mutated gene was placed in a eukaryotic cell, protein synthesis took place but the proteins produced were abnormal. Some of the proteins produced contained extra amino acids, and others contained fewer amino acids than normal.

a. What do these results indicate about how the ribosome recognizes the starting point for translation in eukaryotic cells? Explain your reasoning.

b. If the same experiment had been conducted on bacterial cells, what results would you expect?

c. Explain why some proteins contained extra amino acids while others contained fewer amino acids than normal.

▶ Go to your ⮔LaunchPad to find additional learning resources and the Suggested Readings for this chapter.

16

Control of Gene Expression in Bacteria

The expression of genes in bacteria is often regulated through operons, groups of genes that are transcribed as a unit. Shown here is *Escherichia coli*, a common bacterium found in the intestinal tracts of mammals.

[Pasieka/Science Source.]

Operons and the Noisy Cell

In 2011, geneticists from around the world celebrated the fiftieth anniversary of the operon. An operon is a group of genes that share a common promoter and are transcribed as a unit, producing a single mRNA that encodes several proteins. Typically, these proteins interact in some way; for example, the *trp* operon in the bacterium *Escherichia coli* encodes components of three enzymes that work together to synthesize the amino acid tryptophan. Operons control the expression of genes such as these by regulating their transcription. Genes in bacteria are often organized into operons, but operons are much less common in eukaryotes.

The operon was discovered through the elegant research of Francois Jacob and Jacques Monod, who worked at opposite ends of the attic floor of the Pasteur Institute in Paris. Jacob was studying bacteriophage lambda, a virus that infects *E. coli*; Monod was analyzing the properties of β-galactosidase, an enzyme *E. coli* uses to metabolize the sugar lactose. One summer evening in 1958, Jacob had a flash of inspiration—he saw a connection between the research taking place at each end of the attic. Jacob recognized that the genes that induced phage reproduction were controlled in the same way as the genes that control production of β-galactosidase in *E. coli*. This led to an important collaboration between Jacob and Monod; together they eventually uncovered the structure and function of the *lac* operon, and, in 1965, were awarded the Nobel Prize in Physiology or Medicine along with their collaborator Andre Lwoff.

Following Jacob and Monod's discovery of the *lac* operon, other operons were discovered in bacteria and a great deal of research focused on the mechanism of operon function. Despite extensive research on *how* operons work, much less was known about *why* operons exist: why do prokaryotes have them while most eukaryotes don't? Why are some genes included in operons and others not? These questions intrigued Oleg Igoshin at Rice University, and Christian Ray at the University of Texas MD Anderson Cancer Center. Igoshin and Ray are computational biologists, a new breed of scientists who use complex mathematics to study fundamental problems of biology. Igoshin and Ray took what might seem like an unlikely approach to the question of why operons exist. Instead of growing bacteria, inducing mutations, and examining DNA, they developed a series of mathematical models of gene networks that could be run on the computer. With these models they looked at how genes functioned when grouped into operons and when regulated separately.

Igoshin and Ray knew that random fluctuations occur naturally in the levels of transcription and translation. Because of these fluctuations—noise in the system—the amounts of different proteins can vary widely; the amount of protein produced may be more or less than is optimal for cell growth and survival. Igoshin and Ray hypothesized that coordinating the transcription of several genes through an operon structure might reduce noise in the system and permit more finely tuned control over gene expression.

To test their hypothesis, Igoshin and Ray ran computer models for six different types of interactions between the products of genes that are potentially found in operons. Their models showed that for some types of protein interactions, grouping genes together in an operon decreases the biochemical noise. For other types of protein interactions, grouping the genes in operons actually increases the noise. Thus, operon structure has the potential to increase or decrease noise depending on which genes are grouped together.

Igoshin and Ray then examined genes that are actually found in operons in *E. coli*, and discovered that operons with genes whose interactions decrease noise were more common than expected on a random basis. Conversely, operons whose gene interactions increase noise were less common than expected. They concluded that operons have evolved as a way for the cell to couple together transcription of genes to reduce biochemical noise in the cell, allowing the cell to more finely tune the relative proportions of proteins encoded by the operon. Igoshin and Ray speculated that operons are less common in eukaryotes because the larger cell volume reduces the effect of random fluctuations and, perhaps, because eukaryotes have other mechanisms (such as changes in chromatin structure) that couple the transcription of genes.

This is the first of two chapters about **gene regulation**, the mechanisms and systems that control the expression of genes. In this chapter, we consider systems of gene regulation in bacteria. We begin by considering the necessity for gene regulation, the levels at which gene expression is controlled, and the difference between genes and regulatory elements. We then examine the structure and function of operons, and consider gene regulation in some specific examples of operons, including the *lac* operon studied by Jacob and Monod. Finally, we will discuss several types of bacterial gene regulation that are facilitated by RNA molecules. In Chapter 17 we will discuss mechanisms of gene regulation in eukaryotic genomes.

16.1 The Regulation of Gene Expression Is Critical for All Organisms

A major theme of molecular genetics is the central dogma, which states that genetic information flows from DNA to RNA to proteins (see Figure 10.16). Although the central dogma provided a molecular basis for the connection between genotype and phenotype, it failed to address a critical question: how is the flow of information along the molecular pathway *regulated*?

Consider *E. coli*, a bacterium that resides in your large intestine. Your eating habits completely determine the nutrients available to this bacterium: it can neither seek out nourishment when nutrients are scarce nor move away when confronted with an unfavorable environment. *E. coli* makes up for its inability to alter the external environment by being internally flexible. For example, if glucose is present, *E. coli* uses it to generate ATP; if there's no glucose, it utilizes

lactose, arabinose, maltose, xylose, or any of a number of other sugars. When amino acids are available, *E. coli* uses them to synthesize proteins; if a particular amino acid is absent, *E. coli* produces the enzymes needed to synthesize that amino acid. Thus, *E. coli* responds to environmental changes by rapidly altering its biochemistry. This biochemical flexibility, however, has a high price. Producing all the enzymes necessary for every environmental condition would be energetically expensive. So how does *E. coli* maintain biochemical flexibility while optimizing energy efficiency?

The answer is through gene regulation. Bacteria carry the genetic information for synthesizing many proteins, but only a subset of this genetic information is expressed at any time. When the environment changes, new genes are expressed, and proteins appropriate for the new environment are synthesized. For example, if a carbon source appears in the environment, genes encoding enzymes that take up and metabolize this carbon source are quickly transcribed and translated. When this carbon source disappears, the genes that encode these enzymes are shut off.

Multicellular eukaryotic organisms face a different dilemma. Individual cells in a multicellular organism are specialized for particular tasks. The proteins produced by a nerve cell, for example, are quite different from those produced by a white blood cell. Although they differ in shape and function, a nerve cell and a blood cell still carry the same genetic instructions.

A multicellular organism's challenge is to bring about the specialization of cells that have a common set of genetic instructions (the process of development). This challenge is met through gene regulation: all of an organism's cells carry the same genetic information, but only a subset of genes are expressed in each cell type. Genes needed for other cell types are not expressed. Gene regulation is therefore the key to both unicellular flexibility and multicellular specialization, and it is critical to the success of all living organisms.

In bacteria, gene regulation maintains internal flexibility, turning genes on and off in response to environmental changes. In multicellular eukaryotic organisms, gene regulation brings about cellular differentiation.

The mechanisms of gene regulation were first investigated in bacterial cells, in which the availability of mutants and the ease of laboratory manipulation made it possible to unravel the mechanisms. When the study of these mechanisms in eukaryotic cells began, bacterial gene regulation seemed to clearly differ from eukaryotic gene regulation. However, as more and more information has accumulated about gene regulation, a number of common themes have emerged. Today, many aspects of gene regulation in bacterial and eukaryotic cells are recognized to be similar. Before examining specific elements of bacterial gene regulation (this chapter) and eukaryotic gene regulation (Chapter 17), we will briefly consider some themes of gene regulation common to all organisms.

Genes and Regulatory Elements

In considering gene regulation in both bacteria and eukaryotes, we must distinguish between the DNA sequences that are transcribed and the DNA sequences that regulate the expression of other sequences. **Structural genes** encode proteins that are used in metabolism or biosynthesis or that play a structural role in the cell. **Regulatory genes** are genes whose products, either RNA or proteins, interact with other DNA sequences and affect the transcription or translation of those sequences. In many cases, the products of regulatory genes are DNA-binding proteins (although RNA molecules also affect gene expression). Bacteria and eukaryotes use regulatory genes to control the expression of many of their structural genes. However, a few structural genes, particularly those that encode essential cellular functions, are expressed continually and are said to be **constitutive**. Constitutive genes are therefore not regulated.

We will also encounter DNA sequences that are not transcribed at all but still play a role in regulating genes and other DNA sequences. These **regulatory elements** affect the expression of sequences to which they are physically linked. Regulatory elements are common in both bacterial and eukaryotic cells, and much of gene regulation in both types of organisms takes place through the action of proteins produced by regulatory genes that recognize and bind to regulatory elements.

The regulation of gene expression can be through processes that stimulate gene expression, termed *positive control*, or through processes that inhibit gene expression, termed *negative control*. Bacteria and eukaryotes use both positive and negative control mechanisms to regulate their genes. However, negative control is more important in bacteria, whereas eukaryotes are more likely to use positive control mechanisms.

Regulatory elements are DNA sequences that are not transcribed but affect the expression of genes. Positive control mechanisms stimulate gene expression, whereas negative control inhibits gene expression.

✓ **CONCEPT CHECK 1**

What is a constitutive gene?

Levels of Gene Regulation

In both bacteria and eukaryotes, genes can be regulated at a number of levels along the pathway of information flow from genotype to phenotype (**Figure 16.1**). First, regulation can be through the alteration of DNA or chromatin structure; this type of gene regulation takes place primarily in eukaryotes.

16.1 **Gene expression can be controlled at multiple levels.**

Modifications to DNA or its packaging can help to determine which sequences are available for transcription or the rate at which sequences are transcribed. DNA methylation and changes in chromatin are two processes that play a pivotal role in gene regulation.

A second point at which a gene can be regulated is at the level of transcription. For the sake of cellular economy, limiting the production of a protein early in the process makes sense, and transcription is an important point of gene regulation in both bacterial and eukaryotic cells. A third potential point of gene regulation is mRNA processing. Eukaryotic mRNA is extensively modified before it is translated: a 5′ cap is added, the 3′ end is cleaved and polyadenylated, and introns are removed (see Chapter 14). These modifications determine the stability of the mRNA, the movement of the mRNA into the cytoplasm, whether the mRNA can be translated, the rate of translation, and the amino acid sequence of the protein produced. There is growing evidence that a number of regulatory mechanisms in eukaryotic cells operate at the level of mRNA processing.

A fourth point for the control of gene expression is the regulation of RNA stability. The amount of protein produced depends not only on the amount of mRNA synthesized, but also on the rate at which the mRNA is degraded. A fifth point of gene regulation is at the level of translation, a complex process requiring a large number of enzymes, protein factors, and RNA molecules (see Chapter 15). All of these factors, as well as the availability of amino acids, affect the rate at which proteins are produced and therefore provide points at which gene expression can be controlled. Translation can also be affected by sequences in mRNA.

Finally, many proteins are modified after translation (see Chapter 15), and these modifications affect whether the proteins become active; genes can be regulated through processes that affect posttranslational modification. Gene expression can be affected by regulatory activities at any or all of these points.

CONCEPTS

Gene expression can be controlled at any of a number of levels along the molecular pathway from DNA to protein, including DNA or chromatin structure, transcription, mRNA processing, RNA stability, translation, and posttranslational modification.

✓ **CONCEPT CHECK 2**

Why is transcription a particularly important level of gene regulation in both bacteria and eukaryotes?

DNA-Binding Proteins

Much of gene regulation in bacteria and eukaryotes is accomplished by proteins that bind to DNA sequences and affect their expression. These regulatory proteins generally have discrete functional parts—called **domains**, typically consisting of 60 to 90 amino acids—that are responsible for binding to DNA. Within a domain, only a few amino acids actually make contact with the DNA. These amino acids (most commonly asparagine, glutamine, glycine, lysine, and arginine) often form hydrogen bonds with the bases or interact with the sugar–phosphate backbone of the DNA. Many regulatory proteins have additional domains that can bind other molecules such as other regulatory proteins. By physically attaching to DNA, these proteins can affect the expression of a gene. Most DNA-binding proteins bind dynamically, which means that they are transiently binding and unbinding DNA and other regulatory proteins. Thus, although they may spend most of their time bound to DNA, they are never permanently attached. This dynamic nature means that other molecules can compete with DNA-binding proteins for regulatory sites on the DNA.

DNA-binding proteins can be grouped into several distinct types on the basis of a characteristic structure, called a motif, found within the binding domain. Motifs are simple structures, such as alpha helices, that can fit into the major groove of the DNA. For example, the helix-turn-helix motif (**Figure 16.2a**), consisting of two alpha helices connected

(a) Helix-turn-helix

(b) Zinc fingers

(c) Leucine zipper

16.2 DNA-binding proteins can be grouped into several types on the basis of their structures, or motifs.

Table 16.1	Common DNA-binding motifs		
Motif	Location	Characteristics	Binding Site in DNA
Helix-turn-helix	Bacterial regulatory proteins; related motifs in eukaryotic proteins	Two alpha helices	Major groove
Zinc-finger	Eukaryotic regulatory and other proteins	Loop of amino acids with zinc at base	Major groove
Steroid receptor	Eukaryotic proteins	Two perpendicular alpha helices with zinc surrounded by four cysteine residues	Major groove and DNA backbone
Leucine-zipper	Eukaryotic transcription factors	Helix of leucine residues and a basic arm; two leucine residues interdigitate	Two adjacent major grooves
Helix-loop-helix	Eukaryotic proteins	Two alpha helices separated by a loop of amino acids	Major groove
Homeodomain	Eukaryotic regulatory proteins	Three alpha helices	Major groove

by a turn, is common in bacterial regulatory proteins. The zinc-finger motif (**Figure 16.2b**), common to many eukaryotic regulatory proteins, consists of a loop of amino acids containing a zinc ion. The leucine zipper (**Figure 16.2c**) is another motif found in a variety of eukaryotic binding proteins. These common DNA-binding motifs and others are summarized in **Table 16.1**.

CONCEPTS

Regulatory proteins that bind DNA have common motifs that interact with sequences in the DNA.

✔ CONCEPT CHECK 3

How do amino acids in DNA-binding proteins interact with DNA?
a. By forming covalent bonds with DNA bases
b. By forming hydrogen bonds with DNA bases
c. By forming covalent bonds with DNA sugars

16.2 Operons Control Transcription in Bacterial Cells

A significant difference between bacterial and eukaryotic gene control lies in the organization of functionally related genes. As discussed in the introduction to this chapter, many bacterial genes that have related functions are clustered and under the control of a single promoter. These genes are often transcribed together into a single mRNA. A group of bacterial structural genes that are transcribed together (along with their promoter and additional sequences that control transcription) is called an **operon**. The operon regulates the expression of the structural genes by controlling transcription, which, in bacteria, is usually the most important level of gene regulation.

Operon Structure

The organization of a typical operon is illustrated in **Figure 16.3**. At one end of the operon is a set of structural genes,

16.3 An operon is a single transcriptional unit that includes a series of structural genes, a promoter, and an operator.

shown in Figure 16.3 as gene *a*, gene *b*, and gene *c*. These structural genes are transcribed into a single mRNA, which is translated to produce enzymes A, B, and C. These enzymes carry out a series of biochemical reactions that convert precursor molecule X into product Y. The transcription of structural genes *a*, *b*, and *c* is under the control of a promoter, which lies upstream of the first structural gene. RNA polymerase binds to the promoter and then moves downstream, transcribing the structural genes.

A **regulator gene** helps to control the transcription of the structural genes of the operon. Although it affects operon function, the regulator gene is not considered part of the operon. The regulator gene has its own promoter and is transcribed into a short mRNA, which is translated into a small protein. This **regulator protein** can bind to a region of the operon called the **operator** and affect whether transcription can take place. The operator usually overlaps the 3′ end of the promoter and sometimes the 5′ end of the first structural gene (see Figure 16.3).

CONCEPTS

Functionally related genes in bacterial cells are frequently clustered together as a single transcriptional unit termed an operon. A typical operon includes several structural genes, a promoter for the structural genes, and an operator site to which the product of a regulator gene binds.

✓ **CONCEPT CHECK 4**

What is the difference between a structural gene and a regulator gene?
a. Structural genes are transcribed into mRNA, but regulator genes aren't.
b. Structural genes have complex structures; regulator genes have simple structures.
c. Structural genes encode proteins that function in the structure of the cell; regulator genes carry out metabolic reactions.
d. Structural genes encode proteins; regulator genes control the transcription of structural genes.

Negative and Positive Control: Inducible and Repressible Operons

There are two types of transcriptional control: **negative control**, in which a regulatory protein is a repressor, binding to DNA and inhibiting transcription; and **positive control**, in which a regulatory protein is an activator, stimulating transcription. Operons can also be either inducible or repressible. **Inducible operons** are those in which transcription is normally off (not taking place); something must happen to induce transcription, or turn it on. **Repressible operons** are those in which transcription is normally on (taking place); something must happen to repress transcription, or turn it off. In the next sections, we will consider several varieties of these basic control mechanisms.

NEGATIVE INDUCIBLE OPERONS In a negative inducible operon, the regulator gene encodes an active *repressor* that readily binds to the operator (**Figure 16.4a**). Because the operator site overlaps the promoter site, the binding of this protein to the operator physically blocks the binding of RNA polymerase to the promoter and prevents transcription. For transcription to take place, something must happen to prevent the binding of the repressor at the operator site. This type of system is said to be *inducible* because transcription is normally off (inhibited) and must be turned on (induced).

Transcription is turned on when a small molecule, an **inducer**, binds to the repressor (**Figure 16.4b**). Regulatory proteins frequently have two binding sites: one that binds to DNA and another that binds to a small molecule such as an inducer. The binding of the inducer (precursor V in Figure 16.4b) alters the shape of the repressor, preventing it from binding to DNA. Proteins of this type, which change shape on binding to another molecule, are called **allosteric proteins**.

When the inducer is absent, the repressor binds to the operator, the structural genes are not transcribed, and enzymes D, E, and F (which metabolize precursor V) are not synthesized (see Figure 16.4a). This mechanism is an adaptive one: because no precursor V is available, synthesis of the enzymes would be wasteful when they have no substrate to metabolize. As soon as precursor V becomes available, some of it binds to the repressor, rendering the repressor inactive and unable to bind to the operator site. RNA polymerase can now bind to the promoter and transcribe the structural genes. The resulting mRNA is then translated into enzymes D, E, and F, which convert substrate V into product W (see Figure 16.4b). So, an operon with negative inducible control regulates the synthesis of the enzymes economically: the enzymes are synthesized only when their substrate (V) is available.

Inducible operons usually control proteins that carry out degradative processes—proteins that break down molecules. For these types of proteins, inducible control makes sense because the proteins are not needed unless the substrate (which is broken down by the proteins) is present.

NEGATIVE REPRESSIBLE OPERONS Some operons with negative control are *repressible*, meaning that transcription *normally* takes place and must be turned off, or repressed. The regulator protein in this type of operon also is a repressor but is synthesized in an *inactive* form that cannot by itself bind to the operator. Because no repressor is bound to the operator, RNA polymerase readily binds to the promoter and transcription of the structural genes takes place (**Figure 16.5a**).

To turn transcription off, something must happen to make the repressor active. A small molecule called a **corepressor** binds to the repressor and makes it capable of binding to the operator. In the example illustrated (see Figure 16.5a), the product (U) of the metabolic reaction is the corepressor. As long as the level of product U is high, it is available to bind to the repressor and activate it, preventing transcription

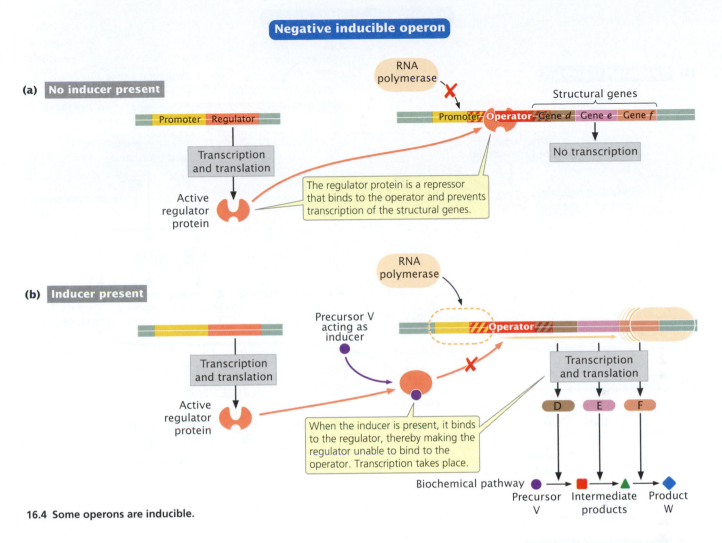

Negative inducible operon

(a) No inducer present

RNA polymerase

Structural genes

Promoter | Regulator

Promoter | **Operator** | Gene *d* | Gene *e* | Gene *f*

Transcription and translation

No transcription

Active regulator protein

The regulator protein is a repressor that binds to the operator and prevents transcription of the structural genes.

(b) Inducer present

RNA polymerase

Precursor V acting as inducer

Operator

Transcription and translation

Transcription and translation

Active regulator protein

D   E   F

When the inducer is present, it binds to the regulator, thereby making the regulator unable to bind to the operator. Transcription takes place.

Biochemical pathway

Precursor V → Intermediate products → Product W

16.4 Some operons are inducible.

(**Figure 16.5b**). With the operon repressed, enzymes G, H, and I are not synthesized, and no more U is produced from precursor T. However, when all of product U is used up, the repressor is no longer activated by product U and cannot bind to the operator. The inactivation of the repressor allows the transcription of the structural genes and the synthesis of enzymes G, H, and I, resulting in the conversion of precursor T into product U. Like inducible operons, repressible operons are economical: the enzymes are synthesized only as needed.

Repressible operons usually control proteins that carry out the biosynthesis of molecules needed in the cell, such as amino acids. For these types of operons, repressible control makes sense because the product produced by the proteins is always needed by the cell. Thus, these operons are normally on and are turned off when there are adequate amounts of the product already present.

Note that both the inducible and the repressible systems that we have considered are forms of negative control, in which the regulatory protein is a repressor. We will now consider positive control, in which a regulator protein stimulates transcription.

POSITIVE CONTROL   With positive control, a regulatory protein is an activator: it binds to DNA (usually at a site other than the operator) and stimulates transcription. Positive control can be inducible or repressible.

In a positive *inducible* operon, transcription is normally turned off because the regulator protein (an activator) is produced in an inactive form. Transcription takes place when an inducer has become attached to the regulatory protein, rendering the regulator active. Logically, the inducer should be the precursor of the reaction controlled by the operon so that the necessary enzymes would be synthesized only when the substrate for their reaction was present.

A positive operon can also be repressible; the regulatory protein is produced in a form that readily binds to DNA, meaning that transcription normally takes place and has to be repressed. Transcription is inhibited when a substance becomes attached to the activator and renders it unable to bind to the DNA so that transcription is no longer stimulated. Here, the product (P) of the reaction controlled by the operon would logically be the repressing substance, because it would be economical for the cell to prevent the transcription of genes that allow the synthesis of P when plenty of P was

Negative repressible operon

(a) No product U present

RNA polymerase

Promoter | Regulator

Structural genes

Promoter—Operator—Gene *g* | Gene *h* | Gene *i*

Transcription and translation → Inactive regulator protein (repressor)

1 The regulator protein is an inactive repressor, unable to bind to the operator.

Transcription and translation

2 Transcription of the structural genes therefore takes place.

Enzymes: G H I

Biochemical pathway: Precursor T → Intermediate products → Product U (corepressor)

(b) Product U present

RNA polymerase

3 Levels of product U build up.

Product U

Transcription and translation → Inactive regulator protein (repressor)

Operator

No transcription

4 Product U binds to the regulator protein,…

5 …making it active and able to bind to the operator…

6 …and thus preventing transcription.

16.5 Some operons are repressible.

already available. The characteristics of positive and negative control in inducible and repressible operons are summarized in **Figure 16.6**. ▶ **TRY PROBLEM 11**

CONCEPTS

There are two basic types of transcriptional control: negative and positive. In negative control, when a regulatory protein (repressor) binds to DNA, transcription is inhibited; in positive control, when a regulatory protein (activator) binds to DNA, transcription is stimulated. Some operons are inducible; transcription is normally off and must be turned on. Other operons are repressible; transcription is normally on and must be turned off.

✔ CONCEPT CHECK 5

In a negative repressible operon, the regulator protein is synthesized as
a. an active activator.
b. an inactive activator.
c. an active repressor.
d. an inactive repressor.

The *lac* Operon of *E. coli*

In 1961, François Jacob and Jacques Monod described the "operon model" for the genetic control of lactose metabolism in *E. coli*. This work and subsequent research on the genetics of lactose metabolism established the operon as the basic unit of transcriptional control in bacteria. Despite the fact that, at the time, no methods were available for determining nucleotide sequences, Jacob and Monod deduced the structure of the operon *genetically* by analyzing the interactions of mutations that interfered with the normal regulation of lactose metabolism. We will examine the effects of some of these mutations after seeing how the *lac* operon regulates lactose metabolism.

LACTOSE METABOLISM Lactose is a major carbohydrate found in milk; it can be metabolized by *E. coli* bacteria that reside in the mammalian gut. Lactose does not easily diffuse across the *E. coli* cell membrane and must be actively transported into the cell by the protein permease (**Figure 16.7**). To utilize lactose as an energy source, *E. coli*

(a) Negative inducible

RNA polymerase

Active repressor

Transcription **off**

Transcription **on**

Substrate makes the repressor inactive.

Substrate   Product

(b) Negative repressible

Inactive repressor

Transcription **on**

Transcription **off**

Product makes the repressor active.

(c) Positive inducible

Inactive activator

Transcription **off**

Transcription **on**

Substrate makes the activator active.

(d) Positive repressible

Active activator

Transcription **on**

Transcription **off**

Product makes the activator inactive.

16.6 A summary of the characteristics of positive and negative control in inducible and repressible operons.

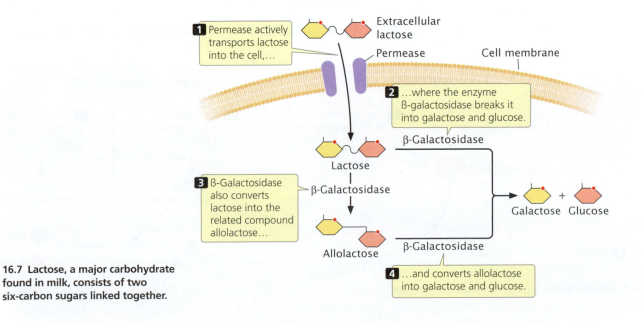

1 Permease actively transports lactose into the cell,...

Extracellular lactose

Permease

Cell membrane

2 ...where the enzyme ß-galactosidase breaks it into galactose and glucose.

β-Galactosidase

Lactose

3 ß-Galactosidase also converts lactose into the related compound allolactose...

β-Galactosidase

β-Galactosidase

Galactose   Glucose

Allolactose

β-Galactosidase

4 ...and converts allolactose into galactose and glucose.

16.7 Lactose, a major carbohydrate found in milk, consists of two six-carbon sugars linked together.

451

must first break it into glucose and galactose, a reaction catalyzed by the enzyme β-galactosidase. This enzyme can also convert lactose into allolactose, a compound that plays an important role in regulating lactose metabolism. A third enzyme, thiogalactoside transacetylase, is also produced by the *lac* operon, but its function in lactose metabolism is not yet clear. One possible function is detoxification, preventing the accumulation of thiogalactosides that are transported into the cell along with lactose by lactose permease.

REGULATION OF THE LAC OPERON The *lac* operon is an example of a negative inducible operon. The enzymes β-galactosidase, permease, and transacetylase are encoded by adjacent structural genes in the *lac* operon of *E. coli* (**Figure 16.8a**) and have a common promoter (*lacP* in Figure 16.8a). β-Galactosidase is encoded by the *lacZ* gene, permease by the *lacY* gene, and transacetylase by the *lacA* gene. When lactose is absent from the medium in which *E. coli* grows, few molecules of each protein are produced. If lactose is added to the medium and glucose

is absent, the rate of synthesis of all three proteins simultaneously increases about a thousandfold within 2 to 3 minutes. This boost in protein synthesis results from the transcription of *lacZ*, *lacY*, and *lacA* and exemplifies **coordinate induction**, the simultaneous synthesis of several proteins, stimulated by a specific molecule, the inducer (**Figure 16.8b**).

Although lactose appears to be the inducer here, allolactose is actually responsible for induction. Upstream of *lacP* is a regulator gene, *lacI*, which has its own promoter (P_I). The *lacI* gene is transcribed into a short mRNA that is translated into a repressor. The repressor consists of four identical polypeptides and has two types of binding sites; one type of site binds to allolactose and the other binds to DNA. In the absence of lactose (and, therefore, allolactose), the repressor binds to the *lac* operator site *lacO* (see Figure 16.8a). Jacob and Monod mapped the operator to a position adjacent to the *lacZ* gene; more-recent nucleotide sequencing has demonstrated that the operator actually overlaps the 3′ end of the promoter and the 5′ end of *lacZ* (**Figure 16.9**)

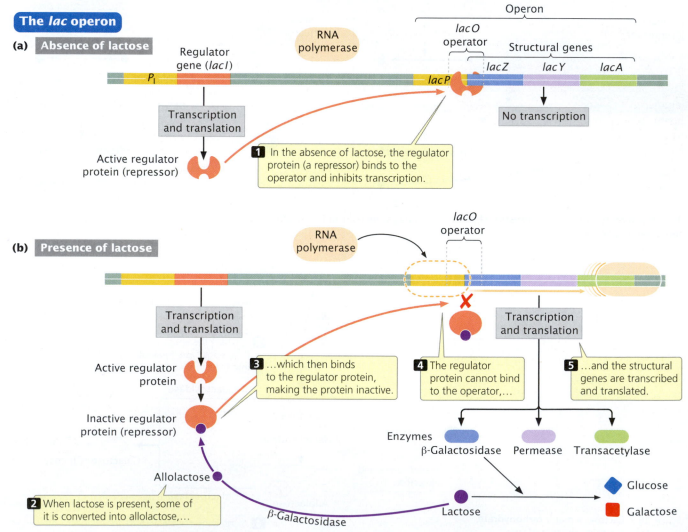

16.8 The *lac* operon regulates lactose metabolism.

16.9 In the *lac* operon, the operator overlaps the promoter and the 5′ end of the first structural gene.

RNA polymerase binds to the promoter and moves down the DNA molecule, transcribing the structural genes. When the repressor is bound to the operator, the binding of RNA polymerase is blocked, and transcription is prevented. When lactose is present, some of it is converted into allolactose, which binds to the repressor and causes the repressor to be released from the DNA. In the presence of lactose, then, the repressor is inactivated, the binding of RNA polymerase is no longer blocked, the transcription of *lacZ*, *lacY*, and *lacA* takes place, and the *lac* proteins are produced.

Have you spotted the flaw in the explanation just given for the induction of the *lac* proteins? You might recall that permease is required to transport lactose into the cell. If the *lac* operon is repressed and no permease is being produced, how does lactose get into the cell to inactivate the repressor and turn on transcription? Furthermore, the inducer is actually allolactose, which must be produced from lactose by β-galactosidase. If β-galactosidase production is repressed, how can lactose metabolism be induced?

The answer is that repression never *completely* shuts down transcription of the *lac* operon. Even with active repressor bound to the operator, there is a low level of transcription and a few molecules of β-galactosidase, permease, and transacetylase are synthesized. When lactose appears in the medium, the permease that is present transports a small amount of lactose into the cell. There, the few molecules of β-galactosidase that are present convert some of the lactose into allolactose, which then induces transcription.

Several compounds related to allolactose also can bind to the *lac* repressor and induce transcription of the *lac* operon. One such inducer is isopropylthiogalactoside (IPTG). Although IPTG inactivates the repressor and allows the transcription of *lacZ*, *lacY*, and *lacA*, this inducer is not metabolized by β-galactosidase; for this reason, IPTG is often used in research to examine the effects of induction, independent of metabolism.

CONCEPTS

The *lac* operon of *E. coli* controls the transcription of three genes needed in lactose metabolism: the *lacZ* gene, which encodes β-galactosidase; the *lacY* gene, which encodes permease; and the *lacA* gene, which encodes thiogalactoside transacetylase. The *lac* operon is negative inducible: a regulator

gene produces a repressor that binds to the operator site and prevents the transcription of the structural genes. The presence of allolactose inactivates the repressor and allows the transcription of the *lac* operon.

✓ CONCEPT CHECK 6

In the presence of allolactose, the *lac* repressor

a. binds to the operator.

b. binds to the promoter.

c. cannot bind to the operator.

d. binds to the regulator gene.

lac Mutations

Jacob and Monod worked out the structure and function of the *lac* operon by analyzing mutations that affected lactose metabolism. To help define the roles of the different components of the operon, they used **partial diploid** strains of *E. coli*. The cells of these strains possessed two different DNA molecules: the full bacterial chromosome and an extra piece of DNA. Jacob and Monod created these strains by allowing conjugation to take place between two bacteria (see Chapter 9). In conjugation, a small circular piece of DNA (the F plasmid, see Chapter 9) is transferred from one bacterium to another. The F plasmid used by Jacob and Monod contained the *lac* operon so the recipient bacterium became partly diploid, possessing two copies of the *lac* operon. By using different combinations of mutations on the bacterial and plasmid DNA, Jacob and Monod determined that some parts of the *lac* operon are cis acting (able to control the expression of genes only when on the same piece of DNA), whereas other parts are trans acting (able to control the expression of genes on other DNA molecules).

STRUCTURAL-GENE MUTATIONS   Jacob and Monod first discovered some mutant strains that had lost the ability to synthesize either β-galactosidase or permease (they did not study in detail the effects of mutations on the transacetylase enzyme, and so transacetylase will not be considered here). The mutations in the mutant strains mapped to the *lacZ* or *lacY* structural genes and altered the amino acid sequences of the proteins encoded by the genes. These mutations clearly affected the *structure* of the proteins but not the regulation of their synthesis.

Through the use of partial diploids, Jacob and Monod were able to establish that mutations at the *lacZ* and *lacY* genes were independent and usually affected only the product of the gene in which the mutation occurred. Partial diploids with *lacZ⁺ lacY⁻* on the bacterial chromosome and *lacZ⁻ lacY⁺* on the plasmid functioned normally, producing β-galactosidase and permease in the presence of lactose. (The genotype of a partial diploid is written by separating the genes on each DNA molecule with a slash: *lacZ⁺ lacY⁻/lacZ⁻ lacY⁺*.) In this partial diploid, a single functional β-galactosidase gene (*lacZ⁺*) is sufficient to produce β-galactosidase; whether the functional β-galactosidase gene is coupled to a functional (*lacY⁺*) or a defective (*lacY⁻*) permease gene makes no difference. The same is true of the *lacY⁺* gene.

REGULATOR-GENE MUTATIONS   Jacob and Monod also isolated mutations that affected the *regulation* of protein production. Mutations in the *lacI* gene affect the production of both β-galactosidase and permease because genes for both proteins are in the same operon and are regulated coordinately.

Some of these mutations were constitutive, causing the *lac* proteins to be produced all the time, whether lactose was

present or not. Such mutations in the regulator gene were designated *lacI⁻*. The construction of partial diploids demonstrated that a *lacI⁺* gene is dominant over a *lacI⁻* gene; a single copy of *lacI⁺* (genotype *lacI⁺/lacI⁻*) was sufficient to bring about normal regulation of protein production. Furthermore, *lacI⁺* restored normal control to an operon even if the operon was located on a different DNA molecule, showing that *lacI⁺* can be trans acting. A partial diploid with genotype *lacI⁺ lacZ⁻/lacI⁻ lacZ⁺* functioned normally, synthesizing β-galactosidase only when lactose was present (**Figure 16.10**). In this strain, the *lacI⁺* gene on the bacterial chromosome was functional, but the *lacZ⁻* gene was defective; on the plasmid, the *lacI⁻* gene was defective, but the *lacZ⁺* gene was functional. The fact that a *lacI⁺* gene could regulate a *lacZ⁺* gene located on a different DNA molecule indicated to Jacob and Monod that the *lacI⁺* gene product was able to operate on either the plasmid or the chromosome.

Some *lacI* mutations isolated by Jacob and Monod prevented transcription from taking place even in the presence of lactose. These mutations were referred to as superrepressors (*lacIˢ*), because they produced defective repressors that

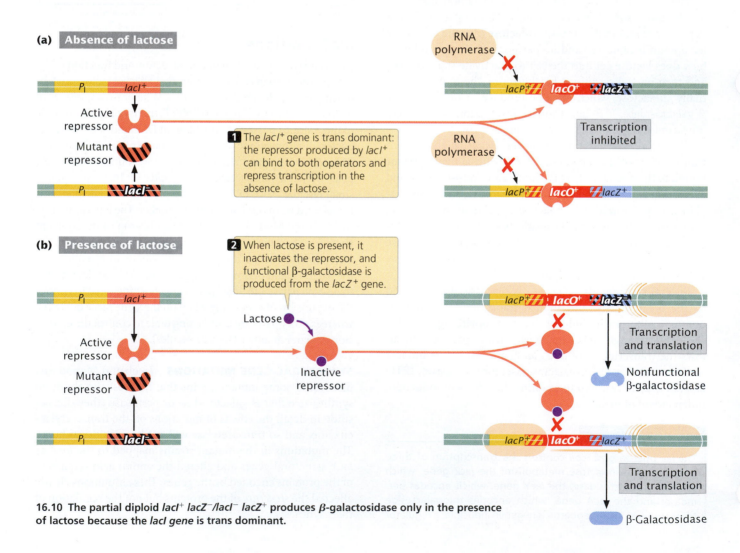

16.10 The partial diploid *lacI⁺ lacZ⁻/lacI⁻ lacZ⁺* produces β-galactosidase only in the presence of lactose because the *lacI* gene is trans dominant.

1 The *lacI*ˢ gene produces a superrepressor that does not bind lactose.

Super-repressor

Active repressor

Lactose → Inactive repressor

RNA polymerase ✗

RNA polymerase ✗

P_I *lacI*ˢ

P_I *lacI*⁺

lacP⁺ *lacO*⁺ *lacZ*⁺

lacP⁺ *lacO*⁺ *lacZ*⁺

2 The *lacI*ˢ gene is trans dominant: the superrepressor binds both operators and prevents transcription in the presence and absence of lactose.

16.11 The partial diploid *lacI*ˢ *lacZ*⁺/*lacI*⁺ *lacZ*⁺ fails to produce β-galactosidase in the presence and absence of lactose because the *lacI*ˢ gene encodes a superrepressor.

could not be inactivated by an inducer. The *lacI*ˢ mutations produced a repressor with an altered inducer-binding site, which made the inducer unable to bind to the repressor; consequently, the repressor was always able to attach to the operator site and prevent transcription of the *lac* genes. Superrepressor mutations were dominant over *lacI*⁺; partial diploids with genotype *lacI*ˢ *lacZ*⁺/*lacI*⁺ *lacZ*⁺ were unable to synthesize either β-galactosidase or permease, whether or not lactose was present (**Figure 16.11**).

OPERATOR MUTATIONS Jacob and Monod mapped another class of constitutive mutants to a site adjacent to *lacZ*. These mutations occurred at the operator site and were referred to as *lacO*ᶜ (*O* stands for operator and "c" for constitutive). The *lacO*ᶜ mutations altered the sequence of DNA at the operator so that the repressor protein was no longer able to bind. A partial diploid with genotype *lacI*⁺ *lacO*ᶜ *lacZ*⁺/*lacI*⁺ *lacO*⁺ *lacZ*⁺ exhibited constitutive synthesis of β-galactosidase, indicating that *lacO*ᶜ is dominant over *lacO*⁺.

Analysis of other partial diploids showed that the *lacO* gene is cis acting, affecting only genes on the same DNA molecule. For example, a partial diploid with genotype *lacI*⁺ *lacO*⁺ *lacZ*⁻/*lacI*⁺ *lacO*ᶜ *lacZ*⁺ was constitutive, producing β-galactosidase in the presence or absence of lactose (**Figure 16.12a**), but a partial diploid with genotype *lacI*⁺ *lacO*⁺ *lacZ*⁺/*lacI*⁺ *lacO*ᶜ *lacZ*⁻ produced β-galactosidase only in the presence of lactose (**Figure 16.12b**). In the constitutive partial diploid (*lacI*⁺ *lacO*⁺ *lacZ*⁻/*lacI*⁺ *lacO*ᶜ *lacZ*⁺; see Figure 16.12a), the *lacO*ᶜ mutation and

the functional *lacZ*⁺ gene are present on the same DNA molecule; but, in *lacI*⁺ *lacO*⁺ *lacZ*⁺/*lacI*⁺ *lacO*ᶜ *lacZ*⁻ (see Figure 16.12b), the *lacO*ᶜ mutation and the functional *lacZ*⁺ gene are on different molecules. The *lacO* mutation affects only genes to which it is physically connected, as is true of all operator mutations. They prevent the binding of a repressor protein to the operator and thereby allow RNA polymerase to transcribe genes on the same DNA molecule. However, they cannot prevent a repressor from binding to normal operators on other DNA molecules. Watch **Animation 16.1** to observe the effects of different combinations of *lacI* and *lacO* mutations on the expression of the *lac* operon. ▶ TRY PROBLEM 21

PROMOTER MUTATIONS Mutations affecting lactose metabolism have also been isolated at the promoter site; these mutations are designated *lacP*⁻, and they interfere with the binding of RNA polymerase to the promoter. Because this binding is essential for the transcription of the structural genes, *E. coli* strains with *lacP*⁻ mutations don't produce *lac* proteins either in the presence or in the absence of lactose. Like operator mutations, *lacP*⁻ mutations are cis acting and thus affect only genes on the same DNA molecule. The partial diploid *lacI*⁺ *lacP*⁺ *lacZ*⁺/*lacI*⁺ *lacP*⁻ *lacZ*⁺ exhibits normal synthesis of β-galactosidase, whereas *lacI*⁺ *lacP*⁻ *lacZ*⁺ / *lacI*⁺ *lacP*⁺ *lacZ*⁻ fails to produce β-galactosidase whether or not lactose is present. The different types of mutations that occur in the lac operon are summarized in **Table 16.2**.

Table 16.2	Characteristics of *lac* operon mutations		
Type	**Location**	**Cis/Trans**	**Effect**
Structural gene mutations	*lacZ*, *lacY*	Only affect *lacZ* or *lacY*	Alter amino acid sequence of protein encoded by gene in which mutation occurs
Regulator gene mutations	*lacI*	Trans	Affect transcription of structural genes
Operator mutations	*lacO*	Cis	Affect transcription of structural genes
Promoter mutations	*lacP*	Cis	Affect transcription of structural genes

(a) Partial diploid *lacI⁺ lacO⁺ lacZ⁻/lacI⁺ lacOᶜ lacZ⁺*

(b) Partial diploid *lacI⁺ lacO⁺ lacZ⁺/lacI⁺ lacOᶜ lacZ⁻*

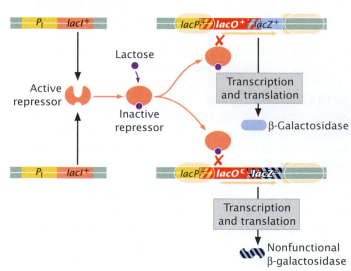

16.12 Mutations in *lacO* are constitutive and cis acting. (a) The partial diploid *lacI⁺lacO⁺ lacZ⁻/lacI⁺ lacOᶜ lacZ⁺* is constitutive, producing β-galactosidase in the presence and absence of lactose. (b) The partial diploid *lacI⁺ lacO⁺ lacZ⁺/lacI⁺ lacOᶜ lacZ⁻* is inducible (produces β-galactosidase only when lactose is present), demonstrating that the *lacO* gene is cis acting.

WORKED PROBLEM

For *E. coli* strains with the following *lac* genotypes, use a plus sign (+) to indicate the synthesis of β-galactosidase and permease and a minus sign (−) to indicate no synthesis of the proteins when lactose is absent and when it is present.

Genotype of strain

a. *lacI⁺ lacP⁺ lacO⁺ lacZ⁺ lacY⁺*
b. *lacI⁺ lacP⁺ lacOᶜ lacZ⁻ lacY⁺*
c. *lacI⁺ lacP⁻ lacO⁺ lacZ⁺ lacY⁻*
d. *lacI⁺ lacP⁺ lacO⁺ lacZ⁻ lacY⁻/lacI⁻ lacP⁺ lacO⁺ lacZ⁺ lacY⁺*

Solution Strategy

What information is required in your answer to the problem?
An indication of whether or not β-galactosidase and permease are produced by each genotype when lactose is present and when lactose is absent, by placing a plus sign (+) or minus sign (−) for each enzyme and condition in the table.

What information is provided to solve the problem?
The genotype of each strain.

Solution Steps

	Lactose absent		Lactose present	
Genotype of strain	β-Galactosidase	Permease	β-Galactosidase	Permease
a. *lacI⁺ lacP⁺ lacO⁺ lacZ⁺ lacY⁺*	−	−	+	+
b. *lacI⁺ lacP⁺ lacOᶜ lacZ⁻ lacY⁺*	−	+	−	+
c. *lacI⁺ lacP⁻ lacO⁺ lacZ⁺ lacY⁻*	−	−	−	−
d. *lacI⁺ lacP⁺ lacO⁺ lacZ⁻ lacY⁻/lacI⁻ lacP⁺ lacO⁺ lacZ⁺ lacY⁺*	−	−	+	+

a. All the genes possess normal sequences so the *lac* operon functions normally: when lactose is absent, the regulator protein binds to the operator and inhibits the transcription of the structural genes, and so β-galactosidase and permease are not produced. When lactose is present, some of it is converted into allolactose, which binds to the repressor and makes it inactive; the repressor does not bind to the operator, and so the structural genes are transcribed and β-galactosidase and permease are produced.

b. The structural *lacZ* gene is mutated so β-galactosidase will not be produced under any conditions. The *lacO* gene has a constitutive mutation, which means that the repressor is unable to bind to *lacO*, and so transcription takes place at all times. Therefore, permease will be produced in both the presence and the absence of lactose.

c. In this strain, the promoter is mutated, and so RNA polymerase is unable to bind and transcription does not take place. Therefore, β-galactosidase and permease are not produced under any conditions.

d. This strain is a partial diploid, which consists of two copies of the *lac* operon—one on the bacterial chromosome and the other on a plasmid. The *lac* operon represented in the upper part of the genotype has mutations in both the *lacZ* and the *lacY* genes, and so it is not capable of encoding β-galactosidase or permease under any conditions. The *lac* operon in the lower part of the genotype has a defective regulator gene, but the normal regulator gene in the upper operon produces a diffusible repressor (trans acting) that binds to the lower operon in the absence of lactose and inhibits transcription. Therefore, no β-galactosidase or permease is produced when lactose is absent. In the presence of lactose, the repressor cannot bind to the operator, and so the lower operon is transcribed and β-galactosidase and permease are produced.

▶ Now try your own hand at predicting the outcome of different *lac* mutations by working Problem 19 at the end of the chapter.

Positive Control and Catabolite Repression

E. coli and many other bacteria metabolize glucose preferentially in the presence of lactose and other sugars. They do so because glucose enters glycolysis without further modification and therefore requires less energy to metabolize than do other sugars. When glucose is available, genes that participate in the metabolism of other sugars are repressed, in a phenomenon known as **catabolite repression**. For example, the efficient transcription of the *lac* operon takes place only if lactose is present and glucose is absent. But how is the expression of the *lac* operon influenced by glucose? What brings about catabolite repression?

Catabolite repression results from positive control in response to glucose. (This regulation is in addition to the negative control brought about by the repressor binding at the operator site of the *lac* operon when lactose is absent.) Positive control is accomplished through the binding of a dimeric protein called the **catabolite activator protein** (CAP) to a site that is about 22 nucleotides long and is located within or slightly upstream of the promoter of the *lac* genes (**Figure 16.13**). RNA polymerase does not bind efficiently to many promoters unless CAP is first bound to the DNA. Before CAP can bind to DNA, it must form a complex with a modified nucleotide called **adenosine-3′, 5′-cyclic monophosphate** (cyclic AMP, or cAMP), which is important in cellular signaling processes in both bacterial and eukaryotic cells. In *E. coli*, the concentration of cAMP is regulated so that its concentration is inversely proportional to the level of available glucose. A high concentration of glucose within the cell lowers the amount of cAMP so little cAMP–CAP complex is available to bind to the DNA. Consequently, RNA polymerase has poor affinity for the *lac* promoter, and little transcription of the *lac* operon takes place. Low concentrations of glucose stimulate high levels of cAMP, resulting in increased cAMP–CAP binding to DNA. This increase enhances the binding of RNA polymerase to the promoter and increases transcription of the *lac* genes by approximately 50-fold.

The catabolite activator protein exerts positive control in more than 20 operons of *E. coli*. The response to CAP varies among these promoters; some operons are activated by low levels of CAP, whereas others require high levels.

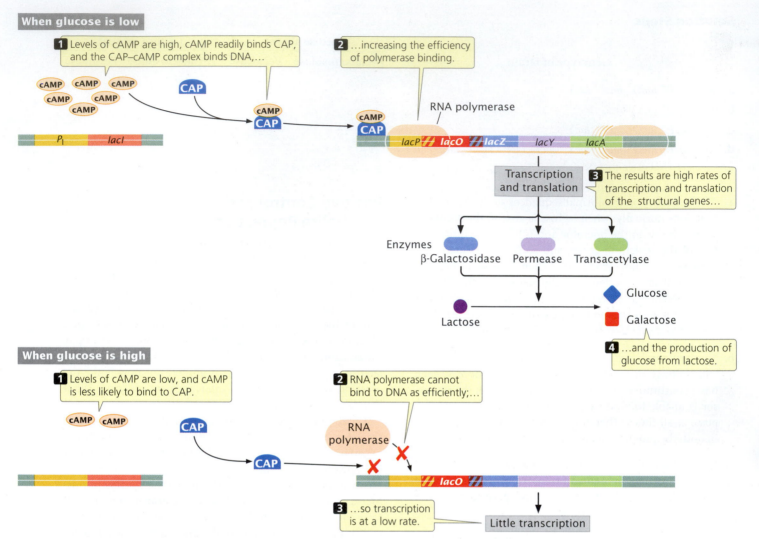

When glucose is low

1 Levels of cAMP are high, cAMP readily binds CAP, and the CAP–cAMP complex binds DNA,...

2 ...increasing the efficiency of polymerase binding.

cAMP cAMP cAMP
cAMP cAMP
cAMP

CAP

cAMP
CAP

cAMP
CAP

RNA polymerase

P_I lacI

lacP lacO lacZ lacY lacA

Transcription and translation

3 The results are high rates of transcription and translation of the structural genes...

Enzymes β-Galactosidase Permease Transacetylase

Lactose

Glucose

Galactose

4 ...and the production of glucose from lactose.

When glucose is high

1 Levels of cAMP are low, and cAMP is less likely to bind to CAP.

2 RNA polymerase cannot bind to DNA as efficiently;...

cAMP cAMP

CAP

CAP

RNA polymerase

lacO

3 ...so transcription is at a low rate.

Little transcription

16.13 The catabolite activator protein (CAP) binds to the promoter of the *lac* operon and stimulates transcription. CAP must complex with adenosine-3′, 5′-cyclic monophosphate (cAMP) before binding to the promoter of the *lac* operon. The binding of cAMP–CAP to the promoter activates transcription by facilitating the binding of RNA polymerase. Levels of cAMP are inversely related to glucose: low glucose stimulates high cAMP; high glucose stimulates low cAMP.

CAP contains a helix-turn-helix DNA-binding motif and, when it binds at the CAP site on DNA, it causes the DNA helix to bend (**Figure 16.14**). The bent helix enables CAP to facilitate the binding of RNA polymerase at the promoter and the initiation of transcription. ▶ TRY PROBLEM 14

CONCEPTS

In spite of its name, catabolite repression is a type of positive control in the *lac* operon. The catabolite activator protein (CAP), complexed with cAMP, binds to a site near the promoter and stimulates the binding of RNA polymerase. Cellular levels of cAMP are controlled by glucose; a low glucose level increases the abundance of cAMP and enhances the transcription of the *lac* structural genes.

✓ CONCEPT CHECK 7

What is the effect of high levels of glucose on the *lac* operon?

a. Transcription is stimulated.
b. Little transcription takes place.
c. Transcription is not affected.
d. Transcription may be stimulated or inhibited, depending on the levels of lactose.

The *trp* Operon of *E. coli*

The *lac* operon just discussed is an inducible operon, one in which transcription does not normally take place and must be turned on. Other operons are repressible; transcription in these operons is normally turned on and must be repressed.

(a)

(b)

16.14 The binding of the cAMP–CAP complex to DNA produces a sharp bend in DNA that activates transcription.

The tryptophan (*trp*) operon in *E. coli*, which controls the biosynthesis of the amino acid tryptophan, is an example of a negative repressible operon.

The *trp* operon contains five structural genes (*trpE*, *trpD*, *trpC*, *trpB*, and *trpA*) that produce the components of three enzymes (two of the enzymes consist of two polypeptide chains). These enzymes convert chorismate into tryptophan (**Figure 16.15**). The first structural gene, *trpE*, contains a long 5′ untranslated region (5′ UTR) that is transcribed but does not encode any of these enzymes. Instead, this 5′ UTR plays an important role in another regulatory mechanism, discussed in the next section. Upstream of the 5′ UTR is the *trp* promoter. When tryptophan levels are low, RNA polymerase binds to the promoter and transcribes the five structural genes into a single mRNA, which is then translated into enzymes that convert chorismate into tryptophan.

Some distance from the *trp* operon is a regulator gene, *trpR*, which encodes a repressor that alone cannot bind DNA (see Figure 16.15). Like the *lac* repressor, the tryptophan repressor has two binding sites, one that binds to DNA at the operator site and another that binds to tryptophan (the activator). Binding with tryptophan causes a conformational change in the repressor that makes it capable of binding to DNA at the operator site, which overlaps the promoter. When the operator is occupied by the tryptophan repressor, RNA polymerase cannot bind to the promoter

16.15 The *trp* operon controls the biosynthesis of the amino acid tryptophan in *E. coli*.

and the structural genes cannot be transcribed. Thus, when cellular levels of tryptophan are low, transcription of the *trp* operon takes place and more tryptophan is synthesized; when cellular levels of tryptophan are high, transcription of the *trp* operon is inhibited and the synthesis of more tryptophan does not take place.

Bacterial Enhancers

Another type of regulatory sequence that affects transcription is an enhancer, a DNA element that affects transcription but, in contrast to promoters, is typically found some distance from the gene (see Chapter 17). Enhancers were originally described in eukaryotes, but research now indicates some also occur in bacteria and archaea.

Like enhancers in eukaryotes, bacterial enhancers contain binding sites for proteins that increase the rate of transcription from promoters that are distant from the gene. They do this by causing the DNA between the promoter and enhancer to loop out, so that the transcription factor at the enhancer directly interacts with RNA polymerase at the promoter. Enhancers are also position independent, meaning that they can be moved without affecting their ability to enhance transcription. Most bacterial enhancers are found upstream of genes that utilize a special type of sigma factor (see Chapter 13) known as sigma 54 (σ^{54}). Enhancers will be discussed in more detail in Chapter 17.

CONCEPTS

The *trp* operon is a negative repressible operon that controls the biosynthesis of tryptophan. In a repressible operon, transcription is normally turned on and must be repressed: this is accomplished through the binding of tryptophan to the repressor, which renders the repressor active. The active repressor binds to the operator and prevents RNA polymerase from transcribing the structural genes. Bacterial enhancers increase the rate of transcription at genes that are distant from the enhancer.

✓ CONCEPT CHECK 8

In the *trp* operon, what happens to the *trp* repressor in the absence of tryptophan?
a. It binds to the operator and represses transcription.
b. It cannot bind to the operator and transcription takes place.
c. It binds to the regulator gene and represses transcription.
d. It cannot bind to the regulator gene and transcription takes place.

16.3 Some Operons Regulate Transcription Through Attenuation, the Premature Termination of Transcription

We've now seen several different ways in which a cell regulates the initiation of transcription in an operon. Some operons have an additional level of control that affects the *continuation* of transcription rather than its initiation. In **attenuation**, transcription begins at the start site, but termination takes place prematurely, before the RNA polymerase even reaches the structural genes. Attenuation takes place in a number of operons that encode enzymes participating in the biosynthesis of amino acids.

Attenuation in the *trp* Operon of *E. coli*

We can understand the process of attenuation most easily by looking at one of the best-studied examples, which is found in the *trp* operon of *E. coli*. The *trp* operon is unusual in that it is regulated both by repression and by attenuation. Most operons are regulated by one of these mechanisms but not by both of them.

Attenuation first came to light when Charles Yanofsky and his colleagues made several observations in the early 1970s that indicated that repression at the operator site is not the only method of regulation in the *trp* operon. They isolated a series of mutants that exhibited high levels of transcription, yet control at the operator site was unaffected, suggesting that some mechanism other than repression at the operator site was controlling transcription. Furthermore, they observed that two mRNAs of different sizes were transcribed from the *trp* operon: a long mRNA containing sequences for the structural genes and a much shorter mRNA of only 140 nucleotides. These observations led Yanofsky to propose that a mechanism that caused premature termination of transcription also regulates transcription in the *trp* operon.

Close examination of the *trp* operon reveals a region of 162 nucleotides that corresponds to the long 5′ UTR of the mRNA (mentioned earlier) transcribed from the *trp* operon (**Figure 16.16a**). The 5′ UTR (also called a leader) contains four regions: region 1 is complementary to region 2, region 2 is complementary to region 3, and region 3 is complementary to region 4 (**Figure 16.16b**). These complementarities allow the 5′ UTR to fold into two different secondary structures (**Figure 16.16c**). Only one of these secondary structures causes attenuation.

One of the secondary structures contains one hairpin produced by the base pairing of regions 1 and 2 and another hairpin produced by the base pairing of regions 3 and 4. Notice that a string of uracil nucleotides follows the 3+4 hairpin. Not coincidentally, the structure of a bacterial intrinsic terminator (see Chapter 13) includes a hairpin followed by a string of uracil nucleotides; this secondary structure in the 5′ UTR of the *trp* operon is indeed a terminator and is called an **attenuator**. The attenuator forms when cellular levels of tryptophan are high, causing transcription to be terminated before the *trp* structural genes can be transcribed.

When cellular levels of tryptophan are low, however, the alternative secondary structure of the 5′ UTR is produced by the base pairing of regions 2 and 3 (see Figure 16.16b). This base pairing also produces a hairpin, but this hairpin is not

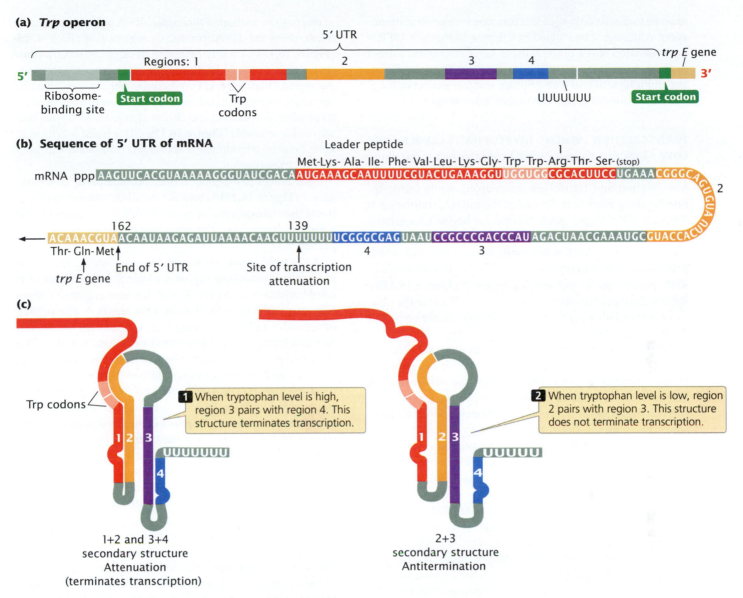

16.16 Two different secondary structures can be formed by the 5′ UTR of the mRNA transcript of the _trp_ operon.

followed by a string of uracil nucleotides; so this structure does *not* function as a terminator. RNA polymerase continues past the 5′ UTR into the coding section of the structural genes, and the enzymes that synthesize tryptophan are produced. Because it prevents the termination of transcription, the 2+3 structure is called an **antiterminator**.

To summarize, the 5′ UTR of the *trp* operon can fold into one of two structures. When the tryptophan level is high, the 3+4 structure forms, transcription is terminated within the 5′ UTR, and no additional tryptophan is synthesized. When the tryptophan level is low, the 2+3 structure forms, transcription continues through the structural genes,

and tryptophan is synthesized. The critical question, then, is: Why does the 3+4 structure arise when the level of tryptophan in the cell is high, whereas the 2+3 structure arises when the level is low?

To answer this question, we must take a closer look at the nucleotide sequence of the 5′ UTR. At the 5′ end, upstream of region 1, is a ribosome-binding site (see Figure 16.16a). Region 1 encodes a small protein. Within the coding sequence for this protein are two UGG codons, which specify the amino acid tryptophan; so tryptophan is required for the translation of this 5′ UTR sequence. The small protein encoded by the 5′ UTR has not been isolated and is presumed to be unstable; its only apparent function is to control attenuation. Although it was stated in Chapter 14 that a 5′ UTR is not translated into a protein, the 5′ UTR of operons subject to attenuation is an exception to this rule. The precise timing and interaction of transcription and translation in the 5′ UTR determine whether attenuation takes place.

TRANSCRIPTION WHEN TRYPTOPHAN LEVELS ARE LOW Let's first consider what happens when intracellular levels of tryptophan are low. Recall that, in prokaryotic cells, transcription and translation are coupled: while transcription is taking place at the 3′ end of the mRNA, translation is initiated at the 5′ end. RNA polymerase begins transcribing the DNA, producing region 1 of the 5′ UTR (**Figure 16.17a**). Closely following RNA polymerase, a ribosome binds to the 5′ UTR and begins to translate the coding region. Meanwhile, RNA polymerase is transcribing region 2 (**Figure 16.17b**). Region 2 is complementary to region 1 but, because the ribosome is translating region 1, the nucleotides in regions 1 and 2 cannot base pair.

RNA polymerase begins to transcribe region 3, and the ribosome reaches the UGG tryptophan codons in region 1. When it reaches the tryptophan codons, the ribosome stalls (**Figure 16.17c**) because the level of tryptophan is low and tRNAs charged with tryptophan are scarce or even unavailable. The ribosome sits at the tryptophan codons, awaiting the arrival of a tRNA charged with tryptophan. Stalling of the ribosome does not, however, hinder transcription; RNA polymerase continues to move along the DNA, and transcription gets ahead of translation.

Because the ribosome is stalled at the tryptophan codons in region 1, region 2 is free to base pair with region 3, forming the 2+3 hairpin (**Figure 16.17d**). This hairpin does not cause termination, and so transcription continues. Because region 3 is already paired with region 2, the 3+4 hairpin (the attenuator) never forms, and so attenuation does not take place and transcription continues. RNA polymerase continues along the DNA, past the 5′ UTR, transcribing all the structural genes into mRNA, which is translated into the enzymes encoded by the *trp* operon. These enzymes then synthesize more tryptophan.

TRANSCRIPTION WHEN TRYPTOPHAN LEVELS ARE HIGH Now let's see what happens when intracellular levels

of tryptophan are high. Once again, RNA polymerase begins transcribing the DNA, producing region 1 of the 5′ UTR (**Figure 16.17e**). Closely following RNA polymerase, a ribosome binds to the 5′ UTR and begins to translate the coding region (**Figure 16.17f**). When the ribosome reaches the two UGG tryptophan codons, it doesn't slow or stall, because tryptophan is abundant and tRNAs charged with tryptophan are readily available (**Figure 16.17g**). This point is critical to note: because tryptophan is abundant, translation can keep up with transcription.

As it moves past region 1, the ribosome partly covers region 2 (**Figure 16.17h**); meanwhile, RNA polymerase completes the transcription of region 3. Although regions 2 and 3 are complementary, the ribosome physically blocks their pairing.

RNA polymerase continues to move along the DNA, eventually transcribing region 4 of the 5′ UTR. Region 4 is complementary to region 3, and, because region 3 cannot base pair with region 2, it pairs with region 4. The pairing of regions 3 and 4 (see Figure 16.17h) produces the attenuator and transcription terminates just beyond region 4. The structural genes are not transcribed, no tryptophan-producing enzymes are translated, and no additional tryptophan is synthesized. Important events in the process of attenuation are summarized in **Table 16.3**. Try pausing the ribosome at the *trp* codon for different lengths of time in **Animation 16.2** and see what effect the pause has on transcription.

A key factor controlling attenuation is the number of tRNA molecules charged with tryptophan, because their availability is what determines whether the ribosome stalls at the tryptophan codons. A second factor concerns the synchronization of transcription and translation, which is critical to attenuation. Synchronization is achieved through a pause site located in region 1 of the 5′ UTR. When this site is transcribed, the RNA folds into a secondary structure that inhibits further transcription. Thus, the RNA polymerase stops temporarily at the pause site, allowing time for a ribosome to bind to the 5′ end of the mRNA. As the ribosome approaches the secondary structure in the RNA, the ribosome disrupts it and allows transcription to continue. Translation then closely follows transcription. It is important to point out that ribosomes do not traverse the convoluted hairpins of the 5′ UTR to translate the structural genes. Ribosomes that attach to the 5′ end of region 1 of the mRNA encounter a stop codon at the end of region 1.

Table 16.3 Events in the process of attenuation

Intracellular Level of Tryptophan	Ribosome Stalls at Trp Codons	Position of Ribosome When Region 3 Is Transcribed	Secondary Structure of 5′ UTR	Termination of Transcription of *trp* Operon
High	No	Covers region 2	3+4 hairpin	Yes
Low	Yes	Covers region 1	2+3 hairpin	No

When tryptophan level is low

(a)

RNA polymerase begins transcribing the DNA, producing region 1 of the 5′ UTR.

(b)

A ribosome attaches to the 5′ end of the 5′ UTR and translates region 1 while region 2 is being transcribed.

(c)

The ribosome stalls at the Trp codons in region 1 because tryptophan is low. Because the ribosome is stalled, region 2 is not covered by the ribosome when region 3 is transcribed.

(d)

When region 3 is transcribed, it pairs with region 2. When region 4 is transcribed, it cannot pair with region 3, because region 3 is already paired with region 2; the attenuator never forms, and transcription continues.

When tryptophan level is high

(e)

RNA polymerase begins transcribing DNA, producing region 1 of the 5′ UTR.

(f)

A ribosome binds to the 5′ end of the 5′ UTR and translates region 1 while region 2 is being transcribed.

(g)

RNA polymerase transcribes region 3. The ribosome does not stall at the Trp codons, because tryptophan is abundant.

(h)

The ribosome covers part of region 2, preventing it from pairing with region 3. Region 4 is transcribed and pairs with region 3, producing the attenuator that terminates transcription.

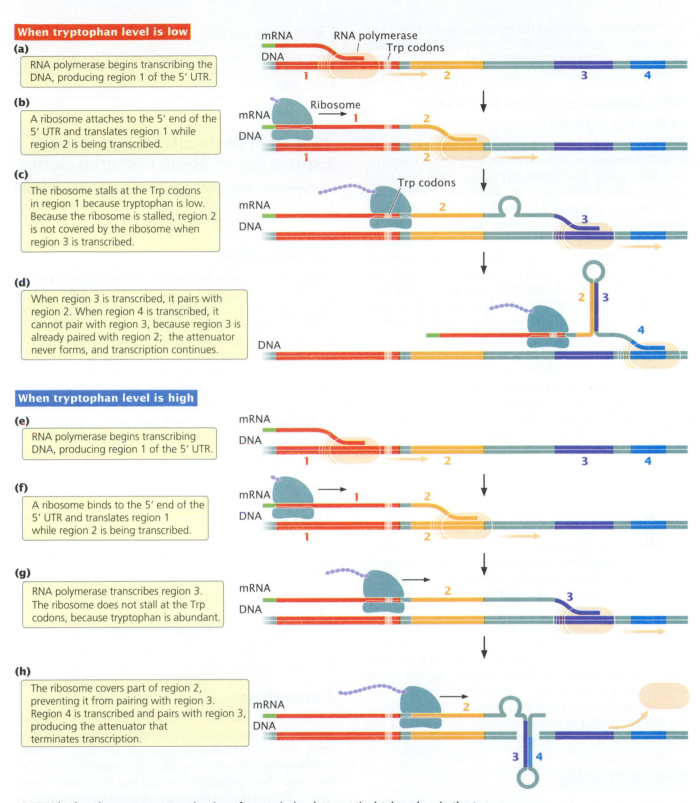

16.17 Whether the premature termination of transcription (attenuation) takes place in the *trp* operon depends on the cellular level of tryptophan.

New ribosomes translating the structural genes attach to a different ribosome-binding site located near the beginning of the *trpE* gene. ▶ TRY PROBLEM 27

Why Does Attenuation Take Place in the *trp* Operon?

Why do bacteria need attenuation in the *trp* operon? Shouldn't repression at the operator site prevent transcription from taking place when tryptophan levels in the cell are high? Why does the cell have two types of control? Part of the answer is that repression is never complete; some transcription is initiated even when the *trp* repressor is active; repression reduces transcription only as much as 70-fold. Attenuation can further reduce transcription another 8- to 10-fold, so together the two processes are capable of reducing transcription of the *trp* operon more than 600-fold. Both mechanisms provide *E. coli* with a much finer degree of control over tryptophan synthesis than either could achieve alone.

Another reason for the dual control is that attenuation and repression respond to different signals: repression responds to the cellular levels of tryptophan, whereas attenuation responds to the number of tRNAs charged with tryptophan. There may be times when a cell's ability to respond to these different signals is advantageous. Finally, the *trp* repressor affects several operons other than the *trp* operon. At an earlier stage in the evolution of *E. coli*, the *trp* operon may have been controlled only by attenuation. The *trp* repressor may have evolved primarily to control the other operons and only incidentally affects the *trp* operon.

Attenuation is a difficult process to grasp because you must simultaneously visualize how two dynamic processes—transcription and translation—interact, and it's easy to confuse them. Remember that attenuation refers to the early termination of *transcription*, not translation (although events in translation bring about the termination of transcription). Attenuation often causes confusion because we know that transcription must precede translation. We're comfortable with the idea that transcription might affect translation, but it's harder to imagine that the effects of translation could influence transcription, as they do in attenuation. The reality is that transcription and translation are closely coupled in prokaryotic cells, and events in one process can easily affect the other.

CONCEPTS

In attenuation, transcription is initiated but terminates prematurely. When tryptophan levels are low, the ribosome stalls at the tryptophan codons and transcription continues. When tryptophan levels are high, the ribosome does not stall at the tryptophan codons, and the 5′ UTR adopts a secondary structure that terminates transcription before the structural genes can be copied into RNA (attenuation).

✔ CONCEPT CHECK 9

Attenuation results when which regions of the 5′ UTR pair?
a. 1 and 3   c. 2 and 4
b. 2 and 3   d. 3 and 4

16.4 RNA Molecules Control the Expression of Some Bacterial Genes

All the regulators of gene expression that we have considered so far have been proteins. Several examples of RNA regulators also have been discovered.

Antisense RNA

Some RNA molecules are complementary to particular sequences on mRNAs and are called **antisense RNA**. They control gene expression by binding to sequences on mRNA and inhibiting translation. Translational control by antisense RNA is seen in the regulation of the *ompF* gene of *E. coli* (**Figure 16.18a**). This gene encodes an outer-membrane protein that functions as a channel for the passive diffusion of small polar molecules, such as water and ions, across the cell membrane. Under most conditions, the *ompF* gene is transcribed and translated and the OmpF protein is synthesized. However, when the osmolarity of the medium increases, the cell depresses the production of OmpF protein to help maintain cellular osmolarity. A regulator gene named *micF*—for mRNA-interfering complementary RNA—is activated and *micF* RNA is produced (**Figure 16.18b**). The *micF* RNA, an antisense RNA, binds to a complementary sequence in the 5′ UTR of the *ompF* mRNA and inhibits the binding of the ribosome. This inhibition reduces the amount of translation (see Figure 16.18b), which results in fewer OmpF proteins in the outer membrane and thus reduces the detrimental movement of substances across the membrane owing to the changes in osmolarity. Recent research has found many antisense RNA molecules in a wide range of bacterial species. ▶ TRY PROBLEM 30

Riboswitches

We have seen that operons of bacteria contain DNA sequences (promoters and operator sites) where the binding of small molecules induces or represses transcription. Some mRNA molecules contain regulatory sequences called **riboswitches**, where molecules can bind and affect gene expression by influencing the formation of secondary structures in the mRNA (**Figure 16.19**). Riboswitches were first discovered in 2002 and now appear to be common in bacteria, regulating about 4% of all bacterial genes. They are also present in archaea, fungi, and plants.

Riboswitches are typically found in the 5′ UTR of the mRNA and fold into compact RNA secondary structures

16.18 Antisense RNA can regulate translation.

with a base stem and several branching hairpins. In some cases, a small regulatory molecule binds to the riboswitch and stabilizes a terminator, which causes premature termination of transcription. In other cases, the binding of a regulatory molecule stabilizes a secondary structure that masks the ribosome-binding site, preventing the initiation of translation. When not bound by the regulatory molecule, the riboswitch assumes an alternative structure that eliminates the premature terminator or makes the ribosome-binding site available.

A RIBOSWITCH CONTROLS THE SYNTHESIS OF VITAMIN B12 An example of a riboswitch is seen in bacterial genes that encode enzymes having roles in the synthesis of

vitamin B12. The genes for these enzymes are transcribed into an mRNA molecule that has a riboswitch. When the activated form of vitamin B12—called coenzyme B12—is present, it binds to the riboswitch and the mRNA folds into a secondary structure that obstructs the ribosome-binding site. Consequently, no translation of the mRNA takes place. In the absence of coenzyme B12, the mRNA assumes a different secondary structure. This secondary structure does not obstruct the ribosome-binding site, and so translation is initiated, the enzymes are synthesized, and more vitamin B12 is produced. For some riboswitches, the regulatory molecule acts as a repressor (as just described) by inhibiting transcription or translation; for others, the regulatory molecule acts as an inducer by causing the formation of a

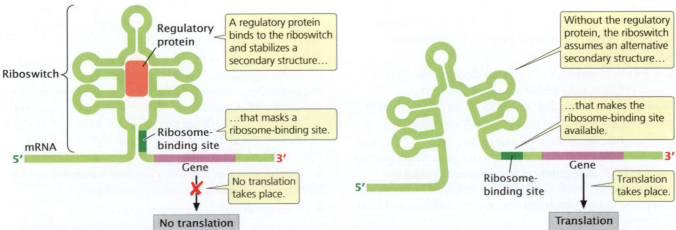

16.19 Riboswitches are RNA sequences in mRNA that affect gene expression.

secondary structure that allows transcription or translation to take place.

RNA-Mediated Repression Through Ribozymes

Another type of gene control is carried out by mRNA molecules called ribozymes, which possess catalytic activity (see Chapter 14). Termed RNA-mediated repression, this type of control has been demonstrated in the *glmS* gene of the bacterium *Bacillus subtilis*. Transcription of this gene produces an mRNA molecule that encodes the enzyme glutamine-fructose-6-phosphate amidotransferase (**Figure 16.20**), which helps synthesize a small sugar called glucosamine-6-phosphate (GlcN6P). Within the 5′ UTR of the *glmS* mRNA are about 75 nucleotides that act as a ribozyme. When GlcN6P is absent, the *glmS* gene is transcribed and translated to produce the enzyme, which synthesizes more GlcN6P. However, when sufficient GlcN6P is present, it binds to the ribozyme part of the mRNA, which then induces self-cleavage of the mRNA—the ribozyme breaks the sugar-phosphate backbone of the RNA. This cleavage then prevents translation of the mRNA.

CONCEPTS

Antisense RNA is complementary to other RNA or DNA sequences. In bacterial cells, it can inhibit translation by binding to sequences in the 5′ UTR of mRNA and preventing the attachment of the ribosome. Riboswitches are sequences in mRNA molecules that bind regulatory molecules and induce changes in the secondary structure of the mRNA that affect gene expression. In RNA-mediated repression, a ribozyme sequence on the mRNA induces self-cleavage and degradation of the mRNA when bound by a regulatory molecule.

16.20 Ribozymes, when bound with small regulatory molecules, can induce the cleavage and degradation of mRNA.

CONCEPTS SUMMARY

- Gene expression can be controlled at different levels, including the alteration of DNA or chromatin structure, transcription, mRNA processing, RNA stability, translation, and posttranslational modification. Much of gene regulation is through the action of regulatory proteins binding to specific sequences in DNA.

- Genes in bacterial cells are typically clustered into operons—groups of functionally related structural genes and the sequences that control their transcription. Structural genes in an operon are transcribed together as a single mRNA molecule.

- In negative control, a repressor protein binds to DNA and inhibits transcription. In positive control, an activator protein binds to DNA and stimulates transcription. In inducible operons, transcription is normally off and must be turned on; in repressible operons, transcription is normally on and must be turned off.

- The *lac* operon of *E. coli* is a negative inducible operon. In the absence of lactose, a repressor binds to the operator

and prevents the transcription of genes that encode β-galactosidase, permease, and transacetylase. When lactose is present, some of it is converted into allolactose, which binds to the repressor and makes it inactive, allowing the structural genes to be transcribed.

- Positive control in the *lac* and other operons is through catabolite repression. When complexed with cyclic AMP, the catabolite activator protein binds to a site in or near the promoter and stimulates the transcription of the structural genes. Levels of cAMP are inversely correlated with glucose; so low levels of glucose stimulate transcription and high levels inhibit transcription.

- The *trp* operon of *E. coli* is a negative repressible operon that controls the biosynthesis of tryptophan.

- Attenuation causes premature termination of transcription. It takes place through the close coupling of transcription and translation and depends on the secondary structure of the 5′ UTR sequence.

■ Antisense RNAs are complementary to sequences in mRNA and may inhibit translation by binding to these sequences, thereby preventing the attachment or progress of the ribosome.

■ When bound by a regulatory molecule, riboswitches in mRNA molecles induce changes in the secondary structure of the mRNA, which affects gene expression. Some mRNAs possess ribozyme sequences that induce self-cleavage and degradation when bound by a regulatory molecule.

IMPORTANT TERMS

gene regulation (p. 444)
structural gene (p. 445)
regulatory gene
   (p. 445)
constitutive gene (p. 445)
regulatory element
   (p. 445)
domain (p. 446)
operon (p. 447)
regulator gene (p. 448)

regulator protein
   (p. 448)
operator (p. 448)
negative control
   (p. 448)
positive control (p. 448)
inducible operon
   (p. 448)
repressible operon (p. 448)
inducer (p. 448)

allosteric protein
   (p. 448)
corepressor (p. 448)
coordinate induction
   (p. 452)
partial diploid (p. 453)
catabolite repression
   (p. 457)
catabolite activator protein
   (CAP) (p. 457)

adenosine-3′, 5′-cyclic monophosphate (cAMP)
   (p. 457)
attenuation (p. 460)
attenuator (p. 460)
antiterminator (p. 461)
antisense RNA (p. 464)
riboswitch (p. 464)

ANSWERS TO CONCEPT CHECKS

1. A constitutive gene is not regulated and is expressed continually.

2. Because it is the first step in the process of information transfer from DNA to protein. For cellular efficiency, gene expression is often regulated early in the process of protein production.

3. b

4. d

5. d

6. c

7. b

8. b

9. d

WORKED PROBLEMS

Problem 1

The *fox* operon, which has sequences A, B, C, and D (which may represent either structural genes or regulatory sequences), encodes enzymes 1 and 2. Mutations in sequences A, B, C, and D have the following effects, where a plus sign (+) indicates that the enzyme is synthesized and a minus sign (−) indicates that the enzyme is not synthesized.

Mutation in sequence	Fox absent		Fox present	
	Enzyme 1	Enzyme 2	Enzyme 1	Enzyme 2
No mutation	−	−	+	+
A	−	−	−	+
B	−	−	−	−
C	−	−	+	−
D	+	+	+	+

a. Is the *fox* operon inducible or repressible?

b. Indicate which sequence (A, B, C, or D) is part of the following components of the operon. Each sequence should be used only once.

Regulator gene _____

Promoter _____

Structural gene for enzyme 1 _____

Structural gene for enzyme 2 _____

What information is required in your answer to the problem?

a. Whether the *fox* operon is inducible or repressive.
b. Which sequence represents each part of the operon.

What information is provided to solve the problem?

For each mutation, whether enzyme 1 and enzyme 2 are produced in the presence and absence of Fox.

For help with this problem, review:

Section 16.2.

Solution Steps

> **Hint:** Review Figure 16.3 for a summary of operon structure.

a. When no mutations are present, enzymes 1 and 2 are produced in the presence of Fox but not in its absence, indicating that the operon is inducible and Fox is the inducer.

b. The mutation in *A* allows the production of enzyme 2 in the presence of Fox, but enzyme 1 is not produced in the presence or absence of Fox, and so *A* must have a mutation in the structural gene for enzyme 1. With the mutation in *B*, neither enzyme is produced under any conditions, and so this mutation most likely occurs in the promoter and prevents RNA polymerase from binding. The mutation in *C* affects only enzyme 2, which is not produced in the presence or absence of Fox; enzyme 1 is produced normally (only in the presence of Fox), and so the mutation in *C* most likely occurs in the structural gene for enzyme 2. The mutation in *D* is constitutive, allowing the production of enzymes 1 and 2 whether or not Fox is present. This mutation most likely occurs in the regulator gene, producing a defective repressor that is unable to bind to the operator under any conditions.

Regulator gene _D_   Structural gene for enzyme 1 _A_
Promoter   _B_   Structural gene for enzyme 2 _C_

Problem 2

A mutation occurs in the 5′ UTR of the *trp* operon that reduces the ability of region 2 to pair with region 3. What will the effect of this mutation be when the tryptophan level is high? When the tryptophan level is low?.

Solution Strategy

What information is required in your answer to the problem?

The effect of the mutation when tryptophan is high and when it is low.

What information is provided to solve the problem?

■ The mutation occurs in the 5′ UTR of the trp operon.

■ mutation reduces the ability of region 2 to pair with region 3.

For help with this problem, review:

Section 16.3.

Solution Steps

> **Hint:** Review 16.17 for a summary of attenuation.

When the tryptophan level is high, regions 2 and 3 do not normally pair, and therefore the mutation will have no effect. When the tryptophan level is low, however, the ribosome normally stalls at the tryptophan codons in region 1 and does not cover region 2, and so regions 2 and 3 are free to pair, which prevents regions 3 and 4 from pairing and forming a terminator, which would end transcription. If regions 2 and 3 cannot pair, then regions 3 and 4 will pair even when tryptophan is low and attenuation will always take place. Therefore, no more tryptophan will be synthesized even in the absence of tryptophan.

COMPREHENSION QUESTIONS

Section 16.1

1. Why is gene regulation important for bacterial cells?
2. Name six different levels at which gene expression might be controlled.

Section 16.2

3. Draw a picture illustrating the general structure of an operon and identify its parts.

4. What is the difference between positive and negative control? What is the difference between inducible and repressible operons?
5. Briefly describe the *lac* operon and how it controls the metabolism of lactose.
6. What is catabolite repression? How does it allow a bacterial cell to use glucose in preference to other sugars?

Section 16.3

7. What is attenuation? What are the mechanisms by which the attenuator forms when tryptophan levels are high and the antiterminator forms when tryptophan levels are low?

Section 16.4

8. What is antisense RNA? How does it control gene expression?

9. What are riboswitches? How do they control gene expression? How do riboswitches differ from RNA-mediated repression?

> For more questions that test your comprehension of the key chapter concepts, go to **LEARNING**Curve for this chapter.

APPLICATION QUESTIONS AND PROBLEMS

Section 16.1

10. Examine **Figure 16.2b**. Why do you think the motif of the DNA binding protein shown is called a zinc finger protein?

Section 16.2

*11. For each of the following types of transcriptional control, indicate whether the protein produced by the regulator gene will be synthesized initially as an active repressor, inactive repressor, active activator, or inactive activator.

  a. Negative control in a repressible operon
  b. Positive control in a repressible operon
  c. Negative control in an inducible operon
  d. Positive control in an inducible operon

12. A mutation at the operator site prevents the regulator protein from binding. What effect will this mutation have in the following types of operons?

  a. Regulator protein is a repressor in a repressible operon.
  b. Regulator protein is a repressor in an inducible operon.

13. The *blob* operon produces enzymes that convert compound A into compound B. The operon is controlled by a regulatory gene S. Normally, the enzymes are synthesized only in the absence of compound B. If gene S is mutated, the enzymes are synthesized in the presence *and* in the absence of compound B. Does gene S produce a repressor or an activator? Is this operon inducible or repressible?

*14. A mutation prevents the catabolite activator protein (CAP) from binding to the promoter in the *lac* operon. What will the effect of this mutation be on the transcription of the operon?

15. Transformation is a process in which bacteria take up new DNA released by dead cells and integrate it into their own genomes (see p. 247 in Chapter 9). In *Streptococcus pneumonia* (which causes many cases of pneumonia, inner-ear infections, and meningitis) the ability to carry out transformation requires from 105 to 124 genes, collectively termed the *com* regulon. The *com* regulon is activated in response to a protein called competence-stimulating peptide (CSP), which is produced by bacteria and is exported into the surrounding medium. When enough CSP accumulates, it attaches to a receptor on the bacterial cell membrane, which then activates a regulator protein that stimulates the transcription of genes within the *com* regulon and sets in motion a series of reactions that ultimately results in transformation. The *com* regulon in *Streptococcus pneumoniae* appears to be controlled through which type of gene regulation? Explain your answer.

  a. Negative inducible
  b. Negative repressible
  c. Positive inducible
  d. Positive repressible

16. Under which of the following conditions would a *lac* operon produce the greatest amount of β-glactosidase? The least? Explain your reasoning.

	Lactose present	Glucose present
Condition 1	Yes	No
Condition 2	No	Yes
Condition 3	Yes	Yes
Condition 4	No	No

17. A mutant strain of *E. coli* produces β-galactosidase in both the presence *and* the absence of lactose. Where in the operon might the mutation in this strain occur?

18. Examine **Figure 16.8**. What would be the effect of a drug that altered the structure of allolactose so that it was unable to bind to the regulator protein?

*19. For *E. coli* strains with the *lac* genotypes shown on the next page use a plus sign (+) to indicate the synthesis of β-galactosidase and permease and a minus sign (−) to indicate no synthesis of the proteins.

	Genotype of strain	Lactose absent		Lactose present	
		β-Galactosidase	Permease	β-Galactosidase	Permease
	$lacI^+$ $lacP^+$ $lacO^+$ $lacZ^+$ $lacY^+$				
	$lacI^-$ $lacP^+$ $lacO^+$ $lacZ^+$ $lacY^+$				
	$lacI^+$ $lacP^+$ $lacO^c$ $lacZ^+$ $lacY^+$				
	$lacI^-$ $lacP^+$ $lacO^+$ $lacZ^+$ $lacY^-$				
	$lacI^-$ $lacP^-$ $lacO^+$ $lacZ^+$ $lacY^+$				
	$lacI^+$ $lacP^+$ $lacO^+$ $lacZ^-$ $lacY^+$ / $lacI^-$ $lacP^+$ $lacO^+$ $lacZ^+$ $lacY^-$				
	$lacI^-$ $lacP^+$ $lacO^c$ $lacZ^+$ $lacY^+$ / $lacI^+$ $lacP^+$ $lacO^+$ $lacZ^-$ $lacY^-$				
	$lacI^-$ $lacP^+$ $lacO^+$ $lacZ^+$ $lacY^-$ / $lacI^+$ $lacP^-$ $lacO^+$ $lacZ^-$ $lacY^+$				
	$lacI^+$ $lacP^-$ $lacO^c$ $lacZ^-$ $lacY^+$ / $lacI^-$ $lacP^+$ $lacO^+$ $lacZ^+$ $lacY^-$				
	$lacI^+$ $lacP^+$ $lacO^+$ $lacZ^+$ $lacY^+$ / $lacI^+$ $lacP^+$ $lacO^+$ $lacZ^+$ $lacY^+$				
	$lacI^s$ $lacP^+$ $lacO^+$ $lacZ^+$ $lacY^-$ / $lacI^+$ $lacP^+$ $lacO^+$ $lacZ^-$ $lacY^+$				
	$lacI^s$ $lacP^-$ $lacO^+$ $lacZ^-$ $lacY^+$ / $lacI^+$ $lacP^+$ $lacO^+$ $lacZ^+$ $lacY^+$				

20. Give all possible genotypes of a *lac* operon that produces β-galactosidase and permease under the following conditions. Do not give partial diploid genotypes.

	Lactose absent		Lactose present	
	β-Galactosidase	Permease	β-Galactosidase	Permease
a.	−	−	+	+
b.	−	−	−	+
c.	−	−	+	−
d.	+	+	+	+
e.	−	−	−	−
f.	+	−	+	−
g.	−	+	−	+

*21. Explain why mutations in the *lacI* gene are trans in their effects, but mutations in the *lacO* gene are cis in their effects.

22. Which strand of DNA (upper or lower) in **Figure 16.9** is the template strand? Explain your reasoning.

23. The *mmm* operon, which has sequences A, B, C, and D (which may be structural genes or regulatory sequences), encodes enzymes 1 and 2. Mutations in sequences A, B, C, and D have the following effects, where a plus sign (+) indicates that the enzyme is synthesized and a minus sign (−) indicates that the enzyme is not synthesized.

Mutation in sequence	Mmm absent		Mmm present	
	Enzyme 1	Enzyme 2	Enzyme 1	Enzyme 2
No mutation	+	+	−	−
A	−	+	−	−
B	+	+	+	+
C	+	−	−	−
D	−	−	−	−

a. Is the *mmm* operon inducible or repressible?

b. Indicate which sequence (A, B, C, or D) is part of the following components of the operon:

Regulator gene	_____
Promoter	_____
Structural gene for enzyme 1	_____
Structural gene for enzyme 2	_____

24. Ellis Engelsberg and his coworkers examined the regulation of genes taking part in the metabolism of arabinose, a sugar (E. Engelsberg et al. 1965. *Journal of Bacteriology* 90:946–957). Four structural genes encode enzymes that help metabolize arabinose (genes A, B, D, and E). An additional gene C is linked to genes A, B, and D. These genes are in the order D-A-B-C. Gene E is distant from the other genes. Engelsberg and his colleagues isolated mutations at the C gene that affected the expression of structural genes A, B, D, and E. In one set of experiments, they created various genotypes at the A and C loci and determined whether arabinose isomerase (the enzyme encoded by gene A) was produced in the presence or absence of arabinose (the substrate of arabinose isomerase). Results from this experiment are shown in the following table, where a plus sign (+) indicates that the arabinose isomerase was synthesized and a minus sign (−) indicates that the enzyme was not synthesized.

Genotype	Arabinose absent	Arabinose present
1. $C^+ A^+$	−	+
2. $C^- A^+$	−	−
3. $C^- A^+ / C^+ A^-$	−	+
4. $C^c A^- / C^- A^+$	+	+

a. On the basis of the results of these experiments, is the C gene an operator or a regulator gene? Explain your reasoning.

b. Do these experiments suggest that the arabinose operon is negatively or positively controlled? Explain your reasoning.

c. What type of mutation is C^c?

25. In *E. coli*, three structural genes (*A*, *D*, and *E*) encode enzymes A, D, and E respectively. Gene O is an operator. The genes are in the order *O-A-D-E* on the chromosome. These enzymes catalyze the biosynthesis of valine. Mutations were isolated at the *A*, *D*, *E*, and *O* genes to study the production of enzymes A, D, and E when cellular levels of valine were low (T. Ramakrishnan and E. A. Adelberg. 1965. *Journal of Bacteriology* 89:654–660). Levels of the enzymes produced by partial-diploid *E. coli* with various combinations of mutations are shown in the following table.

Amount of enzyme produced

Genotype	E	D	A
1. $E^+ D^+ A^+ O^+ / E^+ D^+ A^+ O^+$	2.40	2.00	3.50
2. $E^+ D^+ A^+ O^- / E^+ D^+ A^+ O^+$	35.80	38.60	46.80
3. $E^+ D^- A^+ O^- / E^+ D^+ A^- O^+$	1.80	1.00	47.00
4. $E^+ D^+ A^- O^- / E^+ D^- A^+ O^+$	35.30	38.00	1.70
5. $E^- D^+ A^+ O^- / E^+ D^- A^+ O^+$	2.38	38.00	46.70

a. Is the regulator protein that binds to the operator of this operon a repressor (negative control) or an activator (positive control)? Explain your reasoning.

b. Are genes *A*, *D*, and *E* all under the control of operator O? Explain your reasoning.

c. Propose an explanation for the low level of enzyme E produced in genotype 3.

CHALLENGE QUESTION

Section 16.3

31. Would you expect to see attenuation in the *lac* operon and other operons that control the metabolism of sugars? Why or why not?

Section 16.3

26. At which level of gene regulation shown in **Figure 16.1** does attenuation occur?

*27. Listed in parts *a* through *g* are some mutations that were found in the 5′ UTR region of the *trp* operon of *E. coli*. What will the most likely effect of each of these mutations be on the transcription of the *trp* structural genes?

a. A mutation that prevents the binding of the ribosome to the 5′ end of the mRNA 5′ UTR

b. A mutation that changes the tryptophan codons in region 1 of the mRNA 5′ UTR into codons for alanine

c. A mutation that creates a stop codon early in region 1 of the mRNA 5′ UTR

d. Deletions in region 2 of the mRNA 5′ UTR

e. Deletions in region 3 of the mRNA 5′ UTR

f. Deletions in region 4 of the mRNA 5′ UTR

g. Deletion of the string of adenine nucleotides that follows region 4 in the 5′ UTR

28. Some mutations in the *trp* 5′ UTR region increase termination by the attenuator. Where might these mutations occur and how might they affect the attenuator?

29. Some of the mutations mentioned in Problem 28 have an interesting property. They prevent the formation of the antiterminator that normally takes place when the tryptophan level is low. In one of the mutations, the AUG start codon for the 5′ UTR peptide has been deleted. How might this mutation prevent antitermination from taking place?

Section 16.4

*30. Several examples of antisense RNA regulating translation in bacterial cells have been discovered. Molecular geneticists have also used antisense RNA to artificially control transcription in both bacterial and eukaryotic genes. If you wanted to inhibit the transcription of a bacterial gene with antisense RNA, what sequences might the antisense RNA contain?

Go to your **LaunchPad** to find additional learning resources and the Suggested Readings for this chapter.

Control of Gene Expression in Eukaryotes

Genetic Differences That Make Us Human

Changes in a relatively small number of regulatory sequences help produce the large phenotypic differences between humans and chimpanzees. Illustrated here is a network of interacting genes for transcription factors that are differentially expressed in the brains of humans and chimpanzees and control the expression of other genes. Red circles represent transcription factors that are more highly expressed in the human brain; green circles represent transcription factors that are more highly expressed in the chimpanzee brain. [Edwin Hadley, University of Illinois.]

Over 140 years ago, Charles Darwin proposed that humans shared a recent common ancestor with the African great apes. Today there is significant evidence that our closest living relative is the chimpanzee. Fossil evidence indicates that humans and chimpanzees diverged genetically only 5 to 7 million years ago—a mere blink of the eye in evolutionary time. Yet humans and chimps differ in a large number of anatomical, physiological, behavioral, and cognitive traits. For example, there are numerous differences in the structure of the backbone, pelvis, skull, jaw, teeth, hands, and feet of humans and chimpanzees. The size of the human brain is more than twice that of the chimpanzee; and humans exhibit complex language and cultural characteristics not seen in chimpanzees. Indeed, the degree of phenotypic difference between chimpanzees and humans is so large that scientists place them into entirely different primate families (humans in the family Hominidae and chimpanzees in the family Pongidae).

In spite of the large phenotypic gulf between humans and chimpanzees, sequencing of their genomes reveals that their DNA is remarkably similar. Only about 1% of individual base pairs differ between the two species, along with a 3% difference in insertions and deletions. Thus, 96% of the DNA of humans and chimpanzees is identical. But clearly humans are not chimpanzees. How then, did humans and chimpanzees come to be so different? Where are the genes that make us human?

A possible answer to this paradox was proposed by geneticists Mary-Claire King and A. C. Wilson in 1975. Using the limited techniques available at the time (comparison of amino acid in proteins and DNA hybridization studies), King and Wilson concluded that humans and chimpanzees differed at only about 1% of their DNA sequences. To explain how these very small genetic changes could account for the extensive physical and behavioral differences between humans and chimpanzees, King and Wilson suggested that the genetic variations that make us human are concentrated in regulatory sequences—those parts of the genome that control the expression of other genes. In this way, small genetic changes might influence the expression of numerous other genes and affect the phenotypes of many traits simultaneously. Unfortunately, there were no techniques available at the time to examine regulatory sequences and to test their hypothesis.

Fast-forward to 2009. Using cutting-edge techniques of genomic research and bioinformatics, Katja Nowick and her colleagues at the University of Illinois and the Norwegian University of Science and Technology identified a group of 90 transcription factors whose expression differed significantly between humans and chimpanzees. As discussed in Chapter 13, transcription factors are proteins that bind to DNA and facilitate or repress the synthesis of RNA, the first step in the process of information transfer from genotype to phenotype. Each transcription factor may affect the expression of multiple genes, so a small genetic change affecting the expression of a single transcription factor can influence many additional genes. The differences that Nowick and her colleagues found in the expression of transcription factors were particularly pronounced in brain tissue, where they may account for the large differences in neural and cognitive function of humans and chimpanzees.

Many of the transcription factors that Nowick and colleagues identified were Krüppel-associated box domain zinc finger proteins (KRAB-ZFPs), transcription factors that bind to specific DNA sequences and bring about changes in chromatin structure. As we will discuss in this chapter, changes in chromatin structure are often involved in regulation of gene expression. Other studies have demonstrated that KRAB-ZFPs have evolved rapidly in humans, probably because they were favored by natural selection. The transcription factors identified by Nowick and colleagues clustered into two distinct but interconnected regulatory networks, which control energy metabolism, transcription, vesicular transport, and protein modification.

These results support the idea first proposed by King and Wilson: that changes in a relatively small number of regulatory sequences affect expression of numerous genes in humans and chimpanzees and help produce the large differences we see in anatomy, brain size, cognition, and behavior.

This chapter is about gene regulation in eukaryotic cells, the very changes that help to make humans unique. Gene regulation typically takes place at multiple levels. We begin by considering how gene expression is influenced by changes in chromatin structure, which can be altered by several different mechanisms. We next consider the initiation of transcription in eukaryotes, which is controlled by a complex interaction of several types of proteins and the DNA regulatory elements to which they bind. We examine several ways in which gene expression is controlled through the processing, degradation, and translation of mRNA. We end by revisiting some of the similarities and differences in gene regulation in bacteria and eukaryotes.

nuclear membrane in eukaryotic cells separates transcription and translation in time and space. Therefore, the regulation of gene expression in eukaryotic cells is characterized by a greater diversity of mechanisms that act at different points in the transfer of information from DNA to protein.

Eukaryotic gene regulation is less well understood than bacterial regulation, partly owing to the larger genomes in eukaryotes, their greater sequence complexity, and the difficulty of isolating and manipulating mutations that can be used in the study of gene regulation. Nevertheless, great advances in our understanding of the regulation of eukaryotic genes have been made in recent years.

17.1 Eukaryotic Cells and Bacteria Have Many Features of Gene Regulation in Common, but They Differ in Several Important Ways

As discussed in Chapter 16, many features of gene regulation are common to both bacterial and eukaryotic cells. For example, in both types of cells, DNA-binding proteins influence the ability of RNA polymerase to initiate transcription. However, there are also some differences. First, many prokaryotic genes are organized into operons and are transcribed into a single RNA molecule. Although some operon-like gene clusters have been found in worms and even some primitive chordates, most eukaryotic genes have their own promoters and are transcribed separately. Second, chromatin structure affects gene expression in eukaryotic cells; DNA must unwind from the histone proteins before transcription can take place. Third, the presence of the

17.2 Changes in Chromatin Structure Affect the Expression of Genes

One type of gene control in eukaryotic cells is accomplished through the modification of chromatin structure. In the nucleus, histone proteins associate to form octamers, around which helical DNA tightly coils to create chromatin (see Figure 11.4). In a general sense, this chromatin structure represses gene expression. For a gene to be transcribed, transcription factors, other regulator proteins, and RNA polymerase must bind to the DNA. How can these events take place with DNA wrapped tightly around histone proteins? The answer is that, before transcription, chromatin structure changes and the DNA becomes more accessible to the transcriptional machinery.

DNase I Hypersensitivity

Several types of changes are observed in chromatin structure when genes become transcriptionally *active*. As genes become transcriptionally active, regions around the genes become highly sensitive to the action of DNase I (see Chapter 11). These regions, called **DNase I hypersensitive sites**, frequently develop about 1000 nucleotides upstream of the start site of transcription, suggesting that the chromatin in these regions adopts a more open configuration during transcription. This relaxation of the chromatin structure allows regulatory proteins access to binding sites on the DNA. Indeed, many DNase I hypersensitive sites correspond to known binding sites for regulatory proteins. At least three different processes affect gene regulation by altering chromatin structure: (1) chromatin remodeling; (2) the modification of histone proteins; and (3) DNA methylation. Each of these mechanisms will be discussed in the sections that follow.

CONCEPTS

Sensitivity to DNase I digestion indicates that transcribed DNA assumes an open configuration before transcription.

Chromatin Remodeling

Some transcription factors and other regulatory proteins alter chromatin structure without altering the chemical structure of the histones directly. These proteins are called **chromatin-remodeling complexes**. They bind directly to particular sites on DNA and reposition the nucleosomes, allowing transcription factors and RNA polymerase to bind to promoters and initiate transcription (**Figure 17.1**).

One of the best-studied examples of a chromatin-remodeling complex is SWI–SNF, which is found in yeast, humans, *Drosophila*, and other organisms. This complex utilizes energy derived from the hydrolysis of ATP to reposition nucleosomes, exposing promoters in the DNA to the action of other regulatory proteins and RNA polymerase.

Evidence suggests at least two mechanisms by which remodeling complexes reposition nucleosomes. First, some remodeling complexes cause the nucleosome to slide along the DNA, allowing DNA that was wrapped around the nucleosome to occupy a position in between nucleosomes, where it is more accessible to proteins affecting gene expression (see Figure 17.1). Second, some complexes cause conformational changes in the DNA, in nucleosomes, or in both so that DNA that is bound to the nucleosome assumes a more exposed configuration.

Chromatin-remodeling complexes are targeted to specific DNA sequences by transcriptional activators or repressors that attach to a remodeling complex and then bind to the promoters of specific genes. There is also evidence that chromatin-remodeling complexes work together with enzymes that alter histones, such as acetyltransferase enzymes (see next section), to change chromatin structure and expose DNA for transcription.

Histone Modification

Histones in the octamer core of the nucleosome have two domains: (1) a globular domain that associates with other histones

17.1 Chromatin remodeling complexes reposition the nucleosomes, allowing transcription factors and RNA polymerase to bind to promoters and initiate transcription.

and the DNA and (2) a positively charged tail domain that interacts with the negatively charged phosphate groups on the backbone of DNA. The tails of histone proteins are often modified by the addition or removal of phosphate groups, methyl groups, or acetyl groups. Another modification of histones is ubiquitination, in which small molecules called ubiquitin are added or removed from the histones. All of these modifications have sometimes been called the **histone code**, because they encode information that affects how genes are expressed. The histone code affects gene expression by altering chromatin structure directly or, in some cases, by serving as recognition sites for proteins that bind to DNA and that then regulate transcription.

METHYLATION OF HISTONES One type of histone modification is the addition of methyl groups to the tails of histone proteins. These modifications can bring about either the activation or the repression of transcription, depending on which particular amino acids in the histone tail are methylated. Enzymes called histone methyltransferases add methyl groups to specific amino acids (usually lysine or arginine) of histone proteins. Other enzymes, called histone demethylases, remove methyl groups from histones. Many of the enzymes and proteins that modify histones, such as methyltransferases and demethylases, do not bind to specific DNA sequences, and must be recruited to specific chromatin sites. Sequence-specific binding proteins, pre-existing histone modifications, and RNA molecules serve to recruit histone-modifying enzymes to specific sites.

A common modification is the addition of three methyl groups to lysine 4 in the tail of the H3 histone protein, abbreviated H3K4me3 (K is the abbreviation for lysine). Histones containing the H3K4me3 modification are frequently found near promoters of transcriptionally active genes in eukaryotes. Studies have identified proteins that recognize and bind to H3K4me3, including nucleosome remodeling factor (NURF). NURF and other proteins that recognize H3K4me3 have a common protein-binding domain that binds to the H3 histone tail and then alters chromatin packing, allowing transcription to take place. Research has also demonstrated that some transcription factors, which are necessary for the initiation of transcription (see Chapter 13 and Section 17.3), directly bind to H3K4me3.

ACETYLATION OF HISTONES   Another type of histone modification that affects chromatin structure is acetylation, the addition of acetyl groups (CH_3CO) to histone. The acetylation of histones usually stimulates transcription. For example, the addition of a single acetyl group to lysine 16 in the tail of the H4 histone prevents the formation of the 30-nm chromatin fiber (see Figure 11.4), causing the chromatin to be in an open configuration and available for transcription. In general, acetyl groups destabilize chromatin structure, allowing transcription to take place (**Figure 17.2**). Acetyl groups are added to histone proteins by acetyltransferase enzymes; other enzymes called deacetylases strip acetyl groups from histones and restore chromatin structure, which represses transcription. Certain transcription factors (see Chapter 13) and other proteins that regulate transcription either have acetyltransferase activity or attract acetyltransferases to DNA.

THE ACETYLATION OF HISTONES CONTROLS FLOWER-ING IN *ARABIDOPSIS*   The importance of histone acetylation in gene regulation is demonstrated by the control of flowering in *Arabidopsis thaliana*, a plant with a number of characteristics that make it an excellent genetic model for plant systems (see further information on it in the Reference Guide to Model Genetic Organisms at the end of the book). The time at which flowering takes place is critical to the life of a plant; if flowering is initiated at the wrong time of year, pollinators may not be available to fertilize the flowers or environmental conditions may be unsuitable for the survival and germination of the seeds. Consequently, flowering time in most plants is carefully regulated in response to multiple internal and external cues, such as plant size, photoperiod, and temperature.

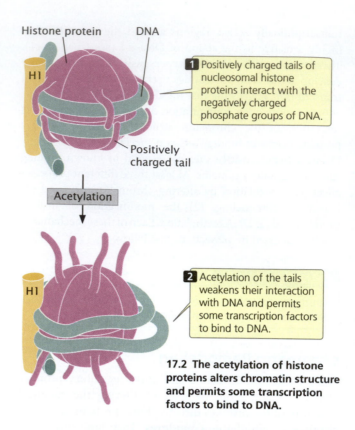

1 Positively charged tails of nucleosomal histone proteins interact with the negatively charged phosphate groups of DNA.

Acetylation

2 Acetylation of the tails weakens their interaction with DNA and permits some transcription factors to bind to DNA.

17.2 The acetylation of histone proteins alters chromatin structure and permits some transcription factors to bind to DNA.

Among the many genes that control flowering in *Arabidopsis* is *flowering locus C* (*FLC*), which plays an important role in suppressing flowering until after an extended period of coldness (a process called vernalization). The *FLC* gene encodes a regulator protein that represses the activity of other genes that affect flowering (**Figure 17.3**). As long as *FLC* is active, flowering remains suppressed. The activity of *FLC* is controlled by another locus called *flowering locus D* (*FLD*), the key role of which is to stimulate flowering by repressing the action of *FLC*. In essence, flowering is stimulated because *FLD* represses the repressor. How does *FLD* repress *FLC*?

FLD encodes a deacetylase enzyme, which removes acetyl groups from histone proteins in the chromatin surrounding *FLC* (see Figure 17.3). The removal of acetyl groups from histones alters chromatin structure and inhibits transcription. The inhibition of transcription prevents *FLC* from being transcribed and removes its repression on flowering. In short, *FLD* stimulates flowering in *Arabidopsis* by deacetylating the chromatin that surrounds *FLC*, thereby removing its inhibitory effect on flowering.

Changes in chromatin structure that affect gene expression, such as the mechanisms just described (chromatin remodeling, histone modification, and DNA methylation), are examples of the phenomenon of epigenetics. Epigenetics is defined as alterations to DNA and chromatin structure that affect traits and are passed on to other cells or generations but are not caused by changes in the DNA base sequences. Epigenetic changes and their affect on gene regulation will be discussed in more detail in Chapter 21. ▶ **TRY PROBLEM 18**

Chromatin   FLC   Acetylated chromatin

1 Acetyl groups on histone proteins destabilize chromatin structure.

Transcription

mRNA

2 Flowering locus C (*FLC*) encodes a regulator protein that represses flowering.

Translation

Regulator protein

3 No flowering takes place.

Repression of flowering

FLD

Transcription

mRNA

Translation

Deacetylase enzyme

4 Flowering locus D (*FLD*) encodes a deacetylase enzyme...

5 ...that removes acetyl groups and restores chromatin structure.

FLC

Restoration of chromatin

No transcription

6 No transcription of *FLC* takes place.

7 Flowering is not suppressed, and so flowering takes place.

No repression of flowering

17.3 Flowering in *Arabidopsis* is controlled in part by *FLD*, a gene that encodes a deacetylase enzyme. This enzyme removes acetyl groups from histone proteins in chromatin surrounding *FLC*, a gene that suppresses flowering. The removal of the acetyl groups from the histones restores chromatin structure and represses the transcription of *FLC*, thereby allowing the plant to flower.

CHROMATIN IMMUNOPRECIPITATION Our understanding of how changes in chromatin structure are associated with gene expression, as well as how DNA binding proteins affect transcription, have been greatly advanced by the use of a technique called chromatin immunoprecipitation (ChIP). This technique allows researchers to determine the specific locations within the genome where proteins interact with DNA. Those proteins might be histones that have undergone modifications, transcription factors, or other proteins that bind to promoters and enhancers (sequences that affect the transcription of distant genes, see Chapter 13).

The basic idea of ChIP is that a particular protein and the DNA to which it is bound are isolated, the protein and DNA are then separated, and the DNA sequence to which the protein was formally bound is identified. The technique has provided a powerful means of determining the genome-wide locations of modified histones and the binding sites for transcription factors and other proteins that affect transcription.

One version of ChIP, called crosslinked ChIP (XChIP) is used for identifying the binding sites of transcription factors and other proteins that bind to DNA. In this procedure (**Figure 17.4**), the protein and associated DNA are temporarily crosslinked, which means that they are treated with formaldehyde or UV light to create covalent bonds between the DNA and protein. The crosslinking holds the DNA and protein together so that the DNA to which the protein is

bound will separate along with the protein. After crosslinking, the cell is lysed and the chromatin is broken into pieces by digestion with an enzyme or by mechanical shearing. Antibodies specific for a particular protein—such as a specific transcription factor—are then applied. The antibodies attach to the protein–DNA complexes and cause them to precipitate. The crosslink is then reversed, separating the DNA and protein. The protein is removed by an enzyme that digests protein but not DNA, leaving fragments of the DNA to which the protein was bound. These fragments can then be sequenced or identified with other methods. The result is the information about the genomic locations of binding sites for the specific protein.

A version of this technique, called native ChIP (nChIP) does not utilize crosslinking. It is often used for finding the locations of modified histone proteins. In this case, crosslinking is not required because the DNA and histone proteins are naturally linked by the nucleosome structure. The chromatin is isolated from the cell and fragmented, and antibodies to a particular protein (usually a specific modified histone) are used to precipitate the protein–DNA complexes. The protein and DNA are separated, the protein digested, and the DNA fragments to which the modified histones were attached are sequenced or otherwise identified.

ChIP analysis has been used to determine the locations of modified histones that activate or repress transcription.

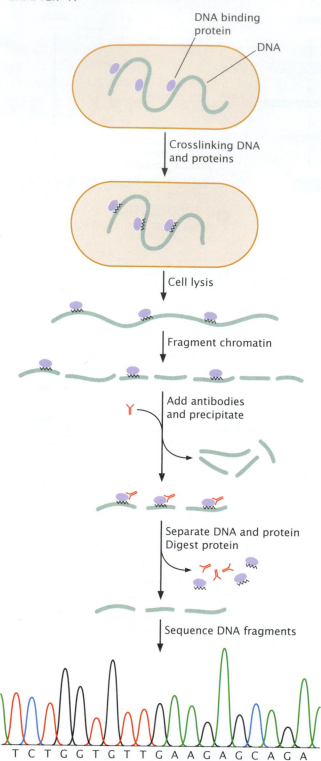

17.4 Chromatin immunoprecipitation (ChIP) can be used to identify DNA binding sites of a specific protein and the locations of modified histone proteins. Shown here is the method for crosslinked chromatin immunoprecipitation (XChIP).

As mentioned earlier, the H3K4me3 histone mark is associated with promoters of active genes. nChIP analysis has successfully identified locations of this modified histone in the human genome, helping to identify active promoters in different tissues.

CONCEPTS

The tails of histone proteins are often modified by the addition or removal of phosphate groups, methyl groups, or acetyl groups, or other changes. These modifications alter chromatin structure and affect the transcription of genes.

DNA Methylation

Another change in chromatin structure associated with transcription is the methylation of cytosine bases, which yields 5-methylcytosine (see Figure 10.19). The methylation of cytosine in DNA is distinct from the methylation of histone proteins mentioned earlier. Heavily methylated DNA is associated with the repression of transcription in vertebrates and plants, whereas transcriptionally active DNA is usually unmethylated in these organisms. Abnormal patterns of methylation are also associated with some types of cancer.

DNA methylation is most common on cytosine bases adjacent to guanine nucleotides (CpG, where p represents the phosphate group in the DNA backbone); so two methylated cytosines sit diagonally across from each other on opposing strands:

$$5' - \overset{m}{C} \, G - 3'$$
$$3' - G \, \underset{m}{C} - 5'$$

DNA regions with many CpG sequences are called **CpG islands** and are commonly found near transcription start sites. While genes are not being transcribed, these CpG islands are often methylated, but the methyl groups are removed before the initiation of transcription. CpG methylation is also associated with long-term gene repression, such as on the inactivated X chromosome of female mammals (see Chapter 4).

Evidence indicates that an association exists between DNA methylation and the deacetylation of histones, both of which repress transcription. Certain proteins that bind tightly to methylated CpG sequences form complexes with other proteins that act as histone deacetylases. In other words, methylation appears to attract deacetylases, which remove acetyl groups from the histone tails, stabilizing the nucleosome structure and repressing transcription. The demethylation of DNA allows acetyltransferases to add acetyl groups, disrupting nucleosome structure and permitting transcription. ▶ **TRY PROBLEM 19**

CONCEPTS

Chromatin structure can be altered by methylation of DNA. In eukaryotes, DNA methylation consists of 5-methylctyosine occurring at CpG dinucleotides. DNA methylation is usually associated with repression of transcription.

✔ **CONCEPT CHECK 1**

What are some of different processes that affect gene regulation by altering chromatin structure?

17.3 The Initiation of Transcription Is Regulated by Transcription Factors and Regulator Proteins

We just considered one level at which gene expression is controlled—the alteration of chromatin and DNA structure. We now turn to another important level of control—control through the binding of proteins to DNA sequences that affect transcription. Transcription is an important level of control in eukaryotic cells, and this control requires a number of different types of proteins and regulatory elements. The initiation of eukaryotic transcription was discussed in detail in Chapter 13. Recall that general transcription factors and RNA polymerase assemble into a *basal transcription apparatus*, the complex of RNA polymerase, transcription factors, and other proteins that carry out transcription. The basal transcription apparatus binds to a *core promoter* located immediately upstream of a gene and is capable of minimal levels of transcription; *transcriptional regulator proteins* are required to bring about normal levels of transcription. These proteins bind to a regulatory promoter, which is located upstream of the core promoter (**Figure 17.5**), and to *enhancers*, which may be located some distance from the gene. Some transcriptional regulator proteins are activators, stimulating transcription. Others are repressors, inhibiting transcription.

Transcriptional Activators and Coactivators

Transcriptional activator proteins stimulate and stabilize the basal transcription apparatus at the core promoter. The activators may interact directly with the basal transcription apparatus or indirectly through protein coactivators. Some activators and coactivators, as well as the general transcription factors, also have acetyltransferase activity and so further stimulate transcription by altering chromatin structure.

Transcriptional activator proteins have two distinct functions (see Figure 17.5). First, they are capable of binding DNA at a specific base sequence, usually a consensus sequence in a regulatory promoter or enhancer; for this function, most transcriptional activator proteins contain one or more DNA-binding motifs, such as the helix-turn-helix, zinc finger, and leucine zipper (see Chapter 16). A second function is the ability to interact with other components of the transcriptional apparatus and influence the rate of transcription.

Within the regulatory promoter are typically several different consensus sequences to which different transcriptional activators can bind. Among different promoters, activator-binding sites are mixed and matched in different combinations (**Figure 17.6**), and so each promoter is regulated by a unique combination of transcriptional activator proteins.

Transcriptional activator proteins bind to the consensus sequences in the regulatory promoter and affect the assembly or stability of the basal transcription apparatus at the core promoter. One of the components of the basal transcription apparatus is a complex of proteins called the **mediator** (see Figure 17.6). Transcriptional activator proteins binding to sequences in the regulatory promoter (or enhancer, see next section) make contact with the mediator and affect the rate at which transcription is initiated. Some regulatory promoters also contain sequences that are bound by proteins that lower the rate of transcription through inhibitory interactions with the mediator.

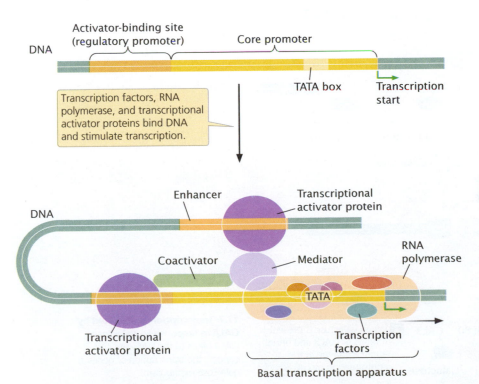

17.5 Transcriptional activator proteins bind to sites on DNA and stimulate transcription. Most act by stimulating or stabilizing the assembly of the basal transcription apparatus.

Regulatory promoter | Core promoter

SV40 early promoter

GC GC GC GC GC GC

Transcription start site

Thymidine kinase promoter

OCT GC CAAT GC TATA

Histone H2B promoter

OCT CAAT CAAT OCT TATA

−120 −100 −80 −60 −40 −20

| | TATA box | | GC box |
| | CAAT box | | Octamer (OCT) element |

17.6 The consensus sequences in the promoters of three eukaryotic genes illustrate the principle that different sequences can be mixed and matched in different combinations. A different transcriptional activator protein binds to each consensus sequence, so each promoter responds to a unique combination of activator proteins.

REGULATION OF GALACTOSE METABOLISM THROUGH GAL4
An example of a transcriptional activator protein is GAL4, which regulates the transcription of several yeast

genes whose products metabolize galactose. Like the genes in the *lac* operon, the genes that control galactose metabolism are inducible: when galactose is absent, these genes are not transcribed and the proteins that break down galactose are not produced; when galactose is present, the genes are transcribed and the enzymes are synthesized. GAL4 contains several zinc fingers and binds to a DNA sequence called UAS$_G$ (upstream activating sequence for GAL4). UAS$_G$ exhibits the properties of an enhancer—a regulatory sequence that may be some distance from the regulated gene and is independent of the gene in position and orientation (see Chapter 13). When bound to UAS$_G$, GAL4 activates the transcription of yeast genes needed for metabolizing galactose. GAL4 and a number of other transcriptional activator proteins contain multiple amino acids with negative charges that form an *acidic activation domain*. These acidic activators stimulate transcription by enhancing the assembly of the basal transcription apparatus.

A particular region of GAL4 binds another protein called GAL80, which regulates the activity of GAL4 in the presence of galactose. When galactose is absent, GAL80 binds to GAL4, preventing GAL4 from activating transcription (**Figure 17.7**). When galactose is present, however, it binds to another protein called GAL3, which interacts with GAL80, causing a conformational change in GAL80 so that it can no longer bind GAL4. The GAL4 protein is then free to activate the transcription of the genes, whose products metabolize galactose.

GAL4 UAS$_G$

1 In the absence of galactose, GAL80 blocks GAL4 from activating transcription.

GAL80

TATA

Basal transcription apparatus

GAL3 ⋯ Galactose

GAL4

TATA

3 GAL4 can now interact with basal transcription apparatus and stimulate transcription.

GAL80 GAL3 **2** When galactose is present, it binds to GAL3 and brings about a change in the conformation of GAL80.

Galactose

17.7 Transcription is activated by GAL4 in response to galactose. GAL4 binds to the UAS$_G$ site and controls the transcription of genes in galactose metabolism.

1 Enhancer I can stimulate the transcription of gene *A*, but its effect on gene *B* is blocked by the insulator.

2 Enhancer II can stimulate the transcription of gene *B*, but its effect on gene *A* is blocked by the insulator.

17.8 An insulator blocks the action of an enhancer on a promoter when the insulator lies between the enhancer and the promoter.

Transcriptional Repressors

Some regulatory proteins in eukaryotic cells act as repressors, inhibiting transcription. These repressors bind to sequences in the regulatory promoter or to distant sequences called *silencers*, which, like enhancers, are position and orientation independent. Unlike repressors in bacteria, most eukaryotic repressors do not directly block RNA polymerase. These repressors may compete with activators for DNA binding sites: when a site is occupied by an activator, transcription is activated, but, if a repressor occupies that site, there is no activation. Alternatively, a repressor may bind to sites near an activator site and prevent the activator from contacting the basal transcription apparatus. A third possible mechanism of repressor action is direct interference with the assembly of the basal transcription apparatus, thereby blocking the initiation of transcription.

CONCEPTS

Transcriptional regulatory proteins in eukaryotic cells can influence the initiation of transcription by affecting the stability or assembly of the basal transcription apparatus. Some regulatory proteins are activators and stimulate transcription; others are repressors and inhibit the initiation of transcription.

✓ CONCEPT CHECK 2

Most transcriptional activator proteins affect transcription by interacting with

a. introns.
b. the basal transcription apparatus.
c. DNA polymerase.
d. the terminator.

Enhancers and Insulators

Enhancers are capable of affecting transcription at distant promoters. For example, an enhancer that regulates the gene encoding the alpha chain of the T-cell receptor is located 69,000 bp downstream of the gene's promoter. Furthermore, the exact position and orientation of an enhancer relative to the promoter can vary. How can an enhancer affect the initiation of transcription taking place at a promoter that is tens of thousands of base pairs away? In many cases, regulator proteins bind to the enhancer and cause the DNA between the enhancer and the promoter to loop out, bringing the promoter and enhancer close to

each other, and so the transcriptional regulator proteins are able to directly interact with the basal transcription apparatus at the core promoter (see Figure 17.5). Some enhancers may be attracted to promoters by proteins that bind to sequences in the regulatory promoter and "tether" the enhancer close to the core promoter. Enhancers may also affect transcription by undergoing modifications that alter chromatin structure. A typical enhancer is about 500 bp in length and contains 10 binding sites for proteins that regulate transcription.

Recent research demonstrates that many enhancers are themselves transcribed into short RNA molecules called enhancer RNAs (eRNAs). Evidence suggests that transcription of an enhancer is often associated with transcription at the promoters that the enhancers affect. How transcription at the enhancer might affect transcription occurring at a distant promoter is not clear. Enhancers might recruit RNA polymerase, which is then transferred to the promoter when the enhancer interacts with the promoter. Alternatively, transcription of the enhancer might allow the chromatin to adopt a more open configuration, which then facilitates transcription at nearby promoters.

Most enhancers are capable of stimulating any promoter in their vicinities. Their effects are limited, however, by **insulators** (also called boundary elements), which are DNA sequences that block or insulate the effect of enhancers in a position-dependent manner. If the insulator lies between the enhancer and the promoter, it blocks the action of the enhancer; but, if the insulator lies outside the region between the two, it has no effect (**Figure 17.8**). Specific proteins bind to insulators and play a role in their blocking activity. Some insulators also limit the spread of changes in chromatin structure that affect transcription. Some enhancer-like elements are found in prokaryotes. ▶TRY PROBLEM 23

CONCEPTS

Some regulatory proteins bind to enhancers, which are regulatory elements that are distant from the gene whose transcription they stimulate. Insulators are DNA sequences that block the action of enhancers.

✓ CONCEPT CHECK 3

How does the binding of regulatory proteins to enhancers affect transcription at genes that are thousands of base pairs away?

Regulation of Transcriptional Stalling and Elongation

Transcription in eukaryotes is often regulated through factors that affect the initiation of transcription, including changes in chromatin structure, transcription factors, and transcriptional regulatory proteins. Research indicates that transcription may also be controlled through factors that affect stalling and elongation of RNA polymerase after transcription has been initiated.

The basal transcription apparatus—consisting of RNA polymerase, transcription factors, and other proteins—assembles at the core promoter. When the initiation of transcription has taken place, RNA polymerase moves downstream, transcribing the structural gene and producing an RNA product. At some genes, RNA polymerase initiates transcription and transcribes from 24 to 50 nucleotides of RNA but then pauses or stalls. For example, stalling is observed at genes that encode **heat-shock proteins** in *Drosophila*—proteins that help to prevent damage from stressing agents such as extreme heat. Heat-shock proteins are produced by a large number of different genes. During times of environmental stress, the transcription of all the heat-shock genes is greatly elevated. RNA polymerase initiates transcription at heat-shock genes in *Drosophila* but, in the absence of stress, stalls downstream of the transcription initiation site. Stalled polymerases are released when stress is encountered, allowing rapid transcription of the genes and the production of heat-shock proteins that facilitate adaptation to the stressful environment.

Stalling was formerly thought to take place at only a small number of genes, but research now indicates that stalling is widespread throughout eukaryotic genomes. For example, stalled RNA polymerases were found at hundreds of genes in *Drosophila*. Several factors that promote stalling have been identified; one of these is a protein called negative elongation factor (NELF), which binds to RNA polymerase and causes it to stall after initiation. Another protein called positive transcription elongation factor b (P-TEFb) relieves stalling and promotes elongation by phosphorylating NELF and RNA polymerase, perhaps by causing NELF to dissociate from the polymerase.

CONCEPTS

At some genes, RNA polymerase may pause or stall downstream of the promoter. Regulatory factors affect stalling and the elongation of transcription.

Coordinated Gene Regulation

Although most eukaryotic cells do not possess operons, several eukaryotic genes may be activated by the same stimulus. Groups of bacterial genes are often coordinately expressed (turned on and off together) because they are physically clustered as an operon and have the same promoter, but coordinately expressed genes in eukaryotic cells are not clustered. How, then, is the transcription of eukaryotic genes coordinately controlled if they are not organized into an operon?

Genes that are coordinately expressed in eukaryotic cells are able to respond to the same stimulus because they have short regulatory sequences in common in their promoters or enhancers. For example, different eukaryotic heat-shock genes possess a common regulatory element upstream of their start sites. Such DNA regulatory sequences are called **response elements**; they are short sequences that typically contain consensus sequences (**Table 17.1**) at varying distances from the gene being regulated. The response elements are binding sites for transcriptional activators. A transcriptional activator protein binds to the response element and elevates transcription. The same response element may be present in different genes, allowing multiple genes to be activated by the same stimulus.

A single eukaryotic gene may be regulated by several different response elements. For example, the metallothionein gene protects cells from the toxicity of heavy metals by encoding a protein that binds to heavy metals and removes them from cells. The basal transcription apparatus assembles around the TATA box, just upstream of the transcription start site for the metallothionein gene, but the apparatus alone is capable of only low rates of transcription.

Other response elements found upstream of the metallothionein gene contribute to increasing its rate of transcription. For example, several copies of a metal response element (MRE) are upstream of the metallothionein gene (**Figure 17.9**). Heavy metals stimulate the binding of activator proteins to MREs, which elevates the rate of transcription of the metallothionein gene. Because there are multiple copies of the MRE, high rates of transcription are induced by metals. Two enhancers also are located in the upstream region of the metallothionein gene; one enhancer contains a response element known as TRE, which stimulates transcription in the presence of an activated protein called AP1. A third response element called GRE is located approximately 250 nucleotides upstream of the metallothionein gene and stimulates transcription in response to certain hormones.

Table 17.1	Some response elements found in eukaryotic cells	
Response Element	**Responds to**	**Consensus Sequence**
Heat-shock element	Heat and other stress	CNNGAANNTCCNNG
Glucocorticoid response element	Glucocorticoids	TGGTACAAATGTTCT
Phorbol ester response element	Phorbal esters	TGACTCA
Serum response element	Serum	CCATATTAGG

Source: After B. Lewin, *Genes IV* (Oxford University Press, 1994), p. 880.

17.9 Multiple response elements (MREs) are found in the upstream region of the metallothionein gene. The basal transcription apparatus binds near the TATA box. In response to heavy metals, activator proteins bind to several MREs and stimulate transcription. The TRE response element is the binding site for transcription factor AP1, which is stimulated by phorbol esters. In response to glucocorticoid hormones, steroid-receptor proteins bind to the GRE response element located approximately 250 nucleotides upstream of the metallothionein gene and stimulate transcription.

This example illustrates a common feature of eukaryotic transcriptional control: a single gene may be activated by several different response elements found in both promoters and enhancers. Multiple response elements allow the same gene to be activated by different stimuli. At the same time, the presence of the same response element in different genes allows a single stimulus to activate multiple genes. In this way, response elements allow complex biochemical responses in eukaryotic cells.

17.4 Some Genes Are Regulated by RNA Processing and Degradation

In bacteria, transcription and translation take place simultaneously. In eukaryotes, transcription takes place in the nucleus and the pre-mRNAs are then processed before moving to the cytoplasm for translation, allowing opportunities for gene control after transcription. Consequently, posttranscriptional gene regulation assumes an important role in eukaryotic cells. A common level of gene regulation in eukaryotes is RNA processing and degradation.

Gene Regulation Through RNA Splicing

Alternative splicing allows a pre-mRNA to be spliced in multiple ways, generating different proteins in different tissues or at different times in development (see Chapter 14). Many eukaryotic genes undergo alternative splicing, and the regulation of splicing is an important means of controlling gene expression in eukaryotic cells.

ALTERNATIVE SPLICING IN THE T-ANTIGEN GENE The T-antigen gene of the mammalian virus SV40 is a well-studied example of alternative splicing. This gene is capable of encoding two different proteins, the large T and small t antigens. Which of the two proteins is produced depends on which of two alternative 5′ splice sites is used in RNA splicing (**Figure 17.10,**). The use of one 5′ splice site produces mRNA that encodes the large T antigen, whereas the use of

the other 5′ splice site (which is farther downstream) produces an mRNA encoding the small t antigen.

A protein called splicing factor 2 (SF2) enhances the production of mRNA encoding the small t antigen (see Figure 17.10). Splicing factor 2 has two binding domains: one domain is an RNA-binding region and the other has alternating serine and arginine amino acids. These two domains are typical of **SR** (serine- and arginine-rich) **proteins**, which often play a role in regulating splicing. Splicing factor 2 stimulates the binding of small nuclear ribonucleoproteins (snRNPs) to the 5′ splice site, one of the earliest steps in RNA splicing (see Chapter 14). The precise mechanism by which SR proteins

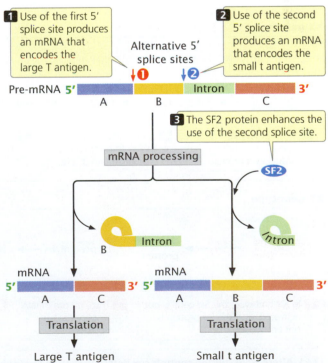

17.10 Alternative splicing leads to the production of the small t antigen and the large T antigen in the mammalian virus SV40.

influence the choice of splice sites is poorly understood. One model suggests that SR proteins bind to specific splice sites on mRNA and stimulate the attachment of snRNPs, which then commit the site to splicing.

ALTERNATIVE SPLICING IN *DROSOPHILA* SEXUAL DEVELOPMENT   Another example of alternative mRNA splicing that regulates gene expression is the determination of whether a fruit fly develops as male or female. Sex differentiation in *Drosophila* arises from a cascade of gene regulation. When two X-chromosomes are present, a female-specific promoter is activated early in development and stimulates the transcription of the *sex-lethal* (*Sxl*) gene (**Figure 17.11**). The protein encoded by *Sxl* regulates the splicing of the pre-mRNA transcribed from another gene called *transformer* (*tra*). The splicing of *tra* pre-mRNA results in the production of the Tra protein (see Figure 17.11). Together with another protein (Tra-2), Tra stimulates the female-specific splicing of pre-mRNA from yet another gene called *doublesex* (*dsx*). This event produces a female-specific Dsx protein, which causes the embryo to develop female characteristics.

In male embryos, which have a single X chromosome (see Figure 17.11), the promoter that transcribes the *Sxl* gene in females is inactive, so no Sxl protein is produced. In the absence of Sxl protein, *tra* pre-mRNA is spliced at a different 3′ splice site to produce a nonfunctional form of Tra protein (**Figure 17.12**). In turn, the presence of this nonfunctional Tra in males causes *dsx* pre-mRNAs to be spliced differently from that in females, and a male-specific Dsx protein is produced (see Figure 17.11). This event causes the development of male-specific traits.

In summary, the Tra, Tra-2, and Sxl proteins regulate alternative splicing that produces male and female phenotypes in *Drosophila*. Exactly how these proteins regulate alternative splicing is not yet known, but the Sxl protein (produced only in females) possibly blocks the upstream splice site on the *tra* pre-mRNA. This blockage would force the spliceosome to use the downstream 3′ splice site, which causes the production of Tra protein and eventually results in female traits (see Figure 17.12). ▶ TRY PROBLEM 24

> ### CONCEPTS
>
> Eukaryotic genes can be regulated through the control of mRNA processing. The selection of alternative splice sites leads to the production of different proteins.

The Degradation of RNA

The amount of a protein that is synthesized depends on the amount of corresponding mRNA available for translation. The amount of available mRNA, in turn, depends on both the rate of mRNA synthesis and the rate of mRNA degradation. Eukaryotic mRNAs are generally more stable than bacterial mRNAs, which typically last only a few minutes before being degraded. Nonetheless, there is great variability in the stability of eukaryotic mRNA: some mRNAs persist for only a few minutes; others last for hours, days, or even months. These variations can produce large differences in the amount of protein that is synthesized.

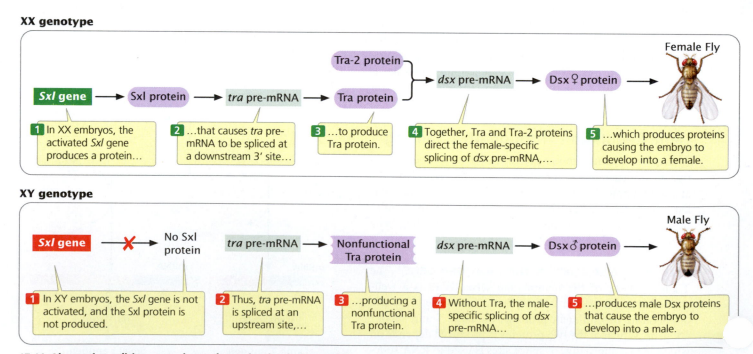

17.11 Alternative splicing controls sex determination in *Drosophila*.

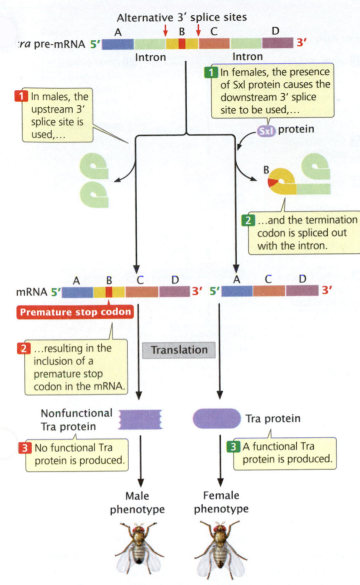

17.12 Alternative splicing of *tra* pre-mRNA. Two alternative 3′ splice sites are present.

Cellular RNA is degraded by ribonucleases, enzymes that specifically break down RNA. Most eukaryotic cells contain 10 or more types of ribonucleases, and there are several different pathways of mRNA degradation. In one pathway, the 5′ cap is first removed, followed by 5′→3′ removal of nucleotides. A second pathway begins at the 3′ end of the mRNA and removes nucleotides in the 3′→5′ direction. In a third pathway, the mRNA is cleaved at internal sites.

Messenger RNA degradation from the 5′ end is most common and begins with the removal of the 5′ cap. This pathway is usually preceded by the shortening of the poly(A) tail. Poly(A)-binding proteins (PABPs) normally bind to the poly(A) tail and contribute to its stability-enhancing effect. The presence of these proteins at the 3′ end of the mRNA protects the 5′ cap. When the poly(A) tail has been shortened below a critical limit, the 5′ cap is removed, and nucleases

then degrade the mRNA by removing nucleotides from the 5′ end. These observations suggest that the 5′ cap and the 3′ poly(A) tail of eukaryotic mRNA physically interact with each other, most likely by the poly(A) tail bending around so that the PABPs make contact with the 5′ cap (see Figure 15.18).

Much of RNA degradation takes place in specialized complexes called P bodies. However, P bodies appear to be more than simply destruction sites for RNA. Evidence suggests that P bodies can temporarily store mRNA molecules, which may later be released and translated. Thus, P bodies help control the expression of genes by regulating which RNA molecules are degraded and which are sequestered for later release. RNA degradation facilitated by small interfering RNAs (siRNAs) also may take place within P bodies (see next section).

Other parts of eukaryotic mRNA, including sequences in the 5′ untranslated region (5′ UTR), the coding region, and the 3′ UTR, also affect mRNA stability. Some short-lived eukaryotic mRNAs have one or more copies of a consensus sequence consisting of 5′-AUUUAUAA-3′, referred to as the AU-rich element, in the 3′ UTR. Messenger RNAs containing AU-rich elements are degraded by a mechanism in which microRNAs take part (see next section). ▶ **TRY PROBLEM 26**

CONCEPTS

The stability of mRNA influences gene expression by affecting the amount of mRNA available to be translated. The stability of mRNA is affected by the 5′ cap, the poly(A) tail, the 5′ UTR, the coding section, and sequences in 3′ UTR.

✓ CONCEPT CHECK 4

How does the poly(A) tail affect stability?

17.5 RNA Interference Is an Important Mechanism of Gene Regulation

The expression of a number of eukaryotic genes is controlled through RNA interference, also known as RNA silencing and posttranscriptional gene silencing (see Chapter 14). Research suggests that as much as 30% of human genes are regulated by RNA interference. RNA interference is widespread in eukaryotes, existing in fungi, plants, and animals. This mechanism is also widely used as a powerful technique for artificially regulating gene expression in genetically engineered organisms (see Chapter 19).

Small Interfering RNAs and MicroRNAs

RNA interference is triggered by microRNAs (miRNAs) and small interfering RNAs (siRNAs), depending on their origin and mode of action (see Chapter 14). An enzyme called Dicer cleaves and processes double-stranded RNA to

17.13 RNA silencing leads to the degradation of mRNA or to the inhibition of translation or transcription. (a) Small interfering RNAs (siRNAs) degrade mRNA by cleavage. (b) MicroRNAs (miRNAs) lead to the inhibition of translation. (c) Some small interfering RNAs (siRNAs) methylate histone proteins or DNA, inhibiting transcription.

produce single-stranded siRNAs or miRNAs that are from 21 to 25 nucleotides in length (**Figure 17.13**) and pair with proteins to form an RNA-induced silencing complex (RISC). The RNA component of RISC then pairs with complementary base sequences of specific mRNA molecules, most often with sequences in the 3′ UTR of the mRNA. Small interfering RNAs tend to base pair perfectly with the mRNAs, whereas miRNAs often form less-than-perfect pairings.

Mechanisms of Gene Regulation by RNA Interference

Small interfering RNAs and microRNAs regulate gene expression through at least four distinct mechanisms: (1) cleavage of mRNA, (2) inhibition of translation, (3) transcriptional silencing, or (4) degradation of mRNA.

RNA CLEAVAGE RISCs that contain an siRNA (and some that contain an miRNA) pair with mRNA molecules and cleave the mRNA near the middle of the bound siRNA (see Figure 17.13a). This cleavage is carried out by a protein that is sometimes referred to as "Slicer." After cleavage, the mRNA is further degraded. Thus, the presence of siRNAs and

miRNAs increase the rate at which mRNAs are broken down and decrease the amount of protein produced.

INHIBITION OF TRANSLATION Some miRNAs regulate genes by inhibiting the translation of their complementary mRNAs (see Figure 17.13b). For example, an important gene in flower development in *Arabidopsis thaliana* is *APETALA2*. The expression of this gene is regulated by an miRNA that base pairs with nucleotides in the coding region of *APETALA2* mRNA and inhibits its translation.

The exact mechanism by which miRNAs repress translation is not known, but some research suggests that it can inhibit both the initiation step of translation and steps after translation initiation such as those causing premature termination. Many mRNAs have multiple miRNA-binding sites, and translation is most efficiently inhibited when multiple miRNAs are bound to the mRNA.

TRANSCRIPTIONAL SILENCING Other siRNAs silence transcription by altering chromatin structure. These siRNAs combine with proteins to form a complex called RITS (for RNA transcriptional silencing; see Figure 17.13c), which is analogous to RISC. The siRNA component of RITS then binds to its complementary sequence in DNA or an RNA molecule in the process of being transcribed and represses transcription by attracting enzymes that methylate the tails of histone proteins. The addition of methyl groups to the histones causes them to bind DNA more tightly, restricting the access of proteins and enzymes necessary to carry out transcription (see earlier section on histone modification). Some miRNAs bind to complementary sequences in DNA and attract enzymes that methylate the DNA directly, which also leads to the suppression of transcription (see earlier section on DNA methylation).

SLICER-INDEPENDENT DEGRADATION OF mRNA A final mechanism by which miRNAs regulate gene expression is by triggering the decay of mRNA in a process that does not require Slicer activity. For example, a short-lived mRNA with an AU-rich element in its $3'$ UTR is degraded by an RNA-silencing mechanism. Researchers have identified an miRNA with a sequence that is complementary to the consensus sequence in the AU-rich element. This miRNA binds to the AU-rich element and, in a way that is not yet fully understood, brings about the degradation of the mRNA in a process that requires Dicer and RISC.

The Control of Development by RNA Interference

Much of development in multicellular eukaryotes is through gene regulation: different genes are turned on and off at specific times (see Chapter 22). In fact, when miRNAs were first discovered, researchers noticed that a mutation in an miRNA in *C. elegans* caused a developmental defect. Research now demonstrates that miRNA molecules are key factors in controlling development in plants, animals, and humans. For example, the vertebrate heart develops through the programmed differentiation and proliferation of cardiomyocytes, which are controlled by a specific miRNA termed miR-1-1.

Recent studies demonstrate that, through their effects on gene expression, miRNAs play important roles in many diseases and disorders. For example, a genetic form of hearing loss has been associated with a mutation in the gene that encodes an miRNA. Other miRNAs are associated with heart disease. One miRNA called miR-1-2 is highly expressed in heart muscle. Mice genetically engineered to express only 50% of the normal amount of miR-1-2 frequently have holes in the wall that separates their left and right ventricles, a common congenital heart defect seen in newborn humans. Overexpression of another miRNA called miR-1 in the hearts of adult mice causes cardiac arrhythmia—irregular electrical activity of the heart that can be life-threatening in humans. Changes in the expression of other miRNAs have been associated with cancer.

CONCEPTS

RNA silencing is initiated by double-stranded RNA molecules that are cleaved and processed. The resulting siRNAs or miRNAs combine with proteins to form complexes that bind to complementary sequences in mRNA or DNA. The siRNAs and miRNAs affect gene expression by cleaving mRNA, inhibiting translation, altering chromatin structure, or triggering RNA degradation.

✓ **CONCEPT CHECK 5**

In RNA silencing, siRNAs and miRNAs usually bind to which part of the mRNA molecules that they control?

a. $5'$ UTR
b. Segment that encodes amino acids.
c. $3'$ poly(A) tail.
d. $3'$ UTR.

17.6 Some Genes Are Regulated by Processes That Affect Translation or by Modifications of Proteins

Ribosomes, aminoacyl tRNAs, initiation factors, and elongation factors are all required for the translation of mRNA molecules. The availability of these components affects the rate of translation and therefore influences gene expression. For example, the activation of T lymphocytes (T cells) is critical to the development of immune responses to viruses (see Chapter 22). T cells are normally in the G_0 stage of the cell cycle and not actively dividing. On exposure to viral antigens, however, specific T cells become activated and undergo rapid proliferation (**Figure 17.14**). Activation includes a 7- to 10-fold increase in protein synthesis that causes cells to enter the cell cycle and proliferate. This burst of protein synthesis does not require an increase in mRNA synthesis. Instead, a global increase in protein synthesis is due to the increased availability of initiation factors taking part in translation—initiation factors that allow ribosomes to bind to mRNA and begin translation. This increase in initiation factors leads to more translation from the existing mRNA molecules, increasing the overall amount of protein synthesized. Similarly, insulin stimulates the initiation of overall protein synthesis by increasing the availability of initiation factors. Initiation factors exist in inactive forms and, in response to various cell signals, can be activated by chemical modifications of their structure, such as phosphorylation.

Mechanisms also exist for the regulation of translation of specific mRNAs. The initiation of translation in some mRNAs is regulated by proteins that bind to an mRNA's $5'$ UTR and inhibit the binding of ribosomes, similar to the way in which repressor proteins bind to operators and prevent the transcription

1 T cells are normally in G_0 and not actually dividing.

2 Exposure to an antigen, such as a virus,…

Virus

Exposure to antigen

T cell

3 …causes an increase in availability of initiation factors,…

Initiation factors

Increase in availability

4 …which allows ribosomes to bind to mRNA and carry out translation.

Ribosomal subunits

+

mRNA

Protein

5 Protein synthesis increases 7- to 10-fold,…

Increase in protein synthesis

Activated T cell

6 …which causes the T cell to undergo proliferation.

17.14 The expression of some eukaryotic genes is regulated by the availability of components required for translation. In this example, exposure to an antigen stimulates an increased availability of initiation factors and a subsequent increase in protein synthesis, leading to T-cell proliferation.

of structural genes. The translation of some mRNAs is affected by the binding of proteins to sequences in the 3′ UTR.

Many eukaryotic proteins are extensively modified after translation by the selective cleavage and trimming of amino acids from the ends, by acetylation, or by the addition of phosphate groups, carboxyl groups, methyl groups, or carbohydrates to the protein. These modifications affect the transport, function, and activity of the proteins.

CONCEPTS

The initiation of translation may be affected by proteins that bind to specific sequences at the 5′ end of mRNA. The availability of ribosomes, tRNAs, initiation and elongation factors, and other components of the translational apparatus may affect the rate of translation. Translation of some mRNAs is regulated by proteins that bind to the 5′ and 3′ untranslated regions of the mRNA.

CONNECTING CONCEPTS

A Comparison of Bacterial and Eukaryotic Gene Control

Now that we have considered the major types of gene regulation in bacteria (Chapter 16) and eukaryotes (this chapter), let's consider some of the similarities and differences in bacterial and eukaryotic gene control.

1. Much of gene regulation in bacterial cells is at the level of transcription (although it does exist at other levels). Gene regulation in eukaryotic cells takes place at multiple levels, including chromatin structure, transcription, mRNA processing, RNA stability, RNA interference, and posttranslational control.

2. Complex biochemical and developmental events in bacterial and eukaryotic cells may require a cascade of gene regulation, in which the activation of one set of genes stimulates the activation of another set.

3. Much of gene regulation in both bacterial and eukaryotic cells is accomplished through proteins that bind to specific sequences in DNA. Regulatory proteins come in a variety of types, but most can be characterized according to a small set of DNA-binding motifs.

4. Chromatin structure plays a role in eukaryotic (but not bacterial) gene regulation. In general, condensed chromatin represses gene expression; chromatin structure must be altered before transcription can take place. Chromatin structure is altered by the chromatin-remodeling proteins, modification of histone proteins, and DNA methylation.

5. Modifications to chromatin structure in eukaryotes may lead to epigenetic changes, which are changes that affect gene expression and are passed on to other cells or future generations.

6. In bacterial cells, genes are often clustered in operons and are coordinately expressed by transcription into a single mRNA molecule. In contrast, many eukaryotic genes have their own promoters and are transcribed independently. Coordinated regulation in eukaryotic cells often takes place through common response elements, present in the promoters and enhancers of the genes. Different genes that have the same response element in common are influenced by the same regulatory protein.

7. Regulatory proteins that affect transcription exhibit two basic types of control: *repressors* inhibit transcription (negative control); *activators* stimulate transcription (positive control). Both negative control and positive control are found in bacterial and eukaryotic cells.

8. The initiation of transcription is a relatively simple process in bacterial cells, and regulatory proteins function by blocking or stimulating the binding of RNA polymerase to DNA. In contrast, eukaryotic transcription requires complex machinery that includes RNA polymerase, general transcription factors, and transcriptional activators, which allows transcription to be influenced by multiple factors.

9. Some eukaryotic transcriptional regulatory proteins function at a distance from the gene by binding to enhancers, causing the formation of a loop in the DNA, which brings the promoter and enhancer into close proximity. Some distant-acting sequences analogous to enhancers have been described in bacterial cells, but they appear to be less common.

10. The greater time lag between transcription and translation in eukaryotic cells compared with that in bacterial cells allows mRNA stability and mRNA processing to play larger roles in eukaryotic gene regulation.

11. RNA molecules (antisense RNA) may act as regulators of gene expression in bacteria. Regulation by siRNAs and miRNAs, which is extensive in eukaryotes, is absent from bacterial cells.

These similarities and differences in gene regulation of bacteria and eukaryotes are summarized in **Table 17.2**.

Table 17.2   **Comparison of gene expression in bacteria and eukaryotes**

Characteristic	Bacterial Gene Control	Eukaryotic Gene Control
Levels of regulation	Primarily transcription	Many levels
Cascades of gene regulation	Present	Present
DNA binding proteins	Important	Important
Role of chromatin structure	Absent	Important
Presence of operons	Common	Uncommon
Negative and positive control	Present	Present
Initiation of transcription	Relatively simple	Relatively complex
Enhancers	Less common	More common
Transcription and translation	Occur simultaneously	Occur separately
Regulation by small RNAs	Rare	Common

CONCEPTS SUMMARY

■ Eukaryotic cells differ from bacteria in several ways that affect gene regulation, including, in eukaryotes, the absence of operons, the presence of chromatin, and the presence of a nuclear membrane.

■ In eukaryotic cells, chromatin structure represses gene expression. In transcription, chromatin structure may be altered by repositioning of nucleosomes and the modification of histone proteins, including acetylation, phosphorylation, and methylation. The methylation of DNA also affects transcription.

■ The initiation of eukaryotic transcription is controlled by general transcription factors that assemble into the basal transcription apparatus and by transcriptional regulator proteins that stimulate or repress normal levels of transcription by binding to regulatory promoters and enhancers.

■ Enhancers affect the transcription of distant genes. Regulatory proteins bind to enhancers and interact with the basal transcription apparatus by causing the intervening DNA to loop out.

■ DNA sequences called insulators limit the action of enhancers by blocking their action in a position-dependent manner.

■ Some regulatory factors cause RNA polymerase to stall downstream of the promoter.

■ Coordinately controlled genes in eukaryotic cells respond to the same factors because they have common response elements that are stimulated by the same transcriptional activator.

■ Gene expression in eukaryotic cells can be influenced by RNA processing.

■ Gene expression can be regulated by changes in RNA stability. The 5′ cap, the coding sequence, the 3′ UTR, and the poly(A) tail are important in controlling the stability of eukaryotic mRNAs. Proteins binding to the 5′ and 3′ ends of eukaryotic mRNA can affect its translation.

■ RNA silencing plays an important role in eukaryotic gene regulation. Small RNA molecules (siRNAs and miRNAs) cleaved from double-stranded DNA combine with proteins and bind to sequences on mRNA or DNA. These complexes cleave RNA, inhibit translation, affect RNA degradation, and silence transcription.

■ Control of the posttranslational modification of proteins may play a role in gene expression.

IMPORTANT TERMS

DNase I hypersensitive site (p. 475)

histone code (p. 475)

chromatin-remodeling complex (p. 475)

CpG island (p. 478)

mediator (p. 479)

insulator (p. 481)

heat-shock protein (p. 482)

response element (p. 482)

SR protein (p. 483)

ANSWERS TO CONCEPT CHECKS

1. Three general processes are chromatin remodeling, the modification of histone proteins (e.g. methylation and acetylation of histones), and DNA methylation.

2. b

3. The DNA between the enhancer and the promoter loops out, so regulatory proteins bound to the enhancer are able to interact directly with the transcription apparatus.

4. The poly(A) tail stabilizes the 5′ cap, which must be removed before the mRNA molecule can be degraded from the 5′ end.

5. d

WORKED PROBLEM

Problem

What would be the effect of a mutation that caused poly(A)-binding proteins to be nonfunctional?

Solution Strategy

What information is required in your answer to the problem?

A statement of the effect of a mutation that eliminated the function of the poly(A) binding protein.

What information is provided to solve the problem?

- A mutation occurs in the gene that encodes poly(A) binding protein.

- The mutation causes the poly(A) protein to be nonfunctional.

For help with this problem, review:

The Degradation of RNA in Section 17.4.

The Addition of the Poly(A) Tail in Section 14.2 (Chapter 14).

Solution Steps

Messenger RNA can be degraded from the 5′ end, the 3′ end, or through internal cleavage. Degradation from the 5′ end requires the removal of the 5′ cap and is usually preceded by the shortening of the poly(A) tail. Poly(A)-binding proteins bind to the poly(A) tail and prevent it from being shortened. Thus, the presence of these proteins on the poly(A) tail protects the 5′ cap, which prevents RNA degradation. If the gene for poly(A)-binding proteins were mutated in such a way that nonfunctional poly(A) proteins were produced, the proteins would not bind to the poly(A) tail. The tail would be shortened prematurely, the 5′ cap removed, and mRNA degraded more easily. The end result would be less mRNA and thus less protein synthesis.

Recall: The poly(A) tail affects the stability of the mRNA.

COMPREHENSION QUESTIONS

Introduction

1. How similar are the genomes of humans and chimpanzees? What genetic changes might be responsible for the large differences in the anatomy, physiology, and behavior of humans and chimpanzees?

Section 17.1

2. List some important differences between bacterial and eukaryotic cells that affect the way in which genes are regulated.

Section 17.2

3. Where are DNase I hypersensitivity sites found and what do they indicate about the nature of chromatin?

4. What changes take place in chromatin structure and what role do these changes play in eukaryotic gene regulation?

5. What is the histone code?

6. How is chromatin immunoprecipitation used to determine the locations of histone modifications in the genome?

Section 17.3

7. Briefly explain how transcriptional activator proteins and repressors affect the level of transcription of eukaryotic genes.

8. What is an enhancer? How does it affect the transcription of distant genes?

9. What is an insulator?

10. What is a response element? How do response elements bring about the coordinated expression of eukaryotic genes?

Section 17.4

11. Outline the role of alternative splicing in the control of sex differentiation in *Drosophila*.

12. What role does RNA stability play in gene regulation? What controls RNA stability in eukaryotic cells?

Section 17.5

13. Briefly list some of the ways in which siRNAs and miRNAs regulate genes.

Section 17.6

14. How does bacterial gene regulation differ from eukaryotic gene regulation? How are they similar?

> ➤ For more questions that test your comprehension of the key chapter concepts, go to **LEARNING**Curve for this chapter.

APPLICATION QUESTIONS AND PROBLEMS

Section 17.2

15. Malaria, one of the most pervasive and destructive of all infectious diseases, is caused by protozoan parasites of the genus *Plasmodium,* which are transmitted from person to person by mosquitoes. *Plasmodium* parasites are able to evade the host immune system by constantly altering the expression of their *var* genes, which encode *Plasmodium* surface antigens (L. H. Freitas-Junior et al. 2005. *Cell* 121:25–36). Individual *var* genes are expressed when chromatin structure is disrupted by chemical changes in histone proteins. What type of chemical changes in the histone proteins might be responsible for these changes in gene expression?

16. A geneticist is trying to determine how many genes are found in a 300,000-bp region of DNA. Analysis shows that four H3K4me3 modifications are found in this piece of DNA. What might their presence suggest about the number of genes located there?

17. In a line of human cells grown in culture, a geneticist isolates a temperature-sensitive mutation at a locus that encodes an acetyltransferase enzyme; at temperatures above 38°C the mutant cells produce a nonfunctional form of the enzyme. What would be the most likely effect of this mutation when the cells are grown at 40°C?

*18. What would be the most likely effect of deleting *flowering locus D* (FLD) in *Arabidopsis thalania*?

*19. X31b is an experimental compound that is taken up by rapidly dividing cells. Research has shown that X31b stimulates the methylation of DNA. Some cancer researchers are interested in testing X31b as a possible drug for treating prostate cancer. Offer a possible explanation for why X31b might be an effective anticancer drug.

Section 17.3

20. How do repressors that bind to silencers in eukaryotes differ from repressors that bind to operators in bacteria?

21. Examine **Figure 17.7**. What would be the effect on transcription if a mutation occurred in the gene that encodes GAL3, so that no functional GAL3 was produced?

22. What would be the effect of moving the insulator shown in **Figure 17.8** to a position between Enhancer II and the promoter for gene B?

*23. An enhancer is surrounded by four genes (*A, B, C,* and *D*), as shown in the adjoining diagram. An insulator lies between gene *C* and gene *D*. On the basis of the positions of the genes, the enhancer, and the insulator, the transcription of which genes is most likely to be stimulated by the enhancer? Explain your reasoning.

| Gene A | Gene B | Enhancer | Gene C | Insulator | Gene D |

Section 17.4

*24. What will be the effect on sexual development in newly fertilized *Drosophila* embryos if the following genes are deleted?

  a. *sex lethal*

  b. *transformer*

  c. *double-sex*

25. Examine **Figure 17.12**. What would be the effect of a mutation that eliminated the second 3′ splice site at the end of exon B in the *tra* pre-mRNA?

*26. Some eukaryotic mRNAs have an AU-rich element in the 3′ untranslated region. What would be the effect on gene expression if this element were mutated or deleted?

27. A strain of *Arabidopsis thaliana* possesses a mutation in the *APETALA2* gene, in which much of the 3′ untranslated region of mRNA transcribed from the gene is deleted. What is the most likely effect of this mutation on the expression of the *APETALA2* gene?

28. What will be the effect of a mutation that destroys the ability of poly(A)-binding protein (PABP) to attach to a poly(A) tail?

Section 17.5

29. Suppose a geneticist introduced a small interfering RNA (siRNA) that was complementary to the FLC mRNA in **Figure 17.3**. What would be the effect on flowering of *Arabidopsis*? Explain your answer.

CHALLENGE QUESTIONS

Section 17.3

30. The yeast gene *SER3*, which has a role in serine biosynthesis, is repressed during growth in nutrient-rich medium, so little transcription takes place and little SER3 enzyme is produced. In an investigation of the repression of the *SER3* gene, a region of DNA upstream of *SER3* was found to be heavily transcribed when *SER3* is repressed (J. A. Martens, L. Laprade, and F. Winston. 2004. *Nature* 429:571–574). Within this upstream region is a promoter that stimulates the transcription of an RNA molecule called *SRG1* RNA (for *SER3* regulatory gene 1). This RNA molecule has none of the sequences necessary for translation. Mutations in the promoter for *SRG1* result in the disappearance of *SRG1* RNA, and these mutations remove the repression of *SER3*. When RNA polymerase binds to the *SRG1* promoter, the polymerase is found to travel downstream, transcribing the *SGR1* RNA, and to pass through and transcribe the promoter for *SER3*. This activity leads to the repression of *SER3*. Propose a possible explanation for how the transcription of *SGR1* might repress the transcription of *SER3*. (Hint: Remember that the *SGR1* RNA does not encode a protein.)

Section 17.5

31. A common feature of many eukaryotic mRNAs is the presence of a rather long 3′ UTR, which often contains consensus sequences. Creatine kinase B (CK-B) is an enzyme important in cellular metabolism. Certain cells—termed U937D cells—have lots of CK-B mRNA, but no CK-B enzyme is present. In these cells, the 5′ end of the CK-B mRNA is bound to ribosomes, but the mRNA is apparently not translated. Something inhibits the translation of the CK-B mRNA in these cells.

Researchers introduced numerous short segments of RNA containing only 3′ UTR sequences into U937D cells. As a result, the U937D cells began to synthesize the CK-B enzyme, but the total amount of CK-B mRNA did not increase. The introduction of short segments of other RNA sequences did not stimulate the synthesis of CK-B; only the 3′ UTR sequences turned on the translation of the enzyme.

On the basis of these results, propose a mechanism for how CK-B translation is inhibited in the U937D cells. Explain how the introduction of short segments of RNA containing the 3′ UTR sequences might remove the inhibition.

Go to your **LaunchPad** to find additional learning resources and the Suggested Readings for this chapter.

18

Gene Mutations and DNA Repair

People with a mutation in the *tinman* gene, named after Tin Man in *The Wizard of Oz*, often have congenital heart defects. [Mary Evans Picture Library/Alamy.]

A Fly Without A Heart

The heart of a fruit fly is a simple organ, an open-ended tube that rhythmically contracts, pumping fluid—rather inefficiently—around the body of the fly. Although simple and inelegant, the fruit fly's heart is nevertheless essential. Remarkably, a few rare mutant fruit flies never develop a heart and die (not surprisingly) at an early embryonic stage. Geneticist Rolf Bodmer analyzed these mutants in the 1980s and made an important discovery—a gene that specifies the development of a heart. He named the gene *tinman*, after the character in *The Wizard of Oz* who also lacked a heart. Bodmer's research revealed that *tinman* encodes a transcription factor that binds to DNA and turns on other genes that are essential for the normal development of a heart. In the mutant flies this gene was lacking, the transcription factor was never produced, and the heart never developed. Findings from subsequent research revealed the existence of a human gene (called *Nkx2.5*) with a sequence similar to that of *tinman*, but the function of the human gene was unknown.

Then, in the 1990s, physicians Jonathan and Christine Seidman began studying people born with abnormal hearts, such as those with a hole in the septum that separates the chambers on the left and right sides of the heart. Such defects result in abnormal blood flow through the heart, causing the heart to work harder than normal and mixing oxygenated and deoxygenated blood. Congenital heart defects are not uncommon; they're found in about 1 of every 125 babies. Some of the defects heal on their own, but others require corrective surgery. Although surgery is often successful in reversing congenital heart

problems, many of these patients begin to have irregular heartbeats in their 20s and 30s. The Seidmans and their colleagues found several families in which congenital heart defects and irregular heartbeats were inherited together in an autosomal dominant fashion. Detailed molecular analysis of one of these families revealed that the gene responsible for the heart problems was located on chromosome 5, at a spot where the human *tinman* gene (*Nkx2.5*) had been previously mapped. All members of this family who inherited the heart defects also inherited a mutation in the *tinman* gene. Subsequent studies with additional patients found that many people with congenital heart defects have a mutation in the *tinman* gene. The human version of this gene, like its counterpart in flies, encodes a transcription factor that controls heart development. Despite tremendous differences in size, anatomy, and physiology, humans and flies use the same gene to make a heart.

The story of *tinman* illustrates the central importance of studying mutations: the analysis of mutants is often a source of key insights into important biological processes. This chapter focuses on gene mutations—how errors arise in genetic instructions and how those errors are studied. We begin with a brief examination of the different types of mutations, including their phenotypic effects, how they can be suppressed, and mutation rates. The next section explores how mutations spontaneously arise during and after the course of DNA replication, as well as how chemicals and radiation induce mutations. After discussing the analysis of mutations, we turn to transposable elements, DNA sequences that are capable of moving within the genome and which often produce mutations when they move. Finally, we take a look at DNA repair and some of the diseases that arise when DNA repair is defective.

18.1 Mutations Are Inherited Alterations in the DNA Sequence

DNA is a highly stable molecule that is replicated with amazing accuracy (see Chapters 10 and 12), but changes in DNA structure and errors of replication do take place. A **mutation** is defined as an inherited change in genetic information; the descendants may be cells or organisms.

The Importance of Mutations

Mutations are both the sustainer of life and the cause of great suffering. On the one hand, mutation is the source of all genetic variation, the raw material of evolution. The ability of organisms to adapt to environmental change depends on the presence of genetic variation in natural populations, and genetic variation is produced by mutation. On the other hand, many mutations have detrimental effects and mutation is the source of many diseases and disorders.

Much of the study of genetics focuses on how variants produced by mutation are inherited; genetic crosses are meaningless if all individual members of a species are identically homozygous for the same alleles. Much of Gregor Mendel's

success in unraveling the principles of inheritance can be traced to his use of carefully selected variants of the garden pea. Similarly, Thomas Hunt Morgan and his students discovered many basic principles of genetics by analyzing mutant fruit flies.

Mutations are also useful for examining fundamental biological processes. Finding or creating mutations that affect different components of a biological system and studying their effects can often lead to an understanding of the system. This method, referred to as genetic dissection, is analogous to figuring out how an automobile works by breaking different parts of a car and observing the effects; for example, smash the radiator and the engine overheats, revealing that the radiator cools the engine. The use of mutations to disrupt function can likewise be a source of insight into biological processes. For example, geneticists have begun to unravel the molecular details of development by studying mutations, such as *tinman*, that interrupt various embryonic stages in *Drosophila* (see Chapter 22). Scientists also used analysis of mutations to reveal the different parts of the *lac* operon (discussed in Chapter 16) and how they function in gene regulation. Although breaking "parts" to determine their function might seem like a crude approach to understanding a system, it is actually very powerful and has been used extensively in biochemistry, developmental biology, physiology, and behavioral science. But this method is *not* recommended for learning how your car works!

CONCEPTS

Mutations are heritable changes in DNA. They are essential to the study of genetics and are useful in many other biological fields.

Categories of Mutations

In multicellular organisms, we can distinguish between two broad categories of mutations: somatic mutations and germline mutations. **Somatic mutations** arise in somatic tissues, which do not produce gametes (**Figure 18.1**). When a somatic cell with a mutation divides (mitosis), the mutation

1 Somatic mutations occur in nonreproductive cells…

2 …and are passed to new cells through mitosis, creating a clone of cells having the mutant gene.

Population of mutant cells

Somatic tissue

Somatic mutation

Mitosis

Mutant cell

Germ-line tissue

Germ-line mutation

Sexual reproduction

All cells carry mutation

No cells carry mutation

3 Germ-line mutations occur in cells that give rise to gametes.

4 Meiosis and sexual reproduction allow germ-line mutations to be passed to approximately half the members of the next generation,…

5 …who will carry the mutation in all their cells.

18.1 The two basic classes of mutations are somatic mutations and germ-line mutations.

is passed on to the daughter cells, leading to a population of genetically identical cells (a clone). The earlier in development that a somatic mutation takes place, the larger the clone of cells will be that contain the mutation.

Because of the huge number of cells present in a typical eukaryotic organism, somatic mutations are numerous. For example, there are about 10^{14} cells in the human body. Typically, a mutation arises once in every million cell divisions, so hundreds of millions of somatic mutations must arise in each person. Many somatic mutations have no obvious effect on the phenotype of the organism, because the function of the mutant cell is replaced by that of normal cells or the mutant cell dies and is replaced by normal cells. However, cells with a somatic mutation that stimulates cell division can increase in number and spread; this type of mutation can give rise to cells with a selective advantage and is the basis for cancers (see Chapter 23).

Germ-line mutations arise in cells that ultimately produce gametes. A germ-line mutation can be passed to future generations, producing individual organisms that carry the mutation in all their somatic and germ-line cells (see Figure 18.1). When we speak of mutations in multicellular organisms, we're usually talking about germ-line mutations.

Historically, mutations have been partitioned into those that affect a single gene, called *gene mutations*, and those that affect the number or structure of chromosomes, called *chromosome mutations*. This distinction arose because chromosome mutations could be observed directly, by looking at chromosomes with a microscope, whereas gene mutations could be detected only by observing their phenotypic effects. Now, DNA sequencing allows direct observation of gene mutations, and chromosome mutations are distinguished from gene mutations somewhat arbitrarily on the basis of the size of the DNA lesion. Nevertheless, it is practical to use *chromosome mutation* for a large-scale genetic alteration that affects chromosome structure or the number of chromosomes and to use **gene mutation** for a relatively small DNA lesion that affects a single gene. This chapter focuses on gene mutations; chromosome mutations were discussed in Chapter 8.

Types of Gene Mutations

There are a number of ways to classify gene mutations. Some classification schemes are based on the nature of the pheno-

typic effect, others are based on the causative agent of the mutation, and still others focus on the molecular nature of the defect. Here, we will categorize mutations primarily on the basis of their molecular nature, but we will also encounter some terms that relate the causes and the phenotypic effects of mutations.

BASE SUBSTITUTIONS The simplest type of gene mutation is a **base substitution**, the alteration of a single nucleotide in the DNA (**Figure 18.2a**). There are two types of base substitutions. In a **transition**, a purine is replaced by a different purine or, alternatively, a pyrimidine is replaced by a different pyrimidine (**Figure 18.3**). In a **transversion**, a purine is replaced by a pyrimidine or a pyrimidine is replaced by a purine. The number of possible transversions (see Figure 18.3) is twice the number of possible transitions, but transitions arise more frequently because transforming a purine into different purine or a pyrimidine into different pyrimidine is easier than transforming a purine into pyrimidine, or vice versa. ▶ **TRY PROBLEM 18**

Original DNA sequence   GGG AGT GTA GAT CGT

(a) Base substitution   GGG AGT G**C**A GAT CGT

A base substitution alters a single codon.

One codon changed

(b) Base insertion   GGG AGT GT**T** AGA TCG T

(c) Base deletion   GGG AGT GAG ATC GT

An insertion or a deletion alters the reading frame and may change many codons.

18.2 Three basic types of gene mutations are base substitutions, insertions, and deletions.

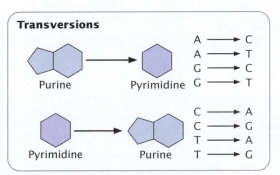

Transitions		Possible base changes
Purine → Purine		A → G G → A
Pyrimidine → Pyrimidine		T → C C → T

Transversions	
Purine → Pyrimidine	A → C A → T G → C G → T
Pyrimidine → Purine	C → A C → G T → A T → G

18.3 A transition is the substitution of a purine for a purine or of a pyrimidine for a pyrimidine; a transversion is the substitution of a pyrimidine for a purine or of a purine for a pyrimidine.

INSERTIONS AND DELETIONS   Another class of gene mutations contains **insertions** and **deletions** (collectively called indels)—the addition or removal, respectively, of one or more nucleotide pairs (**Figure 18.2b** and **c**). Although base substitutions are often assumed to be the most common type of mutation, molecular analysis has revealed that insertions and deletions are often more frequent. Insertions and deletions within sequences that encode proteins may lead to **frameshift mutations**, changes in the reading frame (see p. 421 in Chapter 15) of the gene. Frameshift mutations usually alter all amino acids encoded by nucleotides following the mutation, and so they generally have drastic effects on the phenotype. Some frameshifts also introduce premature stop codons, terminating protein synthesis early and resulting in a shortened (truncated) protein. Not all insertions and deletions lead to frameshifts, however; insertions and deletions consisting of any multiple of three nucleotides will leave the reading frame intact, although the addition or removal of one or more amino acids may still affect the phenotype. Indels not affecting the reading frame are called **in-frame insertions** and **in-frame deletions**.

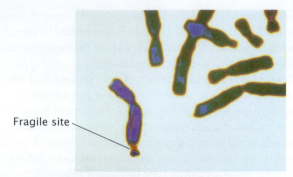

Fragile site

18.4 The fragile-X chromosome is associated with a characteristic constriction (fragile site) on the long arm.
[©2013 Custom Medical Stock Photo.]

CONCEPTS

Gene mutations consist of changes in a single gene and can be base substitutions (a single pair of nucleotides is altered) or insertions or deletions (nucleotides are added or removed). A base substitution can be a transition (substitution of like bases) or a transversion (substitution of unlike bases). Insertions and deletions often lead to a change in the reading frame of a gene.

✓ CONCEPT CHECK 1

Which of the following changes is a transition base substitution?
a.   Adenine is replaced by thymine.
b.   Cytosine is replaced by adenine.
c.   Guanine is replaced by adenine.
d.   Three nucleotide pairs are inserted into DNA.

EXPANDING NUCLEOTIDE REPEATS   Mutations in which the number of copies of a set of nucleotides increases are called **expanding nucleotide repeats**. This type of mutation was first observed in 1991 in a gene called *FMR-1*, which causes fragile-X syndrome, the most common hereditary cause of intellectual disability. The disorder is so named because, in specially treated cells from persons having the condition, the tip of each long arm of the X chromosome is attached by a slender-appearing part of the chromosome (**Figure 18.4**). The normal *FMR-1* allele (not containing the mutation) has 60 or fewer copies of CGG but, in persons with fragile-X syndrome, the allele may harbor hundreds or even thousands of copies.

Expanding nucleotide repeats have been found in almost 30 human diseases, several of which are listed in **Table 18.1**. Most of these diseases are caused by the expansion of a set of three nucleotides (called a trinucleotide), most often CNG, where N can be any nucleotide. However, some diseases are caused by repeats of four, five, and even twelve nucleotides. The number of copies of the nucleotide repeat often correlates with the severity or age of onset of

Table 18.1   Examples of human genetic diseases caused by expanding nucleotide repeats

Disease	Repeated Sequence	Number of Copies of Repeat	
		Normal Range	Disease Range
Spinal and bulbar muscular atrophy	CAG	11–33	40–62
Fragile-X syndrome	CGG	6–54	50–1500
Jacobsen syndrome	CGG	11	100–1000
Spinocerebellar ataxia (several types)	CAG	4–44	21–130
Autosomal dominant cerebellar ataxia	CAG	7–19	37–220
Myotonic dystrophy	CTG	5–37	44–3000
Huntington disease	CAG	9–37	37–121
Friedreich ataxia	GAA	6–29	200–900
Dentatorubral-pallidoluysian atrophy	CAG	7–25	49–75
Myoclonus epilepsy of the Unverricht–Lundborg type	CCCCGCCCCGCG	2–3	12–13

the disease. The number of copies of the repeat also correlates to the instability of nucleotide repeats: when more repeats are present, the probability of expansion to even more repeats increases. This association between the number of copies of nucleotide repeats, the severity of the disease, and the probability of expansion leads to a phenomenon known as anticipation (see Chapter 5), in which diseases caused by nucleotide-repeat expansions become more severe in each generation. Less commonly, the number of nucleotide repeats may decrease within a family. Expanding nucleotide repeats have also been observed in some microbes and plants.

Increases in the number of nucleotide repeats can produce disease symptoms in different ways. In several of the diseases (e.g., Huntington disease), the nucleotide expands within the coding part of a gene, producing a toxic protein that has extra glutamine residues (the amino acid encoded by CAG). In other diseases, the repeat is outside the coding region of a gene and affects its expression. In fragile-X syndrome, additional copies of the nucleotide repeat cause the DNA to become methylated, which turns off the transcription of an essential gene.

Some evidence suggests that expansion of nucleotide repeats occurs in the course of DNA replication and appears to be related to the formation of hairpins and other special DNA structures that form in single-stranded DNA consisting of nucleotide repeats. Such structures may interfere with normal replication by causing strand slippage, misalignment of the sequences, or stalling of replication. One model of how repeat hairpins might result in repeat expansion is shown in **Figure 18.5**. Watch **Animation 18.1** to help you understand how copies of nucleotide repeats increase in number. Other models of repeat expansion that occur through transcription and DNA repair have also been proposed. Many aspects of this phenomenon are unknown, including why repeat expansion occurs in some people and not in others.

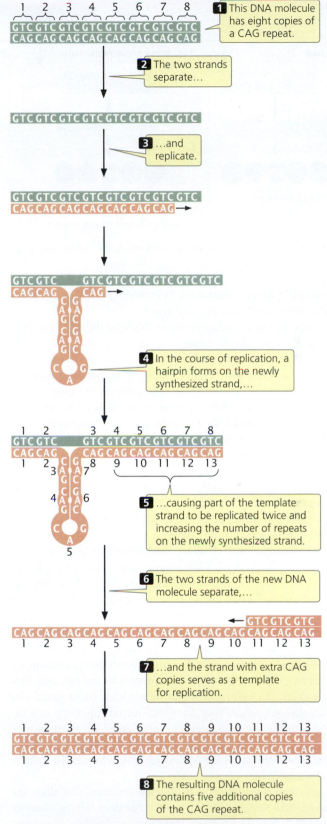

18.5 **A model of how the number of copies of a nucleotide repeat may increase in replication.**

CONCEPTS

Expanding nucleotide repeats are regions of DNA that consist of repeated copies of sets of nucleotides. Increased numbers of nucleotide repeats are associated with several human genetic diseases.

Phenotypic Effects of Mutations

Another way that mutations are classified is on the basis of their phenotypic effects. At the most-general level, we can distinguish a mutation on the basis of its phenotype compared with the wild-type phenotype. A mutation that alters the wild-type phenotype is called a **forward mutation**,

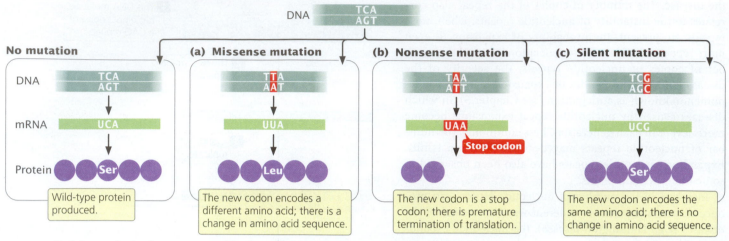

18.6 Base substitutions can cause (a) missense, (b) nonsense, and (c) silent mutations.

whereas a **reverse mutation** (a *reversion*) changes a mutant phenotype back into the wild type.

Geneticists use other terms to describe the effects of mutations on protein structure. A base substitution that results in a different amino acid in the protein is referred to as a **missense mutation** (**Figure 18.6a**). A **nonsense mutation** changes a sense codon (one that specifies an amino acid) into a nonsense codon (one that terminates translation), as shown in **Figure 18.6b**. If a nonsense mutation occurs early in the mRNA sequence, the protein will be truncated and usually nonfunctional.

Because of the redundancy of the genetic code, some different codons specify the same amino acid. A **silent mutation** changes a codon to a synonymous codon that specifies the same amino acid (**Figure 18.6c**), altering the DNA sequence without changing the amino acid sequence of the protein. Not all silent mutations, however, are truly silent: some do have phenotypic effects. For example, silent mutations may have phenotypic effects when different tRNAs (called isoaccepting tRNAs, see Chapter 15) are used for different synonymous codons. Because some isoaccepting tRNAs are more abundant than others, which synonymous codon is used may affect the rate of protein synthesis. The rate of protein synthesis can influence the phenotype by affecting the amount of protein present in the cell and, in a few cases, the folding of the protein. Other silent mutations can alter sequences near the exon–intron junctions that affect splicing (see Chapter 14). Still other silent mutations can influence the binding of miRNAs to complementary sequences in the mRNA, which determine whether the mRNA is translated (see Chapter 14).

A **neutral mutation** is a missense mutation that alters the amino acid sequence of the protein but does not significantly change its function. Neutral mutations occur when one amino acid is replaced by another that is chemically similar or when the affected amino acid has little influence on protein function. For example, neutral mutations occur in the genes that encode hemoglobin; although these mutations alter the amino acid sequence of hemoglobin, they do not affect its ability to transport oxygen.

Loss-of-function mutations cause the complete or partial absence of normal protein function. A loss-of-function mutation so alters the structure of the protein that the protein no longer works correctly—or the mutation can occur in regulatory regions that affect the transcription, translation, or splicing of the protein. Loss-of-function mutations are frequently recessive: an individual diploid organism must be homozygous for a loss-of-function mutation before the effects of the loss of the functional protein can be exhibited. The mutations that cause cystic fibrosis are loss-of-function mutations: these mutations produce a nonfunctional form of the cystic fibrosis transmembrane conductance regulator protein, which normally regulates the movement of chloride ions into and out of the cell (see Chapter 5).

In contrast, a **gain-of-function mutation** causes the cell to produce a protein or gene product whose function is not normally present. This could be an entirely new gene product or one produced in an inappropriate tissue or at an inappropriate time in development. For example, a mutation in a gene that encodes a receptor for a growth factor might cause the mutated receptor to stimulate growth all the time, even in the absence of the growth factor. Gain-of-function mutations are frequently dominant in their expression, because a single copy of the mutation leads to the presence of a new gene product. Still other types of mutations are **conditional mutations**, which are expressed only under certain conditions. For example, some conditional mutations affect the phenotype only at elevated temperatures. Others are **lethal mutations**, causing premature death (Chapter 5). ▶ TRY PROBLEM 22

Suppressor Mutations

A **suppressor mutation** is a genetic change that hides or suppresses the effect of another mutation. This type of mutation is different from a reverse mutation, in which the mutated site changes back into the original wild-type

1 A forward mutation changes the wild type into a mutant phenotype.

2 A reverse mutation restores the wild-type gene and the phenotype.

3 A suppressor mutation occurs at a site different from that of the original mutation...

4 ...and produces an individual that has both the original mutation and the suppressor mutation...

5 ...but has the wild-type phenotype.

Genotype: Wild type $A^+ B^+$ — Forward mutation A^- / Reverse of mutation A^- → Mutation $A^- B^+$ — Suppressor mutation B^- → Mutations $A^- B^-$

Red eyes   White eyes   Red eyes

18.7 Relation of forward, reverse, and suppressor mutations.

sequence (**Figure 18.7**). A suppressor mutation occurs at a site that is distinct from the site of the original mutation; thus, an individual with a suppressor mutation is a double mutant, possessing both the original mutation and the suppressor mutation but exhibiting the phenotype of an unmutated wild type. Geneticists distinguish between two classes of suppressor mutations: intragenic and intergenic.

INTRAGENIC SUPPRESSOR MUTATIONS An **intragenic suppressor mutation** takes place in the same gene as that containing the mutation being suppressed and may work in several ways. The suppressor may change a second nucleotide in the same codon altered by the original mutation, producing a codon that specifies the same amino acid as that specified by the original, unmutated codon (**Figure 18.8**). Intragenic suppressors may also work by suppressing a frameshift mutation. If the original mutation is a one-base deletion, then the addition of a single base elsewhere in the gene will restore the former reading frame.

Consider the following nucleotide sequence on the template stand of DNA and the amino acids that it encodes:

DNA   3′—AAA TCA CTT GGC GTA CAA—5′

mRNA   5′—UUU AGU GAA CCG CAU GUU—3′

Amino acids   Phe   Ser   Glu   Pro   His   Val

Suppose that a one-base deletion occurs in the first nucleotide of the second codon. This deletion shifts the reading frame by one nucleotide and alters all the amino acids that follow the mutation.

One-nucleotide deletion

DNA   3′—AAA X̶CAC TTG GCG TAC AA—5′

mRNA   5′—UUU GUG AAC CGC AUG UU—3′

Amino acids   Phe   Val   Asn   Arg   Met

If a single nucleotide is added to the third codon (the suppressor mutation), the reading frame is restored, although two of the amino acids differ from those specified by the original sequence.

One-nucleotide insertion

DNA   3′—AAA CAC TTT GGC GTA CAA—5′

mRNA   5′—UUU GUG AAA CCG CAU GUU—3′

Amino acids   Phe   Val   Lys   Pro   His   Val

Similarly, a mutation due to an insertion may be suppressed by a subsequent deletion in the same gene.

A third way in which an intragenic suppressor may work is by making compensatory changes in the protein. A first missense mutation can alter the folding of a polypeptide chain by changing the way in which amino acids in the protein interact with one another. A second missense mutation at a different

1 A missense mutation alters a single codon.

2 A second mutation at a different site in the same gene...

3 ...may restore the original amino acid.

DNA   TTA/AAT → Mutation → TTT/AAA → Intragenic supressor mutation → CTT/GAA

mRNA   UUA → UUU → CUU

Protein   Leu → Phe → Leu

18.8 An intragenic suppressor mutation occurs in the gene containing the mutation being suppressed.

499

(a) Wild-type sequence

DNA
TTG
AAC

→ Transcription →

mRNA
UUG

→ Translation →

Ribosome

Leu

AAC
UUG

Leu is incorporated into a protein.

Full-length, functional protein

(b) Base substitution

TAG
ATC

→ Transcription →

Stop codon
UAG

→ Translation →

UAG

Protein synthesis is halted, resulting in a nonfunctional protein.

Termination of translation

Shortened, nonfunctional protein

(c) Base substitution at a second site

Site 1 (first mutation)   Site 2

At site 2 is a gene encoding tyrosine-tRNA.

TAG   ATA
ATC   TAT

tRNA
AUA

Second base-substitution mutation

Normal transcription produces a tRNA with an anticodon AUA (which would pair with the tyrosine codon UAU in translation).

TAG   CTA
ATC   GAT

→ Transcription →

Introduction of an incorrect base (G), results in a mutant tRNA that has anticodon AUC (instead of AUA),…

UAG   tRNA
            AUC

Tyr

→ Translation →

Tyr

…which can pair with the stop codon UAG.

AUC
UAG

Translation continues past the stop codon, and Tyr is incorporated into the protein.

Full-length, functional protein

18.9 An intergenic suppressor mutation occurs in a gene other than the one bearing the original mutation. (a) The wild-type sequence produces a full-length, functional protein. (b) A base substitution at a site in the same gene produces a premature stop codon, resulting in a shortened, nonfunctional protein. (c) A base substitution at a site in another gene, which in this case encodes tRNA, alters the anticodon of tRNA^{Tyr}; tRNA^{Tyr} can pair with the stop codon produced by the original mutation, allowing tyrosine to be incorporated into the protein and translation to continue.

site (the suppressor) can recreate the original folding pattern by restoring interactions between the amino acids.

INTERGENIC SUPPRESSOR MUTATIONS An **intergenic suppressor mutation**, in contrast, occurs in a gene other than the one bearing the original mutation. These suppressors sometimes work by changing the way that the mRNA is translated. In the example illustrated in **Figure 18.9a**, the original DNA sequence is AAC (UUG in the mRNA) and specifies leucine. This sequence mutates to ATC (UAG in mRNA), a termination codon (**Figure 18.9b**). The ATC nonsense mutation could be suppressed by a second mutation in a different gene that encodes a tRNA; this second mutation would result in a codon capable of pairing with the UAG termination codon (**Figure 18.9c**). For example, the gene that encodes the tRNA for tyrosine (tRNA^{Tyr}), which has the anticodon AUA, might

be mutated to have the anticodon AUC, which will then pair with the UAG stop codon. Instead of translation terminating at the UAG codon, tyrosine would be inserted into the protein and a full-length protein would be produced, although tyrosine would now substitute for leucine. The effect of this change would depend on the role of this amino acid in the overall structure of the protein, but the effect of the suppressor mutation is likely to be less detrimental than the effect of the nonsense mutation, which would halt translation prematurely.

Because cells in many organisms have multiple copies of tRNA genes, other unmutated copies of tRNA^{Tyr} would remain available to recognize tyrosine codons in the transcripts of the mutant gene in question and other genes being expressed concurrently. We might expect that the tRNAs that have undergone the suppressor mutation just described would also suppress the normal termination codons at the ends of other

Table 18.2   Characteristics of different types of mutations

Type of Mutation	Definition
Base substitution	Changes the base of a single DNA nucleotide
Transition	Base substitution in which a purine replaces a purine or a pyrimidine replaces a pyrimidine
Transversion	Base substitution in which a purine replaces a pyrimidine or a pyrimidine replaces a purine
Insertion	Addition of one or more nucleotides
Deletion	Deletion of one or more nucleotides
Frameshift mutation	Insertion or deletion that alters the reading frame of a gene
In-frame deletion or insertion	Deletion or insertion of a multiple of three nucleotides that does not alter the reading frame
Expanding nucleotide repeats	Repeated sequence of a set of nucleotides in which the number of copies of the sequence increases
Forward mutation	Changes the wild-type phenotype to a mutant phenotype
Reverse mutation	Changes a mutant phenotype back to the wild-type phenotype
Missense mutation	Changes a sense codon into a different sense codon, resulting in the incorporation of a different amino acid in the protein
Nonsense mutation	Changes a sense codon into a nonsense (stop) codon, causing premature termination of translation
Silent mutation	Changes a sense codon into a synonymous codon, leaving unchanged the amino acid sequence of the protein
Neutral mutation	Changes the amino acid sequence of a protein without altering its ability to function
Loss-of-function mutation	Causes a complete or partial loss of function
Gain-of-function mutation	Causes the appearance of a new trait or function or causes the appearance of a trait in inappropriate tissue or at an inappropriate time
Lethal mutation	Causes premature death
Suppressor mutation	Suppresses the effect of an earlier mutation at a different site
Intragenic suppressor mutation	Suppresses the effect of an earlier mutation within the same gene
Intergenic suppressor mutation	Suppresses the effect of an earlier mutation in another gene

coding sequences, resulting in the production of longer-than-normal proteins, but this event does not usually take place.

Intergenic suppressors can also work through genic interactions (see Chapter 5). Polypeptide chains that are produced by two genes may interact to produce a functional protein. A mutation in one gene may alter the encoded polypeptide such that the interaction between the two polypeptides is destroyed, in which case a functional protein is not produced. A suppressor mutation in the second gene may produce a compensatory change in its polypeptide, therefore restoring the original interaction. Characteristics of some of the different types of mutations are summarized in **Table 18.2**.

CONCEPTS

A suppressor mutation overrides the effect of an earlier mutation at a different site. An intragenic suppressor mutation occurs within the *same* gene as that containing the original mutation; an intergenic suppressor mutation occurs in a *different* gene.

✓ **CONCEPT CHECK 2**

How is a suppressor mutation different from a reverse mutation?

WORKED PROBLEM

A gene encodes a protein with the following amino acid sequence:

Met-Arg-Cys-Ile-Lys-Arg

A mutation of a single nucleotide alters the amino acid sequence to:

Met-Asp-Ala-Tyr-Lys-Gly-Glu-Ala-Pro-Val

A second single-nucleotide mutation occurs in the same gene and suppresses the effects of the first mutation (an intragenic suppressor). With the original mutation and the intragenic suppressor present, the protein has the following amino acid sequence:

Met-Asp-Gly-Ile-Lys-Arg

What is the nature and location of the first mutation and of the intragenic suppressor mutation?

Solution Strategy

What information is required in your answer to the problem?

The type and location of the first mutation and the intragenic suppressor.

What information is provided to solve the problem?

- The amino acid sequence of the protein encoded by the original unmutated gene.

- The amino acid sequence of the protein encoded by the mutated gene.

- The amino acid sequence of the protein encoded by the mutated gene and the intragenic suppressor.

Solution Steps

The first mutation alters the reading frame, because all amino acids after Met are changed. Insertions and deletions affect the reading frame; the original mutation consists of a single-nucleotide insertion or deletion in the second codon. The intragenic suppressor restores the reading frame; the intragenic suppressor also is most likely a single-nucleotide insertion or deletion. If the first mutation is an insertion, the suppressor must be a deletion; if the first mutation is a deletion, then the suppressor must be an insertion. Notice that the protein produced by the suppressor still differs from the original protein at the second and third amino acids, but the second amino acid produced by the suppressor is the same as that in the protein produced by the original mutation. Thus, the suppressor mutation must have occurred in the third codon, because the suppressor does not alter the second amino acid.

▶ For more practice with analyzing mutations, try working Problem 23 at the end of the chapter.

Mutation Rates

The frequency with which a wild-type allele at a locus changes into a mutant allele is referred to as the **mutation rate** and is generally expressed as the number of mutations per biological unit, which may be mutations per cell division, per gamete, or per round of replication. For example, achondroplasia is a type of hereditary dwarfism in humans that results from a dominant mutation. On average, about four achondroplasia mutations arise in every 100,000 gametes, and so the mutation rate is $^4/_{100,000}$, or 0.00004 mutations per gamete. The mutation rate provides information about how often a mutation arises.

FACTORS AFFECTING MUTATION RATES   Calculations of mutation rates are affected by three factors. First, they depend on the frequency with which a change takes place in DNA. A change in the DNA can arise from spontaneous

molecular changes in DNA or it can be induced by chemical, biological, or physical agents in the environment.

The second factor influencing the mutation rate is the probability that when a change takes place, it will be repaired. Most cells possess a number of mechanisms for repairing altered DNA, so most alterations are corrected before they are replicated. If these repair systems are effective, mutation rates will be low; if they are faulty, mutation rates will be elevated. Some mutations increase the overall rate of mutation at other genes; these mutations usually occur in genes that encode components of the replication machinery or DNA-repair enzymes.

The third factor is the probability that a mutation will be detected. When DNA is sequenced, all mutations are potentially detectable. In practice, however, mutations are usually detected by their phenotypic effects. Some mutations may appear to arise at a higher rate simply because they are easier to detect.

VARIATION IN MUTATION RATES   We can draw several general conclusions about mutation rates, though they vary among genes and species (**Table 18.3**). First, spontaneous mutation rates are low for all organisms studied. Typical mutation rates for bacterial genes range from about 1 to 100 mutations per 10 billion cells (from 1×10^{-8} to 1×10^{-10}). The mutation rates for most eukaryotic genes are a bit higher, from about 1 to 10 mutations per million gametes (from 1×10^{-5} to 1×10^{-6}). These higher values in eukaryotes may be due to the fact that the rates are calculated per gamete, and several cell divisions are required to produce a gamete, whereas mutation rates in prokaryotic cells are calculated *per cell division*.

The differences in mutation rates among species may be due to differing abilities to repair mutations, unequal exposures to mutagens, or biological differences in rates of spontaneously arising mutations. Even within a single species, spontaneous rates of mutation vary among genes. The reason for this variation is not entirely understood, but some regions of DNA are known hotspots for mutations.

Recent research suggests that fewer mutations occur in DNA sequences that are occupied by nucleosomes (see Chapter 11). Reduced mutation rates may occur in these sequences because DNA associated with nucleosomes is less exposed to mutagens, but it could also be explained by the effect of nucleosomes on DNA repair, recombination, or replication, all of which influence the rate of mutation.

Several recent studies have measured mutation rates directly by sequencing genes of organisms before and after a number of generations. These new studies suggest that mutation rates are often higher than those previously measured on the basis of changes in phenotype. In one study, geneticists sequenced randomly chosen stretches of DNA in the nematode worm *Caenorhabditis elegans* and found about 2.1 mutations per genome per generation, which was 10 times as high as previous estimates based on phenotypic

Table 18.3	Mutation rates of different genes in different organisms		
Organism	**Mutation**	**Rate**	**Unit**
Bacteriophage T2	Lysis inhibition	1×10^{-8}	Per replication
	Host range	3×10^{-9}	
Escherichia coli	Lactose fermentation	2×10^{-7}	Per cell division
	Histidine requirement	2×10^{-8}	
Neurospora crassa	Inositol requirement	8×10^{-8}	Per asexual spore
	Adenine requirement	4×10^{-8}	
Corn	Kernel color	2.2×10^{-6}	Per gamete
Drosophila	Eye color	4×10^{-5}	Per gamete
	Allozymes	5.14×10^{-6}	
Mouse	Albino coat color	4.5×10^{-5}	Per gamete
	Dilution coat color	3×10^{-5}	
Human	Huntington disease	1×10^{-6}	Per gamete
	Achondroplasia	1×10^{-5}	
	Neurofibromatosis (Michigan)	1×10^{-4}	
	Hemophilia A (Finland)	3.2×10^{-5}	
	Duchenne muscular dystrophy (Wisconsin)	9.2×10^{-5}	

changes. The researchers found that about half of the mutations were insertions and deletions.

Recent genome sequencing has also provided more accurate information about mutation rates in humans. Several sequencing studies suggest that the overall rate of base substitutions in humans is about 1×10^{-8} mutations per base pair per generation. Other research suggests that each person carries approximately 100 loss-of-function germ-line mutations.

ADAPTIVE MUTATION As will be discussed in Chapters 24 through 26, genetic variation is critical for evolutionary change that brings about adaptation to new environments. New genetic variants arise primarily through mutation. For many years, genetic variation was assumed to arise randomly and at rates that are independent of the need for adaptation. However, some evidence suggests that stressful environments—where adaptation may be necessary to survive—can induce more mutations in bacteria, a process that has been termed **adaptive mutation**. The idea of adaptive mutation has been intensely debated; critics counter that most mutations are expected to be deleterious, and so increased mutagenesis would likely be harmful most of the time.

CONCEPTS

Mutation rate is the frequency with which a specific mutation arises. Rates of mutations are generally low and are affected by environmental and genetic factors.

✓ **CONCEPT CHECK 3**

What three factors affect mutation rates?

18.2 Mutations Are Potentially Caused by a Number of Different Factors

Mutations result from both internal and external factors. Those that occur under normal conditions are termed **spontaneous mutations**, whereas those that result from changes caused by environmental chemicals or radiation are **induced mutations**.

Spontaneous Replication Errors

Replication is amazingly accurate: less than one error in a billion nucleotides arises in the course of DNA synthesis (see Chapter 12). However, spontaneous replication errors do occasionally occur.

TAUTOMERIC SHIFTS The primary cause of spontaneous replication errors was formerly thought to be tautomeric shifts, in which the positions of protons in the DNA bases change. Purine and pyrimidine bases exist in different chemical forms called tautomers. The two tautomeric forms of each base are in dynamic equilibrium, although one form is more common than the other. The standard Watson-and-Crick base pairings—adenine with thymine, and cytosine with guanine—occur between the common forms of the bases, but, if the bases are in their rare tautomeric forms, other base pairings are possible. For example, the common form of cytosine pairs with guanine, but the rare tautomer of cytosine pairs with adenine.

Watson and Crick proposed that tautomeric shifts might produce mutations, and, for many years, their proposal was the

Non-Watson-and-Crick base pairing

Thymine–guanine wobble

Cytosine–adenine protonated wobble

18.10 Nonstandard base pairings can occur as a result of the flexibility in DNA structure. Thymine and guanine can pair through wobble between normal bases. Cytosine and adenine can pair through wobble when adenine is protonated (has an extra hydrogen atom).

accepted model for spontaneous replication errors. However, there has never been convincing evidence that the rare tautomers are the cause of spontaneous mutations. Furthermore, research now shows little evidence of tautomers in DNA.

MISPAIRING DUE TO OTHER STRUCTURES Mispairings often arise through wobble (see Chapter 15), in which normal, protonated, and other forms of the bases are able to pair because of flexibility in the DNA helical structure (**Figure 18.10**). These structures have been detected in DNA molecules and are now thought to be responsible for many of the mispairings in replication.

INCORPORATION ERRORS AND REPLICATED ERRORS When a mismatched base has been incorporated into a newly synthesized nucleotide chain, an **incorporated error** is said to have occurred. Suppose that, in replication, thymine (which normally pairs with adenine) mispairs with guanine through wobble (**Figure 18.11**). In the next round of replication, the two mismatched bases separate, and each serves as a

template for the synthesis of a new nucleotide strand. This time, thymine pairs with adenine, producing another copy of the original DNA sequence. On the other strand, however, the incorrectly incorporated guanine serves as the template and pairs with cytosine, producing a new DNA molecule that has an error—a C • G pair in place of the original T • A pair (a T • A → C • G base substitution). The original incorporated error leads to a **replicated error**, which creates a permanent mutation because all the base pairings are correct and there is no mechanism for repair systems to detect the error.

CAUSES OF DELETIONS AND INSERTIONS Mutations due to small insertions and deletions also arise spontaneously in replication and crossing over. **Strand slippage** can occur when one nucleotide strand forms a small loop (**Figure 18.12**). If the looped-out nucleotides are on the newly synthesized strand, an insertion results. At the next round of replication, the insertion will be replicated and both strands will contain the insertion. If the looped-out nucleotides are on the template strand, then the newly replicated strand will have a deletion, and this deletion will be perpetuated in subsequent rounds of replication.

Another process that produces insertions and deletions is unequal crossing over. In normal crossing over, the homologous sequences of the two DNA molecules align, and crossing over produces no net change in the number of nucleotides in either molecule. Misaligned pairing can cause **unequal crossing over**, which results in one DNA molecule with an insertion and the other with a deletion (**Figure 18.13**).

> **CONCEPTS**
>
> Spontaneous replication errors arise from altered-base structures and from wobble-base pairing. Small insertions and deletions can occur through strand slippage in replication and through unequal crossing over.

Spontaneous Chemical Changes

In addition to spontaneous mutations that arise in replication, mutations also result from spontaneous chemical changes in DNA. One such change is **depurination**, the loss of a purine base from a nucleotide. Depurination results when the covalent bond connecting the purine to the 1′-carbon atom

1 DNA strands separate for replication.

2 Thymine on the original template strand base pairs with guanine through wobble, leading to an incorporated error.

DNA TTCG / AAGC   Wild type

TTCG

AAGC

TTCG / AGGC

TTCG / AAGC   Wild type

TTCG → TTCG / AAGC   Wild type

AGGC → TCCG / AGGC   Mutant

3 At the next round of replication, the guanine nucleotide pairs with cytosine, leading to a transition mutation.

18.11 Wobble base pairing leads to a replicated error.

Newly synthesized strand 5′ TACGGACTGAAAA 3′
Template strand 3′ ATGCCTGACTTTTTGCGAAG 5′

1 Newly synthesized strand loops out,…

5′ ACGGACTGAAA A 3′
3′ TGCCTGACTTTTTGCGAA 5′

2 …resulting in the addition of one nucleotide on the new strand.

5′ ACGGACTGAA AAACGCTT 3′
3′ TGCCTGACTTTTTGCGAA 5′

3 Template strand loops out,…

5′ ACGGACTGAA A A 3′
3′ TGCCTGACTT TTGCGAA 5′

4 …resulting in the omission of one nucleotide on the new strand.

5′ ACGGACTGAAA AACGCTT 3′
3′ TGCCTGACTT TTGCGAA 5′

18.12 Insertions and deletions may result from strand slippage.

1 If homologous chromosomes misalign during crossing over,…

Unequal crossing over

2 …one crossover product contains an insertion…

3 …and the other has a deletion.

18.13 Unequal crossing over produces insertions and deletions.

of the deoxyribose sugar breaks (**Figure 18.14a**), producing an apurinic site, a nucleotide that lacks its purine base. An apurinic site cannot act as a template for a complementary base in replication. In the absence of base-pairing constraints, an incorrect nucleotide (most often adenine) is incorporated into the newly synthesized DNA strand opposite the apurinic site (**Figure 18.14b**), frequently leading to an incorporated error. The incorporated error is then transformed into a replicated error at the next round of replication. Depurination is a common cause of spontaneous mutation; a mammalian cell in culture loses approximately 10,000 purines every day. Loss of pyrimidine bases also occurs, but at a much lower rate than depurination.

Another spontaneously occurring chemical change that takes place in DNA is **deamination**, the loss of an amino

group (NH_2) from a base. Deamination can be spontaneous or be induced by mutagenic chemicals.

Deamination can alter the pairing properties of a base: the deamination of cytosine, for example, produces uracil (**Figure 18.15a**), which pairs with adenine in replication. After another round of replication, the adenine will pair with thymine, creating a T • A pair in place of the original C • G pair (C • G → U • A → T • A); this chemical change is a transition mutation. This type of mutation is usually prevented by enzymes that remove uracil whenever it is found in DNA. The ability to recognize the product of cytosine deamination may explain why thymine, not uracil, is found in DNA. In mammals, including humans, some cytosine bases in DNA are naturally methylated and exist in the form of 5-methylcytosine

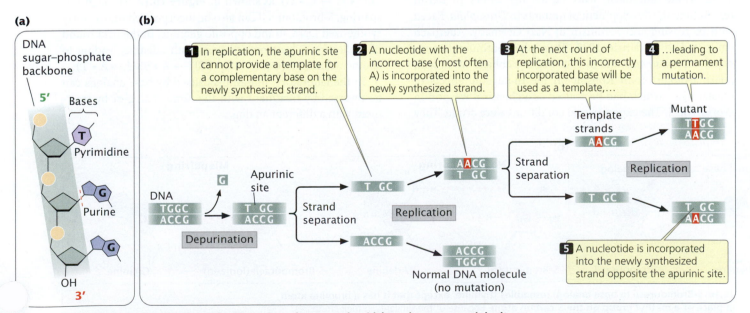

18.14 Depurination (the loss of a purine base from a nucleotide) produces an apurinic site.

18.15 Deamination alters DNA bases.

(5mC; see Figure 10.19). When deaminated, 5mC becomes thymine (**Figure 18.15b**). Because thymine pairs with adenine in replication, the deamination of 5-methylcytosine changes an original C • G pair to T • A (C • G → 5mC • G → T • G → T • A). Consequently, C • G → T • A transitions are frequent in mammalian cells, and 5mC sites are mutation hotspots in humans. ▶ TRY PROBLEM 27

CONCEPTS

Some mutations arise from spontaneous alterations in DNA structure, such as depurination and deamination, which can alter the pairing properties of the bases and cause errors in subsequent rounds of replication.

Chemically Induced Mutations

Although many mutations arise spontaneously, a number of environmental agents are capable of damaging DNA, including certain chemicals and radiation. Any environmental agent that significantly increases the rate of mutation above the spontaneous rate is called a **mutagen**.

The first discovery of a chemical mutagen was made by Charlotte Auerbach, who started her career in Berlin researching the development of mutants in *Drosophila*. Faced with increasing anti-Semitism in Nazi Germany, Auerbach emigrated to Britain in 1933. There she continued her research on *Drosophila* and in 1940 began a collaboration with pharmacologist John Robson on the mutagenic effects of mustard gas, which had been used as a chemical weapon in World War I. The experimental conditions were crude. They

heated liquid mustard gas over a Bunsen burner on the roof of the pharmacology building, and the flies were exposed to the gas in a large chamber. After developing serious burns on her hands from the gas, Auerbach let others carry out the exposures, and she analyzed the flies. Auerbach and Robson showed that mustard gas is indeed a powerful mutagen, reducing the viability of gametes and increasing the numbers of mutations seen in the offspring of exposed flies. Because the research was part of the secret war effort, publication of their findings was delayed until 1947.

BASE ANALOGS One class of chemical mutagens consists of **base analogs**, chemicals with structures similar to that of any of the four standard bases of DNA. DNA polymerases cannot distinguish these analogs from the standard bases; so, if base analogs are present during replication, they may be incorporated into newly synthesized DNA molecules. For example, 5-bromouracil (5BU) is an analog of thymine; it has the same structure as that of thymine except that it has a bromine (Br) atom on the 5-carbon atom instead of a methyl group (**Figure 18.16a**). Normally, 5-bromouracil pairs with adenine just as thymine does, but it occasionally mispairs with guanine (**Figure 18.16b**), leading to a transition (T • A → 5BU • A → 5BU • G → C • G), as shown in **Figure 18.17**. Through mispairing, 5-bromouracil can also be incorporated into a newly synthesized DNA strand opposite guanine. In the next round of replication 5-bromouracil pairs with adenine, leading to another transition (G • C → G • 5BU → A • 5BU → A • T).

In the laboratory, mutations caused by base analogs can be reversed by treatment with the same analog or by treatment with a different analog.

18.16 5-Bromouracil (a base analog) resembles thymine, except that it has a bromine atom in place of a methyl group on the 5-carbon atom. Because of the similarity in their structures, 5-bromouracil may be incorporated into DNA in place of thymine. Like thymine, 5-bromouracil normally pairs with adenine but, when ionized, it may pair with guanine through wobble.

1 In replication, 5-bromouracil may become incorporated into DNA in place of thymine, producing an incorporated error.

2 5-Bromouracil may mispair with guanine in the next round of replication.

3 In the next replication, this guanine nucleotide pairs with cytosine, leading to a permanent mutation.

4 If 5-bromouracil pairs with adenine, no replicated error occurs.

18.17 5-Bromouracil can lead to a replicated error.

Conclusion: Incorporation of bromouracil followed by mispairing leads to a TA ⟶ CG transition mutation.

ALKYLATING AGENTS   Alkylating agents are chemicals that donate alkyl groups, such as methyl (CH_3) and ethyl (CH_3-CH_2) groups, to nucleotide bases. For example, ethylmethylsulfonate (EMS) adds an ethyl group to guanine, producing O^6-ethylguanine, which pairs with thymine (**Figure 18.18a**). Thus, EMS produces C • G → T • A transitions. EMS is also capable of adding an ethyl group to thymine, producing 4-ethylthymine, which then pairs with guanine, leading to a T • A → C • G transition. Because EMS produces both C • G → T • A and T • A → C • G transitions, mutations produced by EMS can be reversed by additional treatment with EMS. Mustard gas is another alkylating agent.

DEAMINATION   In addition to its spontaneous occurrence (see Figure 18.15), deamination can be induced by some chemicals. For instance, nitrous acid deaminates cytosine, creating uracil, which in the next round of replication pairs with adenine (**Figure 18.18b**), producing a C • G → T • A transition mutation. Nitrous acid changes adenine into hypoxanthine, which pairs with cytosine, leading to a T • A → C • G transition. Nitrous acid also deaminates guanine, producing xanthine, which pairs with cytosine just as guanine does; however, xanthine can also pair with thymine, leading to a C • G → T • A transition. Nitrous acid produces exclusively transition mutations and, because both C • G → T • A and T • A → C • G transitions

18.18 Chemicals may alter DNA bases. Shown here are some examples of mutations produced by chemical agents.

	Original base	Mutagen	Modified base	Pairing partner	Type of mutation
(a)	Guanine	EMS Alkylation	O^6-Ethylguanine	Thymine	CG ⟶ TA TA ⟶ CG
(b)	Cytosine	Nitrous acid (HNO_2) Deamination	Uracil	Adenine	CG ⟶ TA TA ⟶ CG
(c)	Cytosine	Hydroxylamine (NH_2OH) Hydroxylation	Hydroxylamino-cytosine	Adenine	CG ⟶ TA

18.19 Oxidative radicals convert guanine into 8-oxy-7, 8-dihydrodeoxyguanine.

are produced, these mutations can be reversed with nitrous acid.

HYDROXYLAMINE Hydroxylamine is a very specific base-modifying mutagen that adds a hydroxyl group to cytosine, converting it into hydroxylaminocytosine (**Figure 18.18c**). This conversion increases the frequency of a rare tautomer that pairs with adenine instead of guanine and leads to $C \cdot G \rightarrow T \cdot A$ transitions. Because hydroxylamine acts only on cytosine, it will not generate $T \cdot A \rightarrow C \cdot G$ transitions; thus, hydroxylamine will not reverse the mutations that it produces. **TRY PROBLEM 25**

OXIDATIVE REACTIONS Reactive forms of oxygen (including superoxide radicals, hydrogen peroxide, and hydroxyl radicals) are produced in the course of normal aerobic metabolism, as well as by radiation, ozone, peroxides, and certain drugs. These reactive forms of oxygen damage DNA and induce mutations by bringing about chemical changes in DNA. For example, oxidation converts guanine into 8-oxy-7,8-dihydrodeoxyguanine (**Figure 18.19**), which frequently mispairs with adenine instead of cytosine, causing a $G \cdot C \rightarrow T \cdot A$ transversion mutation.

INTERCALATING AGENTS Proflavin, acridine orange, ethidium bromide, and dioxin are **intercalating agents** (**Figure 18.20a**), which produce mutations by sandwiching themselves (intercalating) between adjacent bases in DNA, distorting the three-dimensional structure of the helix and causing single-nucleotide insertions and deletions in replication (**Figure 18.20b**). These insertions and deletions frequently produce frameshift mutations, and so the mutagenic effects of intercalating agents are often severe. Because intercalating agents generate both additions and deletions, they can reverse the effects of their own mutations.

CONCEPTS

Chemicals can produce mutations by a number of mechanisms. Base analogs are incorporated into DNA and frequently pair with the wrong base. Alkylating agents, deaminating chemicals, hydroxylamine, and oxidative radicals change the structure of DNA bases, thereby altering their pairing properties. Intercalating agents wedge between the bases and cause single-base insertions and deletions in replication.

18.20 Intercalating agents. Proflavin and acridine orange (a) insert themselves between adjacent bases in DNA, distorting the three-dimensional structure of the helix (b).

✓ CONCEPT CHECK 4

Base analogs are mutagenic because of which characteristic?
a. They produce changes in DNA polymerase that cause it to malfunction.
b. They distort the structure of DNA.
c. They are similar in structure to the normal bases.
d. They chemically modify the normal bases.

Radiation

In 1927, Hermann Muller demonstrated that mutations in fruit flies could be induced by X-rays. The results of subsequent studies showed that X-rays greatly increase mutation rates in all organisms. Because of their high energies, X-rays, gamma rays, and cosmic rays are all capable of penetrating tissues and damaging DNA. These forms of radiation, called ionizing radiation, dislodge electrons from the atoms that they encounter, changing stable molecules into free radicals and reactive ions, which then alter the structures of bases and break phosphodiester bonds in DNA. Ionizing radiation also frequently results in double-strand breaks in DNA. Attempts to repair these breaks can produce chromosome mutations (discussed in Chapter 8).

Ultraviolet (UV) light has less energy than does ionizing radiation and does not eject electrons but is nevertheless highly mutagenic. Purine and pyrimidine bases readily absorb UV light, resulting in the formation of chemical bonds between adjacent pyrimidine molecules on the same strand of DNA and in the creation of **pyrimidine dimers** (**Figure 18.21a**). Pyrimidine dimers consisting of two thymine bases (called thymine dimers) are most frequent, but cytosine dimers and thymine–cytosine dimers also can form. Dimers distort the configuration of DNA (**Figure 18.21b**) and often block replication. Most pyrimidine dimers are immediately repaired by mechanisms discussed later in this chapter, but some escape repair and inhibit replication and transcription.

(a)   **(b)**

18.21 Pyrimidine dimers result from ultraviolet light.
(a) Formation of thymine dimer. (b) Distorted DNA.

When pyrimidine dimers block replication, cell division is inhibited and the cell usually dies; for this reason, UV light kills bacteria and is an effective sterilizing agent. For a mutation—a hereditary error in the genetic instructions—to occur, the replication block must be overcome. Bacteria can sometimes circumvent replication blocks produced by pyrimidine dimers and other types of DNA damage by means of the **SOS system.** This system allows replication blocks to be overcome but, in the process, makes numerous mistakes and greatly increases the rate of mutation. Indeed, the very reason that replication can proceed in the presence of a block is that the enzymes in the SOS system do not strictly adhere to the base-pairing rules. The trade-off is that replication can continue and the cell survives, but only by sacrificing the normal accuracy of DNA synthesis.

CONCEPTS

Ionizing radiation such as X-rays and gamma rays damages DNA by dislodging electrons from atoms; these electrons then break phosphodiester bonds and alter the structure of bases. Ultraviolet light causes mutations primarily by producing pyrimidine dimers that disrupt replication and transcription. The SOS system enables bacteria to overcome replication blocks but introduces mistakes in replication.

18.3 Mutations Are the Focus of Intense Study by Geneticists

Because mutations often have detrimental effects, they are frequently studied by geneticists. These studies have included the development of tests to determine the mutagenic properties of chemical compounds and the investigation of human populations tragically exposed to high levels of radiation.

Detecting Mutations with the Ames Test

People in industrial societies are surrounded by a multitude of artificially produced chemicals: more than 50,000 different chemicals are in commercial and industrial use today, and from 500 to 1000 new chemicals are introduced each year. Some of these chemicals are potential carcinogens, and many natural products are also potentially carcinogenic. One

method for testing the cancer-causing potential of substances is to administer them to laboratory animals (rats or mice) and compare the incidence of cancer in the treated animals with that of control animals. Unfortunately, these tests are time-consuming and expensive. Furthermore, the ability of a substance to cause cancer in rodents is not always indicative of its effect on humans. After all, we aren't rats!

In 1974, Bruce Ames developed a simple test for evaluating the potential of chemicals to cause cancer. The **Ames test** is based on the principle that both cancer and mutations result from damage to DNA, and the results of experiments have demonstrated that 90% of known carcinogens are also mutagens. Ames proposed that mutagenesis in bacteria could serve as an indicator of carcinogenesis in humans.

The Ames test uses different auxotrophic strains of the bacterium *Salmonella typhimurium* that have defects in the lipopolysaccharide coat, which normally protects the bacteria from chemicals in the environment. Furthermore, the DNA-repair system in these strains has been inactivated, enhancing their susceptibility to mutagens.

The most-recent version of the test (called Ames II) uses several auxotrophic strains that detect different types of base-pair substitutions. Other strains detect different types of frameshift mutations. Each strain carries a *his⁻* mutation, which renders it unable to synthesize the amino acid histidine, and the bacteria are plated onto medium that lacks histidine (**Figure 18.22**). Only bacteria that have undergone a reverse mutation of the histidine gene (*his⁻→his⁺*) are able to synthesize histidine and grow on the medium, which makes these mutations easy to detect. Different dilutions of a chemical to be tested are added to plates inoculated with the bacteria, and the number of mutated bacterial colonies that appear on each plate is compared with the number that appear on control plates with no chemical (i.e., that arose through spontaneous mutation). Any chemical that significantly increases the number of colonies appearing on a treated plate is mutagenic and probably also carcinogenic.

Some compounds are not active carcinogens but can be converted into cancer-causing compounds in the body. To make the Ames test sensitive for such *potential* carcinogens, a compound to be tested is first incubated in mammalian liver extract that contains metabolic enzymes.

The Ames test has been applied to thousands of chemicals and commercial products. An early demonstration of its usefulness was the discovery, in 1975, that many hair dyes sold in the United States contained compounds that were mutagenic to bacteria. These compounds were then removed from most hair dyes.

CONCEPTS

The Ames test uses *his⁻* strains of bacteria to test chemicals for their ability to produce *his⁻→his⁺* mutations. Because mutagenic activity and carcinogenic potential are closely correlated, the Ames test is widely used to screen chemicals for their cancer-causing potential.

Experiment

Question: How can chemicals be quickly screened for their ability to cause cancer?

Methods

his⁻ bacteria

1 Bacterial *his⁻* strains are mixed with liver enzymes (which have the ability to convert compounds into potential mutagens).

2 Some of the bacterial strains are also mixed with the chemical to be tested for mutagenic activity.

3 The bacteria are then plated on medium that lacks histidine.

Incubate

Incubate

Results

Control plate (no chemical)

Treatment plate (chemical added)

4 Bacterial colonies that appear on the plates have undergone a *his⁻* → *his⁺* mutation.

Conclusion: Any chemical that significantly increases the number of colonies appearing on the treatment plate is mutagenic and therefore probably also carcinogenic.

18.22 The Ames test is used to identify chemical mutagens.

Radiation Exposure in Humans

People are routinely exposed to low levels of radiation from cosmic, medical, and environmental sources, but there have also been tragic events that produced exposures of much higher degree.

On August 6, 1945, a high-flying American airplane dropped a single atomic bomb on the city of Hiroshima, Japan. The explosion devastated an area of the city measuring 4.5 square miles, killed from 90,000 to 140,000 people, and injured almost as many (**Figure 18.23**). Three days later, the United States dropped an atomic bomb on the city of Nagasaki, this time destroying an area measuring 1.5 square miles and killing between 60,000 and 80,000 people. Huge amounts of radiation were released during these explosions, and many people were exposed.

After the war, a joint Japanese–U.S. effort was made to study the biological effects of radiation exposure on the survivors of the atomic blasts and their children. Somatic mutations were examined by studying radiation sickness and cancer among the survivors; germ-line mutations were assessed by looking at birth defects, chromosome abnormalities, and gene mutations in children born to people that had been exposed to radiation.

Geneticist James Neel and his colleagues examined almost 19,000 children of people who were within 2000 meters (1.2 miles) of the center of the atomic blast at Hiroshima or Nagasaki, along with a similar number of children whose parents did not receive radiation exposure. Radiation doses were estimated for a child's parents on the basis of careful assessment of the parents' location, posture, and position at the time of the blast. A blood sample was collected from each child, and gel electrophoresis was used to investigate amino acid substitutions in 28 proteins. When rare variants were detected, blood samples from the child's parents also were analyzed to establish whether the variant was inherited or a new mutation.

18.23 Hiroshima was destroyed by an atomic bomb on August 6, 1945. The atomic explosion produced many somatic mutations among the survivors. [© Bettmann/Corbis.]

Of a total of 289,868 genes examined by Neel and his colleagues, only one mutation was found in the children of exposed parents; no mutations were found in the control group. From these findings, a mutation rate of 3.4×10^{-6} was estimated for the children whose parents were exposed to the blast, which is within the range of spontaneous mutation rates observed for other eukaryotes. Neel and his colleagues also examined the frequency of chromosome mutations, sex ratios of children born to exposed parents, and frequencies of chromosome aneuploidy. There was no evidence in any of these assays for increased mutations among the children of the people who were exposed to radiation from the atomic explosions, suggesting that germ-line mutations were not elevated.

Animal studies clearly show that radiation causes germ-line mutations; so why was there no apparent increase in germ-line mutations among the inhabitants of Hiroshima and Nagasaki? The exposed parents did exhibit an increased incidence of leukemia and other types of cancers; so somatic mutations were clearly induced. The answer to the question is not known, but the lack of germ-line mutations may be due to the fact that those persons who received the largest radiation doses died soon after the blasts. Additional insight into the genetic effects of radiation have come from studies of people exposed to radiation in the Chernobyl nuclear accident in 1986 and other nuclear accidents, as well as exposure to radiation used in medicine and industry.

18.4 Transposable Elements Cause Mutations

Transposable elements—sequences that can move about in the genome—are also often a cause of mutations. These mobile DNA elements have been given a variety of names, including transposons, transposable genetic elements, movable genes, controlling elements, and jumping genes. They are found in the genomes of all organisms and are abundant in many: for example, they make up at least 45% of human DNA. Most transposable elements are able to insert at many different locations, relying on mechanisms that are distinct from homologous recombination. Through their movement (transposition), transposable elements often cause mutations, either by inserting into another gene and disrupting it or by promoting DNA rearrangements such as chromosome deletions, duplications, and inversions (see Chapter 8).

General Characteristics of Transposable Elements

There are many different types of transposable elements: some have simple structures, encompassing only those sequences necessary for their own transposition, whereas others have complex structures and encode a number of functions not directly related to transposition. Despite this variation, many transposable elements have certain features in common.

Short **flanking direct repeats** from 3 to 12 bp long are present on both sides of most transposable elements. The sequences of these repeats vary, but the length is constant for each type of transposable element. These repeats are not a part of a transposable element and do not travel with it. Rather, they are generated in the process of transposition, at the point of insertion. During transposition, flanking repeats are created when staggered cuts are made in the target DNA, as shown in **Figure 18.24**. The staggered cuts leave short, single-stranded pieces of DNA on either side of the transposable element. Replication of the single-stranded DNA then creates the flanking direct repeats.

At the ends of many, but not all, transposable elements are **terminal inverted repeats**—sequences from 9 to 40 bp in length that are inverted complements of one another. For example, the following sequences are inverted repeats:

$$5'–ACAGTTCAG \ldots CTGAACTGT–3'$$
$$3'–TGTCAAGTC \ldots GACTTGACA–5'$$

On the same strand, the two sequences are not simple inversions, as their name might imply; rather, they are both inverted and complementary. (Notice that the sequence from left to right in the top strand is the same as the

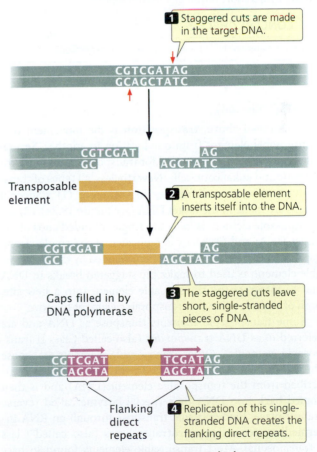

① Staggered cuts are made in the target DNA.

CGTCGATAG
GCAGCTATC

CGTCGAT   AG
GC   AGCTATC

Transposable element

② A transposable element inserts itself into the DNA.

CGTCGAT   AG
GC   AGCTATC

Gaps filled in by DNA polymerase

③ The staggered cuts leave short, single-stranded pieces of DNA.

CGTCGAT   TCGATAG
GCAGCTA   AGCTATC

Flanking direct repeats

④ Replication of this single-stranded DNA creates the flanking direct repeats.

18.24 Flanking direct repeats are generated when a transposable element inserts into DNA.

(a)

Transposable element

```
TGCAA ATCGCA                TGCGAT TGCAA
ACGTT TAGCGT                ACGCTA ACGTT
```

Terminal inverted repeat

Flanking direct repeat

(b)

Transposable element

Terminal inverted repeat

Flanking direct repeat

18.25 Many transposable elements have common characteristics. (a) Most transposable elements generate flanking direct repeats on each side of the point of insertion into target DNA. Many transposable elements also possess terminal inverted repeats. (b) These representations of direct and indirect repeats are used in illustrations throughout this chapter.

sequence from right to left in the bottom strand.) Terminal inverted repeats are recognized by enzymes that catalyze transposition and are required for transposition to take place. **Figure 18.25** summarizes the general characteristics of transposable elements. ▶TRY PROBLEM 33

CONCEPTS

Transposable elements are mobile DNA sequences that often cause mutations. There are many different types of transposable elements; most generate short flanking direct repeats at the target sites as they insert. Many transposable elements also possess short terminal inverted repeats.

✓ **CONCEPT CHECK 5**

How are flanking direct repeats created in transposition?

Transposition

As mentioned above, **transposition** is the movement of a transposable element from one location to another. Several different mechanisms are used for transposition in both prokaryotic and eukaryotic cells. Nevertheless, all types of transposition have several features in common: (1) staggered breaks are made in the target DNA (see Figure 18.24); (2) the transposable element is joined to single-stranded ends of the target DNA; and (3) DNA is replicated at the single-strand gaps. A **transposase** enzyme, often encoded by the transposable element, is used to make the staggered breaks in DNA and to integrate the transposable element into a new site, both of which enable the transposable element to move.

Some transposable elements transpose as DNA and are referred to as **DNA transposons** (also called Class II transposable elements). Other transposable elements transpose through an RNA intermediate. In this case, RNA is transcribed from the transposable element (DNA) and is then copied back into DNA by a special enzyme called reverse transcriptase. Elements that transpose through an RNA intermediate are called **retrotransposons** (also called Class I transposons). Most transposable elements found in bacteria are DNA transposons. Both DNA transposons and retrotransposons are found in eukaryotes, although retrotransposons are more common.

Among DNA transposons, transposition may be replicative or nonreplicative. In **replicative transposition** (also called copy-and-paste transposition), a new copy of the transposable element is introduced at a new site while the old copy remains behind at the original site, and so the number of copies of the transposable element increases as a result of transposition. In **nonreplicative transposition** (cut-and-paste transposition), the transposable element excises from the old site and inserts at a new site without any increase in the number of its copies. Nonreplicative transposition requires the replication of only the few nucleotides that constitute the direct repeats. Retrotransposons use replicative transposition only.

CONTROL OF TRANSPOSITION Many organisms limit transposition by methylating the DNA in regions where transposons are common. DNA methylation usually suppresses transcription (see Chapter 17), preventing the production of the transposase enzyme necessary for transposition. Alterations of chromatin structure also are used to prevent the transcription of transposons. In other cases, translation of the transposase mRNA is controlled. Some animals use small RNA molecules called Piwi-interacting RNAs (piRNAs, see Chapter 14) to silence transposons; piRNAs combine with Piwi proteins and inhibit the expression of transposons sequences.

TRANSPOSITION IN HUMANS About 45% of the human genome is comprised of sequences that are related to transposable elements, mostly retrotransposons (some research suggests that almost two-thirds of the human genome consists of transposable elements). Researchers previously assumed that most of these transposable elements are inactive and that little transposition occurs today, although it clearly took place extensively during past evolution. More recently, however, researchers have begun to map copies of transposons across the genome and have discovered that people often differ in the number and location of transposons. This suggests that recent transpositions are more common than previously thought. The L1 transposon, for example, is estimated to undergo one transposition event about every 100 human births.

Research has also demonstrated that some cancer cells have elevated levels of transposition, probably because patterns of DNA methylation that normally inhibit transposition are disrupted in these cells.

CONCEPTS

Transposition may take place through DNA or an RNA intermediate. In replicative transposition, a new copy of the transposable element inserts in a new location and the old copy stays behind; in nonreplicative transposition, the old copy excises from the old site and moves to a new site. Transposition through an RNA intermediate requires reverse transcription to integrate into the target site. Many cells regulate transposition by a variety of mechanisms.

The Mutagenic Effects of Transposition

Because transposable elements can insert into other genes and disrupt their function, transposition is generally mutagenic. In fact, more than half of all spontaneously occurring mutations in *Drosophila* result from the insertion of a transposable element in or near a functional gene.

A number of cases of human genetic disease have been traced to the insertion of a transposable genetic element into a vital gene. For example, insertion of the L1 transposable element into the gene for blood clotting factor VIII has caused hemophilia. Although most mutations resulting from transposition are detrimental, transposition may occasionally activate a gene or change the phenotype of the cell in a beneficial way. For instance, bacterial transposable elements sometimes carry genes that encode antibiotic resistance, and several transposable elements have created mutations that confer insecticide resistance in insects.

A dramatic example of the mutagenic effect of transposable elements is seen in the color of grapes, which come in black, red, and white varieties (**Figure 18.26**). Black and red grapes result from the production of red pigments (anthocyanins) in the skin, which are lacking in white grapes. White grapes resulted from a mutation in black grapes that turned off the production of anthocyanin pigments. This mutation consisted of the insertion of a 10,422-bp retrotransposon called *Gret1* near a gene that promotes the production of anthocyanins. The *Gret1* retrotransposon apparently disrupted sequences that regulate the gene, effectively shutting down pigment production and producing a white grape with no anthocyanins. Interestingly, red grapes resulted from a second mutation taking place in the white grapes (see Figure 18.26). This mutation (probably resulting from faulty recombination) removed most but not all of the retrotransposons, switching pigment production back on, but not as intensely as in the original black grapes.

Because transposition entails the exchange of DNA sequences and recombination, it often leads to DNA rearrangements. Homologous recombination between multiple copies of transposons can lead to duplications, deletions, and inversions, as shown in **Figure 18.27**. The *Bar* mutation in *Drosophila* (see Figure 8.6) is a tandem duplication thought to have arisen through homologous recombination between two copies of a transposable element present in different locations on the X chromosome. Similarly,

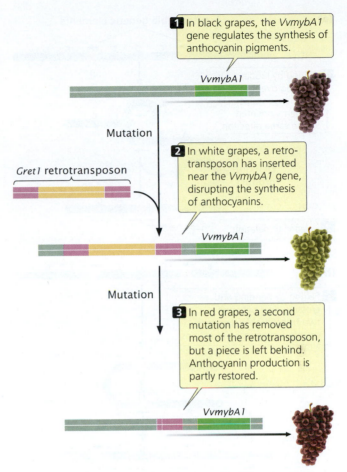

18.26 Red and white color in grapes resulted from insertion and deletion of a retrotransposon.

recombination between copies of the transposable element *Rider* caused a duplication that results in elongated fruit in tomatoes.

DNA rearrangements can also be caused by the excision of transposable elements in a cut-and-paste transposition. If the broken DNA is not repaired properly, a chromosome rearrangement can be generated.

Because most transposable elements insert randomly into DNA sequences, they provide researchers with a powerful tool for inducing mutations throughout the genome, allowing them to determine the functions of genes, study genetic phenomena, and map genes. Furthermore, because the transposable element being used has a known sequence, it can serve as a "tag" for locating the gene in which the mutation has occurred. For example, researchers engineered a transposable element named *Sleeping Beauty* to induce mutations in mice and used it to search for genes that cause cancer. *Sleeping Beauty* was introduced into a strain of mice that produce the transposase needed for transposition, and the transposable element inserted randomly into different locations in the genome. Occasionally, it inserted into a gene that protects against cancer and destroyed its function. By looking for the location of the *Sleeping Beauty* sequence in the DNA from the tumor cells that subsequently developed, geneticists identified a number of genes that protect against cancer.

(a)

Transposable genetic elements

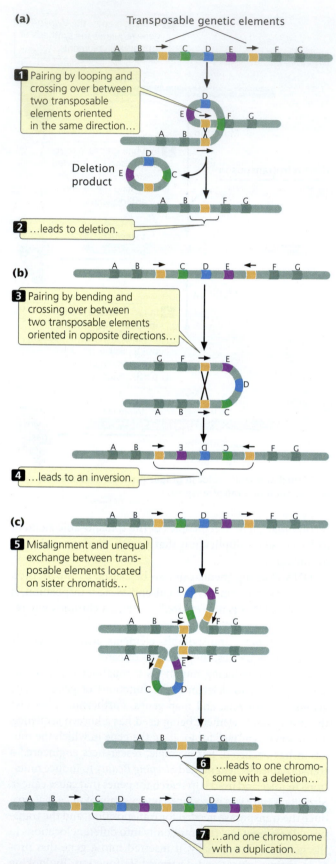

1 Pairing by looping and crossing over between two transposable elements oriented in the same direction...

Deletion product

2 ...leads to deletion.

(b)

3 Pairing by bending and crossing over between two transposable elements oriented in opposite directions...

4 ...leads to an inversion.

(c)

5 Misalignment and unequal exchange between transposable elements located on sister chromatids...

6 ...leads to one chromosome with a deletion...

7 ...and one chromosome with a duplication.

18.27 Many chromosomal rearrangements are generated by transposition.

Bacteria and eukaryotic organisms possess a number of different types of transposable elements, the structures of which vary extensively. In the next two sections, we consider the structure and types of transposable elements in bacteria and eukaryotes.

CONCEPTS

Transposable elements frequently cause mutations and DNA rearrangements.

✔ **CONCEPT CHECK 6**

Briefly explain how transposition causes mutations and chromosome rearrangements.

Transposable Elements in Bacteria

The DNA transposons found in bacteria (there are no retrotransposons in bacteria) constitute two major groups: (1) simple transposable elements, called insertion sequences, that carry only the information required for movement and (2) more-complex transposable elements, called composite transposons, which contain DNA sequences not directly related to transposition.

INSERTION SEQUENCES   The simplest type of transposable element in bacterial chromosomes and plasmids is an **insertion sequence** (IS). This type of element carries only the genetic information necessary for its movement. Insertion sequences are common constituents of bacteria; they can also infect plasmids and viruses and, in this way, can be passed from one cell to another. Geneticists designate each type of insertion sequence with *IS* followed by an identifying number. For example, *IS1* is a common insertion sequence found in *E. coli*.

A number of different insertion sequences have been found in bacteria. They are typically from 800 to 2000 bp in length and possess the two hallmarks of transposable elements: terminal inverted repeats and the generation of flanking direct repeats at the site of insertion. Most insertion sequences contain one or two genes that encode transposase. *IS1*, a typical insertion sequence, is shown in **Figure 18.28.** ▶ **TRY PROBLEM 36**

IS1 (768 bp)

Transposase gene

23-bp terminal inverted repeat

9-bp flanking direct repeat

18.28 Insertion sequences are simple transposable elements found in bacteria.

18.29 *Tn10* is a composite transposon in bacteria.

18.30 Mu is a transposing bacteriophage.

COMPOSITE TRANSPOSONS Any segment of DNA that becomes flanked by two copies of an insertion sequence may itself transpose and is called a **composite transposon**. Each composite transposon is designated by the abbreviation *Tn*, followed by a number. The composite transposon *Tn10*, for example, consists of about 9300 bp that carries a gene for tetracycline resistance between two *IS10* insertion sequences (**Figure 18.29**). The insertion sequences have terminal inverted repeats, so the composite transposon also ends in inverted repeats. Composite transposons also generate flanking direct repeats at their sites of insertion (see Figure 18.29). The insertion sequences at the ends of a composite transposon may be in the same orientation or they may be inverted relative to one other (as in *Tn10*).

The insertion sequences at the ends of a composite transposon are responsible for transposition. The DNA between the insertion sequences is not required for movement and may carry additional information (such as antibiotic resistance). Presumably, composite transposons evolve when one insertion sequence transposes to a location close to another of the same type. The transposase produced by one of the insertion sequences catalyzes the transposition of both insertions sequences, allowing them to move together and carry along the DNA that lies between them. In some composite transposons (such as *Tn10*) one of the insertion sequences may be defective, so its movement depends on the transposase produced by the other.

NONCOMPOSITE TRANSPOSONS Some transposable elements in bacteria lack insertion sequences and are referred to as noncomposite transposons. Noncomposite transposons possess a gene for transposase and have inverted repeats at their ends. For instance, the noncomposite transposon *Tn3* carries genes for transposase and resolvase (an enzyme that functions in recombination), plus a gene that encodes the enzyme β-lactamase, which provides resistance to the antibiotic ampicillin.

A few bacteriophage genomes reproduce by transposition and use transposition to insert themselves into a bacterial chromosome in their lysogenic cycle; the best studied transposing bacteriophage is Mu (**Figure 18.30**). Although Mu does not possess terminal inverted repeats, it does generate short (5-bp) flanking direct repeats when it inserts randomly into DNA. Mu replicates through transposition and causes mutations at the site of insertion, properties characteristic of transposable elements.

CONCEPTS

Insertion sequences are prokaryotic transposable elements that carry only the information needed for transposition. A composite transposon consists of two insertion sequences plus intervening DNA. Noncomposite transposons in bacteria lack insertion sequences but have terminal inverted repeats and carry information not related to transposition. All of these transposable elements generate flanking direct repeats at their points of insertion.

✔ **CONCEPT CHECK 7**

Which type of transposable element possesses terminal inverted repeats?
a. Insertion sequence.
b. Composite transposons.
c. Noncomposite transposon *Tn3*.
d. All the above.

Transposable Elements in Eukaryotes

Eukaryotic transposable elements can be divided into two groups. One group is structurally similar to transposable elements in bacteria, typically ending in short inverted repeats and transposing as DNA: examples include the *P* elements in *Drosophila* and the *Ac* and *Ds* elements in maize (corn). The other group comprises retrotransposons; they use RNA intermediates, and many are similar in structure and movement to retroviruses (see Chapter 9). On the basis of their structure, function, and genomic sequences, some retrotransposons are clearly evolutionarily related to retroviruses. Although their mechanism of movement is fundamentally different from that of other transposable elements, retrotransposons also generate direct repeats at the point of insertion. Retrotransposons include the *Ty* elements in yeast, the *copia* elements in *Drosophila,* and the *Alu* sequences in humans.

Ac **AND** *Ds* **ELEMENTS IN MAIZE** Transposable elements were first identified in maize more than 50 years ago by Barbara McClintock (**Figure 18.31**). McClintock spent much of her long career studying their properties, and her work stands among the landmark discoveries of genetics. Her results, however, were misunderstood and ignored for many years. Not until molecular techniques were developed in the late 1960s and 1970s did the importance of transposable

18.31 Barbara McClintock was the first to discover transposable elements. [Topham/The Image Works.]

18.32 Variegated (multicolored) kernels in corn are caused by mobile genes. The study of variegated corn led Barbara McClintock to discover transposable elements. [Matt Meadows/Getty Images.]

elements become widely accepted. The significance of McClintock's early discoveries was finally recognized in 1983, when she was awarded the Nobel Prize in physiology or medicine.

McClintock's discovery of transposable elements had its genesis in the early work of Rollins A. Emerson on the maize genes that caused variegated (multicolored) kernels. Most corn kernels are either wholly pigmented or colorless (yellow), but Emerson noted that some yellow kernels had spots or streaks of color (**Figure 18.32**). He proposed that these kernels resulted from an unstable mutation: a mutation in the wild-type gene for pigment produced a colorless kernel but, in some cells, the mutation reverted back to the wild type, causing a spot of pigment. However, Emerson didn't know why these mutations were unstable.

McClintock discovered that the cause of the unstable mutation was a gene that moved. She noticed that chromosome breakage in maize often occurred at a gene that she called *Dissociation* (*Ds*) but only if another gene, the *Activator* (*Ac*), also was present. Occasionally, the genes moved together to a different chromosomal location. McClintock called these moving genes controlling elements, because they controlled the expression of other genes.

Since the significance of McClintock's work was recognized, *Ac* and *Ds* elements in maize have been examined in detail. They are DNA transposons that possess terminal inverted repeats and generate flanking direct repeats at the points of insertion (**Figure 18.33a**). Each *Ac* element contains a single gene that encodes a transposase enzyme. Thus *Ac* elements are *autonomous*—able to transpose. *Ds* elements are *Ac* elements with one or more deletions that have inactivated the transposase gene (**Figure 18.33b**). Unable to transpose on their own (*nonautonomous*), *Ds* elements can transpose in the presence of *Ac* elements because they still possess terminal inverted repeats recognized by *Ac* transposase.

Each kernel in an ear of corn is an individual offspring, originating as an ovule fertilized by a pollen grain. A kernel's pigment pattern is determined by several loci. A pigment-encoding allele at one of these loci can be designated *C*, and an allele at the same locus that does not confer pigment can be designated *c*. A kernel with genotype *cc* will be colorless—that is, yellow or white (**Figure 18.34a**); a kernel with genotype *CC* or *Cc* will produce pigment and be purple (**Figure 18.34b**).

A *Ds* element, transposing under the influence of a nearby *Ac* element, may insert into the *C* allele, destroying its ability to produce pigment (**Figure 18.34c**). An allele inactivated by a transposable element is designated by a subscript "t"; so, in this case, the allele would be C_t.

18.33 *Ac* and *Ds* are transposable elements in maize.

If a kernel is initially heterozygous with genotype Cc, then after the transposition of Ds into the C allele, the kernel cell has genotype $C_t c$. This kernel will be colorless (white or yellow), because neither the C_t allele nor the c allele confers pigment. As the original one-celled maize embryo develops and divides by mitosis, additional transpositions may take place in some cells. In any cell in which the transposable element excises from the C_t allele and moves to a new location, the C allele may be rendered functional again: all cells derived from those in which this event has taken place will have the genotype Cc and be purple. The presence of these pigmented cells, surrounded by the colorless ($C_t c$) cells, produces a purple spot or streak (called a sector) in the otherwise yellow kernel (**Figure 18.34d**). The size of the sector varies, depending on when the excision of the transposable element from the C_t allele takes place. If excision is early in development, then many cells will contain the functional C allele and the pigmented sector will be large; if excision is late in development, few cells will have the functional C allele and the pigmented sector will be small.   ▶ **TRY PROBLEM 42**

TRANSPOSABLE ELEMENTS IN *DROSOPHILA*   A number of different transposable elements are found in *Drosophila*. One family of *Drosophila* transposable elements comprises the P elements. Most functional P elements are about 2900 bp long, although shorter P elements containing deletions also exist. Each P element possesses terminal inverted repeats and generates flanking direct repeats at the site of insertion. Like transposable elements in bacteria, P elements are DNA transposons. Each element encodes both a transposase and a repressor of transposition.

The role of this repressor in controlling transposition is demonstrated dramatically in **hybrid dysgenesis**, which is the sudden appearance of numerous mutations, chromosome aberrations, and sterility in the offspring of a cross between a P^+ male fly (*with P elements*) and a P^- female fly

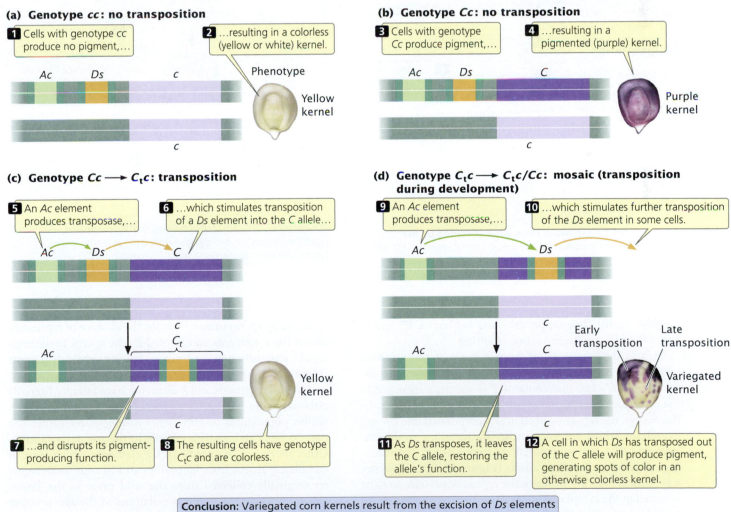

(a) Genotype *cc*: no transposition

1 Cells with genotype cc produce no pigment,…

2 …resulting in a colorless (yellow or white) kernel.

Ac   *Ds*   *c*

Phenotype

Yellow kernel

c

(b) Genotype *Cc*: no transposition

3 Cells with genotype Cc produce pigment,…

4 …resulting in a pigmented (purple) kernel.

Ac   *Ds*   *C*

Purple kernel

c

(c) Genotype *Cc* ⟶ *C_t c*: transposition

5 An Ac element produces transposase,…

6 …which stimulates transposition of a Ds element into the C allele…

Ac   *Ds*   *C*

c
C_t

Ac

Yellow kernel

c

7 …and disrupts its pigment-producing function.

8 The resulting cells have genotype $C_t c$ and are colorless.

(d) Genotype *C_t c* ⟶ *C_t c/Cc*: mosaic (transposition during development)

9 An Ac element produces transposase,…

10 …which stimulates further transposition of the Ds element in some cells.

Ac   *Ds*

c

Early transposition   Late transposition

Ac   *C*

Variegated kernel

c

11 As Ds transposes, it leaves the C allele, restoring the allele's function.

12 A cell in which Ds has transposed out of the C allele will produce pigment, generating spots of color in an otherwise colorless kernel.

Conclusion: Variegated corn kernels result from the excision of Ds elements from genes controlling pigment production during development.

18.34 Transposition results in variegated maize kernels.

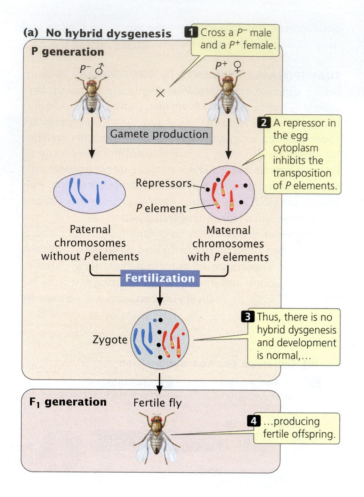

(a) No hybrid dysgenesis

P generation

1 Cross a P⁻ male and a P⁺ female.

$P^- \male$ ✕ $P^+ \female$

Gamete production

2 A repressor in the egg cytoplasm inhibits the transposition of P elements.

Repressors

P element

Paternal chromosomes without P elements

Maternal chromosomes with P elements

Fertilization

Zygote

3 Thus, there is no hybrid dysgenesis and development is normal,…

F₁ generation   Fertile fly

4 …producing fertile offspring.

18.35 Hybrid dysgenesis in *Drosophila* is caused by the transposition of P elements. [After W. Y. Chooi, *Genetics* 68:213–230, 1971.]

(b) Hybrid dysgenesis

P generation

1 Cross a P⁺ male and a P⁻ female.

$P^+ \male$ ✕ $P^- \female$

Gamete production

2 There is no repressor in the egg cytoplasm.

Paternal chromosomes with P elements

Maternal chromosomes without P elements

Fertilization

Zygote

3 Thus, P elements on the paternal chromosomes undergo a burst of transposition—hybrid dysgenesis—…

F₁ generation   Sterile fly

4 …resulting in mutations, chromosome aberrations, and sterile offspring.

Conclusion: Only the cross between a P⁺ male and a P⁻ female causes hybrid dysgenesis, because the sperm does not contribute repressor.

(*without* them). The reciprocal cross between a P^+ female and a P^- male produces normal offspring.

Hybrid dysgenesis arises from a burst of transposition when P elements are introduced into a cell that does not possess them. In a cell that contains P elements, a repressor in the cytoplasm inhibits transposition. When a P^+ female produces eggs, the repressor protein is incorporated into the egg cytoplasm, which prevents further transposition in the embryo and thus prevents mutations from arising. The resulting offspring are fertile as adults (**Figure 18.35a**). However, a P^- female does not produce the repressor protein; so none is stored in the cytoplasm of her eggs. Sperm contain little or no cytoplasm, so a P^+ male does not contribute the repressor protein to his offspring. When eggs from a P^- female are

fertilized by sperm from a P^+ male, the absence of repression allows the P elements contributed by the sperm to undergo rapid transposition in the embryo, causing hybrid dysgenesis (**Figure 18.35b**).

Hybrid dysgenesis and P elements have attracted geneticists' attention because P elements appear to have arisen within populations of *Drosophila melanogaster* within the past 50 years and may play a role in the species evolution. Other species of *Drosophila* lack P elements, which are also completely absent from laboratory strains of *D. melanogaster* originally collected from the wild prior to the 1960s. However, today most wild populations of the species have P elements. Laboratory strains collected during the 1970s are mixed: some have P elements and some do not. This suggests that *P* elements arose sometime during the past century and spread quickly throughout all wild populations of *D. melanogaster*. Because crosses between males with *P* elements and females without them cause sterility, *P* elements and similar transposable elements have the potential to serve as reproductive isolating mechanisms between populations and may play a role in bringing about speciation (see Chapter 26). These observations support the idea that transposable elements play important roles in evolution. ▶ **TRY PROBLEM 38**

TRANSPOSABLE ELEMENTS IN HUMANS   One of the most common transposable elements in the human genome is *Alu*. Every human cell contains more than 1 million related but not identical copies of *Alu* in its chromosomes. *Alu* sequences are similar to the gene that encodes the 7S RNA molecule, which transports newly synthesized proteins across the endoplasmic reticulum. *Alu* sequences create short flanking direct repeats when they insert into DNA and have characteristics that suggest that they have transposed through an RNA intermediate.

Alu belongs to a class of repetitive sequences found frequently in mammalian and some other genomes. These sequences are collectively referred to as short interspersed elements (SINEs) and constitute about 11% of the human genome. Most SINEs are copies of transposable elements that have been shortened at the 5′ end, probably because the reverse-transcription process used in their transposition terminated before the entire sequence was copied. SINEs have been identified as the cause of mutations in more than 20 cases of human genetic disease.

The human genome also has many transposons classified as long interspersed elements (LINEs), which are somewhat more similar in structure to retroviruses. Like SINEs, most LINEs in the human genome have been shortened at the 5′ end. The longest LINEs are usually about 6000 bp but, because most copies are shortened, the average LINE is only about 900 bp. There are approximately 900,000 copies of LINEs in the human genome, collectively constituting 21% of the total human DNA.

CONCEPTS

A great variety of transposable elements exist in eukaryotes. Some resemble transposable elements in prokaryotes, having terminal inverted repeats, and transpose as DNA. Others are retrotransposons with long direct repeats at their ends and transpose through an RNA intermediate.

✓ **CONCEPT CHECK 8**

Hybrid dysgenesis results when

a. a male fly with *P* elements (P^+) mates with a female fly that lacks *P* elements (P^-).

b. a P^- male mates with a P^+ female.

c. a P^+ male mates with a P^+ female.

d. a P^- male mates with a P^- female.

CONNECTING CONCEPTS

Types of Transposable Elements

Now that we have looked at some examples of transposable elements, let's review their major types (**Table 18.4**).

Transposable elements can be divided into two major classes on the basis of structure and movement. Class I comprises the retrotransposons, which possess terminal direct repeats and transpose through RNA intermediates. They generate flanking direct repeats at their points of insertion when they transpose into DNA. Retrotransposons do not encode transposase, but some types are similar in structure to retroviruses and carry sequences that produce reverse transcriptase. Transposition takes place when transcription produces an RNA intermediate, which is then transcribed into DNA by reverse transcriptase and inserted into the target site. Examples of retrotransposons include *Ty* elements in yeast and *Alu* sequences in humans. Retrotransposons are not found in prokaryotes.

Class II consists of DNA transposons that possess terminal inverted repeats and transpose as DNA. Like Class I transposons, they all generate flanking direct repeats at their points of insertion into DNA. Unlike Class I transposons, all active forms of Class II transposable elements encode transposase, which is required for their movement. Some also encode resolvase, repressors, and other proteins. Their transposition may be replicative or nonreplicative, but they never use RNA intermediates. Examples of transposable elements in this class include insertion sequences and all complex transposons in bacteria, *Ac* and *Ds* elements in maize, and *P* elements in *Drosophila*.

Table 18.4   **Characteristics of two major classes of transposable genetic elements**

	Structure	Genes Encoded	Transposition	Examples
Class I (retrotransposon)	Long terminal direct repeats; short flanking direct repeats at target site	Reverse-transcriptase gene (and sometimes others)	By RNA intermediate	*Ty* (yeast) *copia* (Drosophila) *Alu* (human)
Class II	Short terminal inverted repeats; short flanking direct repeats at target site	Transposase gene (and sometimes others)	Through DNA (replicative or nonreplicative)	*IS1* (E. coli) *Tn3* (E. coli) *Ac, Ds* (maize) *P* elements (Drosophila)

Transposable Elements Have Played an Important Role in Genome Evolution

Transposable elements have clearly played an important role in shaping the genomes of many organisms. Much of the tremendous variation in genome size found among eukaryotic organisms is due to differences in numbers of transposable elements. Approximately 45% of the human genome consists of remnants of transposable elements and about 50% of all spontaneous mutations in *Drosophila* are due to transposition. Homologous recombination between copies of transposable elements has been an important force in producing gene duplications and other chromosome rearrangements. Furthermore, some transposable elements may carry extra DNA with them when they transpose to a new site, providing the potential to move DNA sequences that regulate genes to new sites, where they may alter the expression of genes.

TRANSPOSABLE ELEMENTS AS GENOMIC PARASITES

As we have seen, many transposable elements leave a copy behind when they transpose to a new location (copy-and-paste transposition) and therefore increase in number within a genome with the passage of time. This ability to replicate and spread means that many transposable elements may serve no purpose for the cell; they exist simply because they are capable of replicating and spreading. The insertion of transposable elements into a gene will often destroy its function, with harmful consequences for the cell. Furthermore, the time and energy required to replicate large numbers of transposable elements are likely to place a metabolic burden on the cell. Thus, transposable elements can be thought of as genomic parasites that provide no benefit to the cell and may even be harmful.

DOMESTICATION OF TRANSPOSABLE ELEMENTS

Although many transposable elements may be genomic parasites, some have clearly evolved to serve useful purposes for their host cells. These transposons are sometimes referred to as domesticated, implying that their parasitic tendencies have been replaced by properties useful to the cell. For example, the mechanism that generates antibody diversity in the immune systems of vertebrates (see Chapter 22) probably evolved from a transposable element. Immune cells called lymphocytes have the ability to unite several DNA segments that encode antigen-recognition proteins. This mechanism may have arisen from a transposable element that inserted into the germ line of a vertebrate ancestor some 450 million years ago.

Transposable elements have also been played an important role in the evolution of maize. Maize (modern corn) was domesticated from teosinte in Central America more than 8000 years ago. One of the important genetic differences between teosinte and corn involves a gene called *tb1*, which encodes a transcriptional regulator that represses the growth of side branches. In corn, transcription of *tb1* is elevated compared with that of teosinte, with the result that modern corn is more upright and less branched than teosinte. Recent research has demonstrated that the elevated transcription of *tb1* in corn is the result of a transposable element called *Hopscotch* that inserted into regulatory sequences that control *tb1* transcription. This insertion was probably present in teosinte as a variant and was selected by humans during the domestication of corn because it produced a more desirable plant shape. Transposable elements may also play a role in speciation, the process by which new species arise (see earlier section on hybrid dysgenesis).

CONCEPTS

Many transposable elements appear to be genomic parasites, existing in large numbers because of their ability to efficiently increase in copy number. Increases in copy number of transposable elements have contributed to the large size of many eukaryotic genomes. In several cases, transposable elements have been adopted for specific cellular functions.

18.5 A Number of Pathways Repair Changes in DNA

The integrity of DNA is under constant assault from radiation, chemical mutagens, and spontaneously arising changes. In spite of these damaging agents, the rate of mutation remains remarkably low, thanks to the efficiency with which DNA is repaired.

There are a number of complex pathways for repairing DNA, but several general statements can be made about DNA repair. First, most DNA-repair mechanisms require two nucleotide strands of DNA because most replace whole nucleotides, and a template strand is needed to specify the base sequence.

A second general feature of DNA repair is redundancy, meaning that many types of DNA damage can be corrected by more than one pathway of repair. This redundancy illustrates the extreme importance of DNA repair to the survival of the cell: if a mistake escapes one repair system, it's likely to be repaired by another system, ensuring that almost all mistakes are corrected.

We will consider several general mechanisms of DNA repair: mismatch repair, direct repair, base-excision repair, nucleotide-excision repair, and repair of double-strand breaks.

Mismatch Repair

Replication is extremely accurate: each new copy of DNA has less than one error per billion nucleotides. However, in the process of replication, mismatched bases are incorporated into the new DNA with a frequency of about 10^{-4} to 10^{-5}; so most of the errors that initially arise are corrected and never become permanent mutations. Some of these corrections are made in proofreading (see pp. 339–340 in Chapter 12) by the DNA polymerases.

(a)

1 In DNA replication, a mismatched base was added to the new strand.

New DNA

Old (template) DNA

2 Methylation at GATC sequences allows old and newly synthesized nucleotide strands to be differentiated: a lag in methylation means that, immediately after replication, the old strand will be methylated but the new strand will not.

GATC
CTAG

Methyl group

Mismatch-repair complex

(b)

Nick

Methyl group

GATC
CTAG

Mismatch-repair complex

3 The mismatch-repair complex brings the mismatched bases close to the methylated GATC sequence, and the new strand is identified.

T
G

(c)

5′ GATC
CTAG

Methyl group

4 Exonucleases remove nucleotides on the new strand between the GATC sequence and the mismatch.

T
3′

DNA bases

(d)

GATC
CTAG

Methyl group

5 DNA polymerase then replaces the nucleotides, correcting the mismatch, and DNA ligase seals the nick in the sugar–phosphate backbone.

T
A

18.36 Many incorrectly inserted nucleotides that escape proofreading are corrected by mismatch repair.

Many incorrectly inserted nucleotides that escape detection by proofreading are corrected by *mismatch repair*. Incorrectly paired bases are detected and corrected by mismatch-repair enzymes. In addition, the mismatch-repair system corrects small unpaired loops in the DNA, such as those caused by strand slippage in replication (see Figure 18.12). Some nucleotide repeats may form secondary structures on the unpaired strand (see Figure 18.5), allowing them to escape detection by the mismatch-repair system.

After the incorporation error has been recognized, mismatch-repair enzymes cut out a section of the newly synthesized strand and fill the gap with new nucleotides by using the original DNA strand as a template. For this strategy to work, mismatch repair must have some way of distinguishing between the old and the new strands of the DNA so that the incorporation error, but not part of the original strand, is removed.

The proteins that carry out mismatch repair in *E. coli* differentiate between old and new strands by the presence of methyl groups on special sequences of the old strand. After replication, adenine nucleotides in the sequence GATC are methylated. The process of methylation is delayed and so, immediately after replication, the old strand is methylated and the new strand is not (**Figure 18.36a**). The mismatch-repair complex brings an unmethylated GATC sequence in close proximity to the mismatched bases. It nicks the unmethylated strand at the GATC site (**Figure 18.36b**), and degrades the strand between the nick and the mismatched bases (**Figure 18.36c**). DNA polymerase and DNA ligase fill in the gap on the unmethylated strand with correctly paired nucleotides (**Figure 18.36d**).

Mismatch repair in eukaryotic cells is similar to that in *E. coli*, but how the old and new strands are recognized in eukaryotic cells is not known. In some eukaryotes, such as yeast and fruit flies, there is no detectable methylation of DNA, and yet mismatch repair still takes place. Humans who possess mutations in mismatch-repair genes often exhibit elevated somatic mutations and are frequently susceptible to colon cancer.

CONCEPTS

Mismatched bases and other DNA lesions are corrected by mismatch repair. Enzymes cut out a section of the newly synthesized strand of DNA and replace it with new nucleotides.

✔ CONCEPT CHECK 9

Mismatch repair in *E. coli* distinguishes between old and new strands of DNA on the basis of

a. differences in base composition of the two strands.

b. modification of histone proteins.

c. base analogs on the new strand.

d. methyl groups on the old strand.

Direct Repair

Direct repair does not replace altered nucleotides but, instead, changes them back into their original (correct) structures. One of the best understood direct-repair mechanisms is the photoreactivation of UV-induced pyrimidine dimers (see Figure 18.21). *E. coli* and some eukaryotic cells possess an enzyme called photolyase, which uses energy captured from light to break the covalent bonds that link the pyrimidines in a dimer.

Direct repair also corrects O^6-methylguanine, an alkylation product of guanine that pairs with adenine, producing $G \cdot C \rightarrow T \cdot A$ transversions. An enzyme called O^6-methylguanine-DNA methyltransferase removes the methyl group from O^6-methylguanine, restoring the base to guanine (**Figure 18.37**).

O^6-Methylguanine   Guanine

18.37 Direct repair changes nucleotides back into their original structures.

Base-Excision Repair

In **base-excision repair**, a modified base is first excised and then the entire nucleotide is replaced. The excision of modified bases is catalyzed by a set of enzymes called DNA glycosylases, each of which recognizes and removes a specific type of modified base by cleaving the bond that links that base to the 1′-carbon atom of deoxyribose sugar (**Figure 18.38a**). Uracil glycosylase, for example, recognizes and removes uracil produced by the deamination of cytosine. Other glycosylases recognize hypoxanthine, 3-methyladenine, 7-methylguanine, and other modified bases.

After the base has been removed, an enzyme called AP (apurinic or apyrimidinic) endonuclease cuts the phosphodiester bond, and other enzymes remove the deoxyribose sugar (**Figure 18.38b**). DNA polymerase then adds one or more new nucleotides to the exposed 3′-OH group (**Figure 18.38c**), replacing a section of nucleotides on the damaged strand. The nick in the phosphodiester backbone is sealed by DNA ligase (**Figure 18.38d**), and the original intact sequence is restored (**Figure 18.38e**).

Bacteria use DNA polymerase I to replace excised nucleotides, but eukaryotes use DNA polymerase β, which has no proofreading ability and tends to make mistakes. On average, DNA polymerase β makes one mistake per 4000 nucleotides inserted. About 20,000 to 40,000 base modifications per day

18.38 Base-excision repair excises modified bases and then replaces one or more nucleotides.

are repaired by base excision, and so DNA polymerase β may introduce as many as 10 mutations per day into the human genome. How are these errors corrected? Recent research results show that some AP endonucleases have the ability to proofread. When DNA polymerase β inserts a nucleotide with the wrong base into the DNA, DNA ligase cannot seal the nick in the sugar–phosphate backbone, because the 3′-OH and 5′-P groups of adjacent nucleotides are not in the correct orientation for ligase to connect them. In this case, AP endonuclease 1 detects the mispairing and uses its 3′ → 5′ exonuclease activity to excise the incorrectly paired base. DNA polymerase β then uses its polymerase activity to fill in the missing nucleotide. In this way, the fidelity of base-excision repair is maintained.

CONCEPTS

Direct-repair mechanisms change altered nucleotides back into their correct structures. In base-excision repair, glycosylase enzymes recognize and remove specific types of modified bases. The entire nucleotide is then removed and a section of the polynucleotide strand is replaced.

✓ CONCEPT CHECK 10

How do direct-repair mechanisms differ from mismatch repair and base-excision repair?

Nucleotide-Excision Repair

Another repair pathway is **nucleotide-excision repair**, which removes bulky DNA lesions (such as pyrimidine dimers) that distort the double helix. Nucleotide-excision repair can repair many different types of DNA damage and is found in cells of all organisms from bacteria to humans.

The process of nucleotide excision is complex; in humans, a large number of genes take part. First, a complex of enzymes scans DNA, looking for distortions of its three-dimensional configuration (**Figure 18.39a** and **b**). When a distortion is detected, additional enzymes separate the two nucleotide strands at the damaged region, and single-strand-binding proteins stabilize the separated strands (**Figure 18.39c**). Next, the sugar–phosphate backbone of the damaged strand is cleaved on both sides of the damage (**Figure 18.39d**). Part of the damaged strand is peeled away by helicase enzymes (**Figure 18.39e**), and the gap is filled in by DNA polymerase and sealed by DNA ligase (**Figure 18.39f**).

CONCEPTS

Nucleotide-excision repair removes and replaces many types of damaged DNA that distort the DNA structure. The two strands of DNA are separated, a section of the DNA containing the distortion is removed, DNA polymerase fills in the gap, and DNA ligase seals the filled-in gap.

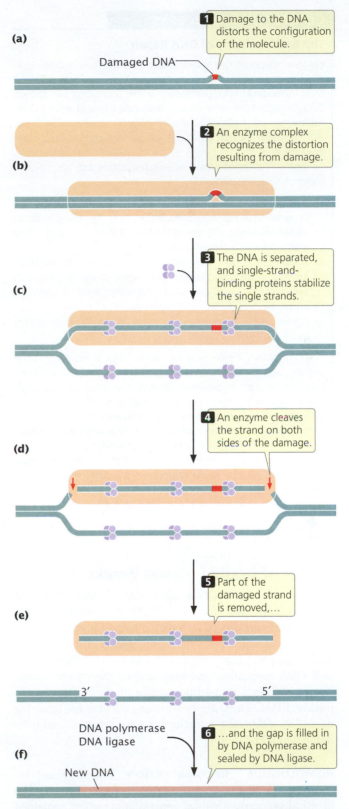

(a) Damaged DNA

1 Damage to the DNA distorts the configuration of the molecule.

(b) **2** An enzyme complex recognizes the distortion resulting from damage.

(c) **3** The DNA is separated, and single-strand-binding proteins stabilize the single strands.

(d) **4** An enzyme cleaves the strand on both sides of the damage.

(e) **5** Part of the damaged strand is removed,…

3′   5′

DNA polymerase
DNA ligase

(f) **6** …and the gap is filled in by DNA polymerase and sealed by DNA ligase.

New DNA

18.39 Nucleotide-excision repair removes bulky DNA lesions that distort the double helix.

The Basic Pathway of DNA Repair

We have now examined several different mechanisms of DNA repair. What do these methods have in common? How are they different? Most methods of DNA repair depend on the presence of two strands, because nucleotides in the damaged area are removed and replaced. Nucleotides are replaced in mismatch repair, base-excision repair, and nucleotide-excision repair but are not replaced by direct-repair mechanisms.

Repair mechanisms that include nucleotide removal utilize a common four-step pathway:

1. Detection: The damaged section of the DNA is recognized.
2. Excision: DNA-repair endonucleases nick the phosphodiester backbone on one or both sides of the DNA damage and one or more nucleotides are removed.
3. Polymerization: DNA polymerase adds nucleotides to the newly exposed 3′-OH group by using the other strand as a template and replacing damaged (and frequently some undamaged) nucleotides.
4. Ligation: DNA ligase seals the nicks in the sugar–phosphate backbone.

The primary differences in the mechanisms of mismatch, base-excision, and nucleotide-excision repair are in the details of detection and excision. In base-excision and mismatch repair, a single nick is made in the sugar–phosphate backbone on one side of the damage; in nucleotide-excision repair, nicks are made on both sides of the DNA lesion. In base-excision repair, DNA polymerase displaces the old nucleotides as it adds new nucleotides to the 3′ end of the nick; in mismatch repair, the old nucleotides are degraded; and, in nucleotide-excision repair, nucleotides are displaced by helicase enzymes. All three mechanisms use DNA polymerase and ligase to fill in the gap produced by the excision and removal of damaged nucleotides.

Repair of Double-Strand Breaks

A common type of DNA damage is a double-strand break, in which both strands of the DNA helix are broken. Double-strand breaks are caused by ionizing radiation, oxidative free radicals, and other DNA-damaging agents. These types of breaks are particularly detrimental to the cell because they stall DNA replication and may lead to chromosome rearrangements, such as deletions, duplications, inversions, and translocations. There are two major pathways for repairing double-strand breaks: homologous recombination and nonhomologous end joining.

HOMOLOGOUS RECOMBINATION Homologous recombination repairs a broken DNA molecule by using the identical or nearly identical genetic information contained in another DNA molecule, usually a sister chromatid—the same mechanism employed in the process of homologous recombination that is responsible for crossing over (see Chapter 12). Homologous recombination begins with the removal of some nucleotides at the broken ends, followed by strand invasion, displacement, and replication (see Figure 12.20). Many of the same enzymes that carry out crossing over are utilized in the repair of double-strand breaks by homologous recombination: two such enzymes are BRCA1 and BRCA2. The genes that code for these proteins are frequently mutated in breast cancer cells.

NONHOMOLOGOUS END JOINING Nonhomologous end joining repairs double-strand breaks without using a homologous template. This pathway is often used when the cell is in G_1 and a sister chromatid is not available for repair through homologous recombination. Nonhomologous end joining uses proteins that recognize the broken ends of DNA, bind to the ends, and then joins them together. Nonhomologous end joining is more error prone than homologous recombination and often leads to deletions, insertions, and translocations. Different types of DNA repair are summarized in **Table 18.5**.

Translesion DNA Polymerases

As discussed in Chapter 12, the high-fidelity DNA polymerases that normally carry out replication operate at high speed and, like a high-speed train, require a smooth track—an undistorted template. Some mutations, such as pyrimidine

Table 18.5	Summary of common DNA repair mechanisms	
Repair System	**Type of Damage Repaired**
Mismatch	Replication errors, including mispaired bases and strand slippage
Direct	Pyrimidine dimers; other specific types of alterations
Base excision	Abnormal bases, modified bases, and pyrimidine dimers
Nucleotide excision	DNA damage that distorts the double helix, including abnormal bases, modified bases, and pyrimidine dimers
Homologous recombination	Double-strand breaks
Nonhomologous end joining	Double-strand breaks

dimers, produce distortions in the three-dimensional structure of the DNA helix, blocking replication by the high-speed polymerases. When distortions of the template are encountered, specialized translesion DNA polymerases take over replication and bypass the lesions.

The translesion polymerases are able to bypass bulky lesions but, in the process, often make errors. Thus, the translesion polymerases allow replication to proceed at the cost of introducing mutations into the sequence. Some of these mutations are corrected by DNA-repair systems, but others escape detection.

An example of a translesion DNA polymerase is polymerase η (eta), which bypasses pyrimidine dimers in eukaryotes. Polymerase η inserts AA opposite a pyrimidine dimer. This strategy seems to be reasonable because about two-thirds of pyrimidine dimers are thymine dimers. However, the insertion of AA opposite a CT dimer results in a $C \cdot G \rightarrow T \cdot A$ transversion. Polymerase η therefore tends to introduce mutations into the DNA sequence.

18.40 Xeroderma pigmentosum results from defects in DNA repair. The disease is characterized by frecklelike spots on the skin (shown here) and a predisposition to skin cancer. [©Stephane AUDRAS/REA/Redux.]

CONCEPTS

Two major pathways exist for the repair of double-strand breaks in DNA: homologous recombination and nonhomologous end joining. Special translesion DNA polymerases allow replication to proceed past bulky distortions in the DNA but often introduce errors as they bypass the distorted region.

Genetic Diseases and Faulty DNA Repair

Several human diseases are connected to defects in DNA repair. These diseases are often associated with high incidences of specific cancers, because defects in DNA repair lead to increased rates of mutation. This concept is discussed further in Chapter 23.

Among the best studied of the human DNA-repair diseases is xeroderma pigmentosum (**Figure 18.40**), a rare autosomal recessive condition that includes abnormal skin pigmentation and acute sensitivity to sunlight. Persons who have this disease also have a strong predisposition to skin cancer, with an incidence ranging from 1000 to 2000 times that found in unaffected people.

Sunlight includes a strong UV component, so exposure to sunlight produces pyrimidine dimers in the DNA of skin cells. Although human cells lack photolyase (the enzyme that repairs pyrimidine dimers in bacteria), most pyrimidine dimers in humans can be corrected by nucleotide-excision repair (see Figure 18.39). However, the cells of most people with xeroderma pigmentosum are defective in

nucleotide-excision repair, and many of their pyrimidine dimers go uncorrected and may lead to cancer.

Xeroderma pigmentosum can result from defects in several different genes. Some persons with xeroderma pigmentosum have mutations in a gene encoding the protein that recognizes and binds to damaged DNA; others have mutations in a gene encoding helicase. Still others have defects in the genes that play a role in cutting the damaged strand on the 5′ or 3′ sides of the pyrimidine dimer. Some persons have a slightly different form of the disease (xeroderma pigmentosum variant) owing to mutations in the gene encoding polymerase η, the translesion DNA polymerase that bypasses pyrimidine dimers.

Another genetic disease caused by faulty DNA repair is an inherited form of colon cancer called hereditary nonpolyposis colon cancer (HNPCC). This is one of the most common hereditary cancers, accounting for about 15% of colon cancers. Research findings indicate that HNPCC arises from mutations in the proteins that carry out mismatch repair (see Figure 18.36). Some genetic diseases associated with defective DNA repair are summarized in **Table 18.6**. ▶ TRY PROBLEM 44

CONCEPTS

Defects in DNA repair are the underlying cause of several genetic diseases. Many of these diseases are characterized by a predisposition to cancer.

✔ **CONCEPT CHECK 11**

Why are defects in DNA repair often associated with increases in cancer?

Table 18.6   Genetic diseases associated with defects in DNA-repair systems

Disease	Symptoms	Genetic Defect
Xeroderma pigmentosum	Frecklelike spots on skin, sensitivity to sunlight, predisposition to skin cancer	Defects in nucleotide-excision repair
Cockayne syndrome	Dwarfism, sensitivity to sunlight, premature aging, deafness, intellectual disability	Defects in nucleotide-excision repair
Trichothiodystrophy	Brittle hair, skin abnormalities, short stature, immature sexual development, characteristic facial features	Defects in nucleotide-excision repair
Hereditary nonpolyposis colon cancer	Predisposition to colon cancer	Defects in mismatch repair
Fanconi anemia	Increased skin pigmentation, abnormalities of skeleton, heart, and kidneys, predisposition to leukemia	Possibly defects in the repair of interstrand cross-links
Li-Fraumeni syndrome	Predisposition to cancer in many different tissues	Defects in DNA damage response
Werner syndrome	Premature aging, predisposition to cancer	Defect in homologous recombination

CONCEPTS SUMMARY

- Mutations are heritable changes in genetic information. Somatic mutations occur in somatic cells; germ-line mutations occur in cells that give rise to gametes.

- The simplest type of mutation is a base substitution, a change in a single base pair of DNA. Transitions are base substitutions in which purines are replaced by purines or pyrimidines are replaced by pyrimidines. Transversions are base substitutions in which a purine replaces a pyrimidine or a pyrimidine replaces a purine.

- Insertions are the addition of nucleotides, and deletions are the removal of nucleotides; these mutations often change the reading frame of the gene.

- Expanding nucleotide repeats are mutations in which the number of copies of a set of nucleotides increases with the passage of time; they are responsible for several human genetic diseases.

- A missense mutation alters the coding sequence so that one amino acid substitutes for another. A nonsense mutation changes a codon that specifies an amino acid into a termination codon. A silent mutation produces a synonymous codon that specifies the same amino acid as the original sequence, whereas a neutral mutation alters the amino acid sequence but does not change the functioning of the protein. A suppressor mutation reverses the effect of a previous mutation at a different site and may be intragenic (within the same gene as the original mutation) or intergenic (within a different gene).

- The mutation rate is the frequency with which a particular mutation arises in a population. Mutation rates are influenced by both genetic and environmental factors.

- Some mutations occur spontaneously. These mutations include the mispairing of bases in replication and spontaneous depurination and deamination.

- Insertions and deletions can arise from strand slippage in replication or from unequal crossing over.

- Base analogs can become incorporated into DNA in the course of replication and pair with the wrong base in subsequent replication events. Alkylating agents and hydroxylamine modify the chemical structure of bases and lead to mutations. Intercalating agents insert into the DNA molecule and cause single-nucleotide additions and deletions. Oxidative reactions alter the chemical structures of bases.

- Ionizing radiation is mutagenic, altering base structures and breaking phosphodiester bonds. Ultraviolet light produces pyrimidine dimers, which block replication. Bacteria use the SOS response to overcome replication blocks produced by pyrimidine dimers and other lesions in DNA.

- The Ames test uses bacteria to assess the mutagenic potential of chemical substances.

- Transposable elements are mobile DNA sequences that insert into many locations within a genome and often cause mutations and DNA rearrangements.

- Most transposable elements have two common characteristics: terminal inverted repeats and the generation of short direct repeats in DNA at the point of insertion.

- Transposition may take place through a DNA molecule or through the production of an RNA molecule that is then reverse transcribed into DNA. Transposition may be replicative, in which the transposable element is copied and the copy moves to a new site, or nonreplicative, in which the transposable element excises from the old site and moves to a new site.

- Retrotransposons transpose through RNA molecules that undergo reverse transcription to produce DNA.

■ Insertion sequences are small bacterial transposable elements that carry only the information needed for their own movement. Composite transposons in bacteria are more complex elements that consist of DNA between two insertion sequences. Some complex transposable elements in bacteria do not contain insertion sequences.

■ DNA transposons in eukaryotic cells are similar to those found in bacteria, ending in short inverted repeats and producing flanking direct repeats at the point of insertion. Others are retrotransposons, similar in structure to retroviruses and transposing through RNA intermediates.

■ Transposons have played an important role in genome evolution.

■ Most damage to DNA is corrected by DNA-repair mechanisms. These mechanisms include mismatch repair, direct repair, base-excision repair, nucleotide-excision repair, and other repair pathways. Most require two strands of DNA and exhibit some overlap in the types of damage repaired.

■ Double-strand breaks are repaired by homologous recombination and nonhomologous end joining. Special translesion DNA polymerases allow replication to proceed past bulky distortions in the DNA.

■ Defects in DNA repair are the underlying cause of several genetic diseases.

IMPORTANT TERMS

mutation (p. 494)
somatic mutation (p. 494)
germ-line mutation (p. 495)
gene mutation (p. 495)
base substitution (p. 495)
transition (p. 495)
transversion (p. 495)
insertion (p. 496)
deletion (p. 496)
frameshift mutation (p. 496)
in-frame insertion (p. 496)
in-frame deletion (p. 496)
expanding nucleotide repeat
   (p. 496)
forward mutation (p. 497)
reverse mutation (reversion)
   (p. 498)
missense mutation (p. 498)
nonsense mutation (p. 498)

silent mutation (p. 498)
neutral mutation (p. 498)
loss-of-function mutation
   (p. 498)
gain-of-function mutation
   (p. 498)
conditional mutation
   (p. 498)
lethal mutation (p. 498)
suppressor mutation (p. 498)
intragenic suppressor
   mutation (p. 499)
intergenic suppressor
   mutation (p. 500)
mutation rate (p. 502)
adaptive mutation (p. 503)
spontaneous mutation
   (p. 503)
induced mutation (p. 503)

incorporated error (p. 504)
replicated error (p. 504)
strand slippage (p. 504)
unequal crossing over
   (p. 504)
depurination (p. 504)
deamination (p. 505)
mutagen (p. 506)
base analog (p. 506)
intercalating agent (p. 508)
pyrimidine dimer (p. 508)
SOS system (p. 509)
Ames test (p. 509)
transposable element
   (p. 511)
flanking direct repeat
   (p. 511)
terminal inverted repeat
   (p. 511)

transposition (p. 512)
transposase (p. 512)
DNA transposon (p. 512)
retrotransposon (p. 512)
replicative transposition
   (p. 512)
nonreplicative transposition
   (p. 512)
insertion sequence (IS)
   (p. 514)
composite transposon
   (p. 515)
hybrid dysgenesis (p. 517)
direct repair (p. 522)
base-excision repair
   (p. 522)
nucleotide-excision repair
   (p. 523)

ANSWERS TO CONCEPT CHECKS

1. c

2. A reverse mutation restores the original phenotype by changing the DNA sequence back to the wild-type sequence. A suppressor mutation restores the phenotype by causing an additional change in the DNA at a site that is different from that of the original mutation.

3. The frequency with which changes arise in DNA, how often these changes are repaired by DNA-repair mechanisms, and our ability to detect the mutation

4. c

5. In transposition, staggered cuts are made in DNA and the transposable element inserts into the cut. Later, replication of the single-stranded pieces of DNA creates short flanking direct repeats on either side of the inserted transposable element.

6. Transposition often results in mutations because the transposable element inserts into a gene, destroying its

function. Chromosome rearrangements arise because transposition includes the breaking and exchange of DNA sequences. Additionally, multiple copies of a transposable element may undergo homologous recombination, producing chromosome rearrangements.

7. d

8. a

9. d

10. Direct-repair mechanisms return an altered base to its correct structure without removing and replacing nucleotides. Mismatch repair and base-excision repair remove and replace nucleotides.

11. Changes in DNA structure may not undergo repair in people with defects in DNA-repair mechanisms. Consequently, increased numbers of mutations occur at all genes, including those that predispose to cancer. This observation indicates that cancer arises from mutations in DNA.

WORKED PROBLEMS

Problem 1

The mutations produced by the following compounds are reversed by the substances shown. What conclusions can you make about the nature of the mutations originally produced by these compounds?

	Mutations produced by compound	5-Bromouracil	EMS	Hydroxyl amine	Acridine orange
a.	1	Yes	Yes	No	No
b.	2	Yes	Yes	Some	No
c.	3	No	No	No	Yes
d.	4	Yes	Yes	Yes	Yes

Solution Strategy

What information is required in your answer to the problem?

Conclusions you can make about the types of mutations produced by each compound based on the substances that reverse the mutations.

What information is provided to solve the problem?

Which substances reverse the mutations produced by each compound.

For help with this problem, review:

Chemically Induced Mutations in Section 18.2.

Solution Steps

Hint: The ability of various compounds to produce reverse mutations reveals important information about the nature of the original mutation.

a. Mutations produced by compound 1 are reversed by 5-bromouracil, which produces both A • T → G • C and G • C → A • T transitions, which tells us that compound 1 produces single-base substitutions that may include the generation of either A • T or G • C pairs. The mutations produced by compound 1 are also reversed by EMS, which, like 5-bromouracil, produces both A • T → G • C and G • C → A • T transitions; so no additional information is provided here. Hydroxylamine does not reverse the mutations produced by compound

1. Because hydroxylamine produces only C • G → T • A transitions, we know that compound 1 does not generate C • G base pairs. Acridine orange, an intercalating agent that produces frameshift mutations, also does not reverse the mutations, revealing that compound 1 produces only single-base-pair substitutions, not insertions or deletions. In summary, compound 1 appears to cause single-base substitutions that generate T • A but not G • C base pairs.

b. Compound 2 generates mutations that are reversed by 5-bromouracil and EMS, indicating that it may produce G • C or A • T base pairs. Some of these mutations are reversed by hydroxylamine, which produces only C • G → T • A transitions, indicating that some of the mutations produced by compound 2 are C • G base pairs. None of the mutations are reversed by acridine orange; so compound 2 does not induce insertions or deletions. In summary, compound 2 produces single-base substitutions that generate both G • C and A • T base pairs.

c. Compound 3 produces mutations that are reversed only by acridine orange; so compound 3 appears to produce only insertions and deletions.

d. Compound 4 is reversed by 5 bromouracil, EMS, hydroxylamine, and acridine orange, indicating that this compound produces single-base substitutions, which include G • C and A • T base pairs, insertions, and deletions.

Problem 2

Certain repeated sequences in eukaryotes are flanked by short direct repeats, suggesting that they originated as transposable elements. These same sequences lack introns and possess a string of thymine nucleotides at one ends. Have these elements transposed through DNA or RNA sequences? Explain your reasoning.

Solution Strategy

What information is required in your answer to the problem?

Whether the transposable elements transpose through a DNA or an RNA intermediate and why you made this conclusion.

What information is provided to solve the problem?

- The element is flanked by short direct repeats.

- The element lacks introns and has a string of Ts at one end.

For help with this problem, review:

Transposition in Section 18.4.

Solution Steps

The absence of introns and the string of thymine nucleotides (which would be complementary to adenine nucleotides in RNA) at one end are characteristics of processed RNA. These similarities to RNA suggest that the element was originally transcribed into mRNA, processed to remove the introns and add a poly(A) tail, and then reverse transcribed into a complementary DNA that was inserted into the chromosome.

Recall: PremRNAs in eukaryotes are processed: a 5′ cap is added, introns are removed and a 3′ poly(A) tail is added.

COMPREHENSION QUESTIONS

Section 18.1

1. What is the difference between a transition and a transversion? Which type of base substitution is usually more common?

2. Briefly describe expanding nucleotide repeats. How do they account for the phenomenon of anticipation?

3. What is the difference between a missense mutation and a nonsense mutation? Between a silent mutation and a neutral mutation?

4. Briefly describe two different ways in which intragenic suppressors can reverse the effects of mutations.

Section 18.2

5. How do insertions and deletions arise?

6. How do base analogs lead to mutations?

7. How do alkylating agents, nitrous acid, and hydroxylamine produce mutations?

Section 18.3

8. What is the purpose of the Ames test? How are *his⁻* bacteria used in this test?

Section 18.4

9. What general characteristics are found in many transposable elements?

10. How does a retrotransposon move?

11. Draw the structure of a typical insertion sequence and identify its parts.

12. Explain how *Ac* and *Ds* elements produce variegated corn kernels.

13. What are some differences between class I and class II transposable elements?

14. Why are transposable elements often called genomic parasites?

Section 18.5

15. List at least three different types of DNA repair and briefly explain how each is carried out.

16. What are two major mechanisms for the repair of double-strand breaks? How do they differ?

> For more questions that test your comprehension of the key chapter concepts, go to **LEARNING**Curve for this chapter.

APPLICATION QUESTIONS AND PROBLEMS

Section 18.1

17. A codon that specifies the amino acid Gly undergoes a single-base substitution to become a nonsense mutation. In accord with the genetic code given in **Figure 15.10**, is this mutation a transition or a transversion? At which position of the codon does the mutation occur?

*18 **a.** If a single transition occurs in a codon that specifies Phe, what amino acids can be specified by the mutated sequence?

b. If a single transversion occurs in a codon that specifies Phe, what amino acids can be specified by the mutated sequence?

c. If a single transition occurs in a codon that specifies Leu, what amino acids can be specified by the mutated sequence?

d. If a single transversion occurs in a codon that specifies Leu, what amino acids can be specified by the mutated sequence?

19. Hemoglobin is a complex protein that contains four polypeptide chains. The normal hemoglobin found in adults—called adult hemoglobin—consists of two alpha and two beta polypeptide chains, which are encoded by different loci. Sickle-cell hemoglobin, which causes sickle-cell anemia, arises from a mutation in the beta chain of adult hemoglobin. Adult hemoglobin and sickle-cell hemoglobin differ in a single amino acid: the sixth amino acid from one end in adult hemoglobin is glutamic acid, whereas sickle-cell hemoglobin has valine at this position. After consulting the genetic code provided in **Figure 15.10**, indicate the type and location of the mutation that gave rise to sickle-cell anemia.

20. The following nucleotide sequence is found on the template strand of DNA. First, determine the amino acids of the protein encoded by this sequence by using the genetic code provided in **Figure 15.10**. Then, give the altered amino acid sequence of the protein that will be found in each of the following mutations:

Sequence
of DNA
template
↳   3′–TAC TGG CCG TTA GTT GAT ATA ACT–5′
↱   1   24
Nucleotide
number

a. Mutant 1:   A transition at nucleotide 11
b. Mutant 2:   A transition at nucleotide 13
c. Mutant 3:   A one-nucleotide deletion at nucleotide 7
d. Mutant 4:   A T → A transversion at nucleotide 15
e. Mutant 5:   An addition of TGG after nucleotide 6
f. Mutant 6:   A transition at nucleotide 9

21. Draw a hairpin turn like that shown in **Figure 18.5** for the repeated sequence found in Fragile X syndrome (see **Table 18.1**).

*22. A polypeptide has the following amino acid sequence:

Met-Ser-Pro-Arg-Leu-Glu-Gly

The amino acid sequence of this polypeptide was determined in a series of mutants listed in parts *a* through *e*. For each mutant, indicate the type of mutation that occurred in the DNA (single-base substitution, insertion, deletion) and the phenotypic effect of the mutation (nonsense mutation, missense mutation, frameshift, etc.).

a. Mutant 1:   Met-Ser-Ser-Arg-Leu-Glu-Gly
b. Mutant 2:   Met-Ser-Pro
c. Mutant 3:   Met-Ser-Pro-Asp-Trp-Arg-Asp-Lys
d. Mutant 4:   Met-Ser-Pro-Glu-Gly
e. Mutant 5:   Met-Ser-Pro-Arg-Leu-Leu-Glu-Gly

*23. A gene encodes a protein with the following amino acid sequence:

Met-Trp-His-Arg-Ala-Ser-Phe

A mutation occurs in the gene. The mutant protein has the following amino acid sequence:

Met-Trp-His-Ser-Ala-Ser-Phe

An intragenic suppressor restores the amino acid sequence to that of the original protein:

Met-Trp-His-Arg-Ala-Ser-Phe

Give at least one example of base changes that could produce the original mutation and the intragenic suppressor. (Consult the genetic code in **Figure 15.10**.)

24. A gene encodes a protein with the following amino acid sequence:

Met-Lys-Ser-Pro-Ala-Thr-Pro

A nonsense mutation from a single-base-pair substitution occurs in this gene, resulting in a protein with the amino acid sequence Met-Lys. An intergenic suppressor mutation allows the gene to produce the full-length protein. With the original mutation and the intergenic suppressor present, the gene now produces a protein with the following amino acid sequence:

Met-Lys-Cys-Pro-Ala-Thr-Pro

Give the location and nature of the original mutation and of the intergenic suppressor.

Section 18.2

*25. Can nonsense mutations be reversed by hydroxylamine? Why or why not?

26. The following nucleotide sequence is found in a short stretch of DNA:

5′–ATGT–3′
3′–TACA–5′

If this sequence is treated with hydroxylamine, what sequences will result after replication?

*27. The following nucleotide sequence is found in a short stretch of DNA:

5′–AG–3′
3′–TC–5′

a. Give all the mutant sequences that can result from spontaneous depurination in this stretch of DNA.

b. Give all the mutant sequences that can result from spontaneous deamination in this stretch of DNA.

28. In many eukaryotic organisms, a significant proportion of cytosine bases are naturally methylated to 5-methylcytosine. Through evolutionary time, the proportion of AT base pairs in the DNA of these organisms increases. Can you suggest a possible mechanism for this increase?

Section 18.3

*29. A chemist synthesizes four new chemical compounds in the laboratory and names them PFI1, PFI2, PFI3, and PFI4. He gives the PFI compounds to a geneticist friend and asks her to determine their mutagenic potential. The geneticist finds that all four are highly mutagenic. She also tests the capacity of mutations produced by the PFI compounds to be reversed by other known mutagens and obtains the following results. What conclusions can you make about the nature of the mutations produced by these compounds?

	Reversed by			
Mutations produced by	**2-Amino-purine**	**Nitrous acid**	**Hydroxyl-amine**	**Acridine orange**
PFI1	Yes	Yes	Some	No
PFI2	No	No	No	No
PFI3	Yes	Yes	No	No
PFI4	No	No	No	Yes

30. Mary Alexander studied the effects of radiation on mutation rates in the sperm of *Drosophila melanogaster*. She irradiated *Drosophila* larvae with either 3000 roentgens (r) or 3975 r, collected the adult males that developed from irradiated larvae, mated them with unirradiated females, and then counted the number of mutant F_1 flies produced by each male. All mutant flies that appeared were used in subsequent crosses to determine if their mutant phenotypes were genetic. She obtained the following results (M. L. Alexander. 1954. *Genetics* 39:409–428):

Group	Number of offspring	Offspring with a genetic mutation
Control (0 r)	45,504	0
Irradiated (3000 r)	49,512	71
Irradiated (3975 r)	50,159	70

a. Calculate the mutation rates of the control group and the two groups of irradiated flies.

b. On the basis of these data, do you think radiation has any effect on mutation? Explain your answer.

31. What conclusion would you make if the number of bacterial colonies in **Figure 18.22** were the same on the control plate and the treatment plate? Explain your reasoning.

32. A genetics instructor designs a laboratory experiment to study the effects of UV radiation on mutation in bacteria. In the experiment, the students expose bacteria plated on petri plates to UV light for different lengths of time, place the plates in an incubator for 48 hours, and then count the number of colonies that appear on each plate. The plates that have received more UV radiation should have more pyrimidine dimers, which block replication; thus, fewer colonies should appear on the plates exposed to UV light for longer periods of time. Before the students carry out the experiment, the instructor warns them that while the bacteria are in the incubator, the students must not open the incubator door unless the room is darkened. Why should the bacteria not be exposed to light?

Section 18.4

*33. A particular transposable element generates flanking direct repeats that are 4 bp long. Give the sequence that will be found on both sides of the transposable element if this transposable element inserts at the position indicated on each of the following sequences.

a.

5′–ATTCGAACTGACCGATCA–3′

b.

5′–ATTCGAACTGACCGATCA–3′

34. White eyes in *Drosophila melanogaster* result from an X-linked recessive mutation. Occasionally, white-eyed mutants give rise to offspring that possess white eyes with small red spots. The number, distribution, and size of the red spots are variable. Explain how a transposable element could be responsible for this spotting phenomenon.

35. What factor might potentially determine the length of the flanking direct repeats that are produced in transposition?

*36. Which of the following pairs of sequences might be found at the ends of an insertion sequence?

a. 5′–GGGCCAATT–3′ and 5′–CCCGGTTAA–3′
b. 5′–AAACCCTTT–3′ and 5′–AAAGGGTTT–3′
c. 5′–TTTCGAC–3′ and 5′–CAGCTTT–3′
d. 5′–ACGTACG–3′ and 5′–CGTACGT–3′
e. 5′–GCCCCAT–3′ and 5′–GCCCAT–3′

37. Explain why the corn kernel in **Figure 18.34d** is varie-gated, with some areas colored and some areas lacking pigment.

*38. Two different strains of *Drosophila melanogaster* are mated in reciprocal crosses. When strain A males are crossed with strain B females, the progeny are normal. However, when strain A females are crossed with strain B males, there are many mutations and chromosome rearrangements in the gametes of the F_1 progeny and the F_1 generation is effectively sterile. Explain these results.

39. An insertion sequence contains a large deletion in its transposase gene. Under what circumstances would this insertion sequence be able to transpose?

40. Zidovudine (AZT) is a drug used to treat patients with AIDS. AZT works by blocking the reverse transcriptase enzyme used by the human immunodeficiency virus (HIV), the causative agent of AIDS. Do you expect that AZT would have any effect on transposable elements? If so, what type of transposable elements would be affected and what would be the most likely effect?

41. A transposable element is found to encode a reverse transcriptase enzyme. On the basis of this information, what conclusions can you make about the likely structure and method of transposition of this element?

*42. A geneticist examines an ear of corn in which most kernels are yellow, but he finds a few kernels with purple spots, as shown here. Give a possible explanation for the appearance of the purple spots in these otherwise yellow kernels, accounting for their different sizes. (Hint: See the section on *Ac* and *Ds* elements in maize.)

Section 18.5

43. Which DNA repair mechanism would most likely correct the incorporated error labeled by balloon 2 in **Figure 18.11**?

*44. A plant breeder wants to isolate mutants in tomatoes that are defective in DNA repair. However, this breeder does not have the expertise or equipment to study enzymes in DNA-repair systems. How can the breeder identify tomato plants that are deficient in DNA repair? What are the traits to look for?

CHALLENGE QUESTIONS

Section 18.1

45. Robert Bost and Richard Cribbs studied a strain of *E. coli* (*araB14*) that possessed a nonsense mutation in the structural gene that encodes L-ribulokinase, an enzyme that allows the bacteria to metabolize the sugar arabinose (R. Bost and R. Cribbs. 1969. *Genetics* 62:1–8). From the *araB14* strain, they isolated some bacteria that possessed mutations that caused them to revert back to the wild type. Genetic analysis of these revertants showed that they possessed two different suppressor mutations. One suppressor mutation (*R1*) was linked to the original mutation in the L-ribulokinase and probably occurred at the same locus. By itself, this mutation allowed the production of L-ribulokinase, but the enzyme was not as effective in metabolizing arabinose as the enzyme encoded by the wild-type allele. The second suppressor mutation (*Su^B*) was not linked to the original mutation. In conjunction with the *R1* mutation, *Su^B* allowed the production of L-ribulokinase, but *Su^B* by itself was not able to suppress the original mutation.

a. On the basis of this information, are the *R1* and *Su^B* mutations intragenic suppressors or intergenic suppressors? Explain your reasoning.

b. Propose an explanation for how *R1* and *Su^B* restore the ability of *araB14* to metabolize arabinose and why *Su^B* is able to more fully restore this ability.

46. Achondroplasia is an autosomal dominant disorder characterized by disproportionate short stature: the legs and arms are short compared with the head and trunk. The disorder is due to a base substitution in the gene, located on the short arm of chromosome 4, for fibroblast growth factor receptor 3 (FGFR3). Although achondroplasia is clearly inherited as an autosomal dominant trait, more than 80% of the people who have achondroplasia are born to parents with normal stature. This high percentage indicates that most cases are caused by newly arising mutations; these cases (not inherited from an affected parent) are referred to as sporadic. Studies have demonstrated that sporadic cases of achondroplasia are almost always caused by mutations inherited from the father (paternal mutations). In addition, the occurrence of achondroplasia is higher among older fathers; approximately 50% of children with achondroplasia are born to fathers older than 35 years of age. There is no

A family of three who have achondroplasia. [Gail Burton/AP.]

association with maternal age. The mutation rate for achondroplasia (about 4×10^{-5} mutations per gamete) is high compared with those for other genetic disorders. Explain why most spontaneous mutations for achondroplasia are paternal in origin and why the occurrence of achondroplasia is higher among older fathers.

47. Tay–Sachs disease is a severe, autosomal recessive genetic disease that produces deafness, blindness, seizures, and, eventually, death. The disease results from a defect in the *HEXA* gene, which encodes hexosaminidase A. This enzyme normally degrades G_{M2} gangliosides. In the absence of hexosaminidase A, G_{M2} gangliosides accumulate in the brain. The results of recent molecular studies showed that the most common mutation causing Tay–Sachs disease is a 4-bp insertion that produces a downstream premature stop codon. Results of further studies have revealed that the transcription of the *HEXA* gene is normal in people who have Tay–Sachs disease, but the *HEXA* mRNA is unstable. Propose a mechanism to account for how a premature stop codon could cause mRNA instability.

48. *Ochre* and *amber* are two types of nonsense mutations. Before the genetic code was worked out, Sydney Brenner, Anthony O. Stretton, and Samuel Kaplan applied different types of mutagens to bacteriophages in an attempt to determine the bases present in the codons responsible for *amber* and *ochre* mutations. They knew that *ochre* and *amber* mutants were suppressed by different types of mutations, demonstrating that each is a different termination codon. They obtained the following results:

(1) A single-base substitution could convert an *ochre* mutation into an *amber* mutation.

(2) Hydroxylamine induced both *ochre* and *amber* mutations in wild-type phages.

(3) 2-Aminopurine caused *ochre* to mutate to *amber*.

(4) Hydroxylamine did not cause *ochre* to mutate to *amber*.

These data do not allow the complete nucleotide sequence of the *amber* and *ochre* codons to be worked out, but they do provide some information about the bases found in the nonsense mutations.

a. What conclusions about the bases found in the codons of *amber* and *ochre* mutations can be made from these observations?

b. Of the three nonsense codons (UAA, UAG, UGA), which represents the *ochre* mutation?

Section 18.3

49. To determine whether radiation associated with the atomic bombings of Hiroshima and Nagasaki produced recessive germ-line mutations, scientists examined the sex ratio of the children of the survivors of the blasts. Can you explain why an increase in germ-line mutations might be expected to alter the sex ratio?

Section 18.4

50. Marilyn Houck and Margaret Kidwell proposed that *P* elements were carried from *Drosophila willistoni* to *Drosophila melanogaster* by mites that fed on fruit flies (M. A. Houck et al. 1991. *Science* 253:1125–1129). What evidence do you think would be required to demonstrate that *D. melanogaster* acquired *P* elements in this way? Propose a series of experiments to provide such evidence.

Section 18.5

51. Trichothiodystrophy is a human inherited disorder characterized by premature aging, including osteoporosis, osteosclerosis, early graying, infertility, and reduced life span. The results of studies showed that the mutation that causes this disorder occurs in a gene that encodes a DNA helicase. Propose a mechanism for how a mutation in a DNA helicase might cause premature aging. Be sure to relate the symptoms of the disorder to possible functions of the helicase enzyme.

▶ Go to your **LaunchPad** to find additional learning resources and the Suggested Readings for this chapter.

19

Molecular Genetic Analysis and Biotechnology

Blindness affects 45 million people throughout the world. Genetic engineering is now being used to treat patients with Leber congenital amaurosis, a genetic form of blindness. [Paul Doyle/Alamy.]

Helping the Blind to See

Genetic engineering, the manipulation and transfer of genes from one organism to another, has revolutionized the study of genetics and has been successfully used to create a plethora of new products, including bacteria that generate chemicals, pest-resistant crops, and farm animals that secrete pharmaceutical products in their milk. In perhaps its ultimate application, genetic engineering is being used to treat disease in humans, a process known as gene therapy. In a dramatic example of gene therapy, researchers in 2007 transferred genes to four blind people, partly restoring their sight.

Eyesight is the most precious of human senses, enabling us to read, navigate physical obstacles, recognize friends, and enjoy the stunning visual beauty of the natural world. Tragically, millions of people are blind or unable to see well. Eyesight often deteriorates with age; indeed, most blindness is found among the elderly. However, some children are born blind, and others lose their eyesight at an early age. Research suggests that heredity is responsible for about half of the cases of blindness before the age of 45. Because of the complexity of the eye, its associated nerves, and those parts of the brain taking part in visual perception, defects in a large number of genes may lead to blindness.

Gene therapy has been applied to a rare genetic form of blindness known as Leber congenital amaurosis (LCA), in which there is a mutation in one of a number of genes responsible for the development or function of light receptors in the retina. LCA is inherited as an autosomal recessive disorder; in people with two defective copies of a gene, the light-sensing cells die. Children with LCA begin losing sight at birth, and most are completely blind by age 40.

One type of LCA is caused by a defect in the *RPE65* gene, which encodes an enzyme that helps convert vitamin A into rhodopsin, a pigment that absorbs light. Without the enzyme, rhodopsin is not produced and the photoreceptor cells atrophy with the passage of time. In 1998, researchers discovered that some Swedish Briard dogs also have a defect in *RPE65* and become blind. Research on these dogs led to a better understanding of how defects in *RPE65* cause blindness and suggested a possible treatment—gene therapy. In 2001, researchers treated three young dogs with gene therapy. The dogs' vision improved enough that they could avoid objects and navigate a maze.

In 2007, researchers in Pennsylvania and London used gene therapy to treat four young adults who had LCA. Each patient was injected with a genetically modified virus carrying a functional copy of *RPE65*. The results were dramatic: all the patients showed significant improvement in visual perception. Some who had formerly been able to detect only hand motions were able to read several lines on an eye chart. One patient who had not been able to navigate an obstacle course was able to make his way through it. All four patients were still legally blind, but the results indicated that gene therapy can be an effective treatment for LCA. Researchers have now used gene therapy to treat a number of additional people with LCA; the results have been particularly dramatic with younger patients, some of whose vision improved enough that they could play sports.

This chapter introduces some of the techniques used in genetic engineering. We begin by considering molecular genetic technology and some of its effects. We then examine a number of methods used to isolate, study, alter, and recombine DNA sequences and place them back into cells. Finally, we explore some of the applications of molecular genetic analysis.

19.1 Techniques of Molecular Genetics Have Revolutionized Biology

In 1973, a group of scientists produced the first organisms with recombinant DNA molecules. Stanley Cohen at Stanford University and Herbert Boyer at the University of California School of Medicine at San Francisco and their colleagues inserted a piece of DNA from one plasmid into another, creating an entirely new, recombinant DNA molecule. They then introduced the recombinant plasmid into *E. coli* cells. These experiments ushered in one of the most momentous revolutions in the history of science.

Recombinant DNA technology is a set of molecular techniques for locating, isolating, altering, and studying DNA segments. The term *recombinant* is used because frequently the goal is to combine DNA from two distinct sources. Genes from two different bacteria might be joined, for example, or a human gene might be inserted into a viral chromosome. Commonly called **genetic engineering**, recombinant DNA technology now encompasses many molecular techniques that can be used to analyze, alter, and recombine virtually any DNA sequences from any number of sources.

The Molecular Genetics Revolution

The techniques of recombinant DNA technology are just part of a vast array of molecular methods that are now available. These molecular techniques have drastically altered the way in which genes are studied. Previously, information about the structure and organization of genes was gained by examining their phenotypic effects, but molecular genetic analysis allows the nucleotide sequences themselves to be read. This analysis has provided new information about the structure and function of genes and has altered many fundamental concepts of genetics. Our detailed understanding of

genetic processes such as replication, transcription, translation, RNA processing, and gene regulation has been obtained through the use of molecular genetic techniques. These techniques are used in many other fields as well, including biochemistry, microbiology, developmental biology, neurobiology, evolution, and ecology.

Recombinant DNA technology and other molecular techniques are also being used to create a number of commercial products, including drugs, hormones, enzymes, and crops (**Figure 19.1**). A complete industry—**biotechnology**—has grown up around the use of these techniques to develop new products. In medicine, molecular genetics is being used to probe the nature of cancer, diagnose genetic and infectious diseases, produce drugs, and treat hereditary disorders.

CONCEPTS

Molecular genetics and recombinant DNA technology are used to locate, analyze, alter, study, and recombine DNA sequences. These techniques are used to probe the structure and function of genes, address questions in many areas of biology, create commercial products, and diagnose and treat diseases.

Working at the Molecular Level

The manipulation of genes at the molecular level presents a serious challenge, often requiring strategies that may not, at first, seem obvious. The basic problem is that genes are minute and every cell contains thousands of them. Individual nucleotides cannot be seen, and no physical features mark the beginning or the end of a gene.

Let's consider a typical situation faced by a molecular geneticist. Suppose we want to use bacteria to produce large quantities of a human protein. The first and most formidable problem is to find the gene that encodes the desired protein. A haploid human genome consists of 3.2 billion base pairs of DNA. Let's assume that the gene that we want to isolate is 3000 bp long. Our target gene occupies only one-millionth of the genome, so searching for our gene in the huge expanse of genomic DNA is more difficult than looking for the proverbial needle in a haystack. But, even if we are able to locate the gene, how do we separate it from the rest of the DNA?

19.1 Recombinant DNA technology has been used to create genetically modified crops. Genetically engineered corn now constitutes 88% of all corn grown in the United States. [Chris Knapton/Photo Researchers.]

If we succeed in locating and isolating the desired gene, we would next need to insert it into a bacterial cell. Linear fragments of DNA are quickly degraded by bacteria so the gene must be inserted in a stable form. It must also be able to successfully replicate or it will not be passed on when the cell divides. If we succeed in transferring our gene to bacteria in a stable form, we must still ensure that the gene is properly transcribed and translated.

Finally, the methods used to isolate and transfer genes are inefficient and, of a million cells that are subjected to these procedures, only *one* cell might successfully take up and express the human gene. So we must search through many bacterial cells to find the one containing the recombinant DNA. We are back to the problem of the needle in a haystack.

Although these problems might seem insurmountable, molecular techniques have been developed to overcome them. Human genes are routinely transferred to bacterial cells, where the genes are expressed.

CONCEPTS

Molecular genetic analyses require special methods because individual genes make up a tiny fraction of the cellular DNA and they cannot be seen.

19.2 Molecular Techniques Are Used to Isolate, Recombine, and Amplify Genes

A first step in the molecular analysis of a DNA segment or gene is to isolate it from the remainder of the DNA and to make many copies of it so that further analyses can be carried out. The isolation and amplification of DNA frequently requires that it be recombined with other DNA molecules. In the sections that follow, we will examine some of the molecular techniques that are used to isolate, recombine, and amplify DNA segments.

Cutting and Joining DNA Fragments

A key event in the development of molecular genetic methods was the discovery in the late 1960s of **restriction enzymes** (also called **restriction endonucleases**) that recognize and make double-stranded cuts in DNA at specific nucleotide sequences. These enzymes are produced naturally by bacteria and are used in defense against viruses. A bacterium protects its own DNA from a restriction enzyme by modifying the recognition sequence, usually by adding methyl groups to its DNA.

Several different types of restriction enzymes have been isolated from bacteria. Type II restriction enzymes recognize specific sequences and cut the DNA at defined sites within or near the recognition sequence. Virtually all molecular genetics work is done with type II restriction enzymes; discussions of restriction enzymes throughout this book refer to type II enzymes.

More than 800 different restriction enzymes that recognize and cut DNA at more than 100 different sequences have been isolated from bacteria. Many of these enzymes are commercially available; examples of some commonly used restriction enzymes are given in **Table 19.1**. The name of each restriction enzyme begins with an abbreviation that signifies its bacterial origin.

The sequences recognized by restriction enzymes are usually from 4 to 8 bp long; most enzymes recognize a sequence of 4 or 6 bp. Most recognition sequences are palindromic—sequences that read the same ($5'$ to $3'$) on the two complementary DNA strands.

Some of the enzymes make staggered cuts in the DNA. For example, *Hin*dIII recognizes the following sequence:

$$5'-AAGCTT-3'$$
$$3'-TTCGAA-5'$$

*Hin*dIII cuts the sugar–phosphate backbone of each strand at the point indicated by the arrow, generating fragments with short, single-stranded overhanging ends:

$$5'-A\qquad AGCTT-3'$$
$$3'-TTCGA\qquad A-5'$$

Such ends are called **cohesive ends** or sticky ends because they are complementary to each other and can spontaneously pair to connect the fragments. Thus, DNA fragments can be "glued" together: any two fragments cleaved by the same enzyme will have complementary ends and will

Table 19.1	Characteristics of some common type II restriction enzymes used in recombinant DNA technology		
Enzyme	**Microorganism from Which Enzyme Is Produced**	**Recognition Sequence**	**Type of Fragment End Produced**
BamHI	Bacillus amyloliquefaciens	↓ 5′–GGATCC–3′ 3′–CCTAGG–3′ ↑	Cohesive
CofI	Clostridium formicoaceticum	↓ 5′–GCGC–3′ 3′–CGCG–5′ ↑	Cohesive
EcoRI	Escherichia coli	↓ 5′–GAATTC–3′ 3′–CTTAAG–5′ ↑	Cohesive
EcoRII	Escherichia coli	↓ 5′–CCAGG–3′ 3′–GGTCC–5′ ↑	Cohesive
HaeIII	Haemophilus aegyptius	↓ 5′–GGCC–3′ 3′–CCGG–5′ ↑	Blunt
HindIII	Haemophilus influenzae	↓ 5′–AAGCTT–3′ 3′–TTCGAA–5′ ↑	Cohesive
PvuII	Proteus vulgaris	↓ 5′–CAGCTG–3′ 3′–GTCGAC–5′ ↑	Blunt

Note: The first three letters of the abbreviation for each restriction enzyme refer to the bacterial species from which the enzyme was isolated (e.g., *Eco* refers to *E. coli*). A fourth letter may refer to the strain of bacteria from which the enzyme was isolated (the "R" in *Eco*RI indicates that this enzyme was isolated from the RY13 strain of *E. coli*). Roman numerals that follow the letters identify different enzymes from the same species.

pair (**Figure 19.2**). When their cohesive ends have paired, two DNA fragments can be joined together permanently by DNA ligase, which seals nicks between the sugar–phosphate groups of the fragments.

Not all restriction enzymes produce staggered cuts and sticky ends. *Pvu*II cuts in the middle of its recognition site and the cuts on the two strands are directly opposite one another, producing blunt-ended fragments:

↓
5′–CAGCTG–3′
3′–GTCGAC–5′
↑

↓

5′–CAG   CTG–3′
3′–GTC   GAC–5′

Fragments that have blunt ends are usually joined together in other ways.

The sequences recognized by a restriction enzyme are located randomly within the genome. Accordingly, there is a relation between the length of the recognition sequence and the number of times that it is present in a genome: there will be fewer longer recognition sequences than shorter recognition sequences because the probability of the occurrence of a particular sequence consisting of, say, six specific bases is less than the probability of the occurrence of a particular sequence of four specific bases. Consequently, restriction enzymes that recognize longer sequences will cut a given piece of DNA into fewer and longer fragments than will restriction enzymes that recognize shorter sequences.

Restriction enzymes are used whenever DNA fragments must be cut or joined. In a typical restriction reaction, a concentrated solution of purified DNA is placed in a small

(a)

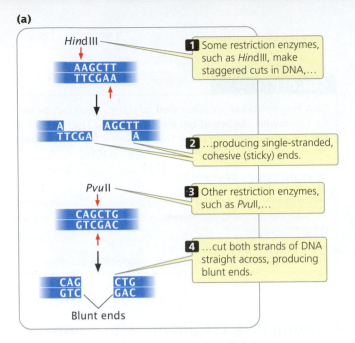

*Hind*III

1 Some restriction enzymes, such as *Hind*III, make staggered cuts in DNA,…

AAGCTT
TTCGAA

A   AGCTT
TTCGA   A

2 …producing single-stranded, cohesive (sticky) ends.

*Pvu*II

3 Other restriction enzymes, such as *Pvu*II,…

CAGCTG
GTCGAC

4 …cut both strands of DNA straight across, producing blunt ends.

CAG   CTG
GTC   GAC

Blunt ends

(b)

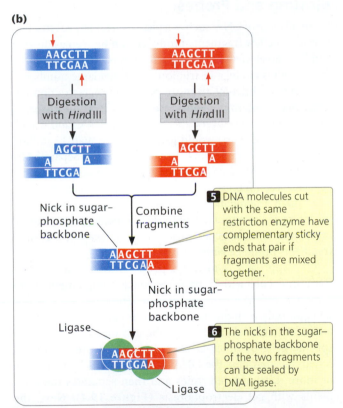

AAGCTT
TTCGAA

AAGCTT
TTCGAA

Digestion with *Hind*III

Digestion with *Hind*III

AGCTT
A
TTCGA

AGCTT
A   A
TTCGA

Nick in sugar–phosphate backbone

Combine fragments

5 DNA molecules cut with the same restriction enzyme have complementary sticky ends that pair if fragments are mixed together.

A AGCTT
TTCGA A

Nick in sugar–phosphate backbone

Ligase

AAGCTT
TTCGAA

6 The nicks in the sugar–phosphate backbone of the two fragments can be sealed by DNA ligase.

Ligase

19.2 Restriction enzymes make double-stranded cuts in DNA, producing cohesive, or sticky, ends.

tube with a buffer solution and a small amount of restriction enzyme. The reaction mixture is then heated at the optimal temperature for the enzyme, usually 37°C. Within a few hours, the enzyme cuts the appropriate restriction sites in all the DNA molecules, producing a set of DNA fragments.

In recent years, geneticists have designed more complex enzymes, termed **engineered nucleases**, which are capable of making double-stranded cuts to the DNA at any predetermined DNA sequence. Engineered nucleases consist of part of a restriction enzyme that cleaves the DNA, coupled with another protein that recognizes and binds to

a specific DNA sequence; the particular sequence to which the protein binds is determined by the protein's amino acid sequence. By altering the amino acid sequence of the binding protein, the engineered nuclease can be custom designed to bind to and cut any particular DNA sequence. Engineered nucleases include zinc finger nucleases (ZFNs), which use a DNA-binding protein called a zinc finger, and transcription activator-like effector nucleases (TALENs), which use a protein that normally binds to sequences in promoters. ▶ TRY PROBLEM 29

CONCEPTS

Type II restriction enzymes cut DNA at specific base sequences that are palindromic. Some restriction enzymes make staggered cuts, producing DNA fragments with cohesive ends; others cut both strands straight across, producing blunt-ended fragments. There are fewer long recognition sequences in DNA than short recognition sequences.

✓ CONCEPT CHECK 1

Where do restriction enzymes come from?

Viewing DNA Fragments

After the completion of a restriction reaction, a number of questions arise. Did the restriction enzyme cut the DNA? Into how many fragments was the DNA cut? What are the sizes of the resulting fragments? Gel electrophoresis provides us with a means of answering these questions.

Electrophoresis is a standard biochemical technique for separating molecules on the basis of their size and electrical charge. There are a number of different types of electrophoresis; to separate DNA molecules, **gel electrophoresis** is used. A porous gel is often made from agarose (a polysaccharide isolated from seaweed), which is melted in a buffer solution and poured into a plastic mold. As it cools, the agarose solidifies, making a gel that looks something like stiff gelatin.

Small wells are made at one end of the gel to hold solutions of DNA fragments (**Figure 19.3a**), and an electrical current is passed through the gel. Because the phosphate group of each DNA nucleotide carries a negative charge, the DNA fragments migrate toward the positive end of the gel. In this migration, the porous gel acts as a sieve, separating the DNA fragments by size. Small DNA fragments migrate more rapidly than do large ones and, with the passage of time, the fragments separate on the basis of their size. Typically, DNA fragments of known length (size standards) are placed in another well. By comparing the migration distance of the unknown fragments with the distance traveled by the size standards, a researcher can determine the approximate size of the unknown fragments (**Figure 19.3b**).

The DNA fragments are still too small to see, so the problem of visualizing the DNA needs to be addressed. Visualization can be accomplished in several ways. The simplest procedure

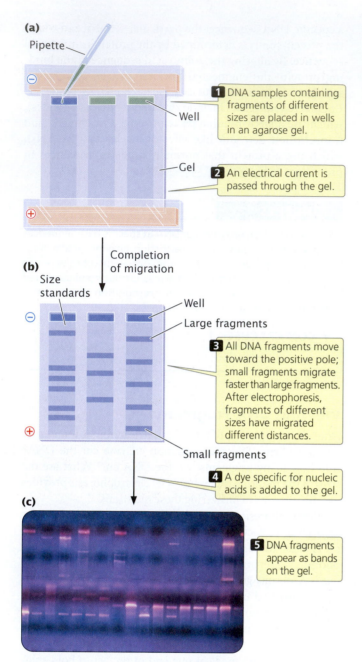

(a)

Pipette

1 DNA samples containing fragments of different sizes are placed in wells in an agarose gel.

Well

Gel

2 An electrical current is passed through the gel.

Completion of migration

(b)

Size standards

Well

Large fragments

3 All DNA fragments move toward the positive pole; small fragments migrate faster than large fragments. After electrophoresis, fragments of different sizes have migrated different distances.

Small fragments

4 A dye specific for nucleic acids is added to the gel.

(c)

5 DNA fragments appear as bands on the gel.

19.3 Gel electrophoresis can be used to separate DNA molecules on the basis of their size and electrical charge.
[Photograph: Klaus Guldbrandsen/Science Source.]

CONCEPTS

DNA fragments can be separated, and their sizes can be determined with the use of gel electrophoresis. The fragments can be viewed by using a dye that is specific for nucleic acids or by labeling the fragments with a chemical tag.

✓ CONCEPT CHECK 2

DNA fragments that are 500 bp, 1000 bp, and 2000 bp in length are separated by gel electrophoresis. Which fragment will migrate farthest in the gel?

a. 2000-bp fragment   c. 500-bp fragment

b. 1000-bp fragment   d. All will migrate equal distances.

Locating DNA Fragments with Southern Blotting and Probes

If a small piece of DNA, such as a plasmid, is cut by a restriction enzyme, the few fragments produced can be seen as distinct bands on an electrophoretic gel. In contrast, if genomic DNA from a cell is cut by a restriction enzyme, a large number of fragments of different sizes are produced. A restriction enzyme that recognizes a four-base sequence would theoretically cut about once every 256 bp. The human genome, with 3.2 billion base pairs, would generate more than 12 million fragments when cut by this restriction enzyme. Separated by electrophoresis and stained, this large set of fragments would appear as a continuous smear on the gel because of the presence of so many fragments of differing size. Usually, researchers are interested in only a few of these fragments, perhaps those carrying a specific gene. How can they locate the desired fragments in such a large pool of DNA?

One approach is to use a **probe**, which is a DNA or RNA molecule with a base sequence complementary to a sequence in the gene of interest. The bases on a probe will pair only with the bases on a complementary sequence and, if suitably labeled, the probe can be used to locate a specific gene or other DNA sequence. To use a probe, a researcher first cuts the DNA into fragments by using one or more restriction enzymes and then separates the fragments with gel electrophoresis (**Figure 19.4**). Next, the separated fragments must be denatured and transferred to a permanent solid medium (such as nitrocellulose or nylon membrane). **Southern blotting** (named after Edwin M. Southern) is one technique used to transfer the denatured, single-stranded fragments from a gel to a permanent solid medium.

After the single-stranded DNA fragments have been transferred, the membrane is placed in a hybridization solution of a labeled probe (see Figure 19.4). The probe will bind to any DNA fragments on the membrane that bear complementary sequences. Often, a probe binds to only a part of the DNA fragment, and so the DNA fragment may contain sequences not found in the probe. The membrane is

---

is to stain the gel with a dye specific for nucleic acids, such as ethidium bromide, which wedges itself tightly (intercalates) between the bases of DNA and fluoresces when exposed to UV light, producing brilliant bands on the gel (**Figure 19.3c**). Alternatively, DNA fragments can be visualized by adding a label to the DNA before it is placed in the gel. For example, chemical labels can be detected by adding antibodies or other substances that carry a dye and will attach to the relevant DNA, which can then be visualized directly. ▶ **TRY PROBLEM 30**

1 DNA fragments are separated by gel electrophoresis.

2 The gel is soaked in an alkali solution to denature the double-stranded DNA and then placed on a platform in a dish containing buffer.

Weight

Blotting paper

3 A membrane is positioned on top of the gel.

Nitrocellulose or other membrane

Gel

Blotting paper

Platform — Alkali solution

4 Buffer drawn up into the top layer of blotting paper passes through the gel, carrying DNA onto the membrane.

Membrane

DNA

5 DNA on the membrane is fixed,…

6 …placed in a hybridization bottle with solution that contains a labeled probe, and gently rotated.

Radioactive probe

7 The probe binds to complementary DNA fragments on the membrane,…

Size standards

8 …and a biochemical method detects fragments with probe attached.

19.4 Southern blotting and hybridization with probes can locate a few specific fragments in a large pool of DNA.

then washed to remove any unbound probe; a biochemical method reveals the presence of the bound probe.

RNA can be transferred from a gel to a solid support by a related procedure called **Northern blotting** (not named after anyone but capitalized to match Southern). The hybridization

of a probe can reveal the size of a particular mRNA molecule, its relative abundance, or the tissues in which the mRNA is transcribed. **Western blotting** is the transfer of protein from a gel to a membrane. Here, the probe is usually an antibody, used to determine the size of a particular protein and the pattern of the protein's expression.

CONCEPTS

Labeled probes, which are sequences of RNA or DNA that are complementary to the sequence of interest, can be used to locate individual genes or DNA sequences. Southern blotting can be used to transfer DNA fragments from a gel to a membrane such as nitrocellulose.

✓ CONCEPT CHECK 3

How do Northern and Western blotting differ from Southern blotting?

Cloning Genes

Many recombinant DNA methods require numerous copies of a specific DNA fragment. One way to amplify a specific piece of DNA is to place the fragment in a bacterial cell and allow the cell to replicate the DNA. This procedure is termed **gene cloning** because identical copies (clones) of the original piece of DNA are produced.

A **cloning vector** is a stable, replicating DNA molecule to which a foreign DNA fragment can be attached for introduction into a cell. An effective cloning vector has three important characteristics (**Figure 19.5**): (1) an origin of replication, which ensures that the vector is replicated within the cell; (2) selectable markers, which enable any cells containing the vector to be selected or identified; and (3) one

1 First, a cloning vector must contain an origin of replication recognized in the host cell so that it is replicated along with the DNA that it carries.

Unique restriction-enzyme cleavage sites

*Bam*HI   *Sal*I

*Pst*I

*Eco*RI

ori (origin of replication)

*Hind*III

Selectable marker

3 Third, a cloning vector needs a single cleavage site for each of one or more restriction enzymes used.

2 Second, it should carry selectable markers—traits that enable cells containing the vector to be selected or identified.

19.5 An idealized cloning vector has an origin of replication, one or more selectable markers, and sites for one or more restriction enzymes.

19.6 The pUC19 plasmid is a typical cloning vector. It contains a cluster of unique restriction sites, an origin of replication, and two selectable markers—an ampicillin-resistance gene and a *lacZ* gene.

or more unique restriction sites into which a DNA fragment can be inserted. The restriction sites used for cloning must be unique; if a vector is cut at multiple recognition sites, several pieces of DNA are generated and getting these back together in the correct order is possible but extremely difficult.

PLASMID VECTORS Plasmids, circular DNA molecules that exist naturally in bacteria (see Chapter 9), are commonly used vectors for cloning DNA fragments in bacteria. They contain origins of replication and are therefore able to replicate independently of the bacterial chromosome. The plasmids typically used in cloning have been constructed from the larger, naturally occurring bacterial plasmids and have multiple restriction sites, an origin of replication site, and selectable markers (**Figure 19.6**).

The easiest method for inserting a gene into a plasmid vector is to cut the foreign DNA (containing the gene) and the plasmid with the same restriction enzyme (**Figure 19.7**). If the restriction enzyme makes staggered cuts in the DNA, complementary sticky ends are produced on the foreign DNA and the plasmid. The DNA and plasmid are then mixed together; some of the foreign DNA fragments will pair with the cut ends of the plasmid. DNA ligase is used to seal the nicks in the sugar–phosphate backbone, creating a recombinant plasmid that contains the foreign DNA fragment. You can learn more about plasmid cloning by viewing **Animation 19.1**.

Sometimes restriction sites are not available at a site where the DNA needs to be cut. In that case, a restriction site can be created with the use of **linkers**, which are small, synthetic DNA fragments that contain one or more restriction sites. Linkers can be attached to the ends of any piece of DNA and are then cut by a restriction enzyme, generating sticky ends that are complementary to sticky ends on the plasmid.

TRANSFORMATION When a gene has been placed inside a plasmid, the plasmid must be introduced into bacterial cells. This task is usually accomplished by *transformation*, in which bacterial cells take up DNA from the external environment (see Chapter 9). Some types of cells undergo transformation naturally; others must be treated chemically or physically before they will undergo transformation. Inside the cell, the plasmids replicate and multiply, and the cells themselves multiply.

SCREENING CELLS FOR RECOMBINANT PLASMIDS Cells bearing recombinant plasmids can be detected with the use of the selectable markers on the plasmid. Common

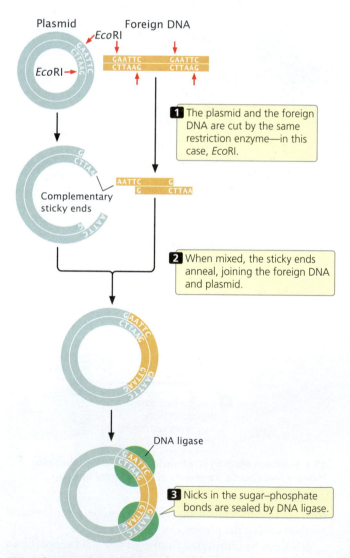

19.7 A foreign DNA fragment can be inserted into a plasmid with the use of restriction enzymes.

selectable markers are genes that confer resistance to an antibiotic: any cell that contains such a plasmid will be able to live in the presence of the antibiotic, which normally kills bacterial cells.

A common way to screen cells for the presence of a recombinant plasmid is to use a plasmid that contains a fragment of the *lacZ* gene—a small part of the front end of the gene (**Figure 19.8**). This partial *lacZ* gene contains a series of unique restriction sites into which a piece of DNA can be inserted and cloned. The bacteria that will do the

1 Foreign DNA is inserted into the partial *lacZ* gene.

2 Bacteria that are *lacZ⁻* are transformed by the plasmid.

3 Bacteria with an original (nonrecombinant) plasmid produce β-galactosidase, which cleaves X-gal and makes the colonies blue.

4 Bacteria with a recombinant plasmid do not synthesize β-galactosidase. Their colonies remain white.

5 Bacteria without a plasmid will not grow.

Conclusion: A white colony consists of bacteria carrying a recombinant plasmid.

19.8 The *lacZ* gene can be used to screen bacteria containing recombinant plasmids. A special plasmid carries a fragment of the *lacZ* gene and an ampicillin-resistance gene. [Photograph courtesy of Edvotek.]

work of cloning have special features that make the presence of the recombinant plasmid evident. The bacteria that are to be transformed by the plasmid are *lacZ⁻*, missing the front end of the *lacZ* gene but containing its back end. Without the plasmid the bacteria are unable to synthesize β-galactosidase. If no foreign DNA has been inserted into the partial *lacZ* gene of the plasmid, the front end of the gene (provided by the plasmid) and back end of the gene (provided by the bacteria) work together within the bacterial cell to produce β-galactosidase. If foreign DNA is successfully inserted into the restriction site, it disrupts the front end of the *lacZ* gene, and β-galactosidase is *not* produced. The plasmid also usually contains a selectable marker, which may be a gene that confers resistance to an antibiotic such as ampicillin.

Bacteria that are *lacZ⁻* are transformed by the plasmids and plated on medium that contains ampicillin. Only cells that have been successfully transformed and contain a plasmid with the ampicillin-resistance gene will survive and grow. Some of these cells will contain an intact plasmid, whereas others possess a recombinant plasmid. The medium also contains the chemical X-gal, which produces a blue substance when cleaved. Bacterial cells with an intact original plasmid—without an inserted fragment—have a functional *lacZ* gene (the front end of the gene provided by the plasmid and the back end provided by the bacteria). These bacteria can synthesize β-galactosidase, which cleaves X-gal and turns the bacteria blue. Bacterial cells with a recombinant plasmid, however, have the front end of the β-galactosidase gene disrupted by the inserted DNA; they do not synthesize β-galactosidase and remain white. (In these experiments, the bacterium's own β-galactosidase gene has been inactivated, and so only bacteria with the plasmid turn blue.) Thus, the color of the bacterial colony allows quick determination of whether a recombinant or intact plasmid is present in the cell. After cells with the recombinant plasmid have been identified, they can be grown in large numbers, replicating the inserted fragment of DNA.

Plasmids make ideal cloning vectors but can hold only DNA less than about 15 kb in size. When large DNA fragments are inserted into a plasmid vector, the plasmid becomes unstable.

OTHER GENE VECTORS   A number of other vectors have been developed for cloning larger pieces of DNA in bacteria. For example, bacteriophage λ, which infects *E. coli*, can be used to clone up to 23,000 bp of foreign DNA; it transfers DNA into bacterial cells with high efficiency. **Cosmids** are plasmids that are packaged into empty viral protein coats and transferred to bacteria by viral infection. They can carry more than twice as much foreign DNA as can a phage vector. **Bacterial artificial chromosomes** (BACs) are vectors originally constructed from the F plasmid (a special plasmid that controls mating and the transfer of genetic material in some bacteria; see Chapter 9) and can hold very large fragments of DNA that can be as long

Table 19.2   Comparison of plasmids, phage lambda vectors, cosmids, and bacterial artificial chromosomes

Cloning Vector	Size of DNA That Can Be Cloned	Method of Propagation	Introduction to Bacteria
Plasmid	As large as 15 kb	Plasmid replication	Transformation
Phage lambda	As large as 23 kb	Phage reproduction	Phage infection
Cosmid	As large as 44 kb	Plasmid reproduction	Phage infection
Bacterial artificial chromosome	As large as 300 kb	Plasmid reproduction	Electroporation

Note: 1 kb = 1000 bp. Electroporation consists of electrical pulses that increase the permeability of a membrane.

as 300,000 bp. **Table 19.2** compares the properties of plasmids, phage λ vectors, cosmids, and BACs.

Sometimes the goal in gene cloning is not just to replicate the gene, but also to produce the protein that it encodes. To ensure transcription and translation, a foreign gene is usually inserted into an **expression vector**, which, in addition to the usual origin of replication, restriction sites, and selectable markers, contains sequences required for transcription and translation in bacterial cells (**Figure 19.9**).

Although manipulating genes in bacteria is simple and efficient, the goal may be to transfer a gene into eukaryotic cells. For example, it might be desirable to transfer a gene conferring herbicide resistance into a crop plant, or to transfer a gene for clotting factor into a person suffering from hemophilia. Many eukaryotic proteins are modified after translation (e.g., sugar groups may be added). Such modifications are essential for proper function, but bacteria do not have the capacity to carry out the modification; thus a functional protein can be produced only in a eukaryotic cell.

A number of cloning vectors have been developed that allow the insertion of genes into eukaryotic cells. Special plasmids have been developed for cloning in yeast, and retroviral vectors have been developed for cloning in mammals.

A **yeast artificial chromosome** (YAC) is a DNA molecule that has a yeast origin of replication, a pair of telomeres, and a centromere. These features ensure that YACs are stable, replicate, and segregate in the same way as yeast chromosomes. YACs are particularly useful because they can carry DNA fragments as large as 600 kb, and some special YACs can carry inserts of more than 1000 kb. YACs have been modified so that they can be used in eukaryotic organisms other than yeast.

The soil bacterium *Agrobacterium tumefaciens*, which invades plants through wounds and induces crown galls (tumors), has been used to transfer genes to plants. This bacterium contains a large plasmid called the **Ti plasmid**, part of which is transferred to a plant cell when *A. tumefaciens* infects a plant. In the plant, part of the Ti plasmid DNA integrates into one of the plant chromosomes, where it is transcribed and translated to produce several enzymes that help support the bacterium (**Figure 19.10a**).

Geneticists have engineered a vector that contains the flanking sequences required to transfer DNA (called TL and TR, see Figure 19.10a), a selectable marker, and restriction sites into which foreign DNA can be inserted (**Figure 19.10b**). When placed in *A. tumefaciens* with a helper Ti plasmid (which carries sequences allowing the bacterium to infect the plant cell, which the artificial vector lacks), this vector will transfer the foreign DNA that it carries into a plant cell, where it will integrate into a plant chromosome. This vector has been used to transfer genes that confer economically significant attributes such as resistances to herbicides, plant viruses, and insect pests. ▶TRY PROBLEM 31

1 Expression vectors contain operon sequences that allow inserted DNA to be transcribed and translated.

2 They also include sequences that regulate—turn on or turn off—the desired gene.

Transcription-initiation sequences

Restriction sites

Transcription-termination sequence

Operator (*O*)

Bacterial promoter (*P*) sequences

Ribosome-binding site

Gene encoding repressor that binds *O* and regulates *P*

ori

Selectable genetic marker (e.g., antibiotic resistance)

19.9 To ensure transcription and translation, a foreign gene may be inserted into an expression vector—in this example, an *E. coli* expression vector.

CONCEPTS

DNA fragments can be inserted into cloning vectors, stable pieces of DNA that will replicate within a cell. A cloning vector must have an origin of replication, one or more unique restriction sites, and selectable markers. An expression vector contains sequences that allow a cloned gene to be transcribed and translated. Special cloning vectors have been developed for introducing genes into eukaryotic cells.

✓ CONCEPT CHECK 4

How is a gene inserted into a plasmid cloning vector?

(a)

Plant chromosome

TL

Bacterial chromosome   Ti plasmid

TR

Agrobacterium tumefaciens

Plant cell

Infection

1 In natural gene transfer, *Agrobacterium* invades the plant at a wound.

2 Part of the Ti plasmid is transferred to the plant cell,...

3 ...where it integrates into one of the plant chromosomes.

(b)

Plasmid vector

Restriction site

Sequences required for transfer

ori

Selectable marker

Foreign DNA

Foreign DNA

3 The helper Ti plasmid is required for infection.

Helper Ti plasmid

Bacterial chromosome   Plasmid vector

Infection

1 Foreign DNA is inserted into a plasmid vector...

2 ...and transferred to *Agrobacterium tumefaciens* with the Ti plasmid.

4 The plasmid vector, along with any foreign DNA that it carries,...

5 ...is then transferred to a plant cell, where it integrates into a plant chromosome.

19.10 The Ti plasmid can be used to transfer genes into plants. Flanking sequences TL and TR are required for the transfer of the DNA segment from bacteria to the plant cell.

Application: The Genetic Engineering of Plants with Pesticides

An early and economically important use of molecular techniques for transferring genes to other organisms is the genetic engineering of plants that produce their own insecticides. The bacterium *Bacillus thuringiensis* naturally produces a protein, known as the Bt toxin, that is lethal to many insects. The Bt toxin is particularly attractive as an insecticide because it is specific to particular insects, breaks down quickly in the environment, and is not toxic to humans and other animals.

In 1987, Mark Vaeck and his colleagues at Plant Genetic Systems in Belgium genetically engineered tobacco plants to express the Bt toxin. Other researchers had isolated the gene that encodes the Bt toxin from *B. thuringiensis* and cloned it into a plasmid of *E. coli* so that the gene sequences could be easily manipulated (**Figure 19.11**). Vaeck and his colleagues used restriction enzymes to cleave the *Bt* gene sequences from these plasmids into fragments of different sizes. The

Bt gene fragments were attached to a gene (*neo*) that confers resistance to the antibiotic kanamycin, which is toxic to plants and other eukaryotes, providing a selectable marker for cells containing the *Bt* gene. These synthetic sequences (called genetic constructs) contained fragments of the *Bt* gene of different lengths linked to a *neo* gene (see Figure 19.11). The constructs were inserted into an expression vector that contained a promoter, to ensure that the introduced sequences would be transcribed. The expression vector also contained poly(A) consensus sequences, which ensured that the mRNA produced from the fused *Bt* and *neo* genes would be processed properly and translated in the plant cells.

Agrobacterium tumefaciens bacteria were then transformed with the expression vectors. After the vectors were inside the *Agrobacterium*, sequences on the vector recombined with sequences on a Ti plasmid, transferring the gene constructs to the Ti plasmid. Small discs of leaves from a tobacco plant were infected with the genetically engineered *Agrobacterium*, which transferred the Ti plasmids into the

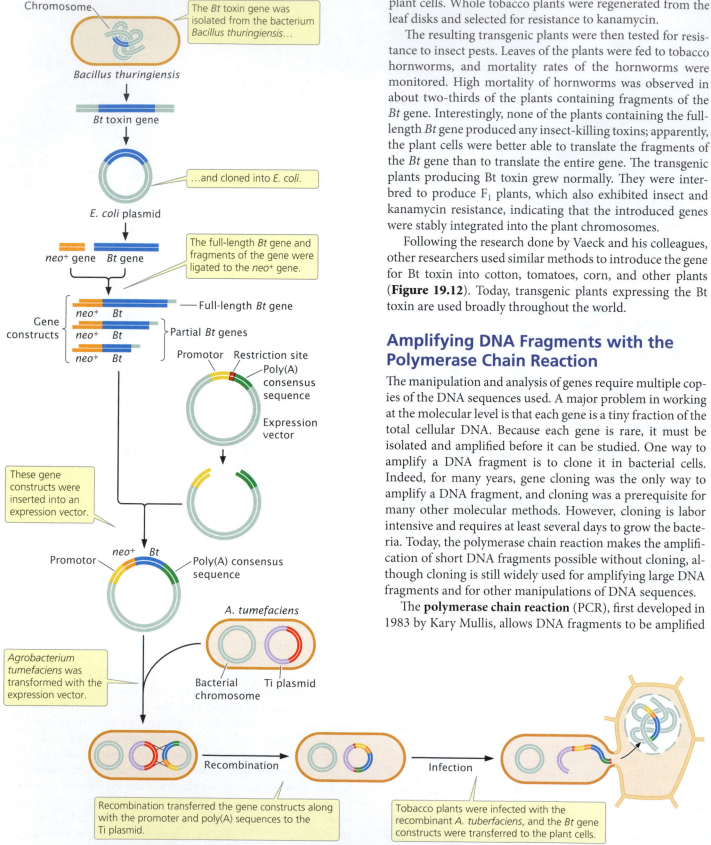

Chromosome

The *Bt* toxin gene was isolated from the bacterium *Bacillus thuringiensis*…

Bacillus thuringiensis

Bt toxin gene

…and cloned into *E. coli.*

E. coli plasmid

neo+ gene   *Bt* gene

The full-length *Bt* gene and fragments of the gene were ligated to the *neo+* gene.

Gene constructs

neo+   *Bt* — Full-length *Bt* gene

neo+   *Bt*

neo+   *Bt* } Partial *Bt* genes

Promotor   Restriction site

Poly(A) consensus sequence

Expression vector

These gene constructs were inserted into an expression vector.

Promotor

neo+   *Bt*   Poly(A) consensus sequence

A. tumefaciens

Agrobacterium tumefaciens was transformed with the expression vector.

Bacterial chromosome   Ti plasmid

Recombination

Recombination transferred the gene constructs along with the promoter and poly(A) sequences to the Ti plasmid.

Infection

Tobacco plants were infected with the recombinant *A. tuberfaciens*, and the *Bt* gene constructs were transferred to the plant cells.

19.11 The *Bt* toxin gene, which encodes an insecticide, was isolated from bacteria and transferred to tobacco plants.

plant cells. Whole tobacco plants were regenerated from the leaf disks and selected for resistance to kanamycin.

The resulting transgenic plants were then tested for resistance to insect pests. Leaves of the plants were fed to tobacco hornworms, and mortality rates of the hornworms were monitored. High mortality of hornworms was observed in about two-thirds of the plants containing fragments of the *Bt* gene. Interestingly, none of the plants containing the full-length *Bt* gene produced any insect-killing toxins; apparently, the plant cells were better able to translate the fragments of the *Bt* gene than to translate the entire gene. The transgenic plants producing Bt toxin grew normally. They were interbred to produce F_1 plants, which also exhibited insect and kanamycin resistance, indicating that the introduced genes were stably integrated into the plant chromosomes.

Following the research done by Vaeck and his colleagues, other researchers used similar methods to introduce the gene for Bt toxin into cotton, tomatoes, corn, and other plants (**Figure 19.12**). Today, transgenic plants expressing the Bt toxin are used broadly throughout the world.

Amplifying DNA Fragments with the Polymerase Chain Reaction

The manipulation and analysis of genes require multiple copies of the DNA sequences used. A major problem in working at the molecular level is that each gene is a tiny fraction of the total cellular DNA. Because each gene is rare, it must be isolated and amplified before it can be studied. One way to amplify a DNA fragment is to clone it in bacterial cells. Indeed, for many years, gene cloning was the only way to amplify a DNA fragment, and cloning was a prerequisite for many other molecular methods. However, cloning is labor intensive and requires at least several days to grow the bacteria. Today, the polymerase chain reaction makes the amplification of short DNA fragments possible without cloning, although cloning is still widely used for amplifying large DNA fragments and for other manipulations of DNA sequences.

The **polymerase chain reaction** (PCR), first developed in 1983 by Kary Mullis, allows DNA fragments to be amplified used in PCR to replicate the templates are short fragments of DNA, typically from 17 to 25 nucleotides long, that are complementary to known sequences on the template.

THE POLYMERASE CHAIN REACTION   To carry out PCR, researchers begin with a solution that includes the target DNA, DNA polymerase, all four deoxyribonucleoside triphosphates (dNTPs—the substrates for DNA polymerase), primers, and magnesium ions and other salts that are necessary for the reaction to proceed. A typical polymerase chain reaction includes three steps (**Figure 19.13**). In step 1, a starting solution of DNA is heated to between 90° and 100°C to break the hydrogen bonds between the strands and thus produce the necessary single-stranded templates. The reaction mixture is held at this temperature for only a minute or two. In step 2, the DNA solution is cooled quickly to between 30° and 65°C and held at this temperature for a minute or less. During this short interval, the DNA strands will not have a chance to reanneal, but the primers will be able to attach to the template strands. In step 3, the solution is heated for a minute or less to 72°C, the temperature at which DNA polymerase can synthesize new DNA strands. At the end of the cycle, two new double-stranded DNA molecules are produced for each original molecule of target DNA.

The whole cycle is then repeated. With each cycle, the amount of target DNA doubles; so the amount of target DNA increases geometrically. One molecule of DNA increases to more than 1000 molecules in 10 PCR cycles, to more than 1 million molecules in 20 cycles, and to more than 1 billion molecules in 30 cycles. Each cycle is completed within a few minutes, so a large amplification of DNA can be achieved within a few hours. To see how the polymerase

19.12 Transgenic crops are broadly cultivated throughout the world. Shown here is Bt corn and non-genetically modified corn growing in alternate strips, which helps prevent the development of resistance of pests to the genetically engineered Bt toxin. [John L. Obermeyer, Purdue Extension Entomology.]

a billionfold within just a few hours. It can be used with extremely small amounts of original DNA, even a single molecule. The polymerase chain reaction has revolutionized molecular biology and is now one of the most widely used molecular techniques. The basis of PCR is replication catalyzed by a DNA polymerase. Replication in this case has two essential requirements: (1) a DNA template from which a new DNA strand can be copied and (2) a pair of primers with a 3′-OH group to which new nucleotides can be added.

Because a DNA molecule consists of two nucleotide strands, each of which can serve as a template, the amount of DNA doubles with each replication event. The primers

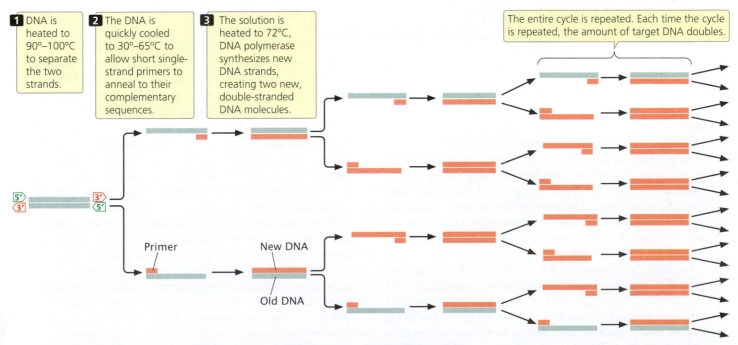

19.13 The polymerase chain reaction is used to amplify even very small samples of DNA.

chain reaction quickly increases the number of copies of a DNA fragment, view **Animation 19.2**.

Two key innovations facilitated the use of PCR in the laboratory. The first was the discovery of a DNA polymerase that is stable at the high temperatures used in step 1 of PCR. The DNA polymerase from *E. coli* that was originally used in PCR denatures at 90°C. For this reason, fresh enzyme had to be added to the reaction mixture in *each* cycle, slowing the process considerably. This obstacle was overcome when DNA polymerase was isolated from the bacterium *Thermus aquaticus*, which lives in the boiling springs of Yellowstone National Park. This enzyme, dubbed *Taq* **polymerase**, is remarkably stable at high temperatures and is not denatured in the strand-separation step of PCR; it can be added to the reaction mixture at the beginning of the PCR process and will continue to function through many cycles.

The second key innovation was the development of automated thermal cyclers—machines that bring about the rapid temperature changes necessary for the different steps of PCR. Originally, tubes containing reaction mixtures were moved by hand among water baths set at the different temperatures required for the three steps of each cycle. In automated thermal cyclers, the reaction tubes are placed in a metal block that changes temperature rapidly according to a computer program.

In addition to amplifying DNA, PCR can be used to amplify sequences corresponding to RNA. This amplification is accomplished by first converting RNA into complementary DNA (cDNA) with the use of the enzyme reverse transcriptase. The cDNA can then be amplified by the usual PCR. This technique is referred to as **reverse-transcription PCR**.

LIMITATIONS OF PCR   The polymerase chain reaction is now often used in place of gene cloning, but it has several limitations. First, the use of PCR requires prior knowledge of at least part of the sequence of the target DNA to allow the construction of the primers. Second, the capacity of PCR to amplify extremely small amounts of DNA makes contamination a significant problem. Minute amounts of DNA from the skin of laboratory workers or small particles in the air can enter a reaction tube and be amplified along with the target DNA. Careful laboratory technique and the use of controls are necessary to circumvent this problem.

A third limitation of PCR is accuracy. Unlike other DNA polymerases, *Taq* polymerase does not have the capacity to proofread (see pp. 339–340 in Chapter 12) and, under standard PCR conditions, it incorporates an incorrect nucleotide about once every 20,000 bp. New heat-stable DNA polymerases with proofreading capacity have been isolated, giving more accurate PCR results.

A fourth limitation of PCR is that the size of the fragments that can be amplified by standard *Taq* polymerase is usually less than 2000 bp. By using a combination of *Taq* polymerase and a DNA polymerase with proofreading capacity, and by modifying the reaction conditions, investigators have been successful in extending PCR amplification to larger fragments, but even these larger fragments are limited in length to 50,000 bp or smaller.

APPLICATIONS OF PCR   In spite of its limitations, PCR has become one of the most widely used tools within molecular biology. The primers used in PCR are specific for known DNA sequences, and so PCR can be used to detect the presence of a particular DNA sequence in a sample. For example, PCR is often used to detect the presence of viruses in blood samples by adding primers complementary to known viral DNA sequences to the reaction. If viral DNA is present, the primers will attach to it, and PCR will amplify a DNA fragment of known length. The presence of a DNA fragment of the proper length on a gel indicates the presence of viral DNA in the blood sample. Modern diagnostic tests for infection with HIV, the causative agent of AIDS (see Chapter 9), use this type of PCR amplification of HIV sequences.

Another common application of PCR is to identify genetic variation in natural populations. PCR can be used to amplify specific segments of DNA, which are then analyzed with other molecular tools that identify differences in nucleotide sequences. Also, primers specific to a particular DNA variant can be used to determine whether that variant is present in the genome of an individual organism.

PCR has also allowed the isolation of DNA from ancient sources, such as DNA from Neanderthals (discussed in the introduction to Chapter 10). It is commonly used to amplify small amounts of DNA from crime scenes and identify the source of a DNA sample through detection of microsatellite variation (see section on DNA fingerprinting later in this chapter).

PCR can be used to introduce new sequences into a fragment of DNA. Although the 3′ end of the primers used in PCR must be complementary to the template, there can be flexibility in the sequences at the 5′ end of the primer. Primers can be designed so that they contain new restriction sites or other desirable sequences at their 5′ ends. These sequences will then be copied by the reaction and thus added to the ends of the DNA amplified by PCR.

A modification of PCR, known as **real-time PCR**, can be used to quantitatively determine the amount of starting nucleic acid. In this procedure, normal PCR is used to amplify a specific DNA fragment, and a sensitive instrument is used to accurately determine the amount of DNA that is present in solution after each PCR cycle. A probe that fluoresces and is specific to the DNA sequence of interest is frequently used in the reaction, and so only the DNA of interest is measured. The technique is called real-time PCR because the amount of DNA amplified is measured as the reaction proceeds.

Often, real-time PCR is combined with reverse-transcription PCR to measure the amount of mRNA in a sample, allowing biologists to determine the level of gene expression in different cells and under different conditions. For example, researchers interested in how gene expression changes in response to the administration of a drug often use real-time PCR to quantitatively measure the amount of mRNA produced by specific genes in cells exposed to the drug and compare it with the amount of mRNA produced by the genes in control cells with no drug exposure.

CONCEPTS

The polymerase chain reaction is an enzymatic, in vitro method for rapidly amplifying DNA. In this process, DNA is heated to separate the two strands, short primers attach to the target DNA, and DNA polymerase synthesizes new DNA strands from the primers. Each cycle of PCR doubles the amount of DNA. PCR has a number of important applications in molecular biology.

✓ **CONCEPT CHECK 5**

Why is the use of a heat-stable DNA polymerase important to the success of PCR?

19.3 Molecular Techniques Can Be Used to Find Genes of Interest

To analyze a gene or to transfer it to another organism, the gene must first be located and isolated. For instance, if we wanted to transfer a human gene for growth hormone to bacteria, we must first find the human gene that encodes growth hormone and separate it from the 3.2 billion bp of human DNA. So far, in our consideration of gene cloning we've glossed over the problem of finding the DNA sequence to be cloned; the solution to this problem has been purposely delayed until now because, paradoxically, researchers must often clone a gene to find it.

This approach—to clone first and search later—is called "shotgun cloning," because it is like hunting with a shotgun: the pellets spray widely in the general direction of the quarry, with a good chance that one or more of the pellets will hit the intended target. In shotgun cloning, a researcher first clones a large number of DNA fragments, knowing that one or more contains the DNA of interest, and then searches for the fragment of interest among the clones.

Gene Libraries

A collection of clones containing all the DNA fragments from one source is called a **DNA library**. For example, we might isolate genomic DNA from human cells, break it into fragments, insert the fragments into vectors, and clone them in bacterial cells. The set of bacterial colonies or phages containing these fragments is a human **genomic library**, containing all the DNA sequences found in the human genome.

CREATING A GENOMIC LIBRARY To create a genomic library, cells are collected and disrupted, which causes them to release their DNA and other cellular contents into an aqueous solution, and the DNA is extracted from the solution. After the DNA has been isolated, it is cut into fragments by a restriction enzyme for a limited amount of time (a partial digestion) so that only *some* of the restriction sites in each DNA molecule are cut. Because the

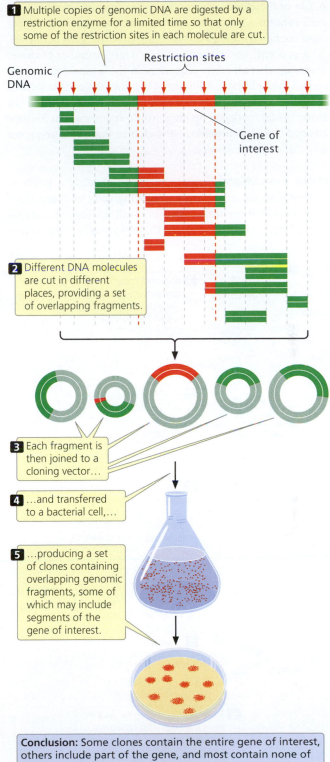

1 Multiple copies of genomic DNA are digested by a restriction enzyme for a limited time so that only some of the restriction sites in each molecule are cut.

Restriction sites

Genomic DNA

Gene of interest

2 Different DNA molecules are cut in different places, providing a set of overlapping fragments.

3 Each fragment is then joined to a cloning vector…

4 …and transferred to a bacterial cell,…

5 …producing a set of clones containing overlapping genomic fragments, some of which may include segments of the gene of interest.

Conclusion: Some clones contain the entire gene of interest, others include part of the gene, and most contain none of the gene of interest.

19.14 A genomic library contains all of the DNA sequences found in an organism's genome.

cutting of sites is random, different DNA molecules will be cut in different places and a set of overlapping fragments will be produced (**Figure 19.14**). The fragments are then joined to vectors, which can be transferred to bacteria.

A few of the clones contain the entire gene of interest (if the gene is not too large) and a few contain parts of the gene, but most contain fragments that have no part of the gene of interest.

A genomic library must contain a large number of clones to ensure that all DNA sequences in the genome are represented in the library. A library of the human genome formed by using cosmids, each carrying a random DNA fragment from 35,000 to 44,000 bp long, would require about 350,000 cosmid clones to provide a 99% chance that every sequence is included in the library.

cDNA LIBRARIES   An alternative to creating a genomic library is to create a library consisting only of those DNA sequences that are transcribed into mRNA (called a **cDNA library** because all the DNA in this library is *complementary*

to mRNA). Much of eukaryotic DNA consists of repetitive (and other DNA) sequences that are not transcribed into mRNA (see pp. 308–309 in Chapter 11), and these sequences are not represented in a cDNA library.

A cDNA library has two additional advantages. First, it is enriched with fragments from actively transcribed genes. Second, introns do not interrupt the cloned sequences; introns would pose a problem when the goal is to produce a eukaryotic protein in bacteria, because most bacteria have no means of removing the introns.

The disadvantage of a cDNA library is that it contains only sequences that are present in mature mRNA. Sometimes, researchers are interested in sequences that are not transcribed, such as those in promoters and enhancers, which are important for transcription but are not themselves transcribed.

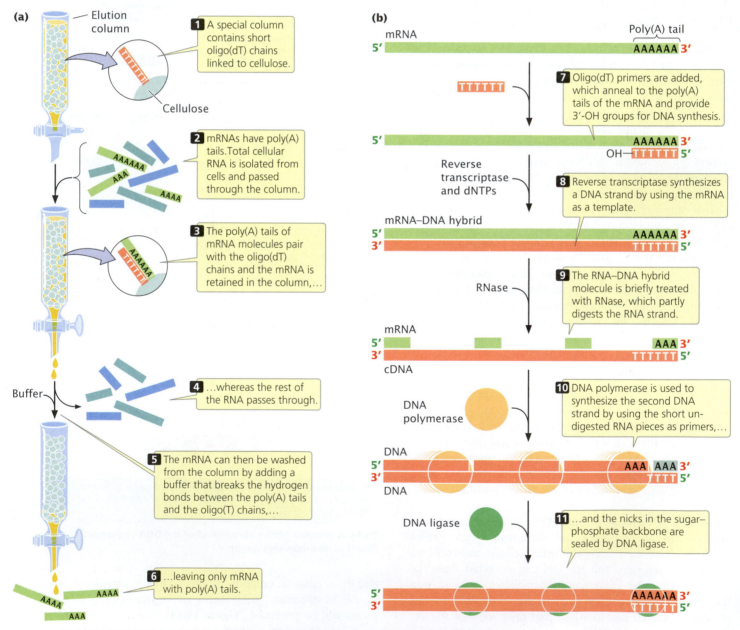

19.15 A cDNA library contains only those DNA sequences that are transcribed into mRNA.

These sequences are not present in a cDNA library. Furthermore, a cDNA library contains only those gene sequences expressed in the tissue from which the RNA was isolated, and the frequency of a particular DNA sequence in a cDNA library depends on the abundance of the corresponding mRNA in the given tissue. So, if a particular gene is not expressed, or is expressed only at low frequency in a particular tissue, it may be absent in a cDNA library prepared from that tissue. In contrast, almost all genes are present at the same frequency in a genomic DNA library.

To create a cDNA library, messenger RNA must first be separated from other types of cellular RNA (tRNA, rRNA, snRNA, etc.). Most eukaryotic mRNAs possess a poly(A) tail, which provides a convenient hook for separating eukaryotic mRNA from the other types. Total cellular RNA is isolated from cells and poured through a column packed with short fragments of DNA consisting entirely of thymine nucleotides—that is, oligo(dT) chains (**Figure 19.15a**). As the RNA moves through the column, the poly(A) tails of mRNA molecules pair with the oligo(dT) chains and are retained in the column, whereas the rest of the RNA passes through it. The mRNA can then be washed from the column by the addition of a buffer that breaks the hydrogen bonds between poly(A) tails and oligo(dT) chains.

The mRNA molecules are then copied into cDNA. Reverse transcriptase, an enzyme isolated from retroviruses (see Chapter 9), synthesizes single-stranded complementary DNA from the RNA template (**Figure 19.15b**). The resulting RNA–DNA hybrid molecule is finally converted into a double-stranded cDNA molecule by DNA polymerase.

▶ **TRY PROBLEM 35**

CONCEPTS

One method of finding a gene is to create and screen a DNA library. A genomic library is created by cutting genomic DNA into overlapping fragments and cloning each fragment in a separate bacterial cell. A cDNA library is created from mRNA that is converted into cDNA and cloned in bacteria.

SCREENING DNA LIBRARIES Creating a genomic or cDNA library is relatively easy compared with screening the library to find clones that contain the gene of interest. The screening procedure used depends on what is known about the gene.

The first step in screening is to plate the clones of the library. If a plasmid or cosmid vector was used to construct the library, the cells are diluted and plated so that each bacterium grows into a distinct colony. If a phage vector was used, the phages are allowed to infect a lawn of bacteria on a petri plate. Each plaque or bacterial colony contains a single, cloned DNA fragment that must be screened for the gene of interest.

A common way to screen libraries is with probes. To use a probe, replicas of the plated colonies or plaques in the library must first be made. **Figure 19.16** illustrates this procedure for a cosmid library.

How is a probe obtained when the gene has not yet been isolated? One option is to use a similar gene from another organism as the probe. For example, if we wanted to screen a human genomic library for the growth-hormone gene and the gene had already been isolated from rats, we could use a purified rat-gene sequence as the probe to find the human gene for growth hormone. Successful hybridization does not require perfect complementarity between the probe and the target sequence, so a related sequence can often be used as a probe.

Alternatively, synthetic probes can be created if the protein produced by the gene has been isolated and its amino acid sequence has been determined. With the use of the genetic code and the amino acid sequence of the protein, possible nucleotide sequences of a small region of the gene can be deduced. Although only one sequence in the gene encodes a particular protein, the presence of synonymous codons means that the same protein could be produced by several different nucleotide sequences, and it is impossible to know which is correct. To overcome this problem, a mixture of all the possible nucleotide sequences is used as a probe. To minimize the number of sequences

1 A disc of nitrocellulose or other membrane is laid on top of the bacterial colonies.

2 A few cells from each colony adhere to the nitrocellulose filter.

3 The cells are disrupted, and their DNA is denatured and fixed to the filter.

4 A labeled probe hybridizes with any complementary DNA.

5 Excess probe is washed off and the membrane is overlaid with X-ray film,...

6 ...which detects the presence of the probe.

7 Comparison of the membrane with the master plate reveals which bacterial colonies have the DNA of interest.

Bacterial colonies on master plate

Nitrocellulose filter

Replica filter

³²P-labeled probe

Membrane   X-ray film

19.16 Genomic and cDNA libraries can be screened with a probe to find the gene of interest.

required in the mixture, a region of the protein is selected with relatively little degeneracy in its codons. When part of the DNA sequence of the gene has been determined, a set of DNA probes can be synthesized chemically by using an automated machine known as an oligonucleotide synthesizer.

Yet another method of screening a library is to look for the protein product of a gene. This method requires that the DNA library be cloned in an expression vector. The clones can be tested for the presence of the protein by using an antibody that recognizes the protein or by using a chemical test for the protein product. This method depends on the existence of a test for the protein produced by the gene. Gene libraries can also be screened using PCR or by sequencing.

CONCEPTS

A DNA library can be screened for a specific gene with the use of complementary probes that hybridize to the gene. Alternatively, the library can be cloned into an expression vector, and the gene can be located by examining the clones for the protein product of the gene.

✓ CONCEPT CHECK 6

Briefly explain how synthetic probes are created to screen a DNA library when the protein encoded by the gene is known.

In Situ Hybridization

DNA probes can be used to determine the chromosomal location of a gene in a process called **in situ hybridization**. The name is derived from the fact that DNA (or RNA) is visualized while it is in the cell (in situ). This technique requires that the cells be fixed and the chromosomes be spread on a microscope slide and denatured. A labeled probe is then applied to the slide, just as it can be applied to a gel. Many probes carry attached fluorescent dyes that can be seen directly with the microscope (**Figure 19.17a**). Several probes with different colored dyes can be used simultaneously to investigate different sequences or chromosomes. Fluorescence in situ hybridization (FISH) has been widely used to identify the chromosomal location of human genes.

In situ hybridization can also be used to determine the tissue distribution of specific mRNA molecules, serving as a source of insight into how gene expression differs among cell types (**Figure 19.17b**). A labeled DNA probe complementary to a specific mRNA molecule is added to tissue, and the location of the probe is determined with the use of radioactive or fluorescent tags. Determining where a gene is expressed often helps define its function. For example, finding that a gene is highly expressed only in brain tissue might suggest that the gene has a role in neural function.

Positional Cloning

For many genes with important functions, no associated protein product is yet known. The biochemical bases of many human genetic diseases, for example, are still unknown. How can these genes be isolated? One approach is to first determine the general location of the gene on the chromosome by using recombination frequencies derived from crosses or pedigrees (see Chapter 7). After the chromosomal region where the gene is found has been identified, genes in this region can be cloned and identified. Then other techniques can be used to identify which of the "candidate" genes might be the one that causes the disease. This approach—to isolate genes on the basis of their position on a gene map—is called **positional cloning**.

In the first step of positional cloning, geneticists use mapping studies (see Chapter 7) to establish linkage between molecular markers and a phenotype of interest, such as a human disease or a desirable physical trait in a plant or animal. Demonstration of linkage between the phenotype and one or more molecular markers would provide information about which chromosome carries the locus that codes for the phenotype and its general location on that chromosome.

19.17 With in situ hybridization, DNA probes are used to determine the cellular or chromosomal location of a gene or its product. (a) A probe with green fluorescence is specific to chromosome 7, revealing a deletion on one copy of chromosome 7. (b) In situ hybridization is used to detect the presence of mRNA from the *tailless* gene in a *Drosophila* embryo. [Part a: Addenbrookes Hospital/Science Source. Part b: Courtesy of L. Tsuda.]

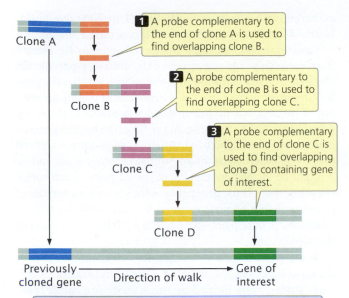

1 A probe complementary to the end of clone A is used to find overlapping clone B.

Clone A

Clone B

2 A probe complementary to the end of clone B is used to find overlapping clone C.

Clone C

3 A probe complementary to the end of clone C is used to find overlapping clone D containing gene of interest.

Clone D

Previously cloned gene   Direction of walk   Gene of interest

Conclusion: By making probes complementary to areas of overlap between cloned fragments in a genomic library, we can connect a gene of interest to a previously mapped, linked gene.

19.18 In chromosome walking, neighboring genes are used to locate a gene of interest.

identified, they can be evaluated to determine which is most likely to be the gene of interest. The expression pattern of the gene—where and when it is transcribed—can often provide clues about its function. For example, genes for neurological disease would likely be expressed in the brain. Geneticists often look in the coding region of the gene for mutations among people with the disease. More will be said about determining the function of genes in sections that follow and in Chapter 20.

CONCEPTS

Positional cloning allows researchers to isolate a gene without having knowledge of its biochemical basis. Linkage studies are used to map the locus producing a phenotype of interest to a particular chromosome region. Chromosome walking and jumping can be used to progress from molecular markers to clones containing sequences that cover the chromosome region. Candidate genes within the region are then evaluated to determine if they encode the phenotype of interest.

✓ **CONCEPT CHECK 7**

How are candidate genes that are identified by positional cloning evaluated to determine whether they encode the phenotype of interest?

Application: Isolating the Gene for Cystic Fibrosis

The first gene responsible for a human genetic disease that was isolated entirely by positional cloning was the gene for cystic fibrosis (CF). Cystic fibrosis is an autosomal recessive disorder characterized by chronic lung infections, insufficient production of pancreatic enzymes that are necessary for digestion, and increased salt concentration in sweat (**Figure 19.19**). It is among the most common genetic diseases in Caucasians,

The next step is to more precisely locate the locus by using additional molecular markers clustered in the chromosomal region where the locus resides. After the gene has been placed on a chromosome map, clones that cover the region can be isolated from a genomic library. With the use of a technique called **chromosome walking** (**Figure 19.18**), it is possible to progress from neighboring genes to linked clones, one of which might contain the gene of interest. The basis of chromosome walking is the fact that a genomic library consists of a set of *overlapping* DNA fragments (see Figure 19.14). We start with a cloned gene marker that is close to the new gene of interest so that the "walk" will be as short as possible. One end of the clone of a neighboring marker (clone A in Figure 19.18) is used to make a complementary probe. This probe is used to screen the genomic library to find a second clone (clone B) that overlaps with the first and extends in the direction of the gene of interest. This second clone is isolated and purified and a probe is prepared from its end. The second probe is used to screen the library for a third clone (clone C) that overlaps with the second. In this way, one can systematically "walk" toward the gene of interest, one clone at a time.

A related technique called chromosome jumping allows one to move from more distantly linked markers to clones that contain a sequence of interest. After clones that cover the delineated region have been obtained by chromosome walking or jumping, all genes located within the region are identified. Genes can be distinguished from other sequences by the presence of characteristic features, such as consensus sequences in the promoter, and a start codon and a stop codon within the same reading frame. After "candidate" genes have been

19.19 Cystic fibrosis was the first genetic disease for which the causative gene was isolated entirely by positional cloning. New treatments have greatly helped patients with cystic fibrosis. This girl wears a "smart vest," which shakes her chest to help break up mucous in her lungs, and inhales from a nebulizer that contains enzymes and salt water, which also help break up mucous. [Jeffrey Sauger Photography.]

Linkage analysis was carried out on families with cystic fibrosis…

Linkage analysis

Chromosome 7

q31–32

…and an association was found between the inheritance of molecular markers from chromosome 7 and CF.

MET   D7S8

Further linkage analysis

Linkage studies with additional markers indicated that the CF locus is close to markers D7S340 and D7S122.

MET   D7S340   D7S122   D7S8

Clones isolated

Overlapping clones

Chromosome jump

Clones from the region were isolated by chromosome walking and jumping.

Linked marker gene

Genes

Analysis of DNA sequences within clones revealed four candidate genes.

Four candidate genes

Further studies of linkage and expression

Additional studies eliminated three of the candidate genes.

Potential CF gene

Sequencing of the gene

DNA sequencing revealed the presence of a 3-bp deletion in the gene of the CF patient.

Person with CF   3-bp deletion

TATCAT   TGGTGTTTCCTA
TATCATCTTTGGTGTTTCCTA

Healthy person

Sequence from coding region of gene

19.20 The gene for cystic fibrosis was located by positional cloning.

occurring with a frequency of about 1 in 2000 live births. Nearly 5% of all Caucasians are carriers of the CF mutation.

Geneticists attempting to isolate the gene for CF faced a formidable task. The symptoms of the disease, especially the elevated salt concentration in sweat, suggested that the gene for CF somehow takes part in the movement of ions into and out of the cell, but no information was available about the protein encoded by the gene. At the time, the human genome had not yet been sequenced. Analyses of pedigrees showed that CF is inherited as an autosomal recessive trait, and so it might be located on any one of the 22 pairs of autosomal chromosomes. Thus, geneticists were seeking an unknown gene—probably encompassing a few thousand or tens of thousands of base pairs—among the 3.2 billion base pairs of the human genome.

Researchers began by looking for associations between the inheritance of CF and that of other traits (**Figure 19.20**). Early studies were limited by the scarcity of genetic traits that varied and could be used for gene-mapping studies, but in the 1980s, advances in molecular biology provided a large number of molecular markers that could be used for linkage analysis (see p. 190 in Chapter 7). Geneticists collected pedigrees of families in which several members had CF. They compared the inheritance of CF with that of molecular markers among the members of these families, looking for evidence of linkage. The gene for CF was found to be closely linked to two markers, MET and D7S8, located on the long arm of chromosome 7. MET and D7S8 are separated by about 1.5 map units (see Chapter 7). In the human genome, each map unit roughly corresponds to 1 million base pairs; so the gene for CF is located somewhere within a stretch of 1.5 million base pairs of DNA, a huge expanse of sequence.

Further linkage studies with additional markers were carried out to more precisely delineate where in the 1.5-million-base-pair region the CF gene lies. Researchers selected additional molecular markers from the region surrounding MET and D7S8 and performed linkage studies between these new markers and CF (see Figure 19.20). These studies identified two additional markers, D7S122 and D7S340, which are closely linked to CF. Furthermore, they showed that the order of the four markers is MET-D7S340-D7S122-D7S8 and that the CF gene lies very close to D7S122 and D7S340. This finding narrowed the region in which the gene for CF lies to about 500,000 bp.

At this stage, geneticists began isolating clones of sequences from the delineated region. Starting from the molecular markers, they used a combination of chromosome walking and chromosome jumping to identify clones from human genomic libraries that completely covered the region of interest (see Figure 19.20). An examination of sequences within these clones revealed the presence of four genes in the region encompassed by the linked markers (see Figure 19.20). Additional studies were then carried out to better characterize these candidate genes. Three of the candidate genes were eventually eliminated, either because linkage

The mRNA is strongly expressed in the pancreas, a tissue known to be affected by CF.

Lanes

19.21 A candidate for the cystic fibrosis gene is expressed in pancreatic, respiratory, and sweat-gland tissues—tissues that are affected by the disease. Shown is a Northern blot of mRNA produced by the candidate gene in different tissues. These data provided evidence that the candidate gene is in fact the gene that causes cystic fibrosis. [From J.R. Riordan et al., *Science* 245:1066–1073, 1989. Reprinted with permission from AAAS.]

studies suggested that they were not closely linked with the inheritance of CF or because analysis of the sequences or their expression patterns suggested they were not the gene for CF.

Hybridization studies were carried out with the one remaining gene to determine where it was expressed. Messenger RNA was isolated from different organ tissues and probed with sequences from the candidate gene. The gene showed high levels of expression in the pancreas, lungs, and sweat glands (**Figure 19.21**), tissues known to be affected by CF.

Copies of the candidate gene from a healthy person and from a person with CF were then sequenced, and the sequence data were examined for differences that might be a mutation causing CF. The findings revealed that the person with CF had a 3-bp deletion in the coding region of the gene, while the healthy person did not have this deletion. The deletion resulted in the absence of a phenylalanine amino acid from the protein encoded by the candidate gene. Then, for a large number of patients with CF, geneticists used PCR to amplify the region of the gene where the deletion was found; 68% of the CF patients had this deletion. Subsequent studies demonstrated that the remaining CF patients possessed mutations at other locations within the candidate gene, thus proving that the candidate gene was indeed the locus that caused CF.

Researchers eventually demonstrated that the gene for CF encodes a membrane protein that controls the movement of chloride into and out of cells and is known today as the *cystic fibrosis transmembrane conductance regulator* (*CFTR*) gene. Patients with CF have two mutated forms of *CFTR*, which cause the chloride channels to remain closed. Chloride ions build up in the cell, leading to the formation of thick mucus and the symptoms of the disease.

19.4 DNA Sequences Can Be Determined and Analyzed

In addition to cloning and amplifying DNA, molecular techniques are used to analyze DNA molecules through the study and determination of their sequences.

Restriction Fragment Length Polymorphisms

A significant contribution of molecular genetics has been to provide numerous genetic markers that can be used in gene mapping. We learned how these markers are essential to the success of positional cloning. One group of such markers comprises **restriction fragment length polymorphisms** (RFLPs), which are variations (polymorphisms) in the patterns of fragments produced when DNA molecules are cut with the same restriction enzyme (**Figure 19.22**). These differences are

1 DNA sequence had two *Hae*III restriction sites.

2 A mutation creates a polymorphism. Some copies have both restriction sites and others only one.

3 When DNA from two persons is digested by *Hae*III,...

4 ... two different patterns appear.

5 Bob's DNA is cut into three bands because his chromosomes possess both restriction sites.

6 Joe's DNA is cut into only two bands because his chromosomes possess only one of the two sites.

7 This example assumes that Bob is homozygous for the A pattern and Joe is homozygous for the B pattern. A person heterozygous for the RFLP would display bands seen in both the A and the B patterns.

19.22 Restriction fragment length polymorphisms are genetic markers that can be used in mapping.

Every child who inherits the *C* allele has the disease. Thus, the RFLP is closely linked to the disease gene.

19.23 Restriction fragment length polymorphisms can be used to detect linkage. The letter *H* refers to the Huntington disease allele; *h* refers to the normal allele. *A*, *B*, and *C* refer to RFLP alleles. Every child who inherits the *C* RFLP allele has Huntington disease, indicating that the genes that encode the RFLP and Huntington disease are closely linked.

inherited and can be used in mapping, similar to the way in which allelic differences are used to map conventional genes.

To illustrate mapping with RFLPs, consider Huntington disease, an autosomal dominant disorder. In the family shown in **Figure 19.23**, the father is heterozygous both for Huntington disease (*Hh*) and for a restriction pattern (*AC*). From the father, each child inherits either a Huntington-disease allele (*H*) or a normal allele (*h*); any child inheriting the *H* allele develops the disease, because it is an autosomal dominant disorder. The child also inherits one of the two RFLP alleles from the father, either *A* or *C*, which produces the corresponding RFLP pattern. In Figure 19.23, every child who inherits the *C* pattern from the father also inherits Huntington disease (and therefore the *H* allele), because the locus for the RFLP is closely linked to the locus for the disease-causing gene. If we had observed no correspondence between the inheritance of the RFLP pattern and the inheritance of the disease, the lack of correspondence would indicate that the genes encoding the RFLP and Huntington disease are assorting independently and are not linked. In recent years, the availability of inexpensive DNA sequencing technology (see below) has decreased the use of RFLPs in gene mapping and genetic diagnosis. Today, the diagnosis of Huntington disease is routinely carried out by amplifying a portion of the Huntington disease gene with PCR. ▶ TRY PROBLEM 41

CONCEPTS

Restriction fragment length polymorphisms are variations in the pattern of fragments produced by restriction enzymes, which reveal variations in DNA sequences. They are used extensively in gene mapping.

DNA Sequencing

A powerful molecular method for analyzing DNA is a technique known as **DNA sequencing**, which quickly determines the sequence of bases in DNA. Sequencing allows the genetic

information in DNA to be read, providing an enormous amount of information about gene structure and function. In the mid-1970s, Frederick Sanger and his colleagues created the dideoxy-sequencing method based on the elongation of DNA by DNA polymerase; at about the same time, Allan Maxam and Walter Gilbert developed a second method based on chemical degradation of DNA. The Sanger method quickly became the standard procedure for sequencing any purified fragment of DNA.

The Sanger, or dideoxy, method of DNA sequencing is based on replication. The fragment to be sequenced is used as a template to make a series of new DNA molecules. In the process, replication is sometimes (but not always) terminated when a specific base is encountered, producing DNA strands of different lengths, each of which ends in the same base.

The method relies on the use of a special substrate for DNA synthesis. Normally, DNA is synthesized from deoxyribonucleoside triphosphates (dNTPs), which have an OH group on the 3′-carbon atom (**Figure 19.24a**). In the Sanger method, a special nucleotide, called a **dideoxyribonucleoside triphosphate** (ddNTP; **Figure 19.24b**), is used as one of the substrates. The ddNTPs are identical to dNTPs, except that they lack a 3′-OH group. In the course of DNA synthesis, ddNTPs are incorporated into a growing DNA strand. However, after a ddNTP has been incorporated into the DNA strand, no more nucleotides can be added, because there is no 3′-OH group to form a phosphodiester bond with an incoming nucleotide. Thus, ddNTPs terminate DNA synthesis.

Although the sequencing of a single DNA molecule is technically possible, most sequencing procedures in use today require a considerable amount of DNA; any DNA

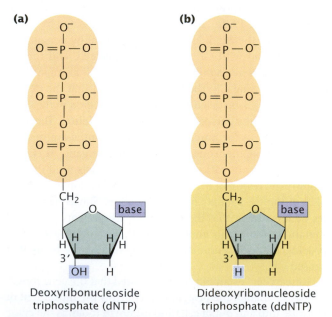

19.24 The dideoxy sequencing reaction requires a special substrate for DNA synthesis. (a) Structure of deoxyribonucleoside triphosphate, the normal substrate for DNA synthesis. (b) Structure of dideoxyribonucleoside triphosphate, which lacks an OH group on the 3′-carbon atom.

1 Each of four reactions contains: single-stranded target DNA to be sequenced,...

Template — **3'** CTAAGCTCGACT **5'**

Primer — **5'** ⬛—OH **3'**

2 ...a primer,...

dCTP dTTP
dATP dGTP
+ DNA polymerase

3 ...all four deoxyribonucleoside triphosphates, DNA polymerase,...

4 ...and one type of dideoxyribonucleoside triphosphate (ddNTP).

+ **ddATP**   + **ddCTP**   + **ddGTP**   + **ddTTP**

5 Nucleotides are added to the 3' end of the primer, with the target DNA being used as a template.

6 When a dideoxynucleotide is incorporated into the growing chain, synthesis terminates because the dideoxynucleotide lacks a 3' OH.

Template CTAAGCTCGACT   CTAAGCTCGACT   CTAAGCTCGACT   CTAAGCTCGACT
         GATTCGAGCTGA   GATTCGAGC   GATTCGAGCTG   GATTCGAGCT
         GATTCGA   GATTC   GATTC   G   GATT
         GA   GATTCG   GAT
                                            G

7 Synthesis terminates at different positions on different strands, which generates a set of DNA fragments of various lengths, each ending in a dideoxynucleotide with the same base.

8 The fragments produced in each reaction are separated by gel electrophoresis.

9 The sequence can be read directly from the bands that appear on the autoradiograph of the gel, starting from the bottom.

10 The sequence obtained is the complement of the original template strand.

A   C   G   T

3' 5'
A   T
G   C
T   A
C   G
G   C
A   T
G   C
A   T
G   C
C   G
T   A
T   A
A   T
G   C
5' 3'

Autoradiogram of electrophoresis gel

Sequence of complementary strand

Sequence of original template strand

19.25 The dideoxy method of DNA sequencing is based on the termination of DNA synthesis.

fragment to be sequenced must first be amplified by PCR or by cloning in bacteria. Copies of the target DNA are isolated and split into four parts (**Figure 19.25**). Each part is placed in a different tube, to which are added:

1. many copies of a primer that is complementary to one end of the target DNA strand;

2. all four types of deoxyribonucleoside triphosphates, the normal precursors of DNA synthesis;

3. a small amount of one of the four types of dideoxyribonucleoside triphosphates, which will terminate DNA synthesis as soon as it is incorporated into any growing chain (each of the four tubes received a different ddNTP); and

4. DNA polymerase.

Either the primer or one of the dNTPs is radioactively or chemically labeled so that newly produced DNA can be detected.

Within each of the four tubes, the DNA polymerase enzyme synthesizes DNA. Let's consider the reaction in one of the four tubes; the one that received ddATP. Within this tube, each of the single strands of target DNA serves as a template for DNA synthesis. The primer pairs to its complementary sequence at one end of each template strand, providing a 3'-OH group for the initiation of DNA synthesis. DNA polymerase elongates a new strand of DNA from this primer. Wherever DNA polymerase encounters a T on the template strand, it uses at random either a dATP or a ddATP to introduce an A in the newly synthesized strand. Because there is more dATP than ddATP in the reaction mixture, dATP is incorporated most often, allowing DNA synthesis to continue. Occasionally, however, ddATP is incorporated into the strand and synthesis terminates. The incorporation of ddA into the new strand occurs randomly at different positions in different copies, producing a set of

1 A single-stranded DNA fragment whose base sequence is to be determined (the template) is isolated.

2 Each of the four ddNTPs is tagged with a different fluorescent dye, and the Sanger sequencing reaction is carried out.

3 The fragments that end in the same base have the same colored dye attached.

4 The products are denatured, and the DNA fragments produced by the four reactions are mixed and loaded into a single well on an electrophoresis gel. The fragments migrate through the gel according to size,…

5 …and the fluorescent dye on the DNA is detected by a laser beam.

6 Each fragment appears as a peak on the computer printout; the color of the peak indicates which base is present.

7 The sequence information is read directly into the computer.

19.26 The dideoxy sequencing method can be automated.

DNA chains of different lengths (12, 7, and 2 nucleotides long in the example illustrated in Figure 19.25), each ending in a nucleotide that contains adenine.

Equivalent reactions take place in the other three tubes, except that synthesis is terminated at nucleotides with a different base in each tube. After the completion of the polymerization reactions, all of the DNA in the tubes is denatured, and the single-strand products of each reaction are separated by gel electrophoresis.

The contents of the four tubes are separated side by side on an acrylamide gel so that DNA strands differing in length by only a single nucleotide can be distinguished. After electrophoresis, the locations, and therefore the sizes, of the DNA strands in the gel are revealed by the use of radioactive or chemical tags.

Reading the DNA sequence is the simplest and shortest part of the procedure. In Figure 19.25, you can see that the band closest to the bottom of the gel is from the tube that contained the ddGTP reaction, which means that the first

nucleotide synthesized had guanine (G). The next band up is from the tube that contained ddATP; so the next nucleotide in the sequence is adenine (A), and so forth. In this way, the sequence is read from the bottom to the top of the gel, with the nucleotides near the bottom corresponding to the 5′ end of the newly synthesized DNA strand and those near the top corresponding to the 3′ end. Keep in mind that the sequence obtained is not that of the target DNA but that of its *complement*. To see dideoxy sequencing in action, view **Animation 19.3**.

▶ **TRY PROBLEM 39**

For many years, DNA sequencing was done largely by hand and was laborious and expensive. Today, sequencing is usually carried out by automated machines that use fluorescent dyes and laser scanners to sequence thousands of base pairs in a few hours (**Figure 19.26**). The dideoxy reaction is also used here, but the ddNTPs used in the reaction are labeled with fluorescent dyes, and a different colored dye is used for each type of dideoxynucleotide. In this case, the four sequencing reactions can take place in the same test tube and can be placed in the same well during electrophoresis. The most recently developed sequencing machines carry out electrophoresis in gel-containing capillary tubes. The different-size fragments produced by the sequencing reaction separate within a tube and migrate past a laser beam and detector. As the fragments pass the laser, their fluorescent dyes are activated and the resulting fluorescence is detected by an optical scanner. Each colored dye emits fluorescence of a characteristic wavelength, which is read by the optical scanner. The information is fed into a computer for interpretation, and the results are printed out as a set of peaks on a graph (see Figure 19.26). Automated sequencing machines may contain 96 or more capillary tubes, allowing from 50,000 to 60,000 bp of sequence to be read in a few hours.

CONCEPTS

The dideoxy sequencing method uses ddNTPs, which terminate DNA synthesis at specific bases.

✓ CONCEPT CHECK 8

In the dideoxy sequencing reaction, what terminates DNA synthesis at a particular base?

a. The absence of a base on the ddNTP halts the DNA polymerase.

b. The ddNTP causes a break in the sugar–phosphate backbone.

c. DNA polymerase will not incorporate a ddNTP into the growing DNA strand.

d. The absence of a 3′-OH group on the ddNTP prevents the addition of another nucleotide.

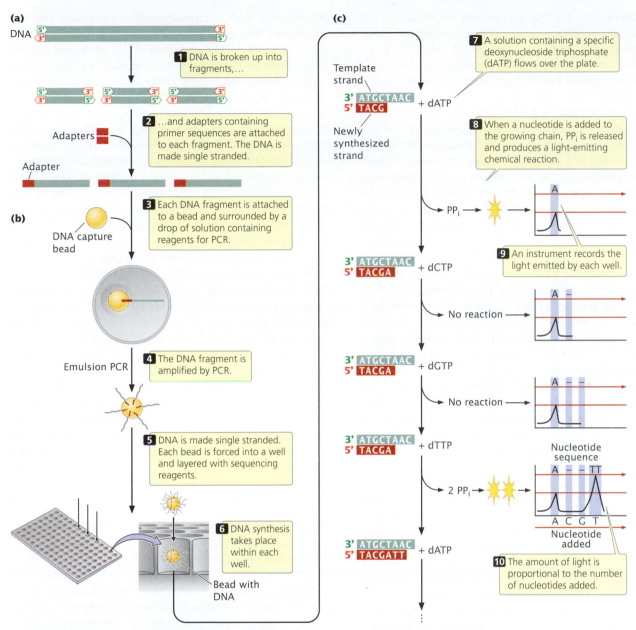

19.27 Next-generation sequencing methods are able to simultaneously determine the sequence of hundreds of thousands or millions of DNA fragments. Pyrosequencing is illustrated here.

Next-Generation Sequencing Technologies

Newer methods, called **next-generation sequencing technologies**, have made sequencing hundreds of times faster and less expensive than the traditional Sanger sequencing method. Most next-generation sequencing technologies do sequencing in parallel, which means that hundreds of thousands or even millions of DNA fragments are simultaneously sequenced.

PYROSEQUENCING One type of next generation sequencing, called pyrosequencing, is based on DNA synthesis:

nucleotides are added one at a time in the order specified by template DNA and the addition of a particular nucleotide is detected with a flash of light, which is generated as the nucleotide is added.

To carry out pyrosequencing, DNA to be sequenced is first fragmented (**Figure 19.27a**). Adaptors, consisting of a short string of nucleotides, are added to each fragment. The adaptor provides a known sequence to prime a PCR reaction. The DNA fragments are then made single stranded. In one version of pyrosequencing, each fragment is then attached to a separate bead and surrounded by a droplet of solution (**Figure 19.27b**). The bead is used to hold the DNA and later deposit it on a plate for the sequencing reaction (see below).

Within the droplet, the fragment is then amplified by PCR and the copies of DNA remain attached to the bead. After amplification by PCR, the beads are mixed with DNA polymerase and are deposited on a plate containing more than a million wells (small holes). Each bead is deposited into a different well.

The sequencing reaction takes place in each well and is based on DNA synthesis. Recall from Chapter 12 that the substrate for DNA synthesis is a deoxynucleoside triphosphate, consisting of a deoxyribose sugar attached to a base and three phosphates. In the process of DNA synthesis, two phosphates (pyrophosphate) are cleaved off, and the resulting nucleotide is attached to the 3′ end of the growing DNA chain. A solution containing one particular type of deoxynucleoside triphosphate—say, deoxyadenosine triphosphate—is passed across the wells (**Figure 19.27c**). If the template within a particular well specifies an adenine nucleotide in the next position of the growing chain, then pyrophosphate is cleaved from the nucleoside triphosphate and the adenine nucleotide is added. A chemical reaction uses the pyrophosphate produced in the reaction to generate a flash of light, which is measured by an optical detector. The amount of light emitted in each well is proportional to the number of nucleotides added: if the template in a well specifies three successive adenine nucleotides (As), then three nucleotides are added and three times more light is emitted than if a single A were added. If the position in the template specifies a base other than adenine, no nucleotide is added, no pyrophosphate is produced, and no light is emitted.

As mentioned, the first solution passed over the plate contains adenosine triphosphate and allows adenine nucleotides to be added to the template. Each well with a template that specifies adenine in the next position will generate a flash of light. Then, a solution with a different type of nucleoside triphosphate—say, deoxyguanosine triphosphate—is passed across the wells. Any fragment that specifies a G in the next position of its growing chain will add a guanine nucleotide and emit a flash of light. The nucleotide triphosphates are passed across the wells in a predetermined order, and the light emitted by each well is measured. In this way, hundreds of thousands or millions of fragments of DNA are sequenced simultaneously on the basis of the order in which nucleotides are added to the 3′ end of the growing chain. Pyrosequencing determines only the sequence of the fragments in each well; it does not, by itself, allow the sequences of these fragments to be reassembled into the sequence of the entire original piece of DNA. The reassembly of sequenced fragments into a continuous stretch of DNA is a general problem in the sequencing of genomes and will be explained in Chapter 20.

ILLUMINA SEQUENCING Several other forms of next-generation sequencing are widely used. Illumina sequencing employs a technology similar to that of the Sanger dideoxy method. Special nucleotides are used that have a fluorescent tag attached, with a different colored tag for each type of nucleotide. Each nucleotide also has a chemical group (a terminator) that, once incorporated into the growing DNA chain, prevents the incorporation of any additional nucleotides. This is similar to termination caused by dideoxynucleotides in Sanger sequencing. However, here the terminator is reversible—it can be chemically removed. To carry out sequencing, the DNA is first fragmented into millions of short overlapping fragments. The fragments are attached to a slide and then amplified, creating clusters of up to 1000 copies of each fragment in close proximity on the slide. The fragments are then denatured and a solution of primers, DNA polymerase, and the special nucleotides are added. The primer attaches to each DNA template and the first nucleotide is incorporated into the newly synthesized strand. The solution is washed away, and the tag on the incorporated nucleotide is excited with a laser, which causes it to fluoresce. As mentioned above, each type of nucleotide (A, T, G, or C) has a different colored fluorescent tag, so the color of the light produced reveals the type of the nucleotide just added. The terminator and the fluorescent tag are then chemically removed and the process is repeated. As the nucleotides are added one at a time, the sequence is read as a series of flashes of colored light from each cluster of DNA. Hundreds of thousands of DNA clusters, each consisting of copies of a different DNA fragment, are sequenced simultaneously, allowing large amounts of DNA to be sequenced in a short time.

Most next-generation sequencing techniques read shorter DNA fragments than the Sanger sequencing reactions do, but because hundreds of thousands or millions of fragments are sequenced simultaneously, these methods are much faster than traditional Sanger sequencing technology.

THIRD-GENERATION SEQUENCING TECHNOLOGY Even more advanced and rapid sequencing methods, typically called third-generation sequencing, are currently under development. For example, nanopore sequencing determines the sequence of individual molecules of DNA. In this method, a single strand of DNA is passed through a tiny hole—a nanopore—in a membrane. As the molecule passes through the nanopore, it disrupts an electrical current in the membrane, and the nature of the disruption is affected by the shape of the molecule passing through the nanopore. Each of the four bases of DNA causes a characteristic disruption, so the sequence of DNA can be read by analyzing the membrane current as the strand passes, one nucleotide at a time, through the nanopore. Hundreds of thousands of nanopores can be created on a single chip, so that many DNA fragments can be read simultaneously. One of the goals of third-generation sequencing technology is to develop a method that can sequence an entire human genome for less than $1000.

CONCEPTS

New next- and third-generation sequencing methods sequence many DNA fragments simultaneously and provide a much faster and less-expensive determination of a DNA base sequence than does the Sanger sequencing method.

Question: How can we identity people based on differences in their DNA?

Methods

Individual 1

DNA samples are collected...

DNA

Microsatellite sequence — 8 repeats of CA

...and subjected to PCR.

Template DNA

Primer

Primer — Template DNA

The length of the DNA fragment produced by PCR depends on the number of copies of the microsatellite sequence.

Individual 2

DNA

2 repeats

Results

The fragments are separated by gel electrophoresis. Different-size fragments appear as different bands.

Results of one STR locus

Conclusion: The patterns of DNA fragments produced by individuals differ.

19.28 DNA fingerprinting can be used to identify people.

The primers used in the PCR reaction are tagged with a fluorescent label so that the resulting DNA fragments can be detected with a laser. Primers for different STR loci are labeled with different colored primers, so that similar-sized products of different loci can be differentiated. After PCR, the fragments are separated on a gel or by a capillary electrophoresis machine. In capillary electrophoresis, the presence of each fragment is detected as it migrates past a laser, and a computer then calculates the size of each fragment based on its rate of migration. The fragments are represented as peaks on a graph; the distance on the horizontal axis represents the size of the fragment, while the height of the peak represents the amount of DNA (**Figure 19.29**). Homozygotes for an STR allele have a single tall peak; heterozygotes have two shorter peaks. When several different microsatellite loci are examined, the probability that two people have the same set of patterns becomes vanishingly small, unless they are identical twins.

The Federal Bureau of Investigation has developed a system using 13 STR loci (**Table 19.3**) that are commonly utilized for identifying people and in solving crimes. These loci make up the Combined DNA Index System (CODIS). Each STR locus in CODIS has a large number of alleles and is located on a different human chromosome,

DNA Fingerprinting

The use of DNA sequences to identify individual persons is called **DNA fingerprinting** or DNA profiling. Because some parts of the genome are highly variable, each person's DNA sequence is unique and, like a traditional fingerprint, provides a distinctive characteristic that allows identification.

Today, most DNA fingerprinting utilizes **microsatellites**, or **short tandem repeats** (STRs), which are very short DNA sequences repeated in tandem (see Chapter 11). These repeated sequences are found at many loci throughout the human genome. People vary in the number of copies of repeat sequences that they possess at each locus. The STRs are typically detected with PCR, using primers flanking the microsatellite repeats so that a DNA fragment containing the repeated sequences is amplified (**Figure 19.28**). The length of the amplified segment depends on the number of repeats; DNA from a person with more repeats will produce a longer amplified segment than will DNA from a person with fewer repeats.

This person is homozygous at the D8S119 locus...

...and heterozygous at the D21S11 locus.

19.29 A DNA profile represents the pattern of DNA fragments produced following PCR of the STR loci. This profile shows the results from four STR loci (D8S1179, D21S11, D7S820, and CSF1P0). The number below each peak represents the number of STR repeats in that DNA fragment.

Table 19.3	Characteristics of 13 STR loci used in CODIS for DNA fingerprinting.		
Locus Name	Chromosome	Number of Repeats	Number of Alleles*
CSF1PO	5	5-17	10
FGA	4	12-51	23
TH01	11	3-14	8
TPOX	2	4-16	8
VWA	12	10-25	10
D3S1358	3	6-26	10
D5S818	5	4-29	9
D7S820	7	5-16	11
D8S1179	8	6-20	11
D13S317	13	5-17	8
D16S539	16	4-17	7
D18S51	18	5-40	19
D21S11	21	12-43	22

*US population.
Source: J. M. Butler and C. R. Hill, *Forensic Science Review* 24:15–26, 2012.

and so variation at each locus assorts independently. When all 13 CODIS loci are used together, the probability of two randomly selected people having the same DNA profile is less than 1 in 10 billion.

In a typical application, DNA fingerprinting might be used to confirm that a suspect was present at the scene of a crime. A sample of DNA from blood, semen, hair, or other body tissue is collected from the crime scene. If the sample is very small, PCR can be used to amplify it so that enough DNA is available for testing. Additional DNA samples are collected from one or more suspects. The pattern of DNA fragments produced by DNA fingerprinting from the sample is then compared with the patterns produced by DNA fingerprinting of the DNA from the suspect. A match between the samples can provide evidence that the suspect was present at the scene of the crime (**Figure 19.30**).

Since its introduction in the 1980s, DNA fingerprinting has helped convict a number of suspects in murder and rape cases. Suspects in other cases have been proved innocent when their DNA failed to match that from the crime scenes. Initially, calculating the odds of a match (the probability that two people could have the same pattern) was controversial, and there were concerns about quality control (such as the accidental contamination of samples and the reproducibility of results) in laboratories where DNA analysis is done. Today, DNA fingerprinting has become an important tool in forensic investigations. In addition to its application in solving crimes, DNA fingerprinting is used to assess paternity,

19.30 DNA fingerprinting can be used to determine the presence of a suspect at a crime scene. The DNA profile of suspect 2 matches that of DNA evidence collected at the crime scene. Shown here are results from 4 STP loci.

study genetic relationships among individual organisms in natural populations, identify specific strains of pathogenic bacteria, and identify human remains.

CONCEPTS

DNA fingerprinting detects genetic differences among people by analyzing highly variable regions of chromosomes.

✔ **CONCEPT CHECK 9**

How are microsatellites detected?

Application: Identifying People Who Died in the Collapse of the World Trade Center

On the morning of September 11, 2001, terrorists hijacked and flew two passenger planes into the World Trade Center towers in New York City. The catastrophic damage and ensuing fire led, within a few hours, to the complete collapse of all 110 floors of both towers, killing almost 3000 building occupants and rescue personnel. The tremendous destructive force generated by the collapse of the towers, with their 425,000 cubic yards of concrete and 200,000 tons of steel, pulverized many of the bodies beyond recognition.

In the days immediately following the World Trade Center collapse, forensic scientists began the task of identifying the remains of those who perished. The goal was to provide evidence for the ongoing criminal investigation of the attack and to identify the remains for families and friends of the victims. This task was unprecedented in scope and difficulty. There was no complete list of victims (such as a passenger list in an airline crash) with which investigators could match the remains. In all, almost 20,000 individual remains were found, varying from whole bodies to tiny fragments of charred bone. The remains were subjected to fires with temperatures exceeding 1000°C that burned for more than 3 months. The collapse of the buildings intermixed many victims' remains, and many body fragments were not recovered for months, during which time they were exposed to dust, water, bacteria, and decay.

The usual means of victim identification—personal items, fingerprints, dental records—were of little use for most of the World Trade Center remains. Identification of the majority of the remains was made with the use of DNA fingerprinting (**Figure 19.31**).

DNA was first extracted from the tissue samples by using sterile techniques to prevent cross-contamination between samples. After the DNA had been extracted, PCR was used to amplify STR loci in the CODIS system (see previous section). The DNA fingerprint generated from each body sample was compared with that of DNA extracted from reference samples, such as the victims' toothbrushes and blood samples, provided by families and friends. If the DNA from a body part had the same alleles at all 13 loci as in the reference sample, then a positive identification was made. When no reference sample was available, investigators collected DNA from family members and tried to match the DNA profiles of remains to those of relatives; in this case, some, but not all STR alleles would match.

Unfortunately, many of the remains were so badly degraded that little DNA remained and one or more of the STR loci could not be amplified. For these remains, DNA fingerprinting was also carried out on mitochondrial DNA (see Chapter 11). Because there are many mitochondria per cell and each mitochondrion contains multiple DNA molecules, there are many more copies of mitochondrial DNA per cell than nuclear DNA; mitochondrial DNA has been successfully extracted and analyzed from ancient remains, such as Neanderthals (see introduction to Chapter 10). Alone, fingerprinting from mitochondrial DNA was insufficient to provide identification with a high degree of confidence (because there are not as many sequences that

19.31 DNA fingerprinting was used to identify the remains of people who died in the collapse of the World Trade Center. The office of the Chief Medical Examiner of New York City used automated DNA fingerprinting to match DNA from human remains recovered from the World Trade Center with DNA extracted from reference samples, such as blood samples and toothbrushes, of possible victims. [Scott Gries/Getty Images.]

vary among people as in the CODIS loci) but, when it was used in conjunction with data from at least some STR loci, a positive identification could often be made.

Used in combination, these techniques allowed the remains of many victims to be positively identified. However, despite the heroic efforts of hundreds of molecular geneticists, forensic anthropologists, and medical examiners, no positively identified remains were recovered for almost half of the people who are thought to have died in the disaster.

19.5 Molecular Techniques Are Increasingly Used to Analyze Gene Function

In the preceding sections, we learned about powerful molecular techniques for isolating, recombining, and analyzing DNA sequences. Although these techniques provide a great deal of information about the organization and nature of gene sequences, the ultimate goal of many molecular studies is to better understand the function of the sequences. In this section, we will explore some advanced molecular techniques that are frequently used to determine gene function and to better understand the genetic processes that these sequences undergo.

Forward and Reverse Genetics

The traditional approach to the study of gene function begins with the isolation of mutant organisms. For example, suppose a geneticist is interested in genes that affect cardiac function in mammals. A first step would be to find individuals—perhaps mice—that have hereditary defects in heart function. The mutations causing the cardiac problems in the mice could then be mapped, and the implicated genes could be isolated and sequenced. The proteins produced by the genes could then be predicted from the gene sequences and isolated. Finally, the biochemistry of the proteins could be studied and their role in heart function discerned. This approach, which begins with a phenotype (a mutant individual) and proceeds to a gene that encodes the phenotype, is called **forward genetics**.

An alternative approach is to begin with a genotype—a DNA sequence—and proceed to the phenotype by altering the sequence or inhibiting its expression. A geneticist might begin with a gene of unknown function, induce mutations in it, and then look to see what effect these mutations have on the phenotype of the organism. This approach is called **reverse genetics**. Today, both forward and reverse genetic approaches are widely used in analyses of gene function.

Creating Random Mutations

Forward genetics depends on the identification and isolation of random mutations that affect a phenotype of interest. Early in the study of genetics, geneticists were forced to rely on naturally occurring mutations, which are usually rare and can be detected only if large numbers of organisms are examined. The discovery of mutagenic agents—environmental factors that increase the rate of mutation (see Chapter 18)—provided a means of increasing the number of mutants in experimental populations of organisms. One of the first examples of experimentally created mutations was Hermann Muller's use of X-rays in 1927 to induce X-linked mutations in *Drosophila melanogaster*.

Today, radiation, chemical mutagens, and transposable elements are all used to create mutations for genetic analysis. To determine all genes that might affect a phenotype, it is desirable to create mutations in as many genes as possible—that is, to saturate the genome with mutations. This procedure is done with a mutagenic screen, which will be described in Chapter 20.

Site-Directed Mutagenesis

Reverse genetics depends on the ability to create mutations, not at random, but in particular DNA sequences, and then to study the effects of these mutations on the organism. Mutations are induced at specific locations through a process called **site-directed mutagenesis**.

A number of different strategies have been developed for site-directed mutagenesis. One strategy that is often used in bacteria is to cut out a short sequence of nucleotides with restriction enzymes and replace it with a short, synthetic oligonucleotide that contains the desired mutated sequence.

The success of this method depends on the availability of restriction sites flanking the sequence to be altered.

If appropriate restriction sites are not available, **oligonucleotide-directed mutagenesis** can be used (**Figure 19.32**). In this method, a single-stranded oligonucleotide is produced that differs from the target sequence by one or a few bases. Because they differ in only a few bases, the target DNA and the oligonucleotide will pair. When successfully paired with the target DNA, the oligonucleotide can act as a primer to initiate DNA synthesis, which produces a double-stranded molecule with a mismatch in the primer region. When this DNA is transferred to bacterial cells, the mismatched bases will be repaired by bacterial enzymes. About half of the time the normal bases will be changed into mutant bases, and

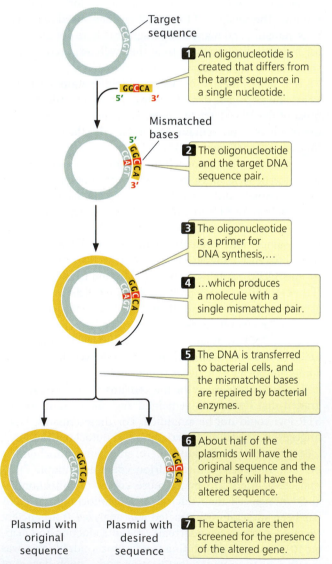

1 An oligonucleotide is created that differs from the target sequence in a single nucleotide.

2 The oligonucleotide and the target DNA sequence pair.

3 The oligonucleotide is a primer for DNA synthesis,…

4 …which produces a molecule with a single mismatched pair.

5 The DNA is transferred to bacterial cells, and the mismatched bases are repaired by bacterial enzymes.

6 About half of the plasmids will have the original sequence and the other half will have the altered sequence.

7 The bacteria are then screened for the presence of the altered gene.

19.32 Oligonucleotide-directed mutagenesis is used to study gene function when appropriate restriction sites are not available.

about half of the time the mutant bases will be changed into normal bases. The bacteria are then screened for the presence of the mutant gene.

Transgenic Animals

Another way that gene function can be analyzed is by adding DNA sequences of interest to the genome of an organism that normally lacks such sequences and then seeing the effect that the introduced sequence has on the organism's phenotype. This method is a form of reverse genetics. An organism that has been permanently altered by the addition of a DNA sequence to its genome is said to be *transgenic*, and the foreign DNA that it carries is called a **transgene** (**Figure 19.33**). Here, we consider techniques for the creation of transgenic mice, which are often used in the study of the function of human genes because they can be genetically manipulated in ways that are impossible with humans and, as mammals, they are more similar to humans than are fruit flies, fish, and other model genetic organisms.

The oocytes of mice and other mammals are large enough that DNA can be injected into them directly. Immediately after penetration by a sperm, a fertilized mouse egg contains two pronuclei, one from the sperm and one from the egg; these pronuclei later fuse to form the nucleus of

the embryo. Mechanical devices can manipulate extremely fine, hollow glass needles to inject DNA directly into one of the pronuclei of a fertilized egg (**Figure 19.34**). Typically,

19.33 The genome of a transgenic organism has been permanently altered by genetic engineering. Left: a transgenic mouse in which a gene affecting hair growth has been knocked out. Right: a normal mouse. [Maggie Bartlett, NHGRI.]

19.34 Transgenic animals have genomes that have been permanently altered through recombinant DNA technology. In the photograph, a mouse embryo is being injected with DNA. [Photograph: Chad Davis/PhotoDisc.]

a few hundred copies of cloned, linear DNA are injected into a pronucleus, and, in a few of the injected eggs, copies of the cloned DNA integrate randomly into one of the chromosomes through a process called nonhomologous recombination. After injection, the embryos are implanted in a pseudopregnant female—a surrogate mother that has been physiologically prepared for pregnancy by mating with a vasectomized male.

Only about 10% to 30% of the embryos survive and, of those that do survive, only a few have a copy of the cloned DNA stably integrated into a chromosome. Nevertheless, if several hundred embryos are injected and implanted, there is a good chance that one or more mice whose chromosomes contain the foreign DNA will be born. Moreover, because the DNA was injected at the one-cell stage of the embryo, these mice usually carry the cloned DNA in every cell of their bodies, including their reproductive cells, and will therefore pass the foreign DNA on to their progeny. Through interbreeding, a strain of mice that carry the foreign gene can be created.

Transgenic mice have proved useful in the study of gene function. For example, proof that the *SRY* gene (see Chapter 4) is the male-determining gene in mice was obtained by injecting a copy of the *SRY* gene into XX embryos and observing that these mice developed as males. In addition, researchers have created a number of transgenic mouse strains that serve as experimental models for human genetic diseases.

Knockout Mice

A useful variant of the transgenic approach is to produce mice in which a normal gene has been not just mutated, but fully disabled. These animals, called **knockout mice**, are particularly helpful in determining the function of a gene: the phenotype of the knockout mouse often gives a good indication of the function of the missing gene.

The creation of knockout mice begins when a normal gene is cloned in bacteria and then "knocked out," or disabled. There are a number of ways to disable a gene, but a common method is to insert a gene called *neo*, which confers resistance to the antibiotic G418, into the middle of the target gene (**Figure 19.35**). The insertion of *neo* both disrupts the target gene and provides a convenient marker for finding copies of the disabled gene. In addition, a second gene, usually the herpes simplex viral thymidine kinase (*tk*) gene, is linked to the disrupted gene. The disabled gene is then transferred to cultured embryonic mouse cells, where it may exchange places with the normal copy on the mouse chromosome through homologous recombination.

19.35 Knockout mice possess a genome in which a gene has been disabled.

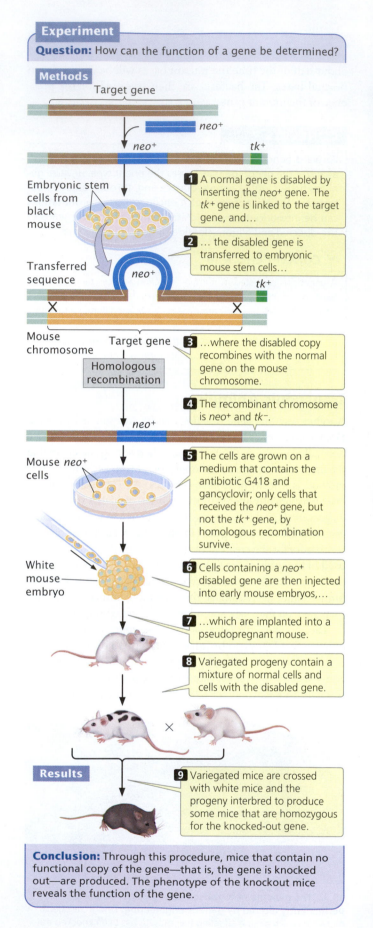

Experiment

Question: How can the function of a gene be determined?

Methods

Target gene

neo⁺

neo⁺   *tk⁺*

Embryonic stem cells from black mouse

1 A normal gene is disabled by inserting the *neo⁺* gene. The *tk⁺* gene is linked to the target gene, and...

Transferred sequence   *neo⁺*

2 ... the disabled gene is transferred to embryonic mouse stem cells...

tk⁺

Mouse chromosome   Target gene

3 ...where the disabled copy recombines with the normal gene on the mouse chromosome.

Homologous recombination

4 The recombinant chromosome is *neo⁺* and *tk⁻*.

neo⁺

Mouse *neo⁺* cells

5 The cells are grown on a medium that contains the antibiotic G418 and gancyclovir; only cells that received the *neo⁺* gene, but not the *tk⁺* gene, by homologous recombination survive.

White mouse embryo

6 Cells containing a *neo⁺* disabled gene are then injected into early mouse embryos,...

7 ...which are implanted into a pseudopregnant mouse.

8 Variegated progeny contain a mixture of normal cells and cells with the disabled gene.

Results

9 Variegated mice are crossed with white mice and the progeny interbred to produce some mice that are homozygous for the knocked-out gene.

Conclusion: Through this procedure, mice that contain no functional copy of the gene—that is, the gene is knocked out—are produced. The phenotype of the knockout mice reveals the function of the gene.

After the disabled gene has been transferred to the embryonic cells, the cells are screened by adding the antibiotic G418 to the medium. Only cells with the disabled gene containing the *neo* insert will survive. Because the frequency of nonhomologous recombination is higher than that of homologous recombination and because the intact target gene is replaced by the disabled copy only through homologous recombination, a means to select for the rarer homologous recombinants is required. The presence of the viral *tk* gene makes the cells sensitive to gancyclovir. Thus, transfected cells that grow on medium containing G418 and gancyclovir will contain the *neo* gene (disabled target gene) but not the adjacent *tk* gene, because the *tk* gene will be eliminated in the double-crossover event. These cells contain the desired homologous recombinants. The nonhomologous recombinants (random insertions) will contain both the *neo* and the *tk* genes, and these transfected cells will die on the selection medium owing to the presence of gancyclovir. The surviving cells are injected into an early-stage mouse embryo, which is then implanted into a pseudopregnant mouse. Cells in the embryo carrying the disabled gene and normal embryonic cells carrying the wild-type gene will develop together, producing a chimera—a mouse that is a genetic mixture of the two cell types. The production of chimeric mice is not itself desirable, but replacing all the cells of the embryo with injected cells is impossible.

Chimeric mice can be easily identified if the injected embryonic cells came from a black mouse and the embryos into which they are injected came from a white mouse; the resulting chimeras will have variegated black and white fur. Some of the chimeras may have the knockout gene in their germline cells and can transmit it to the next generation. The chimeras are crossed to white mice; any black progeny are heterozygous for the knockout. The black progeny can then be intercrossed to produce some progeny that are homozygous for the knockout gene. The effects of disabling a particular gene can be observed in these homozygous mice. Three scientists who helped develop the techniques for creating knockout mice—Mario Capecchi, Oliver Smithies, and Martin Evans—were awarded the Nobel Prize in physiology or medicine in 2007 for their work.

Additional techniques have been developed for knocking out genes only in certain tissues or at certain times; these are called conditional knockouts. A variant of the knockout procedure is to insert in mice a particular DNA sequence into a known chromosome location. For example, researchers might insert the sequence of a human disease-causing allele into the same locus in mice, creating a precise mouse model of the human disease. Mice that carry inserted sequences at specific locations are called **knock-in mice**.

CONCEPTS

A transgenic mouse is produced by the injection of cloned DNA into the pronucleus of a fertilized egg, followed by implantation of the egg into a female mouse. In knockout mice, the injected DNA contains a mutation that disables a gene. Inside the mouse embryo, the disabled copy of the gene can exchange with the normal copy of the gene through homologous recombination.

✔ **CONCEPT CHECK 11**

What is the advantage of using the *neo* gene to disrupt the function of a gene in knockout mice?
a. The *neo* gene produces an antibiotic that kills unwanted cells.
b. The *neo* gene is the right size for disabling other genes.
c. The *neo* gene provides a selectable marker for finding cells that contain the disabled gene.
d. The *neo* gene produces a toxin that inhibits transcription of the target gene.

Silencing Genes with RNAi

In the preceding sections, we considered the analysis of gene function by introducing mutations or new DNA sequences into the genome and analyzing the resulting phenotype to provide information about the function of the altered or introduced DNA. We could also analyze gene function by temporarily turning a gene off and seeing what effect the absence of the gene product has on the phenotype. For many years, there was no method for selectively affecting gene expression. However, the discoveries of siRNAs (small interfering RNAs) and miRNAs (microRNAs; see Chapters 14 and 17) provided powerful tools for controlling the expression of individual genes.

Recall that siRNAs and miRNAs are small RNA molecules that combine with proteins to form the RNA-induced silencing complex (RISC). In a process called RNA interference or RNAi, the RISC pairs with complementary sequences on mRNA and either cleaves the mRNA or prevents the mRNA from being translated. Molecular geneticists have exploited this natural machinery for turning off the expression of specific genes. Studying the effect of silencing a gene with siRNA can often be a source of insight into the gene's function.

The first step in using RNAi technology is to design the siRNAs such that they will be recognized and cleaved by Dicer (the protein that processes siRNAs; see Chapter 14). The complementary sequence must be unique to the target mRNA and not be found on other mRNAs so that the siRNA will not inhibit nontarget mRNAs. Computer programs are often used to design optimal siRNAs.

After the siRNA sequence has been designed, it must be synthesized. One way to create an siRNA is to use an oligonucleotide synthesizer to synthesize a DNA fragment corresponding to the siRNA sequence. The synthesized oligonucleotide can be cloned into a plasmid expression vector between two strong promoters (**Figure 19.36**). *Escherichia coli* are then transformed with the plasmid. Within the bacteria, transcription from the two promoters will proceed in both directions, producing two complementary RNA molecules that will pair to form a double-stranded RNA molecule recognized by Dicer. Alternatively, double-stranded RNA sequences can be synthesized directly with a gene synthesizer.

The next task is to deliver the double-stranded siRNA to the cells. Delivery can be done in a variety of ways, depending on the cell type. The model genetic organism *Caenorhabditis elegans* (see p. A6) can be fed *E. coli* (their natural food) containing the expression vector. Transcription within the bacteria produces double-stranded RNA, that the worms ingest and incorporate into their cells. Alternatively, double-stranded siRNA can be injected directly into cells or the body cavity. Yet another approach is to synthesize a short sequence of DNA that has internal complementarity so that, when transcribed, it folds up into a short hairpin RNA (shRNA) with a double-stranded section. Within a cell, the shRNAs are processed by Dicer to produce siRNAs that bring about gene silencing. DNA sequences containing siRNAs can be introduced into a vector with the use of standard cloning techniques, and the vector can be used to deliver the DNA into a cell. An advantage of this approach is that, with the addition of a DNA sequence, the RNAi sequence has the potential to become a permanent part of the cell's genome and be passed on to progeny.

One of the major advantages of siRNA for controlling gene expression is that it acts in trans—that is, a single copy of an siRNA gene will shut down expression of both copies of the target gene. Another advantage is that the target gene remains intact and therefore the silencing effects are reversible. ▶ **TRY PROBLEM 43**

19.36 Small interfering RNAs can be produced by cloning DNA sequences corresponding to the siRNAs between two strong promoters. When cloned into an expression vector, both DNA strands will be transcribed and the complementary RNA molecules will anneal to form double-stranded RNA that will be processed into siRNA by Dicer.

Application: Using RNAi to Treat Human Disease

In addition to its value in determining gene function, RNAi holds potential as a therapeutic agent for the future treatment of human diseases. This potential includes using siRNAs against RNA viruses, such as HIV, as well as using them to treat genetic diseases and cancer.

TREATMENT OF HIGH CHOLESTEROL WITH RNAI

Research has examined the potential of RNAi for the treatment of cholesterol metabolism disorders. Although cholesterol is essential for life, too much cholesterol is unhealthy: high blood cholesterol is a major contributor to heart disease, the leading cause of death in the United States. Cholesterol is normally transported throughout the body in the form of small particles called lipoproteins, which consist of a core of lipids surrounded by a shell of phospholipids and proteins (see Figure 6.6 in Chapter 6). The ApoB protein is an essential part of lipoproteins. Some people possess genetic mutations that cause elevated levels of ApoB, which predisposes them to coronary artery disease. Findings from studies suggest that lowering the amount of ApoB can reduce the number of lipoproteins and lower blood cholesterol in these people, as well as in people who have elevated cholesterol for other reasons.

In 2006, Tracy Zimmermann and her colleagues at Alnylam Pharmaceuticals and Protiva Biotherapeutics demonstrated that RNAi could be used to reduce the levels of ApoB and blood cholesterol in nonhuman primates. The investigators first created siRNAs that targeted *apoB* gene expression (apoB-siRNAs) on the basis of the known sequence of the gene. The apoB-siRNAs were synthesized in the laboratory and consisted of 21 nucleotides on the sense strand (the strand that was complementary to the *apoB* mRNA) and 23 nucleotides on the complementary antisense strand, with a two-nucleotide overhang.

The next task was to get the apoB-siRNA into the cell. Although siRNAs are readily taken up by the cells of invertebrates such as *C. elegans*, most siRNAs will not readily pass through the membranes of mammalian cells in a form that is still effective in gene silencing. In addition, siRNAs are rapidly removed from circulation. To overcome these problems, Zimmermann and her colleagues encapsulated the apoB-siRNAs in lipids, creating stable nucleic-acid–lipid particles (SNALPs). The SNALPs greatly increased the time spent by the siRNAs in circulation and enhanced their uptake by the cell.

The researchers then tested the effects of the apoB-siRNAs on reducing the synthesis of the ApoB protein and on cholesterol levels. They injected cynomolgus monkeys (**Figure 19.37a**) with SNALPs containing apoB-siRNA at two doses: 1 mg/kg and 2.5 mg/kg. In addition, they injected a third group of monkeys with saline as a control. They found that the apoB-siRNA clearly silenced the *apoB* gene: 48 hours after treatment, *apoB* mRNA in the liver was reduced by 68%

(a)

(b)

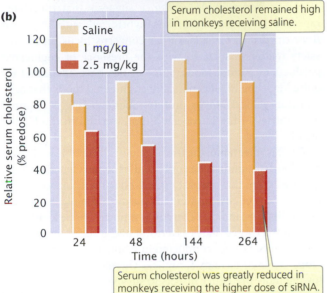

Serum cholesterol remained high in monkeys receiving saline.

Serum cholesterol was greatly reduced in monkeys receiving the higher dose of siRNA.

19.37 Transfer of siRNAs for the ApoB protein into cynomolgus monkeys significantly lowered blood-cholesterol levels.
(a) Cynomolgus monkey. (b) Different groups of monkeys were given saline (control), a 1-mg/kg dose, or a 2.5-mg/kg dose of siRNAs.
[Part a: Tony Camacho/Science Source. Part b: After T. S. Zimmerman, *Nature*, 441:112, 2006, Figure 3b.]

for monkeys receiving the 1-mg dose and 90% for monkeys receiving the 2.5-mg dose.

When the researchers examined serum cholesterol levels in the monkeys, they found that monkeys receiving the apoB-siRNA had a significant reduction in blood-cholesterol levels (**Figure 19.37b**). Importantly, they observed no negative effects of the siRNA treatment. Although preliminary, this study suggests that siRNAs have potential for future treatment of human diseases.

19.6 Biotechnology Harnesses the Power of Molecular Genetics

In addition to providing valuable new information about the nature and function of genes, molecular genetics techniques have many practical applications, including the production

of pharmaceutical products and other chemicals, specialized bacteria, agriculturally important plants, and genetically engineered farm animals. The technology is also used extensively in medical testing and, in a few cases, is even used to correct human genetic defects. Hundreds of companies now specialize in developing products through genetic engineering, and many large multinational corporations have invested enormous sums of money in molecular genetics research. As discussed earlier, the analysis of DNA is also used in criminal investigations and for the identification of human remains.

Pharmaceutical Products

The first commercial products to be developed with the use of genetic engineering were pharmaceutical products used in the treatment of human diseases and disorders. In 1979, the Eli Lilly corporation began selling human insulin produced with the use of recombinant DNA technology. The gene for human insulin was inserted into plasmids and transferred to bacteria that then produced human insulin. Previously, insulin was isolated from pig and cow pancreases; a few diabetics developed allergic reactions to this foreign protein. Recombinant insulin has the advantage of being the same as that produced in the human body. Other pharmaceutical products produced through recombinant DNA technology include human growth hormone (for children with growth deficiencies), clotting factors (for hemophiliacs), and tissue plasminogen activator (used to dissolve blood clots in heart-attack patients).

Specialized Bacteria

Bacteria play an important role in many industrial processes, including the production of ethanol from plant material, the leaching of minerals from ore, and the treatment of sewage and other wastes. The bacteria used in these processes are modified by genetic engineering so that they work more efficiently. New strains of technologically useful bacteria are being developed that will break down toxic chemicals and pollutants, enhance oil recovery, increase nitrogen uptake by plants, and inhibit the growth of pathogenic bacteria and fungi.

Agricultural Products

Recombinant DNA technology has had a major effect on agriculture, where it is now used to create crop plants and domestic animals with valuable traits. For many years, plant pathologists had recognized that plants infected with mild strains of viruses are resistant to infection by virulent strains. Using this knowledge, geneticists created viral resistance in plants by transferring genes for viral proteins to the plant cells. A genetically engineered squash, called Freedom II, carries genes from the watermelon mosaic virus 2 and the zucchini yellow mosaic virus, which protect the squash against viral infections.

Another objective has been to genetically engineer pest resistance into plants to reduce dependence on chemical pesticides. As discussed earlier in the chapter, a gene from the bacterium *Bacillus thuringiensis* that produces an insecticidal toxin has been transferred into corn, tomato, potato, cotton, and other plants. These Bt crops are now grown worldwide. Other genes that confer resistance to viruses and herbicides have been introduced into a number of crop plants. During 2011, 16.7 million farmers worldwide planted 160 million hectares of genetically engineered crops. In the United States, 88% of all corn, 94% of all cotton, and 93% of all soybeans grown in 2012 were genetically engineered.

Recombinant DNA techniques are also applied to domestic animals. For example, the gene for growth hormone was isolated from cattle and cloned in *E. coli*; these bacteria produce large quantities of bovine growth hormone, which is administered to dairy cattle to increase milk production. Transgenic animals are being developed to carry genes that encode pharmaceutical products; some eukaryotic proteins must be modified after translation, and only other eukaryotes (but not bacteria) are capable of carrying out the modifications. For example, a gene for human clotting factor VIII was attached to the regulatory region of the sheep gene for β-lactoglobulin, a milk protein. The fused gene was injected in sheep embryos, creating transgenic sheep that produced in their milk the human clotting factor, which is used to treat hemophiliacs. Transgenic salmon have been created that carry a foreign growth hormone gene and promoter; the transgenic fish grow year round instead of just during warm months, reaching market size more quickly and with less feed than wild salmon. And through genetic engineering scientists have created transgenic chickens that express a small RNA that blocks infection of avian influenza virus.

The genetic engineering of agricultural products is controversial. One area of concern focuses on the potential effects of releasing novel organisms produced by genetic engineering into the environment. There are many examples in which nonnative organisms released into a new environment have caused ecological disruption because they are free of predators and other natural control mechanisms. Genetic engineering normally transfers only small sequences of DNA, relative to the large genetic differences that often exist between species, but even small genetic differences may alter ecologically important traits that might affect the ecosystem.

Another area of concern is the effect of genetically engineered crops on biodiversity. In the largest field test of genetically engineered plants ever conducted, scientists cultivated beets, corn, and rapeseed that were genetically engineered to resist herbicide along with traditional crops on 200 test plots throughout the United Kingdom. They then measured the

biodiversity of native plants and animals in the agricultural fields. They found that the genetically engineered plants were highly successful in the suppression of weeds; however, plots with genetically engineered beets and rapeseed have significantly fewer insects that feed on weeds. For example, plots with genetically engineered rapeseed had 24% fewer butterflies than did plots with traditional crops.

There is also concern that transgenic organisms may hybridize with native organisms and transfer their genetically engineered traits. For example, herbicide resistance engineered into crop plants might be transferred to weeds, which would then be resistant to the herbicides that are now used for their control. Some studies have detected hybridization between genetically engineered crops and wild populations of plants. For example, evidence suggests that transgenic rapeseed (*Brassica napus*) has hybridized with the weed *Brassica rapa* in Canada. Other concerns focus on health-safety matters associated with the presence of engineered products in natural foods; some critics have advocated required labeling of all genetically engineered foods that contain transgenic DNA or protein. Such labeling is required in countries of the European Union but not in the United States.

On the other hand, the use of genetically engineered crops and domestic animals has potential benefits. Genetically engineered crops that are pest resistant have the potential to reduce the use of environmentally harmful chemicals, and research findings indicate that lower amounts of pesticides are used in the United States as a result of the adoption of transgenic plants. Studies conducted in China show that when Bt crops are used farmers spray less chemical insecticides, allowing more predatory insects to survive, and creating more natural pest control. Transgenic crops also increase yields, providing more food per acre, which reduces the amount of land that must be used for agriculture. Genetically engineered plants offer the potential for greater yields that may be necessary to feed the world's future population.

CONCEPTS

Recombinant DNA technology is used to create a wide range of commercial products, including pharmaceutical products, specialized bacteria, genetically engineered crops, and transgenic domestic animals.

✓ CONCEPT CHECK 12

What are some of the concerns about genetically engineered crops?

Genetic Testing

The identification and cloning of many important disease-causing human genes have allowed the development of probes for detecting disease-causing mutations. Prenatal testing is already available for many genetic disorders (see

Chapter 6). Additionally, presymptomatic genetic tests for adults and children are available for an increasing number of disorders. A number of genetic tests are now being offered directly to consumers, without requiring the participation of a health-care provider. Usually offered over the Internet, these direct-to-consumer genetic tests are available for testing a large and growing array of genetic conditions, everything from single-gene disorders such as cystic fibrosis to multifactorial conditions such as obesity, cardiovascular disease, athletic performance, and predisposition to nicotine addiction

The growing availability of genetic tests raises a number of ethical and social questions. For example, is it ethical to test for genetic diseases for which there is no cure or treatment? Other ethical and legal questions concern the confidentiality of test results. Who should have access to the results of genetic testing? Should relatives who also might be at risk be informed of the results of genetic testing?

Another set of concerns is related to the accuracy of genetic tests. For many genetic diseases, the only predictive tests available are those that identify a *predisposing* mutation in DNA, but many genetic diseases may be caused by dozens or hundreds of different mutations. Probes that detect common mutations can be developed, but they won't detect rare mutations and may give a false negative result. Short of sequencing the entire gene—which is expensive and time consuming—there is no way to identify all predisposed persons. These questions and concerns are currently the focus of intense debate by ethicists, physicians, scientists, and patients.

Gene Therapy

Perhaps the ultimate application of recombinant DNA technology is **gene therapy**, the direct transfer of genes into humans to treat disease. Today, thousands of patients have received gene therapy, and many clinical trials are underway. Gene therapy has been used as an experimental treatment for genetic diseases, cancer, heart disease, and even some infectious diseases such as AIDS. A number of different methods for transferring genes into human cells are currently under development. Commonly used vectors include genetically modified retroviruses, adenoviruses, and adeno-associated viruses (**Table 19.4**).

In spite of the growing number of clinical trials for gene therapy, significant problems remain in transferring foreign genes into human cells, getting them expressed, and limiting immune responses to the gene products and the vectors used to transfer the genes to the cells. There are also concerns about safety. In 1999, a patient participating in a gene-therapy trial had a fatal immune reaction after he was injected with a viral vector carrying a gene to treat his metabolic disorder. In addition, five children who underwent gene therapy for severe combined immunodeficiency disease developed leukemia that appeared to be directly related to the insertion of the retroviral gene vectors into cancer-causing genes. Despite these setbacks, gene-therapy research has moved ahead. Unequivocal results demonstrating positive benefits from gene therapy for several different diseases have now been published (see the introduction to this chapter).

Table 19.4	Vectors used in gene therapy	
Vector	**Advantages**	**Disadvantages**
Retrovirus	Efficient transfer	Transfers DNA only to dividing cells, inserts randomly; risk of producing wild-type viruses
Adenovirus	Transfers to nondividing cells	Causes immune reaction
Adeno-associated virus	Does not cause immune reaction	Holds small amount of DNA; hard to produce
Herpes virus	Can insert into cells of nervous system; does not cause immune reaction	Hard to produce in large quantities
Lentivirus	Can accommodate large genes	Safety concerns
Liposomes and other lipid-coated vectors	No replication; does not stimulate immune reaction	Low efficiency
Direct injection	No replication; directed toward specific tissues	Low efficiency; does not work well within some tissues
Pressure treatment	Safe, because tissues are treated outside the body and then transplanted into the patient	Most efficient with small DNA molecules
Gene gun (DNA coated on small gold particles and shot into tissue)	No vector required	Low efficiency

Source: After E. Marshall, Gene therapy's growing pains, *Science* 269:1050–1055, 1995.

Gene therapy conducted to date has targeted only non-reproductive, or somatic, cells. Correcting a genetic defect in these cells (termed *somatic gene therapy*) may provide positive benefits to patients but will not affect the genes of future generations. Gene therapy that alters reproductive, or germ-line, cells (termed *germ-line gene therapy*) is technically possible but raises a number of significant ethical issues because it has the capacity to alter the gene pool of future generations.

CONCEPTS

Gene therapy is the direct transfer of genes into humans to treat disease. Gene therapy was first successfully implemented in 1990 and is now being used to treat genetic diseases, cancer, and infectious diseases.

✓ **CONCEPT CHECK 13**

What is the difference between somatic gene therapy and germ-line gene therapy?

CONCEPTS SUMMARY

- Restriction endonucleases are enzymes that make double-stranded cuts in DNA at specific base sequences.

- DNA fragments can be separated with the use of gel electrophoresis and visualized by staining the gel with a dye that is specific for nucleic acids or by labeling the fragments with a radioactive or chemical tag.

- In gene cloning, a gene or a DNA fragment is placed into a bacterial cell, where it will be multiplied as the cell divides.

- Plasmids, small circular pieces of DNA, are often used as vectors to ensure that a cloned gene is stable and replicated within the recipient cells. Expression vectors contain sequences necessary for foreign DNA to be transcribed and translated.

- The polymerase chain reaction is a method for amplifying DNA enzymatically without cloning. A solution containing DNA is heated, so that the two DNA strands separate, and then quickly cooled, allowing primers to attach to the template DNA. The solution is then heated again, and DNA polymerase synthesizes new strands from the primers. Each time the cycle is repeated, the amount of DNA doubles.

- Genes can be isolated by creating a DNA library—a set of bacterial colonies or viral plaques that each contain a different cloned fragment of DNA. A genomic library contains the entire genome of an organism; a cDNA library contains DNA fragments complementary to all the different mRNAs in a cell.

- In situ hybridization can be used to determine the chromosomal location of a gene and the distribution of the mRNA produced by a gene.

- Positional cloning uses linkage relations to determine the location of genes without any knowledge of their products.

- The Sanger (dideoxy) method of DNA sequencing uses special substrates for DNA synthesis (dideoxynucleoside triphosphates, ddNTPs) that terminate synthesis after they are incorporated into the newly made DNA. Next-generation and third generation sequencing methods sequence many DNA fragments simultaneously, providing a much faster and less-expensive determination of a DNA sequence.

- Short tandem repeats (STRs) and microsatellites are used to identify people by their DNA sequences (DNA fingerprinting).

- Forward genetics begins with a phenotype and conducts analyses to locate the responsible genes. Reverse genetics starts with a DNA sequence and conducts analyses to determine its phenotypic effect.

- Site-directed mutagenesis can be used to produce mutations at specific sites in DNA, allowing genes to be tailored for a particular purpose.

- Transgenic animals, produced by injecting DNA into fertilized eggs, contain foreign DNA that is integrated into a chromosome. Knockout mice are mice in which a normal gene is disabled.

- RNA interference is used to silence the expression of specific genes.

- Techniques of molecular genetics are being used to create products of commercial importance, to develop diagnostic tests, and to treat diseases.

- In gene therapy, diseases are being treated by altering the genes of human cells.

IMPORTANT TERMS

recombinant DNA
   technology (p. 536)
genetic engineering (p. 536)
biotechnology (p. 536)
restriction enzyme (p. 537)

restriction endonuclease
   (p. 537)
cohesive end (p. 537)
engineered nuclease (p. 539)
gel electrophoresis (p. 539)

probe (p. 540)
Southern blotting (p. 540)
Northern blotting (p. 541)
Western blotting (p. 541)
gene cloning (p. 541)

cloning vector (p. 541)
linker (p. 542)
cosmid (p. 543)
bacterial artificial
   chromosome (BAC) (p. 543)

expression vector (p. 544)
yeast artificial
   chromosome (YAC) (p. 544)
Ti plasmid (p. 544)
polymerase chain
   reaction (PCR) (p. 546)
Taq polymerase (p. 548)
reverse-transcription PCR
   (p. 548)
real-time PCR (p. 548)

DNA library (p. 549)
genomic library (p. 549)
cDNA (complementary
   DNA) library (p. 550)
in situ hybridization (p. 552)
positional cloning (p. 552)
chromosome walking (p. 553)
restriction fragment length
   polymorphism (RFLP)
   (p. 555)

DNA sequencing (p. 556)
   dideoxyribonucleoside
   triphosphate (ddNTP)
   (p. 556)
next-generation sequencing
   technologies (p. 559)
DNA fingerprinting (p. 561)
microsatellite (p. 561)
short tandem repeat
   (STR) (p. 561)

forward genetics (p. 564)
reverse genetics (p. 564)
site-directed mutagenesis
   (p. 564)
oligonucleotide-directed
   mutagenesis (p. 564)
transgene (p. 565)
knockout mice (p. 566)
knock-in mice (p. 567)
gene therapy (p. 571)

ANSWERS TO CONCEPT CHECKS

1. Restriction enzymes exist naturally in bacteria, which use them to prevent the entry of viral DNA.

2. c

3. Southern blotting is used to transfer DNA from a gel to a solid medium. Northern blotting is used to transfer RNA from a gel to a solid medium, and Western blotting is used to transfer protein from a gel to a solid medium.

4. The gene and plasmid are cut with the same restriction enzyme and mixed together. DNA ligase is used to seal nicks in the sugar–phosphate backbone.

5. A heat-stable DNA polymerase enzyme is important to the success of PCR because the first step of the reaction requires that the solution be heated to between 90° and 100°C to separate the two DNA strands. Most enzymes are denatured at this temperature. With the use of a heat-stable polymerase, the enzyme can be added at the beginning of the reaction and will function throughout multiple cycles.

6. With the use of the genetic code and the amino acid sequence of the protein, possible nucleotide sequences that cover a small region of the gene can be deduced. A mixture of all the possible nucleotide sequences that might encode the protein, taking into consideration

synonymous codons, is used to probe the library. To minimize the number of sequences required, a region of the protein that has relatively little degeneracy in its codons is selected.

7. The expression pattern of the gene can be examined, and the coding region of copies of the gene from individuals with the mutant phenotype can be compared with the coding region of wild-type individuals.

8. d

9. By using PCR with primers that flank the region containing tandem repeats.

10. b

11. c

12. Possible concerns include: (a) ecological damage caused by introducing novel organisms into the environment; (b) negative effects of transgenic organisms on biodiversity; (c) possible spread of transgenes to native organisms by hybridization; and (d) health effects of eating genetically modified foods.

13. Somatic gene therapy modifies genes only in somatic tissue, and these modifications cannot be inherited. Germ-line gene therapy alters genes in germ-line cells and will be inherited.

WORKED PROBLEMS

Problem 1

A molecule of double-stranded DNA that is 5 million base pairs long has a base composition that is 62% G + C. How many times, on average, are the following restriction sites likely to be present in this DNA molecule?

a. *Bam*HI (recognition sequence is GGATCC)

b. *Hin*dIII (recognition sequence is AAGCTT)

c. *Hpa*II (recognition sequence is CCGG)

Solution Strategy

What information is required in your answer to the problem?

The number of restriction sites likely to be present in the DNA molecule for each of the specified restriction enzymes.

What information is provided to solve the problem?

- The size of the DNA molecule.
- The G + C base composition of the DNA molecule.
- The recognition sequences for each restriction enzyme.

For help with this problem, review:

Cutting and Joining DNA Fragments in Section 19.2.

Solution Steps

The percentages of G and C are equal in double-stranded DNA; so, if G + C = 62%, then %G = %C = 62%/2 = 31%.

The percentage of A + T = (100% − G − C) = 38%, and %A = %T = 38%/2 = 19%. To determine the probability of finding a particular base sequence, we use the multiplication rule, multiplying together the probably of finding each base at a particular site.

a. The probability of finding the sequence GGATCC = 0.31 × 0.31 × 0.19 × 0.19 × 0.31 × 0.31 = 0.0003333. To determine the average number of recognition sequences in a 5-million-base-pair piece of DNA, we multiply 5,000,000 bp × 0.00033 = 1666.5 recognition sequences.

b. The number of AAGCTT recognition sequences is 0.19 × 0.19 × 0.31 × 0.31 × 0.19 × 0.19 × 5,000,000 = 626 recognition sequences.

c. The number of CCGG recognition sequences is 0.31 × 0.31 × 0.31 × 0.31 × 5,000,000 = 46,176 recognition sequences.

> **Hint:** If you know the percentage of any base in the DNA, you can determine the ages of all the bases because ___ and A = T.

> **Recall:** The multiplication rule states that the probability of two or more independent events is calculated by multiplying their independent probabilities.

Problem 2

You are given the following DNA fragment to sequence: 5′–GCTTAGCATC–3′. You first clone the fragment in bacterial cells to produce sufficient DNA for sequencing. You isolate the DNA from the bacterial cells and carry out the dideoxy-sequencing method. You then separate the products of the polymerization reactions by gel electrophoresis. Draw the bands that should appear on the gel from the four sequencing reactions.

Solution Strategy

What information is required in your answer to the problem?

The positions of the bands on the sequencing gel.

What information is provided to solve the problem?

The base sequence of the DNA fragment to be sequenced.

For help with this problem, review:

DNA Sequencing in Section 19.4.

Solution Steps

> **Recall:** In dideoxy sequencing, a new DNA strand is synthesized and that strand is what is sequenced. Thus, the bands that appear on the gel represent the complement of the original sequence.

The first task is to write out the sequence of the newly synthesized fragment, which will be complementary and antiparallel to the original fragment. The original sequence is 5′–GCTTAGCATC–3′, so the newly synthesized sequence will be:

Original (template) sequence: 5′–GCTTAGCATC–3′

Newly synthesized sequence: 3′–CGAATCGTAG–5′

Thus, the sequence of the newly synthesized strand, written 5′→3′ is: 5′–GATGCTAAGC–3′. Bands representing this sequence will appear on the gel, with the bands representing nucleotides near the 5′ end of the molecule at the bottom of the gel.

> **Hint:** Small fragments, those nearer the 5′ end of the newly synthesized strand, will migrate faster and will appear near the bottom of the gel.

COMPREHENSION QUESTIONS

Section 19.1

1. List some of the effects and practical applications of molecular genetic analyses.

Section 19.2

2. What feature is commonly seen in the sequences recognized by type II restriction enzymes?

3. What normal role do restriction enzymes play in bacteria? How do bacteria protect their own DNA from the action of restriction enzymes?

4. Explain how gel electrophoresis is used to separate DNA fragments of different lengths.

5. After DNA fragments have been separated by gel electrophoresis, how can they be visualized?

6. What is the purpose of Southern blotting? How is it carried out?

7. Give three important characteristics of cloning vectors.

8. Briefly describe two different methods for inserting foreign DNA into plasmids, giving the strengths and weaknesses of each method.

9. Briefly explain how an antibiotic-resistance gene and the *lacZ* gene can be used to determine which cells contain a particular plasmid.

10. Briefly explain how the polymerase chain reaction is used to amplify a specific DNA sequence. What are some of the limitations of PCR?

11. What is real-time PCR?

Section 19.3

12. How does a genomic library differ from a cDNA library?

13. How are probes used to screen DNA libraries? Explain how a synthetic probe can be prepared when the protein product of a gene is known.

14. Briefly explain in situ hybridization, giving some applications of this technique.

15. Briefly explain how a gene can be isolated through positional cloning.

16. Explain how chromosome walking can be used to find a gene.

Section 19.4

17. What is the purpose of the dideoxynucleoside triphosphate in the dideoxy sequencing reaction?

18. What is DNA fingerprinting? What types of sequences are examined in DNA fingerprinting?

Section 19.5

19. How does a reverse genetics approach differ from a forward genetics approach?

20. Briefly explain how site-directed mutagenesis is carried out.

21. What are knockout mice, how are they produced, and for what are they used?

22. How is RNA interference used in the analysis of gene function?

Section 19.6

23. What is gene therapy?

> For more questions that test your comprehension of the key chapter concepts, go to **LEARNING**Curve for this chapter.

APPLICATION QUESTIONS AND PROBLEMS

Introduction

24. The introduction to this chapter explains how gene therapy is used to treat Leber congenital amaurosis (LCA, a type of severe blindness). What characteristics of blindness that affect the retina (such as LCA) make it an attractive candidate for treatment by gene therapy?

Section 19.2

*25. Suppose that a geneticist discovers a new restriction enzyme in the bacterium *Aeromonas ranidae*. This restriction enzyme is the first to be isolated from this bacterial species. Using the standard convention for abbreviating restriction enzymes, give this new restriction enzyme a name (for help, see footnote to Table 19.1).

26. How often, on average, would you expect a type II restriction endonuclease to cut a DNA molecule if the recognition sequence for the enzyme had 5 bp? (Assume that the four types of bases are equally likely to be found in the DNA and that the bases in a recognition sequence are independent.) How often would the endonuclease cut the DNA if the recognition sequence had 8 bp?

*27. A microbiologist discovers a new type II restriction endonuclease. When DNA is digested by this enzyme, fragments that average 1,048,500 bp in length are produced. What is the most likely number of base pairs in the recognition sequence of this enzyme?

28. Will restriction sites for an enzyme that has 4 bp in its restriction site be closer together, farther apart, or similarly spaced, on average, compared with those of an enzyme that has 6 bp in its restriction site? Explain your reasoning.

*29. About 60% of the base pairs in a human DNA molecule are AT. If the human genome has 3.2 billion base pairs of DNA, about how many times will the following restriction sites be present?

  a. *Bam*HI (restriction site is 5′–GGATCC–3′)

  b. *Eco*RI (restriction site is 5′–GAATTC–3′)

  c. *Hae*III (restriction site is 5′–GGCC–3′)

*30. Restriction mapping of a linear piece of DNA reveals the following *Eco*RI restriction sites.

```
          EcoRI site 1      EcoRI site 2
              ↓                  ↓
   |  2 kb    |     4 kb         |      5 kb        |
```

  a. This piece of DNA is cut by *Eco*RI, the resulting fragments are separated by gel electrophoresis, and the gel is stained with ethidium bromide. Draw a picture of the bands that will appear on the gel.

  b. If a mutation that alters *Eco*RI site 1 occurs in this piece of DNA, how will the banding pattern on the gel differ from the one that you drew in part *a*?

  c. If mutations that alter *Eco*RI sites 1 and 2 occur in this piece of DNA, how will the banding pattern on the gel differ from the one that you drew in part *a*?

  d. If 1000 bp of DNA were inserted between the two restriction sites, how would the banding pattern on the gel differ from the one that you drew in part *a*?

  e. If 500 bp of DNA between the two restriction sites were deleted, how would the banding pattern on the gel differ from the one that you drew in part *a*?

*31. Which vectors (plasmid, phage λ, cosmid, bacterial artificial chromosome) can be used to clone a continuous fragment of DNA with the following lengths?

  a. 4 kb   c. 35 kb

  b. 20 kb   d. 100 kb

32. A geneticist uses a plasmid for cloning that has the *lacZ* gene and a gene that confers resistance to penicillin. The geneticist inserts a piece of foreign DNA into a restriction site that is located within the *lacZ* gene and uses the plasmid to transform bacteria. Explain how the geneticist can identify bacteria that contain a copy of a plasmid with the foreign DNA.

33. In **Figure 19.11**, what is the purpose of the *neo*⁺ gene that is attached to the *Bt* gene?

Section 19.3

34. Suppose that you have just graduated from college and have started working at a biotechnology firm. Your first job assignment is to clone the pig gene for the hormone prolactin. Assume that the pig gene for prolactin has not yet been isolated, sequenced, or mapped; however, the mouse gene for prolactin has been cloned, and the amino acid sequence of mouse prolactin is known. Briefly explain two different strategies that you might use to find and clone the pig gene for prolactin.

*35. A molecular biologist wants to isolate a gene from a scorpion that encodes the deadly toxin found in its stinger, with the ultimate purpose of transferring this gene to bacteria and producing the toxin for use as a commercial pesticide. Isolating the gene requires a DNA library. Should the molecular biologist create a genomic library or a cDNA library? Explain your reasoning.

[Sahara Nature/Fotalia.com.]

*36. A protein has the following amino acid sequence:

Met-Tyr-Asn-Val-Arg-Val-Tyr-Lys-Ala-Lys-
Trp-Leu-Ile-His-Thr-Pro

You wish to make a set of probes to screen a cDNA library for the sequence that encodes this protein. Your probes should be at least 18 nucleotides in length.

  a. Which amino acids in the protein should be used to construct the probes so that the least degeneracy results? (Consult the genetic code in Figure 15.10.)

  b. How many different probes must be synthesized to be certain that you will find the correct cDNA sequence that specifies the protein?

Section 19.4

37. In **Figure 19.22** Bob and Joe are each homozygous for different restriction fragment patterns. How many bands would you expect to see on the gel if a person was heterozygous for the A and B patterns? Explain your reasoning.

38. Suppose that you want to sequence the following DNA fragment:

5′–TCCCGGGAAA-primer site–3′

You first use PCR to amplify the fragment, so that there is sufficient DNA for sequencing. You then separate the products of the polymerization reactions by gel electrophoresis. Draw the bands that should appear on the gel from the four sequencing reactions.

Reaction containing

ddATP   ddTTP   ddCTP   ddGTP   Origin

*39. Suppose that you are given a short fragment of DNA to sequence. You amplify the fragment with PCR and set up a series of four dideoxy reactions. You then separate the products of the reactions by gel electrophoresis and obtain the following banding pattern:

Reaction containing

ddATP   ddTTP   ddCTP   ddGTP   Origin

Write out the base sequence of the original fragment that you were given.

Original sequence: 5′– _____ –3′

 40. The picture below is a sequencing gel from the original study that first sequenced the cystic fibrosis gene (J. R. Riordan et al. 1989. *Science* 245:1066–1073). From the picture, determine the sequence of the normal copy of the gene and the sequence of the mutated copy of the gene. Identify the location of the mutation that causes cystic fibrosis (CF). Hint: The CF mutation is a 3 bp deletion.

CTAG CTAG

DNA from a healthy person ⎵   DNA from a person with CF ⎵

[From J.R. Riordan et al. *Science* 245:1066–1073. Reprinted with permission from AAAS.]

*41. A hypothetical disorder called G syndrome is an autosomal dominant disease characterized by visual, skeletal, and cardiovascular defects. The disorder appears in middle age. Because its symptoms are variable, the disorder is difficult to diagnose. Early diagnosis is important, however, because the cardiovascular defects can be treated if the disorder is recognized early. The gene for G syndrome is known to reside on chromosome 7, and it is closely linked to two RFLPs on the same chromosome, one at the *A* locus and one at the *C* locus. The *G*, *A*, and *C* loci are very close together, and there is little crossing over between them. The following RFLP alleles are found at the *A* and *C* loci:

> *A* locus: *A1, A2, A3*
> *C* locus: *C1, C2, C3*

Sally, shown in the following pedigree, is concerned that she might have G syndrome. Her deceased mother had G syndrome, and she has a brother with the disorder. Her other brother is middle-aged and does not have the disease; so assume that he does not carry genes for it. A geneticist genotypes Sally and her immediate family for the *A* and *C* loci and obtains the genotypes shown on the pedigree.

A1 A1
C2 C3

A1 A3   *A1 A2*   *A1 A2* ← Sally
C1 C3   *C2 C3*   *C2 C2*

a. Assume that there is no crossing over between the *A*, *C*, and *G* loci. Does Sally carry the gene that causes G syndrome? Explain why or why not.

b. Draw the arrangement of the *A*, *C*, and *G* alleles on the chromosomes for all members of the family.

Section 19.5

42. You have discovered a gene in mice that is similar to a gene in yeast. How might you determine whether this gene is essential for development in mice?

*43. Andrew Fire, Craig Mello, and their colleagues were among the first to examine the effects of double-stranded RNA on gene expression (A. Fire et al. 1998. *Nature* 391:806–811). In one experiment, they used a transgenic strain of *C. elegans* into which a gene (*gfp*) for a green fluorescent pigment had been introduced. They injected some worms with double-stranded RNA complementary to coding sequences of the *gfp* gene and injected other worms with double-stranded RNA complementary to the coding region of a different gene (*unc22A*) that encodes a muscle protein. The photographs on the next page show larvae and adult

progeny of the injected worms. Green fluorescent
pigment appears as bright spots in the photographs.

	Double-stranded *unc22A* RNA	Double-stranded *gfp* RNA
Larvae		
Adult		

[Reprinted by permission from Macmillan Publishers Ltd: *NATURE* 391:806–811
(19 February 1998), copyright 1998.]

a. Explain these results.

b. Fire and Mello conducted another experiment in which
they injected double-stranded RNA complementary
to the introns and promoter sequences of the *gfp* gene.
What results would you expect with this experiment?
Explain your answer.

CHALLENGE QUESTIONS

Section 19.5

44. Suppose that you are hired by a biotechnology firm to
produce a giant strain of fruit flies by using recombinant
DNA technology so that genetics students will not be
forced to strain their eyes when looking at tiny flies. You
go to the library and learn that growth in fruit flies is
normally inhibited by a hormone called shorty substance
P (SSP). You decide that you can produce giant fruit flies
if you can somehow turn off the production of SSP. Shorty
substance P is synthesized from a compound called XSP
in a single-step reaction catalyzed by the enzyme runtase:

$$XSP \xrightarrow{\text{Runtase}} SSP$$

A researcher has already isolated cDNA for runtase and
has sequenced it, but the location of the runtase gene
in the *Drosophila* genome is unknown. In attempting to
devise a strategy for turning off the production of SSP
and producing giant flies by using standard recombinant
DNA techniques, you discover that deleting, inactivating,
or otherwise mutating this DNA sequence in *Drosophila*
turns out to be extremely difficult. Therefore you must
restrict your genetic engineering to gene augmentation

(adding new genes to cells). Describe the methods that
you will use to turn off SSP and produce giant flies by
using recombinant DNA technology.

Section 19.6

45. Much of the controversy over genetically engineered
foods has centered on whether special labeling should be
required on all products made from genetically modified
crops. Some people have advocated labeling that identifies
the product as having been made from genetically
modified plants. Others have argued that labeling should
be required only to identify the ingredients, not the
process by which they were produced. Choose a side in
this issue and justify your stand.

> ▶ Go to your **⌷LaunchPad** to find additional learning
> resources and the Suggested Readings for this chapter.

20 | Genomics and Proteomics

The genome of the honeybee, *Apis mellifera*, was sequenced in 2006, providing new information about honeybee communication, behavior, ecology, and evolution. [K. Wothe/age fotostock.]

Decoding the Waggle Dance: The Genome of the Honeybee

The honeybee, *Apis mellifera*, is one of the world's most amazing animals. Honeybees are highly social, living in complex colonies in which individual bees cooperate and assume responsibility for specialized tasks that benefit the entire hive. In spite of a tiny brain that contains only a million neurons (the human brain contains 100 billion neurons), bees are able to perform intricate behaviors and communicate effectively with thousands of other bees that make up the hive. Each bee recognizes the odor of its own hive, as well as odors that signal alarm, indicate tasks to be performed, and distinguish the caste, age, and sex of other bees. Worker bees possess an extraordinary memory. On finding a clump of flowers that provide a rich supply of nectar, a worker bee returns to the hive and performs the waggle dance, in which it runs a figure eight on the side of the comb. The dance conveys information to other bees about the location of the food source, including the direction in reference to the angle of the sun and the distance from the hive to the source. A bee is able to recall as many as five different flower locations, including the direction and distance of each from the hive, landmarks along the way, and the time of day when the flower produces the most nectar.

Honeybees are not just objects of biological curiosity; they play a critical role in pollinating many important food crops, as well as providing honey and wax. Because of its agricultural and scientific importance, the honeybee was selected for sequencing by the National Genome Research Institute of the U.S. National Institutes of Health. The complete genomic sequence was published in 2006 and has provided a wealth of information about the behavior, neural function, ecology, and evolution of honeybees.

The honeybee genome consists of 10,157 genes that encompass about 236 million base pairs of DNA. On the basis of the genomic information, 36 genes have been identified that encode neuropeptides—brain proteins that affect behavior and memory. More than 3000 genes are active in a bee's brain, and researchers have located regulatory regions that control the expression of many of these genes, including some that are important in the development of foraging behavior. Compared with other insects, honeybees also have more genes that encode olfactory receptors, which is consistent with their sophisticated chemical communication system.

The honeybee has long been associated with humans, but the precise evolutionary origin of honeybees was unknown until recently. With information from the genomic sequence, honeybee geneticists used single-nucleotide polymorphisms (SNPs, sites that vary in the base present at a single nucleotide) to study the evolutionary relationships between bees from different parts of the world. This study revealed that *A. mellifera* originated, not in Asia as was formerly thought, but in Africa. African populations subsequently expanded into Europe and Asia in two waves, one wave populating western Europe and another wave expanding into Asia and eastern Europe.

European honeybees were introduced into North America by early European settlers, but some have been replaced in recent years by descendants of African "killer bees," known for their aggressive stinging behavior. These bees were introduced into Brazil in 1956 and have subsequently spread northward to the United States. Analysis of SNPs of bees collected in the United States before 1990 showed that they possessed only European genes but, from 1993 to 2001, bees from southern Texas showed a transition to genes of predominately African ancestry.

The sequencing of the honeybee genome illustrates the diverse ways that genomic information is being used today. **Genomics** is the field of genetics that attempts to understand the content, organization, function, and evolution of genetic information contained in whole genomes. The field of genomics is at the cutting edge of modern biology; information resulting from research in this field has made significant contributions to human health, agriculture, and numerous other areas. It has provided gene sequences necessary for producing medically important proteins through recombinant DNA technology, and comparisons of genome sequences from different organisms are leading to a better understanding of evolution and the history of life.

We begin this chapter by examining genetic and physical maps and methods for sequencing entire genomes. Next, we explore functional genomics—how genes are identified in genomic sequences and how their functions are defined. The sequence of a genome, by itself, is of limited use, and, now that sequencing genomes has become routine, much of genomics is currently focused on deciphering the function of the sequences that are obtained. In the final part of the chapter, we consider proteomics, the study of the complete set of proteins found in a cell.

20.1 Structural Genomics Determines the DNA Sequences of Entire Genomes

Structural genomics concerns the organization and sequence of genetic information contained within a genome. Often, an early step in characterizing a genome is to prepare genetic and physical maps of its chromosomes. These maps provide information about the relative locations of genes, molecular markers, and chromosome segments, which are often essential for positioning chromosome segments and aligning stretches of sequenced DNA into a whole-genome sequence.

Genetic Maps

Everyone has used a map at one time or another. Maps are indispensable for finding a new friend's house, the way to an unfamiliar city in your state, or the location of a country. Each of these examples requires a map with a different scale. To find a friend's house, you would probably use a city street map; to find your way to an unknown city, you might pick up a state highway map; to find a country such as Kazakhstan, you would need a world atlas. Similarly, navigating a genome requires maps of different types and scales.

CONSTRUCTION OF GENETIC MAPS Genetic maps (also called linkage maps) provide a rough approximation of the locations of genes relative to the locations of other known genes (**Figure 20.1**). These maps are based on the genetic function of recombination (hence the name genetic map). The basic principles of constructing genetic maps are discussed in detail in Chapter 7. In short, individual organisms of known genotype are crossed, and the frequency of recombination between loci is determined by examining the progeny. If the recombination frequency between two loci is 50%, then the loci are found on different chromosomes or are far apart on the same chromosome. If the recombination frequency is less than 50%, the loci are found close together on the same chromosome (they belong to the same linkage group). For linked genes, the rate of recombination is proportional to the physical distance between the loci. Distances on genetic maps are measured in percent recombination (centimorgans, cM), or map units (m.u.). Data from multiple two-point or three-point crosses can be integrated into linkage maps for whole chromosomes.

For many years, genes could be detected only by observing their influence on a trait (the phenotype), and the construction of genetic maps was limited by the availability of single-locus traits that could be examined for evidence of recombination. Eventually, this limitation was overcome by the development of molecular techniques, such as the analysis of restriction fragment length polymorphisms, the polymerase chain reaction, and DNA sequencing (see Chapter 19), which are able to provide molecular markers that can be used to construct and refine genetic maps.

LIMITATIONS OF GENETIC MAPS Genetic maps have several limitations, the first of which is resolution, or detail. The human genome includes 3.2 billion base pairs of DNA

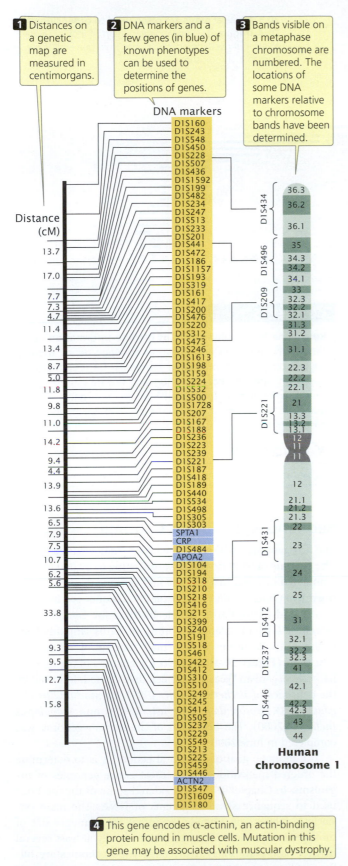

1 Distances on a genetic map are measured in centimorgans.

2 DNA markers and a few genes (in blue) of known phenotypes can be used to determine the positions of genes.

3 Bands visible on a metaphase chromosome are numbered. The locations of some DNA markers relative to chromosome bands have been determined.

DNA markers

Distance (cM)

Human chromosome 1

4 This gene encodes α-actinin, an actin-binding protein found in muscle cells. Mutation in this gene may be associated with muscular dystrophy.

20.1 Genetic maps are based on rates of recombination. Shown here is a genetic map of human chromosome 1.

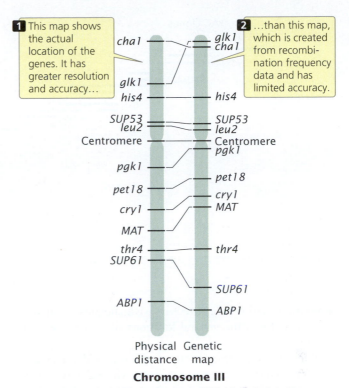

1 This map shows the actual location of the genes. It has greater resolution and accuracy…

2 …than this map, which is created from recombination frequency data and has limited accuracy.

Physical distance   Genetic map

Chromosome III

20.2 Genetic and physical maps may differ in relative distances and even in the position of genes on a chromosome. Genetic and physical maps of yeast chromosome III reveal such differences.

and has a total genetic distance of about 4000 cM, an average of 800,000 bp/cM. Even if a marker were present every centimorgan (which is unrealistic), the resolution in regard to the physical structure of the DNA would still be quite low. In other words, the detail of the map is very limited. A second problem with genetic maps is that they do not always accurately correspond to physical distances between genes. Genetic maps are based on rates of crossing over, which vary somewhat from one part of a chromosome to another; so the distances on a genetic map are only approximations of real physical distances along a chromosome. **Figure 20.2** compares the genetic map of chromosome III of yeast with a physical map determined by DNA sequencing. There are some discrepancies between the distances and even among the positions of some genes. In spite of these limitations, genetic maps have been critical to the development of physical maps and the sequencing of whole genomes.

Physical Maps

Physical maps are based on the direct analysis of DNA, and they place genes in relation to distances measured in number of base pairs, kilobases, or megabases. A common type of physical map is one that connects isolated pieces of genomic DNA that have been cloned in bacteria or yeast (**Figure 20.3**). Physical maps generally have higher resolution and are more accurate than genetic maps. A physical map is analogous to a neighborhood map that shows the location of every house

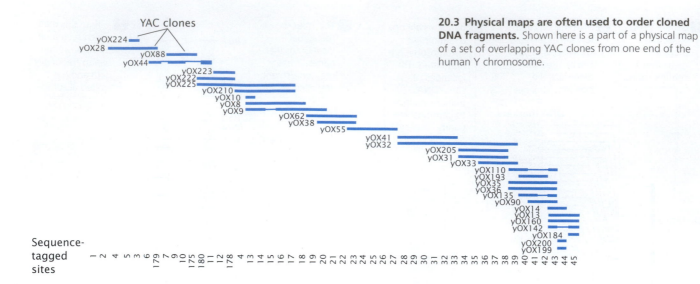

20.3 Physical maps are often used to order cloned DNA fragments. Shown here is a part of a physical map of a set of overlapping YAC clones from one end of the human Y chromosome.

along a street, whereas a genetic map is analogous to a highway map that shows the general locations of major towns and cities.

One of the techniques that has been used for creating physical maps is restriction mapping, which determines the position of restriction sites on DNA. When a piece of DNA is cut with a restriction enzyme and the fragments are separated by gel electrophoresis, the number of restriction sites in the DNA and the distances between them can be determined by the number and positions of bands on the gel (see p. 555 in Chapter 19). However, this information does not tell us the order or the precise location of the restriction sites. To map restriction sites, a sample of the DNA is cut with one restriction enzyme, and another sample is cut with a different restriction enzyme. A third sample is cut with both restriction enzymes together (a double digest). The DNA fragments produced by these restriction digests are then separated by gel electrophoresis, and their sizes are compared. Overlap in size of fragments produced by the digests can be used to position the restriction sites on the original DNA molecule (see the Worked Problem at the end of this chapter). Most restriction mapping is done with several restriction enzymes, used alone and in various combinations, producing many restriction fragments. With long pieces of DNA (greater than 30 kb), computer programs are used to determine the restriction maps, and restriction mapping may be facilitated by tagging one end of a large DNA fragment with radioactivity or by identifying the end with the use of a probe.

Physical maps, such as restriction maps of DNA fragments or even whole chromosomes, are often created for genomic analysis. These lengthy maps are created by combining maps of shorter, overlapping genomic fragments. A number of additional techniques exist for creating physical maps including: sequence-tagged site (STS) mapping, which locates the positions of short unique sequences of DNA on a chromosome; in situ hybridization, by which markers can be visually mapped to locations on chromosomes; and DNA sequencing.

CONCEPTS

Both genetic and physical maps provide information about the relative positions and distances between genes, molecular markers, and chromosome segments. Genetic maps are based on rates of recombination and are measured in percent recombination (map units) or centimorgans. Physical maps are based on physical distances and are measured in base pairs.

✓ **CONCEPT CHECK 1**

What are some of the limitations of genetic maps?

Sequencing an Entire Genome

The first genomes to be sequenced were small virus genomes. The genome of bacteriophage λ, consisting of 49,000 bp, was completed in 1982. In 1995, the first genome of a free-living organism (*Haemophilus influenzae*) was sequenced by Craig Venter and Claire Fraser of The Institute for Genomic Research (TIGR) and Hamilton Smith of Johns Hopkins University. This bacterium has a small genome of 1.8 million base pairs (**Figure 20.4**). By 1996, the genome of the first eukaryotic organism (yeast) had been determined, followed by the genomes of *Escherichia coli* (1997), *Caenorhabditis elegans* (1998), *Drosophila melanogaster* (2000), and *Arabidopsis thaliana* (2000). The first draft of the human genome was completed in June 2000.

The ultimate goal of structural genomics is to determine the ordered nucleotide sequences of entire genomes of organisms. In Chapter 19, we considered some of the methods used to sequence small fragments of DNA. The main obstacle to sequencing a whole genome is the immense size of most genomes. Bacterial genomes are usually at least several million base pairs long; many eukaryotic genomes are billions of base pairs long and are distributed among dozens of chromosomes. Furthermore, for technical reasons, sequencing

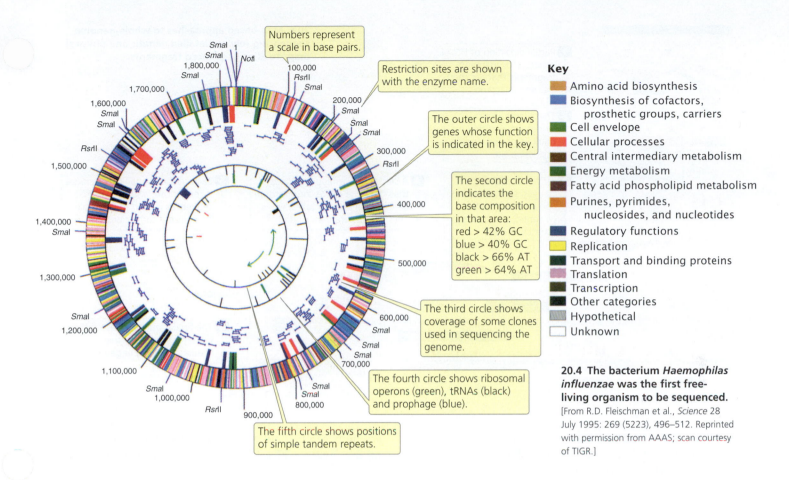

Numbers represent a scale in base pairs.

Restriction sites are shown with the enzyme name.

The outer circle shows genes whose function is indicated in the key.

The second circle indicates the base composition in that area:
red > 42% GC
blue > 40% GC
black > 66% AT
green > 64% AT

The third circle shows coverage of some clones used in sequencing the genome.

The fourth circle shows ribosomal operons (green), tRNAs (black) and prophage (blue).

The fifth circle shows positions of simple tandem repeats.

Key
- Amino acid biosynthesis
- Biosynthesis of cofactors, prosthetic groups, carriers
- Cell envelope
- Cellular processes
- Central intermediary metabolism
- Energy metabolism
- Fatty acid phospholipid metabolism
- Purines, pyrimides, nucleosides, and nucleotides
- Regulatory functions
- Replication
- Transport and binding proteins
- Translation
- Transcription
- Other categories
- Hypothetical
- Unknown

20.4 The bacterium *Haemophilas influenzae* was the first free-living organism to be sequenced. [From R.D. Fleischman et al., *Science* 28 July 1995: 269 (5223), 496–512. Reprinted with permission from AAAS; scan courtesy of TIGR.]

cannot begin at one end of a chromosome and continue straight through to the other end; only small fragments of DNA—usually no more than 500 to 700 nucleotides—can be sequenced at one time. Therefore, determining the sequence for an entire genome requires that the DNA be broken into thousands or millions of smaller fragments that can then be sequenced. The difficulty lies in putting these short sequences back together in the correct order. Two different approaches have been used to assemble the short sequenced fragments into a complete genome: map-based sequencing and whole-genome shotgun sequencing. We will consider these two approaches in the context of the Human Genome Project.

The Human Genome Project

By 1980, methods for mapping and sequencing DNA fragments had been sufficiently developed that geneticists began seriously proposing that the entire human genome could be sequenced. An international collaboration was planned to undertake the Human Genome Project (**Figure 20.5**); initial estimates suggested that 15 years and $3 billion would be required to accomplish the task.

The Human Genome Project officially began in October 1990. Initial efforts focused on developing new and automated methods for cloning and sequencing DNA and on generating detailed physical and genetic maps of the human

genome. The effort was a public project consisting of the international collaboration of 20 research groups and hundreds of individual researchers who formed the International Human Genome Sequencing Consortium. This group used a map-based strategy for sequencing the human genome.

20.5 Craig Venter (left), President of Celera Genomics, and Francis Collins (right), Director of the National Human Genome Research Institute, NIH, announce the completion of a rough draft of the human genome at a press conference in Washington, D.C., on June 26, 2000. [Alex Wong/ Newsmakers/ Getty Images.]

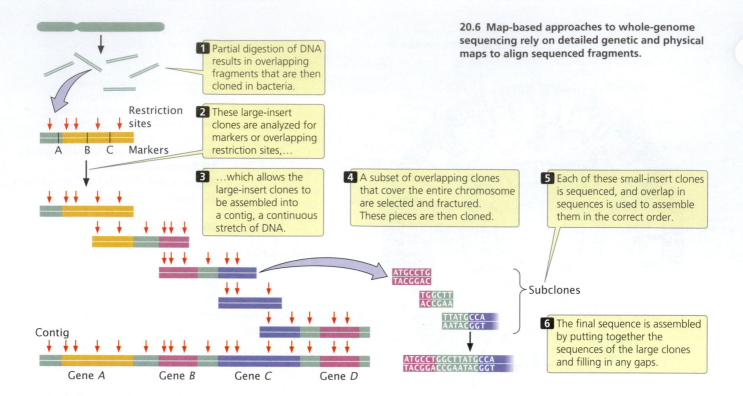

20.6 Map-based approaches to whole-genome sequencing rely on detailed genetic and physical maps to align sequenced fragments.

1 Partial digestion of DNA results in overlapping fragments that are then cloned in bacteria.

2 These large-insert clones are analyzed for markers or overlapping restriction sites,...

3 ...which allows the large-insert clones to be assembled into a contig, a continuous stretch of DNA.

4 A subset of overlapping clones that cover the entire chromosome are selected and fractured. These pieces are then cloned.

5 Each of these small-insert clones is sequenced, and overlap in sequences is used to assemble them in the correct order.

6 The final sequence is assembled by putting together the sequences of the large clones and filling in any gaps.

Restriction sites

A B C Markers

Contig

Gene A   Gene B   Gene C   Gene D

ATGCCTG
TACGGAC

TGGCTT
ACCGAA

TTATGCCA
AATACGGT

Subclones

ATGCCTGGCTTATGCCA
TACGGACCGAATACGGT

MAP-BASED SEQUENCING In **map-based sequencing,** short sequenced fragments are assembled into a whole-genome sequence by first creating detailed genetic and physical maps of the genome, which provide known locations of genetic markers (restriction sites, other genes, or known DNA sequences) at regularly spaced intervals along each chromosome. These markers are later used to help align the short sequenced fragments into their correct order.

After the genetic and physical maps are available, chromosomes or large pieces of chromosomes are separated by pulsed-field gel electrophoresis (PFGE) or by flow cytometry. Standard gel electrophoresis cannot separate very large pieces of DNA, such as whole chromosomes, but PFGE can separate large molecules of DNA or whole chromosomes in a gel by periodically alternating the orientation of an electrical current. In flow cytometry, chromosomes are sorted optically by size.

Each chromosome (or sometimes the entire genome) is then cut up by partial digestion with restriction enzymes (**Figure 20.6**). Partial digestion means that the restriction enzymes are allowed to act for only a limited time so that not all restriction sites in every DNA molecule are cut. Thus, partial digestion produces a set of large overlapping DNA fragments, which are then cloned with the use of cosmids, yeast artificial chromosomes (YACs), or bacterial artificial chromosomes (BACs; see Chapter 19).

Next, these large-insert clones are put together in their correct order on the chromosome (see Figure 20.6). This

assembly can be done in several ways. One method relies on the presence of a high-density map of genetic markers. A complementary DNA probe is made for each genetic marker, and a library of the large-insert clones is screened with the probe, which will hybridize to any colony containing a clone with the marker. The library is then screened for neighboring markers. Because the clones are much larger than the markers used as probes, some clones will have more than one marker. For example, clone A might have markers M1 and M2, clone B markers M2, M3, and M4, and clone C markers M4 and M5. Such a result would indicate that these clones contain areas of overlap, as shown here:

A set of two or more overlapping DNA fragments that form a contiguous stretch of DNA is called a **contig**. This approach was used in 1993 to create a contig consisting of 196 overlapping YAC clones (see Figure 20.3) of the human Y chromosome.

The order of clones can also be determined without the use of preexisting genetic maps. For example, each clone can be cut with a series of restriction enzymes and the resulting fragments then separated by gel electrophoresis. This method generates a unique set of restriction fragments, called a fingerprint, for each clone. The restriction patterns for the clones are stored in a database. A computer program is then used to examine the restriction patterns of all the clones and look for areas of overlap. The overlap is then used to arrange the clones in order, as shown here:

Other genetic markers can be used to help position contigs along the chromosome. ▶ **TRY PROBLEM 27**

When the large-insert clones have been assembled into the correct order on the chromosome, a subset of overlapping clones that efficiently cover the entire chromosome can be chosen for sequencing; the goal is to select the minimum number of clones that are necessary to represent the chromosome. Each of the selected large-insert clones is fractured into smaller overlapping fragments, which are themselves cloned (see Figure 20.6). These smaller clones (called small-insert clones) are then sequenced. The sequences of the small-insert clones are examined for overlap, which allows them to be correctly assembled to give the sequence of the large-insert clones. Enough overlapping small-insert clones are usually sequenced to ensure that the entire genome is sequenced several times. Finally, the whole genome is assembled by putting together the sequences of all overlapping contigs (see Figure 20.6). Often, gaps in the genome map still exist and must be filled in by using other methods.

The International Human Genome Sequencing Consortium used a similar map-based approach to sequencing the human genome. Many copies of the human genome were cut up into fragments of about 150,000 bp each, which were inserted into bacterial artificial chromosomes. Yeast artificial chromosomes and cosmids had been used in early stages of the project but did not prove to be as stable as the BAC clones, although YAC clones were instrumental in putting together some of the larger contigs. Restriction fingerprints were used to assemble the BAC clones into contigs, which were positioned on the chromosomes with the use of genetic markers and probes. The individual BAC clones were sheared into smaller overlapping fragments and sequenced, and the whole genome was assembled by putting together the sequence of the BAC clones.

WHOLE-GENOME SHOTGUN SEQUENCING In 1998, Craig Venter announced that he would lead a company called Celera Genomics in a private effort to sequence the human genome. He proposed using a shotgun sequencing approach, which he suggested would be quicker than the map-based approach employed by the International Human Genome Sequencing Consortium. In **whole-genome shotgun sequencing** (**Figure 20.7**), small-insert clones are prepared directly from genomic DNA and sequenced. Powerful computer programs then assemble the entire genome by examining overlap among the small-insert clones. One advantage of shotgun sequencing is that the small-insert clones can be placed into plasmids, which are simple and easy to manipulate. The requirement for overlap means that most of the genome will be sequenced multiple (often from 10 to 15) times. The average number of times a nucleotide in the genome is sequenced is called the sequencing coverage. For example, 10x coverage means an average nucleotide in the genome has been sequenced 10 times.

Shotgun sequencing was initially used for assembling small genomes such as those of bacteria. When Venter proposed the use of this approach for sequencing the human genome, it was not at all clear that the approach could successfully assemble a complex genome consisting of billions of base pairs such as the human genome. Today, virtually all genomes are sequenced using the whole-genome shotgun approach.

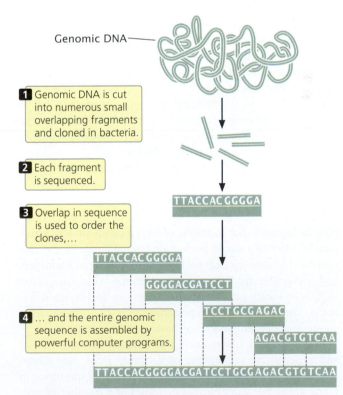

Genomic DNA

1 Genomic DNA is cut into numerous small overlapping fragments and cloned in bacteria.

2 Each fragment is sequenced.

3 Overlap in sequence is used to order the clones,…

4 … and the entire genomic sequence is assembled by powerful computer programs.

TTACCACGGGGA

TTACCACGGGGA
GGGGACGATCCT
TCCTGCGAGAC
AGACGTGTCAA

TTACCACGGGGACGATCCTGCGAGACGTGTCAA

20.7 Whole-genome shotgun sequencing utilizes sequence overlap to align sequenced fragments.

CONCEPTS

Sequencing a genome requires breaking it up into small overlapping fragments of which the DNA sequences can be determined in a sequencing reaction. In map-based sequencing, sequenced fragments are ordered into the final genome sequence with the use of genetic and physical maps. In whole-genome shotgun sequencing, the genome is assembled by comparing overlap in the sequences of small fragments.

✓ CONCEPT CHECK 2

A contig is
a. a set of molecular markers used in genetic mapping.
b. a set of overlapping fragments that form a continuous stretch of DNA.
c. a set of fragments generated by a restriction enzyme.
d. a small DNA fragment used in sequencing.

RESULTS AND IMPLICATIONS OF THE HUMAN GENOME PROJECT

In the summer of 2000, both public and private sequencing projects announced the completion of a rough draft that included most of the sequence of the human genome, 5 years ahead of schedule. The human genome sequence was declared completed in the spring of 2003, although some gaps still remain. For most chromosomes, the finished sequence is 99.999% accurate, with less than one base-pair error per 100,000 bp, an accuracy rate 10 times that of the initial goal.

With the first human genome determined, sequencing additional genomes is much easier. It is now possible to sequence an entire human genome in a single day, and genomes from thousands of different people have now been sequenced. The cost of sequencing a complete human genome has also dropped dramatically and will continue to fall as sequencing technology improves. A number of commercial companies are racing to develop the technology to sequence a complete human genome for $1,000 or less.

The availability of the complete sequence of the human genome is proving to be of great benefit. The sequence has provided tools for detecting and mapping genetic variants across the human genome, greatly facilitating gene mapping in humans. For example, several million sites at which people differ in a single nucleotide (called single-nucleotide polymorphisms; see next section) have now been identified, and these sites are being used in genome-wide association studies to locate genes that affect diseases and traits in humans. The sequence is also providing important information about development and many basic cellular processes.

Next-generation sequencing techniques that allow rapid and inexpensive sequencing of genomic DNA (see Chapter 19) are being used to address fundamental questions in many areas. For example, the genomes of a number of cancer cells have now been completely sequenced and compared with sequences of healthy cells from the same person, allowing complete determination of all the mutations that lead to tumor formation and cancer progression. The complete genome of an unborn baby was recently sequenced from fetal DNA isolated from its mother's blood. The 1000 Genomes Project is sequencing and comparing the genomes of several thousand people from different ethnic groups, with the goal of detecting most of the common variations that exist in the human species. Sequencing of the complete genomes of parents and their children has allowed a direct estimate of mutation rates.

DNA has been extracted from the bones of ancient humans, including Neanderthals and Denisovans (a little-known group of humans that appear to be closely related to Neanderthals), and completely sequenced. Comparisons of the modern human genome with these and other species are adding to our understanding of human evolution, as well as our knowledge of the evolution and the history of all life.

In spite of these benefits and successes, some people have been disappointed by the lack of tangible results from the Human Genome Project. At the time it was proposed, there was speculation that the sequencing of the human genome would immediately revolutionize the practice of medicine, leading to new insights for treating common diseases and resulting in the development of powerful new drugs. Although genomic sequence data have produced numerous new and exciting research findings and has led to a better understanding of many diseases, the data are still seldom used by practicing physicians in the treatment of patients. Undoubtedly, genomic information will be important for medicine in the future, both for tailoring treatment to individual patients (personalized medicine) and for drug discovery, but when this will take place is presently uncertain.

Along with the many potential benefits of having complete sequence information, there are concerns about it being misused. With the knowledge gained from genomic sequencing, many more genes for diseases, disorders, and behavioral and physical traits will be identified, increasing the number of genetic tests that can be performed to make predictions about the future phenotype and health of a person. There is concern that information from genetic testing might be used to discriminate against people who are carriers of disease-causing genes or who might be at risk for some future disease. This matter has been addressed to some extent in the United States with the passage of the Genetic Information Nondiscrimination Act (see Chapter 6), which prohibits health insurers and employers from using genetic information to make decisions about health-insurance coverage and employment (although it does not apply to life, disability, and long-term care insurance). Questions also arise about who should have access to a person's genome sequence. What about relatives, who have similar genomes and might also might be at risk for some of the same diseases? There are also questions about the use of this information to select for specific traits in future offspring. All of these concerns are legitimate and must be addressed if we are to use the information from genome sequencing responsibly.

Single-Nucleotide Polymorphisms

Since the completion of the sequencing of the human genome, sequencers have focused much of their effort on mapping differences among people in their genomic sequences.

Imagine that you are riding in an elevator with a random stranger. How much of your genome do you have in common with this person? Studies of variation in the human genome indicate that you and the stranger will be identical at about 99.9% of your DNA sequences. The difference between you and the stranger is very small in *relative* terms but, because the human genome is so large (3.2 billion base pairs), you and the stranger will be different at more than 3 million base pairs of your genomic DNA. These differences are what make each of us unique, and they greatly affect our physical features, our health, and possibly even our intelligence and personality.

A site in the genome where individual members of a species differ in a single base pair is called a **single-nucleotide polymorphism** (SNP, pronounced "snip"). Arising through mutation, SNPs are inherited as allelic variants (just as are alleles that produce phenotypic differences, such as blood types), although SNPs do not usually produce a phenotypic difference. Single-nucleotide polymorphisms are numerous and are present throughout genomes. In a comparison of the same chromosome from two different people, a SNP can be found approximately every 1000 bp.

HAPLOTYPES AND LINKAGE DISEQUILIBRIUM Most SNPs present within a population arose once from a single mutation that occurred on a particular chromosome and subsequently spread throughout the population. Thus, each SNP is initially associated with other SNPs (as well as other types of genetic variants or alleles) that were present on the particular chromosome on which the mutation arose. The specific set of SNPs and other genetic variants observed on a single chromosome or part of a chromosome is called a **haplotype** (**Figure 20.8**). SNPs within a haplotype are physically linked and therefore tend to be inherited together. New haplotypes can arise through mutation or crossing over, which breaks up the particular set of SNPs in a haplotype. Because the rate of crossing over is proportional to the physical distances between genes, SNPs and other genetic variants that are located close together on the chromosome will be strongly associated as haplotypes. The nonrandom association between genetic variants within a haplotype is called **linkage disequilibrium**. Because the SNPs in a haplotype

20.8 A haplotype is a specific set of SNPs and other genetic variants observed on a single chromosome or part of a chromosome. Chromosomes 1a, 1b, 1c, and 1d represent different copies of a chromosome that might be found in a population.

are inherited together, a haplotype consisting of thousands of SNPs can be identified with the use of only a few SNPs. The few SNPs used to identify a haplotype are called **tag SNPs**. There are about 10 million SNPs in the human population but, because of linkage disequilibrium, these SNPs are present in a much smaller number of haplotypes. Therefore, a relatively small number of SNPs—perhaps only 100,000—can be used to identify most of the haplotypes in humans.

THE USE OF SINGLE-NUCLEOTIDE POLYMORPHISMS

Because of their variability and widespread occurrence throughout the genome, SNPs are valuable as markers in linkage studies. When a SNP is physically close to a disease-causing locus, it will tend to be inherited along with the disease-causing allele. People with the disease will tend to have different SNPs from those of healthy people. A comparison of SNP haplotypes in people with a disease and in healthy people can reveal the presence of genes that affect the disease; because the disease gene and the SNP are closely linked, the location of the disease-causing gene can be determined from the location of associated SNPs. This approach is the same as that used in gene mapping with RFLPs (see Chapter 19), but there are many more SNPs than RFLPs, providing a dense set of variable markers covering the entire genome that can be used more effectively in mapping.

An international effort, called the International HapMap Project, was begun in 2002 with the goal of cataloging and mapping SNPs and other genetic variants that could be used to identify common haplotypes in human populations for use in linkage and family studies. Phase I of the project, completed in 2005, cataloged more than 1 million SNPs in the genomes of 269 people from four diverse human populations

(African, Japanese, Chinese, and European). These SNPs are spread over all 23 human chromosomes at a distance of about one SNP per 5000 bp. Phase II of the project, completed in 2006, cataloged a total of 4.6 million SNPs.

Alleles from most of the common SNPs are found in all four human ethnic groups, although the frequencies of the alleles vary considerably among human populations. The greatest genetic diversity of SNPs is found within Africans, which is consistent with many other studies that suggest that humans first evolved in Africa.

Data from the HapMap Project have provided important information about the function and evolution of the human genome. For example, studies of linkage disequilibrium with the use of SNPs have determined that recombination does not take place randomly across the chromosomes: there are numerous recombination hotspots, where more recombination takes place than expected merely by chance. The distribution of SNPs is also providing information about regions of the human genome that have evolved quickly in the recent past, providing clues to how humans have responded to natural selection.

GENOME-WIDE ASSOCIATION STUDIES Many common diseases are caused by complex interactions among multiple genes: the availability of SNPs has greatly facilitated the search for these genes. **Genomewide association studies** use numerous SNPs scattered across the genome to find genes of interest. Soon after the completion of phase I of the HapMap Project, researchers used the SNP data to conduct a genome-wide association study for the presence of a major gene that causes age-related macular degeneration, one of the leading causes of blindness among the elderly. In this study, researchers genotyped 96 people with macular degeneration and 50 people with healthy eyes for more than 100,000 SNPs that blanketed the genome. The study revealed a strong association between the disease and a gene on chromosome 1 that encodes complement factor H, which has a role in immune function. This finding suggests that macular degeneration might be treated with drugs that affect complement proteins. Another study uncovered a major gene associated with Crohn disease, a common inflammatory disorder of the gastrointestinal tract. Other studies using genome-wide scans of SNPs have identified genes that contribute to heart disease, bone density, prostate cancer, and diabetes.

In one successful application of SNPs for finding disease associations, researchers genotyped 17,000 people in the United Kingdom for 500,000 SNPs in 2007. They detected strong associations between 24 genes and chromosome segments and the incidence of seven common diseases, including coronary artery disease, Crohn disease, rheumatoid arthritis, bipolar disorder, hypertension, and two types of diabetes. The importance of this study is its demonstration that genome-wide association studies utilizing SNPs could successfully locate genes that contribute to complex diseases caused by multiple genetic and environmental factors.

Within the past few years, SNPs have been used in genome-wide association studies to successfully locate genes that influence many additional traits, such as the age of puberty and menopause in women, variation in facial features, skin pigmentation, eye color, body weight, bone density, glaucoma, and even susceptibility to infectious diseases such as meningococcal disease and tuberculosis. Unfortunately, the genes identified often explain only a small proportion of the genetic influence on the trait. For example, one huge genome-wide association study combined data from over 100,000 human subjects in an attempt to locate genes coding for blood lipids and cardiovascular disease. Although the study identified 95 different loci associated with lipid traits, these genes only correspond to 25-30% of the total genetic variation in these traits.

DNA sequences that encode the majority of the missing genetic variation in such traits—sometimes called the "dark matter of the genome"—have, this far, remained largely undetected. The low percentage of variation explained by most current genome-wide association studies means that the genes identified are not, by themselves, useful predictors of the risk of inheriting the disease or trait. Nevertheless the identification of specific genes that influence a disease or trait can lead to a better understanding of the biological processes that produce the phenotype.

Copy-Number Variations

A diploid person normally possesses two copies of every gene, one inherited from the mother and one inherited from the father. Nevertheless, studies of the human genome have revealed differences among people in the number of copies of large DNA sequences (greater than 1000 bp); these variations are called **copy-number variations** (CNVs). Copy-number variations may include deletions, causing some people to have only a single copy of a sequence, or duplications, causing some people to have more than two copies. A study of CNVs in 270 people from four populations (those studied in the HapMap Project) identified a surprising number of these types of variants: more than 1447 genomic regions varied in copy number, encompassing 12% of the human genome. But only a small part of the human genome was surveyed in this study, and so the CNVs detected are only a small subset of all that exist. Many of the CNVs encompassed large regions of DNA sequence, often several hundred thousand base pairs in length. Thus, people differ not only at millions of individual SNPs, but also in the number of copies of many larger segments of the genome.

Most CNVs contain multiple genes and potentially affect the phenotype by altering gene dosage and by changing the position of sequences, which may affect the regulation of nearby genes. Indeed, several studies have found associations between CNVs and disease, and even between CNVs and normal phenotypic variability in human populations. For example, variations in the number of copies of the *UGT2B17* gene contribute to differences in testosterone

Sequence-Tagged Sites and Expressed-Sequence Tags

A **sequence-tagged site** (STS) is a short (200-500 bp) sequence of DNA, present only once in the genome, whose chromosomal location has been determined. STSs are often used to determine the genomic location of a DNA clone. Researchers have developed PCR primers for each STS, allowing the amplification of the STS by PCR. In this way, PCR can be used to detect the presence of an STS in a set of clones. Because the chromosomal location of the STS is known, the presence of the STS in a particular clone provides information about where in the genome that clone originated.

Another type of identified sequence is an **expressed sequence tag** (EST). These are sequences corresponding to DNA that is transcribed into mRNA. In most eukaryotic organisms, only a small percentage of the DNA encodes proteins; for example, in humans, about 1.5% of the DNA encodes proteins. If only protein-encoding genes are of interest, an examination of mRNA rather than the entire DNA genomic sequence is often more efficient. Expressed-sequence tags are obtained by isolating RNA from a cell and subjecting it to reverse transcription, producing a set of cDNA fragments that correspond to RNA molecules from the cell. Short stretches from the ends of these cDNA fragments are then sequenced, and the sequence obtained (called a tag) provides a marker that identifies the DNA fragment. Expressed-sequence tags can be used to find active genes in a particular tissue or at a particular point in development.

CONCEPTS

In addition to collecting genomic-sequence data, genomic projects are collecting databases of nucleotides that vary among individual organisms (single-nucleotide polymorphisms, SNPs), variations in the number of copies of sequences (copy-number variations), unique sequences whose chromosomal location has been mapped (sequence-tagged sites, STSs), and markers associated with transcribed sequences (expressed-sequence tags, ESTs).

✓ **CONCEPT CHECK 3**

What was the goal of the HapMap Project?

Bioinformatics

Complete genome sequences have now been determined for numerous organisms, with many additional projects underway. These studies are producing tremendous quantities of sequence data. Cataloging, storing, retrieving, and analyzing this huge data set are major challenges of modern genetics.

Bioinformatics is an emerging field consisting of molecular biology and computer science that centers on the development of databases, computer-search algorithms, gene-prediction software, and other analytical tools that are used to make sense of DNA-, RNA-, and protein-sequence data. Bioinformatics develops and applies these tools to "mine the data," extracting the useful information from sequencing projects. The development and use of algorithms and computer software for analyzing DNA- and protein-sequence data have helped to make molecular biology a more-quantitative field. Sequence data in publicly available online databases enable scientists and students throughout the world to access this tremendous resource.

A number of databases have been established for the collection and analysis of DNA- and protein-sequence information. Primary databases contain the sequence information, along with information that describes the source of the sequence and its determination. Secondary databases contain the results of analyses carried out on the primary sequence data, such as information about particular sequence patterns, variations, mutations, and evolutionary relationships.

After a genome has been sequenced, one of the first tasks is to identify potential genes within the sequence. Unfortunately, there are no universal characteristics that mark the beginning and end of a gene. The enormous amount of DNA in a genome and the complexities of gene structure make finding genes within the sequence a difficult task. Computer programs have been developed to look for specific sequences in DNA that are associated with certain genes. There are two general approaches to finding genes. The *ab initio* approach scans the sequence looking for features that are usually within a gene. For example, protein-encoding genes are characterized by an open reading frame, which includes a start codon and a stop codon in the same reading frame. Specific sequences mark the splice sites at the beginning and end of introns; other specific sequences are present in promoters immediately upstream of start codons. The comparative approach looks for similarity between a new sequence and sequences of all known genes. If a match is found, then the new sequence is assumed to be a similar gene. Some of these computer programs are capable of examining databases of EST and protein sequences to see if there is evidence that a potential gene is expressed.

It is important to note that the programs that have been developed to identify genes on the basis of DNA sequence are not perfect. Therefore, the numbers of genes reported in most genome projects are estimates. The presence of multiple introns, alternative splicing, multiple copies of some genes, and much noncoding DNA between genes makes accurate identification and counting of genes difficult.

After a gene has been identified, it must be **annotated**, which means linking its sequence information to other information about its function and expression, the protein that it encodes, and information on similar genes in other species. There are a number of methods of probing a gene's function, which will be discussed in the Section 20.2 on functional genomics. Computer programs are available for determining whether similar sequences have already been found, either in the same species or in different species. The most widely used of these programs is the Basic Local Alignment Search Tool (BLAST). To conduct a BLAST search, a researcher submits a query sequence and the program searches the database for any other sequences that have regions of high similarity to the query sequence. The program returns all sequences in the database that are similar, along with information about the degree of similarity and the significance of the match (how likely the similarity would be to occur by chance alone).

Additional databases contain information on sequence diversity—how and where a genome varies among individual organisms. The human genome has now been mapped for millions of SNPs, and information about the frequency of these variants in different populations is being collected. Additional databases have been developed to catalog all known mutations causing particular diseases; the Human Variome Project is an effort to collect and make available all genetic variations that affect human health. Important information about the expression patterns of the thousands of genes found in genomes also is being compiled.

▶ TRY PROBLEM 29

CONCEPTS

Bioinformatics is an interdisciplinary field that combines molecular biology and computer science. It develops databases of DNA and protein sequences and tools for analyzing those sequences.

✓ CONCEPT CHECK 4

The *ab initio* approach finds genes by looking for
a. common sequences found in most genes.
b. similarity in sequence with known genes.
c. mRNA with the use of in situ hybridization.
d. mutant phenotypes.

Metagenomics

Advances in sequencing technology, which have made sequencing faster and less expensive, now provide the possibility of sequencing not just the genomes of individual species but the genomes of entire communities of organisms. **Metagenomics** is an emerging field in which the genome sequences of an entire group of organisms that inhabit a common environment are sampled and determined.

Thus far, metagenomics has been applied largely to microbial communities. Traditionally, bacteria have been studied by growing and analyzing them in the laboratory. However, many bacteria cannot be cultured with the use of laboratory techniques. Metagenomics analyzes microbial communities by extracting DNA from the environment, determining its sequences, and reconstructing community composition and function on the basis of genome sequences. This technique allows the identification and genetic analysis of species that cannot be grown in the laboratory and have never been studied by traditional microbiological methods. The entire genomes of some dominant species have been reconstructed from environmental samples, providing scientists with a great deal of information on the biology of these microbes.

An early metagenomic study analyzed the microbial community found in acid drainage from a mine and determined that this community consisted of only a few dominant bacterial species. Another study, called the Global Ocean Sampling Expedition, followed the route of Darwin's voyage on the H.M.S. *Beagle* in the 1800s. Scientists collected ocean samples and used metagenomic methods to determine their microbial communities. In this study, scientists cataloged sequences for more than 6 million proteins, including more than 1700 new protein families. Some important results have already emerged from metagenomic studies. Analyses of bacterial genomes found in ocean samples led to the discovery of proteorhodopsin proteins, which are light-driven proton pumps. Subsequent research demonstrated that these proteins are found in diverse microbial groups and are a major source of energy flux in the world's oceans.

Other metagenomic studies have examined the genes of bacteria that inhabit the human intestinal tract. These bacteria, along with those that inhabit the skin and other parts of the human body, are termed the human **microbiome**. The microbiome of a typical person includes over 100 trillion cells—more than 10 times as many as human cells—and contains 100 times as many genes as the human genome.

Research is demonstrating that the human microbiome plays an important role in human health. One study examined the gut microflora of obese and lean people. Two groups of bacteria are common in the human gut: Bacteroidetes and Firmicutes. Researchers discovered that obese people have relatively more Firmicutes than do lean people and that the proportion of Firmicutes decreases in obese people who lose weight on a low-calorie diet. These same results were observed in obese and lean mice. In an elegant experiment, researchers transferred bacteria from obese to lean mice. The mice that received bacteria from obese mice extracted more calories from their food and stored more fat, suggesting that gut microflora might play some role in obesity.

Ecological communities have traditionally been characterized on the basis of the species that they contain. Metagenomics affords the possibility of characterizing communities on the basis of their component genes, an approach that has been termed gene-centric. A gene-centric approach leads to new questions: Are certain types of genes more common in some communities than others? Are some genes essential for energy flow and nutrient recycling within a community? Because of the larger sizes of their genomes, eukaryotic communities have not yet been the focus of these approaches, but many researchers predict that they will be in the future.

CONCEPTS

Metagenomic studies examine the genomes of communities of organisms that inhabit a common environment. This approach has been applied to microbial communities and allows the composition and genetic makeup of the community to be determined without cultivating and isolating individual species. Metagenomic studies are a source of important new insights into microbial communities.

Synthetic Biology

The ability to sequence and study whole genomes, coupled with an increased understanding of the genetic information required for basic biological processes, now provides the possibility of creating—entirely from scratch—novel organisms that have never before existed. Synthetic biology is a new field that seeks to design organisms that might provide useful functions, such as microbes that provide clean energy or break down toxic wastes.

Synthetic biologists have already mixed and matched parts from different organisms to synthesize microbes. In 2002, geneticists recreated the poliovirus by joining together pieces of DNA that were synthesized in the laboratory. Even more impressively, in 2010 Daniel Gibson and his colleagues synthesized from scratch the complete 1.08-million-base-pair genome of the bacterium *Mycoplasma mycoides*. They started with a thousand pieces of DNA that were synthesized in the laboratory and then joined them together in successively larger pieces until they had assembled a complete copy of the genome. Within their synthetic genome they included a set of DNA sequences that spelled out—in code—an email address, the names of the researchers who participated in the project, and several well-known quotations. Finally, the researchers transplanted the artificial genome into a cell of a different bacterial species, *M. capricolum*, whose original genome had been removed. The new cell then began expressing the traits specified by the synthetic genome.

Synthetic biology has also been extended to eukaryotic cells. In 2011, geneticists replaced 90,000 bp of DNA of yeast chromosome 9 and 30,000 bp of yeast chromosome 6 with synthetic DNA. These partially synthetic yeast cells grew normally and exhibited only minor differences in gene expression. The ultimate goal of these experiments is to replace the entire 12 million bp genome of yeast with human-designed sequences.

These types of experiments have raised a number of concerns. The ability to tailor-make novel genomes and mix and match parts from different organisms creates the potential for the synthesis of dangerous microbes that might create ecological havoc if they escape from the laboratory or that could be used in biological warfare or bioterrorism. Ongoing discussions among geneticists, ethicists, security experts, and politicians are addressing these concerns and whether synthetic genomes can be safely made and used.

20.2 Functional Genomics Determines the Function of Genes by Using Genomic-Based Approaches

A genomic sequence is, by itself, of limited use. Merely knowing the sequence would be like having a huge set of encyclopedias without being able to read: you could recognize the different letters but the text would be meaningless. **Functional genomics** characterizes what the sequences do—their function. The goals of functional genomics include the identification of all the RNA molecules transcribed from a genome, called the **transcriptome** of that genome, and all the proteins encoded by the genome, called the **proteome**. Functional genomics uses both bioinformatics and laboratory-based experimental approaches in its search to define the function of DNA sequences.

Chapter 19 considered several methods for identifying genes and assessing their functions, including in situ hybridization, experimental mutagenesis, and the use of transgenic animals and knockouts. These methods can be applied to individual genes and can provide important information about the locations and functions of genetic information. In this section, we will focus primarily on methods that rely on knowing the sequences of other genes or methods that can be applied to large numbers of genes simultaneously.

Predicting Function from Sequence

The nucleotide sequence of a gene can be used to predict the amino acid sequence of the protein that it encodes. The protein can then be synthesized or isolated and its properties studied to determine its function. However, this biochemical approach to understanding gene function is both time consuming and expensive. A major goal of functional genomics has been to develop computational methods that allow gene function to be identified from DNA sequence alone, bypassing the laborious process of isolating and characterizing individual proteins.

HOMOLOGY SEARCHES One computational method (often the first employed) for determining gene function is to conduct a homology search, which relies on comparisons of DNA and protein sequences from the same organism and from different organisms. Genes that are evolutionarily related are said to be **homologous**. Homologous genes found in different species that evolved from the same gene in a common ancestor are called **orthologs** (**Figure 20.9**). For example, both mouse and human genomes contain a gene that encodes the alpha subunit of hemoglobin; the mouse and human alpha-hemoglobin genes are said to be orthologs, because both genes evolved from an alpha-hemoglobin gene in a common mammalian ancestor. Homologous genes in the same organism (arising by duplication of a single gene in the evolutionary past) are called **paralogs** (see Figure 20.9).

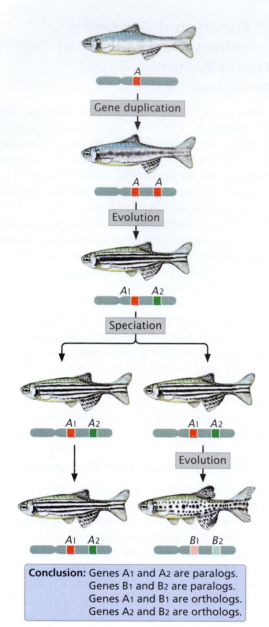

Conclusion: Genes A₁ and A₂ are paralogs.
Genes B₁ and B₂ are paralogs.
Genes A₁ and B₁ are orthologs.
Genes A₂ and B₂ are orthologs.

20.9 Homologous genes are evolutionarily related. Orthologs are homologous genes found in different species; paralogs are homologous genes in the same species.

Within the human genome is a gene that encodes the alpha subunit of hemoglobin and another, homologous gene that encodes the beta subunit of hemoglobin. These paralogs arose because an ancestral gene underwent duplication and the resulting two genes diverged through evolutionary time, giving rise to the alpha- and beta-subunit genes (see Figure 26.18). Homologous genes (both orthologs and paralogs) often have the same or related functions, so, after a function has been assigned to a particular gene, it can provide a clue to the function of a homologous gene.

Databases containing genes and proteins found in a wide array of organisms are available for homology searches. Powerful computer programs, such as BLAST, have been developed for scanning these databases to look for particular sequences. Suppose a geneticist sequences a genome and locates a gene that encodes a protein of unknown function. A homology search conducted on databases containing the DNA or protein sequences of other organisms may identify one or more orthologous sequences. If a function is known for a protein encoded by one of these sequences that could provide information about the function of the newly discovered protein. Sometimes searches do not identify homologous genes; in that case, researchers may look for parts of a gene that are similar to parts of other genes with known function. ▶ TRY PROBLEM 33

PROTEIN DOMAINS Complex proteins often contain regions that have specific shapes or functions called **protein domains**. For example, certain DNA-binding proteins attach to DNA in the same way; these proteins have in common a domain that provides the DNA-binding function. Each protein domain has an arrangement of amino acids common to that domain. There are probably a limited, though large, number of protein domains, which have mixed and matched through evolutionary time to yield the protein diversity seen in present-day organisms.

Many protein domains have been characterized, and their molecular functions have been determined. The sequence from a newly identified gene can be scanned against a database of known domains. If the gene sequence encodes one or more domains whose functions have been previously determined, the function of the domain can provide important information about a possible function of the new gene.

CONCEPTS

The function of an unknown gene can sometimes be determined by finding genes with a similar sequence whose function is known. A gene's function may also be determined by identifying functional domains in the protein that it encodes.

✓ CONCEPT CHECK 5

What is the difference between orthologs and paralogs?
a. Orthologs are homologous sequences; paralogs are analogous sequences.
b. Orthologs are more similar than paralogs.
c. Orthologs are in the same organism; paralogs are in different organisms.
d. Orthologs are in different organisms; paralogs are in the same organism.

Gene Expression and Microarrays

Many important clues about gene function come from knowing when and where the genes are expressed. The development of microarrays has allowed the expression of thousands of genes to be monitored simultaneously.

Microarrays rely on nucleic acid hybridization (see Chapter 19), in which a known DNA fragment is used as a probe to find complementary sequences (**Figure 20.10**). Numerous known DNA fragments are fixed to a solid support in an orderly pattern, or array, usually as a series of dots. These DNA fragments (the probes) usually correspond to known genes from a particular organism.

After the microarray has been constructed, mRNA, DNA, or cDNA isolated from experimental cells is labeled with fluorescent nucleotides and applied to the array. Any of the DNA or RNA molecules that are complementary to probes on the array will hybridize with them and emit fluorescence, which can be detected by an automated scanner. An array containing tens of thousands of probes can be applied to a

Experiment

Question: Can variation in gene expression, detected by microarrays, be used to predict the recurrence of breast cancer?

Methods Microarray chip with DNA probes

1 A microarray consists of DNA probes fixed to a solid support, such as a nylon membrane or glass slide.

2 Each spot consists of a different DNA probe.

Cancer cells   RNA   cDNA with fluorescent bases

Noncancer cells

Hybridization

3 Cancer and noncancer cells removed from 78 women with breast cancer.

4 Messenger RNA from the cells…

5 …is converted into cDNA and labeled with red (cancer cells) or green (noncancer cells) fluorescent nucleotides.

6 The cDNAs are mixed…

7 …and hybridized to DNA probes on a chip.

8 The chip is scanned spot by spot. Yellow fluorescence (red + green) indicates equal expression of the gene in both types of cells; red indicates more expression in cancer cells; and green indicates more expression in noncancer cells.

Results Each spot represents the expression of one gene in one patient's tumor compared with the expression of that gene in her noncancer cells.

Tumors above the solid yellow line came primarily from patients who remained cancer free for at least 5 years.

Tumors below the solid yellow line came primarily from patients in whom the cancer spread within 5 years of diagnosis.

Conclusion: Seventy genes were identified whose expression patterns accurately predicted the recurrence of breast cancer within 5 years of treatment.

20.10 Microarrays can be used to examine gene expression associated with disease progression. [Reprinted by permission from Macmillan Publishers Ltd: van'T Veer, Laura J., et al., "Gene expression profiling predicts clinical outcome of breast cancer," NATURE 405:532, copyright 2002.]

glass slide or silicon wafer just a few square centimeters in size.

THE USE OF MICROARRAYS   Used with cDNA, microarrays can provide information about the expression of thousands of genes, enabling scientists to study which genes are active in particular tissues. They can also be used to investigate how gene expression changes in the course of biological processes such as development or disease progression. For example, breast cancer affects 1 of 10 women in the United States, and half of those women die from it. Current treatment depends on a number of factors, including a woman's age, the size of the tumor, the characteristics of the tumor cells, and whether the cancer has already spread to nearby lymph nodes. Many women whose cancer has not spread are treated by removal of the tumor and radiation therapy, yet the cancer later reappears in some of the women thus treated. These women might benefit from more-aggressive treatment when the cancer is first detected.

Using microarrays, researchers examined the expression patterns of 25,000 genes from primary tumors of 78 young women who had breast cancer (Figure 20.10). Messenger RNA from cancer cells and noncancer cells was converted into cDNA and labeled with red fluorescent nucleotides and with green fluorescent nucleotides, respectively. The labeled cDNAs were mixed and hybridized to a DNA chip, which contained DNA probes from different genes. Hybridization of the red (cancer) and green (noncancer) cDNAs was proportional to the relative amounts of mRNA in the samples. The fluorescence of each spot was assessed with microscopic scanning and appeared as a single color. Red indicates the overexpression of a gene in the cancer cells relative to that in the noncancer cells (more red-labeled cDNA hybridizes), whereas green indicates the underexpression of a gene in the cancer cells relative to that in the noncancer cells (more green-labeled cDNA hybridizes). Yellow indicates equal expression in both types of cells (equal hybridization of red- and green-labeled cDNAs), and no color indicates no expression in either type of cell.

In 34 of the 78 patients, the cancer later spread to other sites; the other 44 patients remained free of breast cancer for 5 years after their initial diagnoses. The researchers identified a subset of 70 genes for which the expression patterns in the initial tumors accurately predicted whether the cancer would later spread (see Figure 20.10). This degree of prediction was much higher than that of traditional predictive measures, which are based on the size and histology of the tumor.

Researchers have also used microarrays to examine the expression of microRNAs (miRNAs) in human cancers. Recent research indicates that miRNAs are frequently expressed abnormally in cancerous tissue and may contribute to the progression of cancer (see Chapter 23). For example, one study using microarrays found that several miRNAs

20.11 **Microarrays have been used to compare the expression of miRNAs in cancerous cervical cells compared to normal cervical cells.** [Adapted from X. Wang, S. Tang, S-Y. Le, R. Lu, J.S. Rader et al. (2008) Aberrant Expression of Oncogenic and Tumor-Suppressive Micro RNAs in Cervical Cancer Is Required for Cancer Cell Growth. *PLoSONE* 3(7): e2557. doi:10.1371/journal.]

were overexpressed in cancerous cervical tissue, compared to normal cervical tissue, while other miRNAs were underexpressed (**Figure 20.11**). Other studies using microarrays have demonstrated that miRNA expression is associated with resistance of tumors to chemotherapy and radiation and that miRNA expression can be used to predict the response of some tumors to anticancer treatment. Results such as these suggest that gene-expression data obtained from microarrays can be a powerful tool in determining the nature of cancer research and treatment. The products of the genes that show differences in expression are being examined as possible targets for drug therapy. ▶ TRY PROBLEM 34

Microarrays that allow the detection of specific alleles, SNPs, and even particular proteins also have been created. Importantly, not all DNA molecules bind equally to microarrays, and so microarrays may sometimes overestimate or underestimate the expression of specific genes. Thus, verification of the results of microarrays through other methods is desirable.

CONCEPTS

Microarrays, consisting of DNA probes attached to a solid support, can be used to determine which RNA and DNA sequences are present in a mixture of nucleic acids. They are capable of determining which RNA molecules are being synthesized and can thus be used to examine changes in gene expression.

Gene Expression and Reporter Sequences

Patterns of gene expression can also be determined visually by using a reporter sequence. In this approach, genomic fragments are first cloned in BACs or other vectors that are capable of holding the coding region of a gene plus its regulatory sequences. The coding region of a gene whose expression is to be studied is then replaced with a reporter sequence, which encodes an easily observed product. For example, a commonly used reporter sequence encodes a green fluorescent protein (GFP) from jellyfish. The BAC is then inserted into an embryo, creating a transgenic organism. The regulatory sequence of the cloned gene ensures that it is expressed at the appropriate time and in the appropriate tissue within the transgenic organism. The product of the gene expression is a fluorescent green pigment, which is easily observed (**Figure 20.12**).

This technique is being used to study the expression patterns of genes that affect brain function. In the Gene Expression Nervous System Atlas (GENSAT) project, scientists are systematically replacing the coding regions of hundreds of genes with the GFP reporter sequence and observing their patterns of expression in transgenic mice. The goal is to produce a comprehensive atlas of gene expression in the mouse brain. This project has already shed light on where in the brain several genes having roles in inherited

20.12 A reporter sequence can be used to examine expression of a gene. Expression of the neural-specific β-tubulin gene in the brain of a tadpole is revealed by fluorescent green pigment. [Photo by Miranda Gomperts, courtesy Enrique Amaya, The Amaya Lab.]

neurological disorders are expressed. Another functional genomics project, the Allen Brain Atlas, has compiled information on the expression patterns of 20,000 genes in the mouse brain.

Genome-Wide Mutagenesis

As discussed in Chapter 18, one of the best methods for determining the function of a gene is to examine the phenotypes of individual organisms that possess a mutation in the gene. Traditionally, genes encoding naturally occurring variations in a phenotype were mapped, the causative genes were isolated, and their products were studied. But this procedure was limited by the number of naturally occurring mutations and the difficulty of mapping genes with a limited number of chromosomal markers. The number of naturally occurring mutations can be increased by exposure to mutagenic agents, and the accuracy of mapping is increased dramatically by the availability of mapped molecular markers, such as RFLPs, microsatellites, STSs, ESTs, and SNPs. These two methods—random inducement of mutations on a genome-wide basis, and mapping with molecular markers—are coupled and automated in a **mutagenesis screen**.

Genome-wide mutagenesis screens can be used to search for all genes affecting a particular function or trait. For example, mutagenesis screens of mice are being used to identify genes having roles in cardiovascular function. When these genes are located in mice, homology searches are carried out to determine if similar genes exist in humans. Those genes can then be studied to better understand cardiac disease in humans.

To conduct a mutagenesis screen, random mutations are induced in a population of organisms, creating new phenotypes. The mutations are induced by exposing the organisms to radiation, a chemical mutagen (see Chapter 18), or transposable elements (DNA sequences that insert randomly into the DNA; see Chapter 18). The procedure for a typical mutagenesis screen is illustrated in **Figure 20.13**. Here, male zebrafish are treated with ethylmethylsulfonate (EMS), a chemical that induces germ-line mutations in their sperm. The treated males are mated with wild-type female fish. A few of the offspring will be heterozygous for mutations induced by EMS; the offspring are screened for any mutant phenotypes that might be the products of dominant mutations expressed in these heterozygous fish. Recessive mutations will not be expressed in the F_1 progeny but can be revealed with further breeding.

The fish with variant phenotypes undergo further breeding experiments to verify that each variant phenotype is, in fact, due to a single-gene mutation, although the gene in which the mutation occurs is not known. The gene can be located, however, by positional cloning (see Chapter 19).

Mutagenesis screens have been used to study genes that control vertebrate development. A team of developmental

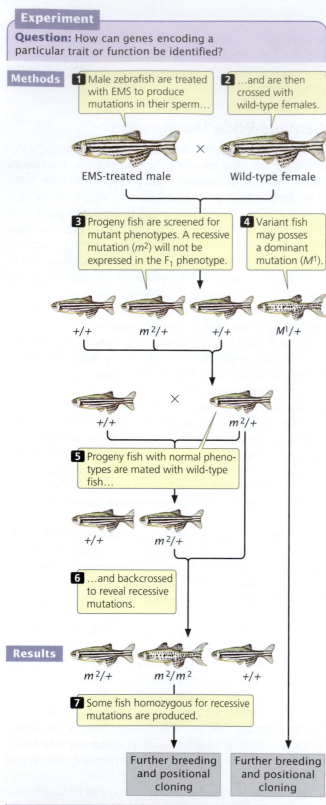

Experiment

Question: How can genes encoding a particular trait or function be identified?

Methods

1 Male zebrafish are treated with EMS to produce mutations in their sperm...

2 ...and are then crossed with wild-type females.

EMS-treated male × Wild-type female

3 Progeny fish are screened for mutant phenotypes. A recessive mutation (m^2) will not be expressed in the F_1 phenotype.

4 Variant fish may posses a dominant mutation (M^1).

+/+ m^2/+   +/+ M^1/+

+/+   × m^2/+

5 Progeny fish with normal phenotypes are mated with wild-type fish...

+/+ m^2/+

6 ...and backcrossed to reveal recessive mutations.

Results

m^2/+ m^2/m^2 +/+

7 Some fish homozygous for recessive mutations are produced.

Further breeding and positional cloning

Further breeding and positional cloning

Conclusion: The mutagenesis screen produces fish with a mutation affecting the trait. These fish are further analyzed to identify the mutation.

20.13 Genes affecting a particular characteristic or function can be identified by a genome-wide mutagenesis screen. In this illustration, M^1 represents a dominant mutation and m^2 represents a recessive mutation.

geneticists have produced thousands of mutations in the zebrafish that affect development and are systematically locating and characterizing the loci where the mutations occur. Zebrafish are ideal genetic models for this type of study because they reproduce quickly, are easily reared in the laboratory, and have transparent embryos in which developmental deformities are easy to spot. This research has already identified a number of genes that are important in embryonic development, many of which have counterparts in humans.

CONCEPTS

Genome-wide mutagenesis screening coupled with positional cloning can be used to identify genes that affect a specific characteristic or function.

✓ **CONCEPT CHECK 6**

Which is the correct order of steps in a mutagenesis screen?
a. Positional cloning, mutagenesis, identification of mutants, verification of genetic basis.
b. Mutagenesis, positional cloning, identification of mutants, verification of genetic basis.
c. Mutagenesis, identification of mutants, verification of genetic basis, positional cloning.
d. Identification of mutants, positional cloning, mutagenesis, verification of genetic basis.

20.3 Comparative Genomics Studies How Genomes Evolve

Genome-sequencing projects provide detailed information about gene content and organization in different species and even in different members of the same species, allowing inferences about how genes function and genomes evolve. They also provide important information about evolutionary relationships among organisms and about factors that influence the speed and direction of evolution. **Comparative genomics** is the field of genomics that compares similarities and differences in gene content, function, and organization among genomes of different organisms.

Prokaryotic Genomes

More than 2000 prokaryotic genomes have now been sequenced. Most prokaryotic genomes consist of a single circular chromosome. However, there are exceptions, such as *Vibrio cholerae*, the bacterium that causes cholera, which has two circular chromosomes, and *Borrelia burgdorferi*, which has one large linear chromosome and 21 smaller chromosomes.

GENOME SIZE AND NUMBER OF GENES The total amount of DNA in prokaryotic genomes ranges from 490,885 bp in *Nanoarchaeum equitans*, an archaea that lives entirely within another archaea, to 9,105,828 bp in *Bradyrhizobium japonicum*, a soil bacterium (**Table 20.1**). Although this range in genome size might seem extensive, it is much less than the

Table 20.1	Characteristics of representative prokaryotic genomes that have been completely sequenced	
Species	**Size (millions of base pairs)**	**Number of Predicted Genes**
Archaea		
Archaeoglobus fulgidus	2.18	2407
Methanobacterium thermoautotrophicum	1.75	1869
Nanoarchaeum equitans	0.490	536
Eubacteria		
Bacillus subtilis	4.21	4100
Bradyrhizobium japonicum	9.11	8317
Escherichia coli	4.64	4289
Haemophilus influenzae	1.83	1709
Mycobacterium tuberculosis	4.41	3918
Mycoplasma genitalium	0.58	480
Staphylococcus aureus	2.88	2697
Vibrio cholerae	4.03	3828

Source: Data from the Genome Atlas of the Center for Biological Sequence Analysis, www.cbs.dtu.dk/services/GenomeAtlas.

enormous range of genome sizes seen in eukaryotes, which can vary from a few million base pairs to hundreds of billions of base pairs. *Escherichia coli*, the most widely used bacterium for genetic studies, has a fairly typical genome size at 4.6 million base pairs. Archaea and bacteria are similar in their ranges of genome size. Surprisingly, genome size also varies extensively within some species; for example, different strains of *E. coli* vary in genome size by more than 1 million base pairs.

Among prokaryotes, the number of genes typically varies from 1000 to 2000, but some species have as many as 6700 and others as few as 480. Interestingly, the density of genes is rather constant across all species, with an average of about one gene per 1000 bp. Thus, prokaryotes with larger genomes will have more genes, in contrast with eukaryotes, for which there is little association between genome size and number of genes (see the next section). The evolutionary factors that determine the size of prokaryotic genomes (as well as eukaryotic genomes) are still largely unknown. However, the time required for cell division is frequently longer in organisms with larger genomes because of the time required to replicate the DNA before division. Thus, selection may favor smaller genomes in organisms that occupy environments where rapid reproduction is advantageous.

Prokaryotes with the smallest genomes tend to be in species that occupy restricted habitats, such as bacteria that live inside other organisms. The constant environment and metabolic functions supplied by the host organism may allow these bacteria to survive with fewer genes. Bacteria with the larger genomes tend to occupy highly complex and variable environments, such as the soil or the root nodules of plants. In these environments, genes that are needed only occasionally may be useful. In complex environments where resources are abundant, there may be little need for rapid division (which favors smaller genomes).

An example of a large-genome bacterium occupying a complex environment is *Streptomyces coeliocolor*, a filamentous bacterium with a genome size of 8,677,507 bp. This bacterium has been called the "Boy Scout" bacterium because it has a diverse set of genes and therefore follows the Scout motto: be prepared. In addition to the usual housekeeping genes for genetic functions such as replication, transcription, and translation, its genome contains a number of genes for the breakdown of complex carbohydrates, allowing it to consume decaying matter from plants, animals, insects, fungi, and other bacteria. It has genes for a large number of proteins that produce secondary metabolites (breakdown products) that can function as antibiotics and protect against desiccation and low temperatures. In this species, a large genome with lots of genes appears to be beneficial in the complex environment that it inhabits.

HORIZONTAL GENE TRANSFER   Bacteria possess a number of mechanisms by which they can gain and lose DNA. DNA can be lost through simple deletion and can be gained by gene duplication and through the insertion of transposable genetic elements. Another mechanism for gaining new genetic information is horizontal gene transfer, a process by which both closely and distantly related bacterial species periodically exchange genetic information over evolutionary time. Such exchange may take place through the bacterial uptake of DNA in the environment (transformation), through the exchange of plasmids, and through viral vectors (see Chapter 9). Horizontal gene transfer has been recognized for some time,

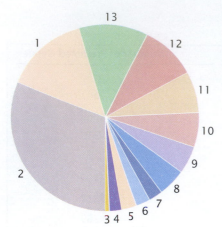

1. Metabolism
2. Unknown
3. Ionic homeostasis
4. Protein synthesis
5. Energy
6. Transport facilitation
7. Cellular biogenesis
8. Intracellular transport
9. Protein destination
10. Cellular communication and signal transduction
11. Cell rescue, defense, cell death, and aging
12. Cell growth, cell division, and DNA synthesis
13. Transcription

20.14 The functions of many genes in prokaryotes cannot be determined by comparison with genes in other prokaryotes. Proportion of the circle occupied by each color represents the proportion of genes affecting various known and unknown functions in *E. coli*.

but analyses of many microbial genomes now indicate that it is more extensive than was formerly recognized. The widespread occurrence of horizontal transfer has caused some biologists to question whether distinct species even exist in bacteria (see Chapter 9 for more discussion of this matter).

FUNCTION OF GENES   Only about half of the genes identified in prokaryotic genomes can be assigned a function. Almost a quarter of the genes have no significant sequence similarity to any known genes in other bacteria. The number of genes that encode biological functions such as transcription and translation tends to be similar among species, even when their genomes differ greatly in size. This similarity suggests that these functions are encoded by a basic set of genes that does not vary among species. In contrast, the number of genes taking part in biosynthesis, energy metabolism, transport, and regulatory functions varies greatly among species and tends to be higher in species with larger genomes. The functions of predicted genes (i.e., genes identified by computer programs) and known genes in *E. coli* are presented in **Figure 20.14**. A substantial part of the "extra" DNA found in the larger bacterial genomes is made up of paralogous genes that have arisen by duplication.

CONCEPTS

Comparative genomics compares the content and organization of whole genomic sequences from different organisms. Prokaryotic genomes are small, usually ranging from 1 million to 3 million base pairs of DNA, with several thousand genes. Prokaryotes with the smallest genomes tend to occupy restricted habitats, whereas those with the largest genomes are usually found in complex environments. Horizontal gene transfer has played a major role in bacterial genome evolution.

✔ CONCEPT CHECK 7

What is the relation between genome size and gene number in prokaryotes?

Eukaryotic Genomes

The genomes of more than 150 eukaryotic organisms have been completely sequenced, including a number of fungi and protists, several insects, almost 30 species of plants, and numerous vertebrates. Sequenced eukaryotes include papayas, corn, rice, sorghum, grapevine, silkworms, several fruit flies, aphids, mosquitoes, anemones, mice, rats, dogs, cows, horses, orangutans, chimpanzees, and humans. Even the genomes of some extinct organisms have now been sequenced, including those of the wooly mammoth and Neanderthals. Hundreds of additional eukaryotic genomes are in the process of being sequenced. It is important to note that, even though the genomes of these organisms have been "completely sequenced," many of the final assembled sequences contain gaps, and regions of heterochromatin may not have been sequenced at all. Thus, the sizes of eukaryotic genomes are often estimates, and the number of base pairs given for the genome size of a particular species may vary. Predicting the number of genes that are present in a genome also is difficult and may vary, depending on the assumptions made and the particular gene-finding software used.

GENOME SIZE AND NUMBER OF GENES   The genomes of eukaryotic organisms (**Table 20.2**) are larger than those of prokaryotes, and, in general, multicellular eukaryotes have more DNA than do simple, single-celled eukaryotes such as yeast (see p. 308 in Chapter 11). However, there is no close relation between genome size and complexity among the multicellular eukaryotes. For example, the mosquito (*Anopheles gambiae*) and fruit fly (*Drosophila melanogaster*) are both insects with similar structural complexity, yet the mosquito has 60% more DNA than the fruit fly.

In general, eukaryotic genomes also contain more genes than do prokaryotes (but some large bacteria have more genes than single-celled yeasts have), and the genomes of multicellular eukaryotes have more genes than do the

Table 20.2 Characteristics of representative eukaryotic genomes that have been completely sequenced

Species	Genome Size (millions of base pairs)	Number of Predicted Genes
Saccharomyces cerevisiae (yeast)	12	6,144
Physcomitrella patens (moss)	480	38,354
Arabidopsis thaliana (mustard plant)	125	25,706
Zea mays (corn)	2,400	32,000
Hordeum vulgare (barley)	5,100	26,159
Caenorhabditis elegans (roundworm)	103	20,598
Drosophila melanogaster (fruit fly)	170	13,525
Anopheles gambiae (mosquito)	278	14,707
Danio rerio (zebrafish)	1,465	22,409
Takifugu rubripes (tiger pufferfish)	329	22,089
Xenopus tropicalis (clawed frog)	1,510	18,429
Anolis carolinesis (anole lizard)	1,780	17,792
Mus musculus (mouse)	2,627	26,762
Pan troglodytes (chimpanzee)	2,733	22,524
Homo sapiens (human)	3,223	~20,000

Source: Ensembl Web site: http://useast.ensembl.org/index.html and plants.ensembl.org/index.html

genomes of single-celled eukaryotes. In contrast with bacteria, there is no correlation between genome size and number of genes in eukaryotes. The *number* of genes among multicellular eukaryotes also is not obviously related to phenotypic complexity: humans have more genes than do invertebrates but only twice as many as fruit flies and fewer than the plant *A. thaliana*. The nematode *C. elegans* has more genes than *D. melanogaster* but is less complex. Additionally, the pufferfish has only about one-tenth the amount of DNA present in humans and mice but about as many genes. Eukaryotic genomes contain multiple copies of many genes, indicating that gene duplication has been an important process in genome evolution.

Most of the DNA in multicellular organisms is noncoding, and many genes are interrupted by introns. In the more complex eukaryotes, both the number and the length of the introns are greater.

SEGMENTAL DUPLICATIONS AND MULTIGENE FAMILIES Many eukaryotic genomes, especially those of multicellular organisms, are filled with **segmental duplications**, regions greater than 1000 bp that are almost identical in sequence. For example, about 4% of the human genome consists of segmental duplications. In most segmental duplications, the two copies are found on the same chromosome (an intrachromosomal duplication), but, in others, the two copies are found on different chromosomes (an interchromosomal duplication). In the human genome, the average size of segmental duplications is 15,000 bp.

Segmental duplications arise from processes that generate chromosome duplications, such as unequal crossing over (see Chapter 8). After a segmental duplication arises, it promotes further duplication by the occurrence of misalignment among the duplicated regions. Segmental duplication plays an important role in evolution by giving rise to new genes. After a segmental duplication arises, the original copy of the gene can continue its function while the new copy undergoes mutation. These changes may eventually lead to new function. The importance of gene duplication in genome evolution is demonstrated by the large number of multigene families that exist in many eukaryotic genomes. A **multigene family** is a group of evolutionarily related genes that arose through repeated evolution of an ancestral gene. For example, the globin gene family in humans consists of 13 genes that encode globinlike molecules, most of which produce proteins that carry oxygen. An even more spectacular example is the human olfactory multigene family, which consists of about 1000 genes that encode olfactory receptor molecules used in our sense of smell.

NONCODING DNA Most eukaryotic organisms contain vast amounts of DNA that do not encode proteins. For example, only about 1.5 percent of the human genome consists of DNA that directly specifies the amino acids of proteins. The function of the remainder of DNA sequences, called noncoding DNA, has long been in question. Some research has suggested that much of this DNA has no function. For example, Marcelo Nobrega and his colleagues genetically engineered mice that were missing a large chromosomal region with no protein-encoding genes (called a **gene desert**). In one experiment, they created mice that were missing a 1,500,000-bp gene desert from mouse chromosome 3; in another, they created mice missing an 845,000-bp gene desert from chromosome 19. Remarkably, these mice appeared healthy and were indistinguishable from control mice. The researchers concluded that large regions of the mammalian genome can be deleted without major phenotypic effects and may, in fact, be superfluous.

However, other research suggested that gene deserts may contain sequences that have a functional role. For example, genome-wide association studies demonstrated that DNA sequences contained within a gene desert on human chromosome 9 are associated with coronary artery disease, and subsequent studies have demonstrated the presence of 33 enhancers in this gene desert.

In 2002 the Encyclopedia of DNA Elements (ENCODE) project was undertaken to determine if noncoding DNA had any function. Researchers cataloged all nucleotides within the genome that provide some function, including sequences encoding proteins and RNA molecules, and those serving as control sites for gene expression. This 10-year project was carried out by a team of over 400 scientists from around the world. In a series of papers published in 2012, ENCODE concluded that at least 80% of the human genome is involved in some type of gene function. Many of the functional sequences consisted of sites where proteins bind and influence the expression of genes. Prior to this study, much of the genome was considered "junk DNA" with no function, but the ENCODE study has greatly altered this view and suggests that there is little nonfunctional DNA in the human genome.

TRANSPOSABLE ELEMENTS A substantial part of the genomes of most multicellular organisms consists of moderately and highly repetitive sequences (see Chapter 11), and the percentage of repetitive sequences is usually higher in those species with larger genomes (**Table 20.3**). Most of these repetitive sequences appear to have arisen through transposition. In the human genome, 45% of the DNA is derived from transposable elements, many of which are

Table 20.3 **Percentage of genome consisting of interspersed repeats derived from transposable elements**

Organism	Percentage of Genome
Arabidopsis thaliana (plant)	10.5
Zea mays (corn)	85.0
Caenorhabditis elegans (worm)	6.5
Drosophila melanogaster (fly)	3.1
Takifugu rubripes (tiger pufferfish)	2.7
Homo sapiens (human)	44.4

Table 20.4 **Number of estimated protein domains encoded by some eukaryotic genomes**

Species	Number of Predicted Protein Domains
Saccharomyces cerevisiae (yeast)	851
Arabidopsis thaliana (plant)	1012
Caenorhabditis elegans (roundworm)	1014
Drosophila melanogaster (fruit fly)	1035
Homo sapiens (human)	1262

Source: Number of genes and protein-domain families from the International Human Genome Sequencing Consortium, Initial sequencing and analysis of the human genome, *Nature* 409:860–921, Table 23, 2001.

defective and no longer able to move. In corn, 85% of the genome is derived from transposable elements.

PROTEIN DIVERSITY In spite of only a modest increase in gene number, vertebrates have considerably more protein diversity than do invertebrates. One way to measure the amount of protein diversity is by counting the number of protein domains, which are characteristic parts of proteins that are often associated with a function. Vertebrate genomes do not encode significantly more protein domains than do invertebrate genomes; for example, there are 1262 domains in humans compared with 1035 in fruit flies (**Table 20.4**). However, the existing domains in humans are assembled into more combinations, leading to many more types of proteins.

HOMOLOGOUS GENES An obvious and remarkable trend seen in eukaryotic genomes is the degree of homology among genes found in even distantly related species. For example, mice and humans have about 99% of their genes in common. About 50% of the genes in fruit flies are homologous to genes in humans, and, even in plants, about 18% of the genes are homologous to those found in humans.

COLLINEARITY BETWEEN RELATED GENOMES One of the features of genome evolution revealed by comparing the gene sequences of different organisms is that many genes are present in the same order in related genomes, a phenomenon that is sometimes termed collinearity. The reason for collinearity among genomes is that they are descended from a common ancestral genome, and evolutionary forces have maintained the same gene order in the genomes of descendants. Genomic studies of grasses—plants in the family Poaceae—illustrate the principle of collinearity.

Grasses comprise more than 10,000 species, including economically important crops such as rice, wheat, barley, corn, millet, oat, and sorghum. Taken together, grasses make up about 60% of the world's food production. The genomes of these species vary greatly in size and chromosome number. For example, chromosome number in grasses ranges from 4 to 266; the rice genome consists of only about 460

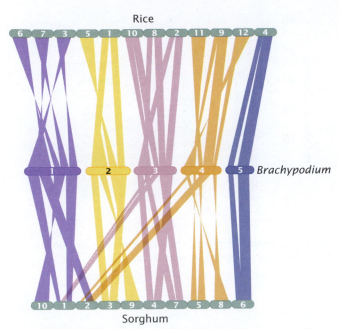

Rice

Brachypodium

Sorghum

20.15 Collinear relationships among blocks of genes found in rice, sorghum, and *Brachypodium* (wild grass). Each colored band represents a block of genes that is collinear between chromosomes of the three species.

million base pairs, whereas the genome of wheat contains 17 billion base pairs. In spite of these large differences in chromosome number and genome size, the position and order of many genes within the genomes are remarkably conserved. Through evolutionary time, regions of DNA between the genes (intergenic regions) have increased, decreased, and undergone rearrangements, whereas the genes themselves have stayed relatively constant in order and content. An example of the collinear relationship of genes between rice, sorghum, and *Brachypodium* (wild grass) is shown in **Figure 20.15**. ▶ TRY PROBLEM 38

CONCEPTS

Genome size varies greatly among eukaryotic species. For multicellular eukaryotic organisms, there is no clear relation between organismal complexity and amount of DNA or gene number. A substantial part of the genome in eukaryotic organisms consists of repetitive DNA, much of which is derived from transposable elements. Many eukaryotic genomes have homologous genes in common, and genes are often in the same order in the genomes of related organisms.

✔ CONCEPT CHECK 8

Segmental duplications play an important role in evolution by
a. giving rise to new genes and multigene families.
b. keeping the number of genes in a genome constant.
c. eliminating repetitive sequences produced by transposition.
d. controlling the base content of the genome.

Comparative *Drosophila* Genomics

The fruit fly *Drosophila melanogaster* is one of the workhorses of genetics. Its genome was sequenced in 2000, the second animal genome to be deciphered. In 2007, researchers published genomic sequences of 10 additional species of *Drosophila*. Combined with the already sequenced genomes of *D. melanogaster* and *D. pseudoobscura*, this effort provided genomic data for 12 species of *Drosophila*, allowing detailed evolutionary analysis of this group of insects.

The 12 *Drosophila* species that were sequenced are found throughout the world, and all diverged from a common ancestor some 60 million years ago (**Figure 20.16**). Genome size in the group varies from 130 million base pairs in *D. mojavensis* to 364 million base pairs in *D. virilis*. The number of protein-encoding genes varies less widely, from 13,733 to 17,325. Many of the genes are found in all 12 of the species; for example, 77% of the 13,733 genes found in *D. melanogaster* have homologs in all of the other species.

Transposable elements have played an important role in the evolution of *Drosophila* genomes. The extent of DNA that consists of the remnants of transposable elements ranges from 2.7% of the genome in *D. simulans* to 25% of the genome of *D. ananassae*. The amount of DNA contributed by transposable elements accounts for much of the difference in genome size among the species. Genomic rearrangements (inversions and translocations) were frequent in the evolution of this group. The rate of evolution varies among different classes of genes: genes controlling olfaction, immunity, and insecticide resistance have evolved at a faster rate compared with other genes.

The Human Genome

The human genome, which is fairly typical of mammalian genomes, has been extensively studied and analyzed because of its importance to human health and evolution. It is 3.2

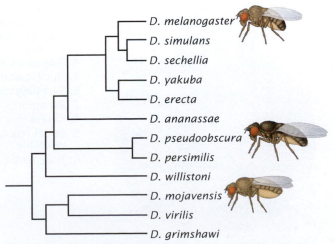

20.16 Twelve *Drosophila* species whose genomes have been determined and analyzed diverged from a common ancestor approximately 60 million years ago.

Table 20.5	Average characteristics of genes in the human genome
Characteristic	**Average**
Number of exons	8.8
Size of internal exon	145 bp
Size of intron	3,365 bp
Size of 5′ untranslated region	300 bp
Size of 3′ untranslated region	770 bp
Size of coding region	1,340 bp
Total length of gene	27,000 bp

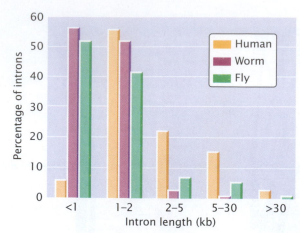

20.17 The introns of genes in humans are generally longer than the introns of genes in worms and flies.

billion base pairs in length, but only about 1.5% encodes proteins. Active genes are often separated by vast regions of noncoding DNA, much of which consists of repeated sequences derived from transposable elements.

The average gene in the human genome is approximately 27,000 bp in length, with about 9 exons (**Table 20.5**). (One exceptional gene has 234 exons.) The introns of human genes are much longer, and there are more of them than in other genomes (**Figure 20.17**). The human genome does not encode substantially more protein domains (see Table 20.4), but the domains are combined in more ways to produce a relatively diverse proteome. Gene functions encoded by the human genome are presented in **Figure 20.18**. A single gene often encodes multiple proteins through alternative splicing; each gene encodes, on the average, two or three different mRNAs.

Gene density varies among human chromosomes; chromosomes 17, 19, and 22 have the highest densities and chromosomes X, Y, 4, 13, and 18 have the lowest densities. Some proteins encoded by the human genome that are not found in other animals include: those affecting immune function; neural development, structure, and function; intercellular and intracellular signaling pathways in development; hemostasis; and apoptosis.

Transposable elements are much more common in the human genome than in worm, plant, and fruit-fly genomes (see Table 20.3). The human genome contains a variety of types of transposable elements, including LINEs, SINEs, retrotransposons, and DNA transposons (see Chapter 18). Most appear to be evolutionarily old and are defective, containing mutations and deletions making them no longer capable of transposition.

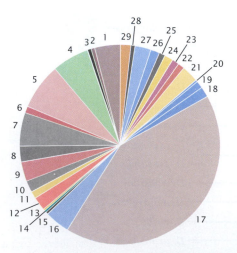

1. Miscellaneous
2. Viral protein
3. Transfer or carrier protein
4. Transcription factor
5. Nucleic acid enzyme
6. Signaling molecule
7. Receptor
8. Kinase
9. Select regulatory molecule
10. Transferase
11. Synthase and synthetase
12. Oxidoreductase
13. Lyase
14. Ligase
15. Isomerase
16. Hydrolase
17. Molecular function unknown
18. Transporter
19. Intracellular transporter
20. Select calcium-binding protein
21. Protooncogene
22. Structural protein of muscle
23. Motor
24. Ion channel
25. Immunoglobulin
26. Extracellular matrix
27. Cytoskeletal structural protein
28. Chaperone
29. Cell adhesion

20.18 Functions for many human genes have yet to be determined. Proportion of the circle occupied by each color represents the proportion of genes affecting various known and unknown functions.

20.4 Proteomics Analyzes the Complete Set of Proteins Found in a Cell

DNA sequence data offer tremendous insight into the biology of an organism, but they are not the whole story. Many genes encode proteins, and proteins carry out the vast majority of the biochemical reactions that shape the phenotype of an organism. Although proteins are encoded by DNA, many proteins undergo modifications after translation and, in more-complex eukaryotes there are many more proteins than genes. Thus, in recent years, molecular biologists have turned their attention to analysis of the protein content of cells. The ultimate goal is to determine the proteome, the complete set of proteins found in a given cell. The study of the proteome is termed **proteomics**.

Plans are underway to identify and characterize all proteins in the human body, an effort that has been called the **Human Proteome Project**. The project would catalog which proteins are present in which cell types, where each protein is located within the cell, and which other proteins each interacts with. Many researchers feel that this information will be of immense benefit in identifying drug targets, understanding the biological basis of disease, and understanding the molecular basis of many biological processes.

Determination of Cellular Proteins

The basic procedure for characterizing the proteome is first to separate the proteins found in a cell and then to identify and quantify the individual proteins. One method for separating proteins is **two-dimensional polyacrylamide gel electrophoresis** (2D-PAGE), in which the proteins are separated in one dimension by charge, separated in a second dimension by mass, and then stained (**Figure 20.19a**). This procedure separates the different proteins into spots, with the size of each spot proportional to the amount of protein present. A typical 2D-PAGE gel may contain several hundred to several thousand spots (**Figure 20.19b**).

Because 2D-PAGE does not detect some proteins in low abundance and is difficult to automate, researchers have turned to liquid chromatography for separating proteins. In liquid chromatography, a mixture of molecules is dissolved in a liquid and passed through a column packed with solid particles. Different affinities for the liquid and solid phases cause some components of the mixture to travel through the column more slowly than others, resulting in separation of the components of the mixture.

The traditional method for identifying a protein is to remove its amino acids one at a time and determine the identity of each amino acid removed. This method is far too slow and labor intensive for analyzing the thousands of proteins present in a typical cell. Today, researchers use **mass spectrometry**, which is a method for precisely determining the molecular mass of a molecule. In mass spectrometry, a

(a)
1 A mixture of proteins is separated by charge in one dimension…

2 …and by mass in a second dimension,…

3 …producing a series of spots on a gel.

Charge →
Molecular mass

(b)

20.19 Two-dimensional acrylamide gel electrophoresis (2D-PAGE) can be used to separate cellular proteins. [After G. Gibson and S. Muse. 2004. *A Primer of Genome Science*, 2e. Sinauer Associates, Inc. p. 274, Fig. 5.4.]

molecule is ionized and its migration rate in an electrical field is determined. Because small molecules migrate more rapidly than larger molecules, the migration rate can accurately determine the mass of the molecule.

To analyze proteins with mass spectrometry, a protein is first digested with the enzyme trypsin, which cleaves the protein into smaller peptide fragments, each containing several amino acids (**Figure 20.20a**). Mass spectrometry is then used to separate the peptides on the basis of their mass-to-charge ratio (**Figure 20.20b**). This separation produces a profile of peaks, in which each peak corresponds to the mass-to-charge ratio of one peptide (**Figure 20.20c**). A computer program then searches through a database of proteins to find a match between the profile generated and the profile expected with a known protein (**Figure 20.20d**), allowing the protein in the sample to be identified. Using bioinformatics, the computer creates "virtual digests" and predicts the profiles of all proteins found in a genome, given the DNA sequences of the protein-encoding genes.

Mass spectrometric methods can also be used to measure the amount of each protein identified. With recent advances, researchers now carry out "shotgun" proteomics, which eliminates most of the initial protein-separation stage. In this procedure, a complex mixture of proteins (such as those from a tissue sample) is digested and analyzed with mass spectrometry. The computer program then sorts out the proteins present in the original sample from the peptide profiles.

Mary Lipton and her colleagues used this approach to study the proteome of *Deinococcus radiodurans*, an exceptional bacterium that is able to withstand high doses of ionizing radiation that are lethal to all other organisms. The genome of *D. radiodurans* had already been sequenced. Lipton and her colleagues extracted proteins from the bacteria, digested them with trypsin, separated the fragments with liquid chromatography, and then determined the proteins from the peptide fragments with mass spectrometry. They were able to identify 1910 proteins, which is more than 60% of the proteins predicted on the basis of the genome sequence.

Deciphering the proteome of even a single cell is a challenging task. Every cell contains a complete sequence of genes, but different cells express vastly different proteins. Each gene may produce a number of different proteins through alternative processing (see Chapter 14) and post-translational protein processing (see Chapter 15). A typical human cell contains as many as 100,000 different proteins that vary greatly in abundance, and no technique such as PCR can be used to easily amplify proteins.

Affinity Capture

Proteomics concerns not just the identification of all proteins in a cell, but also an understanding of how these proteins interact and how their expression varies with the passage of time. Researchers have developed a number of techniques for identifying proteins that interact within the cell. In **affinity capture**, an antibody (see Figure 22.20) to a specific protein is used to capture one protein from a complex mixture of proteins. The protein captured will "pull down" with it any proteins with which the captured protein physically interacts. The pulled-down mixture of proteins can then be analyzed by mass spectrometry to identify the proteins. Various modifications of affinity capture and other techniques can be used to determine the complete set of protein interactions in a cell, termed the **interactome**.

Protein Microarrays

Protein–protein interactions can also be analyzed with **protein microarrays** (**Figure 20.21**), which are similar to the microarrays used for examining gene expression. With this technique, a large number of different proteins are applied to a glass slide as a series of spots, with each spot containing a different protein. In one application, each spot is an antibody for a different protein, labeled with a tag that fluoresces when bound. An extract of tissue is applied to the protein microarray. A spot of fluorescence

(a) A protein is treated with the enzyme trypsin,…

…which breaks it into short peptides.

(b) Accelerator — Mass spectrometer — Detector

The peptides are analyzed with a mass spectrometer, which determines their mass-to-charge ratio.

(c) Counts / Mass (*m/z*)

A profile of peaks is produced.

(d) A computer program compares the profile with those of known and predicted proteins.

GIVPPTKVWYRAITNDEKTSLAFR

A match identifies the protein.

20.20 Mass spectrometry is used to identify proteins.

(a) **(b)**

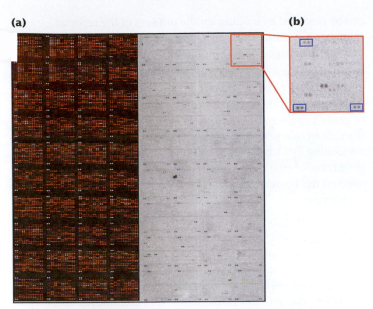

20.21 Protein microarrays can be used to examine interactions among proteins. (a) A microarray containing 4400 proteins found in yeast. (b) The array was probed with an enzyme that phosphylates proteins to determine which proteins serve as substrate for the enzyme. Dark spots represent proteins that were phosphylated by the enzyme. Proteins that phosphylate themselves (autophosphorylate) are included in each block of the microarray (shown in blue boxes) to serve as reference points. [From D. Hall, J. Ptacek, and M. Snyder, 2006. *Mechanisms of Ageing and Development* 128 (2007) 161–167. © 2006, with permission from Elsevier.]

appears when a protein in the extract binds to an antibody, indicating the presence of that particular protein in the tissue.

Structural Proteomics

The high-resolution structure of a protein provides a great deal of useful information. It is often a source of insight into the function of an unknown protein; it may also suggest the location of active sites and provide information about other molecules that interact with the protein. Knowledge of a protein's structure often suggests targets for potential drugs that might interact with the protein. Because structure often provides information about function, a goal of proteomics is to determine the structure of every protein found in a cell.

Two procedures are currently used to solve the structures of complex proteins: (1) X-ray crystallography, in which crystals of the protein are bombarded with X-rays and the diffraction patterns of the X-rays are used to determine the structure (see Chapter 10) and (2) nuclear magnetic resonance (NMR), which provides information on the position of specific atoms within a molecule by using the magnetic properties of nuclei.

Both X-ray crystallography and NMR require human intervention at many stages and are too slow for determining the structure of thousands of proteins that may exist within a cell. Because the structures of hundreds of thousands of proteins are required for studies of the proteome, researchers ultimately hope to be able to predict the structure of a protein from its amino acid sequence. This method is not possible at the present time, but the hope is that, if enough high-resolution structures are solved, it may be possible in the future to model the structure from the amino acid sequence alone. As scientists work on automated methods that will speed the structural determination of proteins, bioinformaticists are developing better computer programs for predicting protein structure from sequence.

> ## CONCEPTS
>
> The proteome is the complete set of proteins found in a cell. Techniques of protein separation and mass spectrometry are used to identify the proteins present within a cell. Affinity capture and microarrays are used to determine sets of interacting proteins. Structural proteomics attempts to determine the structure of all proteins.
>
> ✔ **CONCEPT CHECK 9**
>
> Why is knowledge of a protein's structure important?

CONCEPTS SUMMARY

■ Genomics is the field of genetics that attempts to understand the content, organization, and function of genetic information contained in whole genomes.

■ Genetic maps position genes relative to other genes by determining rates of recombination and are measured in percent recombination. Physical maps are based on the physical distances between genes and are measured in base pairs.

■ The Human Genome Project was an effort to determine the entire sequence of the human genome. The project began officially in 1990; rough drafts of the human genome sequence were completed in 2000. The final draft of the human genome sequence was completed in 2003.

■ Sequencing a whole genome requires breaking it into small overlapping fragments whose DNA sequences can be determined in sequencing reactions. The individual sequences can be ordered into a whole-genome sequence using a map-based approach, in which fragments are assembled in order by using previously created genetic and physical maps, or with the use of a whole-genome shotgun approach, in which overlap between fragments is used to assemble them into a whole-genome sequence. Today, almost all genomes are sequenced using whole-genome shotgun sequencing.

■ Single-nucleotide polymorphisms are single-base differences in DNA between individual organisms and are valuable as markers in linkage studies. Individual organisms may also differ in the number of copies of DNA sequences, called copy-number variations.

■ Sequence-tagged sites are unique DNA sequences whose chromosomal location has been determined. Expressed-sequence tags are markers associated with expressed (transcribed) DNA sequences and can be used to find the genes expressed in a genome.

■ Bioinformatics is a synthesis of molecular biology and computer science that develops tools to store, retrieve, and analyze DNA-, cDNA-, and protein-sequence data.

■ Metagenomics studies the genomes of entire groups of organisms. Synthetic biology is developing techniques for creating genomes and organisms.

■ Homologous genes are evolutionarily related. Orthologs are homologous sequences found in different organisms, whereas paralogs are homologous sequences found in the same organism. Gene function may be determined by looking for homologous sequences (both orthologs and paralogs) whose function has been previously determined.

■ A microarray consists of DNA fragments fixed in an orderly pattern to a solid support, such as a nylon filter or glass slide. When a solution containing a mixture of DNA or RNA is applied to the array, any nucleic acid that is complementary to the probe being used will bind to the probe. Microarrays can be used to monitor the expression of thousands of genes simultaneously.

■ By linking a reporter sequence with the regulatory sequences of a gene, the expression pattern of the gene can be observed by looking for the product of the reporter sequence. Genes affecting a particular function can also be identified through whole-genome mutagenesis screens.

■ Most prokaryotic species have between 1 million and 3 million base pairs of DNA and from 1000 to 2000 genes. Compared with that of eukaryotic genomes, the density of genes in prokaryotic genomes is relatively uniform, with about one gene per 1000 bp. There is relatively little noncoding DNA between prokaryotic genes. Horizontal gene transfer (the movement of genes between different species) has been an important evolutionary process in prokaryotes.

■ Eukaryotic genomes are larger and more variable in size than prokaryotic genomes. There is no clear relation between organismal complexity and the amount of DNA or number of genes among multicellular organisms. Much of the genomes of eukaryotic organisms consist of repetitive DNA. Transposable elements are very common in most eukaryotic genomes.

■ Proteomics determines the protein content of a cell and the functions of those proteins. Proteins within a cell can be separated and identified with the use of mass spectrometry. Structural proteomics attempts to determine the three-dimensional shape of proteins.

IMPORTANT TERMS

genomics (p. 580)
structural genomics (p. 580)
genetic (linkage) map (p. 580)
physical map (p. 581)
map-based sequencing (p. 584)
contig (p. 584)
whole-genome shotgun
   sequencing (p. 585)
single-nucleotide polymorphism
   (SNP) (p. 587)
haplotype (p. 587)
linkage disequilibrium (p. 587)
tag single-nucleotide polymorphism
   (tag-SNP) (p. 587)
genome-wide association study (p. 588)

copy-number variation
   (CNV) (p. 588)
sequence-tagged site (STS) (p. 589)
expressed-sequence tag (EST) (p. 589)
bioinformatics (p. 589)
annotation (p. 589)
metagenomics (p. 590)
microbiome (p. 590)
functional genomics (p. 591)
transcriptome (p. 591)
proteome (p. 591)
homologous genes (p. 591)
orthologous genes (p. 591)
paralogous genes (p. 591)
protein domain (p. 592)

microarray (p. 593)
mutagenesis screen (p. 595)
comparative genomics (p. 596)
segmental duplication (p. 599)
multigene family (p. 599)
gene desert (p. 600)
proteomics (p. 603)
Human Proteome Project (p. 603)
two-dimensional polyacrylamide
   gel electrophoresis (2D-PAGE)
   (p. 603)
mass spectrometry (p. 603)
affinity capture (p. 604)
interactome (p. 604)
protein microarray (p. 604)

ANSWERS TO CONCEPT CHECKS

1. Accuracy and resolution

2. b

3. To catalog and map SNPs and other human genetic variants

4. a

5. d

6. c

7. Species with larger genomes generally have more genes than species with smaller genomes, and so gene density is quite constant.

8. a

9. Structure often provides important information about how a protein functions and the types of proteins with which it is likely to interact.

WORKED PROBLEM

Problem

A linear piece of DNA that is 30 kb long is first cut with *Bam*HI, then with *Hpa*II, and, finally, with both *Bam*HI and *Hpa*II together. Fragments of the following sizes were obtained from this reaction:

    *Bam*HI: 20-kb, 6-kb, and 4-kb fragments

    *Hpa*II: 21-kb and 9-kb fragments

    *Bam*HI and Hpa*II*: 20-kb, 5-kb, 4-kb, and 1-kb fragments

Draw a restriction map of the 30-kb piece of DNA, indicating the locations of the *Bam*HI and *Hpa*II restriction sites.

Solution Strategy

What information is required in your answer to the problem?

A map that includes the number and relative locations of restriction sites for *Bam*HI and *Hpa*II and the distances in bp between the sites.

What information is provided to solve the problem?

- The piece of DNA is 30 kb long.
- The sizes of the fragments produced when the DNA is cut with *Bam*HI, with *Hpa*II, and with both enzymes together.

For help with this problem, review:

Physical Maps in Section 20.1, along with the Worked Problem in Section 20.1.

Solution Steps

> **Note:** This problem can be solved correctly through a variety of approaches; this solution applies one possible approach.

When cut by *Bam*HI alone, the linear piece of DNA is cleaved into three fragments; so there must be two *Bam*HI restriction sites. When cut with *Hpa*II alone, a clone of the same piece of DNA is cleaved into only two fragments; so there is a single *Hpa*II site.

> **Hint:** For linear DNA, the number of restriction sites is one less than the number of fragments produced.

Let's begin to determine the location of these sites by examining the *Hpa*II fragments. Notice that the 21-kb fragment produced when the DNA is cut by *Hpa*II is not present in the fragments produced when the DNA is cut by *Bam*HI and *Hpa*II together (the double digest); this result indicates that the 21-kb *Hpa*II fragment has within it a *Bam*HI site. If we examine the fragments produced by the double digest, we see that the 20-kb and 1-kb fragments sum to 21 kb; so a *Bam*HI site must be 20 kb from one end of the fragment and 1 kb from the other end.

> **Hint:** Look for fragments in the double digest that sum to the length of a fragment present in the single digest.

Similarly, we see that the 9-kb *Hpa*II fragment does not appear in the double digest and that the 5-kb and 4-kb fragments in the double digest add up to 9 kb; so another *Bam*HI site must be 5 kb from one end of this fragment and 4 kb from the other end.

Now, let's examine the fragments produced when the DNA is cut by *Bam*HI alone. The 20-kb and 4-kb fragments are also present in the double digest; so neither of these fragments contains an *Hpa*II site. The 6-kb fragment, however, is not present in the double digest, and the 5-kb and 1-kb fragments in the double digest sum to 6 kb; so this fragment contains an *Hpa*II site that is 5 kb from one end and 1 kb from the other end.

We have accounted for all the restriction sites, but we must still determine the order of the sites on the original 30-kb fragment.

Notice that the 5-kb fragment must be adjacent to both the 1-kb and the 4-kb fragments; so it must be in between these two fragments.

We have also established that the 1-kb and 20-kb fragments are adjacent; because the 5-kb fragment is on one side, the 20-kb fragment must be on the other, completing the restriction map:

> **Hint:** Two fragments in the double digest that were produced by cutting a fragment in the single digest must be adjacent to one another.

COMPREHENSION QUESTIONS

Section 20.1

1. What is the difference between a genetic map and a physical map? Which generally has higher resolution and accuracy and why?

2. What is the difference between a map-based approach to sequencing a whole genome and a whole-genome shotgun approach?

3. What are some of the ethical concerns arising out of the information produced by the Human Genome Project?

4. What is a single-nucleotide polymorphism (SNP)? How are SNPs used in genomic studies?

5. What is a haplotype? How do different haplotypes arise?

6. What is linkage disequilibrium? How does it result in haplotypes?

7. How is a genome-wide association study carried out?

8. What is copy-number variation? How does it arise?

9. (a) What is an expressed-sequence tag (EST)? (b) How are ESTs created? (c) How are ESTs used in genomics studies?

10. How are genes recognized within genomic sequences?

Section 20.2

11. What are homologous sequences? What is the difference between orthologs and paralogs?

12. Describe several different methods for inferring the function of a gene by examining its DNA sequence.

13. What is a microarray? How can it be used to obtain information about gene function?

14. Explain how a reporter sequence can be used to provide information about the expression pattern of a gene.

15. Briefly outline how a mutagenesis screen is carried out.

Section 20.3

16. What is the relation between genome size and gene number in prokaryotes?

17. What is horizontal gene transfer? How might it take place between different species of bacteria?

18. DNA content varies considerably among different multicellular organisms. Is this variation closely related to the number of genes and the complexity of the organism? If not, what accounts for some of these differences?

19. More than half of the genome of *Arabidopsis thaliana* consists of duplicated sequences. What mechanisms are thought to have been responsible for these extensive duplications?

20. What is a segmental duplication?

21. The human genome does not encode substantially more protein domains than do invertebrate genomes, yet it encodes many more proteins. How are more proteins encoded when the number of domains does not differ substantially?

22. (a) What is genomics and how does structural genomics differ from functional genomics? (b) What is comparative genomics?

Section 20.4

23. How does proteomics differ from genomics?

24. How is mass spectrometry used to identify proteins in a cell?

> ➤ For more questions that test your comprehension of the key chapter concepts, go to **LEARNING**Curve for this chapter.

APPLICATION QUESTIONS AND PROBLEMS

Section 20.1

*25. A 22-kb piece of DNA has the following restriction sites:

A batch of this DNA is first fully digested by *Hpa*I alone, then another batch is fully digested by *Hin*dIII alone, and, finally, a third batch is fully digested by both *Hpa*I and *Hin*dIII together. The fragments resulting from each of the three digestions are placed in separate wells of an agarose gel, separated by gel electrophoresis, and stained by ethidium bromide. Draw the bands as they would appear on the gel.

*26. A linear piece of DNA that is 14 kb long is cut first by *Eco*RI alone, then by *Sma*I alone, and, finally, by both *Eco*RI and *Sma*I together. The following results are obtained:

Digestion by *Eco*RI alone	Digestion by *Sma*I alone	Digestion by both *Eco*RI and *Sma*I
3-kb fragment	7-kb fragment	2-kb fragment
5-kb fragment	7-kb fragment	3-kb fragment
6-kb fragment		4-kb fragment
		5-kb fragment

Draw a map of the *Eco*RI and *Sma*I restriction sites on this 14-kb piece of DNA, indicating the relative positions of the restriction sites and the distances between them.

*27. The presence (+) or absence (−) of six sequence-tagged sites (STSs) in each of five bacterial artificial chromosome (BAC) clones (A–E) is indicated in the following table. Using these markers, put the BAC clones in their correct order and indicate the locations of the STS sites within them.

	STSs					
BAC clone	**1**	**2**	**3**	**4**	**5**	**6**
A	+	−	−	−	+	−
B	−	−	−	+	−	+
C	−	+	+	−	−	−
D	−	−	+	−	+	−
E	+	−	−	+	−	−

28. A linear piece of DNA was broken into random, overlapping fragments and each fragment was sequenced. The sequence of each fragment is shown below.

Fragment 1: 5′–TAGTTAAAAC–3′

Fragment 2: 5′–ACCGCAATACCCTAGTTAAA–3′

Fragment 3: 5′–CCCTAGTTAAAAC–3′

Fragment 4: 5′–ACCGCAATACCCTAGTT–3′

Fragment 5: 5′–ACCGCAATACCCTAGTTAAA–3′

Fragment 6: 5′–ATTTACCGCAAT–3′

On the basis of overlap in sequence, create a contig sequence of the original fragment.

*29. How does the density of genes found on chromosome 22 compare with the density of genes found on chromosome 21, two similar-sized chromosomes? How does the number of genes on chromosome 22 compare with the number found on the Y chromosome?

To answer these questions, go to www.ensembl.org. Under the heading *Species*, select *Human*. On the left-hand side of the next page click on *Karyotype*. Pictures of the human chromosomes will appear. Click on chromosome 22 and select *Chromosome Summary*. You will be shown a picture of this chromosome and a histogram illustrating the densities of total genes (uncolored bars) and of known genes (colored bars). The total numbers of genes, along with the chromosome length in base pairs are given at the bottom of the diagram. Write down the total length of the chromosome and the number of protein-coding genes.

Now go to chromosome 21 by selecting it from the Change Chromosome drop-down. Examine the total length and total number of protein-coding genes for chromosome 21. Now do the same for the Y chromosome. Calculate the gene density (number of genes/length) for chromosomes 22, 21, and Y.

a. Which chromosome has the highest density and greatest number of genes? Which has the fewest?

b. Examine in more detail the genes at the tip of the short arm of the Y chromosome by clicking on the top bar in the histogram of genes. Jump to location view. A more detailed view will be shown. What known genes are found in this region? How many protein-coding genes are there in this region?

Human chromosomes 21 and 22. [Leonard Lessin/Science Source.]

30. In recent years, honeybee colonies throughout North America have been decimated by the rapid death of worker bees, a disorder termed colony collapse disorder (CCD). First noticed by beekeepers in 2004, the disorder has been responsible for the loss of 50% to 90% of beekeeping operations in the United States. Evidence suggests that CCD is caused by a pathogen. Diana Cox-Foster and her colleagues (2007. *Science* 318:283–287) used a metagenomic approach to try to identify the causative agent of CCD by isolating DNA from normal honeybee hives and from hives that had experienced CCD. A number of different bacteria, fungi, and viruses were identified in the metagenomic analysis. The following table gives the percentage of CCD hives and non-CCD hives that tested positive for four potential pathogens identified in the metagenomic analysis. On the basis of these data, which potential pathogen appears most likely to be responsible for CCD? Explain your reasoning. Do these data prove that this pathogen is the cause of CCD? Explain.

Virus	**CCD hives infected** ($n = 30$)	**Non-CCD hives infected** ($n = 21$)
Israeli acute paralysis virus	83.3%	4.8%
Kashmir bee virus	100%	76.2%
Nosema apis	90%	47.6%
Nosema cernae	100%	80.8%

31. James Noonan and his colleagues (2005. *Science* 309:597–599) set out to study genome sequences of an extinct species of cave bear. They extracted DNA from 40,000-year-old bones from a cave bear

and used a metagenomic approach to isolate, identify, and sequence the cave-bear DNA. Why did they use a metagenomic approach when their objective was to sequence the genome of one species (the cave bear)?

[Larry Miller/Photo Researchers.]

Section 20.2

32. In **Figure 20.9**, explain why genes A2 and B2 are orthologs and not paralogs.

*33. Examine **Figure 26.18**. Are the epsilon (ε) and beta (β) genes on chromosome 11 orthologs or paralogs? Explain your answer.

*34. Microarrays can be used to determine levels of gene expression. In one type of microarray, hybridization of the red (experimental) and green (control) cDNAs is proportional to the relative amounts of mRNA in the samples. Red indicates the overexpression of a gene and green indicates the underexpression of a gene in the experimental cells relative to the control cells, yellow indicates equal expression in experimental and control cells, and no color indicates no expression in either experimental or control cells.

In one experiment, mRNA from a strain of antibiotic-resistant bacteria (experimental cells) is converted into cDNA and labeled with red fluorescent nucleotides; mRNA from a nonresistant strain of the same bacteria (control cells) is converted into cDNA and labeled with green fluorescent nucleotides. The cDNAs from the resistant and nonresistant cells are mixed and hybridized to a chip containing spots of DNA from genes 1 through 25. The results are shown in the adjoining illustration. What conclusions can you make about which genes might be implicated in antibiotic resistance in these bacteria? How might this information be used to design new antibiotics that are less vulnerable to resistance?

35. For the genes in the microarray shown in the lower part of **Figure 20.10**, are most of these genes over expressed or under expressed in tumors from patients that remained cancer free for at least five years? Explain your reasoning.

36. What does the photograph in **Figure 20.12** reveal about the expression of β-tubulin?

Section 20.3

37. *Dictyostelium discoideum* is a soil-dwelling, social ameba: much of the time, the organism consists of single, solitary cells, but, during times of starvation, amebae come together to form aggregates that have many characteristics of multicellular organisms. Biologists have long debated whether *D. discoideum* is a unicellular or multicellular organism; in 2005 its genome was completely sequenced. The table below lists some genomic characteristics of *D. discoideum* and other eukaryotes (L. Eichinger et al. 2005. *Nature* 435:43–57).

Table for Problem 37

Feature	D. discoideum	P. falciparium	S. cerevisiae	A. thaliana	D. melanogaster	C. elegans	H. sapiens
Organism	ameba	malaria parasite	yeast	plant	fruit fly	worm	human
Cellularity	?	uni	uni	multi	multi	multi	multi
Genome size (millions of base pairs)	34	23	13	125	180	103	2,851
Number of genes	12,500	5,268	5,538	25,498	13,676	19,893	22,287
Average gene length (bp)	1,756	2,534	1,428	2,036	1,997	2,991	27,000
Genes with introns (%)	69	54	5	79	38	5	85
Mean number of introns	1.9	2.6	1.0	5.4	4.0	5.0	8.1
Mean intron size (bp)	146	179	nd*	170	nd*	270	3,365
Mean G + C (exons)	27%	24%	28%	28%	55%	42%	45%

*nd = not determined

a. On the basis of the organisms listed in the table other than *D. discoideum*, what are some differences in genome characteristics between unicellular and multicellular organisms?

b. On the basis of these data, do you think the genome of *D. discoideum* is more like those of other unicellular eukaryotes or more like those of multicellular eukaryotes? Explain your answer.

Dictyostelium discoideum.
[David Scharf/Science Source.]

*38. How do the following genomic features of prokaryotic organisms compare with those of eukaryotic organisms? How do they compare among eukaryotes?

a. Genome size

b. Number of genes

c. Gene density (bp/gene)

d. Number of exons

39. A group of scientists sequenced the genomes of 12 species of *Drosophila* (*Drosophila* 12 Genomes Consortium. 2007. *Nature* 450:203–218). Data on genome size and number of protein-encoding genes from this study are given in the accompanying table. Plot the number of protein-encoding genes as a function of genome size for 12 species of *Drosophila*. Is there a relation between genome size and number of genes in fruit flies? How does this compare with the relation between genome size and number of genes across all eukaryotes?

Characteristics of 12 *Drosophila* species genomes

Species	Genome Size (millions of base pairs)	Number of Protein-Encoding Genes
D. melanogaster	200	13,733
D. simulans	162	15,983
D. sechellia	171	16,884
D. yakuba	190	16,423
D. erecta	135	15,324
D. ananassae	217	15,276
D. pseudoobscura	193	16,363
D. persimilis	193	17,325
D. willistoni	222	15,816
D. virilis	364	14,680
D. mojavensis	130	14,849
D. grimshawi	231	15,270

Section 20.4

40. A scientist determines the complete genomes and proteomes of a liver cell and a muscle cell from the same person. Would you expect bigger differences in the genomes or proteomes of these two cell types? Explain your answer.

CHALLENGE QUESTIONS

Section 20.1

41. The genome of *Drosophila melanogaster,* a fruit fly, was sequenced in 2000. However, this "completed" sequence did not include most heterochromatin regions. The heterochromatin was not sequenced until 2007 (R. A. Hoskins et al. 2007. *Science* 316:1625–1628). Most completed genome sequences do not include heterochromatin. Why is heterochromatin usually not sequenced in genomic projects? (Hint: See Chapter 11 for a more detailed discussion of heterochromatin.)

42. In metagenomic studies, a comparison of ribosomal RNA sequences is often used to determine the number of different species present. What are some characteristics of ribosomal sequences that make them useful for determining what species are present?

*43. Some synthetic biologists have proposed creating an entirely new, free-living organism with a minimal genome, the smallest set of genes that allows for replication of the organism in a particular environment. This genome could be used to design and create, from "scratch," novel organisms that might perform specific tasks such as the breakdown of toxic materials in the environment.

a. How might the minimal genome required for life be determined?

b. What, if any, social and ethical concerns might be associated with the construction of an entirely new organism with a minimal genome?

Go to your **LaunchPad** to find additional learning resources and the Suggested Readings for this chapter.

21

Epigenetics

Through epigenetics, crop abundance and failure can have health effects that persist for several generations. [*Harvest*, by Hugh Cameron (1835–1918); Private Collection/Bridgeman Art Library Ltd.]

How Your Grandfather's Diet Could Affect Your Health

Överkalix parish in northern Sweden lies above the Arctic Circle, with an inhospitable climate characterized by long, cold winters and short summers. The community has always been small; even today there are fewer than 4000 residents. During the 1800s and early 1900s, there were few roads in the region, and transportation during the winter was limited by the ice and snow. Because of its extreme northern location, growing crops was a constant challenge. When crops failed, the remoteness of the region limited the import of food and people starved. Crops failed frequently throughout the 1800s (for example, in 1800, 1812, 1821, and 1829) and the period of 1831–1836 was marked by total crop failure and extreme hardship. However, because of the unpredictability of weather, years of successful harvest and food abundance often followed these bad years.

In the 1980s, researchers became interested in the effects of feast and famine on the long-term health of the people of northern Sweden. They wanted to determine whether food shortages that people experienced in their childhood affected the future health of their descendants. By combing through harvest statistics, grain prices, and other historical facts, the researchers were able to determine the availability of food in the area throughout the nineteenth and early twentieth centuries. They also scrutinized the health records of the inhabitants; this was made possible by the availability of centralized medical records that Sweden maintains on all its citizens.

The researchers focused on the health of three groups of people, born in 1890, 1905, and 1920. They examined the life span of these individuals and looked at their risk of dying from cardiovascular disease and diabetes. They then tracked down their parents and grandparents, determined the availability of food when they were children, and looked for correlations between the diet of the parents and grandparents, and the health of their descendants.

What the researchers found was startling. Individuals whose parents and grandparents were exposed to food shortages as children lived longer than individuals whose ancestors were exposed to food excesses. On the other hand, people whose ancestors grew up during times of abundant food died at an earlier age and were more likely to die from cardiovascular disease and diabetes. For example, if a paternal grandfather had access to excess food as a child, his grandchildren were four times more likely to die from diabetes than the grandchildren of people who were not exposed to excess food during childhood.

How can the amount of food available to a person during childhood affect the health of his or her children and grandchildren living 20 to 60 years later? One of the principles of modern genetics is that our genes are stable (except for rare mutations) and not altered by the environment, so how can diet influence the traits of descendants for two generations? The effect of the grandfather's diet on later generations was particularly remarkable. Mothers provide their offspring with the cytoplasm of the egg and a uterine environment, as well as genes, but through their sperm fathers contribute only a set of paternal genes to their offspring.

The researchers proposed that the effect they observed occurred through epigenetics: changes to chromatin and DNA that are heritable but do not involve alteration of the base sequence of the DNA. Epigenetic inheritance was not envisioned by Mendel nor, until recently, by most modern geneticists, but epigenetic processes appear to play an important role in the inheritance of many phenotypes. Today the study of epigenetics is the focus of intensive research.

Another question that this study raises is why would the adverse conditions of famine during childhood *lower* the risk of dying from cardiovascular disease and diabetes in future generations, while conditions of excess food *increase* the risk? One might expect just the opposite, that nutritional stress during childhood would increase the risk of dying, while excess food would lower it. Evolutionary biologists have proposed an explanation for this relationship, which has also been observed in other studies. This explanation, termed the *thrifty phenotype hypothesis*, is based on the assumption that information about the parental environment may be useful to the offspring, allowing them to respond in ways that increase their own survival and reproduction. This hypothesis proposes that when environmental conditions are poor for the parent, they are likely to persist and also to be poor for the offspring. Therefore, when the parent experiences hard times, natural selection favors parents that produce metabolically thrifty offspring—offspring that eat as much as possible when food is available, minimize energy expenditure, and hoard calories—because the parent's environment predicts that there will be little food available to the offspring. This strategy was probably advantageous in the distant past, before agriculture, but it often backfires in modern society. Eating all you can, minimizing energy expenditure, and hoarding calories when food is plentiful often leads to obesity, heart disease, and diabetes, as was seen in the children and grandchildren of the people of Överkalix.

This chapter is about epigenetics, the explanation proposed for the effect of diet on the health of the residents of Överkalix. We begin by discussing the origin of the term epigenetics, and what the term encompasses today. We then review the types of changes to chromatin that can occur and the major process that alter chromatin structure. We also take a look at how changes to chromatin structure might be passed on to future cells and future generations. We then look at a number of epigenetic effects, including paramutation, behavioral effects, effects of chemicals, metabolic effects, effects on monozygotic twins, X-inactivation, cell differentiation, and genomic imprinting. We end the chapter by discussing efforts to map the genome-wide location of epigenetic marks—the epigenome.

21.1 What is Epigenetics?

The term epigenetics was first used by Conrad Waddington (**Figure 21.1**) in 1942 to describe how, through the process of development, a genotype produces a phenotype. In coining the term, Waddington combined the words "epigenesis," which is how an embryo develops, with "genetics," the study of genes and heredity. Waddington's goal was to encourage

the merging of genetics and development. However, his use of the term preceded our modern understanding of DNA

21.1 Carl Waddington first used the word epigenetics to refer to how a phenotype develops from a genotype. [Godfrey Argent Studio/The Royal Society.]

and chromosome structure, and today epigenetics has taken on a narrower meaning.

The Greek root "epi" means "over" or "above"; epigenetics has come to represent the inheritance of variation above and beyond differences in DNA sequence. Today epigenetics usually refers to the phenotypes and processes that are transmitted to other cells and sometimes future generations, but are not the result of differences in the DNA base sequence. Often epigenetic effects are caused by changes in gene expression that result from alterations to chromatin structure or other aspects of DNA structure, such as DNA methylation. One definition of an epigenetic trait is: a stably inherited phenotype resulting from changes in chromatin without alterations in the DNA sequence. Some have broadened the definition of epigenetics to refer to any alteration of chromatin or DNA structure that affects gene expression. Here, we will use epigenetics to refer to changes in gene expression and/or a phenotype that are potentially heritable without alteration of the underlying DNA base sequence.

Many epigenetic changes are stable, persisting across cell divisions or even generations. However, epigenetic alterations are also potentially influenced by environmental factors. For example, environmental stress has been shown to alter methylation of the rat *Bdnf* gene, which encodes a growth factor that plays an important role in brain development. DNA methylation has been tied to the expression of genes and the phenotypes they produce. As we will see, altered DNA methylation is capable of being replicated across cell division, resulting in progeny with the new phenotype, although there is no corresponding difference in the DNA base sequence of individuals that "inherit" the new phenotype. The fact that epigenetic traits may be induced by environmental effects and transmitted to future generations has been interpreted by some to mean that genes have memory through epigenetics—that environmental factors acting on individuals can have effects that are transmitted to future generations, as was seen with the effect of diet on life expectancy in the introduction to this chapter. Epigenetics has been called "inheritance, but not as we know it."

Epigenetics is providing an explanation for how changes outside of the DNA sequence can influence the phenotype and how those changes are heritable. It is also making important contributions to the study of behavior, environmental science, cancer, neurobiology, and pharmacology.

▶ TRY PROBLEM 2

CONCEPTS

Epigenetic effects are phenotypes that are passed to other cells and sometimes future generations, but are not the result of differences in the DNA base sequence. The study of epigenetics is making important contributions to many areas of biology.

21.2 Several Molecular Processes Lead to Epigenetic Changes

Epigenetics alters the expression of genes; these alterations are stable enough to be transmitted through mitosis (and sometimes meiosis) but can also be changed. Most evidence suggests that epigenetic effects are brought about by physical changes to chromatin structure. In Chapter 11 we considered a number of chemical changes in DNA and histone proteins that affect chromatin structure, including DNA methylation, modification of histone proteins, and repositioning of nucleosomes. In Chapter 17, we discussed the role that these alterations have on the expression of genes. These chromatin changes are thought to play a role in epigenetic traits. Chapter 14 discussed small RNA molecules, some of which play an important role in bringing about epigenetic changes.

We will consider three types of molecular mechanisms that alter chromatin structure and underlie many epigenetic phenotypes: (1) changes in patterns of DNA methylation; (2) chemical modifications of histone proteins; and (3) RNA molecules that affect chromatin structure and gene expression.

CONCEPTS

Many epigenetic phenotypes are the result of alterations to chromatin structure, mediated through three major processes: DNA methylation, histone modification, and RNA molecules.

DNA Methylation

The best understood mechanism of epigenetic change is methylation of DNA. DNA methylation refers to the addition of methyl groups to the nucleotide bases. In eukaryotes, the predominant type of DNA methylation is the methylation of cytosine to produce 5-methylcytosine (**Figure 21.2a**). As discussed in Chapter 17, DNA methylation is often associated with repression of transcription.

DNA methylation often occurs on cytosine bases that are immediately adjacent to guanine nucleotides, referred to as CpG dinucleotides (where p represents the phosphate group that connects C and G nucleotides). In CpG dinucleotides, cytosine nucleotides on the two DNA strands are diagonally across from one another. Usually both cytosine bases are methylated, so that methyl groups occur on both DNA strands, as shown below and in **Figure 21.2b**.

$$5'-\overset{m}{C} G-3'$$
$$3'-G \underset{m}{C}-5'$$

In plants, DNA methylation also occurs at CpNpG trinucleotides, where N represents a nucleotide with any base.

(a)

Cytosine (C) → 5-Methylcytosine

on opposing strands. Before replication, cytosine bases on both strands are methylated (**Figure 21.3**). Immediately after semiconservative replication, the cytosine base on the template strand will be methylated, but the cytosine base on the newly replicated strand will be unmethylated. Special methyltransferase enzymes recognize the hemimethylated state of CpG dinucleotides and add methyl

(b)

Methyl group

21.2 DNA methylation is a common epigenetic modification of chromatin. (a) Cytosine bases are often modified to form 5-methylcytosine. (b) Three-dimensional structure of DNA showing methylation of CpG dinucleotides.

Some DNA regions have many CpG dinucleotides and are referred to as CpG islands. In mammalian cells, CpG islands are often located in or near the promoters of genes. These CpG islands are usually not methylated when genes are being actively transcribed. However, methylation of CpG islands near a gene leads to repression of transcription.

Cells repress and activate genes by methylating and demethylating cytosine bases. Enzymes called DNA methyltransferases methylate DNA by adding methyl groups to cytosine bases to create 5-methylcytosine. Other enzymes called demethylases remove methyl groups, converting 5-methylcytosine back to cytosine (see Chapter 17).

MAINTENANCE OF METHYLATION The fact that epigenetic changes are passed to other cells and (sometimes) to future generations means that changes in chromatin structure associated with epigenetic phenotypes must be faithfully maintained when chromosomes replicate. How are epigenetic changes retained and replicated through the process of cell division?

Methylation of CpG sequences means that two methylated cytosine bases sit diagonally across from each other

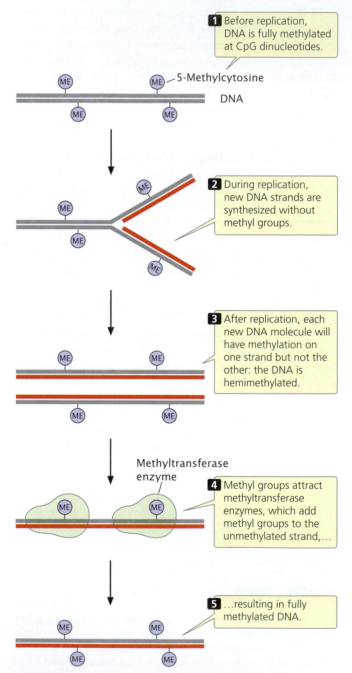

1 Before replication, DNA is fully methylated at CpG dinucleotides.

5-Methylcytosine

DNA

2 During replication, new DNA strands are synthesized without methyl groups.

3 After replication, each new DNA molecule will have methylation on one strand but not the other: the DNA is hemimethylated.

Methyltransferase enzyme

4 Methyl groups attract methyltransferase enzymes, which add methyl groups to the unmethylated strand,...

5 ...resulting in fully methylated DNA.

21.3 DNA methylation is stably maintained through DNA replication.

groups to the unmethylated cytosine bases, creating two new DNA molecules that are fully methylated. In this way, the methylation pattern of DNA is maintained across cell division. ▶ TRY PROBLEM 26

DNA METHYLATION IN HONEYBEES

A remarkable example of epigenetics is seen in honeybees. Queen bees and worker bees are both female, but the resemblance ends there. The queen is large and develops functional ovaries, whereas workers are small and sterile (**Figure 21.4**). The queen goes on a mating flight and spends her entire life reproducing, whereas workers spend all of their time collecting nectar and pollen, tending the queen, and raising her offspring. In spite of these significant differences in anatomy, physiology, and behavior, queens and workers are genetically the same; both develop from ordinary eggs. How they differ is in diet: worker bees produce and feed a few female larvae a special substance called royal jelly, which causes these larvae to develop as queens. Other larvae are fed ordinary bee food, and they develop as workers. This simple difference in diet greatly affects gene expression, causing different genes to be activated in queens and workers and resulting in a very different set of phenotypic traits.

How royal jelly affects gene expression has long been a mystery, but research now suggests that it changes an epigenetic mark. In 2008 Ryszard Kucharski and his colleagues demonstrated that royal jelly silences the expression of a key gene called *Dnmt3,* whose product normally adds methyl groups to DNA. With *Dnmt3* shut down, bee DNA is less methylated and many genes that are normally silenced in workers are expressed, leading to the development of queen characteristics. Kucharski and his coworkers demonstrated the importance of DNA methylation in queen development by injecting into bee larvae small RNA molecules (small interfering RNAs, or siRNAs; see Chapter 14) that specifically inhibited the expression of *Dnmt3.* These larvae had lower levels of DNA methylation and many developed as queens with fully functional ovaries. This experiment demonstrated

21.4 Epigenetic changes are responsible for differences in the phenotypes of honeybee (*Apis mellifera*) queens (left) and workers (right). [WILDLIFE GmbH/Alamy.]

that royal jelly brings about epigenetic changes (less DNA methylation), which are transmitted through cell division and modify developmental pathways, eventually leading to a queen bee. ▶ TRY PROBLEM 25

REPRESSION OF TRANSCRIPTION BY DNA METHYLATION

How does DNA methylation suppress gene expression? The methyl group of 5-methylcytosine sits within the major groove of the DNA, which is recognized by many DNA binding proteins. The presence of the methyl group in the major groove inhibits the binding of transcription factors and other proteins required for transcription to occur. 5-methylcytosine also attracts certain proteins that directly repress transcription. In addition, DNA methylation attracts histone deacetylase enzymes, which remove acetyl groups from the tails of histone proteins, altering chromatin structure in a way that represses transcription (see Chapter 17).

> ### CONCEPTS
>
> Cytosine bases are often methylated to form 5-methylcytosine, which is associated with repression of transcription. DNA methylation is stably maintained through replication by methyltransferase enzymes that recognize the hemimethylated state of CpG dinucleotides and add methyl groups to the unmethylated cytosine bases.
>
> ### ✔ CONCEPT CHECK 1
>
> Which of the following is true of CpG islands?
> a. They are methylated near promoters of actively transcribed genes.
> b. They are unmethylated near promoters of actively transcribed genes.
> c. Acetylation of CpG islands leads to repression of transcription.
> d. CpG islands code for RNA molecules that activate transcription.

Histone Modifications

Epigenetic changes can also occur through modification of histone proteins. In eukaryotic cells, DNA is complexed to histone proteins in the form of nucleosomes, which are the basic repeating units of chromatin structure (see Chapter 11). More than 100 different posttranslational modifications of histone proteins have been detected. Many of these modifications take place in the positively charged tails of the histone proteins, which interact with the DNA and affect chromatin structure. Modifications to histones include the addition of phosphates, methyl groups, acetyl groups, and ubiquitin to their tails. These modifications often alter chromatin structure and affect transcription of genes (see Chapter 17). The modifications may also serve as binding sites for proteins such as transcription factors that are required for transcription.

The addition of acetyl groups to amino acids in the histone tails (histone acetylation) generally destabilizes chromatin structure, causing it to assume a more open configuration and is associated with increased transcription (see Figure 17.2 in Chapter 17). The addition of methyl groups to histones (histone methylation) also alters chromatin structure, but the effect varies depending on the specific amino acid that is methylated; some types of histone methylation are associated with increased transcription and other types are associated with decreased transcription. For example, the addition of three methyl groups to lysine 4 in the H3 histone (H3K4me3, K stands for lysine) is often found near transcriptionally active genes. Methylation of lysine 36 in the H3 histone (H3K36me3) is also associated with increased transcription. On the other hand, the addition of three methyl groups to lysine 9 in H3 (H3K9me3) and to lysine 20 in histone 4 (H4K20me3) is associated with repression of transcription. Many additional histone marks have also been shown to associate with the level of transcription. These types of modifications have been called **epigenetic marks**.

Histone modifications are added and removed by special proteins. The polycomb group (PcG) proteins are a large group of proteins that repress transcription by modifying histones. These modifications alter chromatin structure so that the DNA is not accessible to transcription factors, RNA polymerase, and other proteins required for transcription. For example, polycomb repressive complex 2 (PRC2) adds two or three methyl groups to lysine 27 of histone H3, creating the H3K27me3 epigenetic mark that represses transcription.

Many of the enzymes and proteins that produce epigenetic marks cannot bind to specific DNA sequences by themselves. Thus, they must be recruited to specific targets on the chromosome. Sequence-specific transcription factors, pre-existing chromatin marks, and noncoding RNA molecules serve to recruit histone-modifying enzymes to specific sites.

Research indicates that single histone modifications, such as those mentioned here, do not individually determine the transcriptional activity of a gene. Rather, it is the combined presence of multiple epigenetic marks that determine the activity level. There is also considerable "crosstalk" between epigenetic marks—one histone mark may affect whether additional marks occur nearby and how they function. Crosstalk occurs because histone modifications attract enzymes and proteins that modify other histones. Histone modifications not only affect transcription, but can also influence other molecular processes such as DNA repair and cell cycle checkpoint signaling (see pp. 670–672 in Chapter 23). For example, ubiquitination of histone H2B is required for repair of double-strand breaks in DNA. This modification leads to other histone modifications, such as methylation of lysine 79 of H3 (H3K79me); these modifications alter chromatin structure and allow access of proteins that repair double-strand breaks.

MAINTENANCE OF HISTONE MODIFICATIONS The process by which histone modifications are maintained across cell division is not as well understood as that of DNA methylation. There is no universal mechanism for maintaining histone modifications; different types of modifications are undoubtedly maintained by different mechanisms.

Several models have been proposed to explain how histone modifications are faithfully transmitted to daughter cells. During the process of DNA replication, nucleosomes are disrupted and the original histone proteins are distributed randomly between the two new DNA molecules. Newly synthesized histones are then added to complete the formation of new nucleosomes (see Chapter 12). Most models assume that after replication the epigenetic marks remain on the original histones, and these marks recruit enzymes that make similar changes to the new histones. For example, PRC2 adds the H3K27me3 epigenetic mark to histones. PRC2 preferentially targets histones in chromatin that already contains an H327me3 mark, ensuring that any new nucleosomes that are added after replication will also become methylated. In this way the histone modifications can be maintained across cell division.

CONCEPTS

Modification of histone proteins, including the addition of methyl groups, acetyl groups, phosphates, and ubiquitination alter chromatin structure. Some of these modifications are passed to daughter cells during cell division and to future generations.

Epigenetic Effects by RNA Molecules

Evidence increasingly demonstrates that RNA molecules play an important role in bringing about epigenetic effects. The first discovered and still best understood example of RNA mediating epigenetic change is X inactivation, in which a long, noncoding RNA called *Xist* suppresses transcription on one of the X chromosomes in female mammals. Another example involves paramutation in corn, in which an epigenetically altered allele induces a change in another allele that then gets transmitted to future generations. Paramutation in corn is brought about by siRNAs (see Chapter 14). Both of these examples will be discussed later in this chapter.

Different mechanisms are involved in epigenetic changes through RNA molecules. In the case of X inactivation, the *Xist* RNA coats one X chromosome and then attracts PRC2, which deposits methyl groups on lysine 27 of histone H3, creating the H3K27me3 epigenetic mark that alters chromatin structure and represses transcription.

Other examples of RNA-associated epigenetic phenotypes occur through siRNA molecules that silence genes and transposable elements (see Chapter 18) by directing DNA methylation or histone modifications to specific DNA sequences. In addition, research has demonstrated that epigenetic processes such as methylation and histone modification influence the expression of microRNAs (see Chapter 14) which, in turn, play an important role in regulating other genes. MicroRNAs also control the expression of genes that produce epigenetic effects, such as enzymes that methylate DNA and modify histone proteins. How RNA-based epigenetic changes are maintained across cell divisions is less clear, although some apparently involve small RNAs that are transmitted through the cytoplasm.

CONCEPTS

RNA molecules bring about modification of chromatin by a variety of processes.

✓ CONCEPT CHECK 2

Which is *not* a major mechanism of epigenetic change?
a. DNA methylation.
b. Alteration of a DNA base sequence in a promoter.
c. Histone acetylation.
d. Action of RNA molecules.

21.3 Epigenetic Processes Produce a Diverse Set of Effects

Initially, epigenetic mechanisms were thought to play a role in a small number of unusual phenotypes. However, research during the past 15 years has revealed that epigenetics underlies an impressive and ever-increasing array of biological phenomena. In this section, we will look at some examples of these epigenetic effects.

Paramutation

One of the first examples of epigenetics was a curious phenotype that Alexander Brink described in corn in the 1950s. Brink was studying the *r1* locus, which helps to determine pigmentation in the seeds of corn. The R^r allele at this locus normally produces purple kernels, while the R^{st} allele codes for spotted kernels. Brink observed that when R^{st} was present in a genotype with R^r allele, the R^{st} allele permanently altered the expression of the R^r allele, so that it now also produced spotted seeds. This diminished effect of the altered R^r allele on pigmentation persisted for several generations, even in the absence of the R^{st} allele. Brink called this phenomenon paramutation. Paramutation violates Mendel's principle of segregation, which states that when gametes are formed, each allele separates and is transmitted independently to the next generation.

Today, **paramutation** is defined as an interaction between two alleles that leads to a heritable change in expression of one of the alleles. Surprisingly, paramutation produces these differences in phenotype without any alteration in the DNA base sequence of the converted allele. The phenomenon of paramutation has several important features. First, the newly established expression pattern of the converted allele is transmitted to future generations, even though the allele that brought about the alteration is no longer present with it. Second, the altered allele is now able to convert other alleles to the new phenotype. And third, there are no associated DNA sequence differences in the altered alleles. A number of examples of paramutation have now been discovered in different organisms, and geneticists have begun to unravel the molecular mechanism of this curious phenomenon.

PARAMUTATION IN CORN   A few years after Brink reported paramutation at the *r1* locus in corn, another related example was discovered by Ed Coe, Jr. This case involved interaction between the alleles at the *b1* locus in corn, which also aids in determining pigmentation. Paramutation at the *b1* locus is more straightforward than at the *r1* locus, so we will use *b1* to examine the process and mechanism of paramutation.

The *b1* locus helps to determine the amount of purple anthocyanin that a corn plant produces. The locus actually encodes a transcription factor that regulates genes involved in pigment production. Plants homozygous for the *B-I* allele (*B-I B-I*) show high expression of the *b1* locus and are dark purple (**Figure 21.5**). Plants homozygous for the *B′* allele (*B′B′*) have a lower expression of *b1* and are lightly pigmented.

21.5 In paramutation at the *b1* locus in corn, a copy of the *B′* allele converts the *B-I* allele to *B′**, which has the same phenotype as *B′*. The *B-I B-I* genotype produces a pigmented plant, while *B′B′* and *B′ B′** genotypes are lightly pigmented.

However, the DNA sequences of the *B-I* and *B'* alleles are identical. Genetically identical alleles such as these, which produce heritable differences in phenotypes through epigenetic processes, are referred to as **epialleles**.

In plants that are heterozygous *B-I B'*, the *B-I* allele is converted to *B'*, with the result that the heterozygous plants are lightly pigmented (see Figure 21.5), just like the *B'B'* homozygotes. The newly converted allele is usually designated *B'**. Importantly, there is no functional difference between *B'* and *B'**; the *B'** allele is now fully capable of converting other *B-I* alleles into *B'** alleles in subsequent generations.

Research has demonstrated that one of the features required for paramutation at the *b1* locus is the presence of a series of seven tandemly repeated sequences that are located approximately 100,000 base pairs upstream of the coding sequence for the *b1* locus (**Figure 21.6**). Each repeat consists of 853 bp and does not encode any protein. Both the *B-I* and *B'* alleles have these tandem repeats, but the chromatin structure of the two alleles differs: the *B-I* allele has more open chromatin. The tandem repeats are required for high expression of the *B-I* allele and pigment

production. It has been suggested that the repeats act like an enhancer (see Chapter 17), stimulating transcription at the *b1* locus, but only when the chromatin surrounding the repeats is in an open configuration, as it is in the *B-I* allele. The more closed configuration of the *B'* allele may prevent the repeats from interacting with the promoter of *b1* and stimulating transcription. How the repeats might interact with the *B'* allele is not known.

The different chromatin states of *B-I* and *B'* may explain their different levels of expression, but how does the *B'* allele convert the *B-I* allele to *B'**? Although the mechanism is not completely understood, recent research demonstrates that the communication between *B'* and *B-I* probably occurs through the action of small RNA molecules.

The tandem repeats that are required for paramutation encode 25-nucleotide long siRNAs, (Chapter 14). Some siRNAs are known to modify chromatin structure by directing DNA methylation to specific DNA sequences. Geneticists have isolated several genes in corn that are required for paramutation to take place; inactivating these genes eliminates paramutation. One of these genes is *mop1*, which encodes an RNA-directed RNA polymerase (an enzyme that synthesizes RNA from an RNA template). This gene is required to generate the siRNAs encoded by the tandem repeats, although it does not appear to be the enzyme that actually transcribes the DNA copies of tandem repeats. Another gene required for paramutation, called *rmr1*, encodes a chromatin-remodeling protein.

Thus, the current evidence suggests that siRNA molecules convert *B-I* to *B'**, and this conversion probably involves a change to the chromatin states of the alleles. There are other examples from plants where siRNAs influence DNA methylation and chromatin structure. However, exactly how the siRNA molecules bring about this change in chromatin in paramutation is unknown. The altered chromatin state of the repeats in the *B'* allele probably decreases transcription of the *b1* locus, perhaps by interfering with interaction between the tandem repeats and the promoter of *b1*. It is assumed that the alteration to the chromatin of the *B'* allele is stable and transmitted to future generations. Research also shows that the transcription of the tandem repeats and generation of siRNAs from them are necessary but not sufficient for paramutation, so additional factors must be involved. It is also not clear how the production of the siRNAs is transmitted across generations.

PARAMUTATION IN MICE Several examples of paramutation have also been observed in mice. One involves the *Kit* locus, which encodes a tyrosine kinase receptor and functions in pigment production, germ cell development, and production of blood cells. Geneticists had earlier genetically engineered a mutant *Kit* allele (designated here as *Kit*[t]), which

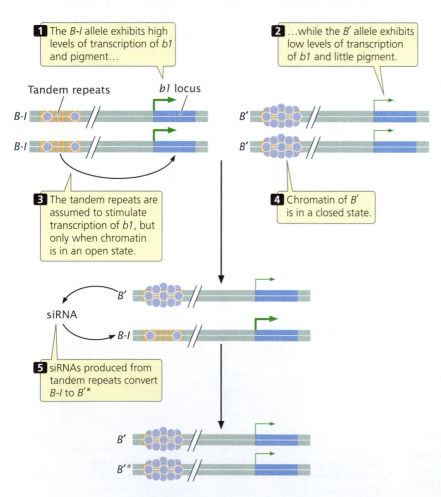

1 The *B-I* allele exhibits high levels of transcription of *b1* and pigment…

2 …while the *B'* allele exhibits low levels of transcription of *b1* and little pigment.

Tandem repeats   *b1* locus

B-I

B-I

B'

B'

3 The tandem repeats are assumed to stimulate transcription of *b1*, but only when chromatin is in an open state.

4 Chromatin of *B'* is in a closed state.

B'

siRNA

B-I

5 siRNAs produced from tandem repeats convert *B-I* to *B'**

B'

*B'**

21.6 Paramutation at the *b1* locus in corn requires the presence of 7 tandem repeats upstream of the *b1* locus.

P generation

Kit⁺ Kit⁺ homozygotes have solid color.

Kit⁺ Kitᵗ heterozygotes have white feet and tail tip.

×

$Kit^+ Kit^+$

$Kit^+ Kit^t$

F₁ generation

Some *Kit⁺ Kit⁺* progeny develop the phenotype of *Kit⁺ Kitᵗ* genotype.

$\frac{1}{2}\ Kit^+ Kit^+$

$\frac{1}{2}\ Kit^+ Kit^t$

Conclusion: A cross between *Kit⁺ Kit⁺* and *Kit⁺ Kitᵗ* produces ¹/₂ *Kit⁺ Kit⁺* and ¹/₂ *Kit⁺ Kitᵗ* progeny, but some *Kit⁺ Kit⁺* develop the phenotype of heterozygotes.

21.7 Paramutation at the *Kit* locus in mice.

carries a 3000 bp portion of the *lacZ* gene (see Chapter 16) inserted into the *Kit* locus. Mice that are homozygous for the wild-type allele (*Kit⁺ Kit⁺*) have normal pigment. Mice that are homozygous for mutant *Kit* alleles (*Kitᵗ Kitᵗ*) die shortly after birth. Mice heterozygous for wild type and mutant alleles (*Kit⁺ Kitᵗ*) have white tail tips and white feet (**Figure 21.7**). When a heterozygous mouse is crossed with a homozygous wild-type mouse, half of the progeny are homozygous (*Kit⁺ Kit⁺*) and half are heterozygous (*Kit⁺ Kitᵗ*), as expected. However, many of the *Kit⁺ Kit⁺* mice develop white tails and feet, the phenotype expected of the heterozygotes. In the presence of the *Kitᵗ* allele in the heterozygote, the *Kit⁺* allele is altered so that it has the same phenotype as *Kitᵗ*. Mice with these altered alleles are designated as *Kit**. The altered *Kit** allele is stably transmitted to future generations, where it continues to produce white tails and feet. Some humans with a white spot in the forehead hair and areas of reduced pigment (called the piebald trait) have mutations in the *Kit* locus; other mutations in *Kit* result in a predisposition to some cancers.

Researchers have demonstrated that this example of paramutation is also mediated through RNA molecules, although the mechanism is likely to be different from that seen in paramutation in corn. In mice, the white tail and feet of the *Kitᵗ* allele appears to be caused by microRNAs (miRNAs) that degrade the *Kit* mRNA, and these miRNAs are transmitted to future generations through the gametes. Researchers observed a two-fold decrease in *Kit* mRNA in both the heterozygous mice and the *Kit** mice, suggesting that the

Table 21.1	Effects of injection of different types of RNA into wild-type mice (*Kit⁺ Kit⁺*)
Type of RNA Injected	**Presence of White Tail Tips and Feet**
Kit⁺ Kit⁺ mRNA	Uncommon
Kit⁺ Kitᵗ mRNA	More common
miRNA to *Kit* mRNA	More common
nonspecific miRNA	Uncommon

white tails and feet of the heterozygotes is due to a reduction in the amount of *Kit* mRNA. To determine whether RNA was responsible for paramutation they injected some wild-type embryos with RNA from *Kit⁺ Kit⁺* homozygotes and injected other wild-type embryos with RNA from *Kit⁺ Kitᵗ* heterozygotes. Among the mice that completed development, they observed white tail tips and feet more frequently in those injected with RNA from heterozygotes, suggesting that RNA from the heterozygotes is capable of altering the *Kit⁺* allele of the wild type mice (**Table 21.1**). The researchers then injected into wild-type embryos miRNAs that degrade *Kit* mRNA. This produced more mice with white tails and feet than when they injected nonspecific miRNAs into the embryos (Table 21.1) The ability to produce the white tails and feet characteristic of the *Kit⁺ Kitᵗ* genotype by injecting miRNA into embryos suggests that this case of paramutation is associated with miRNA molecules that are transferred to the embryo via egg and sperm. However, there are still many unknown aspects of paramutation at the *Kit* locus.

CONCEPTS

Paramutation occurs when one allele creates a heritable alteration of another allele without any change in DNA sequence. Research suggests that paramutation in corn and mice is mediated through small RNA molecules.

✓ CONCEPT CHECK 3

Which is a characteristic of paramutation?
a. One allele is able to alter another allele when both are present in a heterozygote.
b. Altered alleles must be passed on to future generations.
c. Altered alleles must be capable of altering other alleles in future generations.
d. All of the above.

Behavioral Epigenetics

Research has shown that life experiences, especially those early in life, can have long-lasting effects on behavior, in some cases into future generations. Increasingly, researchers are finding that these long-term effects are mediated through epigenetic processes. The number of studies that convincingly demonstrate that life experience alters chromatin

21.8 Young rats exposed to more licking and grooming from their mothers develop different patterns of DNA methylation, which alters the expression of stress-response genes and makes them less fearful as adults. [Eric Isselee/Shutterstock.]

structure is currently small (and some are still controversial), but a number of researchers are actively looking for epigenetic effects of experience and their long-term effects on chromatin structure and behavior.

EPIGENETIC CHANGES INDUCED BY MATERNAL BEHAVIOR   A fascinating example of behavioral epigenetics is seen in the long-lasting effects of maternal behavior in rats. A mother rat licks and grooms her offspring (**Figure 21.8**), usually while she arches her back and nurses them. The offspring of mothers who display more licking and grooming behavior are less fearful as adults and show reduced hormonal responses to stress compared with the offspring of mothers who lick and groom less. These long-lasting differences in the offspring are not due to genetic differences inherited from their mothers—at least not genetic differences in the base sequences of the DNA. Offspring exposed to more licking and grooming develop a different pattern of DNA methylation compared with offspring exposed to less licking and grooming. These differences in DNA methylation affect the acetylation of histone proteins that persist into adulthood and alter the expression of the glucocorticoid receptor gene, which plays a role in hormonal responses to stress. The expression of other stress-response genes also is affected.

To demonstrate the effect of altered chromatin structure on the stress response of the offspring, researchers infused the brains of young rats with a deacetylase inhibitor, which prevents the removal of acetyl groups from the histone proteins. After infusion of the deacetylase inhibitor, differences in DNA methylation and histone acetylation associated with grooming behavior disappeared, as did the difference in responses to fear and stress in the adults. This demonstrates that the mother rat's licking and grooming behavior brings about epigenetic changes in the offspring's chromatin, which causes long-lasting differences in their behavior.

▶ **TRY PROBLEM 28**

EPIGENETIC EFFECTS OF EARLY STRESS IN HUMANS
Numerous studies have demonstrated that stress during childhood and adolescence produce a number of adverse effects that persist into adult life. For example, childhood

abuse increases the probability that the child will experience depression, anxiety, and suicide as an adult. In one study, researchers examined the brains of 24 people who had committed suicide, half of whom had experienced childhood abuse. They found that those who experienced childhood abuse had a greater degree of methylation of the glucocorticoid receptor gene, a gene involved in the stress response, than those who had not experienced abuse. Although the number of brains studied was small, the study suggests that early childhood stress can indeed cause epigenetic modifications to chromatin structure in humans.

Other studies have demonstrated that gene expression is affected by early life experience. For example, researchers found that growing up in a lower socioeconomic environment before the age of 5 altered the expression of over 100 genes related to immune function of adults. The introduction to Chapter 11 discusses the observation that early childhood stress—in the form of growing up in an orphanage—alters telomere length, a type of epigenetic change.

EPIGENETICS IN COGNITION   A number of research studies have shown that abnormalities in DNA methylation are associated with disorders of development and intellectual ability in humans. These findings prompted researchers to look for effects of chromatin structure on learning, memory, and cognitive ability in mice and rats. One study found that training mice to avoid an adverse stimulus at a specific location reduced DNA methylation of the *Bdnf* gene, which encodes a growth factor that stimulates the growth of connections between neurons. When demethylated, the *Bdnf* gene was more active. When researchers injected into the mice's brains a drug that inhibits demethylation, activity of the *Bdnf* gene was decreased, and the mice's memory of where the adverse stimulus occurred also decreased.

Another study found that a drug that promotes the acetylation of histone proteins improved learning and memory in mice that have a disorder similar to Alzheimer disease. Acetylation of histones alters chromatin structure by loosening the association of DNA and histone proteins and stimulates transcription of many genes. Other studies have found that histone acetylation decreases with age in mice, with diminished expression of genes related to learning and memory. When researchers injected mice with a drug that is an inhibitor of deacetylase activity, acetylation of histones increased, transcription of genes involved in memory increased, and memory of the mice improved. These studies suggest that changes in chromatin structure may be involved in memory and learning.

CONCEPTS

Studies are providing evidence that early life experiences can produce epigenetic changes that have long-lasting effects on behavior.

Epigenetic Effects of Environmental Chemicals

Because some chemicals are capable of modifying chromatin structure, researchers have looked for long-term effects of environmental toxicants on chromatin structure and epigenetic traits.

There has been much recent interest in chemicals, called endocrine disruptors, which mimic or interfere with natural hormones. Endocrine disruptors are capable of interfering with processes regulated by natural hormones, such as sexual development and reproduction. One endocrine disruptor is vinclozolin, a common fungicide used to control fungal diseases in grapes, fruits, and vegetables and to treat turf on golf courses. Vinclozolin acts as an antagonist at the androgen receptor—vinclozolin and its metabolites mimic testosterone and bind to the androgen receptor, preventing testosterone from binding. But vinclozolin and its metabolites do not properly activate the receptor and, in this way, vinclozolin inhibits the action of androgens and prevents sperm production.

In one study, researchers found that the exposure of embryonic rats to vinclozolin led to reduced sperm production not only in the treated animals (when they reached puberty), but also in several subsequent generations. Increased DNA methylation was seen in sperm of the males that were exposed to vinclozolin and these patterns of methylation were inherited. This study and others have raised concerns that, through epigenetic changes, environmental exposure to some chemicals might have effects on the health of future generations. ▶ TRY PROBLEM 29

> ### CONCEPTS
>
> Through epigenetic changes, environmental chemicals may have influences that extend to later generations.

Transgenerational Epigenetic Effects on Metabolism

In the introduction to this chapter, we discussed how diet during childhood can have effects on health that can carry across generations. These types of epidemiological studies on humans are supported by laboratory studies of mice and rats. In one study, researchers fed inbred male mice either a normal (control) diet or a diet low in protein. They then bred mice in both groups to control females fed a normal diet. The males were then separated from the females and never had any contact with their offspring; their only contribution to the offspring was a set of paternal genes transferred through the sperm.

The offspring were raised and their lipid and cholesterol levels examined. The offspring of males fed a low protein diet exhibited increased expression of genes involved in lipid and cholesterol metabolism and a corresponding decrease in levels of cholesterol, compared to the offspring of males fed a normal diet. They also observed numerous differences in DNA methylation in the offspring of the two types of fathers, although no differences could be found in the methylation patterns of the sperm of the two groups of fathers. These results suggest that epigenetic changes altered the cholesterol metabolism of the offspring, although how the differences in methylation were transmitted from father to offspring was unclear.

In another study, researchers fed male rats a high-fat diet and, not surprisingly, they gained weight. They then bred these males to females that had been fed a normal diet. The offspring were also fed a normal diet. The daughters of the male rats on the high-fat diet had normal weight, but as adults they developed a diabetes-like condition of impaired glucose tolerance and insulin secretion. The researchers observed that in the insulin-secreting pancreatic islet cells of the daughters the expression of 642 genes involved in insulin secretion and glucose tolerance was altered, demonstrating that the father's diet affected gene expression in his daughters.

Epigenetic Effects in Monozygotic Twins

Monozygotic (identical) twins develop from a single egg fertilized by a single sperm that divides and gives rise to two zygotes (see Chapter 6). Monozygotic twins are genetically identical, in the sense that they possess identical DNA sequences, but they often differ somewhat in appearance, health, and behavior. The nature of these differences in the phenotypes of identical twins is not well understood, but recent evidence suggests that at least some of these differences may be due to epigenetic changes. In one study, Mario Fraga at the Spanish National Cancer Center and his colleagues examined 80 identical twins and compared the degree and location of their DNA methylation and histone acetylation. They found that DNA methylation and histone acetylation in identical twins were similar early in life, but older twins had remarkable differences in their overall content and distribution of DNA methylation and histone acetylation. Furthermore, these differences affected gene expression in the twins. This research suggests that identical twins do differ epigenetically and that phenotypic differences between them may be caused by differential gene expression.

> ### CONCEPTS
>
> Phenotypic differences between genetically identical monozygotic twins may result from epigenetic effects.
>
> ✓ **CONCEPT CHECK 4**
>
> What degree of differences would you expect to see in the DNA base sequences and epigenetic marks of monozygotic twins?
>
> a. Similar differences in DNA base sequence and epigenetic marks.
> b. Greater differences in DNA base sequence than epigenetic marks.
> c. Greater differences in epigenetic marks than DNA base sequence.
> d. No differences in either DNA base sequence or epigenetic marks.

X Inactivation

In female mammals, one X chromosome in each cell is randomly inactivated to provide equal expression of X-linked genes in males and females (see Chapter 4). Through this process, termed X inactivation, many genes on the inactivated X chromosome are permanently silenced and are not transcribed. Once a particular X chromosome is inactivated in a cell, that same X chromosome remains inactivated when the DNA is replicated, and the inactivation mark is passed on to daughter cells through mitosis. This phenomenon is responsible for the patchy distribution of black and orange pigment seen in tortoiseshell cats (see Chapter 4). X inactivation is a type of epigenetic effect because it results in a stable change in gene expression that is passed on to other cells.

A great deal of research has demonstrated that which X chromosome is inactivated within a cell is controlled by a particular segment of the X chromosome called the **X-inactivation center**, which is 100,000 to 500,000 bp in length. Inactivation is initiated at the X-inactivation center and then spreads to the remainder of the inactivated X chromosome. Examination of the X-inactivation center led to the discovery of several genes that play a role in inactivating one X chromosome in each female cell and keeping the other X chromosome active (**Figure 21.9**).

The key player in X inactivation is a gene called *Xist* (for X-inactivation-specific transcript) which encodes a long noncoding RNA (lncRNA) that is 17,000 bp in length (**Figure 21.10**). As its name implies, this RNA molecule does not encode a protein. Instead, *Xist* lncRNA coats the X chromosome from which it was transcribed. *Xist* lncRNA then attracts polycomb repressor complex 2 (PRC2) and eventually polycomb repressor complex 1 (PRC1). These proteins produce epigenetic marks, such as histone 3 lysine 27 trimethylation (H3K27me3) and other histone modifications that repress transcription. Eventually, many CpG dinucleotides are methylated, leading to permanent silencing of the inactivated X chromosome.

In mice, there are two separate inactivation events. Soon after fertilization, when the embryo reaches the 8-cell stage, the X chromosome from the male parent is inactivated, while the maternal X chromosome remains active. In the developing embryo, the paternal X chromosome is then reactivated during blastocyst maturation. Inactivation occurs again in early development, but now which X is inactivated is random—the X from the male parent and the X from the female parent are equally likely to be inactivated. From this point on, whichever X is inactivated remains silenced through subsequent cell divisions. However, some genes on the inactivated X chromosome escape inactivation and continue to be transcribed. How these genes escape X inactivation is not known. Interestingly, in marsupial mammals, the paternal X chromosome is the copy that remains permanently silenced in all cells.

As mentioned, X-inactivation is brought about by the transcription of the *Xist* gene on the inactive X chromosome to produce *Xist* lncRNA, which coats the inactive X chromosome and leads to changes in chromatin structure that silence transcription. But what happens on the active X chromosome? Why isn't it coated by *Xist* RNA and silenced? Although all details of this process are not yet understood, recent research has demonstrated that there are several additional genes in the X-inactivation center that encode other lncRNAs. These lncRNAs help bring about X inactivation of the inactive X, while not silencing the active X (see Figure 21.9). One of these is the *Tsix* gene, which is transcribed on the active X chromosome. *Tsix* is antisense to *Xist*, which means that it overlaps with the *Xist* gene and is transcribed from the opposite strand (see Figure 21.10), producing a *Tsix* lncRNA that is complementary to *Xist* lncRNA. Through several mechanisms, *Tsix* represses the expression of *Xist* on the active X chromosome. Another major player is a gene called *Jpx*, which encodes an lncRNA that stimulates transcription of *Xist* on the inactive X chromosome. Thus, *Xist* is controlled by two parallel switches with opposite effects: (1) *Jpx* stimulates *Xist* expression on the inactive X chromosome, causing *Xist* to be transcribed and leading to X-inactivation; and (2) *Tsix* represses *Xist* on the active X chromosome, causing *Xist* not to be transcribed on that chromosome and preventing inactivation. Several other genes are also involved. A gene called *Xite* encodes a lncRNA that sustains *Tsix* expression

21.9 In X inactivation, the *Xist* gene on the inactive X produces a long noncoding RNA that coats the inactive X chromosome and suppresses transcription.

21.10 Several genes within the X-inactivation center interact to bring about inactivation of one X chromosome while keeping the other X chromosome active.

Table 21.2	Major genes involved in X inactivation	
Gene	**Encodes**	**Action of Gene**
Xist	lncRNA	Coats inactive X chromosome and leads to silencing of transcription of many genes on the inactive X
Tsix	lncRNA	Inhibits transcription of *Xist* on active X chromosome
Jpx	lncRNA	Stimulates transcription of *Xist* on inactive X chromosome
Xite	lncRNA	Sustains *Tsix* expression on active X, which inhibits *Xist* and maintains transcription of genes on active X chromosome

on the active X chromosome. The major genes involved in the process of X-inactivation are summarized in **Table 21.2**.

This complex process ensures that in each female cell one X chromosome is inactivated and one remains active. Scientists have long recognized that X-inactivation also involves some type of mechanism that is capable of counting X chromosomes, because all but one X chromosome in each cell is inactivated. Thus, the single X in the cells of an XY male remains active (no X-inactivation occurs), and two X chromosomes are inactivated in XXX females (see discussion of Barr bodies in Chapter 4). The nature of this counting mechanism is not yet well understood. ▶ **TRY PROBLEM 31**

CONCEPTS

Epigenetic changes underlie X-inactivation, in which one X chromosome in female cells is permanently silenced. X-inactivation occurs through the action of several genes in the X-inactivation center that encode long noncoding RNAs. The products of these genes interact to ensure that one X chromosome is inactive and one remains active in each female cell.

✔ **CONCEPT CHECK 5**

What would be the effect of introducing into a female cell siRNAs that degrade *Xist* RNA?

Epigenetic Changes Associated with Cell Differentiation

All cells in the human body are genetically identical, and yet different cell types exhibit remarkably different phenotypes— a nerve cell is quite distinct in its shape, size, and function from an intestinal cell. These differences in phenotypes are stable and passed from one cell to another, despite the fact that the DNA sequences of all the cells are the same.

Stem cells are undifferentiated cells that are capable of forming every type of cell in an organism, a property referred to as **pluripotency**. As a stem cell divides and gives rise to a more specialized type of cell, the gene-expression program of the cell becomes progressively fixed, so that each particular cell type expresses only those genes necessary to carry out the functions of that cell type. Though the control of these cell-specific expression programs is not well understood, changes in DNA methylation and chromatin structure clearly play important roles in silencing some genes and activating others.

Stem cells provide a potential source of cells for regeneration of tissues, medical treatment, and research. In the past, the only source of stem cells with the capacity to differentiation into adult tissues were cells from embryos, but because of ethical concerns about creating and using human embryos for harvesting stem cells, researchers have long sought the ability to induce adult somatic cells to dedifferentiate and revert to stem cells. Such cells are called **induced pluripotent stem cells** (iPSCs). Researchers have now successfully created iPSCs by treating fibroblasts (fully differentiated human connective tissue cells) in culture with a cocktail of transcription factors (**Figure 21.11**), although less than 1% of the cells that are treated actually revert to iPSCs. Transcription

21.11 Differentiated adult cells can be reprogrammed to form induced pluripotent stem cells (iPSCs), which are capable of differentiating into many different types of cells.

factors that induce pluripotency cause extensive epigenetic reprogramming, altering patterns of DNA methylation and histone modifications that accumulate with cellular differentiation. Recent research has shown, however, that iPSCs retain a memory of their past and are not completely equivalent to embryonic stem cells (those derived from embryos). One study found that although the DNA methylation patterns of iPSCs differ greatly from those of differentiated somatic cells, the iPSCs retained some methylation marks of the somatic cells and that the methylation of iPSCs was not identical with that of embryonic stem cells. Another study compared histone modifications of fibroblasts, iPSCs, and embryonic stem cells. The iPSCs and embryonic stem cells had many fewer H3K27me3 and H3K9me3 marks than did the fibroblasts, but researchers also found significantly more of these marks on iPSCs than on embryonic stem cells.

Genomic Imprinting

Diploid organisms usually possess two alleles at each autosomal locus, one allele inherited from the mother and one allele inherited from the father. For most genes, both alleles are expressed, and the effect of a particular allele on the phenotype is independent of which parent transmitted the allele to the offspring. However, for a few genes, the sex of the parent that contributed the allele influences how the allele is expressed—alleles inherited from the mother and father are not equivalent (**Figure 21.12**). This phenomenon, in which the sex of the parent that transmits the allele determines its expression, is termed genomic imprinting (see Chapter 5). For some imprinted genes, the allele inherited from the male parent is expressed and the allele inherited from the female parent is silent; for other genes, the allele inherited from the female parent is expressed and the allele inherited from the male parent is silent. As discussed in Chapter 5, genomic imprinting is thought to be due to different degrees of methylation of genes inherited from the parents.

Previous research suggested that the number of imprinted genes was limited, but more recent research suggests the number is much higher. A study conducted by Christopher Gregg at Harvard University and his colleagues found that over 1300 genes in the mouse brain exhibit evidence of genomic imprinting. Many of these imprinted genes were not completely silenced: there was biased expression, with one sex transmitting an allele that was more highly expressed than the allele transmitted by the other sex. Gregg and his colleagues also found that imprinting was highly variable; some genes were imprinted only in certain tissues or at certain times of development.

Genomic imprinting has a number of interesting parallels to X-inactivation. Most imprinted genes are located in clusters of 3-12 genes that occur at

a discrete region of a particular chromosome. Each cluster contains genes that encode proteins, as well as genes that produce noncoding RNA. In each of the well-studied examples, there is an imprinting control region that determines imprinting; deletion of this region destroys the ability to imprint. In addition, the imprinting control region exhibits different chromatin modifications between alleles inherited from the male and female parents. Each imprinting cluster contains genes for one or more lncRNAs that play an important role in imprinting and are themselves imprinted. For example, the gene for insulin-like growth factor 2 (*Igf2*) in humans exhibits genomic imprinting; the *Igf2* allele transmitted from the male parent is expressed, while the *Igf2* allele transmitted from the female is silenced (see Figure 5.18 in Chapter 5). Several lncRNAs produced by other genes in the imprinting control region are required for gene silencing of *Igf2* in females, although how they bring about repression of transcription is not clear.

Many of the well-studied clusters of imprinted genes are associated with disorders that result from faulty imprinting. Beckwith-Wiedemann syndrome is one such disorder. Children with Beckwith-Wiedemann syndrome exhibit excessive growth during fetal development and early childhood. They also have unusual embryonic malignant tumors.

21.12 In genomic imprinting, the expression of an allele depends on whether it is inherited from the male or female parent. In this case, the A^1 allele is expressed only when inherited from the male parent, but in other cases, an allele is expressed only when inherited from the female parent.

Beckwith-Wiedemann syndrome is associated with imprinting of a cluster of genes on chromosome 11, including the *Igf2* gene. Individuals with Beckwith-Wiedemann syndrome often have small deletions on chromosome 11 that interfere with the normal process of imprinting. For example, *Igf2* is normally expressed only when inherited from the father, but in some children with Beckwith-Wiedemann syndrome, deletions within the imprinting center lead to expression of alleles from both parents. The result is that too much *Igf2* is produced, leading to excessive growth and cancer. Prader-Willi syndrome and Angelman syndrome are disorders that are due to imprinting defects on chromosome 15 (see Chapter 5).

IMPRINTING AND GENETIC CONFLICT Many genes that are genomically imprinted affect fetal and early embryonic growth. One possible explanation for genomic imprinting is the **genetic-conflict hypothesis**, which suggests that there are different and conflicting evolutionary pressures acting on maternal and paternal alleles for genes (such as *Igf2*) that affect fetal growth. From an evolutionary standpoint, paternal alleles that maximize the size of the offspring are favored, because birth weight is strongly associated with infant mortality and adult health. Thus, it is to the advantage of the male parent to pass on alleles that promote maximum fetal growth of their offspring. In contrast, maternal alleles that cause more-limited fetal growth are favored: committing too many nutrients to any one fetus may limit a mother's ability to reproduce in the future and giving birth to very large babies is also difficult and risky. The genetic-conflict hypothesis predicts that genomic imprinting will evolve: paternal copies of genes that affect fetal growth should be maximally expressed, whereas maternal copies of the same genes should be less actively expressed or even silent. Indeed, *Igf2* follows this pattern: the paternal allele is active and promotes growth; the maternal allele is silent and does not contribute to growth. Recent findings demonstrate that the paternal copy of *Igf2* promotes fetal growth by directing more maternal nutrients to the fetus through the placenta.

CONCEPTS

Genomic imprinting is caused by epigenetic differences in the alleles inherited from male and female parents. The genetic conflict hypothesis suggests that imprinting evolves because of conflicting evolutionary pressures acting on maternal and paternal alleles.

✓ **CONCEPT CHECK 6**

Which is true of genomic imprinting?
a. The sex of the parent that transmits an allele affects the expression of the allele in the offspring.
b. The sex of the offspring affects the expression of an allele inherited from one of the parents.
c. The sex of the parent affects how an allele is transmitted to the offspring.
d. The sex of the offspring affects which allele is inherited from the parent.

21.4 The Epigenome

In 2003 researchers declared that essentially the entire human genome had been sequenced. This monumental achievement provided a wealth of information about how genetic information is encoded within the genome. Yet, the DNA base sequence is only a partial record of the heritable information. As we have discussed, additional epigenetic information is contained within chromatin structure: information that is heritable and affects how the base sequence is expressed. The overall pattern of chromatin modifications possessed by each individual organism has been termed the **epigenome**. Over the past few years, techniques have become available for detecting and describing epigenetic modifications across the genome.

DETECTING DNA METHYLATION A number of techniques have been developed for examining levels of DNA methylation. Some of these rely on restriction endonucleases, enzymes that make double-stranded cuts in the DNA at specific base sequences (see Chapter 19). Some restriction enzymes are sensitive to methylation and will not cut a sequence that contains 5-methylcytosine, whereas other restriction enzymes are insensitive to methylation. By cutting DNA with enzymes that are sensitive to methylation and with enzymes that are not, and then analyzing the resulting fragments, overall patterns of methylation can be determined.

A more precise and widely used technique for analyzing DNA methylation is bisulfite sequencing (**Figure 21.13**). In this technique, genomic DNA is first treated with sodium bisulfite, which chemically converts unmethylated cytosine to uracil. Uracils are then detected as thymine during sequencing. However, 5-methylcytosine is not chemically altered by treatment with bisulfite and is detected as cytosine during sequencing (see Chapter 19 for a discussion of DNA sequencing). By sequencing genomic DNA with and without bisulfite treatment, researchers are able to determine the locations of all copies of 5-methylcytosine in the DNA.

DETECTING HISTONE MODIFICATIONS Histone modifications can be detected by breaking the chromatin into fragments and applying an antibody specific to a particular histone modification, a process called chromatin immunoprecipitation (abbreviated ChIP; see Chapter 17). The antibody causes the chromatin with the histone modification to precipitate and separate from chromatin fragments without the modification. The protein is then removed by digestion with an enzyme that degrades protein but not DNA, and the DNA fragment with which the histone was associated is sequenced. This provides information about where in the genome the histone modifications occur.

GENOME-WIDE EPIGENETIC MARKS Using these techniques, geneticists have compared the epigenomes of different types of cells. For example, researchers determined the distribution of 5-methylcytosine across the

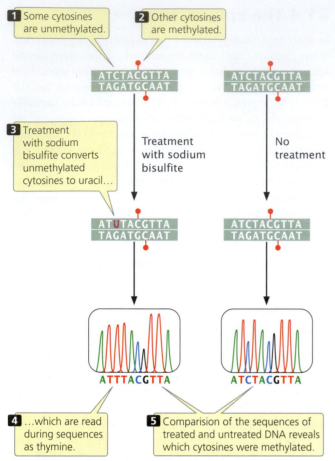

21.13 Bisulfite sequencing can be used to determine the locations of 5-methylcytosines.

entire genome in two cell types: (1) an undifferentiated human stem cell; and (2) a fibroblast. The researchers found widespread differences in the methylation patterns of these cells. The finding that patterns of DNA methylation vary among cell types supports the idea that cytosine methylation provides the means by which cell types stably maintain their differences during development. Other researchers have compared the epigenomes of cancer cells and normal cells and observed distinct epigenetic marks associated with cancer.

Similarly, researchers have mapped the genomic locations of histone modifications in different cell types. These studies detected specific histone modifications associated with promoters and enhancers of active genes. In one study, researchers mapped nine different epigenetic marks in nine different types of human cells. They were able to determine how the chromatin marks varied across cell types and compared the epigenetic marks associated with active and repressed genes. Because specific epigenetic marks are often associated with regulatory elements such as promoters and enhancers, researchers have used the presence of these marks to map the locations of these regulatory elements throughout the genome.

CONCEPTS

The epigenome is the complete set of chromatin modifications possessed by an individual organism.

CONCEPTS SUMMARY

■ The term epigenetics, first coined by Conrad Waddington, today refers to effects by which phenotypes are passed to other cells or future generations but do not include differences in the base sequence of DNA.

■ Many epigenetic phenotypes result from changes to chromatin structure. Epigenetic effects occur through DNA methylation, histone modification, and RNA molecules.

■ Epigenetic changes are stable but are potentially affected by environmental factors.

■ Three molecular mechanisms underlie many epigenetic phenotypes: (1) changes in patterns of DNA methylation; (2) chemical modifications of histone proteins; and (3) RNA molecules that affect chromatin structure and gene expression.

■ Some epigenetic effects result from DNA methylation, in which cytosine bases are methylated to form 5-methylcytosine. Methylation often occurs at CpG dinucleotides. The presence of 5-methylcytosine is associated with repression of transcription.

■ DNA regions with many CpG dinucleotides are referred to as CpG islands. Methylation of CpG islands near a gene often leads to repression of transcription.

■ DNA methylation inhibits transcription by inhibiting the binding of transcription factors and others proteins required for transcription to occur. DNA methylation also attracts proteins that repress transcription and histone deacetylase enzymes that alter chromatin structure.

■ DNA methylation is maintained across cell division by methyltransferase enzymes that recognize methylation of CpG dinucleotides on one strand of DNA and add methyl groups to the unmethylated cytosine bases on the other strand.

■ Modifications of histone proteins alter chromatin structure. Histone modifications may be passed across cell division.

■ RNA molecules bring about modification of chromatin by a variety of processes.

■ Paramutation is a heritable alteration of one allele by another allele, without any change in DNA sequence. Paramutation in corn and mice is mediated through small RNA molecules.

■ Early life experiences can produce epigenetic changes that have long-lasting effects on behavior.

- Environmental chemicals may produce epigenetic effects that are passed to later generations.

- Epigenetic modifications have effects on metabolism that extend across generations.

- Phenotypic differences between genetically identical monozygotic twins may result from epigenetic effects.

- X inactivation occurs when one X chromosome in female cells is permanently silenced. Epigenetic changes bring about X inactivation and require the action of several genes that encode long noncoding RNAs.

- Genomic imprinting occurs when the expression of a gene depends on which parent transmitted the gene. It is caused by epigenetic changes to chromatin structure that are passed to offspring. The genetic conflict hypothesis suggests that imprinting evolves because of conflicting evolutionary pressures acting on maternal and paternal alleles.

- The epigenome is the complete set of chromatin modifications possessed by an individual organism.

IMPORTANT TERMS

epigenetic marks (p. 618)
paramutation (p. 619)
epialleles (p. 620)
X-inactivation center (p. 624)
stem cells (p. 625)
pluripotency (p. 625)
induced pluripotent stem cells (p. 625)
genetic-conflict hypothesis (p. 627)
epigenome (p. 627)

ANSWERS TO CONCEPT CHECKS

1. b

2. b

3. d

4. c

5. No *Xist* RNA would be present to coat the X chromosome, and X inactivation would not occur. Both X chromosomes would remain active.

6. a

WORKED PROBLEM

Problem

The *b1* allele encodes a transcription factor that stimulates production of anthocyanin, a purple pigment in plants. What would be the effect of deleting the seven tandem repeats that are located 100,000 bp upstream of the *b1* locus in maize?

Solution Strategy

What information is required in your answer to the problem?

The effects of deleting the repeats.

What information is provided to solve the problem?

- The *b1* locus encodes a transcription factor that stimulates anthocyanin production in plants.

- The seven tandem repeats are located upstream of the *b1* locus.

For help with this problem, review:

Paramutation in Corn in Section 21.3.

Solution Steps

The information provided in the text indicates that the *B-I B-I* plant normally has high expression of the *b1*

locus, produces anthocyanin, and is purple in color. The tandem repeats are required for high transcription of the *b1* locus, which encodes a transcription factor that stimulates production of anthocyanin, a purple pigment. The tandem repeats act like an enhancer, stimulating transcription of the *b1* locus in the *B-I B-I* genotype. Without the enhancer-like action of the repeats, *b1* will be transcribed at minimal levels and little anthocyanin will be produced. This will result in lightly pigmented *B-I B-I* plants, the same phenotype as is usually seen in plants with genotype *B'B'* plants.

The tandem repeats also encode 25 nucleotide-long siRNAs, which are required for paramutation (conversation of the *B-I* allele into *B'** alleles). Deletion of the tandem repeats would thus eliminate the ability of B' alleles to carry out paramutation.

Recall: Enhancers stimulate transcription at genes that may be distant from the enhancer (see Chapter 17).

COMPREHENSION QUESTIONS

Introduction

1. What is the thrifty phenotype hypothesis? How does it help to explain the long-term effects of diet that were observed among the residents of Överkalix?

Section 21.1

*2. What are the important characteristics of an epigenetic trait?

Section 21.2

3. What three molecular mechanisms alter chromatin structure and are responsible for many epigenetic phenotypes?

4. What is the major form of DNA methylation that is seen in eukaryotes? At what type of DNA sequence is DNA methylation usually found?

5. How does DNA methylation repress transcription?

6. Briefly explain how patterns of DNA methylation are transmitted across cell division.

7. What types of histone modifications are responsible for epigenetic phenotypes?

Section 21.3

8. What is paramutation? What are the key features of this phenomenon?

9. Briefly describe paramutation at the *Kit* locus in mice. What evidence suggests that small RNA molecules play a role in this phenomenon?

10. What evidence suggests that cognition in mice is influenced by epigenetic changes?

11. Explain how vinclozolin acts as an endocrine disrupter.

12. Give an example of a transgenerational epigenetic effect of diet on metabolism.

13. What evidence suggests that differences in monozygotic twins may be caused by epigenetic effects?

14. How is X inactivation an epigenetic phenotype?

15. Briefly describe the molecular processes that cause one X chromosome in each female cell to be active and the other X chromosome to become inactivated.

16. What are induced pluripotent stem cells? How are they derived from adult somatic cells?

17. Define genomic imprinting.

18. What is the genomic conflict hypothesis for the origin of genomic imprinting?

Section 21.4

19. What is the epigenome?

> ▶ For more questions that test your comprehension of the key chapter concepts, go to **LEARNING**Curve for this chapter.

APPLICATION QUESTIONS AND PROBLEMS

Introduction

20. The introduction to this chapter describes the long-term effects of diet on the residents of Överkalix, Sweden.

  a. What evidence suggests that these effects are due to epigenetic effects?

  b. What additional evidence would help to demonstrate that these changes are due to epigenetic changes?

Section 21.1

21. How do epigenetic traits differ from traditional genetic traits, such as the differences in color and shape of peas that Mendel studied?

22. Epigenetics has been described as "inheritance, but not as we know it." Do you think this is a good definition? Why or why not?

Section 21.2

23. What would be required to prove that a phenotype is caused by an epigenetic change?

24. Which honeybee in **Figure 21.4** (the worker or the queen) will have more copies of 5-methylcytosine in its DNA? Explain your answer.

*25. What would be the effect of deleting the *Dnmt3* gene in honeybees?

*26. Much of DNA methylation in eukaryotes occurs at CpG dinucleotides, but some individual cytosine nucleotides are also methylated to form 5-methylcytosine. Considering what you know about the process by which DNA methylation at CpG dinucleotides is maintained across cell division, do you think that methylation at individual C nucleotides would also be maintained by the same process? Explain your reasoning.

Section 21.3

27. A cross between the F_1 individual in **Figure 21.5** and a plant with genotype *B-I B-I* will produce progeny with what phenotype?

*28. A scientist does an experiment in which she removes the offspring of rats from their mother at birth and has her genetics students feed and rear the offspring. Assuming that the students do not lick and groom the baby rats as the mother rats normally do, what long-term behavioral and epigenetic effects would you expect to see in the rats when they grow up?

*29. Pregnant female rats were exposed to a daily dose of 100 or 200 mg/kg of vinclozolin, a fungicide commonly used in the wine industry (M. D. Anway et al. 2005. *Science* 308:1466–1469). The F_1 offspring of the exposed female rates were interbred, producing F_2, F_3, and F_4 rats. None of the F_2, F_3, or F_4 rats were exposed to vinclozolin. Testes from the F_1–F_4 male rats were examined and compared with those of control rats from females that were not exposed to vinclozolin (see adjoining graphs). These effects were seen in more than 90% of the F_1–F_4 male offspring. Furthermore, 8% of the F_1–F_4 males from vinclozolin-exposed females developed complete infertility, compared with 0% of the F_1–F_4 males of control females. Molecular analysis of the testes demonstrated that DNA methylation patterns differed between offspring of vinclozolin-exposed females and offspring of control females. Provide an explanation for the transgenerational effects of vinclozolin on male fertility.

30. Based on the information from studies of the long-term effects of diet on metabolism in mice, what might the epigenetic effects be on the children and grandchildren of people from Överkalix who were exposed to famine as children? Include in your answer the types of epigenetic changes to chromatin you might expect to see and the phenotypic effects on lipid and cholesterol metabolism.

*31. What would be the effect on X inactivation of adding siRNAs that eliminated the products of each of the following genes?

a. *Xist*

b. *Jpx*

Section 21.4

32. A DNA fragment with the following base sequence has some cytosine bases that are methylated (indicated by C*) and others that are unmethylated. To determine the location of methylated and unmethylated cytosines, researchers sequenced this fragment both with and without treatment with sodium bisulfite. Give the sequence of bases that will be read with and without bisulfite treatment.

—ATCGC*GTTAC*ATTGC*ATCA—

33. A geneticist is interested in determining the locations of methylated cytosines within a fragment of DNA. She treats some copies of the fragment with sodium bisulfite and leaves some copies untreated. She then sequences

(a)

(b)

(c)

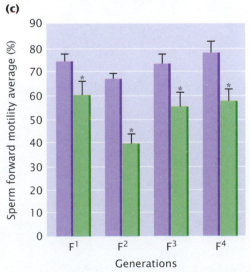

[Graphs after M. D. Anway et al. 2005. *Science* 308:1466–1469.]

the treated and untreated copies of the fragment and obtains the following results. Give the original sequence of the DNA fragment and indicate the presence of methylated cytosines.

Sequence without treatment: —AATTGCCCGATCGATTAAGCCA—

Sequence with treatment: —AATTGTTTGATCGATTAAGCTA—

CHALLENGE QUESTIONS

Section 21.3

34. Would the genomic conflict hypothesis likely explain genomic imprinting for genes involved in adult memory? Why or why not?

35. In recent years, techniques have been developed to clone mammals through a process called nuclear transfer, in which the nucleus of a somatic cell is transferred to an egg cell from which the nuclear material has been removed (see Chapter 22). Research has demonstrated that when a nucleus from a differentiated somatic cell is transferred to an egg cell only a small percent of the resulting embryos complete development and many of those that do die shortly after birth. In contrast, when a nucleus from an undifferentiated embryonic stem cell is transferred into an egg cell, the percentage of embryos that complete development is significantly higher (W. M. Rideout, K. Eggan, and R. Jaenisch. 2001. *Science* 293:1095–1098). Propose a possible reason for why successful development of cloned embryos is higher when the nucleus transferred comes from an undifferentiated embryonic stem cell.

Section 21.4

36. The use of embryonic stem cells has been proposed for replacing cells that are destroyed by disease or injury. Because of ethical concerns about creating and destroying embryos to produce embryonic stem cells, researchers have attempted to create induced pluripotent cells (iPSCs). In this chapter we discussed studies showing that iPSCs retain some epigenetic marks of the differentiated adult cells from which they were derived. What implications might this research have for attempts to use iPSCs to regrow cells and tissue?

▶ Go to your **LaunchPad** to find additional learning resources and the Suggested Readings for this chapter.

22

Developmental Genetics and Immunogenetics

Alterations of key regulatory sequences often bring about major developmental changes. Mutations in an enhancer of the *Pitx1* gene cause the loss of pelvic fins in freshwater populations of threespine sticklebacks (shown here). [Barrett & MacKay/All Canada Photos/Corbis.]

The Origin of Spineless Sticklebacks

Threespine sticklebacks (*Gasterosteus aculeatus*) are curious little fish. In spite of their small size—they reach only about 2 inches (5 cm) in length—they are heavily armored, with protective plates on their back, sides, and belly, and possess three spines on their dorsal surface, thus giving rise to their name. Each fish also possesses two impressive pelvic spines, which are anchored to the pelvic girdle and project out from their sides (**Figure 22.1**). The dorsal and pelvic spines make sticklebacks difficult to swallow, allowing them to survive in an environment where an unprotected 2-inch fish is an easily caught meal for numerous predators.

Most sticklebacks are marine, living in the ocean, but a few isolated populations can be found inland in freshwater lakes. In North America, these freshwater populations originated 10,000 to 20,000 years ago, at the end of the last ice age, when marine sticklebacks invaded the lakes. Many lake populations of threespine sticklebacks have lost their armor and spines, probably because fish predators that might eat them are absent, there is little calcium present to develop the plates and spines, and invertebrate predators found in the lakes catch the fish by grabbing onto their spines. Biologists have long been interested in how marine sticklebacks made the evolutionary transition from the marine to the freshwater environment: how did a heavily armored fish become spineless? Research conducted by developmental geneticists has begun to provide an answer to this question.

In 1998, geneticist David Kingsley from Stanford University began a collaboration with Dolph Schluter, an evolutionary biologist from the University of British Columbia. Their goal was to understand how threespine sticklebacks lost their pelvic spines during the evolutionary transition from marine to freshwater environments. The scientists crossed a female marine stickleback that possessed spines with a male from Paxton Lake, British Columbia that lacked pelvic spines. All of the F_1 fish from this cross possessed pelvic spines. They then crossed two of the F_1 fish, producing a total of 375 F_2 progeny. These F_2 progeny showed a wide range of variation in their pelvic spines: some had fully developed spines, some lacked spines altogether, and others had varying degrees of spine reduction.

Kingsley and his colleagues then examined the association of pelvic spines in the F_2 progeny and the inheritance of genetic markers across the genome. They found that most of the variation in pelvic spines was associated with genetic markers from a particular region on chromosome 7. Interestingly, this same region contains *Pitx1*, a gene that is often expressed in the hind limbs of developing vertebrates. Mice with a mutation in their *Pitx1* gene often have reduced hind limbs, as well as other developmental abnormalities. This observation suggested that mutations within the *Pitx1* gene might be responsible for the absence of pelvic

633

(a) Marine sticklebacks

Pelvic spines

(b) Freshwater sticklebacks

22.1 Marine populations of threespine sticklebacks possess pelvic spines (a) but the spines are reduced or missing in sticklebacks from many freshwater lakes (b). [Adapted by permission from Macmillan Publishers Ltd: Neil H. Shubin and Randall D. Dahn, *Evolutionary Biology: Lost and Found. Nature* 428, 703-704 (15 April 2004), copyright 2004.]

spines in freshwater populations of sticklebacks. But when the researchers examined the DNA sequences of the *Pitx1* gene in marine and freshwater threespine sticklebacks, they found no differences that would alter the amino acid sequence of the protein encoded by the gene. What they did find was that the expression of the *Pitx1* gene differed in marine and freshwater fish. Sticklebacks with pelvic fins expressed *Pitx1* in their pelvic region; sticklebacks without pelvic fins expressed *Pitx1* in other tissues, but the gene was completely inactive in the pelvic region. This suggested that although the *Pitx1* gene itself is not mutated in spineless fish, mutations in regulatory elements that affect the expression of *Pitx1* might be the source of variation in presence or absence of pelvic spines.

In 2010, Kingsley and his team located the mutation that causes the absence of spines in freshwater populations of threespine sticklebacks. They found an enhancer 500 bp upstream of the *Pitx1* gene that controls the tissue-specific expression of *Pitx1*. Enhancers are DNA sequences that promote the transcription of distant genes, often in a tissue-specific manner (see Chapter 17). The researchers determined that spineless fish from Paxton Lake possessed a deletion that removed this enhancer, thus preventing expression of *Pitx1* and ultimately resulting in the loss of pelvic spines. Spineless fish from other lakes in Canada, Alaska, and even Iceland also possessed deletions of the enhancer, but fish from different lakes possessed different deletions. This observation suggests that during the course of evolution, pelvic spines have been lost multiple times through natural selection acting on different mutations that have the same phenotypic effect.

T he story of how the stickleback lost its spines illustrates that major anatomical alterations can occur through small genetic changes in key regulatory sequences that affect development, a theme throughout this chapter on the genetic control of development. The chapter begins with a consideration of how cell differentiation occurs, not through loss of genes but rather through alteration of gene expression. We then discuss the genetic control of early development of *Drosophila* embryos, one of the best-understood developmental systems. We next consider the genetic control of floral structure in plants, another model system that has been well studied, followed by a more detailed look at programmed cell death and the use of development for understanding evolution. At the end of the chapter, we turn to the development of immunity and its genetic control.

22.1 Development Takes Place Through Cell Determination

Every multicellular organism begins life as a unicellular, fertilized egg. This single-celled zygote undergoes repeated cell divisions, eventually producing millions or trillions of cells that constitute a complete adult organism. Initially, each

cell in the embryo is **totipotent**: it has the potential to develop into any cell type. Many cells in plants and fungi remain totipotent, but animal cells usually become committed to developing into specific types of cells after just a few early embryonic divisions. This commitment often comes well before a cell begins to exhibit any characteristics of a particular cell type; when the cell becomes committed, it does not normally reverse its fate and develop into a different cell type. A cell becomes committed by a process called **determination**.

For many years, the work of developmental biologists was limited to describing the changes that take place in the course of development, because techniques for probing the intracellular processes behind these changes were unavailable. But, in recent years powerful genetic and molecular techniques have had a tremendous influence on the study of development; for example, DNA sequencing has provided much information about the nature and organization of DNA sequences that control developmental processes. In some model systems such as *Drosophila* and *Arabidopsis*, the molecular mechanisms underlying developmental change are now beginning to be understood.

If all cells in a multicellular organism are derived from the same original cell, how do different cell types arise? Before

1 Phloem tissue from the carrot is disrupted…

2 …and single cells are isolated.

3 A single cell is placed in a nutritive medium that contains growth hormones…

4 …and eventually gives rise to a complete carrot plant.

22.2 Many kinds of plants can be cloned from isolated single cells. This type of experiment demonstrates that none of the original genetic material is lost during plant development.

the 1950s, two hypotheses were considered. One possibility was that throughout development genes might be selectively lost or altered, causing different cell types to have different genomes. Alternatively, each cell might contain the same genetic information, but different genes might be expressed in each cell type. The results of early cloning experiments helped settle this issue.

Cloning Experiments on Plants

In the 1950s, Frederick Steward developed methods for cloning plants. He disrupted phloem tissue from the root of a carrot by separating and isolating individual cells, and then placed individual cells in a sterile medium that contained nutrients and other substances required for growth. Steward was successful in getting the cells to grow and divide, and eventually he obtained whole edible carrots from single cells (**Figure 22.2**). Because all parts of the plant were regenerated from a specialized phloem cell, he concluded that each phloem cell contained the genetic potential for a whole plant; none of the original genetic material was lost during determination.

Cloning Experiments on Animals

The results of other studies demonstrated that most animal cells also retain a complete set of genetic information during development. In 1952, Robert Briggs and Thomas King removed the nuclei from unfertilized oocytes of the frog *Rana pipiens*. They then isolated nuclei from frog blastulas (an early embryonic stage), injected these nuclei individually into the oocytes, and pricked the eggs with a needle to stimulate them to divide. Although most were damaged in the process, a few eggs developed into complete tadpoles that eventually metamorphosed into frogs.

In the late 1960s, John Gurdon used these methods to successfully clone a few frogs with nuclei isolated from the intestinal cells of tadpoles. This suggested that the differentiated intestinal cells carried the genetic information necessary to encode traits found in all other cells. However, Gurdon's

successful clonings may have resulted from the presence of a few undifferentiated stem cells in the intestinal tissue, which were inadvertently used as the nuclei donors.

In 1997, researchers at the Roslin Institute of Scotland announced that they had successfully cloned a sheep by using the genetic material from a differentiated cell of an adult animal. To perform this experiment, they fused an udder cell from a white-faced Finn Dorset ewe with an enucleated egg cell and stimulated the egg electrically to initiate development. After growing the embryo in the laboratory for a week, they implanted it into a Scottish black-faced surrogate mother. Dolly, the first mammal cloned from an adult cell, was born on July 5, 1996 (**Figure 22.3**). Since then, a number of other animals including sheep, goats, mice, rabbits, cows, pigs, horses, mules, dogs, and cats have been cloned from differentiated adult cells. Importantly, although Dolly and other mammals that have been cloned contain the same *nuclear* genetic material as that of their cloned parent, they are not identical for *cytoplasmic* genes, such as those on the

22.3 In 1996, researchers at the Roslin Institute of Scotland successfully cloned a sheep named Dolly. They used the genetic material from a differentiated cell of an adult animal. [Paul Clements/AP.]

mitochondrial chromosome, because the cytoplasm is donated by both the donor cell and the enucleated egg cell.

The cloning experiments demonstrated that genetic material is not lost or permanently altered during development: development must require the selective expression of genes. But how do cells regulate their gene expression in a coordinated manner to give rise to a complex, multicellular organism? Research has now begun to provide some answers to this important question.

CONCEPTS

The ability to clone plants and animals from single specialized cells demonstrates that genes are not lost or permanently altered during development.

✔ **CONCEPT CHECK 1**

Scientists have cloned some animals by injecting a nucleus from an early embryo into an enucleated egg cell. Does this outcome demonstrate that genetic material is not lost during development? Why or why not?

22.2 Pattern Formation in *Drosophila* Serves As a Model for the Genetic Control of Development

Pattern formation consists of the developmental processes that lead to the shape and structure of complex, multicellular organisms. One of the best-studied systems for the genetic

control of pattern formation is the early embryonic development of *Drosophila melanogaster*. Geneticists have isolated a large number of mutations in fruit flies that influence all aspects of their development, and molecular analysis of these mutations has provided much information about how genes control early development in *Drosophila*.

The Development of the Fruit Fly

An adult fruit fly possesses three basic body parts: head, thorax, and abdomen (**Figure 22.4**). The thorax consists of three segments: the first thoracic segment carries a pair of legs; the second thoracic segment carries a pair of legs and a pair of wings; and the third thoracic segment carries a pair of legs and the halteres (rudiments of the second pair of wings found in most other insects). The abdomen consists of a number of segments.

When a *Drosophila* egg has been fertilized, its diploid nucleus (**Figure 22.5a**) immediately divides nine times without division of the cytoplasm, creating a single, multinucleate cell (**Figure 22.5b**). These nuclei are scattered throughout the cytoplasm but later migrate toward the periphery of the embryo and divide several more times (**Figure 22.5c**). Next, the cell membrane grows inward and around each nucleus, creating a layer of approximately 6000 cells at the outer surface of the embryo (**Figure 22.5d**). Nuclei at one end of the embryo develop into pole cells, which eventually give rise to germ cells. The early embryo then undergoes further development in three distinct stages: (1) the anterior–posterior axis and the dorsal–ventral axis of

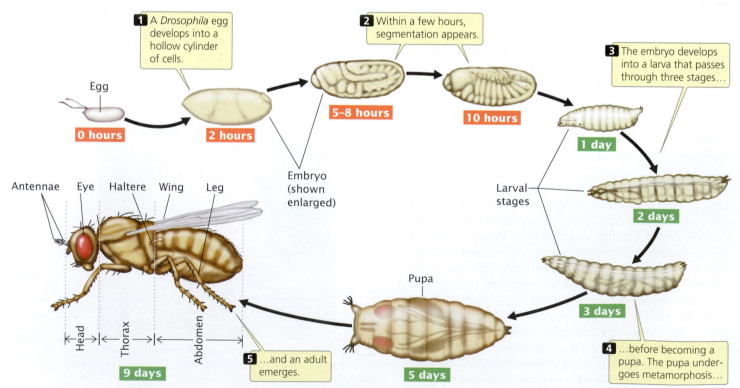

1 A *Drosophila* egg develops into a hollow cylinder of cells.

2 Within a few hours, segmentation appears.

3 The embryo develops into a larva that passes through three stages…

4 …before becoming a pupa. The pupa undergoes metamorphosis…

5 …and an adult emerges.

Egg
0 hours
2 hours
5–8 hours
10 hours
1 day
2 days
3 days
5 days
9 days

Embryo (shown enlarged)
Larval stages
Pupa

Antennae   Eye   Haltere   Wing   Leg
Head   Thorax   Abdomen

22.4 The fruit fly *Drosophila melanogaster* passes through three larval stages and a pupa before developing into an adult fly. The three major body parts of the adult are head, thorax, and abdomen.

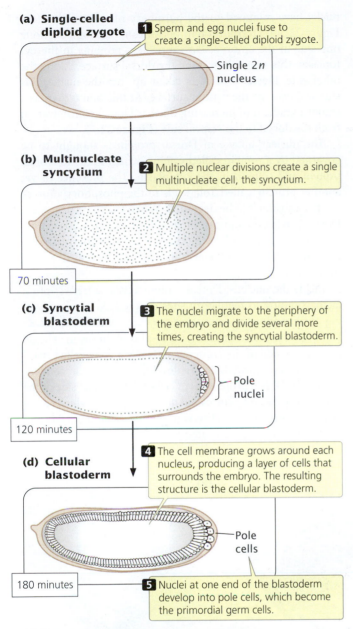

(a) Single-celled diploid zygote

1 Sperm and egg nuclei fuse to create a single-celled diploid zygote.

Single 2n nucleus

(b) Multinucleate syncytium

2 Multiple nuclear divisions create a single multinucleate cell, the syncytium.

70 minutes

(c) Syncytial blastoderm

3 The nuclei migrate to the periphery of the embryo and divide several more times, creating the syncytial blastoderm.

Pole nuclei

120 minutes

(d) Cellular blastoderm

4 The cell membrane grows around each nucleus, producing a layer of cells that surrounds the embryo. The resulting structure is the cellular blastoderm.

Pole cells

180 minutes

5 Nuclei at one end of the blastoderm develop into pole cells, which become the primordial germ cells.

22.5 Early development of a *Drosophila* embryo.

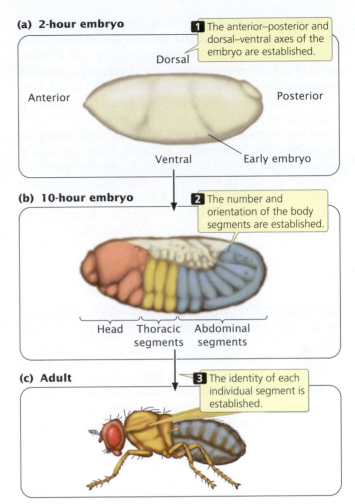

(a) 2-hour embryo

1 The anterior–posterior and dorsal–ventral axes of the embryo are established.

Dorsal

Anterior   Posterior

Ventral   Early embryo

(b) 10-hour embryo

2 The number and orientation of the body segments are established.

Head   Thoracic   Abdominal
        segments   segments

(c) Adult

3 The identity of each individual segment is established.

22.6 In an early *Drosophila* embryo, the major body axes are established, the number and orientation of the body segments are determined, and the identity of each individual segment is established. Different sets of genes control each of these three stages.

Table 22.1	Stages in the early development of fruit flies and the genes that control each stage	
Developmental Stage		**Genes**
Establishment of main body axes		Egg-polarity genes
Determination of number and polarity of body segments		Segmentation genes
Establishment of identity of each segment		Homeotic genes

the embryo are established (**Figure 22.6a**); (2) the number and orientation of the body segments are determined (**Figure 22.6b**); and (3) the identity of each individual segment is established (**Figure 22.6c**). Different sets of genes control each of these three stages (**Table 22.1**).

Egg-Polarity Genes

The **egg-polarity genes** (genes that determine polarity, or direction) play a crucial role in establishing the two main axes of development in fruit flies. You can think of these axes as the longitude and latitude of development: any location in the *Drosophila* embryo can be defined in relation to these two axes.

The egg-polarity genes are transcribed into mRNAs in the course of egg formation in the maternal parent, and these mRNAs become incorporated into the cytoplasm of the egg. After fertilization, the mRNAs are translated into proteins that play an important role in determining the anterior–posterior and dorsal–ventral axes of the embryo. Because the mRNAs of the polarity genes are produced by the female parent and influence the phenotype of the offspring, the traits encoded by them are examples of genetic maternal effects (see Chapter 5).

There are two sets of egg-polarity genes: one set determines the anterior–posterior axis, and the other determines the dorsal–ventral axis. These genes work by setting up concentration gradients of morphogens within the developing embryo. A **morphogen** is a protein that varies in concentration and elicits different developmental responses at different concentrations. Egg-polarity genes function by producing proteins that become asymmetrically distributed in the cytoplasm, giving the egg polarity. This asymmetrical distribution may take place in a couple of ways. An mRNA may be localized to particular regions of the egg cell, leading to an abundance of the protein in those regions when the mRNA is translated. Alternatively, the mRNA may be randomly distributed, but the protein that it encodes may become asymmetrically distributed by a transport system that delivers it to particular regions of the cell, by regulation of its translation, or by its removal from particular regions by selective degradation.

DETERMINATION OF THE DORSAL–VENTRAL AXIS   The dorsal–ventral axis defines the back (dorsum) and belly (ventrum) of a fly (see Figure 22.6). At least 12 different genes determine this axis, one of the most important being a gene called *dorsal*. The *dorsal* gene is transcribed and translated in the maternal ovary, and the resulting mRNA and protein are transferred to the egg during oogenesis. In a newly laid egg, mRNA and protein encoded by the *dorsal* gene are uniformly distributed throughout the cytoplasm but, after the

nuclei have migrated to the periphery of the embryo (see Figure 22.5c), Dorsal protein becomes redistributed. Along one side of the embryo, Dorsal protein remains in the cytoplasm; this side will become the dorsal surface. Along the other side, Dorsal protein is taken up into the nuclei; this side will become the ventral surface. At this point, there is a smooth gradient of increasing nuclear Dorsal concentration from the dorsal to the ventral side (**Figure 22.7**).

The nuclear uptake of Dorsal protein is thought to be governed by a protein called Cactus, which binds to Dorsal protein and traps it in the cytoplasm. The presence of yet another protein, called Toll, leads to the phosphorylation of Cactus, causing it to be degraded. When Cactus is degraded, Dorsal is released and can move into the nucleus. Together, Cactus and Toll regulate the nuclear distribution of Dorsal protein, which in turn determines the dorsal–ventral axis of the embryo.

Inside the nucleus, Dorsal protein acts as a transcription factor, binding to regulatory sites on the DNA and activating or repressing the expression of other genes (**Table 22.2**). High nuclear concentration of Dorsal protein (as in cells on the ventral side of the embryo) activates a gene called *twist*, which causes ventral tissues to develop. Low nuclear concentrations of Dorsal protein (as in cells on the dorsal side of the embryo), activate a gene called *decapentaplegic*, which specifies dorsal structures. In this way, the ventral and dorsal sides of the embryo are determined.

22.7 Dorsal protein in the nuclei helps to determine the dorsal–ventral axis of the *Drosophila* embryo. (a) Relative concentrations of Dorsal protein in the cytoplasm and nuclei of cells in the early *Drosophila* embryo. (b) Micrograph of a cross section of the embryo showing the Dorsal protein, darkly stained, in the nuclei along the ventral surface. [Part b: Max Planck Institute for Developmental Biology.]

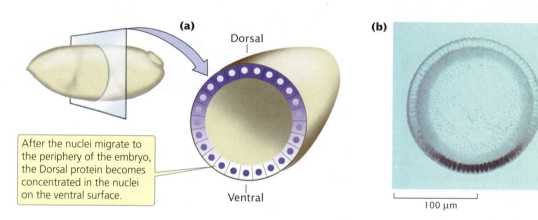

(a)

Dorsal

After the nuclei migrate to the periphery of the embryo, the Dorsal protein becomes concentrated in the nuclei on the ventral surface.

Ventral

(b)

100 μm

Table 22.2	Key genes that control the development of the dorsal–ventral axis in fruit flies and their action	
Gene	**Where Expressed**	**Action of Gene Product**
dorsal	Ovary	Affects the expression of genes such as *twist* and *decapentaplegic*
cactus	Ovary	Traps Dorsal protein in the cytoplasm
toll	Ovary	Leads to the phosphorylation of Cactus, which is then degraded, releasing Dorsal to move into the nuclei of ventral cells
twist	Embryo	Takes part in the development of mesodermal tissues*
decapentaplegic	Embryo	Takes part in the development of gut structures

*One of the three primary tissue layers in the early embryo.

Table 22.3	Some key genes that determine the anterior–posterior axis in fruit flies	
Gene	**Where Expressed**	**Action**
bicoid	Ovary	Regulates expression of genes responsible for anterior structures; stimulates *hunchback*
nanos	Ovary	Regulates expression of genes responsible for posterior structures; inhibits translation of *hunchback* mRNA
hunchback	Embryo	Regulates transcription of genes responsible for anterior structures

DETERMINATION OF THE ANTERIOR–POSTERIOR AXIS

One of the most important early developmental events is the determination of the anterior (head) and posterior (butt) ends of an animal. We will consider several key genes that establish this anterior–posterior axis of the *Drosophila* embryo (**Table 22.3**). An important gene in this regard is *bicoid*, which is first transcribed in the ovary of an adult female during oogenesis. The *bicoid* mRNA becomes incorporated into the cytoplasm of the egg; as it passes into the egg, *bicoid* mRNA becomes anchored to the anterior end of the egg by part of its 3′ end. This anchoring causes *bicoid* mRNA to become concentrated at the anterior end (**Figure 22.8a**). (A number of other genes that are active in the ovary are required for proper localization of *bicoid* mRNA in the egg.) When the egg has been laid, *bicoid* mRNA is translated into Bicoid protein. Because most of the mRNA

is at the anterior end of the egg, Bicoid protein is synthesized there and forms a concentration gradient along the anterior–posterior axis of the embryo, with a high concentration at the anterior end and a low concentration at the posterior end. This gradient is maintained by the continuous synthesis of Bicoid protein and its short half-life.

The high concentration of Bicoid protein at the anterior end induces the development of anterior structures such as the head of the fruit fly. It stimulates the development of anterior structures by binding to regulatory sequences in the DNA and influencing the expression of other genes. One of the most important of the genes stimulated by Bicoid protein is *hunchback*, which is required for the development of the head and thoracic structures of the fruit fly.

The development of the anterior–posterior axis is also greatly influenced by a gene called *nanos*, an egg-polarity

(a) Anterior determinant

Anterior   Posterior

Localized *bicoid* mRNA

1 The *bicoid* mRNA is localized at the anterior end of the egg.

Distribution of *bicoid* mRNA

Distribution of Bicoid protein

2 Bicoid protein forms a gradient with high concentration at the anterior end, which induces development of the anterior structures of a fruit fly.

(b) Posterior determinant

Anterior   Posterior

Localized *nanos* mRNA

3 The *nanos* mRNA is localized at the posterior end of the egg.

Distribution of *nanos* mRNA

Distribution of Nanos protein

4 After fertilization, *nanos* mRNA is translated into Nanos protein, which becomes concentrated at the posterior end of the egg and inhibits formation of anterior structures.

22.8 The anterior–posterior axis in a *Drosophila* embryo is determined by concentrations of Bicoid and Nanos proteins. [Part a and b: From Christiane Nüsslein-Volhard, "Determination of the Embryonic Axes of *Drosophilia*," Development, Suppl. 1, 1991, 1. © Company of Biologists. Part c and d: Courtesy of E. R. Gavis, L. K. Dickenson, and R. Lehman, Massachusetts Institute of Technology.]

Table 22.4   **Segmentation genes and the effects of mutations in them**

Class of Gene	Effect of Mutations	Examples of Genes
Gap genes	Delete adjacent segments	*hunchback, Krüppel, knirps, giant, tailless*
Pair-rule genes	Delete same part of pattern in every other segment	*runt, hairy, fushi tarazu, even-skipped, odd-paired, sloppy paired, odd-skipped*
Segment-polarity genes	Affect polarity of segment; part of segment replaced by mirror image of part of another segment	*engrailed, wingless, gooseberry, cubitus interruptus, patched, hedgehog, disheveled, costal-2, fused*

gene that acts at the posterior end of the axis. The *nanos* gene is transcribed in the adult female, and the resulting mRNA becomes localized at the posterior end of the egg (**Figure 22.8b**). After fertilization, *nanos* mRNA is translated into Nanos protein, which diffuses slowly toward the anterior end. The Nanos protein gradient is opposite that of the Bicoid protein: Nanos is most concentrated at the posterior end of the embryo and is least concentrated at the anterior end. Nanos protein inhibits the formation of anterior structures by repressing the translation of *hunchback* mRNA. The synthesis of the Hunchback protein is therefore stimulated at the anterior end of the embryo by Bicoid protein and is repressed at the posterior end by Nanos protein. This combined stimulation and repression results in a Hunchback protein concentration gradient along the anterior–posterior axis that, in turn, affects the expression of other genes and helps determine the anterior and posterior structures.

CONCEPTS

The major axes of development in early fruit-fly embryos are established as a result of initial differences in the distribution of specific mRNAs and proteins encoded by genes in the female parent (genetic maternal effect). These differences in distribution establish concentration gradients of morphogens, which cause different genes to be activated in different parts of the embryo.

✔ CONCEPT CHECK 2

High concentration of which protein stimulates the development of anterior structures?
a. Dorsal.   c. Bicoid.
b. Toll.   d. Nanos.

Segmentation Genes

Like all insects, the fruit fly has a segmented body plan. When the basic dorsal–ventral and anterior–posterior axes of the fruit-fly embryo have been established, **segmentation genes** control the differentiation of the embryo into individual segments. These genes affect the number and organization of the segments, and mutations in them usually disrupt whole sets of segments. The approximately 25 segmentation genes in *Drosophila* are transcribed after fertilization, so they don't exhibit a genetic maternal effect and their expression is regulated by the Bicoid and Nanos protein gradients.

The segmentation genes fall into three classes as shown in **Table 22.4** and **Figure 22.9**. The three classes act sequentially, affecting progressively smaller regions of the embryo. First, the products of the egg-polarity genes activate or repress **gap genes**, which divide the embryo into broad regions. The gap genes, in turn, regulate **pair-rule genes**, which affect the development of pairs of segments. Finally, the pair-rule genes influence **segment-polarity genes**, which guide the development of individual segments.

Gap genes define large sections of the embryo; mutations in these genes eliminate whole groups of adjacent segments. Mutations in the *Krüppel* gene, for example, cause the absence of several adjacent segments. Pair-rule genes define regional sections of the embryo and affect alternate segments. Mutations in the *even-skipped* gene cause the deletion of even-numbered segments, whereas mutations in the *fushi tarazu* gene cause the absence of odd-numbered segments. Segment-polarity genes affect the orientation of segments. Mutations in these genes cause part of each segment to be deleted and replaced by a mirror image of part or all of an adjacent segment. For example, mutations in the *gooseberry* gene cause the posterior half of each segment to be replaced by the anterior half of an adjacent segment.

CONCEPTS

When the major axes of the fruit-fly embryo have been established, segmentation genes determine the number, orientation, and basic organization of the body segments.

✔ CONCEPT CHECK 3

The correct sequence in which the segmentation genes act is:
a. segment-polarity genes → gap genes → pair-rule genes.
b. gap genes → pair-rule genes → segment-polarity genes.
c. segment-polarity genes → pair-rule genes → gap genes.
d. gap genes → segment-polarity genes → pair-rule genes.

Homeotic Genes in *Drosophila*

After the segmentation genes have established the number and orientation of the segments, **homeotic genes** become active and determine the *identity* of individual segments. Eyes normally arise only on the head segment, whereas legs develop only on the thoracic segments. The products of homeotic genes activate other genes that encode these

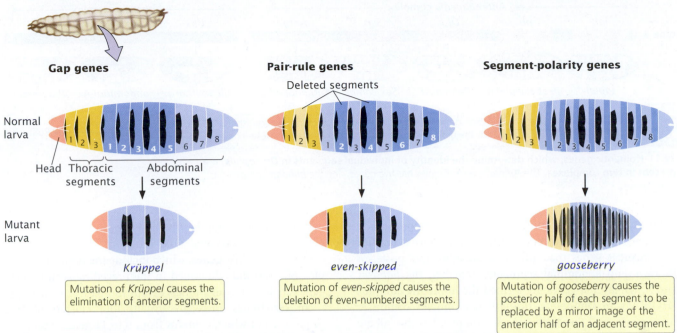

Gap genes   **Pair-rule genes**   **Segment-polarity genes**

Deleted segments

Normal
larva

Head Thoracic Abdominal
      segments segments

Mutant
larva

Krüppel   *even-skipped*   *gooseberry*

Mutation of *Krüppel* causes the elimination of anterior segments.

Mutation of *even-skipped* causes the deletion of even-numbered segments.

Mutation of *gooseberry* causes the posterior half of each segment to be replaced by a mirror image of the anterior half of an adjacent segment.

22.9 Segmentation genes control the differentiation of the *Drosophila* embryo into individual segments. Gap genes affect large sections of the embryo. Pair-rule genes affect alternate segments. Segment-polarity genes affect the orientation of segments.

segment-specific characteristics. Mutations in the homeotic genes cause body parts to appear in the wrong segments.

In the late 1940s, Edward Lewis began to study homeotic mutations in *Drosophila*—mutations that cause bizarre rearrangements of body parts. Mutations in the *Antennapedia* gene, for example, cause legs to develop on the head of a fly in place of the antennae (**Figure 22.10**). Homeotic genes create addresses for the cells of particular segments, telling the cells where they are within the regions defined by the segmentation genes. When a homeotic gene is mutated, the address is wrong and cells in the segment develop as though they were somewhere else in the embryo.

Homeotic genes in *Drosophila* are expressed after fertilization and are activated by specific concentrations of the proteins produced by the gap, pair-rule, and segment-polarity genes. The homeotic gene *Ultrabithorax* (*Ubx*), for example, is activated when the concentration of Hunchback protein (a product of a gap gene) is within certain values. These concentrations exist only in the middle region of the embryo; so *Ubx* is expressed only in these segments.

The homeotic genes in animals encode regulatory proteins that bind to DNA; each gene contains a subset of nucleotides, called a **homeobox**, that are similar in all homeotic genes. The homeobox encodes 60 amino acids that serve as a DNA-binding domain; this domain is related to the helix-turn-helix motif (see Figure 16.2a). Homeoboxes are also present in segmentation genes and other genes that play a role in spatial development.

There are two major clusters of homeotic genes in *Drosophila*. One cluster, the **Antennapedia complex**, affects

(a)   **(b)**

22.10 The homeotic mutation *Antennapedia* substitutes legs for the antennae of a fruit fly.
(a) Normal, wild-type antennae.
(b) *Antennapedia* mutant. [Dr. Walter J. Gehring.]

22.11 Homeotic genes, which determine the identity of individual segments in *Drosophila*, are present in two complexes. The *Antennapedia* complex has five genes, and the *bithorax* complex has three genes.

the development of the adult fly's head and anterior thoracic segments. The other cluster consists of the **bithorax** complex and includes genes that influence the adult fly's posterior thoracic and abdominal segments. Together, the *bithorax* and *Antennapedia* genes are termed the **homeotic complex (HOM-C)**. In *Drosophila*, the *bithorax* complex comprises three genes, and the *Antennapedia* complex has five; all are located on the same chromosome (**Figure 22.11**). In addition to these eight genes, HOM-C contains many sequences that regulate the homeotic genes. Remarkably, the order of the genes in the HOM-C is the same as the order in which the genes are expressed along the anterior–posterior axis of the body. The genes that are expressed in the more-anterior segments are found at one end of the complex, whereas those expressed in the more-posterior end of the embryo are found at the other end of the complex (see Figure 22.11).

CONCEPTS

Homeotic genes help determine the identity of individual segments in *Drosophila* embryos by producing DNA-binding proteins that activate other genes. Each homeotic gene contains a consensus sequence called a homeobox, which encodes the DNA-binding domain.

✔ CONCEPT CHECK 4

Mutations in homeotic genes often cause
a. the deletion of segments.
b. the absence of structures.
c. too many segments.
d. structures to appear in the wrong place.

Homeobox Genes in Other Organisms

After homeotic genes in *Drosophila* had been isolated and cloned, molecular geneticists set out to determine if similar genes exist in other animals; probes complementary to the homeobox of *Drosophila* genes were used to search for homologous genes that might play a role in the development of other animals. The search was hugely successful: homeobox-containing genes have been found in all animals, including nematodes, beetles, sea urchins, frogs, birds, and mammals. Genes with homeoboxes have even been discovered in fungi

and plants, indicating that the homeobox arose early in the evolution of eukaryotes. One group of homeobox genes comprises the **Hox genes**, which include the homeotic complex of *Drosophila* described in the preceding section. *Hox* genes are found in all animals except sponges.

In vertebrates, there are usually four clusters of *Hox* genes, each of which contains from 9 to 11 genes. Mammalian *Hox* genes, like those in *Drosophila*, encode transcription factors that help determine the identity of body regions along an anterior–posterior axis. The *Hox* genes of other organisms often exhibit the same relation between order on the chromosome and order of their expression along the anterior–posterior axis of the embryo as that of *Drosophila* (**Figure 22.12**), but this pattern is not universal. For example, the tunicate *Oikopleura dioica* (a primitive relative of vertebrates) has 9 *Hox* genes, but the genes are scattered throughout the genome, unlike the clustered arrangement seen in most animals. Despite a lack of physical clustering, the *Hox* genes in *O. dioica* are expressed in the same anterior–posterior order as that seen in vertebrates.

The *Hox* genes of vertebrates also exhibit a relation between their order on the chromosome and the timing of their expression: genes at one end of the complex (those expressed at the anterior end) are expressed early in development, whereas genes at the other end (those expressed at the posterior end) are expressed later. If a *Hox* gene is experimentally moved to a new location within the *Hox*-gene complex, it is expressed in the appropriate tissue but the timing of its expression is altered, suggesting that the timing of gene expression is controlled by its physical location within the complex. Although the mechanism of this sequential control is not well understood, evidence suggests that, in mice, it is controlled by a progressive change in the methylation patterns of histone proteins, an epigenetic change that alters chromatin structure and affects transcription (see Chapter 21). Recent studies have also identified microRNA genes (see Chapter 14) within the *Hox*-gene clusters, and evidence suggests that miRNAs play a role in controlling the expression of some *Hox* genes.

Hox genes and their expression are often correlated with anatomical changes in animals, and *Hox* genes have been hypothesized to play an important role in the evolution of animals. For example, the lancet *Branchiostoma* (another

Genes shown in the same color are homologous.

There are four clusters of *Hox* genes in mammals, each cluster containing from 9 to 11 genes.

The mammalian *Hox* genes are similar in sequence to the homeotic genes found in *Drosophila*, and they are in the same order.

22.12 Hox genes in mammals are similar to those found in *Drosophila*. The complexes are arranged so that genes with similar sequences lie in the same column. See Figure 22.11 for the full names of the *Drosophila* genes.

primitive relative of vertebrates) has a simple body form and possesses only 14 *Hox* genes in a single cluster, whereas some fishes, with much more complex body architecture, have as many as 48 *Hox* genes in seven clusters.

CONNECTING CONCEPTS

The Control of Development

Development is a complex process consisting of numerous events that must take place in a highly specific sequence. Studies of fruit flies and other organisms reveal that this process is regulated by a large number of genes. In *Drosophila*, the dorsal–ventral axis and the anterior–posterior axis are established by maternal genes; these genes encode mRNAs and proteins that are localized to specific regions within the egg and cause specific genes to be expressed in different regions of the embryo. The proteins of these genes then stimulate other genes, which in turn stimulate yet other genes in a cascade of control. As might be expected, most of the gene products in the cascade are regulatory proteins, which bind to DNA and activate other genes.

In the course of development, successively smaller regions of the embryo are determined (**Figure 22.13**). In *Drosophila*, first, the major axes and regions of the embryo are established by egg-polarity genes. Next, patterns within each region are determined by the action of segmentation genes: the gap genes define large sections, the pair-rule genes define regional sections of the embryo and affect alternate segments, and the segment-polarity genes affect individual segments. Finally, the homeotic genes provide each segment with a unique identity. Initial gradients in proteins and mRNA stimulate localized gene expression, which produces more finely located gradients that stimulate even more localized gene expression. Developmental regulation thus becomes more and more narrowly defined.

The processes by which limbs, organs, and tissues form (called morphogenesis) are less well understood, although this pattern of generalized-to-localized gene expression is encountered frequently. ▶ TRY PROBLEM 22

Epigenetic Changes in Development

Early development of the fruit fly is controlled in large part by the products of certain key genes selectively activating or repressing the expression of other genes. As discussed in Chapter 21, gene expression in eukaryotes is affected by epigenetic changes, and indeed epigenetics also plays an important role in development.

In Chapter 21, epigenetic changes were defined as heritable alterations to DNA and chromatin structure—alterations that affect gene expression and are passed on to other cells but are not changes to the DNA base sequence. In the course of development, a single-cell zygote divides and gives rise to many cells, which differentiate and acquire the characteristics of specific organs and tissues.

22.13 A cascade of gene regulation establishes the polarity and identity of individual segments of *Drosophila*. In development, successively smaller regions of the embryo are determined.

Each type of cell eventually expresses a different subset of genes, producing the proteins needed for that cell type, and this program of gene expression is passed on when the differentiated cell divides.

The gene-expression program of cells that make up a particular organ or tissue type is often defined by epigenetic marks. As development and differentiation proceed, cells acquire epigenetic changes that turn specific sets of genes on and off. In Chapter 21, we considered several types of epigenetic marks, including DNA methylation, the modification of histone proteins, and changes due to RNA molecules. These epigenetic changes help determine which genes are expressed by a cell. In early stages of development, genes that may be required at later stages of development are often held in a transiently silent state by histone modifications. Histone modifications are generally flexible and easily reversed, and so these genes can be activated later in developmental stages. In the course of development, other genes are permanently silenced; this more-long-term silencing is often accomplished by DNA methylation.

22.3 Genes Control the Development of Flowers in Plants

We have now examined in detail pattern formation in *Drosophila*, which serves as a model system of development. Another model system that has provided important insight into how genes influence patterns of growth and development is the formation of flower parts in angiosperms.

One of the most important developmental events in the life of a plant is the switch from vegetative growth to flowering. The precise timing of this switch is affected by season, day length, plant size, and a number of other factors and is under the control of a large number of different genes. The development of the flower itself also is under genetic control, and homeotic genes play a crucial role in the determination of the floral structures.

Flower Anatomy

A flower is made up of four concentric rings of modified leaves, called whorls. The outermost whorl (whorl 1) consists of the green leaflike sepals. The next whorl (whorl 2) consists of the petals, which typically lack chlorophyll. Whorl 3 consists of the stamens, which bear pollen, and whorl 4 consists of carpels that are often fused to form the stigma bearing the ovules. In wild-type *Arabidopsis*, a model genetic plant (see the Reference Guide to Model Genetic Organisms and **Figure 22.14**), there are four sepals, four white petals, six stamens (four long and two short), and two carpels (**Figure 22.15a**).

Genetic Control of Flower Development

Elliot Meyerowitz and his colleagues conducted a series of experiments to examine the genetic basis of flower development

22.14 The flower produced by *Arabidopsis thaliana* has four sepals, four white petals, six stamens, and two carpels. [Darwin Dale/Photo Researchers,Inc.]

in *Arabidopsis*. They began by isolating and analyzing homeotic mutations in *Arabidopsis*. Homeotic mutations were actually first identified in plants, in 1894, when William Bateson noticed that the floral parts of plants occasionally appeared in the wrong place: he found, for example, flowers in which stamens grew in the place where petals normally grow.

Meyerowitz and his coworkers used these types of mutants to reveal the genes that control flower development. They were able to place the homeotic mutations that they isolated into three groups on the basis of their effect on floral structure. Class A mutants had carpels instead of sepals in the first whorl and stamens instead of petals in the second whorl (**Figure 22.15b**). The third whorl consisted of stamens, and the fourth whorl consisted of carpels, the normal pattern. Class B mutants had sepals in the first and second whorls and carpels in the third and fourth whorls (**Figure 22.15c**). The final group, class C mutants, had sepals and petals in the first and second whorls, respectively, as is normal, but had petals in the third whorl and sepals in the fourth whorl (**Figure 22.15d**)

Meyerowitz and his colleagues concluded that each class of mutants was missing the product of a gene or the products of a set of genes that are critical to proper flower development: class A mutants were missing gene A activity, class B mutants were missing gene B activity, and class C mutants were missing gene C activity. They hypothesized that the class A genes are active in the first and second whorls. Class A gene products alone cause the first whorl to differentiate into sepals, and together with class B gene products they cause the second whorl to develop into petals. Class C gene products together with Class B gene products induce the third whorl to develop into stamens. Class C genes alone cause the fourth whorl to become carpels. The products of the different gene classes and their effects are summarized in the conclusion of Figure 22.15.

Experiment

Question: How do genes control the development of flower structures?

Methods Isolate and analyze homeotic mutants that affect flower development.

Results

(a) Wild-type flower

Stamen Carpel

Sepal

Petal

Whorl 1: Sepals
Whorl 2: Petals
Whorl 3: Stamens
Whorl 4: Carpels

(b) Class A mutants

Whorl 1: Carpels
Whorl 2: Stamens
Whorl 3: Stamens
Whorl 4: Carpels

(c) Class B mutants

Whorl 1: Sepals
Whorl 2: Sepals
Whorl 3: Carpels
Whorl 4: Carpels

(d) Class C mutants

Whorl 1: Sepals
Whorl 2: Petals
Whorl 3: Petals
Whorl 4: Sepals

Conclusion: In wild-type flowers:

Gene Product		Flower whorl affected
Class A gene products	→	sepals in the 1st whorl
Class A + class B gene products	→	petals in the 2nd whorl
Class B + class C gene products	→	stamens in the 3rd whorl
Class C gene products	→	carpels in the 4th whorl

22.15 Analysis of homeotic mutants in *Arabidopsis thaliana* led to an understanding of the genes that determine floral structures in plants.

To explain the results, they also proposed that the genes of some classes affect the activities of others. Where class A is active, class C is repressed, and where class C is active, class A is repressed. Additionally, if a mutation inactivates A, then C becomes active and vice versa. Class A genes are normally expressed in whorls 1 and 2, class B genes are expressed in whorls 2 and 3, and class C genes are expressed in whorls 3 and 4 (**Figure 22.16**).

The interaction of these three classes of genes explains the different classes of mutants in Figure 22.15. For example, class A mutants are lacking class A gene products, and therefore class C genes are active in all tissues because when A is inactivated C becomes active. Therefore whorl 1, with only class C gene products, will consist of carpels; whorl 2, with class C and class B gene products, will produce stamens; whorl 3, with class B and class C gene products, will produce stamens; and whorl 4, with only class C gene activity, will produce carpels (see Figure 22.15b):

Class C (in the absence of class A)
    gene products   → carpels (1st whorl)
Class B + class C (in the absence
    of class A) gene products   → stamens (2nd whorl)
Class B + class C gene products   → stamens (3rd whorl)
Class C gene products   → carpels (4th whorl)

To confirm this explanation, Meyerowitz and his colleagues bred double and triple mutants and predicted the outcome. The resulting flower structures fit their predictions. In subsequent studies, they isolated the genes of each class.

22.16 Expression of class A, B, and C genes varies among the structures of a flower.

There are two class A genes, termed *APETALA1* (*AP1*) and *APETALA2* (*AP2*); two class B genes, termed *APETALA3* (*AP3*) and *PISTILLATA* (*PI*); and one class C gene termed *AGAMOUS* (*AG*). The cloning and sequencing of these genes revealed that all are MADS-box genes—genes that function as transcription factors and affect the expression of other genes. MADS-box genes in plants play a similar role to that of homeobox genes in animals, although MADS-box genes and homeobox genes are not homologous.

The results of other studies have demonstrated the presence of an additional group of genes called *SEPALLATA* (*SEP*) that are expressed in whorls 2, 3, and 4, and they, too, are required for normal floral development. If the *SEP* genes are defective, the flower consists entirely of sepals. Findings from studies of other species have demonstrated that this system of flower development exists not only in *Arabidopsis* but also in other flowering plants. It is important to note that these genes are necessary but not sufficient for proper flower development; other genes also take part in the identity of the different parts of flowers. ▶ **TRY PROBLEM 25**

CONCEPTS

Plant homeotic genes control the development of floral structures in plants. The products of three classes of homeotic genes interact to determine the formation of the four whorls that constitute a complete flower.

✓ CONCEPT CHECK 5

What types of flower structures would you expect to see in whorls 1 through 4 of a mutant plant that failed to produce both class A and class B gene products?

a. Carpels, stamens, stamens, carpels.

b. Sepals, sepals, carpels, carpels.

c. Sepals, sepals, sepals, sepals.

d. Carpels, carpels, carpels, carpels.

CONNECTING CONCEPTS

Comparison of Development in *Drosophila* and Flowers

We have now considered two very different model systems of development: the formation of body form and pattern in fruit flies and the development of flower structures in angiosperms. In spite of their differences, these two systems exhibit similarities in how genes control development.

First, both pattern formation in *Drosophila* and flower development in plants are controlled by numerous genes that interact in complex ways. For example, we saw in *Drosophila* how a large complex of genes successively defines smaller and smaller regions of the fruit fly embryo and how genes at one level stimulate and inhibit genes at other levels. Similarly, flower development is controlled by class A, class B, and class C genes. The action of each class depends on which products of other classes are present, and the individual genes of each class interact in complex ways to control the differentiation of each whorl of a flower.

Another common feature of development in fruit flies and flowers is that many of the genes involved in these processes function by influencing the expression of other genes. In both fruit flies and flowers there is a cascade of development, in which gene products stimulate other genes, which in turn stimulate yet other genes. Many of the gene products are regulatory proteins that bind to DNA and affect the transcription of other genes. For example, *Hox* genes in *Drosophila* and MADS-box genes in flowers encode transcription factors that play an important role in development.

A final similarity is that each system contains homeotic genes, which define the identity of particular structures or segments. Mutation of these homeotic genes produces structures that are in the wrong place, such as legs where antennas are normally found in fruit flies or carpels where sepals usually occur in flowers.

22.4 Programmed Cell Death Is an Integral Part of Development

An important aspect of development is the death of cells. Cell death shapes many body parts in the course of development: it is responsible for the disappearance of a tadpole's tail during metamorphosis and causes the removal of tissue between the digits to produce the human hand. Cell death is also used to eliminate dangerous cells that have escaped normal controls (see the section on mutations in cell-cycle control and cancer in Chapter 23).

APOPTOSIS   Cell death in animals is often initiated by the cell itself in **apoptosis**, or cellular suicide. In this process, a cell's DNA is degraded, its nucleus and cytoplasm shrink, and the cell undergoes phagocytosis by other cells without any leakage of its contents (**Figure 22.17a**). Cells that are injured, on the other hand, die in a relatively uncontrolled manner called *necrosis*. In this process, a cell swells and bursts, spilling its contents over neighboring cells and eliciting an inflammatory response (**Figure 22.17b**). Apoptosis is essential to embryogenesis; most multicellular animals cannot complete development if the process is inhibited.

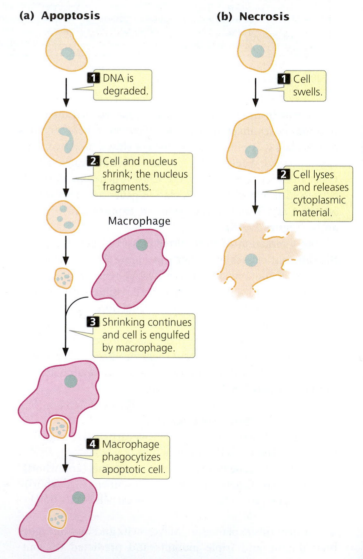

(a) Apoptosis

1 DNA is degraded.

2 Cell and nucleus shrink; the nucleus fragments.

Macrophage

3 Shrinking continues and cell is engulfed by macrophage.

4 Macrophage phagocytizes apoptotic cell.

(b) Necrosis

1 Cell swells.

2 Cell lyses and releases cytoplasmic material.

22.17 Programmed cell death by apoptosis is distinct from uncontrolled cell death through necrosis.

REGULATION OF APOPTOSIS Surprisingly, most cells are programmed to undergo apoptosis and will survive only if the internal death program is continually held in check. Apoptosis is highly regulated and depends on numerous signals inside and outside the cell. Geneticists have identified a number of genes that have roles in various stages of the regulation of apoptosis. Some of these genes encode enzymes called **caspases**, which cleave other proteins at specific sites (after aspartic acid). Each caspase is synthesized as a large, inactive precursor (a procaspase) that is activated by cleavage, often by another caspase. When one caspase is activated, it cleaves other procaspases that trigger even more caspase activity. The resulting cascade of caspase activity eventually cleaves proteins essential to cell function, such as those supporting the nuclear membrane and cytoskeleton. Caspases also cleave a protein that normally keeps an enzyme that degrades DNA (DNase) in an inactive form. Cleavage of this protein activates DNase and leads to the breakdown of cellular DNA, which eventually leads to cell death.

Procaspases and other proteins required for cell death are continuously produced by healthy cells, and so the potential for cell suicide is always present. A number of different signals can trigger apoptosis; for instance, infection by a virus can activate immune cells to secrete substances onto an infected cell, causing that cell to undergo apoptosis. This process is believed to be a defense mechanism designed to prevent the reproduction and spread of viruses. Similarly, DNA damage can induce apoptosis and thus prevent the replication of mutated sequences. Damage to mitochondria and the accumulation of a misfolded protein in the endoplasmic reticulum also stimulate programmed cell death.

APOPTOSIS IN DEVELOPMENT Apoptosis plays a critical role in development. As animals develop, excess cells are often produced and then later culled by apoptosis to produce the proper number of cells required for an organ or a tissue. In some cases, whole structures are created that are later removed by apoptosis. For example, early mammalian embryos develop both male and female reproductive ducts, but the Wolffian ducts degenerate in females and the Mullerian ducts degenerate in males. Apoptosis also plays an important role in the development of immune function (see section 22.6), where lymphocytes that recognize the body's own cells normally undergo apoptosis so they do not attack self tissues. The immune system can also induce apoptosis in cells that are infected with viruses.

During embryonic development in *Drosophila*, large numbers of cells die. Three genes in *Drosophila* activate caspases that are essential for apoptosis: *reaper* (*rpr*), *grim*, and *head involution defective* (*hid*). Embryos possessing a deletion that removes all three genes exhibit no apoptosis and die in the course of embryogenesis with an excess of cells. Numerous other genes also affect the process of apoptosis.

Apoptosis is also crucial in metamorphosis, the process by which larval structures are transformed into adult structures.

For example, the large salivary glands of larval fruit flies regress during metamorphosis. The hormone ecdysone stimulates metamorphosis, including the onset of apoptosis. Ecdysone induces the expression of *rpr* and *hid* and inhibits the expression of other genes, which then leads to apoptosis of salivary gland cells.

APOPTOSIS IN DISEASE The symptoms of many diseases and disorders are caused by apoptosis or, in some cases, its absence. In neurodegenerative disorders such as Parkinson disease and Alzheimer disease, symptoms are caused by a loss of neurons through apoptosis. In heart attacks and stroke, some cells die through necrosis, but many others undergo apoptosis. Cancer is often stimulated by mutations in genes that regulate apoptosis, leading to a failure of apoptosis that would normally eliminate cancer cells (see Chapter 23).

CONCEPTS

Cells are capable of apoptosis (programmed cell death), a highly regulated process that depends on enzymes called caspases. Apoptosis plays an important role in animal development and is associated with a number of diseases.

✓ **CONCEPT CHECK 6**

How does cell death from apoptosis differ from cell death from necrosis?

22.5 The Study of Development Reveals Patterns and Processes of Evolution

"Ontogeny recapitulates phylogeny" is a familiar phrase that was coined in the 1860s by German zoologist Ernst Haeckel to describe his belief—now considered an oversimplification—that during their development (ontogeny) organisms repeat their evolutionary history (phylogeny). According to Haeckel's belief, a human embryo passes through fish, amphibian, reptilian, and mammalian stages before developing human traits. Scientists have long recognized that organisms do not pass through the adult stages of their ancestors during their development, but the embryos of these related organisms often display similarities.

COMMON GENES IN DEVELOPMENTAL PATHWAYS Although ontogeny does not precisely recapitulate phylogeny, many evolutionary biologists today are turning to the study of development for a better understanding of the processes and patterns of evolution. Sometimes called "evo-devo," the study of evolution through the analysis of development is revealing that the same genes often shape developmental pathways in distantly related organisms. Biologists once thought that segmentation in vertebrates and invertebrates was only superficially similar, but we now know that in both

Drosophila and the primitive chordate *Branchiostoma*, the *engrailed* gene divides the embryo into specific segments. A gene called *distalless*, which creates the legs of a fruit fly, plays a role in the development of crustacean branched appendages. This same gene also stimulates body outgrowths of many other organisms, from polycheate worms to starfish. Another example is the role of *Pitx1* in controlling the presence and absence of pelvic spines in sticklebacks, discussed in the introduction to this chapter. This same gene is found in mice, where it also affects hind limb development, and in humans, where mutations of the gene have been associated with the development of clubfoot.

An amazing example of how the same genes in distantly related organisms can shape similar developmental pathways is seen in the development of eyes in fruit flies, mice, and humans. Walter Gehring and his collaborators examined the effect of the *eyeless* gene in *Drosophila*, which is required for proper development of the fruit-fly eye. Gehring and his coworkers genetically engineered cells that expressed *eyeless* in parts of the fly where the gene is not normally expressed. When these flies hatched, they had eyes on their wings, antennae, and legs (**Figure 22.18**). These structures were not just tissue that resembled eyes, but complete eyes with a cornea, cone cells, and photoreceptors that responded to light, although the flies could not use these eyes to see, because they lacked a connection to the nervous system.

The *eyeless* gene has counterparts in mice and humans that affect the development of mammalian eyes. There is a striking similarity between the *eyeless* gene of *Drosophila* and the *Small eye* gene that exists in mice. In mice, a mutation in one copy of *Small eye* causes small eyes; a mouse that is homozygous for the *Small eye* mutation has no eyes. There is also a similarity between the *eyeless* gene in *Drosophila* and the *Aniridia* gene in humans; a mutation in *Aniridia* produces a severely malformed human eye. Similarities in the sequences of *eyeless*, *Small eye*, and *Aniridia* suggest that all three genes evolved

from a common ancestral sequence and a common pathway underlies eye development in flies, mice, and humans. This possibility is surprising, because the eyes of insects and mammals were thought to have evolved independently.

Similar genes may be part of a developmental pathway common to two different species but have quite different effects. For example, a *Hox* gene called *AbdB* helps define the posterior end of a *Drosophila* embryo; a similar group of genes in birds divides the wing into three segments. In another example, the *sog* gene in fruit flies stimulates cells to assume a ventral orientation in the embryo, but the expression of a similar gene called *chordin* in vertebrates causes cells to assume dorsal orientation, exactly the opposite of the situation in fruit flies. In vertebrates, *toll* genes encode proteins called Toll-like receptors that bind to molecules on pathogens and stimulate the immune system. In fruit flies, the *toll* gene similarly functions in immunity, but it also encodes a protein that helps determine the dorsal–ventral axis, as mentioned earlier. The theme emerging from these studies is that a small, common set of genes may underlie many basic developmental processes in many different organisms.

EVOLUTION THROUGH CHANGE IN GENE EXPRESSION

Another concept revealed by studies in evo-devo is that many major evolutionary adaptations are not accomplished through changes in the types of proteins produced but through changes in the expression of genes that encode proteins that regulate development. This is illustrated in the introduction to this chapter, where deletion of an enhancer that stimulates the *Pitx1* gene in sticklebacks is responsible for the evolution of pelvic spine reduction in freshwater populations of the fish.

This concept is also seen in the evolution of blind cavefish. The Mexican tetra, *Astyanax mexicanus*, normally occurs in surface waters of streams and rivers in Texas and northern Mexico. Some 10,000 years ago, a few tetras migrated into caves. In the total darkness of their cave environment, vision was not needed and, with the passage of time, these tetras lost their eyes. Today, some 30 distinct populations of Mexican tetras are totally blind and eyeless (**Figure 22.19**), whereas surface-dwelling populations of the same species have retained normal eye development.

22.18 Expression of the *eyeless* gene causes the development of an eye on the leg of a fruit fly. Genes similar to *eyeless* also control eye development in mice and humans. [U. Kloter and G. Halder/ Biozentrum.]

22.19 Mexican tetras that live in caves have lost their eyes through a developmental change in gene expression. [Mark Smith/ Science Source.]

How did Mexican tetras lose their eyes? Mexican tetra zygotes begin to develop eyes, just like their surface-dwelling cousins, but about 24 hours after fertilization, eye development aborts, and the cells that were destined to become the lens spontaneously die. The absence of the lens prevents other components of the eye, such as the cornea and iris, from developing. The optic cup and retina form, but their growth is retarded and photoreceptor cells never differentiate. The degenerate eye sinks into the orbit and is eventually covered by a flap of skin.

Blind Mexican tetras have the same genes as surface-dwelling Mexican tetras; how they differ is in gene expression. Two genes named *sonic hedgehog* (*shh*) and *tiggy-winkle hedgehog* (*twhh*) are more widely expressed in the eye primordium of blind cavefish than in surface fish. (The original *hedgehog* gene was named for a mutant phenotype in *Drosophila*, in which the mutant embryo is covered with denticles like a hedgehog. The *sonic hedgehog* gene is named after the video game character, and the *tiggy-winkle hedgehog* gene is named for Mrs. Tiggy-Winkle, a character from Beatrix Potter's books.) The expanded expression of *shh* and *twhh* activates the transcription of other genes, which cause lens cells to undergo apoptosis, and the lens degenerates.

When geneticists injected *twhh* or *shh* mRNA into the embryos of surface fish, the development of the lens aborted, and adults that developed from these embryos were missing eyes. When drugs were used to inhibit the expression of *twhh* and *shh* in cavefish embryos, eye development in these fish was partly restored. These results demonstrate that eye development in Mexican tetras is regulated by the precise expression of *shh* and *twhh*. Overexpression of one or both of these genes in the cavefish induces the death of lens cells and aborts normal eye development. A small increase in the transcription of either gene during embryonic development results in a major anatomical change in adult fish, change that has allowed the fish to adapt to the darkness of the cave environment. ▶ TRY PROBLEM 27

Another example of differences in gene expression bringing about evolutionary change is seen in Darwin's finches, a group of 14 closely related birds found on the Galápagos Islands (see Figure 26.9). The birds differ primarily in the size and shape of their beaks: ground finches have deep and wide beaks, cactus finches have long and pointed beaks, and warbler finches have sharp and thin beaks. These differences are associated with diet, and evolutionary changes in beak shape and size have taken place in the past when climate changes brought about shifts in the abundance of food items.

To investigate the underlying genetic basis of these evolutionary changes, Arkaht Abzhanov and his colleagues used microarrays (see Chapter 20) to examine differences in transcription levels of several thousand genes in five species of Darwin's finches. They found differences in the expression of a gene that encodes a protein called calmodulin (CaM); the gene that encodes CaM was more highly expressed in the long and pointed beak of cactus-finch embryos than in the beaks of the other species. CaM takes part in a process called calcium signaling, which is known to affect many aspects of development. When Abzhanov and his coworkers activated calcium signaling in developing chicken embryos, the chickens had longer beaks, like those of the cactus finch. Thus, these researchers were able to reproduce, at least in part, the evolutionary difference that distinguishes cactus finches. This experiment shows that changes in the expression of a single gene in the course of development can produce significant anatomical differences in adults.

In these studies and others, the combined efforts of developmental biologists, geneticists, and evolutionary biologists are sources of important insights into how evolution takes place. Although Haeckel's euphonious phrase "ontogeny recapitulates phylogeny" was incorrect, evo-devo is proving that development can reveal much about the process of evolution.

22.6 The Development of Immunity Is Through Genetic Rearrangement

As we have seen in our consideration of animal and plant development, a basic principle of developmental biology is that every somatic cell carries an identical set of genetic information and that no genes are lost in development. Although this principle holds for most cells, there are some important exceptions, one of which concerns genes that encode immune function in vertebrates. In the development of immunity, individual segments of certain genes are rearranged into different combinations, producing immune cells that contain different genetic information and are each adapted to attack one particular foreign substance. This rearrangement and loss of genetic material is key to the power of our immune systems to protect us against almost any conceivable foreign substance.

The immune system provides protection against infection by bacteria, viruses, fungi, and parasites. The focus of an immune response is an **antigen**, defined as any molecule (usually a protein) that elicits an immune reaction. The immune system is remarkable in its ability to recognize an almost unlimited number of potential antigens. The body is full of proteins, so it is essential that the immune system be able to distinguish between self-antigens and foreign antigens. Occasionally, the ability to make this distinction breaks down, and the body produces an immune reaction to its own antigens, resulting in an **autoimmune disease** (**Table 22.5**).

Table 22.5	Examples of autoimmune diseases
Disease	**Tissues Attacked**
Graves disease, Hashimoto thyroiditis	Thyroid gland
Rheumatic fever	Heart muscle
Systemic lupus erythematosus	Joints, skin, and other organs
Rheumatoid arthritis	Joints
Insulin-dependent diabetes mellitus	Insulin-producing cells in pancreas
Multiple sclerosis	Myelin sheath around nerve cells

The Organization of the Immune System

The immune system contains a number of different components and uses several mechanisms to provide protection against pathogens, but most immune responses can be grouped into two major classes: humoral immunity and cellular immunity. Although it is convenient to think of these classes as separate systems, they interact and influence each other significantly.

HUMORAL IMMUNITY   Immune function is carried out by specialized blood cells called lymphocytes, which are a type of white blood cell. **Humoral immunity** centers on the production of antibodies by specialized lymphocytes called **B cells** (**Figure 22.20**), which mature in the bone marrow. **Antibodies** are proteins that circulate in the blood and other body fluids, binding to specific antigens and marking them for destruction by phagocytic cells. Antibodies also activate a set of proteins called complement that help to lyse cells and attract macrophages.

CELLULAR IMMUNITY   **T cells** (see Figure 22.20), are specialized lymphocytes that mature in the thymus and respond only to antigens found on the surfaces of the body's own cells. These lymphocytes are responsible for the second type of immune response, **cellular immunity**.

After a pathogen such as a virus has infected a host cell, some viral antigens appear on the cell's surface. Proteins, called **T-cell receptors**, on the surfaces of T cells bind to these antigens and mark the infected cell for destruction. T-cell receptors must simultaneously bind a foreign antigen and a self-antigen called a **major histocompatibility complex (MHC) antigen** on the cell surface (discussed later in the chapter). Not all T cells attack cells having foreign antigens; some help regulate immune responses, providing communication among different components of the immune system.

CLONAL SELECTION   How can the immune system recognize an almost unlimited number of foreign antigens? Remarkably, each mature lymphocyte is genetically programmed to attack one and only one specific antigen: each mature B cell produces antibodies against a single antigen, and each T cell is capable of attaching to only one type of foreign antigen.

If each lymphocyte is specific for only one type of antigen, how does an immune response develop? The **theory of clonal selection** states that, initially, there is a large pool of millions of different lymphocytes, each capable of binding only one antigen (**Figure 22.21**), so millions of different foreign antigens can be detected. To illustrate clonal selection, let's imagine that a foreign protein enters the body. Only a few lymphocytes in the pool will be specific for this particular foreign antigen. When one of these lymphocytes encounters the foreign antigen and binds to it, that lymphocyte is stimulated to divide. The lymphocyte proliferates rapidly, producing a large population of genetically identical cells—a clone—each of which is specific for that particular antigen.

This initial proliferation of antigen-specific B and T cells is known as a **primary immune response** (see Figure 22.21); in most cases, the primary response destroys the foreign antigen. Following the primary immune response, most of the lymphocytes in the clone die, but a few continue to circulate in the body. These **memory cells** may remain in circulation for years or even for the rest of a person's life. Should the same antigen reappear at some time in the future, memory cells specific to that antigen become activated and quickly give rise to another clone of cells capable of binding the antigen. The rise of this second clone is termed a **secondary immune response** (see Figure 22.21). The ability to quickly produce a second clone of antigen-specific cells permits the long-lasting immunity that often follows recovery from a disease. For example, people who have chicken pox usually have life-long immunity to the disease. The secondary immune response

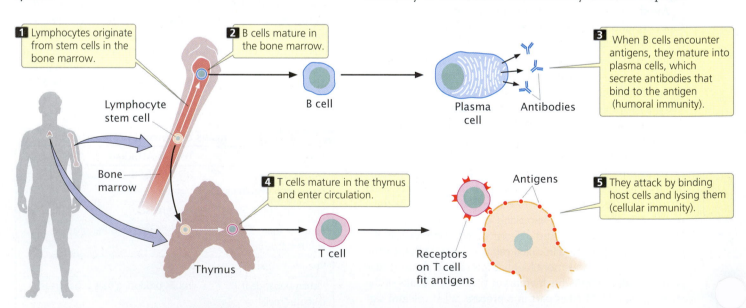

1 Lymphocytes originate from stem cells in the bone marrow.

2 B cells mature in the bone marrow.

3 When B cells encounter antigens, they mature into plasma cells, which secrete antibodies that bind to the antigen (humoral immunity).

Lymphocyte stem cell

B cell

Plasma cell

Antibodies

Bone marrow

4 T cells mature in the thymus and enter circulation.

Antigens

5 They attack by binding host cells and lysing them (cellular immunity).

Thymus

T cell

Receptors on T cell fit antigens

22.20 Immune responses are classified as humoral immunity, in which antibodies are produced by B cells, and cellular immunity, which is produced by T cells.

1 In a large pool of B lymphocytes, each is specific for one antigen.

Antigens

2 When an antigen binds to a B cell, the B cell divides…

Primary immune response

Clone of B cells

3 …and gives rise to a clone of B cells, all specific for the same antigen.

4 This proliferation of lymphocytes is the primary immune response.

Plasma cells

5 Some cells differentiate into antibody-secreting plasma cells.

Antibodies

6 Antibodies are specific for the antigen.

Secondary immune response

Memory cells

7 Memory cells remain in circulation.

Antigen

8 If the same antigen reappears at a later time,…

9 …the antigen binds to the memory cells,…

10 …which rapidly give rise to a secondary immune response.

Plasma cells

Memory cells

Antibodies

22.21 An immune response to a specific antigen is produced through clonal selection.

is also the basis for vaccination, which stimulates a primary immune response to an antigen and results in memory cells that can quickly produce a secondary response if that same antigen appears in the future. Three sets of proteins are used in immune responses: antibodies, T-cell receptors, and the major histocompatibility antigens. The next section explores how the enormous diversity in these proteins is generated.

CONCEPTS

Each B cell and T cell of the immune system is genetically capable of binding one type of foreign antigen. When a lymphocyte binds to an antigen, the lymphocyte undergoes repeated division, giving rise to a clone of genetically identical lymphocytes (the primary response), all of which are specific for that same antigen. Memory cells remain in circulation for long periods of time; if the antigen reappears, the memory cells rapidly proliferate and generate a secondary immune response.

Immunoglobulin Structure

The principal products of the humoral immune response are antibodies—also called immunoglobulins. Each immunoglobulin (Ig) molecule consists of four polypeptide chains—two identical light chains and two identical heavy chains—that form a Y-shaped structure (**Figure 22.22**).

(a)

Antigen-binding site

Heavy chains

Antigen-binding site

Variable region (V)

Joining region (J)

Light chain

Light chain

Constant region (C)

(b)

Antigen-binding site

Light chains

Antigen-binding site

Heavy chains

22.22 Each immunoglobulin molecule consists of four polypeptide chains—two light chains and two heavy chains—that combine to form a Y-shaped structure. (a) Structure of an immunoglobulin. (b) Folded, space-filling model.

Disulfide bonds link the two heavy chains in the stem of the Y and attach a light chain to a heavy chain in each arm of the Y. Binding sites for antigens are at the ends of the two arms.

The light chains of an immunoglobulin are of two basic types: kappa chains and lambda chains. An immunoglobulin molecule can have two kappa chains or two lambda chains, but it cannot have one of each type. Both the light and the heavy chains have a variable region at one end and a constant region at the other end; the variable regions of different immunoglobulin molecules vary in amino acid sequence, whereas the constant regions of different immunoglobulins are similar in sequence. The variable regions of both light and heavy chains make up the antigen-binding regions and specify the type of antigen to which the antibody can bind.

The Generation of Antibody Diversity

The immune system is capable of making antibodies against virtually any antigen that might be encountered in a person's lifetime: each person is capable of making about 10^{15} different antibody molecules. Antibodies are proteins, so the amino acid sequences of all 10^{15} potential antibodies must be encoded in the human genome. However, there are fewer than 1×10^{5} genes in the human genome and, in fact, only 3×10^{9} total base pairs— so how can this huge diversity of antibodies be encoded?

The answer lies in the fact that antibody genes are composed of segments. There are a number of copies of each type of segment, each differing slightly from the others. In the maturation of a lymphocyte, the segments are joined together to create an immunoglobulin gene. The particular copy of each segment used is random and, because there are multiple copies of each type, there are many possible combinations of the segments. A limited number of segments can therefore encode a huge diversity of antibodies.

To illustrate this process of antibody assembly, let's consider the immunoglobulin light chains. Kappa and lambda chains are encoded by separate genes on different chromosomes. Each gene is composed of three types of segments: V, for variable; J, for joining; and C, for constant. The V segments encode most of the variable region of the light chains, the C segment encodes the constant region of the chain, and the J segments encode a short set of nucleotides that join the V and C segments together. The number of V, J, and C segments differs among species. For the human kappa gene, there are from 30 to 35 different functional V gene segments, 5 different J genes, and a single C gene segment, all of which are present in the germ-line DNA (**Figure 22.23a**). The V gene segments, which are about 400 bp in length, are located on the same chromosome and are separated from one another by about 7000 bp. The J gene segments are about 30 bp in length and all together encompass about 1400 bp.

22.23 Antibody diversity is produced by somatic recombination. Shown here is recombination among gene segments that encode the human kappa light chain.

Initially, an immature lymphocyte inherits all of the *V* gene segments and all of the *J* gene segments present in the germ line. In the maturation of the lymphocyte, **somatic recombination** within a single chromosome moves one of the *V* gene segments to a position next to one of the *J* gene segments (**Figure 22.23b**). In Figure 22.23b, V_2 (the second of approximately 35 different *V* gene segments) undergoes somatic recombination, which places it next to J_3 (the third of 5 *J* gene segments); the intervening segments are lost.

After somatic recombination has taken place, the light chain gene is transcribed into pre-mRNA that contains one *V* gene and several *J* genes, along with the *C* gene (**Figure 22.23c**). The resulting pre-mRNA is processed (**Figure 22.23d**) to produce a mature mRNA that contains only sequences for a single *V*, *J*, and *C* segment; this mRNA is translated into a functional light chain (**Figure 22.23e**). In this way, each mature human B cell produces a unique type of kappa light chain, and different B cells produce slightly different kappa chains, depending on the combination of *V* and *J* segments that are joined together.

The gene that encodes the lambda light chain is organized in a similar way but differs from the kappa gene in the number of copies of the different segments. Somatic recombination takes place among the segments in the same way as that in the kappa gene, generating many possible combinations of lambda light chains. The gene that encodes the immunoglobulin heavy chain also is arranged in *V*, *J*, and *C* segments, but this gene possesses *D* (for diversity) segments as well. Thus, many different types of light and heavy chains are possible.

Somatic recombination is brought about by RAG1 and RAG2 proteins, which generate double-strand breaks at specific nucleotide sequences called recombination signal sequences that flank the *V*, *D*, *J*, and *C* gene segments. DNA-repair proteins then process and join the ends of particular segments together.

In addition to somatic recombination, other mechanisms add to antibody diversity. First, each type of light chain can potentially combine with each type of heavy chain to make a functional immunoglobulin molecule, increasing the amount of possible variation in antibodies. Second, the recombination process that joins *V*, *J*, *D*, and *C* gene segments in the developing B cell is imprecise, and a few random nucleotides are frequently lost or gained at the junctions of the recombining segments. This **junctional diversity** greatly enhances variation among antibodies. A third mechanism that adds to antibody diversity is **somatic hypermutation**, a process that leads to a high mutation rate in the antibody genes. This process is initiated when cytosine bases are deaminated, converting them into uracil. The uracil bases are detected and replaced by DNA-repair mechanisms (see Chapter 18) that are error prone and often replace the original cytosine with a different base, leading to a mutation.

Through the processes of somatic recombination, junctional diversity, and somatic hypermutation, each lymphocyte comes to possess a unique set of genetic information (different from that in other lymphocytes) that encodes an antibody specific to a particular antigen. ▶ TRY PROBLEM 28

CONCEPTS

The genes encoding the antibody chains are organized in segments, and germ-line DNA contains multiple versions of each segment. The many possible combinations of *V*, *J*, and D segments permit an immense variety of different antibodies to be generated. This diversity is augmented by the different combinations of light and heavy chains, the random addition and deletion of nucleotides at the junctions of the segments, and the high mutation rates in the immunoglobulin genes.

✔ **CONCEPT CHECK 7**

How does somatic recombination differ from alternative splicing of RNA?

T-Cell-Receptor Diversity

Like B cells, each mature T cell has genetically determined specificity for one type of antigen that is mediated through the cell's receptors. T-cell receptors are structurally similar to immunoglobulins (**Figure 22.24**) and are located on the cell surface; most T-cell receptors are composed of one alpha and one beta polypeptide chain held together by disulfide bonds. One end of each chain is embedded in the cell membrane; the other end projects away from the cell and binds antigens. Like the immunoglobulin chains, each chain of the T-cell receptor possesses a constant region and a variable region (see Figure 22.22); the variable regions of the two chains provide the antigen-binding site.

The genes that encode the alpha and beta chains of the T-cell receptor are organized much like those that encode the heavy and light chains of immunoglobulins: each gene is made up of segments that undergo somatic recombination before the gene is transcribed. For example, the human gene for the

22.24 A T-cell receptor is composed of two polypeptide chains, each having a variable and constant region. Most T-cell receptors are composed of alpha and beta polypeptide chains held together by disulfide bonds. One end of each chain traverses the cell membrane; the other end projects away from the cell and binds antigens.

alpha chain initially consists of 44 to 46 *V* gene segments, 50 *J* gene segments, and a single *C* gene segment. The organization of the gene for the beta chain is similar, except that it also contains *D* gene segments. Alpha and beta chains combine randomly and there is junctional diversity, but there is no evidence for somatic hypermutation in T-cell-receptor genes.

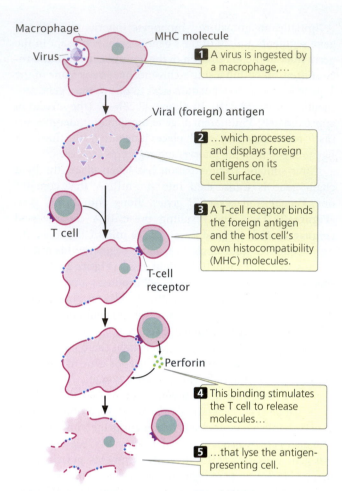

> **CONCEPTS**
>
> Like the genes that encode antibodies, the genes for the T-cell-receptor chains consist of segments that undergo somatic recombination, generating an enormous diversity of antigen-binding sites.

Major Histocompatibility Complex Genes

When tissues are transferred from one species to another or even from one member to another within a species, the transplanted tissues are usually rejected by the host animal. The results of early studies demonstrated that this graft rejection is due to an immune response that takes place when antigens on the surface of cells of the grafted tissue are detected and attacked by T cells in the host organism. The antigens that elicit graft rejection are referred to as histocompatibility antigens, and they are encoded by a cluster of genes called the major histocompatibility complex (MHC).

T cells are activated only when the T-cell receptor simultaneously binds both a foreign antigen and the host cell's own histocompatibility antigen. The reason for this requirement is not clear; it may reserve T cells for action against pathogens that have invaded cells. When a foreign body, such as a virus, is ingested by a macrophage or other cell, partly digested pieces of the foreign body containing antigens are displayed on the cell's surface (**Figure 22.25**). A cell infected with a virus may also express viral antigens on its cell surface. Through their T-cell receptors, T cells bind to both the histocompatibility protein and the foreign antigen and secrete substances that either destroy the antigen-containing cell, activate other B and T cells, or do both.

The MHC genes are among the most variable genes known: there are more than 100 different alleles for some MHC loci. Because each person possesses five or more MHC loci and because many alleles are possible at each locus, no two people (with the exception of identical twins) produce the same set of histocompatibility antigens. The variation in histocompatibility antigens provides each of us with a unique identity for our own cells, which allows our immune systems to distinguish self from nonself. This variation is also the cause of rejection in organ transplants.

> **CONCEPTS**
>
> The MHC genes encode proteins that provide identity to the cells of each individual organism. To bring about an immune response, a T-cell receptor must simultaneously bind both a histocompatibility (self) antigen and a specific foreign antigen.

22.25 T cells are activated by binding both to a foreign antigen and to a histocompatibility antigen on the surface of a self-cell.

Genes and Organ Transplants

For a person with a seriously impaired organ, a transplant operation may offer the only hope of survival. Successful transplantation requires more than the skills of a surgeon; it also requires a genetic match between the patient and the person donating the organ. The fate of transplanted tissue depends largely on the type of antigens present on the surface of its cells. Because foreign tissues are usually rejected by the host, the successful transplantation of tissues between different persons is very difficult. Tissue rejection can be partly inhibited by drugs that interfere with cellular immunity. Unfortunately, this treatment can create serious problems for transplant patients because they may have difficulty fighting off common pathogens and may thus die of infection. The only other option for controlling the immune reaction is to carefully match the donor and the recipient, maximizing the genetic similarities.

The tissue antigens that elicit the strongest immune reaction are the very ones used by the immune system to mark its own cells: those encoded by the major histocompatibility complex. The MHC spans a region of more than 3 million base pairs on human chromosome 6 and has many alleles, providing different MHC antigens on the cells of different people and allowing the immune system to recognize foreign cells. The severity of an immune rejection of a transplanted organ depends on the number of mismatched MHC antigens on the cells of the transplanted tissue. The ABO red-blood-cell antigens also are important because they elicit a strong immune reaction. The ideal donor is the patient's own identical twin, who will have exactly the same MHC and ABO antigens. Unfortunately, most patients don't have an identical twin. The next-best donor is a sibling with the same major MHC and ABO antigens. If a sibling is not available, donors from the general population are considered. An attempt is made to match as many of the MHC antigens of the donor and recipient as possible, and immunosuppressive drugs are used to control rejection due to the mismatches. The long-term success of transplants depends on the closeness of the match.

Survival rates after kidney transplants (the most successful of the major organ transplants) increase from 63% with zero or one MHC match to 90% with four matches.

Scientists have now been successful in inducing adult cells to lose their specialized characteristics and return to an undifferentiated state called a pluripotent stem cell (see Chapter 21), which is capable of developing into many different cell types. In the future it may be possible to create pluripotent stem cells from a person's adult cells, and then grow those cells into tissues or organs that could be transplanted back into the same person. Such cells would have the same MHC antigens as the original cell, avoiding immune rejection that occurs with transplants between different people.

CONCEPTS SUMMARY

- Each multicellular organism begins as a single cell that has the potential to develop into any cell type. As development proceeds, cells become committed to particular fates. The results of cloning experiments demonstrated that this process arises from differential gene expression.

- In the early *Drosophila* embryo, determination is brought about through a cascade of gene control.

- The dorsal–ventral and anterior–posterior axes of the *Drosophila* embryo are established by egg-polarity genes, which are expressed in the female parent and produce RNA and proteins that are deposited in the egg cytoplasm. Initial differences in the distribution of these molecules regulate gene expression in various parts of the embryo. The dorsal–ventral axis is defined by a concentration gradient of the Dorsal protein, and the anterior–posterior axis is defined by concentration gradients of the Bicoid and Nanos proteins.

- Three types of segmentation genes act sequentially to determine the number and organization of the embryonic segments in *Drosophila*. The gap genes establish large sections of the embryo, the pair-rule genes affect alternate segments, and the segment-polarity genes affect the organization of individual segments. Homeotic genes then define the identity of individual *Drosophila* segments.

- Homeotic genes control the development of flower structure. Three sets of genes interact to determine the identity of the four whorls found in a complete flower.

- Apoptosis, or programmed cell death, is a highly regulated process that depends on caspases—proteins that cleave proteins. Apoptosis plays an important role in the development of many animals.

- The immune system is the primary defense network in vertebrates. In humoral immunity, B cells produce antibodies that bind foreign antigens; in cellular immunity, T cells attack cells carrying foreign antigens.

- Each B and T cell is capable of binding only one type of foreign antigen. When a lymphocyte binds to an antigen, the lymphocyte divides and gives rise to a clone of cells, each specific for that same antigen—the primary immune response. A few memory cells remain in circulation for long periods of time and, on exposure to that same antigen, can proliferate rapidly and generate a secondary immune response.

- Immunoglobulins (antibodies) are encoded by genes that consist of several types of gene segments; germ-line DNA contains multiple copies of these gene segments, which differ slightly in sequence. Somatic recombination randomly brings together one version of each segment to produce a single complete gene, allowing many combinations. Diversity is further increased by the random addition and deletion of nucleotides at the junctions of the segments and by a high mutation rate.

- The germ-line genes for T-cell receptors consist of segments with multiple varying copies. Somatic recombination generates many different types of T-cell receptors in different cells. Junctional diversity also adds to T-cell-receptor variability.

- The major histocompatibility complex encodes a number of histocompatibility antigens. The MHC antigen allows the immune system to distinguish self from nonself. Each locus for the MHC contains many alleles.

IMPORTANT TERMS

totipotency (p. 634)
determination (p. 634)
egg-polarity gene (p. 637)
morphogen (p. 638)
segmentation gene (p. 640)

gap gene (p. 640)
pair-rule gene (p. 640)
segment-polarity gene (p. 640)
homeotic gene (p. 640)

homeobox (p. 641)
Antennapedia complex (p. 641)
bithorax complex (p. 642)

homeotic complex (HOM-C) (p. 642)
Hox gene (p. 642)
apoptosis (p. 646)
caspase (p. 647)

antigen (p. 649)
autoimmune disease
    (p. 649)
humoral immunity
    (p. 650)
B cell (p. 650)
antibody (p. 650)

T cell (p. 650)
cellular immunity
    (p. 650)
T-cell receptor (p. 650)
major histocompatibility
    complex (MHC) antigen
    (p. 650)

theory of clonal selection
    (p. 650)
primary immune response
    (p. 650)
memory cell (p. 650)
secondary immune response
    (p. 650)

somatic recombination
    (p. 653)
junctional diversity
    (p. 653)
somatic hypermutation
    (p. 653)

ANSWERS TO CONCEPT CHECKS

1. No, it does *not* prove that genetic material is not lost during development, because differentiation has not yet taken place in an early embryo. The early embryo would likely still contain all its genes and could therefore give rise to a complete animal. The use of specialized cells, such as a cell from an udder, does prove that genes are not lost during development because if they were lost there would be no cloned animal.

2. c

3. b

4. d

5. d

6. In cell death from necrosis, the cell swells and bursts, causing an inflammatory reponse. In cell death through apoptosis, the cell's DNA is degraded, its nucleus and cytoplasm shrink, and the cell is phagocytized, without leakage of cellular contents.

7. Somatic recombination takes place through the rearrangement of DNA segments, so each lymphocyte has a different sequence of nucleotides in its DNA. Alternative splicing (Chapter 14) takes place through the rearrangement of RNA sequences in pre-mRNA; there is no change in the DNA that encodes the pre-mRNA. The generation of antibody diversity requires both the somatic recombination of DNA sequences and the alternative splicing of the pre-mRNA sequences.

WORKED PROBLEMS

Problem 1

If a fertilized *Drosophila* egg is punctured at the anterior end and a small amount of cytoplasm is allowed to leak out, what will be the most likely effect on the development of the fly embryo?

Recall: The *bicoid* gene is an egg-polarity gene that helps determine the anterior-posterior axis of the developing embryo.

Solution Strategy

What information is required in your answer to the problem?

Likely effects on development of removing cytoplasm from the anterior end of a fertilized fly egg.

What information is provided to solve the problem?

Cytoplasm is removed from the anterior end.

For help with this problem, review:

Egg-polarity genes in Section 22.2.

Solution Steps

The egg-polarity genes determine the major axes of development in the *Drosophila* embryo. One of these

genes is *bicoid*, which is transcribed in the maternal parent. As *bicoid* mRNA passes into the egg, the mRNA becomes anchored to the anterior end of the egg. After the egg has been laid, *bicoid* mRNA is translated into Bicoid protein, which forms a concentration gradient along the anterior–posterior axis of the embryo. The high concentration of Bicoid protein at the anterior end induces the development of anterior structures such as the head of the fruit fly. If the anterior end of the egg is punctured, cytoplasm containing high concentrations of Bicoid protein will leak out, reducing the concentration of Bicoid protein at the anterior end. The result will be that the embryo fails to develop head and thoracic structures at the anterior end.

Problem 2

The immunoglobulin molecules of a particular mammalian species have kappa and lambda light chains and heavy chains. The kappa gene consists of 250 *V* and 8 *J* segments. The lambda gene contains 200 *V* and 4 *J* segments. The gene for the heavy chain consists of 300 *V*, 8 *J*, and 4 *D* segments. If just somatic recombination and random combinations of light and heavy chains are taken into consideration, how many different types of antibodies can be produced by this species?

Solution Strategy

What information is required in your answer to the problem?

The number of different types of antibodies that can be produced if somatic recombination and random combinations of light and heavy chains are considered.

What information is provided to solve the problem?

- The kappa gene has 250 *V* and 8 *J* segments.
- The lambda gene has 200 *V* and 4 *J* segments.
- The heavy chain has 300 *V*, 8 *J*, and 4 *D* segments.

For help with this problem, review:

The Generation of Antibody Diversity in Section 22.6.

Solution Steps

For the kappa light chain, there are $250 \times 8 = 2000$ combinations; for the lambda light chain, there are $200 \times 4 = 800$ combinations; so a total of 2800 different types of light chains are possible. For the heavy chains, there are $300 \times 8 \times 4 = 9600$ possible types. Any of the 2800 light chains can combine with any of the 9600 heavy chains; so there are $2800 \times 9600 = 26,880,000$ different types of antibodies possible from somatic recombination and random combination alone. Junctional diversity and somatic hypermutation would greatly increase this diversity.

Hint: The number of each type of light chain consists of the number of V segments x the number of J segments.

Hint: To determine the number of different types of antibodies, multiply the number of possible light chains × the number of possible heavy chains.

COMPREHENSION QUESTIONS

Section 22.1

1. What experiments suggested that genes are not lost or permanently altered in development?

Section 22.2

2. Briefly explain how the Dorsal protein is redistributed in the formation of the *Drosophila* embryo and how this redistribution helps to establish the dorsal–ventral axis of the early embryo.
3. Briefly describe how the *bicoid* and *nanos* genes help to determine the anterior–posterior axis of the fruit fly.
4. List the three major classes of segmentation genes and outline the function of each.
5. What role do homeotic genes play in the development of fruit flies?

Section 22.3

6. How do class A, B, and C genes in plants work together to determine the structures of the flower?

Section 22.4

7. What is apoptosis and how is it regulated?

Section 22.6

8. Explain how each of the following processes contributes to antibody diversity.
   a. Somatic recombination.
   b. Junctional diversity.
   c. Hypermutation.
9. What is the function of the MHC antigens? Why are the genes that encode these antigens so variable?

> For more questions that test your comprehension of the key chapter concepts, go to **LEARNING**Curve for this chapter.

APPLICATION QUESTIONS AND PROBLEMS

Section 22.1

10. If telomeres are normally shortened after each round of replication in somatic cells (see Chapter 12), what prediction would you make about the length of telomeres in Dolly, the first cloned sheep?

Section 22.2

11. A drug causes the degradation of Cactus protein. What would be the effect of administering this drug to developing *Drosophila* embryos?
12. What would be the effect of deleting the *toll* gene in *Drosophila* embryos?

13. Why do mutations in *bicoid* and *nanos* exhibit genetic maternal effects (a mutation in the maternal parent produces a phenotype that shows up in the offspring, see Chapter 5), but mutations in *runt* and *gooseberry* do not? (Hint: See Tables 22.3 and 22.4.)

*14. Give examples of genes that affect development in fruit flies by regulating gene expression at the level of (a) transcription and (b) translation.

15. Using **Figure 22.6**, indicate at which stage segmentation genes, homeotic genes, and egg polarity genes would have an effect on development.

16. What would be the most likely effect on development of puncturing the posterior end of a *Drosophila* egg, allowing a small amount of cytoplasm to leak out, and then injecting that cytoplasm into the anterior end of another egg?

17. Christiane Nüsslein-Volhard and her colleagues carried out several experiments in an attempt to understand what determines the anterior and posterior ends of a *Drosophila* larva (reviewed in C. Nüsslein-Volhard, H. G. Frohnhofer, and R. Lehmann. 1987. *Science* 238:1675–1681). They isolated fruit flies with mutations in the *bicoid* gene (*bcd⁻*). These flies produced embryos that lacked a head and thorax. When they transplanted cytoplasm from the anterior end of an egg from a wild-type female into the anterior end of an egg from a mutant *bicoid* female, normal head and thorax development took place in the embryo. However, transplanting cytoplasm from the posterior end of an egg from a wild-type female into the anterior end of an egg from a *bicoid* female had no effect. Explain these results in regard to what you know about proteins that control the determination of the anterior–posterior axis.

18. What would be the most likely result of injecting *bicoid* mRNA into the posterior end of a *Drosophila* embryo and inhibiting the translation of *nanos* mRNA?

19. What would be the most likely effect of inhibiting the translation of *hunchback* mRNA throughout the embryo?

*20. Molecular geneticists have performed experiments in which they altered the number of copies of the *bicoid* gene in flies, affecting the amount of Bicoid protein produced.

  a. What would be the effect on development of an increased number of copies of the *bicoid* gene?

  b. What would be the effect of a decreased number of copies of *bicoid*? Justify your answers.

21. What would be the most likely effect on fruit-fly development of a deletion in the *nanos* gene?

*22. Give an example of a gene found in each of the categories of genes (egg-polarity, gap, pair-rule, and so forth) listed in **Figure 22.13**.

23. In Chapter 1, we considered preformationism, the early idea about heredity that suggested that inside the egg or sperm is a tiny adult called a homunculus, with all the features of an adult human in miniature. According to this idea, the homunculus simply enlarges during development. What types of evidence presented in this chapter prove that preformationism is false?

Section 22.3

24. Explain how (a) the absence of class B gene expression produces the flower structures seen in class B mutants (see **Figure 22.15c**) and (b) the absence of class C gene product produces the structures seen in class C mutants (see **Figure 22.15d**).

*25. What would you expect a flower to look like in a plant that lacked both class A and class B genes? In a plant that lacked both class B and class C genes?

26. What will be the flower structure of a plant in which expression of the following genes is inhibited?

  a. Expression of class B genes is inhibited in the second whorl, but not in the third whorl.

  b. Expression of class C genes is inhibited in the third whorl, but not in the fourth whorl.

  c. Expression of class A genes is inhibited in the first whorl, but not in the second whorl.

  d. Expression of class A genes is inhibited in the second whorl, but not in the first whorl.

Section 22.5

*27. William Jeffrey and his colleagues crossed surface-dwelling Mexican tetras that had fully developed eyes with cave-dwelling blind Mexican tetras. The progeny from this cross had uniformly small eyes compared with those of surface fish (Y. Yamamoto, D. W. Stock, and W. R. Jeffrey. 2004. *Nature* 431:844–847). What prediction can you make about the expression of *shh* in the embryos of these progeny relative to its expression in the embryos of surface fish?

Section 22.6

*28. In a particular species, the gene for the kappa light chain has 200 *V* gene segments and 4 *J* segments. In the gene for the lambda light chain, this species has 300 *V* segments and 6 *J* segments. If only the variability arising from somatic recombination is taken into consideration, how many different types of light chains are possible?

29. Based on the information provided in **Figure 22.21**, what would be the likely effect of a mutation that prevented the formation of memory cells?

30. In the book *Chromosome 6* by Robin Cook, a biotechnology company genetically engineers individual bonobos (a type of chimpanzee) to serve as future organ donors for clients. The genes of the bonobos are altered so that no tissue rejection takes place when their organs are transplanted into a client. What genes would need to be altered for this scenario to work? Explain your answer.

[Ronald van der Beek/Shutterstock.]

CHALLENGE QUESTIONS

Section 22.2

31. As we have learned in this chapter, the Nanos protein inhibits the translation of *hunchback* mRNA, lowering the concentration of Hunchback protein at the posterior end of a fruit-fly embryo and stimulating the differentiation of posterior characteristics. The results of experiments have demonstrated that the action of Nanos on *hunchback* mRNA depends on the presence of an 11-base sequence that is located in the 3′ untranslated region (3′ UTR) of *hunchback* mRNA. This sequence has been termed the Nanos response element (NRE). There are two copies of NRE in the 3′ UTR of *hunchback* mRNA. If a copy of NRE is added to the 3′ UTR of another mRNA produced by a different gene, the mRNA now becomes repressed by Nanos. The repression is greater if several NREs are added. On the basis of these observations, propose a mechanism for how Nanos inhibits Hunchback translation.

32. Given the distribution of *Hox* genes among animals, what would you predict about the number and type of *Hox* genes in the common ancestor of all animals?

Section 22.6

33. Ataxia-telangiectasis (ATM) is a rare genetic neurodegenerative disease. About 20% of people with ATM develop acute lymphocytic leukemia or lymphoma, cancers of the immune cells. Cells in many of these cancers exhibit chromosome rearrangments, with chromosome breaks occurring at antibody and T-cell-receptor genes (A. L. Bredemeyer et al. 2006. *Nature* 442:466–470). Many people with ATM also have a weakened immune system, making them susceptible to respiratory infections. Research has shown that the locus that causes ATM has a role in the repair of double-strand breaks. Explain why people who have a genetic defect in the repair of double-strand breaks might have a high incidence of chromosome rearrangements in their immune cells and why their immune systems might be weakened.

▶ Go to your **LaunchPad** to find additional learning resources and the Suggested Readings for this chapter.

23

Cancer Genetics

Villa designed by Renaissance architect Andrea Palladio, for whom the *palladin* gene is named. *Palladin* encodes an essential component of a cell's cytoskeleton; when mutated, *palladin* contributes to the spread of pancreatic cancer. [Gianni Dagli Orti/The Art Archive at Art Resource, NY.]

Palladin and the Spread of Cancer

Pancreatic cancer is among the most serious of all cancers. With about 45,000 new cases each year in the United States, it is only the tenth most-common form of the disease, but it is the fourth leading cause of death due to cancer, killing more than 38,000 people each year. Most people with pancreatic cancer survive less than 6 months after they are diagnosed; only 5% survive more than 5 years. A primary reason for pancreatic cancer's lethality is its propensity to spread rapidly to the lymph nodes and other organs. Most symptoms don't appear until the cancer is advanced and has invaded other organs. So what makes pancreatic cancer so likely to spread?

In 2006, researchers identified a key gene that contributes to the development of pancreatic cancer—an important source of insight into the disease's aggressive nature. Geneticists at the University of Washington in Seattle found a unique family in which nine members over three generations were diagnosed with pancreatic cancer (**Figure 23.1**). Nine additional family members had precancerous growths that were likely to develop into pancreatic cancer. In this family, pancreatic cancer was inherited as an autosomal dominant trait.

By using gene-mapping techniques, the geneticists determined that the gene causing pancreatic cancer in the family was located within a region on the long arm of chromosome 4. Unfortunately, this region encompasses 16 million base pairs and includes 250 genes.

To determine which of the 250 genes in the delineated region might be responsible for cancer in the family, researchers designed a unique microarray (see Chapter 20) that contained sequences from the region. They used this microarray to examine gene expression in pancreatic tumors and precancerous growths in family members, as well as in sporadic pancreatic tumors in other people and in normal pancreatic tissue from unaffected people. The researchers reasoned that the cancer gene might be overexpressed or underexpressed in the tumors relative to normal tissue. Data from the microarray revealed that the most-overexpressed gene in the pancreatic tumors and precancerous growths was a gene encoding a critical component of the cytoskeleton—a gene called *palladin*. Sequencing demonstrated that all members of the family with pancreatic cancer had an identical mutation in exon 2 of the *palladin* gene.

The *palladin* gene is named for Renaissance architect Andrea Palladio because it plays a central role in the architecture of the cell. Palladin protein functions as a scaffold for the

23.1 Pancreatic cancer is inherited as an autosomal dominant trait in a family that possesses a mutant *palladin* gene. [After K. L. Pogue et al., *Plos Medicine* 3:2216–2228, 2006.]

binding of the other cytoskeleton proteins that are necessary for maintaining cell shape, movement, and differentiation. The ability of a cancer cell to spread is directly related to its cytoskeleton; cells that spread typically have poor cytoskeleton architecture, enabling them to detach easily from a primary tumor mass and migrate through other tissues. To determine whether mutations in the *palladin* gene affect cell mobility, researchers genetically engineered cells with a mutant copy of the *palladin* gene and tested the ability of these cells to migrate. The cells with mutated *palladin* were 33% more efficient at migrating than were cells with normal *palladin*, demonstrating that the *palladin* gene contributes to pancreatic cancer cells' ability to spread.

The discovery of *palladin*'s link to pancreatic cancer illustrates the power of modern molecular genetics for unraveling the biological nature of cancer. In this chapter, we examine the genetic nature of cancer, a disease that is fundamentally genetic but is often not inherited. We begin by considering the nature of cancer and how multiple genetic alterations are required to transform a normal cell into a cancerous one. We then consider some of the types of genes that contribute to cancer, including oncogenes and tumor-suppressor genes, genes that control the cell cycle, genes encoding DNA-repair systems and telomerase, and genes that, like *palladin*, contribute to the spread of cancer. Next, we take a look at epigenetic changes associated with cancer, and examine how specific genes contribute to the progression of colon cancer. Finally, we discuss chromosome mutations associated with cancer and the role of viruses in some cancers.

23.1 Cancer Is a Group of Diseases Characterized by Cell Proliferation

One of every five people in the United States will die from cancer, and cancer treatments cost billions of dollars per year. Cancer is not a single disease; rather, it is a heterogeneous group of disorders characterized by the presence of cells that do not respond to the normal controls on division. Cancer cells divide rapidly and continuously, creating tumors that crowd out normal cells and eventually rob healthy tissues of nutrients (**Figure 23.2**). The cells of an advanced tumor can separate from the tumor and travel to distant sites in the body, where they may take up residence and develop into new tumors. The most-common cancers in the United States are those of the prostate gland, breast, lung, colon and rectum, and blood (**Table 23.1**).

Tumor Formation

Normal cells grow, divide, mature, and die in response to a complex set of internal and external signals. A normal cell receives both stimulatory and inhibitory signals, and its growth and division are regulated by a delicate balance between these opposing forces. In a cancer cell, one or more of the signals has been disrupted, which causes the cell to proliferate at an abnormally high rate. As they lose their response to the normal controls, cancer cells gradually lose their regular shape and boundaries, eventually forming a distinct mass of abnormal cells—a tumor. If the tumor cells remain localized, the tumor is said to be benign; if the cells invade other tissues, the tumor is said to be **malignant**. Cells that travel to other sites in the body, where they establish secondary tumors, have undergone **metastasis**.

(a)

(b)

Tumor cells   Normal cells

23.2 Abnormal proliferation of cancer cells produces a tumor that crowds out normal cells. (a) Metastatic breast cancer masses (white protrusions) growing in a human liver. (b) A light micrograph of a liver section with tumors. The cancer cells are the light, pale-stained cells; the darker cells are healthy liver cells. [CNRI/Science Source.]

Table 23.1 Estimated incidences of various cancers and cancer mortality in the United States in 2013

Type of Cancer	New Cases per Year	Deaths per Year
Prostate	238,590	29,720
Breast	234,580	40,030
Lung and bronchus	228,190	159,480
Colon and rectum	142,820	50,830
Lymphoma	79,030	20,200
Melanoma	76,690	9,480
Bladder	72,570	15,210
Uterus	49,560	8,190
Leukemias	48,610	23,720
Pancreas	45,220	38,460
Oral cavity and pharynx	41,380	7,890
Liver	30,640	21,670
Brain and nervous system	23,130	14,080
Ovary	22,240	14,030
Stomach	21,600	10,990
Uterine cervix	12,340	4,030
Cancers of soft tissues including heart	11,410	4,390
All cancers	1,660,290	580,350

Source: American Cancer Society, *Cancer Facts and Figures, 2013* (Atlanta: American Cancer Society, 2013), p. 4.

Cancer As a Genetic Disease

Cancer arises as a result of fundamental defects in the regulation of cell division, and its study therefore has significance not only for public health, but also for our basic understanding of cell biology. Through the years, many ideas have been put forth to explain cancer, but we now recognize that most, if not all, cancers arise from defects in DNA.

GENETIC EVIDENCE FOR CANCER Early observations suggested that cancer might result from genetic damage. First, many agents, such as ionizing radiation and chemicals, that cause mutations also cause cancer (are carcinogens, see Chapter 18). Second, some cancers are consistently associated with particular chromosome abnormalities. About 90% of people with chronic myeloid leukemia, for example, have a reciprocal translocation between chromosome 22 and chromosome 9. Third, some specific types of cancers tend to run in families. Retinoblastoma, a rare childhood cancer of the retina, appears with high frequency in a few families and is inherited as an autosomal dominant trait, suggesting that a single gene is responsible for these cases of the disease.

Although these observations hinted that genes play some role in cancer, the theory of cancer as a genetic disease had several significant problems. If cancer is inherited, every cell in the body should receive the cancer-causing gene and therefore every cell should become cancerous. In the types of cancer that run in families, however, tumors typically appear only in certain tissues and often only when the person reaches an advanced age. Finally, many cancers do not run in families at all and, even in regard to those cancers that generally do, isolated cases crop up in families with no history of the disease.

KNUDSON'S MULTISTEP MODEL OF CANCER In 1971, Alfred Knudson proposed a model to explain the genetic basis of cancer. Knudson was studying retinoblastoma, which usually develops in only one eye but occasionally appears in both. Knudson found that, when retinoblastoma appears in both eyes, onset is at an early age, and affected children often have close relatives who also have the disease.

Knudson proposed that retinoblastoma results from two separate genetic defects, both of which are necessary for cancer to develop (**Figure 23.3**). He suggested that, in the cases in which the disease affects just one eye, a single cell in one eye undergoes two successive mutations. Because the chance of these two mutations occurring in a single cell is remote, retinoblastoma is rare and typically develops in only one eye. For the bilateral case, Knudson proposed that the child inherited one of the two mutations required for the cancer, and so every cell contains this initial mutation. In these cases, all that is required for cancer to develop is for one eye cell to undergo the second mutation. Because each eye possesses millions of cells, the probability that the second mutation will occur in at least one cell of each eye is high, producing tumors in both eyes at an early age.

Knudson's proposal suggests that cancer is the result of a multistep process that requires several mutations. If one or more of the required mutations is inherited, fewer additional mutations are required to produce cancer, and the cancer will tend to run in families. Knudson's idea has been called the "two-hit hypothesis" because, for retinoblastoma, only two mutations are necessary to cause a tumor. However, for most cancers more than two mutations are involved in the transformation of normal cells into cancer cells. In the case of retinoblastoma, the two required mutations occur at the same locus (both alleles become mutated), but for many cancers mutations at different loci are required for the development of cancer. The idea that cancer results from multiple mutations turns out to be correct for most cancers.

Knudson's genetic theory for cancer has been confirmed by the identification of genes that, when mutated, cause cancer. Today, we recognize that cancer is fundamentally a genetic disease, although few cancers are actually inherited. Most tumors arise from somatic mutations that accumulate in a person's life span, either through spontaneous mutation or in response to environmental mutagens.

▶ **TRY PROBLEM 24**

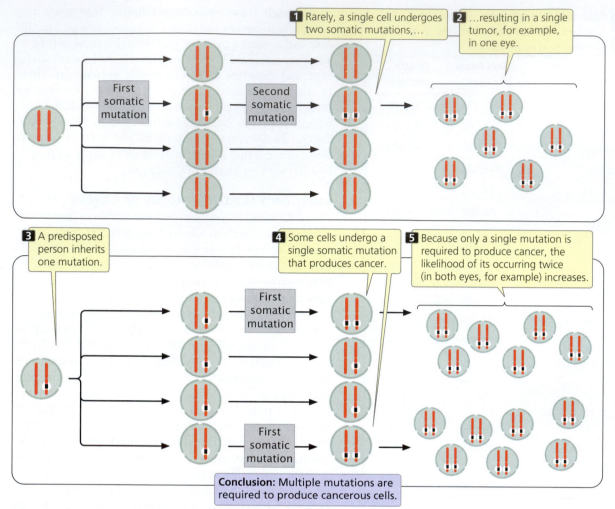

1 Rarely, a single cell undergoes two somatic mutations,...

2 ...resulting in a single tumor, for example, in one eye.

First somatic mutation

Second somatic mutation

3 A predisposed person inherits one mutation.

4 Some cells undergo a single somatic mutation that produces cancer.

5 Because only a single mutation is required to produce cancer, the likelihood of its occurring twice (in both eyes, for example) increases.

First somatic mutation

First somatic mutation

Conclusion: Multiple mutations are required to produce cancerous cells.

23.3 Alfred Knudson proposed that retinoblastoma results from two separate genetic defects, both of which are necessary for cancer to develop.

THE CLONAL EVOLUTION OF TUMORS   Cancer begins when a single cell undergoes a mutation that causes the cell to divide at an abnormally rapid rate. The cell proliferates, giving rise to a clone of cells, each of which carries the same mutation. Because the cells of the clone divide more rapidly than normal, they soon outgrow other cells. An additional mutation that arises in some of the clone's cells may further enhance the ability of those cells to proliferate, and cells carrying both mutations soon become the most-common cells in the clone. Eventually, they may be overtaken by cells that contain yet more mutations that enhance proliferation. In this process, called **clonal evolution**, the tumor cells acquire more mutations that allow them to become increasingly more aggressive in their proliferative properties (**Figure 23.4**).

The rate of clonal evolution depends on the frequency with which new mutations arise. Any genetic defect that allows more mutations to arise will accelerate cancer progression. Genes that regulate DNA repair are often found to have been mutated in the cells of advanced cancers, and inherited disorders of DNA repair are usually characterized by increased incidences of cancer. Because DNA-repair mechanisms normally eliminate many of the mutations that arise, cells with defective DNA-repair systems are more likely to retain mutations than are normal cells, including mutations in genes that regulate cell division. Xeroderma pigmentosum, for example, is a rare disorder caused by a defect in DNA repair (see Chapter 18). People with this condition have elevated rates of skin cancer when exposed to sunlight (which induces mutation). Similarly, breast cancer can be caused by mutations in *BRCA1* and *BRCA2*, two genes that function in DNA repair.

Mutations in genes that affect chromosome segregation also may contribute to the clonal evolution of tumors. Many cancer cells are aneuploid (contain extra or missing copies of individual chromosomes, see Chapter 8) and, clearly, chromosome mutations contribute to cancer progression by duplicating some genes (those on extra chromosomes) and eliminating others (those on

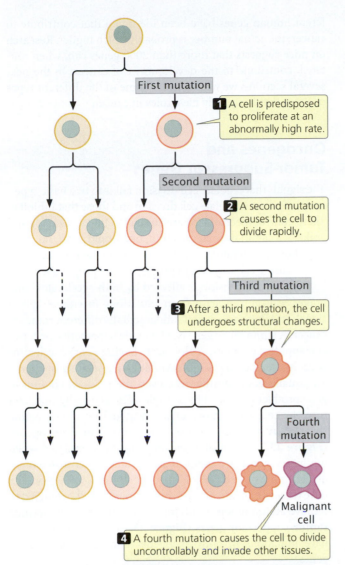

First mutation

1 A cell is predisposed to proliferate at an abnormally high rate.

Second mutation

2 A second mutation causes the cell to divide rapidly.

Third mutation

3 After a third mutation, the cell undergoes structural changes.

Fourth mutation

Malignant cell

4 A fourth mutation causes the cell to divide uncontrollably and invade other tissues.

23.4 Through clonal evolution, tumor cells acquire multiple mutations that allow them to become increasingly more aggressive and proliferative. To conserve space, a dashed arrow is used to represent a second cell of the same type in each case.

deleted chromosomes). Cellular defects that interfere with chromosome separation increase aneuploidy and may therefore accelerate cancer progression.

CONCEPTS

Cancer is fundamentally a genetic disease. Mutations in several genes are usually required to produce cancer. If one of these mutations is inherited, fewer somatic mutations are necessary for cancer to develop, and the person may have a predisposition to cancer. Clonal evolution is the accumulation of mutations in a clone of cells.

✔ CONCEPT CHECK 1

How does the multistep model of cancer explain the observation that sporadic cases of retinoblastoma usually appear in only one eye, whereas inherited forms of the cancer appear in both eyes?

The Role of Environmental Factors in Cancer

Although cancer is a genetic disease, most cancers are not inherited, and many are influenced by environmental factors. The role of environmental factors in cancer is suggested by differences in the incidence of specific cancers throughout the world (**Table 23.2**). The results of studies show that migrant populations typically take on the cancer incidence of their host country. For example, the overall rates of cancer are considerably lower in Japan than in Hawaii. However, within a single generation after migration to Hawaii, Japanese people develop cancer at rates similar to those of native Hawaiians. The increased cancer among the migrants is due to the fact that they are exposed to the same environmental factors as are the natives.

A number of environmental factors contribute to cancer, but those that have the greatest effects include tobacco use, diet, obesity, alcohol, and UV radiation (**Table 23.3**). Other environmental factors that induce cancer are certain types of chemicals, such as benzene (used as an industrial solvent), benzo[a]pyrene (found in cigarette smoke), and polychlorinated biphenyls (PCBs; used in industrial transformers and capacitors). Most environmental factors associated with cancer cause somatic mutations that stimulate cell division or otherwise affect the process of cancer progression.

Environmental factors may interact with genetic predispositions to cancer. For example, lung cancer is clearly associated with smoking, an environmental factor. Genomewide association studies (see Chapter 20) revealed that variation at

Table 23.2	Examples of geographic variation in the incidence of cancer	
Type of Cancer	**Location**	**Incidence Rate***
Lip	Canada (Newfoundland)	15.1
	Brazil (Fortaleza)	1.2
Nasopharynx	Hong Kong	30.0
	United States (Utah)	0.5
Colon	United States (Iowa)	30.1
	India (Mumbai)	3.4
Lung	United States (New Orleans, African Americans)	110.0
	Costa Rica	17.8
Prostate	United States (Utah)	70.2
	China (Shanghai)	1.8
Bladder	United States (Connecticut, Whites)	25.2
	Philippines (Rizal)	2.8
All cancer	Switzerland (Basel)	383.3
	Kuwait	76.3

Source: C. Muir et al., *Cancer Incidence in Five Continents*, vol. 5 (Lyon: International Agency for Research on Cancer, 1987), Table 12-2.
*The incidence rate is the age-standardized rate in males per 100,000 population.

Table 23.3	Percentage of cancer cases in the United Kingdom caused by environmental factors
Factor	Percentage of Cancer Cases
Tobacco	19.4
Diet	9.2
Overweight and obesity	5.5
Alcohol	4.0
Occupation	3.7
Radiation (UV)	3.5
Infections	3.1
Radiation (ionizing)	1.8
All environmental factors	42.7

Source: Parkin, D.M., L. Boyd, and L.C. Walker. 2011. Fraction of cancer attributable to lifestyle and environmental factors in the UK in 2010. *British Journal of Cancer* 105:S77-S81.

several genes predisposes some people to smoking-induced lung cancer. Variants at some of these genes cause people to be more likely to become addicted to smoking. Other predisposing genes encode receptors that bind potential carcinogens in cigarette smoke. ▶ TRY PROBLEM 23

23.2 Mutations in a Number of Different Types of Genes Contribute to Cancer

As we have learned, cancer is a disease caused by alterations in DNA. However, there are many different types of genetic alterations that may contribute to cancer. More than 350 dif-

ferent human genes have been identified that contribute to cancer; the actual number is probably much higher. Research on mice suggests that more than 2000 genes can, when mutated, contribute to the development of cancer. In the next several sections, we will consider some of the different types of genes that frequently have roles in cancer.

Oncogenes and Tumor-Suppressor Genes

The signals that regulate cell division fall into two basic types: molecules that stimulate cell division and those that inhibit it. These control mechanisms are similar to the accelerator and brake of a car. In normal cells (but, one would hope, not your car), both accelerators and brakes are applied at the same time, causing cell division to proceed at the proper speed.

Because cell division is affected by both accelerators and brakes, cancer can arise from mutations in either type of signal, and there are several fundamentally different routes to cancer. A stimulatory gene can be made hyperactive or active at inappropriate times, analogous to having a car's accelerator stuck in the floored position. Mutations in stimulatory genes are usually dominant because even the reduced amount of gene product produced by a single allele is usually sufficient to produce a stimulatory effect. Mutated dominant-acting stimulatory genes that cause cancer are termed **oncogenes** (**Figure 23.5a**). Cell division may also be stimulated when inhibitory genes are made *inactive*, analogous to having a defective brake in a car. Mutated inhibitory genes generally have recessive effects, because both copies must be mutated to remove all inhibition. Inhibitory genes in cancer are termed **tumor-suppressor genes** (**Figure 23.5b**). Many cancer cells have mutations in both oncogenes and tumor-suppressor genes.

(a) Oncogenes

(b) Tumor-suppressor genes

23.5 Both oncogenes and tumor-suppressor genes contribute to cancer but differ in their modes of action and dominance.

Although oncogenes or mutated tumor-suppressor genes or both are required to produce cancer, mutations in DNA-repair genes can increase the likelihood of acquiring mutations in these genes. Having mutated DNA-repair genes is analogous to having a lousy car mechanic who does not make the necessary repairs on a broken accelerator or brake.

ONCOGENES Oncogenes were the first cancer-causing genes to be identified. In 1909, a farmer brought physician Peyton Rous a hen with a large connective-tissue tumor (sarcoma) growing on its breast. When Rous injected pieces of this tumor into other hens, they also developed sarcomas. Rous conducted experiments that demonstrated that the tumors were being transmitted by a virus, which became known as the Rous sarcoma virus. A number of other cancer-causing viruses were subsequently isolated from various animal tissues. These viruses were generally assumed to carry a cancer-causing gene that was transferred to the host cell. The first oncogene, called *src*, was isolated from the Rous sarcoma virus in 1970.

In 1975, Michael Bishop, Harold Varmus, and their colleagues began to use probes for viral oncogenes to search for related sequences in normal cells. They discovered that the genomes of all normal cells carry DNA sequences that are closely related to oncogenes. These normal cellular genes are called **proto-oncogenes**. They are responsible for basic cellular functions in normal cells but, when mutated, become oncogenes that contribute to the development of cancer. When a virus infects a cell, a proto-oncogene may become incorporated into the viral genome through recombination. Within the viral genome, the proto-oncogene may mutate to an oncogene that, when inserted back into a cell, causes rapid cell division and cancer. Because the proto-oncogenes are more likely to undergo mutation or recombination within a virus, viral infection is often associated with the cancer.

Proto-oncogenes can be converted into oncogenes in viruses several different ways. The sequence of the proto-oncogene may be altered or truncated as it is incorporated into the viral genome. This mutated copy of the gene may then produce an altered protein that causes uncontrolled cell proliferation. Alternatively, through recombination a proto-oncogene may end up next to a viral promoter or enhancer, which then causes the gene to be overexpressed. Finally, sometimes the function of a proto-oncogene in the host cell may be altered when a virus inserts its own DNA into the gene, disrupting its normal function. While viruses are capable of converting proto-oncogenes into oncogenes, most proto-oncogenes are mutated to form oncogenes without the involvement of a virus.

Many oncogenes have been identified by experiments in which selected fragments of DNA are added to cells in culture. Some of the cells take up the DNA and, if these cells become cancerous, then the DNA fragment that was added to the culture must contain an oncogene. The fragments can then be sequenced, and the oncogene can be identified. A large number of oncogenes have now been discovered (**Table 23.4**). About 90% of all cancer genes are thought to be dominant oncogenes.

Table 23.4 Some oncogenes and functions of their corresponding proto-oncogenes

Gene	Normal function	Cancer in which gene is mutated
erbB	Part of growth factor receptor	Many types of cancer
fos	Transcription factor	Osteosarcoma and endometrial carcinoma
jun	Transcription factor, cell cycle control	Lung cancer, breast cancer
myc	Transcription factor	Lymphomas, leukemias, neuroblastoma
ras	GTP binding and GTPase	Many types of cancer
sis	Growth factor	Glioblastomas and other cancers
src	Protein tyrosine kinase	Many types of cancer

TUMOR-SUPPRESSOR GENES Tumor-suppressor genes are more difficult to identify than oncogenes because they *inhibit* cancer and are recessive; both alleles must be mutated before the inhibition of cell division is removed. Because the *failure* of their function promotes cell proliferation, tumor-suppressor genes cannot be identified by adding them to cells and looking for cancer. About 10% of cancer-causing genes are thought to be tumor-suppressor genes.

Defects in both copies of a tumor-suppressor gene are usually required to cause cancer; an organism can inherit one defective copy of the tumor-suppressor gene (is heterozygous for the cancer-causing mutation) and not have cancer, because the remaining normal allele produces the tumor-suppressing product. However, these heterozygotes are often predisposed to cancer because the inactivation or loss of the one remaining allele is all that is required to completely eliminate the tumor-suppressor product. Inactivation of the remaining wild-type allele in heterozygotes is referred to as the **loss of heterozygosity**. A common mechanism for the loss of heterozygosity is a deletion on the chromosome that carried the normal copy of the tumor-suppressor gene (**Figure 23.6**).

Among the first tumor-suppressor genes to be identified was the gene causing retinoblastoma. In 1985, Raymond White and Webster Cavenne showed that large segments of chromosome 13 were missing in cells of retinoblastoma tumors, and, later, the tumor-suppressor gene was isolated from these segments. Another example of a tumor-suppressor gene is *BRCA1*, mutations of which are associated with increased risk of breast and ovarian cancer. *BRCA1* produces a protein that normally helps in repair of double-strand breaks in DNA by homologous recombination (see Chapter 18). It also acts as a transcription factor and interacts

This person is heterozygous (*Aa*) for a tumor-suppressor gene.

Loss of the wild-type allele, in this case through a chromosome deletion, causes loss of the tumor-supressor activity.

Chromosome deletion

Conclusion: People heterozygous for a tumor-suppressor gene are predisposed to cancer.

23.6 The loss of heterozygosity often leads to cancer in a person heterozygous for a tumor-suppressor gene.

with histone deacetylase enzymes, which affect transcription. A number of tumor-suppressor genes have now been discovered (**Table 23.5**).

Sometimes the mutation or loss of a single allele of a recessive tumor-suppressor gene is sufficient to cause cancer. This effect—the appearance of the trait in an individual cell or organism that is heterozygous for a normally recessive trait—is called **haploinsufficiency**. This phenomenon is thought to be due to dosage effects: the heterozygote produces only half as much of the product encoded by the tumor-suppressing gene. Normally, this amount is sufficient for the cellular processes that prevent tumor formation, but it is less than the optimal amount, and other factors may sometimes combine with the lowered tumor-suppressor product to cause cancer. ▶ **TRY PROBLEM 25**

Table 23.5	Some tumor-suppressor genes and their normal functions	
Gene	**Normal function**	**Cancer in which gene is mutated**
APC	Scaffold protein, interacts with microtubules	Colorectal
BRCA1	DNA repair, transcription factor	Breast and ovarian
CDKN2A	Regulates cell division	Melanoma
NF1	GTPase activator	Neurofibromatosis
p53	Regulates cell division	Many types of cancer
RB	Regulates cell division	Retinoblastoma

CONCEPTS

Proto-oncogenes are genes that control normal cellular functions; when mutated, they become oncogenes that stimulate cell proliferation. They tend to be dominant in their action. Tumor-suppressor genes normally inhibit cell proliferation; when mutated, they allow cells to proliferate. Tumor-suppressor genes tend to be recessive in their action. Individual organisms that are heterozygous for tumor-suppressor genes are often predisposed to cancer.

✔ **CONCEPT CHECK 2**

Why are oncogenes usually dominant in their action, whereas tumor-suppressor genes are recessive?

Mutations in Genes That Control the Cycle of Cell Division

The cell cycle is the normal process by which cells undergo growth and division. Normally, progression through the cell cycle is tightly regulated so that cells divide only when additional cells are needed, when all the components necessary for division are present, and when the DNA has been replicated without damage. Sometimes, however, errors arise in one or more of the components that regulate the cell cycle. These errors often cause cells to divide at inappropriate times or rates, leading to cancer. Indeed, many proto-oncogenes and tumor-suppressor genes function normally by helping to control the cell cycle. Before considering how errors in this system contribute to cancer, we must first understand how the cell cycle is usually regulated.

CONTROL OF THE CELL CYCLE As discussed in Chapter 2, the cell cycle is the period from one cell division to the next. Cells that are actively dividing pass through the G_1, S, and G_2 phases of interphase and then move directly into the M phase, when cell division takes place. Nondividing cells exit from G_1 into the G_0 stage, in which they are functional but not actively growing or dividing. Progression from one stage of the cell cycle to another is influenced by a number of internal and external signals and is regulated at key points in the cycle called checkpoints.

For many years, the biochemical events that control the progression of cells through the cell cycle were completely unknown, but research has now revealed many of the details of this process. Key events of the cell cycle are controlled by **cyclin-dependent kinases** (CDKs), which are enzymes that add phosphate groups to other proteins. Sometimes phosphorylation activates the other protein and other times it inactivates the protein. As their name implies, CDKs are functional only when they associate with another protein called a **cyclin**. The level of cyclin oscillates in the course of the cell cycle; when bound to a

CDK, cyclin specifies which proteins the CDK will phosphorylate. Each cyclin appears at a specific point in the cell cycle, usually because its synthesis and destruction are regulated by another cyclin. Cyclins and CDKs are called by different names in different organisms; here, we will use the terms applied to these molecules in mammals.

G₁-TO-S TRANSITION Let's begin by looking at the G₁-to-S transition. As mentioned earlier, checkpoints ensure that all cellular components are present and in good working order before the cell proceeds to the next stage of the cycle. The G₁/S checkpoint is in G₁, just before the cell enters into the S phase and replicates its DNA. The cell is prevented from passing through the G₁/S checkpoint by the retinoblastoma (RB) protein (**Figure 23.7**), which binds to another molecule called E2F and keeps it inactive. In G₁, cyclin D and cyclin E continuously increase in concentration and combine with their associated CDKs. Cyclin-D–CDK and cyclin-E–CDK both phosphorylate molecules of RB. Late in G₁, the phosphorylation of RB is completed, which inactivates RB. Without the inhibitory effects of RB, the E2F protein is released. E2F is a transcription factor that stimulates the transcription of genes that produce enzymes necessary for the replication of DNA, and the cell moves into the S stage of the cell cycle.

G₂-TO-M TRANSITION Regulation of the G₂-to-M transition is similar to that of the G₁-to-S transition. In the G₂-to-M transition, cyclin B combines with a CDK to form an inactive complex called *mitosis-promoting factor* (MPF). After

MPF has been formed, it must be activated by the removal of a phosphate group (**Figure 23.8a**). During G₁, cyclin B levels are low; so the amount of MPF also is low. As more cyclin B is produced, it combines with CDK to form increasing amounts of MPF. Near the end of G₂, the amount of active MPF reaches a critical level, which commits the cell to divide. The MPF concentration continues to increase, reaching a peak in mitosis.

The active form of MPF phosphorylates other proteins, which then bring about many of the events associated with mitosis, such as nuclear-membrane breakdown, spindle formation, and chromosome condensation. At the end of metaphase, cyclin B is abruptly degraded, which lowers the amount of MPF and, initiating anaphase, sets in motion a chain of events that ultimately brings mitosis to a close (**Figure 23.8b**). In brief, high levels of active MPF stimulate mitosis, and low levels of MPF bring a return to interphase conditions.

The G₂/M checkpoint is at the end of G₂, before the cell enters mitosis. A number of factors stimulate the synthesis of cyclin B and the activation of MPF, whereas other factors inhibit MPF. Together, these factors ensure that mitosis is not initiated until conditions are appropriate for cell division. For example, DNA damage inhibits the activation of MPF; consequently, the cell is arrested in G₂ and does not undergo division.

SPINDLE-ASSEMBLY CHECKPOINT Yet another checkpoint, called the spindle-assembly checkpoint, is in metaphase. This checkpoint delays the onset of anaphase until all chromosomes are aligned on the metaphase plate and sister kinetochores are attached to spindle fibers from opposite poles. If all chromosomes are not properly aligned, the checkpoint blocks the destruction of cyclin B. The persistence of cyclin B keeps MPF active and maintains the cell in a mitotic state. An additional checkpoint controls the cell's exit from mitosis.

MUTATIONS IN CELL-CYCLE CONTROL AND CANCER
Many cancers are caused by defects in the cell cycle's regulatory machinery. For example, mutations in the gene that encodes the RB protein—which normally holds the cell in G₁ until the DNA is ready to be replicated—are associated with many cancers, including retinoblastoma. When the *RB* gene is mutated, cells pass through the G₁/S checkpoint without the normal controls that prevent cell proliferation. The gene that encodes cyclin D (thus stimulating the passage of cells through the G₁/S checkpoint) is overexpressed in about 50% of all breast cancers, as well as some cases of esophageal and skin cancer. Likewise, the tumor-suppressor gene *p53*, which is mutated in about 75% of all colon cancers, regulates a potent inhibitor of CDK activity.

Some proto-oncogenes and tumor-suppressor genes have roles in **apoptosis**, a process of programmed cell death in which the cell's DNA is degraded, its nucleus and cytoplasm shrink, and the cell undergoes phagocytosis by other cells without the leakage of its contents. Cells have the ability to assess themselves and, when they are abnormal or damaged, they normally undergo apoptosis (see pp. 646–647 in Chapter 22). Cancer cells

23.7 The RB protein helps control the progression through the G₁/S checkpoint by binding transcription factor E2F.

RB protein

RB binds E2F and keeps it inactive.

E2F

Cyclin-D–CDK

Cyclin-E–CDK

Phosphorylation

Increasing concentrations of cyclin-D–CDK and cyclin-E–CDK phosphorylate RB,...

P P

RB E2F

...which inactivates RB, and it releases E2F.

DNA

Promoter   Gene

E2F binds to DNA and stimulates the transcription of genes required for DNA replication.

Transcription

RNA

23.8 Progression through the G₂/M checkpoint is regulated by cyclin B.

frequently have chromosome mutations, DNA damage, and other cellular anomalies that would normally stimulate apoptosis and prevent their proliferation. Often, these cells have mutations in genes that regulate apoptosis, and therefore they do not undergo programmed cell death. The ability of a cell to initiate apoptosis in response to DNA damage, for example, depends on *p53*, which is inactive in many human cancers.

CONCEPTS

Progression through the cell cycle is controlled at checkpoints, which are regulated by interactions between cyclins and cyclin-dependent kinases. Genes that control the cell cycle are frequently mutated in cancer cells.

✔ CONCEPT CHECK 3

What would be the most likely effect of a mutation that causes cyclin B to be unable to bind to CDK?
a. Cells pass through the G₂/M checkpoint and enter mitosis even when DNA has not been replicated.
b. Cells never pass through the G₁/S checkpoint.
c. Cells pass through mitosis more quickly than unmutated cells.
d. Cells fail to pass the G₂/M checkpoint and do not enter into mitosis.

SIGNAL-TRANSDUCTION PATHWAYS   Whether cells pass through the cell cycle and continue to divide is influenced by a large number of internal and external signals. External signals are initiated by hormones and growth factors.

These molecules are often unable to pass through the cell membrane because of their size or charge; they exert their effects by binding to receptors on the cell surface, which triggers a series of intracellular reactions that then carry the message to the nucleus or other site within the cell. This type of system, in which an external signal triggers a cascade of intracellular reactions that ultimately produce a specific response, is called a **signal-transduction pathway**. Defects in signal-transduction pathways are often associated with cancer.

A signal-transduction pathway begins with the binding of an external signaling molecule to a specific receptor that is embedded in the cell membrane. Receptors in signal-transduction pathways usually have three parts: (1) an extracellular domain that protrudes from the cell and binds the signaling molecule; (2) a transmembrane domain that passes across the membrane and conducts the signal to the interior of the cell; and (3) an intracellular domain that extends into the cytoplasm and, on the binding of the signaling molecule, undergoes a chemical or conformational change that is transmitted to molecules of the signal-transduction pathway in the cytoplasm. The binding of a signaling molecule to the membrane-bound receptor activates a protein in the pathway. On activation, this protein activates the next molecule in the pathway, often by adding or removing phosphate groups or causing changes in the conformation of the protein. The newly activated protein activates the next molecule in the pathway, and in this way, the signal is passed along through a cascade of reactions and ultimately produces the response, such as stimulating or inhibiting the cell cycle.

In the past decade, much research has been conducted to determine the pathways by which various signals influence the cell cycle. To illustrate signal transduction, let's consider the Ras signal-transduction pathway, which plays an important role in control of the cell cycle. Each Ras protein cycles between an active form and an inactive form. In the inactive form, the Ras protein is bound to guanosine diphosphate (GDP); in the active form, it is bound to guanosine triphosphate (GTP).

The Ras signal-transduction pathway is activated when a growth factor, such as epidermal growth factor (EGF), binds to a receptor on the cell membrane (**Figure 23.9**). The binding of EGF causes a conformational change in the receptor and the addition of phosphate groups to it. The addition of the phosphate groups allows adaptor molecules to bind to the receptor. These adaptor molecules link the receptor with an inactive molecule of Ras protein. The adaptor molecules stimulate Ras to release GDP and bind GTP, activating Ras. The newly activated Ras protein then binds to an inactive form of another protein called Raf and activates it. After activating Raf, Ras hydrolyzes GTP to GDP, which converts Ras back into the inactive form.

Activated Raf then sets in motion a cascade of reactions, ending in the activation of a protein called MAP kinase. Activated MAP kinase moves into the nucleus and activates a number of transcription factors that stimulate the transcription of genes taking part in the cell cycle. In this way, the original external signal promotes cell division. A number of other transduction pathways have been identified that affect the cell cycle and cell proliferation.

23.9 The Ras signal-transduction pathway conducts signals from growth factors and hormones to the nucleus and stimulates the cell cycle. Mutations in this pathway often contribute to cancer.

Because signal-transduction pathways help control the cell cycle, defects in their components often contribute to cancer. For example, genes that encode Ras proteins are frequently oncogenes, and mutations in these genes are often found in cancer cells: 75% of tumors of the pancreas and 50% of those in the thyroid and colon have mutations in *ras* genes. Mutations in these genes produce mutant Ras proteins that are permanently activated and continuously stimulate cell division.

CONCEPTS

Molecules outside the cell often bring about intracellular responses by binding to a membrane receptor and stimulating a cascade of intracellular reactions, known as a signal-transduction pathway. Many molecules in the pathway are proteins that alternate between active and inactive forms. Defects in signal-transduction pathways are often associated with cancer.

✔ CONCEPT CHECK 4

Ras proteins are activated when they
a. bind GTP.
b. release GTP.
c. bind GDP.
d. undergo acetylation.

DNA-Repair Genes

Cancer arises from the accumulation of multiple mutations in a single cell. Some cancer cells have normal rates of mutation, and multiple mutations accumulate because each mutation gives the cell a replicative advantage over cells without the mutations. Other cancer cells may have higher-than-normal rates of mutation in all of their genes, which leads to more-frequent mutation of oncogenes and tumor-suppressor genes. What might be the source of these high rates of mutation in some cancer cells?

Two processes control the rate at which mutations arise within a cell: (1) the rate at which errors arise during and after the course of replication and (2) the efficiency with which these errors are corrected. The error rate in replication is controlled by the fidelity of DNA polymerases and other proteins in the replication process (see Chapter 12). However, defects in genes encoding replication proteins have not been strongly linked to cancer.

The mutation rate is also strongly affected by whether errors are corrected by DNA-repair systems (see pp. 520–526 in Chapter 18). Defects in genes that encode components of these repair systems have been consistently associated with a number of cancers. People with xeroderma pigmentosum, for example, are defective in nucleotide-excision repair, an important cellular repair system that normally corrects DNA damage caused by a number of mutagens, including

ultraviolet light. Likewise, about 13% of colorectal, endometrial, and stomach cancers have cells that are defective in mismatch repair, another major repair system in the cell.

A particular type of colon cancer called nonpolyposis colorectal cancer is inherited as an autosomal dominant trait. In families with this condition, a person can inherit one mutated and one normal allele of a gene that controls mismatch repair. The normal allele provides sufficient levels of the protein for mismatch repair to function, but it is highly likely that this normal allele will become mutated or lost in at least a few cells. If it does so, there is no mismatch repair, and these cells undergo higher-than-normal rates of mutation, leading to defects in oncogenes and tumor-suppressor genes that cause the cells to proliferate.

Defects in DNA-repair systems may also contribute to the generation of chromosome rearrangements and genomic instability. Many DNA-repair systems make single- and double-strand breaks in the DNA. If these breaks are not repaired properly, then chromosome rearrangements often result.

Genes That Regulate Telomerase

Another factor that may contribute to the progression of cancer is the inappropriate activation of the enzyme telomerase. Telomeres are special sequences at the ends of eukaryotic chromosomes. Recall that the ends of chromosomes cannot be replicated, and telomeres become shorter with each cell division. This shortening eventually leads to the destruction of the chromosome and cell death, so somatic cells are only capable of a limited number of cell divisions.

In germ cells and stem cells, telomerase replicates the chromosome ends (see pp. 344–346 in Chapter 12), thereby maintaining the telomeres, but this enzyme is not normally expressed in somatic cells. In many tumor cells, however, sequences that regulate the expression of the telomerase gene are mutated, allowing the enzyme to be expressed, and the cell is capable of unlimited cell division. This mutation allows cancer cells to divide indefinitely. Although the expression of telomerase appears to contribute to the development of many cancers, its precise role in tumor progression is unknown and under investigation.

Genes That Promote Vascularization and the Spread of Tumors

A final set of factors that contribute to the progression of cancer includes genes that affect the growth and spread of tumors. Oxygen and nutrients, which are essential to the survival and growth of tumors, are supplied by blood vessels, and the growth of new blood vessels (angiogenesis) is important to tumor progression. Angiogenesis is stimulated by growth factors and others proteins encoded by genes whose expression is carefully regulated in normal cells. In tumor cells, genes encoding these proteins are often overexpressed compared with normal cells, and inhibitors of angiogenesis-promoting factors may be inactivated or underexpressed. At least one inherited cancer—von Hippel–Lindau disease, in which people develop multiple types of tumors—is caused by the mutation of a gene that affects angiogenesis.

In the development of many cancers, the primary tumor gives rise to cells that spread to distant sites, producing secondary tumors. This process of metastasis is the cause of death in 90% of human cancer cases; it is influenced by cellular changes induced by somatic mutation. As discussed in the introduction to this chapter, the *palladin* gene, when mutated, contributes to the metastasis of pancreatic tumors. By using microarrays to measure levels of gene expression, researchers have identified other genes that are transcribed at a significantly higher rate in metastatic cells compared with nonmetastatic cells. For example, one study detected a set of 95 genes that were overexpressed or underexpressed in a population of metastatic breast-cancer cells that were strongly metastatic to the lung, compared with a population of cells that were only weakly metastatic to the lung. Genes that contribute to metastasis often encode components of the extracellular matrix and the cytoskeleton. Others encode adhesion proteins, which help hold cells together.

Advances in sequencing technology have now made possible the complete sequencing of the DNA of tumor cells to see how their genomes differ from those of normal cells. In one experiment, researchers sequenced the entire genome of cells from a metastasized breast-cancer tumor and compared it with the genome of noncancer cells from the same person. They also compared the genome of the metastasized tumor with the genome of the primary tumor (from which the metastasis originated), which had been removed from the patient nine years earlier. The researchers found 32 different somatic mutations in the coding regions of genes from the tumor cells, 19 of which were not detected in the primary tumor. This finding suggests that the metastasized tumor underwent considerable genetic changes in its nine-year evolution from the primary tumor. In contrast, another study of a breast-cancer metastasis found only two mutations that were not present in the primary tumor but, in this case, the metastasis had evolved in only one year.

> ### CONCEPTS
>
> Mutations in genes that encode components of DNA-repair systems are often associated with cancer; these mutations increase the rate at which mutations are retained and result in an increased number of mutations in proto-oncogenes, tumor-suppressor genes, and other genes that contribute to cell proliferation. Mutations that allow telomerase to be expressed in somatic cells and those that affect vascularization and metastasis also can contribute to cancer progression.

> ### ✔ CONCEPT CHECK 5
>
> Which type of mutation in telomerase is associated with cancer cells?
> a. Mutations that produce an inactive form of telomerase.
> b. Mutations that decrease the expression of telomerase.
> c. Mutations that increase the expression of telomerase.
> d. All of the above.

MicroRNAs and Cancer

MicroRNAs (miRNAs) are a class of small RNA molecules that pair with complementary sequences on mRNA and degrade the mRNA or inhibit its translation (see Chapter 14). Given the fact that miRNAs are important in controlling gene expression and development, it is not surprising that they are also associated with tumor development. Many tumor cells exhibit widespread reduction in the expression of many miRNAs. Researchers have genetically engineered mouse tumor cells that lacked the machinery to generate miRNAs and found that these cells showed enhanced tumor progression when implanted into mice. Interestingly, this effect was seen only in cells that had already initiated tumor development, suggesting that miRNAs play a role in later stages of tumor progression.

Lowered levels of miRNAs may contribute to cancer by allowing oncogenes that are normally controlled by the miRNAs to be expressed at high levels. For example, let-7 miRNA normally controls the expression of the *ras* oncogene, probably by binding to complementary sequences in the 3′untranslated region of mRNA and inhibiting translation. In lung-cancer cells, levels of let-7 miRNA are often low, allowing the Ras protein to be highly expressed, which then leads to the development of lung cancer.

A transcription factor called c-MYC is often expressed at high levels in cancer cells. Evidence suggests that c-MYC helps to drive cell proliferation and the development of cancer. Among other effects, c-MYC binds to the promoters of miRNA genes and decreases their transcription, decreasing the abundance of the miRNAs. Some of these miRNAs are known to suppress tumor development. Research has shown that, if, through genetic manipulation, the miRNAs are expressed at high levels, the development of tumors decreases. All of these findings suggest that altered expression of miRNAs plays an important role in cancer.

Several miRNAs have been implicated in the process of metastasis. A particular miRNA called miR-10b has been associated with the formation of metastatic breast tumors. In one experiment, investigators manipulated a line of breast-cancer cells so that miR-10b was overexpressed. When the manipulated cells were injected into mice, many of the mice developed metastatic tumors. In contrast with the preceding examples in which the lower expression of miRNA is associated with cancer, here high levels of miRNA appear to promote the spread of cancer cells. Further study revealed that, in humans, the levels of miR-10b are elevated in metastatic tumors compared with tumors in metastasis-free patients. miR-10b regulates the expression of a number of other genes, including some that are known to suppress the spread of tumor cells. Other miRNAs are known to inhibit metastasis.

Cancer Genome Projects

Formed in 2008, the International Cancer Genome Consortium coordinates efforts to determine the genomic sequences of tumors. A goal of the consortium is to completely sequence 500 tumors from each of 50 different types of cancer, along with the genomes of normal tissues from the same persons. This effort is producing important results, revealing the numbers and types of mutations that are associated with particular cancers. The hope is that new cancer-causing genes will be identified, which will lead to a better understanding of the nature of cancer and suggest new targets for cancer treatment. Another research project called The Cancer Genome Atlas (TCGA) project began in the United States in 2005. This project seeks to provide a comprehensive genomic analysis of over 100 different types of cancer cells, including sequencing of all exons across the genome, as well as characterization of mRNA expression, DNA methylation, copy number variations, and microRNAs of the tumor cells.

The genomes of a number of tumors have been sequenced and many more are currently being sequenced. For example, the entire genome of a small-cell lung carcinoma (a type of lung cancer) was sequenced in 2010 and compared with the genome of normal cells from the same person. More than 22,000 base-pair mutations were identified in the tumor, of which 134 were within protein-encoding genes. The tumor also possessed 58 chromosome rearrangements and 334 copy-number variations (see Chapter 20). In another study that was part of TCGA, researchers examined mRNA expression, microRNAs, DNA methylation, and copy number variations in 489 ovarian adenocarcinomas, and sequenced the DNA of exons from 316 of the tumors. Almost all of the tumors contained mutations in *p53*, a tumor-suppressor gene involved in DNA repair and cell cycle control. Mutations in *BRCA1* and *BRCA2*, two tumor-suppressor genes that also occur in breast cancer, occurred in 22% of the tumors. Mutations in seven other genes occurred statistically more often than in normal cells. The results suggested several new approaches for drug treatment of ovarian cancer.

Another series of studies sequenced a number of genes in samples of malignant gliomas, an incurable and deadly form of brain cancer. The researchers examined DNA sequences, copy number variations, DNA methylation, and RNA expression of these tumors. The analyses revealed mutations in several genes that appear to be important in the development of glioma tumors. All of these genomic studies are providing new insight in the genetic basis of cancer.

The large number of mutations found in cancer genomes can be divided into two types: **mutation drivers** and **mutation passengers**. Drivers are mutations that drive the cancer process: they directly contribute to the development of cancer. Drivers include mutations in oncogenes, tumor-suppressor genes, DNA-repair genes, and the other types of cancer genes discussed in this chapter. Passengers are mutations that arise randomly in the process of tumor development and do not contribute to the cancer process. Many passengers are in introns (regions between genes) and other DNA that is not transcribed and translated, but they can also arise within protein-encoding genes. A major challenge is to determine which of the numerous mutations found in tumors are drivers and actually contribute to the development of cancer and which are passengers with no effect.

23.3 Epigenetic Changes Are Often Associated with Cancer

Epigenetic changes—alterations to chromatin structure that affect gene expression (see Chapter 21)—are seen in many cancer cells. Two broad lines of evidence suggest that epigenetic changes play an important role in cancer progression. First, genes encoding proteins that are important regulators of epigenetic changes are often mutated in some cancers. For example, almost 90% of cases of follicular lymphoma exhibit mutations in the *MLL2* gene, which encodes a histone methyltransferase enzyme; this enzyme adds methyl groups to DNA, a type of epigenetic modification that alters chromatin structure and affects transcription. Similarly, the *UTX* gene, which encodes a histone demethylase (enzyme that removes methyl groups from histone proteins) is mutated in a number of different types of cancer.

A second line of evidence suggesting that epigenetic alterations are important in cancer comes from recent genomic studies that have compared chromatin structure of cancer cells and normal cells from the same individual. These studies often find that cancer cells have significant alterations to DNA methylation and histone structure. One type of epigenetic alteration often observed in cancer cells is an overall level of less DNA methylation (hypomethylation). As discussed in Chapter 17, DNA methylation is often associated with repression of transcription. It is assumed that hypomethylation leads to transcription of oncogenes, which then stimulate cancer. Some evidence also suggests that hypomethylation causes chromosome instability, a hallmark of many tumors. Tumor cells from mice that have been genetically engineered to have reduced DNA methylation show increased gains and losses of chromosomes, but how hypomethylation might cause chromosome instability is unclear.

A number of studies have observed that although the overall level of DNA methylation is often lower in cancer cells, some specific CpG islands (see Chapter 17) have extra methylation (are hypermethylated). For example, one study found that 5% to 10% of normally unmethylated CpG islands located at promoters become abnormally methylated in cancer cells. This excess methylation may inhibit transcription of tumor-suppressor genes, thus stimulating the development of cancer. The methylation of the promoter of the *Apaf-1* gene is seen in many malignant melanoma cells. *Apaf-1* helps bring about apoptosis of cells with damaged DNA; methylation of its promoter reduces the expression of *Apaf-1*, interrupting the process of apoptosis and allowing abnormal cancer cells to survive.

Research has also demonstrated that the histone proteins in nucleosomes, the fundamental unit of chromatin, are often abnormally modified in cancer cells. Modification of histone proteins, including methylation and acetylation, alters chromatin structure and affects whether transcription occurs. Global patterns of histone acetylation are often altered in cancer cells. However, acetylation not only affects histones but also a number of other proteins that may stimulate or suppress cell division, so whether the effect of acetylation on cancer occurs through changes to chromatin structure is unclear. Epigenetic processes are receiving increasing attention by cancer researchers because they may be amenable to drug therapy.

CONCEPTS

Epigenetic changes, including DNA methylation and histone modification, are often associated with cancer.

✓ CONCEPT CHECK 6

Hypermethylation is thought to contribute to cancer by
a. inhibiting DNA replication.
b. inhibiting the expression of tumor-suppressor genes.
c. stimulating the translation of oncogenes.
c. stimulating telomerase.

23.4 Colorectal Cancer Arises Through the Sequential Mutation of a Number of Genes

Colorectal cancer is an excellent example of how cancer often arises through the accumulation of successive genetic defects. Mutations that contribute to colorectal cancer have been extensively studied.

Colorectal cancer arises in the cells lining the colon and rectum. More than 143,000 new cases of colorectal cancer are diagnosed in the United States each year, where this cancer is responsible for more than 51,000 deaths annually. If detected early, colorectal cancer can be treated successfully; consequently, there has been much interest in identifying the molecular events responsible for the initial stages of this cancer.

Colorectal cancer is thought to originate as benign tumors called adenomatous polyps (**Figure 23.10**). Initially, these polyps are microscopic but, in time, they enlarge and the cells of

23.10 Mutations in multiple genes contribute to the progression of colorectal cancer.

the polyp acquire the abnormal characteristics of cancer cells. In the later stages of the disease, the tumor may invade the muscle layer surrounding the gut and metastasize. The progression of the disease is slow; from 10 to 35 years may be required for a benign tumor to develop into a malignant tumor.

Most cases of colorectal cancer are sporadic, developing in people with no family history of the disease, but a few families display a clear genetic predisposition to it. In one form of hereditary colon cancer, known as familial adenomatous polyposis coli, hundreds or thousands of polyps develop in the colon and rectum; if these polyps are not removed, one or more almost invariably become malignant.

Because polyps and tumors of the colon and rectum can be easily observed and removed with a colonoscope (a fiber-optic instrument used to view the interior of the rectum and colon), much is known about the progression of colorectal cancer, and some of the genes responsible for its clonal evolution have been identified. Mutations in these genes are responsible for the different steps of colorectal-cancer progression. Among the earliest steps is a mutation that inactivates the *APC* gene, which increases the rate of cell division, leading to polyp formation (see Figure 23.10). A person with familial adenomatous polyposis coli inherits one defective copy of the *APC* gene, and defects in this gene are associated with the numerous polyps that appear in those who have this disorder. Mutations in *APC* are also found in the polyps that develop in people who do not have adenomatous polyposis coli.

Mutations of the *ras* oncogene usually occur later, in larger polyps consisting of cells that have acquired some genetic mutations. As discussed earlier in this chapter, the normal *ras* proto-oncogene is a key player in a signal-transduction pathway that relays signals from growth factors to the nucleus, where the signal stimulates cell division. When *ras* is mutated, the protein that it encodes continually relays a stimulatory signal for cell division even when growth factor is absent.

Mutations in *p53* and other genes appear still later in tumor progression; these mutations are rare in polyps but common in malignant cells. About 75% of colorectal cancers have mutations in tumor-suppressor gene *p53*. Because *p53* prevents the replication of cells with genetic damage and controls proper chromosome segregation, mutations in *p53* can allow a cell to rapidly acquire further gene and chromosome mutations, which then contribute to further proliferation and invasion into surrounding tissues.

The sequence of steps just outlined is not the only route to colorectal cancer, and the mutations need not occur in the order presented here. However, this sequence is a common pathway by which colon and rectal cells become cancerous.

23.5 Changes in Chromosome Number and Structure Are Often Associated with Cancer

Most tumors contain cells with chromosome mutations. For many years, geneticists argued about whether these chromosome mutations were the cause or the result of cancer. Some types of tumors are consistently associated with *specific* chromosome mutations; for example, most cases of chronic myelogenous leukemia are associated with a reciprocal translocation between chromosomes 22 and 9. These types of associations suggest that chromosome mutations contribute to the cause of the cancer. Yet many cancers are not associated with specific types of chromosome abnormalities, and individual *gene* mutations are now known to contribute to many types of cancer. Nevertheless, chromosome instability is a general feature of cancer cells, causing them to accumulate chromosome mutations, which then affect individual genes that may contribute to the cancer process. Thus, chromosome mutations appear to both *cause* cancer and *result* from it.

At least three types of chromosome rearrangements— deletions, inversions, and translocations—are associated with certain types of cancer. Deletions can result in the loss of one or more tumor-suppressor genes. Inversions and translocations contribute to cancer in several ways. First, the chromosomal breakpoints that accompany these mutations can lie within tumor-suppressor genes, disrupting their function and leading to cell proliferation. Second, translocations and inversions can bring together sequences from two different genes, generating a fused protein that stimulates some aspect of the cancer process.

Fusion proteins are seen in most cases of chronic myelogenous leukemia, which affects bone-marrow cells. Most patients with chronic myelogenous leukemia have a reciprocal translocation between the long arm of chromosome 22 and the tip of the long arm of chromosome 9 (**Figure 23.11**). This translocation produces a shortened chromosome 22, called the Philadelphia chromosome because it was first discovered in Philadelphia. At the end of a normal chromosome 9 is a potential cancer-causing gene called c-*ABL*. As a result of the

23.11 A reciprocal translocation between chromosomes 9 and 22 causes chronic myelogenous leukemia.

23.12 A reciprocal translocation between chromosomes 8 and 14 causes Burkitt lymphoma.

translocation, part of the c-*ABL* gene is fused with the *BCR* gene from chromosome 22. The protein produced by this *BCR*–c-*ABL* fusion gene is much more active than the protein produced by the normal c-*ABL* gene; the fusion protein stimulates increased, unregulated cell division and eventually leads to leukemia.

A third mechanism by which chromosome rearrangements can produce cancer is by the transfer of a potential cancer-causing gene to a new location, where it is activated by different regulatory sequences. Burkitt lymphoma is a cancer of the B cells, the lymphocytes that produce antibodies. Many people with Burkitt lymphoma possess a reciprocal translocation between chromosome 8 and chromosome 2, 14, or 22 (**Figure 23.12**). This translocation relocates a gene called c-*MYC* from the tip of chromosome 8 to a position on chromosome 2, 14, or 22 that is next to a gene that encodes an immunoglobulin protein. At this new location, c-*MYC*, a cancer-causing gene, comes under the control of regulatory sequences that normally activate the production of immunoglobulins, and c-*MYC* is expressed in B cells. The c-*MYC* protein stimulates the division of the B cells and leads to Burkitt lymphoma.

CONCEPTS

Many tumors contain a variety of types of chromosome mutations. Some tumors are associated with specific deletions, inversions, and translocations. Deletions can eliminate or inactivate genes that control the cell cycle; inversions and translocations can cause breaks in genes that suppress tumors, fuse genes to produce cancer-causing proteins, or move genes to new locations, where they are under the influence of different regulatory sequences.

Most advanced tumors contain cells that exhibit a dramatic variety of chromosome anomalies, including extra chromosomes, missing chromosomes, and chromosome rearrangements (**Figure 23.13**). Some cancer researchers believe that cancer is initiated when genetic changes take place that cause the genome to become unstable, generating numerous chromosome abnormalities that then alter the expression of oncogenes and tumor-suppressor genes.

A number of genes that contribute to genomic instability and lead to missing or extra chromosomes (aneuploidy) have now been identified. Aneuploidy in somatic cells usually arises when chromosomes do not segregate properly in mitosis. Normal cells have a spindle-assembly checkpoint that monitors the proper assembly of the mitotic spindle (see p. 671); if chromosomes are not properly attached to the microtubules at metaphase, the onset of anaphase is blocked. Some aneuploid cancer cells contain mutant alleles for genes that encode proteins having roles in this checkpoint; in these cells, anaphase is entered despite the improper assembly of the spindle or lack of it, and chromosome abnormalities result. For example, mutations in *RB* increase aneuploidy by increasing the expression of a protein called Mad2, which is a critical component of the spindle-assembly checkpoint.

Mutations in genes that encode parts of the spindle apparatus also may contribute to abnormal segregation and lead

23.13 Cancer cells often possess chromosome abnormalities, including extra chromosomes, missing chromosomes, and chromosome rearrangements. Shown here are chromosomes from a colon-cancer cell, which has numerous chromosome abnormalities. For comparison, see a normal karyotype in Figure 2.6 [Courtesy Dr. Peter Duesberg, UC Berkeley.]

to chromosome abnormalities. *APC* is a tumor-suppressor gene that is often mutated in colon-cancer cells. *APC* has several functions, one of which is to interact with the ends of the microtubules that associate with the kinetochore. Dividing mouse cells that have defective copies of the *APC* gene give rise to cells with many chromosome defects.

The tumor-suppressor gene *p53*, in addition to controlling apoptosis, plays a role in the duplication of the centrosome, which is required for proper formation of the spindle and for chromosome segregation. Normally, the centrosome duplicates once per cell cycle. If *p53* is mutated or missing, however, the centrosome may undergo extra duplications, resulting in the unequal segregation of chromosomes. In this way, mutation of the *p53* gene may generate chromosome mutations that contribute to cancer. The *p53* gene is also a tumor-suppressor gene that prevents cell division when the DNA is damaged. ▶ TRY PROBLEM 32

23.6 Viruses Are Associated with Some Cancers

As mentioned earlier in this chapter, viruses are responsible for a number of cancers in animals, and there is evidence that some viruses contribute to at least a few cancers in humans (**Table 23.6**). For example, about 95% of all women with cervical cancer are infected with **human papilloma viruses** (HPVs). Similarly, infection with the virus that causes hepatitis B increases the risk of liver cancer in some people. The Epstein–Barr virus, which is responsible for mononucleosis, has been linked to several types of cancer that are prevalent in parts of Africa, including Burkitt lymphoma.

RETROVIRUSES AND CANCER   Many of the viruses that cause cancer in animals are retroviruses; earlier, we saw how studies of the Rous sarcoma retrovirus in chickens led to the identification of oncogenes in humans. Retroviruses sometimes cause cancer by mutating and rearranging host genes,

converting proto-oncogenes into oncogenes (**Figure 23.14a**). Another way in which viruses can contribute to cancer is by altering the expression of host genes (**Figure 23.14b**). Retroviruses often contain strong promoters to ensure that their own genetic material is transcribed by the host cell. If the provirus inserts near a proto-oncogene, viral promoters can stimulate high levels of expression of the proto-oncogene, leading to cell proliferation.

There are only a few retroviruses that cause cancer in humans. HTLV-1, the first human retrovirus discovered, is associated with human adult T-cell leukemia. Other human cancers are associated with DNA viruses, which, like retroviruses, integrate into the host chromosome but, unlike retroviruses, do not utilize reverse transcription. For example, HPV is a DNA virus that is strongly associated with cervical cancer.

HUMAN PAPILLOMA VIRUS AND CERVICAL CANCER
Human papilloma virus causes warts and other types of benign tumors of epithelial cells. More than 100 different types of HPV are known, about 30 of which are sexually transmitted and cause genital warts. A few of them are associated with cervical cancer. In the United States, 70% of the cases of cervical cancer are caused by HPV-16 and HPV-18. These viruses cause cervical cancer by producing proteins that attach to and inactivate *RB* and *p53*, two proteins that play key roles in the regulation of the cell cycle. When these proteins are inactivated, cells are stimulated to progress through the cell cycle and divide without the normal controls that prevent cell proliferation.

About 75% of sexually active women in the United States are infected with HPV, but only a small number of these women will ever develop cervical cancer. The risk of infection by HPV can be reduced by limiting the number of sexual partners and by a vaccine that was approved by the U.S. Food and Drug Administration in 2006. This vaccine is highly effective against four of the most common HPVs associated with cervical cancer. In spite of the large number of women infected with HPVs, the incidence of cervical cancer in the United States has declined 75% in the past 40 years. The primary reason for this decline is widespread use of the Pap test, which detects early stages of cervical cancer, as well as cervical dysplasia, a precancerous growth that can be removed before it develops into cancer.

Although rare in the United States, cervical cancer is the second most common cause of cancer in women worldwide, with high incidences in many developing countries such as those of Sub-Saharan Africa, South Asia, and South and Central America. An estimated 510,000 women throughout the world develop cervical cancer each year, and 288,000 die from the disease. The primary reason for the high incidence and death rate in developing regions is lack of access to the cervical-cancer screening procedures such as the Pap test. The availability of the vaccine against cervical cancer in these developing countries could greatly reduce the incidence of cervical cancer.

Table 23.6	Some human cancers associated with viruses
Virus	**Cancer**
Human papilloma viruses (HPVs)	Cervical, penile, and vulvar cancers
Hepatitis B virus	Liver cancer
Human T-cell leukemia virus 1 (HTLV-1)	Adult T-cell leukemia
Human T-cell leukemia virus 2 (HTLV-2)	Hairy-cell leukemia
Epstein–Barr virus	Burkitt lymphoma, nasopharyngeal cancer, Hodgkin lymphoma
Human herpes virus	Kaposi sarcoma
Merkel cell polyomavirus	Merkel cell carcinoma

Note: Some of these associations between cancer and viruses exist only in certain populations and geographic areas.

(a)

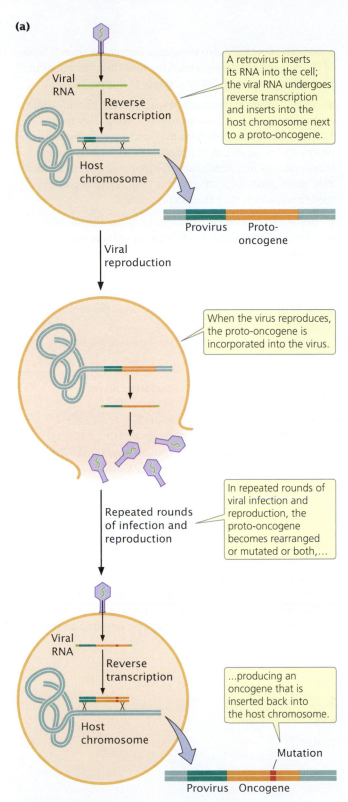

A retrovirus inserts its RNA into the cell; the viral RNA undergoes reverse transcription and inserts into the host chromosome next to a proto-oncogene.

When the virus reproduces, the proto-oncogene is incorporated into the virus.

In repeated rounds of viral infection and reproduction, the proto-oncogene becomes rearranged or mutated or both,...

...producing an oncogene that is inserted back into the host chromosome.

23.14 Retroviruses cause cancer by (a) mutating and rearranging proto-oncogenes or (b) inserting strong promoters near proto-oncogenes.

CONCEPTS

Viruses contribute to a few cancers in humans by mutating and rearranging host genes that then contribute to cell proliferation, or by altering the expression of host genes.

(b)

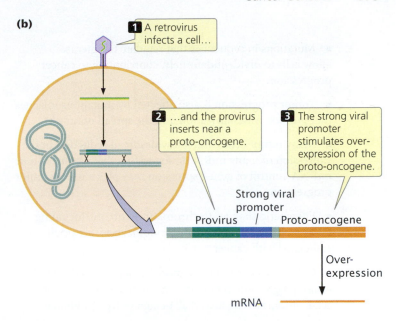

1 A retrovirus infects a cell...

2 ...and the provirus inserts near a proto-oncogene.

3 The strong viral promoter stimulates over-expression of the proto-oncogene.

CONCEPTS SUMMARY

■ Cancer is fundamentally a genetic disorder, arising from somatic mutations in multiple genes that affect cell division and proliferation. If one or more mutations are inherited, then fewer additional mutations are required for cancer to develop.

■ A mutation that allows a cell to divide rapidly provides the cell with a growth advantage; this cell gives rise to a clone of cells having the same mutation. Within this clone, other mutations occur that provide additional growth advantages, and cells with these additional mutations become dominant in the clone. In this way, the clone evolves.

■ Environmental factors play an important role in the development of many cancers by increasing the rate of somatic mutations.

■ Oncogenes are dominant mutated copies of normal genes (proto-oncogenes) that stimulate cell division. Tumor-suppressor genes normally inhibit cell division; recessive mutations in these genes may contribute to cancer. Sometimes, the mutation of a single allele of a tumor-suppressor gene is sufficient to cause cancer, a phenomenon known as haploinsufficiency.

■ The cell cycle is controlled by cyclins and cyclin-dependent kinases. Mutations in genes that control the cell cycle are often associated with cancer.

■ Signal-transduction pathways conduct external signals to intracellular responses. These pathways consist of a series of proteins that are activated and inactivated in a cascade of reactions. Mutations in the components of signal-transduction pathways disrupt the cell cycle and may contribute to cancer.

■ Defects in DNA-repair genes often increase the overall mutation rate of other genes, leading to defects in proto-oncogenes and tumor-suppressor genes that can contribute to cancer progression.

■ Mutations in sequences that regulate telomerase allow cells to divide indefinitely, contributing to cancer progression.

■ Tumor progression is also affected by mutations in genes that promote vascularization and the spread of tumors.

■ Many tumor cells exhibit a widespread reduction in the expression of many miRNAs, suggesting that a reduction in miRNA control of gene expression may play a role in tumor progression.

■ Epigenetic changes to chromatin structure, including DNA methylation and histone modification, are often associated with cancer.

■ Colorectal cancer offers a model system for understanding tumor progression in humans. Initial mutations stimulate cell division, leading to a small benign polyp. Additional mutations allow the polyp to enlarge, invade the muscle layer of the gut, and eventually spread to other sites. Mutations in particular genes affect different stages of this progression.

■ Some cancers are associated with specific chromosome mutations, including chromosome deletions, inversions, and translocations. Deletions may cause cancer by removing or disrupting genes that suppress tumors; inversions and translocations may break tumor-suppressing genes or they may move genes to positions next to different regulatory sequences, which alter their expression.

■ Mutations in some genes cause or allow chromosomes to segregate improperly, leading to aneuploidy that can contribute to cancer.

■ Viruses are associated with some cancers; they contribute to cell proliferation by mutating and rearranging host genes or by altering the expression of host genes.

IMPORTANT TERMS

malignant tumor (p. 662)
metastasis (p. 662)
clonal evolution (p. 664)
oncogene (p. 666)

tumor-suppressor gene (p. 666)
proto-oncogene (p. 667)
loss of heterozygosity (p. 667)
haploinsufficiency (p. 668)

cyclin-dependent kinase (CDK) (p. 668)
cyclin (p. 668)
apoptosis (p. 669)
signal-transduction pathway (p. 671)

mutation driver (p. 674)
mutation passenger (p. 674)
human papilloma virus (HPV) (p. 678)

ANSWERS TO CONCEPT CHECKS

1. Retinoblastoma results from at least two separate genetic defects, both of which are necessary for cancer to develop. In sporadic cases, two successive mutations must occur in a single cell, which is unlikely and therefore typically affects only one eye. In people who have inherited one of the two required mutations, every cell contains this mutation, and so a single additional mutation is all that is required for cancer to develop. Given the millions of cells in each eye, there is a high probability that the second mutation will occur in at least one cell of each eye, producing tumors in both eyes and the inheritance of this type of retinoblastoma.

2. Oncogenes have a stimulatory effect on cell proliferation. Mutations in oncogenes are usually dominant because a mutation in a single copy of the gene is usually sufficient to produce a stimulatory effect. Tumor-suppressor genes inhibit cell proliferation. Mutations in tumor-suppressor genes are generally recessive because both copies must be mutated to remove all inhibition.

3. d

4. a

5. c

6. b

7. d

WORKED PROBLEM

Problem

In some cancer cells, a specific gene has become duplicated many times. Is this gene likely to be an oncogene or a tumor-suppressor gene? Explain your reasoning.

Solution Strategy

What information is required in your answer to the problem?

Whether the gene is likely to be an oncogene or tumor-suppressor gene and why.

What information is provided to solve the problem?

In cancer cells, the gene has been amplified many times.

For help with this problem, review:

Oncogenes and Tumor-Suppressor Genes in Section 23.2.

Solution Steps

The gene is likely to be an oncogene. Oncogenes stimulate cell proliferation and act in a dominant manner. Therefore, extra copies of an oncogene will result in cell proliferation and cancer. Tumor-suppressor genes, on the other hand, suppress cell proliferation and act in a recessive manner; a single copy of a tumor-suppressor gene is sufficient to prevent cell proliferation. Therefore, extra copies of the tumor-suppressor gene will not lead to cancer.

COMPREHENSION QUESTIONS

Section 23.1

1. What types of evidence indicate that cancer arises from genetic changes?

2. How is cancer different from most other types of genetic diseases?

3. Outline Knudson's two-hit hypothesis of retinoblastoma and describe how it helps to explain unilateral and bilateral cases of retinoblastoma.

4. Briefly explain how cancer arises through clonal evolution.

Section 23.2

5. What is the difference between an oncogene and a tumor-suppressor gene? Give some examples of the functions of proto-oncogenes and tumor suppressers in normal cells.

6. What is haploinsufficiency? How might it affect cancer risk?

7. How do cyclins and CDKs differ? How do they interact in controlling the cell cycle?

8. Briefly outline the events that control the progression of cells through the G_1/S checkpoint in the cell cycle.

9. Briefly outline the events that control the progression of cells through the G_2/M checkpoint of the cell cycle.

10. What is a signal-transduction pathway? Why are mutations in components of signal-transduction pathways often associated with cancer?

11. How is the Ras protein activated and inactivated?

12. Why do mutations in genes that encode DNA-repair enzymes often produce a predisposition to cancer?

13. What role do telomeres and telomerase play in cancer progression?

Section 23.3

14. How is an epigenetic change different from a mutation?

15. How is DNA methylation related to cancer?

Section 23.4

16. Briefly outline some of the genetic changes commonly associated with the progression of colorectal cancer.

Section 23.5

*17. Explain how chromosome deletions, inversions, and translocations may cause cancer.

18. Briefly outline how the Philadelphia chromosome leads to chronic myelogenous leukemia.

19. What is genomic instability? Give some ways in which genomic instability may arise.

Section 23.6

*20. How do viruses contribute to cancer?

> For more questions that test your comprehension of the key chapter concepts, go to *LEARNINGCurve* for this chapter.

APPLICATION QUESTIONS AND PROBLEMS

Introduction

21. What characteristics of the pedigree shown in **Figure 23.1** suggest that pancreatic cancer in this family is inherited as an autosomal dominant trait?

Section 23.1

22. If cancer is fundamentally a genetic disease, how might an environmental factor such as smoking cause cancer?

*23. Both genes and environmental factors contribute to cancer. **Table 23.2** shows that prostate cancer is 30 times as common among Caucasians from Utah as among Chinese from Shanghai. Briefly outline how you might determine if these differences in the incidence of prostate cancer are due to differences in the genetic makeup of the two populations or to differences in their environments.

*24. A couple has one child with bilateral retinoblastoma. The mother is free from cancer, but the father has unilateral retinoblastoma and he has a brother who has bilateral retinoblastoma.

a. If the couple has another child, what is the probability that this next child will have retinoblastoma?

b. If the next child has retinoblastoma, is it likely to be bilateral or unilateral?

c. Explain why the father's case of retinoblastoma is unilateral, whereas his son's and brother's cases are bilateral.

Section 23.2

*25. The *palladin* gene, which plays a role in pancreatic cancer (see the introduction to this chapter), is said to be an oncogene. Which of its characteristics suggest that it is an oncogene rather than a tumor-suppressor gene?

26. Mutations in the *RB* gene are often associated with cancer. Explain how a mutation that results in a nonfunctional RB protein contributes to cancer.

27. Cells in a tumor contain mutated copies of a particular gene that promotes tumor growth. Gene therapy can be used to introduce a normal copy of this gene into the tumor cells. Would you expect this therapy to be effective if the mutated gene were an oncogene? A tumor-suppressor gene? Explain your reasoning.

28. What would be the effect on the cell cycle of a drug that inhibited each of the following?

a. MPF

b. cyclin-E-CDK

c. cyclin-D-CDK

29. What would be the effect of a drug that inhibited the breakdown of cyclin B?

Section 23.3

30. David Seligson and his colleagues examined levels of histone protein modification in prostate tumors and their association with clinical outcomes (D. B. Seligson et al. 2005. *Nature* 435:1262–1266). They used antibodies to stain for acetylation at three different sites and for methylation at two different sites on histone proteins. They found that the degree of histone acetylation and methylation helped predict whether prostate cancer would return within 10 years in the patients who had a prostate tumor removed. Explain how acetylation and methylation might be associated with tumor recurrence in prostate cancer. (Hint: See Chapter 17.)

31. Some cancers have been treated with drugs that demethylate DNA. Explain how these drugs might work. Do you think the cancer-causing genes that respond to the demethylation are likely to be oncogenes or tumor-suppressor genes? Explain your reasoning.

Section 23.5

*32. Some cancers are consistently associated with the deletion of a particular part of a chromosome. Does the deleted region contain an oncogene or a tumor-suppressor gene? Explain.

Section 23.6

33. Assume that the provirus in **Figure 23.14** inserts just upstream of a tumor suppressor gene. Would this be likely to cause cancer? Why or why not?

CHALLENGE QUESTIONS

Section 23.2

34. Many cancer cells are immortal (will divide indefinitely) because they have mutations that allow telomerase to be expressed. How might this knowledge be used to design anticancer drugs?

35. Bloom syndrome is an autosomal recessive disease that exhibits haploinsufficiency. A recent survey showed that people heterozygous for mutations at the *BLM* locus are at increased risk of colon cancer. Suppose you are a genetic counselor. A young woman is referred to you whose mother has Bloom syndrome; the young woman's father has no family history of Bloom syndrome. The young woman asks whether she is likely to experience any other health problems associated with her family history of Bloom syndrome. What advice would you give her?

36. Imagine that you discover a large family in which bladder cancer is inherited as an autosomal dominant trait. Briefly outline a series of studies that you might conduct to identify the gene that causes bladder cancer in this family.

Go to your **LaunchPad** to find additional learning resources and the Suggested Readings for this chapter.

24

Quantitative Genetics

Methods of quantitative genetics coupled with molecular techniques have been used to identify a gene that determines oil content in corn.
[Jim Craigmyle/Corbis.]

Corn Oil and Quantitative Genetics

In 2012, the world's population surpassed 7 billion. The United Nations projects that by 2050 it will increase by another 2 billion: feeding those billions of additional people will be a major challenge for agriculture in the next few decades. Crop plants will have to provide most of the calories and nutrients required for the world's future population. Because of dwindling petroleum supplies and concerns about global warming, plants are also increasingly being utilized as sources of biofuels, placing additional demands on crop production.

To help meet the need for increased crop yields, plant breeders are using the latest genetic techniques in their quest to develop higher-yielding, more-efficient crop plants. The power of this approach is demonstrated by research aimed at increasing the oil content of corn. The oil content of corn is inherited, but the inheritance is more complex than that of the characteristics that we have studied so far; it is not a simple single-gene characteristic such as seed shape in peas. Numerous genes and environmental factors contribute to the oil content of corn. For characteristics such as oil content, several loci frequently interact and their expression is affected by environmental factors. Can the inheritance of a complex characteristic such as oil content be studied? Is it possible to predict the oil content of a plant on the basis of its breeding? The answers are yes—at least in part—but these questions cannot be addressed with the methods that we used for simple genetic characteristics. Instead, we must use statistical procedures that have been developed for analyzing complex characteristics. The genetic analysis of complex characteristics such as the oil content of corn is known as **quantitative genetics**.

In 2008, geneticists used a combination of quantitative genetics and molecular techniques to identify a key gene that controls the oil content of corn. First, they conducted crosses between high-oil and low-oil corn plants to identify chromosomal regions that play an important role in determining oil production. Chromosomal regions containing genes that influence a quantitative trait are termed **quantitative trait loci** (QTLs). Through these crosses, the geneticists located several QTLs that affected oil content; one of them was on corn chromosome 6.

Fine-scale genetic mapping further narrowed the QTL down to a small region of 4.2 centiMorgans on chromosome 6. Researchers sequenced DNA from the region and found that it contained five genes, one of which was *DGAT1-2*, a gene known to produce an enzyme that catalyzes the final step in a pathway for triacyglycerol biosynthesis. DNA from the *DGAT1-2* gene in a high-oil-producing strain contained an insertion of a codon that added phenylalanine to the enzyme, an insertion that was missing from a low-oil-producing strain. The researchers confirmed the effect of the additional phenylalanine codon on oil production by producing transgenic corn that contained the extra codon; oil production of these transgenic strains increased compared with that of transgenic strains without the extra codon. Interestingly, the extra phenylalanine codon is present in wild relatives of corn, suggesting that the codon was lost in the process of domestication or subsequent breeding of modern varieties.

This research suggests that the oil content of corn and other plants might be increased by genetically modifying their *DGAT1-2* genes to contain the extra codon for phenylalanine. Other studies that similarly combine quantitative and molecular analyses have led to the identification of genes that increase the vitamin A content of rice and increase sugar production in tomatoes.

This chapter is about the genetic analysis of complex characteristics such as the oil content of corn. We begin by considering the differences between quantitative and qualitative characteristics and why the expression of some characteristics varies continuously. We'll see how quantitative characteristics are often influenced by many genes, each of which has a small effect on the phenotype. Next, we'll examine statistical procedures for describing and analyzing quantitative characteristics. We'll consider how much of phenotypic variation can be attributed to genetic and environmental influences. Finally, we'll look at the effects of selection on quantitative characteristics.

24.1 Quantitative Characteristics Vary Continuously and Many Are Influenced by Alleles at Multiple Loci

Qualitative, or discontinuous, characteristics possess only a few distinct phenotypes (**Figure 24.1a**); these characteristics are the types studied by Mendel (e.g., round and wrinkled peas) and have been the focus of our attention thus far. However, many characteristics vary continuously along a scale of measurement with many overlapping phenotypes (**Figure 24.1b**). They are referred to as *continuous characteristics*, and are also called *quantitative characteristics* because any individual's phenotype must be described by a quantitative measurement. Examples of quantitative characteristics include height, weight, and blood pressure in humans, growth rate in mice, seed weight in plants, and milk production in cattle.

Quantitative characteristics arise from two phenomena. First, many are polygenic: they are influenced by genes at many loci. If many loci take part, many genotypes are possible, each producing a slightly different phenotype. Second, quantitative characteristics often arise when environmental factors affect the phenotype because environmental differences result in a single genotype producing a range of phenotypes. Most continuously varying characteristics are *both* polygenic *and* influenced by environmental factors, and these characteristics are said to be multifactorial.

The Relation Between Genotype and Phenotype

For some discontinuous characteristics, the relation between genotype and phenotype is straightforward. Each genotype produces a single phenotype, and most phenotypes are

(a) Discontinuous characteristic

1 A discontinuous (qualitative) characteristic exhibits only a few, easily distinguished phenotypes.

2 The plants are either tall or dwarf.

Number of individuals

Tall
Dwarf

Phenotype (height)

(b) Continuous characteristic

3 A continuous (quantitative) characteristic exhibits a continuous range of phenotypes.

4 The plants exhibit a wide range of heights.

Number of individuals

Dwarf
Tall

Phenotype (height)

24.1 Discontinuous and continuous characteristics differ in the number of phenotypes exhibited.

encoded by a single genotype. Dominance and epistasis may allow two or three genotypes to produce the same phenotype, but the relation remains simple. This simple relation between genotype and phenotype allowed Mendel to decipher the basic rules of inheritance from his crosses with pea plants; it also permits us both to predict the outcome of genetic crosses and to assign genotypes to individuals.

For quantitative characteristics, the relation between genotype and phenotype is often more complex. If the characteristic is polygenic, many different genotypes are possible, several of which may produce the same phenotype. For instance, consider a plant whose height is determined by three loci (A, B, and C), each of which has two alleles. Assume that one allele at each locus (A^+, B^+, and C^+) encodes a plant hormone that causes the plant to grow 1 cm above its baseline height of 10 cm. The other allele at each locus (A^-, B^-, and C^-) does not encode a plant hormone and thus does not contribute to additional height. If we consider only the two alleles at a single locus, 3 genotypes are possible (A^+A^+, A^+A^-, and A^-A^-). If all three loci are taken into account, there are a total of $3^3 = 27$ possible multilocus genotypes ($A^+A^+ B^+B^+ C^+C^+$, $A^+A^- B^+B^+ C^+C^+$, etc.). Although there are 27 genotypes, they produce only seven phenotypes (10 cm, 11 cm, 12 cm, 13 cm, 14 cm, 15 cm, and 16 cm in height). Some of the genotypes produce the same phenotype (**Table 24.1**); for example, genotypes $A^+A^- B^-B^- C^-C^-$, $A^-A^- B^+B^- C^-C^-$, and $A^-A^- B^-B^- C^+C^-$ all have one gene that encodes a plant hormone. Each of these genotypes produces one dose of the hormone and results in a plant that is 11 cm tall. Even in this simple example of only three loci, the relation between genotype and phenotype is quite complex. The more loci encoding a characteristic, the greater the complexity. As the number of loci encoding a characteristic increases, the number of potential phenotypes increases, and differences between individual phenotypes becomes more difficult to distinguish.

The influence of environment on a characteristic also can complicate the relation between genotype and phenotype. Because of environmental effects, the same genotype can produce a range of potential phenotypes. The phenotypic ranges of different genotypes can overlap, making it difficult to know whether individuals differ in phenotype because of genetic or environmental differences (**Figure 24.2**).

In summary, the simple relation between genotype and phenotype that exists for many qualitative (discontinuous) characteristics is absent for quantitative characteristics, and it is impossible to assign a genotype to an individual on the basis of its phenotype alone. The methods used for analyzing qualitative characteristics (examining the phenotypic ratios of progeny from a genetic cross) will not work for quantitative characteristics. Our goal remains the same: we wish to make predictions about the phenotypes of offspring produced in a genetic cross. We may also want to know how much of the variation of a characteristic results from genetic differences and how much results from environmental differences.

Table 24.1	Hypothetical example of plant height determined by pairs of alleles at each of three loci	
Plant Genotype	**Doses of Hormone**	**Height (cm)**
$A^-A^-\ B^-B^-\ C^-C^-$	0	10
$A^+A^-\ B^-B^-\ C^-C^-$	1	11
$A^-A^-\ B^+B^-\ C^-C^-$		
$A^-A^-\ B^-B^-\ C^-C^+$		
$A^+A^+\ B^-B^-\ C^-C^-$	2	12
$A^-A^-\ B^+B^+\ C^-C^-$		
$A^-A^-\ B^-B^-\ C^+C^+$		
$A^+A^-\ B^+B^-\ C^-C^-$		
$A^+A^-\ B^-B^-\ C^+C^-$		
$A^-A^-\ B^+B^-\ C^+C^-$		
$A^+A^+\ B^+B^-\ C^-C^-$	3	13
$A^+A^+\ B^-B^-\ C^+C^-$		
$A^+A^-\ B^+B^+\ C^-C^-$		
$A^-A^-\ B^+B^+\ C^+C^-$		
$A^+A^-\ B^-B^-\ C^+C^+$		
$A^-A^-\ B^+B^-\ C^+C^+$		
$A^+A^-\ B^+B^-\ C^+C^-$		
$A^+A^+\ B^+B^+\ C^-C^-$	4	14
$A^+A^+\ B^+B^-\ C^+C^-$		
$A^+A^-\ B^+B^+\ C^+C^-$		
$A^-A^-\ B^+B^+\ C^+C^+$		
$A^+A^+\ B^-B^-\ C^+C^+$		
$A^+A^-\ B^+B^-\ C^+C^+$		
$A^+A^+\ B^+B^+\ C^+C^-$	5	15
$A^+A^-\ B^+B^+\ C^+C^+$		
$A^+A^+\ B^+B^-\ C^+C^+$		
$A^+A^+\ B^+B^+\ C^+C^+$	6	16

Note: Each + allele contributes 1 cm in height above a baseline of 10 cm.

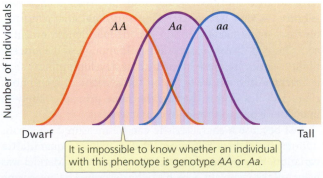

It is impossible to know whether an individual with this phenotype is genotype *AA* or *Aa*.

24.2 For a quantitative characteristic, each genotype may produce a range of possible phenotypes. In this hypothetical example, the phenotypes produced by genotypes *AA*, *Aa*, and *aa* overlap.

To answer these questions, we must turn to statistical methods that allow us to make predictions about the inheritance of phenotypes in the absence of information about the underlying genotypes. ▶ TRY PROBLEM 16

Types of Quantitative Characteristics

Before we look more closely at polygenic characteristics and relevant statistical methods, we need to more clearly define what is meant by a quantitative characteristic. Thus far, we have considered only quantitative characteristics that vary continuously in a population. A *continuous characteristic* can theoretically assume any value between two extremes; the number of phenotypes is limited only by our ability to precisely measure the phenotype. Human height is a continuous characteristic because, within certain limits, people can theoretically have any height. Although the number of phenotypes possible with a continuous characteristic is infinite, we often group similar phenotypes together for convenience; we may say that two people are both 5 feet 11 inches tall, but careful measurement may show that one is slightly taller than the other.

Some characteristics are not continuous but are nevertheless considered quantitative because they are determined by multiple genetic and environmental factors. **Meristic characteristics**, for instance, are measured in whole numbers. An example is litter size: a female mouse can have 4, 5, or 6 pups but not 4.13 pups. A meristic characteristic has a limited number of distinct phenotypes, but the underlying determination of the characteristic can still be quantitative. These characteristics must therefore be analyzed with the same techniques that we use to study continuous quantitative characteristics.

Another type of quantitative characteristic is a **threshold characteristic**, which is simply present or absent. Although threshold characteristics exhibit only two phenotypes, they are considered quantitative because they, too, are determined by multiple genetic and environmental factors. The expression of the characteristic depends on an underlying susceptibility (usually referred to as liability or risk) that varies continuously. When the susceptibility is larger than a threshold value, a specific trait is expressed (**Figure 24.3**). Diseases are often threshold characteristics because many factors, both genetic and environmental, contribute to disease susceptibility. If enough of the susceptibility factors are present, the disease develops; otherwise, it is absent. Although we focus on the genetics of continuous characteristics in this chapter, the same principles apply to many meristic and threshold characteristics.

Just because a characteristic can be measured on a continuous scale does not mean that it exhibits quantitative variation. One of the characteristics studied by Mendel was height of the pea plant, which can be described by measuring the length of the plant's stem. However, Mendel's particular plants exhibited only two distinct phenotypes (some were tall and others short), and these differences were determined by

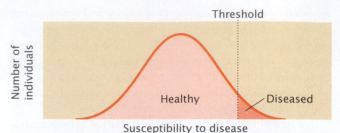

24.3 Threshold characteristics display only two possible phenotypes—the trait is either present or absent—but they are quantitative because the underlying susceptibility to the characteristic varies continuously. When the susceptibility exceeds a threshold value, the characteristic is expressed.

alleles at a single locus. The differences that Mendel studied were therefore discontinuous in nature.

CONCEPTS

Characteristics for which the phenotypes vary continuously are called quantitative characteristics. For most quantitative characteristics, the relation between genotype and phenotype is complex. Some characteristics for which the phenotypes do not vary continuously also are considered quantitative because they are influenced by multiple genes and environmental factors.

Polygenic Inheritance

After the rediscovery of Mendel's work in 1900, questions soon arose about the inheritance of continuously varying characteristics. These characteristics had already been the focus of a group of biologists and statisticians, led by Francis Galton, who used statistical procedures to examine the inheritance of quantitative characteristics such as human height and intelligence. The results of these studies showed that quantitative characteristics were at least partly inherited, although the mechanism of inheritance was not yet known. Some biometricians argued that the inheritance of quantitative characteristics could not be explained by Mendelian principles, whereas others felt that Mendel's principles acting on numerous genes (polygenes) could adequately account for the inheritance of quantitative characteristics.

This conflict began to be resolved by the work of Wilhelm Johannsen, who showed that continuous variation in the weight of beans was influenced by both genetic and environmental factors. George Udny Yule, a mathematician, proposed in 1906 that several genes acting together could produce continuous characteristics. This hypothesis was later confirmed by Herman Nilsson-Ehle, working on wheat and tobacco, and by Edward East, working on corn. The argument was finally laid to rest in 1918, when Ronald Fisher demonstrated that the inheritance of quantitative characteristics could indeed be explained by the cumulative effects of many genes, each following Mendel's rules.

Kernel Color in Wheat

To illustrate how multiple genes acting on a characteristic can produce a continuous range of phenotypes, let us examine one of the first demonstrations of polygenic inheritance. Nilsson-Ehle studied kernel color in wheat and found that the intensity of red pigmentation was determined by three unlinked loci, each of which had two alleles.

NILSSON-EHLE'S CROSS Nilsson-Ehle obtained several homozygous varieties of wheat that differed in color. Like Mendel, he performed crosses between these homozygous varieties and studied the ratios of phenotypes in the progeny. In one experiment, he crossed a variety of wheat that possessed white kernels with a variety that possessed purple (very dark red) kernels and obtained the following results:

P   Plants with white kernels   ×   Plants with purple kernels

F$_1$ Plants with red kernels

F$_2$
- $^1/_{16}$ plants with purple kernels
- $^4/_{16}$ plants with dark-red kernels
- $^6/_{16}$ plants with red kernels
- $^4/_{16}$ plants with light-red kernels
- $^1/_{16}$ plants with white kernels

INTERPRETATION OF THE CROSS Nilsson-Ehle interpreted this phenotypic ratio as the result of the segregation of alleles at two loci (although he found alleles at three loci that affect kernel color, the two varieties used in this cross differed at only two of the loci). He proposed that there were two alleles at each locus: one that produced red pigment and another that produced no pigment. We'll designate the alleles that encoded pigment A^+ and B^+ and the alleles that encoded no pigment A^- and B^-. Nilsson-Ehle recognized that the effects of the genes were additive. Each gene seemed to contribute equally to color; so the overall phenotype could be determined by adding the effects of all the genes, as shown in the following table.

Genotype	Doses of pigment	Phenotype
$A^+A^+\ B^+B^+$	4	Purple
$A^+A^+\ B^+B^-$ $A^+A^-\ B^+B^+$	3	Dark red
$A^+A^+\ B^-B^-$ $A^-A^-\ B^+B^+$ $A^+A^-\ B^+B^-$	2	Red
$A^+A^-\ B^-B^-$ $A^-A^-\ B^+B^-$	1	Light red
$A^-A^-\ B^-B^-$	0	White

Notice that the purple and white phenotypes are each encoded by a single genotype, but other phenotypes may result from several different genotypes.

From these results, we see that five phenotypes are possible when alleles at two loci influence the phenotype and the effects of the genes are additive. When alleles at more than two loci influence the phenotype, more phenotypes are possible, and the color would appear to vary continuously between white and purple. If environmental factors had influenced the characteristic, individuals of the same genotype would vary somewhat in color, making it even more difficult to distinguish between discrete phenotypic classes. Luckily, environment played little role in determining kernel color in Nilsson-Ehle's crosses, and only a few loci encoded color; so Nilsson-Ehle was able to distinguish among the different phenotypic classes. This ability allowed him to see the Mendelian nature of the characteristic.

Let's now see how Mendel's principles explain the ratio obtained by Nilsson-Ehle in his F$_2$ progeny. Remember that Nilsson-Ehle crossed the homozygous purple variety ($A^+A^+ B^+B^+$) with the homozygous white variety ($A^-A^- B^-B^-$), producing F$_1$ progeny that were heterozygous at both loci ($A^+A^- B^+B^-$). This is a dihybrid cross, like those that we worked in Chapter 3 except that both loci encode the same trait. All the F$_1$ plants possessed two pigment-producing alleles that allowed two doses of color to make red kernels. The types and proportions of progeny expected in the F$_2$ can be found by applying Mendel's principles of segregation and independent assortment.

Let's first examine the effects of each locus separately. At the first locus, two heterozygous F$_1$s are crossed ($A^+A^- \times A^+A^-$). As we learned in Chapter 3, when two heterozygotes are crossed, we expect progeny in the proportions $^1/_4\ A^+A^+$, $^1/_2\ A^+A^-$, and $^1/_4\ A^-A^-$. At the second locus, two heterozygotes also are crossed, and, again, we expect progeny in the proportions $^1/_4\ B^+B^+$, $^1/_2\ B^+B^-$, and $^1/_4\ B^-B^-$.

To obtain the probability of combinations of genes at both loci, we must use the multiplication rule of probability (see Chapter 3), the use of which assumes Mendel's principle of independent assortment. The expected proportion of F$_2$ progeny with genotype $A^+A^+ B^+B^+$ is the product of the probability of obtaining genotype A^+A^+ ($^1/_4$) and the probability of obtaining genotype B^+B^+ ($^1/_4$), or $^1/_4 \times ^1/_4 = ^1/_{16}$ (**Figure 24.4**). The probabilities of each of the phenotypes can then be obtained by adding the probabilities of all the genotypes that produce that phenotype. For example, the red phenotype is produced by three genotypes:

Genotype	Phenotype	Probability
$A^+A^+\ B^-B^-$	Red	$^1/_{16}$
$A^-A^-\ B^+B^+$	Red	$^1/_{16}$
$A^+A^-\ B^+B^-$	Red	$^1/_4$

Question: How is a continous trait, such as kernel color in wheat, inherited?

Methods Cross wheat with white kernels and wheat with purple kernels. Intercross the F$_1$ to produce F$_2$.

P generation

$$A^+A^+B^+B^+ \times A^-A^-B^-B^-$$

Purple   White

Results

F$_1$ generation

$$A^+A^-B^+B^- \longrightarrow \text{Red}$$

Break into simple crosses

$$A^+A^- \times A^+A^- \qquad B^+B^- \times B^+B^-$$

$$\tfrac{1}{4}A^+A^+ \quad \tfrac{1}{2}A^+A^- \quad \tfrac{1}{4}A^-A^- \qquad \tfrac{1}{4}B^+B^+ \quad \tfrac{1}{2}B^+B^- \quad \tfrac{1}{4}B^-B^-$$

Combine results

F$_2$ generation — Number of pigment genes — Phenotype

$\tfrac{1}{4}A^+A^+$	$\tfrac{1}{4}B^+B^+ \rightarrow \tfrac{1}{4}\times\tfrac{1}{4}=\tfrac{1}{16}$ $A^+A^+B^+B^+$	4	Purple
	$\tfrac{1}{2}B^+B^- \rightarrow \tfrac{1}{4}\times\tfrac{1}{2}=\tfrac{2}{16}$ $A^+A^+B^+B^-$	3	Dark red
	$\tfrac{1}{4}B^-B^- \rightarrow \tfrac{1}{4}\times\tfrac{1}{4}=\tfrac{1}{16}$ $A^+A^+B^-B^-$	2	Red
$\tfrac{1}{2}A^+A^-$	$\tfrac{1}{4}B^+B^+ \rightarrow \tfrac{1}{2}\times\tfrac{1}{4}=\tfrac{2}{16}$ $A^+A^-B^+B^+$	3	Dark red
	$\tfrac{1}{2}B^+B^- \rightarrow \tfrac{1}{2}\times\tfrac{1}{2}=\tfrac{4}{16}$ $A^+A^-B^+B^-$	2	Red
	$\tfrac{1}{4}B^-B^- \rightarrow \tfrac{1}{2}\times\tfrac{1}{4}=\tfrac{2}{16}$ $A^+A^-B^-B^-$	1	Light red
$\tfrac{1}{4}A^-A^-$	$\tfrac{1}{4}B^+B^+ \rightarrow \tfrac{1}{4}\times\tfrac{1}{4}=\tfrac{1}{16}$ $A^-A^-B^+B^+$	2	Red
	$\tfrac{1}{2}B^+B^- \rightarrow \tfrac{1}{4}\times\tfrac{1}{2}=\tfrac{2}{16}$ $A^-A^-B^+B^-$	1	Light red
	$\tfrac{1}{4}B^-B^- \rightarrow \tfrac{1}{4}\times\tfrac{1}{4}=\tfrac{1}{16}$ $A^-A^-B^-B^-$	0	White

Combine common phenotypes

F$_2$ ratio

Frequency	Number of pigment genes	Phenotype
$\tfrac{1}{16}$	4	Purple
$\tfrac{4}{16}$	3	Dark red
$\tfrac{6}{16}$	2	Red
$\tfrac{4}{16}$	1	Light red
$\tfrac{1}{16}$	0	White

Conclusion: Kernel color in wheat is inherited according to Mendel's principles acting on alleles at two loci.

24.4 Nilsson-Ehle demonstrated that kernel color in wheat is inherited according to Mendelian principles. The ratio of phenotypes in the F$_2$ can be determined by breaking the dihybrid cross into two simple single-locus crosses and combining the results by using the multiplication rule.

Thus, the overall probability of obtaining red kernels in the F$_2$ progeny is $\tfrac{1}{16} + \tfrac{1}{16} + \tfrac{1}{4} = \tfrac{6}{16}$. Figure 24.4 shows that the phenotypic ratio expected in the F$_2$ is $\tfrac{1}{16}$ purple, $\tfrac{4}{16}$ dark red, $\tfrac{6}{16}$ red, $\tfrac{4}{16}$ light red, and $\tfrac{1}{16}$ white. This phenotypic ratio is precisely what Nilsson-Ehle observed in his F$_2$ progeny, demonstrating that the inheritance of a continuously varying characteristic such as kernel color is indeed according to Mendel's basic principles.

CONCLUSIONS AND IMPLICATIONS Nilsson-Ehle's crosses demonstrated that the difference between the inheritance of genes influencing quantitative characteristics and the inheritance of genes influencing discontinuous characteristics is in the *number* of loci that determine the characteristic. When multiple loci affect a characteristic, more genotypes are possible, so the relation between the genotype and the phenotype is less obvious. As the number of loci affecting a characteristic increases, the number of phenotypic classes in the F$_2$ increases (**Figure 24.5**).

Several conditions of Nilsson-Ehle's crosses greatly simplified the polygenic inheritance of kernel color and made it possible for him to recognize the Mendelian nature of the characteristic. First, genes affecting color segregated at only two or three loci. If genes at many loci had been segregating, he would have had difficulty in distinguishing the phenotypic classes. Second, the genes affecting kernel color had strictly additive effects, making the relation between genotype and phenotype simple. Third, environment played almost no role in the phenotype; had environmental factors modified the phenotypes, distinguishing between the five phenotypic classes would have been difficult. Finally, the loci that Nilsson-Ehle studied were not linked, so the genes assorted independently. Nilsson-Ehle was fortunate: for many polygenic characteristics, these simplifying conditions are not present and Mendelian inheritance of these characteristics is not obvious. ▶ **TRY PROBLEM 17**

Determining Gene Number for a Polygenic Characteristic

The proportion of F$_2$ individuals that resemble one of the original parents can be used to estimate the number of genes affecting a polygenic trait. When two individuals homozygous for different alleles at a single locus are crossed ($A^1A^1 \times A^2A^2$) and the resulting F$_1$ are interbred ($A^1A^2 \times A^1A^2$), one-fourth of the F$_2$ should be homozygous like each of the original parents. If the original parents are homozygous for different alleles at two loci, as are those in Nilsson-Ehle's crosses, then $\tfrac{1}{4} \times \tfrac{1}{4} = \tfrac{1}{16}$ of the F$_2$ should resemble one of the original homozygous parents. Generally, $(\tfrac{1}{4})^n$ will be the proportion of individuals in the F$_2$ progeny that should resemble each of the original homozygous parents, where n equals the number of loci with a segregating pair of alleles that affects the characteristic. This equation provides us with a possible means

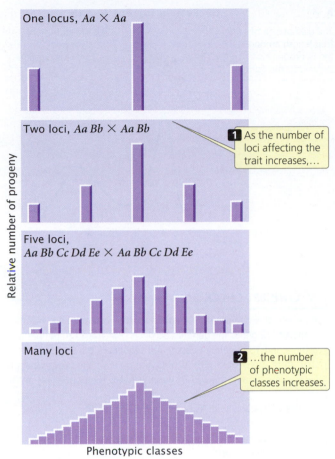

24.5 The results of crossing individuals heterozygous for different numbers of loci affecting a characteristic.

of determining the number of loci influencing a quantitative characteristic.

To illustrate the use of this equation, assume that we cross two different homozygous varieties of pea plants that differ in height by 16 cm, interbreed the F_1, and find that approximately $\frac{1}{256}$ of the F_2 are similar to one of the original homozygous parental varieties. This outcome would suggest that four loci with segregating pairs of alleles ($\frac{1}{256} = (\frac{1}{4})^4$) are responsible for the height difference between the two varieties. Because the two homozygous strains differ in height by 16 cm and there are four loci each of which has two alleles (eight alleles in all), each of the alleles contributes 16 cm/8 = 2 cm in height.

This method for determining the number of loci affecting phenotypic differences requires the use of homozygous strains, which may be difficult to obtain in some organisms. It also assumes that all the genes influencing the characteristic have equal effects, that their effects are additive, and that the loci are unlinked. For many polygenic characteristics, these assumptions are not valid, and so this method of determining the number of genes affecting a characteristic has limited application. ▶ **TRY PROBLEM 19**

CONCEPTS

The same principles determine the inheritance of quantitative and discontinuous characteristics, but more genes take part in the determination of quantitative characteristics.

✓ CONCEPT CHECK 1

Briefly explain how the number of genes influencing a polygenic trait can be determined.

24.2 Statistical Methods Are Required for Analyzing Quantitative Characteristics

Because quantitative characteristics are described by a measurement and are influenced by multiple factors, their inheritance must be analyzed statistically. This section will explain the basic concepts of statistics that are used to analyze quantitative characteristics.

Distributions

Understanding the genetic basis of any characteristic begins with a description of the numbers and kinds of phenotypes present in a group of individuals. Phenotypic variation in a group can be conveniently represented by a **frequency distribution**, which is a graph of the frequencies (numbers or proportions) of the different phenotypes (**Figure 24.6**). In a typical frequency distribution, the phenotypic classes are plotted on the horizontal (x) axis, and the numbers (or proportions) of individuals in each class are plotted on the vertical (y) axis. A frequency distribution is a concise method of summarizing all phenotypes of a quantitative characteristic.

Connecting the points of a frequency distribution with a line creates a curve that is characteristic of the distribution. Many quantitative characteristics exhibit a symmetrical

24.6 A frequency distribution is a graph that displays the number or proportion of different phenotypes. Phenotypic values are plotted on the horizontal axis, and the numbers (or proportions) of individuals in each class are plotted on the vertical axis.

(a) Sugar beet percentage of sucrose

1 This type of symmetrical (bell-shaped) distribution is called a normal distribution.

(b) Squash fruit length

2 The distribution of fruit length among the F$_2$ progeny is skewed to the right.

(c) Earwig forceps length

3 A distribution with two peaks is bimodal.

24.7 Distributions of phenotypes can assume several different shapes.

(bell-shaped) curve called a **normal distribution** (**Figure 24.7a**). Normal distributions arise when a large number of independent factors contribute to a measurement, as is often the case in quantitative characteristics. Two other common types of distributions (skewed and bimodal) are illustrated in **Figure 24.7b** and **c**.

Samples and Populations

Biologists frequently need to describe the distribution of phenotypes exhibited by a group of individuals. We might want to describe the height of students at the University of Texas (UT), but UT has more than 40,000 students and measuring all of them is not practical. Scientists are constantly confronted with this problem: the group of interest, called the **population**, is too large for a complete census. One solution is to measure a smaller collection of individuals, called a **sample**, and use those measurements to describe the population.

To provide an accurate description of the population, a good sample must have several characteristics. First, it must be representative of the whole population—for instance, if our sample consisted entirely of members of the UT basketball team we would probably overestimate the true height of the students. One way to ensure that a sample is representative of the population is to select the members of the sample randomly. Second, the sample must be large enough that chance differences between individuals in the sample and the overall population do not distort the estimate of the population measurements. If we measured only three students at UT and just by chance all three were short, we would underestimate the true height of the student population. Statistics can provide information about how much confidence to have in estimates based on random samples.

CONCEPTS

In statistics, the population is the group of interest; a sample is a subset of the population. The sample should be representative of the population and large enough to minimize chance differences between the population and the sample.

✔ CONCEPT CHECK 2

A geneticist is interested in whether asthma is caused by a mutation in the *DS112* gene. The geneticist collects DNA from 120 people with asthma and 100 healthy people, and sequences the DNA. She finds that 35 of the people with asthma and none of the healthy people have a mutation in the *DS112* gene. What is the population in this study?

a. The 120 people with asthma.
b. The 100 healthy people.
c. The 35 people with a mutation in their gene.
d. All people with asthma.

The Mean

The **mean**, also called the average, provides information about the center of the distribution. If we measured the heights of 10-year-old and 18-year-old boys and plotted a frequency distribution for each group, we would find that both distributions are normal, but the two distributions would be centered at different heights, and this difference would be indicated in their different means (**Figure 24.8**).

Suppose that we have five measurements of height in centimeters: 160, 161, 167, 164, and 165. If we represent a group of measurements as x_1, x_2, x_3, and so forth, then the mean ($\bar{x}$) is calculated by adding all the individual measurements and dividing by the total number of measurements in the sample (n):

$$\bar{x} = \frac{x_1 + x_2 + x_3 + \ldots + x_n}{n} \tag{24.1}$$

In our example, $x_1 = 160$, $x_2 = 161$, $x_3 = 167$, and so forth. The mean height ($\bar{x}$) equals:

$$\bar{x} = \frac{160 + 161 + 167 + 164 + 165}{5} = \frac{817}{5} = 163.4$$

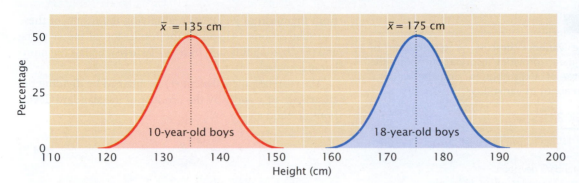

24.8 The mean provides information about the center of a distribution. Both distributions of heights of 10-year-old and 18-year-old boys are normal, but they have different locations along a continuum of height, which makes their means different.

A shorthand way to represent this formula is

$$\bar{x} = \frac{\sum x_i}{n} \qquad (24.2)$$

or

$$\bar{x} = \frac{1}{n}\sum x_i \qquad (24.3)$$

where the symbol $\sum$ means "the summation of" and x_i represents individual x values.

The Variance and Standard Deviation

A statistic that provides key information about a distribution is the **variance**, which indicates the variability of a group of measurements, or how spread out the distribution is. Distributions can have the same mean but different variances (**Figure 24.9**). The larger the variance, the greater the spread of measurements in a distribution about its mean.

The variance (s^2) is defined as the average squared deviation from the mean:

$$s^2 = \frac{\sum (x_i - \bar{x})^2}{n - 1} \qquad (24.4)$$

To calculate the variance, we (1) subtract the mean from each measurement and square the value obtained, (2) add all the squared deviations, and (3) divide this sum by the number of original measurements minus 1.

Another statistic that is closely related to the variance is the **standard deviation** (s), which is defined as the square root of the variance:

$$s = \sqrt{s^2} \qquad (24.5)$$

Whereas the variance is expressed in units squared, the standard deviation is in the same units as the original measurements; so the standard deviation is often preferred for describing the variability of a measurement.

A normal distribution is symmetrical; so the mean and standard deviation are sufficient to describe its shape. The mean plus or minus one standard deviation ($\bar{x} \pm s$) includes approximately 66% of the measurements in a normal distribution; the mean plus or minus two standard deviations ($\bar{x} \pm 2s$) includes approximately 95% of the measurements, and the mean plus or minus three standard deviations ($\bar{x} \pm 3s$) includes approximately 99% of the measurements (**Figure 24.10**). Thus, only 1% of a normally distributed population lies outside the range of ($\bar{x} \pm 3s$). ▶ TRY PROBLEM 22

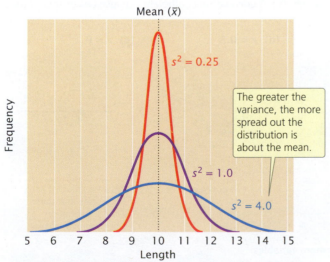

24.9 The variance provides information about the variability of a group of phenotypes. Shown here are three distributions with the same mean but different variances.

24.10 The proportions of a normal distribution occupied by plus or minus one, two, and three standard deviations from the mean.

The mean and the variance describe a distribution of measurements: the mean provides information about the location of the center of a distribution, and the variance provides information about its variability.

✓ CONCEPT CHECK 3

The measurements of a distribution with a higher _____ will be more spread out.
a. mean
c. standard deviation
b. variance
d. variance and standard deviation

Correlation

The mean and the variance can be used to describe an individual characteristic, but geneticists are frequently interested in more than one characteristic. Often, two or more characteristics vary together. For instance, both the number and the weight of eggs produced by hens are important to the poultry industry. These two characteristics are not independent of each other. There is an inverse relation between egg number and weight: hens that lay more eggs produce smaller eggs. This kind of relation between two characteristics is called a **correlation**. When two characteristics are correlated, a change in one characteristic is likely to be associated with a change in the other.

Correlations between characteristics are measured by a **correlation coefficient** (designated r), which measures the strength of their association. Consider two characteristics, such as human height (x) and arm length (y). To determine how these characteristics are correlated, we first obtain the covariance (cov) of x and y:

$$\text{cov}_{xy} = \frac{\sum (x_i - \bar{x})(y_i - \bar{y})}{n - 1} \qquad (24.6)$$

The covariance is computed by (1) taking an x value for an individual and subtracting it from the mean of x ($\bar{x}$); (2) taking the y value for the same individual and subtracting it from the mean of y ($\bar{y}$); (3) multiplying the results of these two subtractions; (4) adding the results for all the xy pairs; and (5) dividing this sum by $n - 1$ (where n equals the number of xy pairs).

The correlation coefficient (r) is obtained by dividing the covariance of x and y by the product of the standard deviations of x and y:

$$r = \frac{\text{cov}_{xy}}{s_x s_y} \qquad (24.7)$$

A correlation coefficient can theoretically range from -1 to $+1$. A positive value indicates that there is a direct association between the variables (**Figure 24.11a**): as one variable increases, the other variable also tends to increase. A positive correlation exists for human height and weight: tall people tend to weigh more. A negative correlation coefficient indicates that there is an inverse relation between the two variables (**Figure 24.11b**): as one variable increases, the other tends to decrease (as is the case for egg number and egg weight in chickens).

The absolute value of the correlation coefficient (the size of the coefficient, ignoring its sign) provides information about the strength of association between the variables. A coefficient of -1 or $+1$ indicates a perfect correlation between the variables, meaning that a change in x is always accompanied by a proportional change in y. Correlation coefficients close to -1 or close to $+1$ indicate a strong association between the variables: a change in x is almost always associated with a proportional increase in y, as seen in **Figure 24.11c**. On the other hand, a correlation coefficient closer to 0 indicates a weak correlation: a change in x is associated with a change in y but not always (**Figure 24.11d**). A correlation of 0 indicates that there is no association between variables (**Figure 24.11e**).

A correlation coefficient can be computed for two variables measured for the same individual, such as height (x) and weight (y). A correlation coefficient can also be computed for a

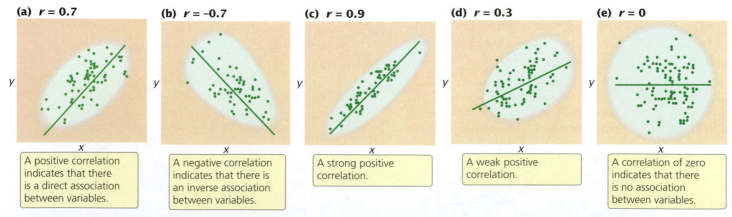

(a) $r = 0.7$

A positive correlation indicates that there is a direct association between variables.

(b) $r = -0.7$

A negative correlation indicates that there is an inverse association between variables.

(c) $r = 0.9$

A strong positive correlation.

(d) $r = 0.3$

A weak positive correlation.

(e) $r = 0$

A correlation of zero indicates that there is no association between variables.

24.11 The correlation coefficient describes the relation between two or more variables.

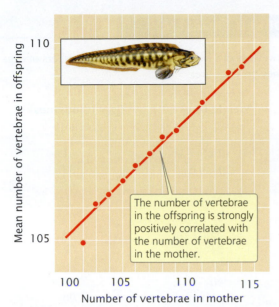

24.12 **A correlation coefficient can be computed for a single variable measured for pairs of individuals.** Here, the numbers of vertebrae in mothers and offspring of the fish *Zoarces viviparus* are compared.

single variable measured for pairs of individuals. For example, we can calculate for fish the correlation between the number of vertebrae of a parent (*x*) and the number of vertebrae of its offspring (*y*), as shown in **Figure 24.12**. This approach is often used in quantitative genetics.

A correlation between two variables indicates only that the variables are associated; it does not imply a cause-and-effect relation. Correlation also does not mean that the values of two variables are the same; it means only that a change in one variable is associated with a proportional change in the other variable. For example, the *x* and *y* variables in the following list are almost perfectly correlated, with a correlation coefficient of 0.99.

x value	y value
12	123
14	140
10	110
6	61
3	32
Average: 9	90

A high correlation is found between these *x* and *y* variables; larger values of *x* are always associated with larger values of *y*. Note that the *y* values are about 10 times as large as the corresponding *x* values; so, although *x* and *y* are correlated, they are not identical. The distinction between correlation and identity becomes important when we consider the effects of heredity and environment on the correlation of characteristics. ▶TRY PROBLEM 25

Regression

Correlation provides information only about the strength and direction of association between variables. However, we often want to know more than just whether two variables are associated; we want to be able to predict the value of one variable given a value of the other.

A positive correlation exists between the body weight of parents and the body weight of their offspring; this correlation exists in part because genes influence body weight, and parents and children have genes in common. Because of this association between parental and offspring phenotypes, we can predict the weight of an individual on the basis of the weights of its parents. This type of statistical prediction is called **regression**. This technique plays an important role in quantitative genetics because it allows us to predict the characteristics of offspring from a given mating, even without knowledge of the genotypes that encode the characteristics.

Regression can be understood by plotting a series of *x* and *y* values. **Figure 24.13** illustrates the relation between the weight of a father (*x*) and the weight of his son (*y*). Each father–son pair is represented by a point on the graph. The overall relation between these two variables is depicted by the regression line, which is the line that best fits all the points on the graph (deviations of the points from the line are minimized). The regression line defines the relation between the *x* and *y* variables and can be represented by

$$y = a + bx \qquad (24.8)$$

In Equation 24.8, *x* and *y* represent the *x* and *y* variables (in this case, the father's weight and the son's weight, respectively). The variable *a* is the *y* intercept of the line, which is the expected value of *y* when *x* is 0. Variable *b* is the slope of the regression line, also called the **regression coefficient**.

Trying to position a regression line by eye is not only very difficult but also inaccurate when there are many

24.13 **A regression line defines the relation between two variables.** Illustrated here is a regression of the weights of fathers against the weights of sons. Each father–son pair is represented by a point on the graph: the *x* value of a point is the father's weight and the *y* value of the point is the son's weight.

The regression coefficient represents the slope of the regression line.

b = 1

When the regression coefficient is 1, a 1-unit increase in *x* is associated with a 1-unit increase in *y*.

b = 0.4

b = 0.2

When the regression coefficient is 0.2, a 1-unit increase in *x* is associated with a 0.2-unit increase in *y*.

24.14 The regression coefficient, *b*, represents the change in *y* per unit change in *x*. Shown here are regression lines with different regression coefficients.

points scattered over a wide area. Fortunately, the regression coefficient and the *y* intercept can be obtained mathematically. The regression coefficient (*b*) can be computed from the covariance of *x* and *y* (cov_{xy}) and the variance of *x* (*s*) by

$$b = \frac{cov_{xy}}{s_x^2} \qquad (24.9)$$

The regression coefficient indicates how much *y* increases, on average, per increase in *x*. Several regression lines with different regression coefficients are illustrated in **Figure 24.14**. Notice that as the regression coefficient increases, the slope of the regression line increases.

After the regression coefficient has been calculated, the *y* intercept can be calculated by substituting the regression coefficient and the mean values of *x* and *y* into the following equation:

$$a = \bar{y} - b\bar{x} \qquad (24.10)$$

The regression equation ($y = a + bx$, Equation 24.8) can then be used to predict the value of any *y* given the value of *x*.

CONCEPTS

A correlation coefficient measures the strength of association between two variables. The sign (positive or negative) indicates the direction of the correlation; the absolute value measures the strength of the association. Regression is used to predict the value of one variable on the basis of the value of a correlated variable.

✔ CONCEPT CHECK 4

In Lubbock, Texas, rainfall and temperature exhibit a significant correlation of −0.7. Which conclusion is correct?
a. There is usually more rainfall when the temperature is high.
b. There is usually more rainfall when the temperature is low.
c. Rainfall is equally likely when the temperature is high or low.

WORKED PROBLEM

Body weights of 11 female fishes and the numbers of eggs that they produce are:

Weight (mg)	Eggs (thousands)
x	*y*
14	61
17	37
24	65
25	69
27	54
33	93
34	87
37	89
40	100
41	90
42	97

What are the correlation coefficient and the regression coefficient for body weight and egg number in these 11 fishes?

Solution Strategy

What information is required in your answer to the problem?
The correlation coefficient (*r*) and the regression coefficient (*b*) for body weight and egg number in the fish.

What information is provided to solve the problem?
Body weights and egg numbers for a sample of 11 fish.

Solution Steps
The computations needed to answer this question are given in the table on the facing page. To calculate the correlation and regression coefficients, we first obtain the sum of all the x_i values ($\sum x_i$) and the sum of all the y_i values ($\sum y_i$); these sums are shown in the last row of the table. We can calculate the means of the two variables by dividing the sums by the number of measurements, which is 11:

$$\bar{x} = \frac{\sum x_i}{n} = \frac{334}{11} = 30.36$$

$$\bar{y} = \frac{\sum y_i}{n} = \frac{842}{11} = 76.55$$

After the means have been calculated, the deviations of each value from the mean are computed; these deviations are

A Weight (mg)	B	C	D Eggs (thousands)	E	F	G
x	$x_i - \bar{x}$	$(x_i - \bar{x})^2$	y	$y_i - \bar{y}$	$(y_i - \bar{y})^2$	$(x_i - \bar{x})(y_i - \bar{y})$
14	−16.36	267.65	61	−15.55	241.80	254.40
17	−13.36	178.49	37	−39.55	1564.20	528.39
24	−6.36	40.45	65	−11.55	133.40	73.46
25	−5.36	28.73	69	−7.55	57.00	40.47
27	−3.36	11.29	54	−22.55	508.50	75.77
33	2.64	6.97	93	16.45	270.60	43.43
34	3.64	13.25	87	10.45	109.20	38.04
37	6.64	44.09	89	12.45	155.00	82.67
40	9.64	92.93	100	23.45	549.90	226.06
41	10.64	113.21	90	13.45	180.90	143.11
42	11.64	135.49	97	20.45	418.20	238.04
$\Sigma x_i = 334$		$\Sigma(x_i - \bar{x})^2 = 932.55$	$\Sigma y_i = 842$		$\Sigma(y_i - \bar{y})^2 = 4188.70$	$\Sigma(x_i - \bar{x})(y_i - \bar{y}) = 1743.84$

Source: R. R. Sokal and F. J. Rohlf, *Biometry*, 2d ed. (San Francisco: W. H. Freeman and Company, 1981).

shown in columns B and E of the table. The deviations are then squared (columns C and F) and summed (last row of columns C and F). Next, the products of the deviation of the *x* values and the deviation of the *y* values (($x_i - \bar{x}$) ($y_i - \bar{y}$)) are calculated; these products are shown in column G, and their sum is shown in the last row of column G.

To calculate the covariance, we use Equation 24.6:

$$\text{cov}_{xy} = \frac{\Sigma(x_i - \bar{x})(y_i - \bar{y})}{n - 1} = \frac{1743.84}{10} = 174.38$$

To calculate the correlation and the regression requires the variances and standard deviations of *x* and y:

$$s_x^2 = \frac{\Sigma(x_i - \bar{x})^2}{n - 1} = \frac{932.55}{10} = 93.26$$

$$s_x = \sqrt{s_x^2} = \sqrt{93.26} = 9.66$$

$$s_y^2 = \frac{\Sigma(y_i - \bar{y})^2}{n - 1} = \frac{4188.70}{10} = 418.87$$

$$s_y = \sqrt{s_y^2} = \sqrt{418.87} = 20.47$$

We can now compute the correlation and regression coefficients as shown here.

Correlation coefficient:

$$r = \frac{\text{cov}_{xy}}{s_x s_y} = \frac{174.38}{9.66 \times 20.47} = 0.88$$

Regression coefficient:

$$b = \frac{\text{cov}_{xy}}{s_x^2} = \frac{174.38}{93.26} = 1.87$$

▶ Practice your understanding of correlation and regression by working Problem 26 at the end of the chapter.

Applying Statistics to the Study of a Polygenic Characteristic

Edward East carried out an early statistical study of polygenic inheritance on the length of flowers in tobacco (*Nicotiana longiflora*). He obtained two varieties of tobacco that differed in flower length: one variety had a mean flower length of 40.5 mm, and the other had a mean flower length of 93.3 mm (**Figure 24.15**). These two varieties had been inbred for many generations and were homozygous at all loci contributing to flower length. Thus, there was no genetic variation in the original parental strains; the small differences in flower length within each strain were due to environmental effects on flower length.

When East crossed the two strains, he found that flower length in the F_1 was about halfway between that in the two parents (see Figure 24.15), as would be expected if the genes determining the differences in the two strains were additive in their effects. The variance of flower length in the F_1 was similar to

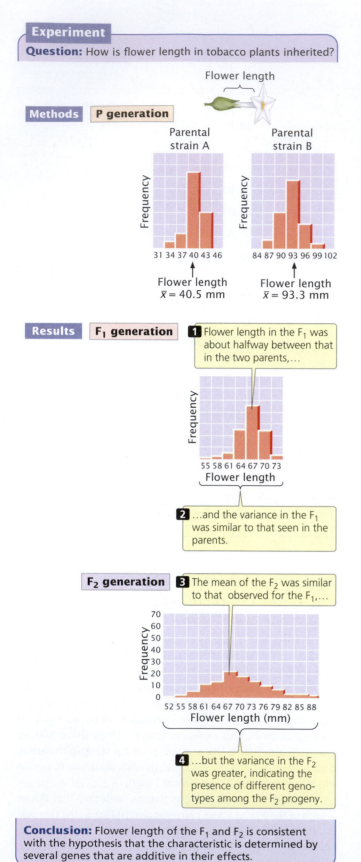

Question: How is flower length in tobacco plants inherited?

Methods   **P generation**

Flower length

Parental strain A

Parental strain B

Frequency

31 34 37 40 43 46

Flower length
$\bar{x} = 40.5$ mm

Frequency

84 87 90 93 96 99 102

Flower length
$\bar{x} = 93.3$ mm

Results   **F₁ generation**

1 Flower length in the F₁ was about halfway between that in the two parents,…

Frequency

55 58 61 64 67 70 73
Flower length

2 …and the variance in the F₁ was similar to that seen in the parents.

F₂ generation

3 The mean of the F₂ was similar to that observed for the F₁,…

Frequency
70
60
50
40
30
20
10
0

52 55 58 61 64 67 70 73 76 79 82 85 88
Flower length (mm)

4 …but the variance in the F₂ was greater, indicating the presence of different genotypes among the F₂ progeny.

Conclusion: Flower length of the F₁ and F₂ is consistent with the hypothesis that the characteristic is determined by several genes that are additive in their effects.

24.15 Edward East conducted an early statistical study of the inheritance of flower length in tobacco.

that seen in the parents, because all the F₁ had the same genotype, as did each parental strain (the F₁ were all heterozygous at the genes that differed between the two parental varieties).

East then interbred the F₁ to produce F₂ progeny. The mean flower length of the F₂ was similar to that of the F₁, but the variance of the F₂ was much greater (see Figure 24.15). This greater variability indicates that not all of the F₂ progeny had the same genotype.

East selected some F₂ plants and interbred them to produce F₃ progeny. He found that flower length of the F₃ depended on flower length in the plants selected as their parents. This finding demonstrated that flower-length differences in the F₂ were partly genetic and were therefore passed to the next generation. None of the 444 F₂ plants raised by East exhibited flower lengths similar to those of the two parental strains. This result suggested that more than four loci with pairs of alleles affected flower length in his varieties, because four allelic pairs are expected to produce 1 of 256 progeny $[(^1/_4)^4 = {}^1/_{256}]$ having one or the other of the original parental phenotypes.

24.3 Heritability Is Used to Estimate the Proportion of Variation in a Trait That Is Genetic

In addition to being polygenic, quantitative characteristics are frequently influenced by environmental factors. Knowing how much of the variation in a quantitative characteristic is due to genetic differences and how much is due to environmental differences is often useful. The proportion of the total phenotypic variation that is due to genetic differences is known as the **heritability**.

Consider a dairy farmer who owns several hundred milk cows. The farmer notices that some cows consistently produce more milk than others. The nature of these differences is important to the profitability of his dairy operation. If the differences in milk production are largely genetic in origin, then the farmer may be able to boost milk production by selectively breeding the cows that produce the most milk. On the other hand, if the differences are largely environmental in origin, selective breeding will have little effect on milk production, and the farmer might better boost milk production by adjusting the environmental factors associated with higher milk production. To determine the extent of genetic and environmental influences on variation in a characteristic, phenotypic variation in the characteristic must be partitioned into components attributable to different factors.

Phenotypic Variance

To determine how much of the phenotypic variation in a population is due to genetic factors and how much is due to environmental factors, we must first have some quantitative measure of the phenotype under consideration. Consider a population of wild plants that differ in size. We could collect a representative sample of plants from the population, weigh each plant in the sample, and calculate the mean and variance of plant weight. This **phenotypic variance** is represented by V_P.

COMPONENTS OF PHENOTYPIC VARIANCE First, some of the phenotypic variance may be due to differences in genotypes among individual members of the population. These differences are termed the **genetic variance** and are represented by V_G.

Second, some of the differences in phenotype may be due to environmental differences among the plants; these differences are termed the **environmental variance**, V_E. Environmental variance includes differences in environmental factors such as the amount of light or water that the plant receives; it also includes random differences in development that cannot be attributed to any specific factor. Any variation in phenotype that is not inherited is, by definition, a part of the environmental variance.

Third, **genetic–environmental interaction variance** (V_{GE}) arises when the effect of a gene depends on the specific environment in which it is found. An example is shown in **Figure 24.16**. In a dry environment, genotype AA produces a plant that averages 12 g in weight, and genotype aa produces a smaller plant that averages 10 g. In a wet environment, genotype aa produces the larger plant, averaging 24 g in weight, whereas genotype AA produces a plant that averages 20 g. In this example, there are clearly differences in the two environments: both genotypes produce heavier plants in the wet environment. There are also differences in the weights of the two genotypes, but the relative performances of the genotypes depend on whether the plants are grown in a wet or a dry environment. In this case, the influences on phenotype cannot be neatly allocated into genetic and environmental components, because the expression of the genotype depends on the environment in which the plant grows. The phenotypic variance must therefore include a component that accounts for the way in which genetic and environmental factors interact.

In summary, the total phenotypic variance can be apportioned into three components:

$$V_P = V_G + V_E + V_{GE} \tag{24.11}$$

COMPONENTS OF GENETIC VARIANCE Genetic variance can be further subdivided into components consisting of different types of genetic effects. First, **additive genetic variance** (V_A) comprises the additive effects of genes on the phenotype, which can be summed to determine the overall effect on the phenotype. For example, suppose that, in a plant, allele A^1 contributes 2 g in weight and allele A^2 contributes 4 g. If the alleles are strictly additive, then the genotypes would have the following weights:

$$A^1A^1 = 2 + 2 = 4 \text{ g}$$
$$A^1A^2 = 2 + 4 = 6 \text{ g}$$
$$A^2A^2 = 4 + 4 = 8 \text{ g}$$

The genes that Nilsson-Ehle studied, which affect kernel color in wheat, were additive in this way. The additive genetic variance primarily determines the resemblance between parents and offspring. For example, if all of the phenotypic variance is due to additive genetic variance, then the average phenotype of the offspring will be exactly intermediate between those of the parents.

Second, there is **dominance genetic variance** (V_D) when some genes have a dominance component. In this case, the alleles at a locus are not additive; rather, the effect of an allele depends on the identity of the other allele at that locus. For example, with a dominant allele (T), genotypes TT and Tt have the same phenotype. Here, we cannot simply add the effects of the alleles together, because the effect of the small t allele is masked by the presence of the large T allele. Instead, we must add a component (V_D) to the genetic variance to account for the way in which alleles interact.

Third, genes at different loci may interact in the same way that alleles at the same locus interact. When this genic interaction takes place, the effects of the genes are not additive. For example, Chapter 5 described how coat color in Labrador retrievers exhibits genic interaction; genotypes $BB\ ee$ and $bb\ ee$ both produce yellow dogs, because the effect of alleles at the B locus are masked when ee alleles are present at the E locus. With genic interaction, we must add a third component, called **genic interaction variance** (V_I), to the genetic variance:

$$V_G = V_A + V_D + V_I \tag{24.12}$$

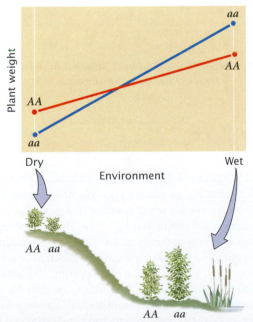

24.16 Genetic–environmental interaction variance is obtained when the effect of a gene depends on the specific environment in which it is found. In this theoretical example, the genotype affects plant weight, but the environmental conditions determine which genotype produces the heavier plant.

SUMMARY EQUATION   We can now integrate these components into one equation to represent all the potential contributions to the phenotypic variance:

$$V_P = V_A + V_D + V_I + V_E + V_{GE} \qquad (24.13)$$

This equation provides us with a model that describes the potential causes of differences that we observe among individual phenotypes. It's importatnt to note that this model deals strictly with the observable *differences* (variance) in phenotypes among individual members of a population; it says nothing about the absolute value of the characteristic or about the underlying genotypes that produce these differences.

Types of Heritability

The model of phenotypic variance that we've just developed can be used to determine how much of the phenotypic variance in a characteristic is due to genetic differences. **Broad-sense heritability** (H^2) represents the proportion of phenotypic variance that is due to genetic variance and is calculated by dividing the genetic variance by the phenotypic variance:

$$\text{broad-sense heritability} = H^2 = \frac{V_G}{V_P} \qquad (24.14)$$

The symbol H^2 represents broad-sense heritability because it is a measure of variance, which is in units squared.

Broad-sense heritability can potentially range from 0 to 1. A value of 0 indicates that none of the phenotypic variance results from differences in genotype and all of the differences in phenotype result from environmental variation. A value of 1 indicates that all of the phenotypic variance results from differences in genotypes. A heritability value between 0 and 1 indicates that both genetic and environmental factors influence the phenotypic variance.

Often, we are more interested in the proportion of the phenotypic variance that results from the additive genetic variance because, as mentioned earlier, the additive genetic variance primarily determines the resemblance between parents and offspring. **Narrow-sense heritability** (h^2) is equal to the additive genetic variance divided by the phenotypic variance:

$$\text{narrow-sense heritability} = h^2 = \frac{V_A}{V_P} \qquad (24.15)$$

▶ TRY PROBLEM 27

Calculating Heritability

Having considered the components that contribute to phenotypic variance and developed a general concept of heritability, we can now ask how we go about estimating these different components and calculating heritability.

There are several ways to measure the heritability of a characteristic. They include eliminating one or more variance components, comparing the resemblance of parents and offspring, comparing the phenotypic variances of individuals with different degrees of relatedness, and measuring the response to selection. The mathematical theory that underlies these calculations of heritability is complex, and so, here, we will focus on developing a general understanding of how heritability is measured.

HERITABILITY BY ELIMINATION OF VARIANCE COMPONENTS   One way of calculating the broad-sense heritability is to eliminate one of the variance components. We have seen that $V_P = V_G + V_E + V_{GE}$. If we eliminate all environmental variance ($V_E = 0$), then $V_{GE} = 0$ (because, if either V_G or V_E is zero, no genetic–environmental interaction can take place), and $V_P = V_G$. In theory, we might make V_E equal to 0 by ensuring that all individuals were raised in exactly the same environment but, in practice, it is virtually impossible. Instead, we could make V_G equal to 0 by raising genetically identical individuals, causing V_P to be equal to V_E. In a typical experiment, we might raise cloned or highly inbred, identically homozygous individuals in a defined environment and measure their phenotypic variance to estimate V_E. We could then raise a group of genetically variable individuals and measure their phenotypic variance (V_P). Using V_E calculated on the genetically identical individuals, we could obtain the genetic variance of the variable individuals by subtraction:

$$V_{G(\text{of genetically varying individuals})}$$
$$= V_{P\,(\text{of genetically varying individuals})} - V_{E\,(\text{of genetically identical individuals})} \qquad (24.16)$$

The broad-sense heritability of the genetically variable individuals would then be calculated as follows:

$$H^2 = \frac{V_{G(\text{of genetically varying individuals})}}{V_{P(\text{of genetically varying individuals})}} \qquad (24.17)$$

Sewall Wright used this method to estimate the heritability of white spotting in guinea pigs. He first measured the phenotypic variance for white spotting in a genetically variable population and found that $V_P = 573$. Then he inbred the guinea pigs for many generations so that they were essentially homozygous and genetically identical. When he measured their phenotypic variance in white spotting, he obtained V_P equal to 340. Because $V_G = 0$ in this group, their $V_P = V_E$. Wright assumed this value of environmental variance for the original (genetically variable) population and estimated their genetic variance:

$$V_P - V_E = V_G$$
$$573 - 340 = 233$$

24.17 The narrow-sense heritability, h^2, equals the regression coefficient, b, in a regression of the mean phenotype of the offspring against the mean phenotype of the parents. (a) There is no relation between the mean parental phenotype and the mean offspring phenotype. (b) The mean offspring phenotype is the same as the mean parental phenotype. (c) Both genes and environment contribute to the differences in phenotype.

He then estimated the broad-sense heritability from the genetic and phenotypic variance:

$$H^2 = \frac{V_G}{V_P}$$

$$H^2 = \frac{233}{573} = 0.41$$

This value implies that 41% of the variation in spotting of guinea pigs in Wright's population was due to differences in genotype.

Estimating heritability with this method assumes that the environmental variance of genetically identical individuals is the same as the environmental variance of the genetically variable individuals, which may not be true. Additionally, this approach can be applied only to organisms for which it is possible to create genetically identical individuals.

▶TRY PROBLEM 32

HERITABILITY BY PARENT–OFFSPRING REGRESSION

Another method for estimating heritability is to compare the phenotypes of parents and offspring. When genetic differences are responsible for phenotypic variance, offspring should resemble their parents more than they resemble unrelated individuals, because offspring and parents have some genes in common that help determine their phenotype. Correlation and regression can be used to analyze the association of phenotypes in different individuals.

To calculate the narrow-sense heritability in this way, we first measure the characteristic on a series of parents and offspring. The data are arranged into families, and the mean parental phenotype is plotted against the mean offspring phenotype (**Figure 24.17**). Each data point in the graph represents one family; the value on the x (horizontal) axis is the mean phenotypic value of the parents in a family, and the

value on the y (vertical) axis is the mean phenotypic value of the offspring for the family.

Let's assume that there is no narrow-sense heritability for the characteristic ($h^2 = 0$), meaning that genetic differences do not contribute to the phenotypic differences among individuals. In this case, offspring will be no more similar to their parents than they are to unrelated individuals, and the data points will be scattered randomly, generating a regression coefficient of zero (see Figure 24.17a). Next, let's assume that all of the phenotypic differences are due to additive genetic differences ($h^2 = 1.0$). In this case, the mean phenotype of the offspring will be equal to the mean phenotype of the parents, and the regression coefficient will be 1 (see Figure 24.17b). If genes and environment both contribute to the differences in phenotype, both heritability and the regression coefficient will lie between 0 and 1 (see Figure 24.17c). The regression coefficient therefore provides information about the magnitude of the heritability.

A complex mathematical proof (which we will not go into here) demonstrates that, in a regression of the mean phenotype of the offspring against the mean phenotype of the parents, narrow-sense heritability (h^2) equals the regression coefficient (b):

$$h^2 = b_{\text{(regression of mean offspring against mean of both parents)}} \qquad (24.18)$$

An example of calculating heritability by regression of the phenotypes of parents and offspring is illustrated in **Figure 24.18**.

Sometimes, only the phenotype of one parent is known. In a regression of the mean offspring phenotype against the phenotype of only one parent, the narrow-sense heritability equals twice the regression coefficient:

$$h^2 = 2b_{\text{(regression of mean offspring against mean of one parent)}} \qquad (24.19)$$

With only one parent, the heritability is twice the regression coefficient because only half the genes of the offspring come

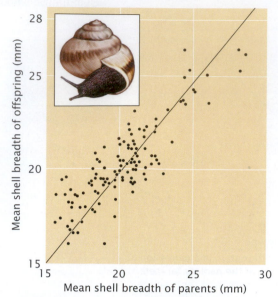

24.18 The heritability of shell breadth in snails can be determined by regression of the mean phenotype of offspring against the mean phenotype of the parents. The regression coefficient, which equals the heritability, is 0.70. [From L. M. Cook, *Evolution* 19:86–94, 1965.]

from one parent; thus, we must double the regression coefficient to obtain the full heritability.

HERITABILITY AND DEGREES OF RELATEDNESS A third method for calculating heritability is to compare the phenotypes of individuals with different degrees of relatedness. This method is based on the concept that the more closely related two individuals are, the more genes they have in common.

Monozygotic (identical) twins have 100% of their genes in common, whereas dizygotic (nonidentical) twins have, on average, 50% of their genes in common. If genes are important in determining variability in a characteristic, then monozygotic twins should be more similar in a particular characteristic than dizygotic twins. By using correlation to compare the phenotypes of monozygotic and dizygotic twins, we can estimate broad-sense heritability. A rough estimate of the broad-sense heritability can be obtained by taking twice the difference of the correlation coefficients for a quantitative characteristic in monozygotic and dizygotic twins:

$$H^2 = 2(r_{MZ} - r_{DZ}) \qquad (24.20)$$

where r_{MZ} equals the correlation coefficient among monozygotic twins and r_{DZ} equals the correlation coefficient among dizygotic twins. For example, suppose we found the correlation of height among the two members of monozygotic twin pairs (r_{MZ}) to be 0.9 and the correlation of height among the two members of dizygotic twins (r_{DZ}) to be 0.5. The broad-sense heritability for height would be $H^2 = 2(0.9 - 0.5) = 2(0.4) = 0.8$. This calculation assumes that the two individuals of a monozygotic twin pair experience environments that are no more similar to each other than those experienced by the two individuals of a dizygotic twin pair. This assumption is often not met when twins have been reared together.

Narrow-sense heritability also can be estimated by comparing the phenotypic variances for a characteristic in

full siblings (who have both parents in common as well as an average of 50% of their genes) and half siblings (who have only one parent in common and thus an average of 25% of their genes).

All estimates of heritability depend on the assumption that the environments of related individuals are not more similar than those of unrelated individuals. This assumption is difficult to meet in human studies, because related people are usually reared together. Heritability estimates for humans should therefore always be viewed with caution.
▶ TRY PROBLEM 36

The Limitations of Heritability

Knowledge of heritability has great practical value because it allows us to statistically predict the phenotypes of offspring on the basis of their parent's phenotype. It also provides useful information about how characteristics will respond to selection (see Section 24.4). In spite of its importance, heritability is frequently misunderstood. It does not provide information about an individual's genes or the environmental factors that control the development of a characteristic, and it says nothing about the nature of differences between groups. This section outlines some limitations and common misconceptions concerning broad- and narrow-sense heritability.

HERITABILITY DOES NOT INDICATE THE DEGREE TO WHICH A CHARACTERISTIC IS GENETICALLY DETERMINED Heritability is the proportion of the phenotypic variance that is due to genetic variance; it says nothing about the degree to which genes determine a characteristic. Heritability indicates only the degree to which genes determine *variation* in a characteristic. The determination of a characteristic and the determination of variation in a characteristic are two very different things.

Consider polydactyly (the presence of extra digits) in rabbits, which can be caused either by environmental factors or by a dominant gene. Suppose that we have a group of rabbits all homozygous for a gene that produces the usual numbers of digits. None of the rabbits in this group carries a gene for polydactyly, but a few of the rabbits are polydactylous because of environmental factors. Broad-sense heritability for polydactyly in this group is zero, because there is no genetic variation for polydactyly; all of the variation is due to environmental factors. However, it would be incorrect for us to conclude that genes play no role in determining the number of digits in rabbits. Indeed, we know that there are specific alleles that can produce extra digits (though these alleles are not present in the group of rabbits under consideration). Heritability indicates nothing about whether genes control the development of a characteristic; it provides information only about causes of the variation in a characteristic within a defined group.

AN INDIVIDUAL DOES NOT HAVE HERITABILITY Broad- and narrow-sense heritabilities are statistical values based on the genetic and phenotypic variances found in a *group* of individuals. Heritability cannot be calculated for an individual, and heritability has no meaning for a specific individual. Suppose that we calculate the narrow-sense heritability of adult body weight for the students in a biology class and obtain a value of 0.6. We could conclude that 60% of the variation in adult body weight among the students in this class is determined by additive genetic variation. We should not, however, conclude that 60% of any particular student's body weight is due to additive genes.

THERE IS NO UNIVERSAL HERITABILITY FOR A CHARACTERISTIC The value of heritability for a characteristic is specific for a given population in a given environment. Recall that broad-sense heritability is genetic variance divided by phenotypic variance. Genetic variance depends on which alleles are present, which often differs between populations. In the example of polydactyly in rabbits, there were no alleles for polydactyly in the group, so the heritability of the characteristic was zero. A different group of rabbits might contain many alleles for polydactyly, and the heritability of the characteristic might then be high.

Environmental differences can also affect heritability because V_P is composed of both genetic and environmental variance. When the environmental differences that affect a characteristic differ between two groups, the heritabilities for the two groups also often differ.

Because heritability is specific to a defined population in a given environment, it is important not to extrapolate heritabilities from one population to another. For example, human height is determined by environmental factors (such as nutrition and health) and by genes. If we measured the heritability of height in a developed country, we might obtain a value of 0.8, indicating that the variation in height in this population is largely genetic. This population has a high heritability because most people have adequate nutrition and health care (V_E is low), so most of the phenotypic variation in height is genetically determined. It would be incorrect for us to assume that height has a high heritability in all human populations. In developing countries, there may be more variation in a range of environmental factors; some people may enjoy good nutrition and health, whereas others may have a diet deficient in protein and suffer from diseases that affect stature. If we measured the heritability of height in such a country, we would undoubtedly obtain a lower value than we observed in the developed country because there is more environmental variation and the genetic variance in height constitutes a smaller proportion of the phenotypic variation, making the heritability lower. The important point to remember is that heritability must be calculated separately for each population and each environment.

EVEN WHEN HERITABILITY IS HIGH, ENVIRONMENTAL FACTORS CAN INFLUENCE A CHARACTERISTIC High heritability does not mean that environmental factors cannot influence the expression of a characteristic. High heritability indicates only that the environmental variation to which the population is *currently* exposed is not responsible for variation in the characteristic. Let's look again at human height. In most developed countries, the heritability of human height is high, indicating that genetic differences are responsible for most of the variation in height. It would be wrong for us to conclude that human height cannot be changed by the alteration of the environment. Indeed, height decreased in several European cities during World War II owing to hunger and disease, and height can be increased dramatically by the administration of growth hormone to children. The absence of environmental variation in a characteristic does not mean that the characteristic will not respond to environmental change.

HERITABILITIES INDICATE NOTHING ABOUT THE NATURE OF POPULATION DIFFERENCES IN A CHARACTERISTIC A common misconception about heritability is that it provides information about population differences in a characteristic. Heritability is specific for a given population in a given environment, so it cannot be used to draw conclusions about why populations differ in a characteristic.

Suppose that we measured heritability for human height in two groups. One group is from a small town in a developed country, where everyone consumes a high-protein diet. Because there is little variation in the environmental factors that affect human height and there is some genetic variation, the heritability of height in this group is high. The second group comprises the inhabitants of a single village in a developing country. The consumption of protein by these people is only 25% of that consumed by those in the first group, so their average adult height is several centimeters less than that in the developed country. Again, there is little variation in the environmental factors that determine height in this group, because everyone in the village eats the same types of food and is exposed to the same diseases. Because there is little environmental variation and there is some genetic variation, the heritability of height in this group also is high.

Thus, the heritability of height in both groups is high, and the average height in the two groups is considerably different. We might be tempted to conclude that the difference in height between the two groups is genetically based—that the people in the developed country are genetically taller than the people in the developing country. This conclusion is obviously wrong, however, because these differences in height are due largely to diet—an environmental factor. Heritability provides no information about the causes of differences between populations.

These limitations of heritability have often been ignored, particularly in arguments about the possible social implications of genetic differences between humans. Soon after Mendel's principles of heredity were rediscovered, some geneticists began to claim that many human behavioral characteristics are determined entirely by genes. This claim led to debates about whether characteristics such as human intelligence are determined by genes or environment. Many of the early claims of genetically based human behavior were based on poor research; unfortunately, the results of these studies were often accepted at face value and led to a number of eugenic laws that discriminated against certain groups of people. Today, geneticists recognize that many behavioral characteristics are influenced by a complex interaction of genes and environment, and separating genetic effects from those of the environment is very difficult.

The results of a number of modern studies indicate that human intelligence as measured by IQ and other intelligence tests has a moderately high heritability (usually from 0.4 to 0.8). On the basis of this observation, some people have argued that intelligence is innate and that enhanced educational opportunities cannot boost intelligence. This argument is based on the misconception that, when heritability is high, changing the environment will not alter the characteristic. In addition, because heritabilities of intelligence range from 0.4 to 0.8, a considerable amount of the variance in intelligence originates from environmental differences.

Another argument based on a misconception about heritability is that ethnic differences in measures of intelligence are genetically based. Because the results of some genetic studies show that IQ has moderately high heritability and other studies find differences in the average IQ of ethnic groups, some people have suggested that ethnic differences in IQ are genetically based. As in the example of the effects of diet on human height, heritability provides no information about causes of differences among groups; it indicates only the degree to which phenotypic variance within a single group is genetically based. High heritability for a characteristic does not mean that phenotypic differences between ethnic groups are genetic. We should also remember that separating genetic and environmental effects in humans is very difficult, so heritability estimates themselves may be unreliable. ▶TRY PROBLEM 35

CONCEPTS

Heritability provides information only about the degree to which *variation* in a characteristic is genetically determined. There is no universal heritability for a characteristic; heritability is specific for a given population in a specific environment. Environmental factors can potentially affect characteristics with high heritability, and heritability says nothing about the nature of population differences in a characteristic.

✔ CONCEPT CHECK 6

Suppose that you just learned that the narrow-sense heritability of blood pressure measured among a group of African Americans in Detroit, Michigan, is 0.4. What does this heritability tell us about genetic and environmental contributions to blood pressure?

Locating Genes That Affect Quantitative Characteristics

The statistical methods described for use in analyzing quantitative characteristics can be used both to make predictions about the average phenotype expected in offspring and to estimate the overall contribution of genes to variation in the characteristic. These methods do not, however, allow us to identify and determine the influence of individual genes that affect quantitative characteristics. As discussed in the introduction to this chapter, chromosome regions with genes that control polygenic characteristics are referred to as quantitative trait loci. Although quantitative genetics has made important contributions to basic biology and to plant and animal breeding, the past inability to identify QTLs and measure their individual effects severely limited the application of quantitative genetic methods.

MAPPING QTLs In recent years, numerous genetic markers have been identified and mapped with the use of molecular techniques, making it possible to identify QTLs by linkage analysis. The underlying idea is simple: if the inheritance of a genetic marker is associated consistently with the inheritance of a particular characteristic (such as increased height), then that marker must be linked to a QTL that affects height. The key is to have enough genetic markers so that QTLs can be detected throughout the genome. With the introduction of restriction fragment length polymorphisms, microsatellite variations, and single-nucleotide polymorphisms (SNPs; see Chapters 19 and 20), variable markers are now available for mapping QTLs in a number of different organisms (**Figure 24.19**).

24.19 QTL mapping is used to identify genes that influence many important quantitative traits, including muscle mass in pigs. [USDA.]

24.20 Mapping quantitative train loci by linkage analysis can help identify genes that help determine differences in quantitative traits. Genotypes at the *C* locus are associated with the inheritance of differences in plant height, indicating that a QTL for height (the *T* locus) is closely linked to the *C* locus.

A common procedure for mapping QTLs is to cross two homozygous strains that differ in alleles at many loci (**Figure 24.20**). The resulting F_1 progeny are then intercrossed or backcrossed to allow the genes to recombine through independent assortment and crossing over. Genes on different chromosomes and genes that are far apart on the same chromosome will recombine freely; genes that are closely linked will be inherited together. The offspring are measured for one or more quantitative characteristics; at the same time, they are genotyped for numerous genetic markers that span the genome. Any correlation between the inheritance of a particular marker allele and a quantitative phenotype indicates that a QTL is linked to that marker. If enough markers are used, the detection of all the QTLs affecting a characteristic is theoretically possible. It is important to recognize that a QTL is not a gene; rather, it is a map location for a chromosome region that is associated with that trait. After a QTL has been identified, it can be studied for the presence of one or more

Table 24.2 Examples of quantitative characteristics for which QTLs have been detected

Organism	Quantitative Characteristic
Tomato	Soluble solids
	Fruit mass
	Fruit pH
	Growth
	Leaflet shape
	Height
Corn	Height
	Leaf length
	Tiller number
	Glume hardness
	Grain yield
	Number of ears
	Thermotolerance
Common bean	Number of nodules
Mung bean	Seed weight
Cow pea	Seed weight
Wheat	Preharvest sprout
Pig	Growth
	Length of small intestine
	Average back fat
	Abdominal fat
Mouse	Epilepsy
Rat	Hypertension

Source: After S. D. Tanksley, Mapping polygenes, *Annual Review of Genetics* 27:218, 1993.

specific genes or other sequences that influence the quantitative trait. The introduction to this chapter describes how this approach was used to identify a major gene that affects oil production in corn. QTL mapping has been used to detect genes affecting a variety of characteristics in plant and animal species (**Table 24.2**).

GENOMEWIDE ASSOCIATION STUDIES The traditional method of identifying QTLs is to carry out crosses between varieties that differ in a quantitative trait and then genotype numerous progeny for many markers. Although effective, this method is slow and labor intensive.

An alternative technique for identifying genes that affect quantitative traits is to conduct genomewide association studies, which were introduced in Chapter 7. Unlike traditional linkage analysis, which examines the association of a trait and gene markers among the *progeny of a cross,* genomewide association studies look for associations between traits and genetic markers in a *biological population,* a group of interbreeding individuals. The presence of an association between genetic markers and a trait indicates that the genetic markers are closely linked to one or more genes that affect variation in the trait. Genomewide association studies have been

facilitated by the identification of single-nucleotide polymorphisms, which are positions in the genome where individual organisms vary in a single base pair (see Chapter 20). Many SNPs have been identified in organisms through genomic sequencing. Individual organisms can often be quickly and inexpensively genotyped for numerous SNPs, providing the genetic markers necessary to conduct genomewide association studies.

Genomewide association studies have been widely used to locate genes that affect quantitative traits in humans, including disease susceptibility, obesity, intelligence, and height. A number of quantitative traits in plants have also been studied, including kernel composition, size, color and taste, disease resistance, and starch quality. Genomewide association studies in domestic animals have identified chromosomal segments affecting body weight, body composition, reproductive traits, hormone levels, hair characteristics, and behaviors.

CONCEPTS

The availability of numerous genetic markers revealed by molecular methods makes it possible to map chromosome segments containing genes that contribute to polygenic characteristics. Genomewide association studies locate genes that affect quantitative traits by detecting associations between genetic markers and a trait within a population of individuals.

24.4 Genetically Variable Traits Change in Response to Selection

Evolution is genetic change taking place among members of a population with the passage of time. Several different forces are potentially capable of bringing about evolution, and we will explore these forces and the process of evolution more fully in Chapter 25. Here, we consider how one of these forces—natural selection—can bring about genetic change in a quantitative characteristic.

Charles Darwin proposed the idea of natural selection in his book *On the Origin of Species* in 1859. **Natural selection** arises through the differential reproduction of individuals with different genotypes. Because of the genes that they possess, some individuals produce more offspring than others. The more-successful reproducers give rise to more offspring, which inherit the genes that confer a reproductive advantage. Thus, the frequencies of the genes that confer a reproductive advantage increase with the passage of time and the population evolves. Natural selection is among the most important of the forces that bring about evolutionary change. Through natural selection, organisms become genetically suited to their environments; as environments change, groups of organisms change in ways that make them better able to survive and reproduce.

For thousands of years, humans have practiced a form of selection by promoting the reproduction of organisms with traits perceived as desirable. This form of selection is **artificial** selection, and it has produced the domestic plants and animals that make modern agriculture possible. The power of artificial selection, the first application of genetic principles by humans, is illustrated by the tremendous diversity of shapes, colors, and behaviors of modern domesticated dogs (**Figure 24.21**).

Predicting the Response to Selection

When a quantitative characteristic is subjected to natural or artificial selection, it frequently changes with the passage of time, provided that there is genetic variation for that characteristic in the population. Suppose that a dairy farmer wants to increase milk production among the cows in his herd. Variation at several loci potentially affects milk production in cows; some alleles at these loci confer high milk production whereas other alleles confer low milk production. The dairy farmer breeds only those cows in his herd that have the highest milk production. If there is genetic variation in milk production (i.e., there are different alleles at the loci that control milk production), the mean milk production in the offspring of the selected cows should be higher than the mean milk production of the original herd. This increased production is due to the fact that the selected cows possess more alleles for high milk production than does the average cow, and these alleles are passed on to the offspring. The offspring of the selected cows possess a higher proportion of alleles for greater milk yield and therefore produce more milk than the average cow in the initial herd.

The extent to which a characteristic subjected to selection changes in one generation is termed the **response to selection**. Suppose that the average cow in a dairy herd produces 80 liters of milk per week. A farmer selects for increased milk production by breeding the highest milk producers, and the progeny of these selected cows produce 100 liters of milk per week on average. The response to selection is calculated by subtracting the mean phenotype of the original population (80 liters) from the mean phenotype of the offspring (100 liters), obtaining a response to selection of 100 − 80 = 20 liters per week.

FACTORS INFLUENCING RESPONSE TO SELECTION

The response to selection is determined primarily by two factors. First, it is affected by narrow-sense heritability, which largely determines the degree of resemblance between parents and offspring. When the narrow-sense heritability is high, offspring will tend to resemble their parents; conversely, when the narrow-sense heritability is low, there will be little resemblance between parents and offspring.

The second factor that determines the response to selection is how much selection there is. If the farmer is very stringent in the choice of parents and breeds only the highest milk producers in the herd (say, the top two cows), then all the offspring will receive genes for high-quality milk

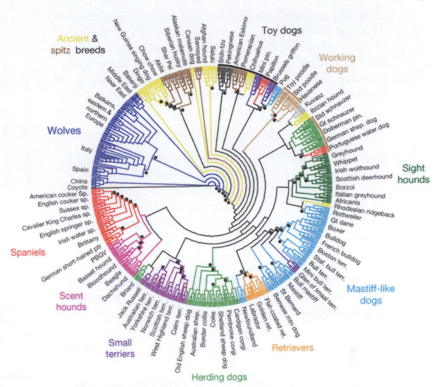

24.21 Artificial selection has produced the tremendous diversity of shape, size, color, and behavior seen today among breeds of domestic dogs. This diagram depicts the evolutionary relationships among wolves and different breeds of dogs from analyses of DNA sequences. [Reprinted by permisson from Macmillan Publishers Ltd.: B.M. von Holdt et al, *Nature*, 464 (7290), 898–902. © 2010.]

production. If the farmer is less selective and breeds the top 20 milk producers in the herd, then the offspring will not carry as many superior genes for high milk production, and on average they will not produce as much milk as the offspring of the top two producers. The response to selection depends on the phenotypic difference of the individuals that are selected as parents; this phenotypic difference is measured by the **selection differential**, defined as the difference between the mean phenotype of the selected parents and the mean phenotype of the original population. If the average milk production of the original herd is 80 liters and the farmer breeds cows with an average milk production of 120 liters, then the selection differential is 120 − 80 = 40 liters.

CALCULATION OF RESPONSE TO SELECTION

The response to selection (R) depends on the narrow-sense heritability (h^2) and the selection differential (S):

$$R = h^2 \times S \qquad (24.21)$$

This equation can be used to predict the magnitude of change in a characteristic when a given selection differential is applied. G. A. Clayton and his colleagues estimated the response to selection that would take place in the abdominal bristle number of *Drosophila melanogaster*. By using several different methods, including parent–offspring regression, they first estimated the narrow-sense heritability of the abdominal bristle number in one population of fruit flies to be 0.52. The mean number of bristles in the original population was 35.3. They selected individual flies with a mean bristle number of 40.6 and intercrossed them to produce the next generation. The selection differential was $40.6 - 35.3 = 5.3$; so they predicted a response to selection to be

$$R = 0.52 \times 5.3 = 2.8$$

The response to selection of 2.8 is the expected increase in the characteristic of the offspring above the mean of the original population. They therefore expected the average number of abdominal bristles in the offspring of their selected flies to be $35.3 + 2.8 = 38.1$. Indeed, they found an average bristle number of 37.9 in these flies.

ESTIMATING HERITABILITY FROM RESPONSE TO SELECTION Rearranging Equation 24.21 provides another way to calculate narrow-sense heritability:

$$h^2 = \frac{R}{S} \qquad (24.22)$$

In this way, h^2 can be calculated by conducting a response-to-selection experiment. First, the selection differential is obtained by subtracting the population mean from the mean of selected parents. The selected parents are then interbred, and the mean phenotype of their offspring is measured. The difference between the mean of the offspring and that of the initial population is the response to selection, which can be used with the selection differential to estimate heritability. Heritability determined by a response-to-selection experiment is usually termed the **realized heritability**. If certain assumptions are met, the realized heritability is identical with the narrow-sense heritability.

One of the longest selection experiments is a study of oil and protein content in corn seeds (**Figure 24.22**). This experiment began at the University of Illinois on 163 ears of corn with an oil content ranging from 4% to 6%. Corn plants with high oil content and those with low oil content were selected and interbred. Response to selection for increased oil content (the upper line in Figure 24.22) reached about 20%, whereas response to selection for decreased oil content reached a lower limit near zero. Genetic analyses of the high- and low-oil-content strains revealed that at least 20 loci take part in determining oil content, one of which we explored in the introduction to this chapter. ▶ TRY PROBLEM 40

CONCEPTS

The response to selection is influenced by narrow-sense heritability and the selection differential.

✓ CONCEPT CHECK 7

The narrow-sense heritability for a trait is 0.4 and the selection differential is 0.5. What is the predicted response to selection?

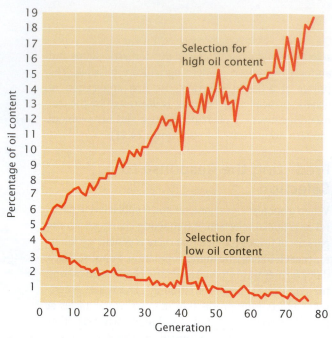

24.22 In a long-term response-to-selection experiment, selection for oil content in corn increased oil content in one line to about 20%, whereas it almost eliminated it in another line.

Limits to Selection Response

When a characteristic has been selected for many generations, the response may eventually level off, and the characteristic no longer responds to selection (**Figure 24.23**). A potential reason for this leveling off is that the genetic variation in the population may be exhausted; at some point, all individuals in the population have become homozygous for alleles that encode the selected trait. When there is no more additive genetic variation, heritability equals zero, and no further response to selection can take place.

Sometimes, the response to selection may level off even while some genetic variation remains in the population. This leveling off takes place because natural selection opposes

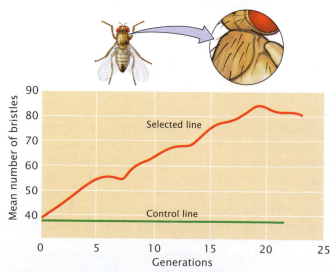

24.23 The response of a population to selection often levels off at some point in time. In a response-to-selection experiment for increased abdominal chaetae bristle number in female fruit flies, the number of bristles increased steadily for about 20 generations and then leveled off.

further change in the characteristic. The response to selection for small body size in mice, for example, eventually levels off because the smallest animals are sterile and cannot pass on their genes for small body size. In this case, artificial selection for small size is opposed by natural selection for fertility, and the population can no longer respond to the artificial selection.

Correlated Responses

Often when a specific trait is selected, other traits change at the same time. This type of associated response is due to the fact that the traits are encoded by the same genes.

Two or more characteristics are often correlated. For example, in many plants, plant size and number of seeds produced exhibit a positive correlation: larger plants, on average, produce more seeds than smaller plants. This correlation is a **phenotypic correlation** because the association is between two phenotypes of the same individual. Phenotypic correlations may be due to environmental or genetic correlations. Environmental correlations refer to two or more characteristics that are influenced by the same environmental factor. Moisture availability, for example, may affect both the size of a plant and the number of seeds produced by the plant. Plants growing in environments with lots of water are large and produce many seeds, whereas plants growing in environments with limited water are small and have few seeds.

Alternatively, a phenotypic correlation may result from a **genetic correlation**, which means that the genes affecting two characteristics are associated. The primary genetic cause of phenotypic correlations is pleiotropy, which is due to the effect of one gene on two or more characteristics (see Chapter 5). In humans, for example, many body structures respond to growth hormone, and there are genes that affect the amount of growth hormone secreted by the pituitary gland. People with certain genes produce high levels of growth hormone, which increases both height and hand size. Others possess genes that produce lower levels of growth hormone, which leads to both short stature and small hands. Height and hand size are therefore phenotypically correlated in humans, and this correlation is due to a genetic correlation—the fact that both characteristics are affected by the same genes that control the amount of growth hormone. Genetically speaking, height and hand size are the same characteristic because they are the phenotypic manifestation of a single set of genes. When two characteristics are influenced by the same genes they are genetically correlated.

Genetic correlations are quite common (**Table 24.3**) and may be positive or negative. A positive genetic correlation between two characteristics means that genes that cause an increase in one characteristic also produce an increase in the other characteristic. Thorax length and wing length in *Drosophila* are positively correlated because the genes that increase thorax length also increase wing length. A negative genetic correlation means that genes that cause an increase in one characteristic produce a decrease in the other characteristic. Milk yield and percentage of butterfat are negatively correlated in cattle: genes that cause higher milk production result in milk with a lower percentage of butterfat.

Table 24.3	Genetic correlations in various organisms	
Organism	**Characteristics**	**Genetic Correlation**
Cattle	Milk yield and percentage of butterfat	−0.38
Pig	Weight gain and back-fat thickness	0.13
	Weight gain and efficiency	0.69
Chicken	Body weight and egg weight	0.42
	Body weight and egg production	−0.17
	Egg weight and egg production	−0.31
Mouse	Body weight and tail length	0.29
Fruit fly	Abdominal bristle number and sternopleural bristle number	0.41

Source: After D. S. Falconer and T. F. C. Mackay. *Introduction to Quantitative Genetics* (Pearson, New York, 1996), p. 314.

Genetic correlations are important in animal and plant breeding because they produce a correlated response to selection, which means that, when one characteristic is selected, genetically correlated characteristics also change. Correlated responses to selection are due to the fact that both characteristics are influenced by the same genes; selection for one characteristic causes a change in the genes affecting that characteristic, and these genes also affect the second characteristic, causing it to change at the same time. Correlated responses may well be undesirable and may limit the ability to alter a characteristic by selection. From 1944 to 1964, domestic turkeys were subjected to intense selection for growth rate and body size. At the same time, fertility, egg production, and egg hatchability all declined. These correlated responses were due to negative genetic correlations between body size and fertility; eventually, these genetic correlations limited the extent to which the growth rate of turkeys could respond to selection. Genetic correlations may also limit the ability of natural populations to respond to selection in the wild and adapt to their environments. ▶ TRY PROBLEM 45

CONCEPTS

Genetic correlations result from pleiotropy. When two characteristics are genetically correlated, selection for one characteristic will produce a correlated response in the other characteristic.

✓ CONCEPT CHECK 8

In a herd of dairy cattle, milk yield and the percentage of butterfat exhibit a genetic correlation of −0.38. If greater milk yield is selected in this herd, what will be the effect on the percentage of butterfat?

CONCEPTS SUMMARY

■ Quantitative genetics focuses on the inheritance of complex characteristics whose phenotypes vary continuously. For many quantitative characteristics, the relation between genotype and phenotype is complex because many genes and environmental factors influence a characteristic.

■ The individual genes that influence a polygenic characteristic follow the same Mendelian principles that govern discontinuous characteristics but, because many genes participate, the expected ratios of phenotypes are obscured.

■ A population is the group of interest, and a sample is a subset of the population used to describe it.

■ A frequency distribution, in which the phenotypes are represented on one axis and the number of individuals possessing each phenotype is represented on the other axis, is a convenient means of summarizing phenotypes found in a group of individuals.

■ The mean and variance provide key information about a distribution: the mean gives the central location of the distribution, and the variance provides information about how the phenotype varies within a group.

■ The correlation coefficient measures the direction and strength of association between two variables. Regression can be used to predict the value of one variable on the basis of the value of a correlated variable.

■ Phenotypic variance in a characteristic can be divided into components that are due to additive genetic variance, dominance genetic variance, genic interaction variance, environmental variance, and genetic–environmental interaction variance.

■ Broad-sense heritability is the proportion of the phenotypic variance due to genetic variance; narrow-sense heritability is the proportion of the phenotypic variance due to additive genetic variance.

■ Heritability provides information only about the degree to which variation in a characteristic results from genetic differences. Heritability is based on the variances present within a group of individuals, and an individual does not have heritability. The heritability of a characteristic varies among populations and among environments. Even if the heritability for a characteristic is high, the characteristic may still be altered by changes in the environment. Heritabilities provide no information about the nature of population differences in a characteristic.

■ Quantitative trait loci (QTL) are chromosome segments containing genes that control polygenic characteristics. QTLs can be mapped by examining the association between the inheritance of a quantitative characteristic and the inheritance of genetic markers. Genes influencing quantitative traits can also be located with the use of genomewide association studies.

■ The amount that a quantitative characteristic changes in a single generation when subjected to selection (the response to selection) is directly related to the selection differential and narrow-sense heritability.

■ A genetic correlation may be present when the same gene affects two or more characteristics (pleiotropy). Genetic correlations produce correlated responses to selection.

IMPORTANT TERMS

quantitative genetics (p. 683)
quantitative trait locus (QTL) (p. 683)
meristic characteristic (p. 686)
threshold characteristic (p. 686)
frequency distribution (p. 689)
normal distribution (p. 690)
population (p. 690)

sample (p. 690)
mean (p. 690)
variance (p. 691)
standard deviation (p. 691)
correlation (p. 692)
correlation coefficient (p. 692)
regression (p. 693)
regression coefficient (p. 693)
heritability (p. 696)
phenotypic variance (p. 697)

genetic variance (p. 697)
environmental variance (p. 697)
genetic–environmental interaction variance (p. 697)
additive genetic variance (p. 697)
dominance genetic variance (p. 697)
genic interaction variance (p. 697)

broad-sense heritability (p. 698)
narrow-sense heritability (p. 698)
natural selection (p. 704)
artificial selection (p. 704)
response to selection (p. 705)
selection differential (p. 705)
realized heritability (p. 706)
phenotypic correlation (p. 707)
genetic correlation (p. 707)

ANSWERS TO CONCEPT CHECKS

1. Cross two individuals that are each homozygous for different genes affecting the traits and then intercross the resulting F_1 progeny to produce the F_2. Determine what proportion of the F_2 progeny resembles one of the original homozygotes in the P generation. This proportion should be $(^1/_4)^n$, where n equals the number of loci with a segregating pair of alleles that affect the characteristic.

2. d

3. d

4. b

5. a

6. It indicates that about 40% of the differences in blood pressure among African Americans in Detroit are due to additive genetic differences. It neither provides information about the heritability of blood pressure in other groups of people nor indicates anything about the nature of differences in blood pressure between African Americans in Detroit and people in other groups.

7. 0.2

8. The percentage of butterfat will decrease.

WORKED PROBLEMS

Problem 1

Seed weight in a particular plant species is determined by pairs of alleles at two loci (a^+a^- and b^+b^-) that are additive and equal in their effects. Plants with genotype $a^-a^-\ b^-b^-$ have seeds that average 1 g in weight, whereas plants with genotype $a^+a^+\ b^+b^+$ have seeds that average 3.4 g in weight. A plant with genotype $a^-a^-\ b^-b^-$ is crossed with a plant of genotype $a^+a^+\ b^+b^+$.

  a. What is the predicted weight of seeds from the F_1 progeny of this cross?

  b. If the F_1 plants are intercrossed, what are the expected seed weights and proportions of the F_2 plants?

Solution Strategy

What information is required in your answer to the problem?

  ▪ The predicted weight of seeds from the F_1 progeny.

  ▪ The expected seed weights and their proportions among the F_2 plants?

What information is provided to solve the problem?

  ▪ Seed weight is determined by pairs of alleles at two loci (a^+a^- and b^+b^-).

  ▪ The alleles are additive and equal in their effects.

  ▪ Plants with genotype $a^-a^-\ b^-b^-$ have seeds that average 1 g in weight.

  ▪ Plants with genotype $a^+a^+\ b^+b^+$ have seeds that average 3.4 g in weight.

  ▪ A plant with genotype $a^-a^-\ b^-b^-$ is crossed with a plant of genotype $a^+a^+\ b^+b^+$.

For help with this problem, review:

Polygenic Inheritance in Section 24.1.

Solution Steps

The difference in average seed weight of the two parental genotypes is 3.4 g − 1 g = 2.4 g. These two genotypes differ in four genes, so each gene difference contributes an additional 2.4 g/4 = 0.6 g of weight to the 1-g weight of a plant ($a^-a^-\ b^-b^-$) that has none of the contributing genes.

> **Hint:** Because the alleles are equal and additive, each allele contributes the same amount to height.

The cross between the two homozygous genotypes produces the F_1 and F_2 progeny shown on next page.

  a. The F_1 are heterozygous at both loci ($a^+a^-\ b^+b^-$) and possess two genes that contribute an additional 0.6 g each to the 1-g weight of a plant that has no contributing genes. Therefore, the seeds of the F_1 should average 1 g + 2(0.6 g) = 2.2 g.

  b. The F_2 will have the following phenotypes and proportions: $^1/_{16}$ 1 g; $^4/_{16}$ 1.6 g; $^6/_{16}$ 2.2 g; $^4/_{16}$ 2.8 g; and $^1/_{16}$ 3.4 g.

P $\qquad a^-a^-\ b^-b^-\qquad\times\qquad a^+a^+\ b^+b^+$
$\qquad\qquad$ 1 g $\qquad\qquad\qquad$ 3.4 g

F$_1$ $\qquad\qquad a^+a^-\ b^+b^-$
$\qquad\qquad\qquad$ 2.2 g

Recall: The probability of each two-locus genotype can be determined by multiplying the probability of the single locus genotypes.

Genotype	Probability	Number of contributing genes	Average seed weight
$a^-a^-\ b^-b^-$	$\frac{1}{4}\times\frac{1}{4}=\frac{1}{16}$	0	$1\text{ g} + (0\times0.6\text{ g}) = 1\text{ g}$
$a^+a^-\ b^-b^-$	$\frac{1}{2}\times\frac{1}{4}=\frac{1}{8}$ $\Big\}\frac{2}{8}=\frac{4}{16}$		
$a^-a^-\ b^+b^-$	$\frac{1}{4}\times\frac{1}{2}=\frac{1}{8}$	1	$1\text{ g} + (1\times0.6\text{ g}) = 1.6\text{ g}$
$a^+a^+\ b^-b^-$	$\frac{1}{4}\times\frac{1}{4}=\frac{1}{16}$		
$a^-a^-\ b^+b^+$	$\frac{1}{4}\times\frac{1}{4}=\frac{1}{16}$ $\Big\}\frac{2}{16}+\frac{1}{4}+=\frac{6}{16}$	2	$1\text{ g} + (2\times0.6\text{ g}) = 2.2\text{ g}$
$a^+a^-\ b^+b^-$	$\frac{1}{2}\times\frac{1}{2}=\frac{1}{4}$		
$a^+a^+\ b^+b^-$	$\frac{1}{4}\times\frac{1}{2}=\frac{1}{8}$ $\Big\}\frac{2}{8}=\frac{4}{16}$		
$a^+a^-\ b^+b^+$	$\frac{1}{4}\times\frac{1}{2}=\frac{1}{8}$	3	$1\text{ g} + (3\times0.6\text{ g}) = 2.8\text{ g}$
$a^+a^+\ b^+b^+$	$\frac{1}{4}\times\frac{1}{4}=\frac{1}{16}$	4	$1\text{ g} + (4\times0.6\text{ g}) = 3.4\text{ g}$

(F$_2$ label appears at left of the middle rows)

Problem 2

A farmer is raising rabbits. The average body weight in his population of rabbits is 3 kg. The farmer selects the 10 largest rabbits in his population, whose average body weight is 4 kg, and interbreeds them. If the heritability of body weight in the rabbit population is 0.7, what is the expected body weight among offspring of the selected rabbits?

Solution Strategy

What information is required in your answer to the problem?

The expected weight of the offspring of the selected rabbits.

What information is provided to solve the problem?

- The average body weight in the population is 3 kg.

- The average body weight of selected rabbits is 4 kg.

- The heritability of body weight is 0.7.

For help with this problem, review:

Predicting the Response to Selection in Section 24.4.

Solution Steps

The farmer has carried out a response-to-selection experiment. The selection differential equals the difference in average weights of the selected rabbits and the entire population: $4\text{ kg} - 3\text{ kg} = 1\text{ kg}$. The narrow-sense heritability is given as 0.7; so the expected response to selection is: $R = h^2 \times S = 0.7 \times 1\text{ kb} = 0.7\text{ kg}$. This value is the increase in weight that is expected in the offspring of the selected parents; so the average weight of the offspring is expected to be: $3\text{ kg} + 0.7\text{ kg} = 3.7\text{ kg}$.

Recall: The response to selection equals the selection differential multiplied by the narrow-sense heritability.

COMPREHENSION QUESTIONS

Section 24.1

1. How does a quantitative characteristic differ from a discontinuous characteristic?

2. Briefly explain why the relation between genotype and phenotype is frequently complex for quantitative characteristics.

3. Why do polygenic characteristics have many phenotypes?

Section 24.2

4. Explain the relation between a population and a sample. What characteristics should a sample have to be representative of the population?

5. What information do the mean and variance provide about a distribution?

6. How is the standard deviation related to the variance?

7. What information does the correlation coefficient provide about the association between two variables?

8. What is regression? How is it used?

Section 24.3

9. List all the components that contribute to the phenotypic variance and define each component.

10. How do broad-sense and narrow-sense heritabilities differ?

11. Briefly outline some of the ways in which heritability can be calculated.

12. Briefly describe common misunderstandings or misapplications of the concept of heritability.

13. Briefly explain how genes affecting a polygenic characteristic are located with the use of QTL mapping.

Section 24.4

14. How is the response to selection related to narrow-sense heritability and the selection differential? What information does the response to selection provide?

15. Why does the response to selection often level off after many generations of selection?

> ▶ For more questions that test your comprehension of the key chapter concepts, go to **LEARNING***Curve* for this chapter.

APPLICATION QUESTIONS AND PROBLEMS

Section 24.1

*16. For each of the following characteristics, indicate whether it would be considered a discontinuous characteristic or a quantitative characteristic. Briefly justify your answer.

a. Kernel color in a strain of wheat, in which two codominant alleles segregating at a single locus determine the color. Thus, there are three phenotypes present in this strain: white, light red, and medium red.

b. Body weight in a family of Labrador retrievers. An autosomal recessive allele that causes dwarfism is present in this family. Two phenotypes are recognized: dwarf (less than 13 kg) and normal (greater than 23 kg).

c. Presence or absence of leprosy. Susceptibility to leprosy is determined by multiple genes and numerous environmental factors.

d. Number of toes in guinea pigs, which is influenced by genes at many loci.

e. Number of fingers in humans. Extra (more than five) fingers are caused by the presence of an autosomal dominant allele.

*17. Assume that plant weight is determined by a pair of alleles at each of two independently assorting loci (A and a, B and b) that are additive in their effects. Further assume that each allele represented by an uppercase letter contributes 4 g to weight and that each allele represented by a lowercase letter contributes 1 g to weight.

a. If a plant with genotype $AA\ BB$ is crossed with a plant with genotype $aa\ bb$, what weights are expected in the F_1 progeny?

b. What is the distribution of weight expected in the F_2 progeny?

18. Assume that three loci, each with two alleles (A and a, B and b, C and c), determine the differences in height between two homozygous strains of a plant. These genes are additive and equal in their effects on plant height. One strain ($aa\ bb\ cc$) is 10 cm in height. The other strain ($AA\ BB\ CC$) is 22 cm in height. The two strains are crossed, and the resulting F_1 are interbred to produce F_2 progeny. Give the phenotypes and the expected proportions of the F_2 progeny.

*19. A farmer has two homozygous varieties of tomatoes. One variety, called Little Pete, has fruits that average only 2 cm in diameter. The other variety, Big Boy, has fruits that average a whopping 14 cm in diameter. The farmer crosses Little Pete and Big Boy; he then intercrosses the F_1 to produce F_2 progeny. He grows 2000 F_2 tomato plants and doesn't find any F_2 offspring that produce fruits as small as Little Pete or as large as Big Boy. If we assume that the differences in fruit size of these varieties are produced by genes with equal and additive effects, what can we conclude about the minimum number of loci with pairs of alleles determining the differences in fruit size of the two varieties?

20. Seed size in a plant is a polygenic characteristic. A grower crosses two pure-breeding varieties of the plant and measures seed size in the F_1 progeny. She then backcrosses the F_1 plants to one of the parental varieties and measures seed size in the backcross progeny. The grower finds that seed size in the backcross progeny has a higher variance than does seed size in the F_1 progeny. Explain why the backcross progeny are more variable.

Section 24.2

21. The following data are the numbers of digits per foot in 25 guinea pigs. Construct a frequency distribution for these data.

4, 4, 4, 5, 3, 4, 3, 4, 4, 5, 4, 4, 3, 2, 4, 4, 5, 6, 4, 4, 3, 4, 4, 4, 5

*22. Ten male Harvard students were weighed in 1916. Their weights are given here in kilograms. Calculate the mean, variance, and standard deviation for these weights.

51, 69, 69, 57, 61, 57, 75, 105, 69, 63

23. Among a population of tadpoles, the correlation coefficient for size at metamorphosis and time required for metamorphosis is −0.74. On the basis of this

correlation, what conclusions can you make about the relative sizes of tadpoles that metamorphose quickly and those that metamorphose more slowly?

24. A researcher studying alcohol consumption in North American cities finds a significant, positive correlation between the number of Baptist preachers and alcohol consumption. Is it reasonable for the researcher to conclude that the Baptist preachers are consuming most of the alcohol? Why or why not?

*25. Body weight and length were measured on six mosquito fish; these measurements are given in the following table. Calculate the correlation coefficient for weight and length in these fish.

[A. Hartl/age fotostock.]

Wet weight (g)	Length (mm)
115	18
130	19
210	22
110	17
140	20
185	21

26. The heights of mothers and daughters are given in the following table:

Height of mother (in)	Height of daughter (in)
64	66
65	66
66	68
64	65
63	65
63	62
59	62
62	64
61	63
60	62

a. Calculate the correlation coefficient for the heights of the mothers and daughters.

b. Using regression, predict the expected height of a daughter whose mother is 67 inches tall.

Section 24.3

*27. Phenotypic variation in the tail length of mice has the following components:

Additive genetic variance (V_A)	= 0.5
Dominance genetic variance (V_D)	= 0.3
Genic interaction variance (V_I)	= 0.1
Environmental variance (V_E)	= 0.4
Genetic−environmental interaction variance (V_{GE})	= 0.0

a. What is the narrow-sense heritability of tail length?

b. What is the broad-sense heritability of tail length?

28. The narrow-sense heritability of ear length in Reno rabbits is 0.4. The phenotypic variance (V_P) is 0.8, and the environmental variance (V_E) is 0.2. What is the additive genetic variance (V_A) for ear length in these rabbits?

29. Assume that human ear length is influenced by multiple genetic and environmental factors. Suppose you measured ear length on three groups of people, in which group A consists of five unrelated persons, group B consists of five siblings, and group C consists of five first cousins.

a. With the assumption that the environment for each group is similar, which group should have the highest phenotypic variance? Explain why.

b. Is it realistic to assume that the environmental variance for each group is similar? Explain your answer.

30. A characteristic has a narrow-sense heritability of 0.6.

a. If the dominance variance (V_D) increases and all other variance components remain the same, what will happen to the narrow-sense heritability? Will it increase, decrease, or remain the same? Explain.

b. What will happen to the broad-sense heritability? Explain.

c. If the environmental variance (V_E) increases and all other variance components remain the same, what will happen to the narrow-sense heritability? Explain.

d. What will happen to the broad-sense heritability? Explain.

31. Flower color in the varieties of pea plants studied by Mendel is controlled by alleles at a single locus. A group of peas homozygous for purple flowers is grown. Careful study of the plants reveals that all their flowers are purple, but there is some variability in the intensity of the purple color. What would the estimated heritability be for this variation in flower color? Explain your answer.

*32. A graduate student is studying a population of bluebonnets along a roadside. The plants in this population are genetically variable. She counts the seeds produced by 100 plants and measures the mean and variance of seed number. The variance is 20. Selecting one plant, the student takes cuttings from it and cultivates them in the greenhouse, eventually producing many genetically identical clones of the same plant. She then transplants these clones into the roadside population, allows them to grow for 1 year, and then counts the number of seeds produced by each of the cloned plants. The student finds that the variance in seed number among these cloned plants is 5. From the phenotypic variance of the genetically variable and genetically identical plants, she calculates the broad-sense heritability.

[Purestock/Getty Images, Inc.]

a. What is the broad-sense heritability of seed number for the roadside population of bluebonnets?

b. What might cause this estimate of heritability to be inaccurate?

33. Many researchers have estimated the heritability of human traits by comparing the correlation coefficients of monozygotic and dizygotic twins (see pp. 148–149). One of the assumptions in using this method is that two monozygotic twins experience environments that are no more similar to each other than those experienced by two dizygotic twins. How might this assumption be violated? Give some specific examples of how the environments of two monozygotic twins might be more similar than the environments of two dizygotic twins.

34. What conclusion can you draw from **Figure 24.18** about the proportion of phenotypic variation in shell breadth that is due to genetic differences? Explain your reasoning.

*35. A genetics researcher determines that the broad-sense heritability of height among Southwestern University undergraduate students is 0.90. Which of the following conclusions would be reasonable? Explain your answer.

a. Since Sally is a Southwestern University undergraduate student, 10% of her height is determined by nongenetic factors.

b. Ninety percent of variation in height among all undergraduate students in the United States is due to genetic differences.

c. Ninety percent of the height of Southwestern University undergraduate students is determined by genes.

d. Ten percent of the variation in height among Southwestern University undergraduate students is determined by variation in nongenetic factors.

e. Because the heritability of height among Southwestern University students is so high, any change in the students' environment will have minimal effect on their height.

*36. The length of the middle joint of the right index finger was measured on 10 sets of parents and their adult offspring. The mean parental lengths and the mean offspring lengths for each family are listed in the following table. Calculate the regression coefficient for regression of mean offspring length against mean parental length and estimate the narrow-sense heritability for this characteristic.

Mean parental length (mm)	Mean offspring length (mm)
30	31
35	36
28	31
33	35
26	27
32	30
31	34
29	28
40	38
33	34

37. Assume that in **Figure 24.14**, x equals the mean phenotype of the parents and y equals the mean phenotype of the offspring. Which line represents the highest heritability? Explain your answer.

38. *Drosophila buzzatii* is a fruit fly that feeds on the rotting fruits of cacti in Australia. Timothy Prout and Stuart Barker calculated the heritabilities of body size, as measured by thorax length, for a natural population of *D. buzzatii* raised in the wild and for a population of *D. buzzatii* collected in the wild but raised in the laboratory (T. Prout and J. S. F. Barker. 1989. *Genetics* 123:803–813). They found the following heritabilities.

Population	Heritability of body size ($\pm$ standard error)
Wild population	0.0595 ± 0.0123
Laboratory-reared population	0.3770 ± 0.0203

Why do you think that the heritability measured in the laboratory-reared population is higher than that measured in the natural population raised in the wild?

39. Mr. Jones is a pig farmer. For many years, he has fed his pigs the food left over from the local university cafeteria, which is known to be low in protein, deficient in vitamins, and downright untasty. However, the food is free, and his pigs don't complain. One day a salesman from a feed company visits Mr. Jones. The salesman claims that his company sells a new, high-protein, vitamin-enriched feed that enhances weight gain in pigs. Although the food is expensive, the salesman claims that the increased weight gain of the pigs will more than pay for the cost of the feed, increasing Mr. Jones's profit. Mr. Jones responds that he took a genetics class at university and that he has conducted some genetic experiments on his pigs; specifically, he has calculated the narrow-sense heritability of weight gain for his pigs and found it to be 0.98. Mr. Jones says that this heritability value indicates that 98% of the variance in weight gain among his pigs is determined by genetic differences, and therefore the new pig feed can have little effect on the growth of his pigs. He concludes that the feed would be a waste of his money. The salesman doesn't dispute Mr. Jones' heritability estimate, but he still claims that the new feed can significantly increase weight gain in Mr. Jones' pigs. Who is correct and why?

Section 24.4

*40. Joe is breeding cockroaches in his dorm room. He finds that the average wing length in his population of cockroaches is 4 cm. He chooses six cockroaches that have the largest wings; the average wing length among these selected cockroaches is 10 cm. Joe interbreeds these selected cockroaches. From earlier studies, he knows that the narrow-sense heritability for wing length in his population of cockroaches is 0.6.

a. Calculate the selection differential and expected response to selection for wing length in these cockroaches.

b. What should be the average wing length of the progeny of the selected cockroaches?

41. Three characteristics in beef cattle—body weight, fat content, and tenderness—are measured, and the following variance components are estimated:

	Body weight	Fat content	Tenderness
V_A	22	45	12
V_D	10	25	5
V_I	3	8	2
V_E	42	64	8
V_{GE}	0	0	1

In this population, which characteristic would respond best to selection? Explain your reasoning.

42. A rancher determines that the average amount of wool produced by a sheep in her flock is 22 kg per year. In an attempt to increase the wool production of her flock, the rancher picks five male and five female sheep with the greatest wool production; the average amount of wool produced per sheep by those selected is 30 kg. She interbreeds these selected sheep and finds that the average wool production among the progeny of the selected sheep is 28 kg. What is the narrow-sense heritability for wool production among the sheep in the rancher's flock?

43. A strawberry farmer determines that the average weight of individual strawberries produced by plants in his garden is 2 g. He selects the 10 plants that produce the largest strawberries; the average weight of strawberries among these selected plants is 6 g. He interbreeds these selected plants. The progeny of these selected plants produce strawberries that weigh 5 g. If the farmer were to select plants that produce an average strawberry weight of 4 g, what would be the predicted weight of strawberries produced by the progeny of these selected plants?

44. Has the response to selection leveled off in the strain of corn selected for high oil content shown in **Figure 24.22**? What does this observation suggest about genetic variation in the strain selected for high oil content?

*45. The narrow-sense heritability of wing length in a population of *Drosophila melanogaster* is 0.8. The narrow-sense heritability of head width in the same population is 0.9. The genetic correlation between wing length and head width is −0.86. If a geneticist selects for increased wing length in these flies, what will happen to head width?

46. Pigs have been domesticated from wild boars. Would you expect to find higher heritability for weight among domestic pigs or wild boars? Explain your answer.

CHALLENGE QUESTIONS

Section 24.1

47. Bipolar illness is a psychiatric disorder with a strong hereditary basis, but the exact mode of inheritance is not known. Research has shown that siblings of patients with bipolar illness are more likely to develop the disorder than are siblings of unaffected persons. Findings from one study demonstrated that the ratio of bipolar brothers to bipolar sisters is higher when the patient is male than when the patient is female. In other words, relatively more brothers of bipolar patients also have the disease when the patient is male than when the patient is female. What does this observation suggest about the inheritance of bipolar illness?

Section 24.3

48. We have explored some of the difficulties in separating the genetic and environmental components of human behavioral characteristics. Considering these difficulties and what you know about calculating heritability, propose an experimental design for accurately measuring the heritability of musical ability.

49. A student who has just learned about quantitative genetics says, "Heritability estimates are worthless! They don't tell you anything about the genes that affect a characteristic. They don't provide any information about the types of offspring to expect from a cross. Heritability estimates measured in one population can't be used for other populations, so they don't even give you any general information about how much of a characteristic is genetically determined. Heritabilities don't do anything but make undergraduate students sweat during tests." How would you respond to this statement? Is the student correct? What good are heritabilities, and why do geneticists bother to calculate them?

Section 24.4

50. Eugene Eisen selected for increased 12-day litter weight (total weight of a litter of offspring 12 days after birth) in a population of mice (E. J. Eisen. 1972. *Genetics* 72:129–142). The 12-day litter weight of the population steadily increased but then leveled off after about 17 generations. At generation 17, Eisen took one family of mice from the selected population and reversed the selection procedure: in this group, he selected for *decreased* 12-day litter weight. This group immediately responded to decreased selection; the 12-day litter weight dropped 4.8 g within one generation and dropped 7.3 g after 5 generations. On the basis of the results of the reverse selection, what is the most likely explanation for the leveling off of 12-day litter weight in the original population?

[J & C Sohns/age fotostock.]

> Go to your ▶LaunchPad to find additional learning resources and the Suggested Readings for this chapter.

25

Population Genetics

Genetic Rescue of Bighorn Sheep

Rocky Mountain bighorn sheep (*Ovis canadensis*). A population of bighorn sheep at the National Bison Range suffered the loss of genetic variation due to genetic drift; the introduction of sheep from other populations dramatically increased genetic variation and the fitness of the sheep. [© Kitchin & Hurst/age fotostock America, Inc.]

Rocky Mountain bighorn sheep (*Ovis canadensis*) are among North America's most-spectacular animals, characterized by the male's magnificent horns that curve gracefully back over the ears, spiraling down and back up beside the face. Two hundred years ago, bighorn sheep were numerous throughout western North America, ranging from Mexico to southern Alberta and from Colorado to California. Meriwether Lewis and William Clark reported numerous sightings of these beautiful animals in their expedition across the western United States from 1804 to 1806. Before 1900, there were about 2 million bighorn sheep in North America.

Unfortunately, settlement of the west by Europeans was not kind to the bighorns. Beginning in the late 1800s, hunting, loss of habitat, competition from livestock, and diseases carried by domestic sheep decimated the bighorns. Today, fewer than 70,000 bighorn sheep remain, scattered across North America in fragmented and isolated populations.

In 1922, wildlife biologists established a population of bighorn sheep at the National Bison Range, an isolated tract of 18,000 acres nestled between the mountains of northwestern Montana. In that year, 12 bighorn sheep—four males (rams) and eight females (ewes)—were trapped at Banff National Park in Canada and transported to the National Bison Range. No additional animals were introduced to this population for the next 60 years.

At first, the population of bighorns at the National Bison Range flourished, protected from hunting and livestock. Within 8 years, the population had grown to 90 sheep but then began to slowly decrease in size. Population size waxed and waned through the years, but the number of sheep had dropped to about 50 by 1985, and the population was in trouble. The amount of genetic variation was low compared with other native populations of bighorn sheep. The reproductive rate of both male and female sheep had dropped, and the size and survival of the sheep were lower than in healthier populations. The population at the National Bison Range was suffering from genetic drift, an evolutionary force operating in small populations that causes random changes in the gene pool and the loss of genetic variation.

715

To counteract the negative effects of genetic drift, biologists added five new rams from other herds in Montana and Wyoming in 1985, mimicking the effects of natural migration among herds. Another 10 sheep were introduced between 1990 and 1994. This influx of new genes had a dramatic effect on the genetic health of the population. Genetic variation among individual sheep increased significantly. Outbred rams (those containing the new genes) were more dominant, more likely to copulate, and more likely to produce offspring. Outbred ewes had more than twice the annual reproductive success of inbred females. Adult survival increased after the introduction of new genes, and, slowly, the population grew in size, reaching 69 sheep by 2003.

The bighorn sheep at the National Bison Range illustrate an important principle of genetics: small populations lose genetic variation with the passage of time through genetic drift, often with catastrophic consequences for survival and reproduction. The introduction of new genetic variation into an inbred population, called **genetic rescue**, often dramatically improves the health of the population and can better ensure its long-term survival. These effects have important implications for wildlife management, as well as for how organisms evolve in the natural world.

This chapter introduces *population genetics*, the branch of genetics that studies the genetic makeup of groups of individuals and how a group's genetic composition changes with time. Population geneticists usually focus their attention on a **Mendelian population**, which is a group of interbreeding, sexually reproducing individuals that have a common set of genes—the **gene pool**. A population evolves through changes in its gene pool; therefore, population genetics is also the study of evolution. Population geneticists study the variation in alleles within and between groups and the evolutionary forces responsible for shaping the patterns of genetic variation found in nature. In this chapter, we will learn how the gene pool of a population is measured and what factors are responsible for shaping it.

(a)

(b)

25.1 All organisms exhibit genetic variation. (a) Extensive variation among humans. (b) Variation in the spotting patterns of Asian lady beetles. [Part a: Michael Dwyer/Alamy.]

25.1 Genotypic and Allelic Frequencies Are Used to Describe the Gene Pool of a Population

An obvious and pervasive feature of life is variability. Students in a typical college class will vary in eye color, hair color, skin pigmentation, height, weight, facial features, blood type, and susceptibility to numerous diseases and disorders. No two students in the class are likely to be the same in appearance.

Humans are not unique in their extensive variability (**Figure 25.1a**); almost all organisms exhibit variation in phenotype. For instance, lady beetles are highly variable in their spotting patterns (**Figure 25.1b**), mice vary in body size, snails have different numbers of stripes on their shells, and plants vary in their susceptibility to pests. Much of this phenotypic variation is hereditary. Recognition of the extent of phenotypic variation led Charles Darwin to the idea of evolution through natural selection. Genetic variation is the basis of all evolution, and the extent of genetic variation within a population affects its potential to adapt to environmental change.

In fact, even more genetic variation exists in populations than is visible in the phenotype. Much variation exists at the molecular level owing, in part, to the redundancy of the genetic code, which allows different codons to specify the same amino acid. Thus, two members of a population can produce the same protein even if their DNA sequences are different. DNA sequences between the genes and introns within genes do not encode proteins; much of the variation in these sequences also has little effect on the phenotype.

An important but frequently misunderstood tool used in population genetics is the mathematical model. Let's take a moment to consider what a model is and how it can be used. A mathematical model usually describes a process as an equation. Factors that may influence the process are represented by variables in the equation; the equation defines the way in which the variables influence the process. Most models are simplified representations of a process, because the simultaneous consideration of all of the influencing factors is impossible; some factors must be ignored in order to examine the effects of others. At first, a model might consider only one or a few factors but, after their effects are understood, the model can be improved by the addition of more details. Importantly, even a simple model can be a source of valuable insight into how a process is influenced by key variables.

Before we can explore the evolutionary processes that shape genetic variation, we must be able to describe the genetic structure of a population. The usual way of describing this structure is to enumerate the types and frequencies of genotypes and alleles in a population.

Calculating Genotypic Frequencies

A frequency is simply a proportion or a percentage, usually expressed as a decimal fraction. For example, if 20% of the alleles at a particular locus in a population are A, we would say that the frequency of the A allele in the population is 0.20. For large populations, for which a determination of the genes of all individual members is impractical, a sample of the population is usually taken and the genotypic and allelic frequencies are calculated for this sample (see Chapter 24). The genotypic and allelic frequencies of the sample are then used to represent the gene pool of the population.

To calculate a **genotypic frequency**, we simply add up the number of individuals possessing the genotype and divide by the total number of individuals in the sample (N). For a locus with three genotypes AA, Aa, and aa, the frequency (f) of each genotype is

$$f(AA) = \frac{\text{number of } AA \text{ individuals}}{N}$$

$$f(Aa) = \frac{\text{number of } Aa \text{ individuals}}{N} \quad (25.1)$$

$$f(aa) = \frac{\text{number of } aa \text{ individuals}}{N}$$

The sum of all the genotypic frequencies always equals 1.

Calculating Allelic Frequencies

The gene pool of a population can also be described in terms of the allelic frequencies. There are always fewer alleles than genotypes, so the gene pool of a population can be described in fewer terms when the allelic frequencies are used. In a sexually reproducing population, the genotypes are only temporary assemblages of the alleles.

As described by Mendel's principle of segregation, the genotypes break down each generation and individual alleles are passed to the next generation through the gametes. Thus, the types and numbers of alleles, rather than genotypes, have real continuity from one generation to the next and make up the gene pool of a population.

Allelic frequencies can be calculated from (1) the numbers or (2) the frequencies of the genotypes. To calculate the **allelic frequency** from the numbers of genotypes, we count the number of copies of a particular allele present in a sample and divide by the total number of all alleles in the sample:

$$\text{frequency of an allele} = \frac{\begin{array}{c}\text{number of copies}\\\text{of the allele}\end{array}}{\begin{array}{c}\text{number of copies of all}\\\text{alleles at the locus}\end{array}} \quad (25.2)$$

For a locus with only two alleles (A and a), the frequencies of the alleles are usually represented by the symbols p and q and can be calculated as follows:

$$p = f(A) = \frac{2n_{AA} + n_{Aa}}{2N}$$

$$q = f(a) = \frac{2n_{aa} + n_{Aa}}{2N} \quad (25.3)$$

where n_{AA}, n_{Aa}, and n_{aa} represent the numbers of AA, Aa, and aa individuals, and N represents the total number of individuals in the sample. To obtain the number of copies of the allele in the numerator of the equation, we add twice the number of homozygotes (because each has two copies of the allele for which the frequency is being calculated) to the number of heterozygotes (because each has a single copy of the allele). We divide by $2N$ because each diploid individual has two alleles at a locus. The sum of the allelic frequencies always equals 1 ($p + q = 1$); so, after p has been obtained, q can be determined by subtraction: $q = 1 - p$.

Alternatively, allelic frequencies can be calculated from the genotypic frequencies. This is useful if the genotypic frequencies have already been calculated and the numbers of the different genotypes are not available. To calculate an allelic frequency from genotypic frequencies, we add the frequency of the homozygote for each allele to half the frequency of the heterozygote (because half of the heterozygote's alleles are of each type):

$$p = f(A) = f(AA) + \tfrac{1}{2}f(Aa)$$

$$q = f(a) = f(aa) + \tfrac{1}{2}f(Aa) \quad (25.4)$$

We obtain the same values of p and q whether we calculate the allelic frequencies from the numbers of genotypes (Equation 25.3) or from the genotypic frequencies (Equation 25.4). A sample calculation of allelic frequencies is provided in the next Worked Problem. ▶ **TRY PROBLEM 16**

LOCI WITH MULTIPLE ALLELES We can use the same principles to determine the frequencies of alleles for loci with more than two alleles. To calculate the allelic frequencies from the numbers of genotypes, we count up the number of copies of an allele by adding twice the number of homozygotes to the number of heterozygotes that possess the allele and divide this sum by twice the number of individuals in the sample. For a locus with three alleles (A^1, A^2, and A^3) and six genotypes (A^1A^1, A^1A^2, A^2A^2, A^1A^3, A^2A^3, and A^3A^3), the frequencies (p, q, and r) of the alleles are

$$p = f(A^1) = \frac{2n_{A^1A^1} + n_{A^1A^2} + n_{A^1A^3}}{2N}$$

$$q = f(A^2) = \frac{2n_{A^2A^2} + n_{A^1A^2} + n_{A^2A^3}}{2N} \quad (25.5)$$

$$r = f(A^3) = \frac{2n_{A^3A^3} + n_{A^1A^3} + n_{A^2A^3}}{2N}$$

Alternatively, we can calculate the frequencies of multiple alleles from the genotypic frequencies by extending Equation 25.4. Once again, we add the frequency of the homozygote to half the frequency of each heterozygous genotype that possesses the allele:

$$p = f(A^1A^1) + \tfrac{1}{2}f(A^1A^2) + \tfrac{1}{2}f(A^1A^3)$$

$$q = f(A^2A^2) + \tfrac{1}{2}f(A^1A^2) + \tfrac{1}{2}f(A^2A^3) \quad (25.6)$$

$$r = f(A^3A^3) + \tfrac{1}{2}f(A^1A^3) + \tfrac{1}{2}f(A^2A^3)$$

X-LINKED LOCI To calculate allelic frequencies for genes at X-linked loci, we apply these same principles. However, we must remember that a female possesses two X chromosomes and therefore has two X-linked alleles, whereas a male has only a single X chromosome and one X-linked allele.

Suppose there are two alleles at an X-linked locus, X^A and X^a. Females may be either homozygous (X^AX^A or X^aX^a) or heterozygous (X^AX^a). All males are hemizygous (X^AY or X^aY). To determine the frequency of the X^A allele (p), we first count the number of copies of X^A: we multiply the number of X^AX^A females by two and add the number of X^AX^a females and the number of X^AY males. We then divide the sum by the total number of alleles at the locus, which is twice the total number of females plus the number of males:

$$p = f(X^A) = \frac{2n_{X^AX^A} + n_{X^AX^a} + n_{X^AY}}{2n_{females} + n_{males}} \quad (25.7a)$$

Similarly, the frequency of the X^a allele is

$$q = f(X^a) = \frac{2n_{X^aX^a} + n_{X^AX^a} + n_{X^aY}}{2n_{females} + n_{males}} \quad (25.7b)$$

The frequencies of X-linked alleles can also be calculated from genotypic frequencies by adding the frequency of the females that are homozygous for the allele, half the frequency of the females that are heterozygous for the allele, and the frequency of males hemizygous for the allele:

$$p = f(X^A) = f(X^AX^A) + \tfrac{1}{2}f(X^AX^a) + f(X^AY)$$

$$q = f(X^a) = f(X^aX^a) + \tfrac{1}{2}f(X^AX^a) + f(X^aY) \quad (25.8)$$

If you remember the logic behind these calculations, you can determine allelic frequencies for any set of genotypes, and it will not be necessary to memorize all of the formulas. ▶ TRY PROBLEM 18

CONCEPTS

Population genetics concerns the genetic composition of a population and how it changes over time. The gene pool of a population can be described by the frequencies of genotypes and alleles in the population.

✓ CONCEPT CHECK 1

What are some advantages of using allelic frequencies to describe the gene pool of a population instead of using genotypic frequencies?

WORKED PROBLEM

The human MN blood-type antigens are determined by two codominant alleles, L^M and L^N (see p. 106 in Chapter 5). The MN blood types and corresponding genotypes of 398 Finns from Karjala are tabulated here.

Phenotype	Genotype	Number
MM	L^ML^M	182
MN	L^ML^N	172
NN	L^NL^N	44

Source: W. C. Boyd, *Genetics and the Races of Man* (Boston: Little, Brown, 1950.)

Calculate the genotypic and allelic frequencies at the MN locus for the Karjala population.

Solution Strategy

What information is required in your answer to the problem?

The genotypic and allelic frequencies of the population.

What information is provided to solve the problem?

The numbers of the different MN genotypes in the sample.

Solution Steps

The genotypic frequencies for the population are calculated with the following formula:

genotypic frequency

$$= \frac{\text{number of individuals with genotype}}{\text{total number of individuals in sample } (N)}$$

$$f(L^ML^M) = \frac{\text{number of } L^ML^M \text{ individuals}}{N} = \frac{182}{398} = 0.457$$

$$f(L^ML^N) = \frac{\text{number of } L^ML^M \text{ individuals}}{N} = \frac{172}{398} = 0.432$$

$$f(L^NL^N) = \frac{\text{number of } L^NL^N \text{ individuals}}{N} = \frac{44}{398} = 0.111$$

The allelic frequencies can be calculated from either the numbers or the frequencies of the genotypes. To calculate allelic frequencies from numbers of genotypes, we add the number of copies of the allele and divide by the number of copies of all alleles at that locus.

$$\text{frequency of an allele} = \frac{\text{number of copies of the allele}}{\text{number of copies all alleles}}$$

$$p = f(L^M) = \frac{(2n_{L^ML^M}) + (n_{L^ML^N})}{2N}$$

$$= \frac{2(182) + 172}{2(398)} = \frac{536}{796} = 0.673$$

$$q = f(L^N) = \frac{(2n_{L^NL^N}) + (n_{L^ML^N})}{2N}$$

$$= \frac{2(44) + 172}{2(398)} = \frac{260}{796} = 0.327$$

To calculate the allelic frequencies from genotypic frequencies, we add the frequency of the homozygote for that genotype to half the frequency of each heterozygote possessing that allele:

$$p = f(L^M) = f(L^ML^M) + \tfrac{1}{2}f(L^ML^N)$$
$$= 0.457 + \tfrac{1}{2}(0.432) = 0.673$$

$$q = f(L^N) = f(L^NL^N) + \tfrac{1}{2}f(L^ML^N)$$
$$= 0.111 + \tfrac{1}{2}(0.432) = 0.327$$

▶ Now try your hand at calculating genotypic and allelic frequencies by working Problem 17 at the end of the chapter.

25.2 The Hardy–Weinberg Law Describes the Effect of Reproduction on Genotypic and Allelic Frequencies

The primary goal of population genetics is to understand the processes that shape a population's gene pool. First, we must ask what effects reproduction and Mendelian principles have on the genotypic and allelic frequencies: How do the segregation of alleles in gamete formation and the combining of alleles in fertilization influence the gene pool? The answer to this question lies in the **Hardy–Weinberg law**, among the most important principles of population genetics.

The Hardy–Weinberg law was formulated independently by G. H. Hardy and Wilhelm Weinberg in 1908 (similar conclusions were reached by several other geneticists at about the same time). The law is actually a mathematical model that evaluates the effect of reproduction on the genotypic and allelic frequencies of a population. It makes several simplifying assumptions

about the population and provides two key predictions if these assumptions are met. For an autosomal locus with two alleles, the Hardy–Weinberg law can be stated as follows:

> **Assumptions** If a population is large, randomly mating, and not affected by mutation, migration, or natural selection, then:
> **Prediction 1** the allelic frequencies of a population do not change; and
> **Prediction 2** the genotypic frequencies stabilize (will not change) after one generation in the proportions p^2 (the frequency of AA), $2pq$ (the frequency of Aa), and q^2 (the frequency of aa), where p equals the frequency of allele A and q equals the frequency of allele a.

The Hardy–Weinberg law indicates that, when the assumptions are met, reproduction alone does not alter allelic or genotypic frequencies and the allelic frequencies determine the frequencies of genotypes.

The statement that genotypic frequencies stabilize after one generation means that they may change in the first generation after random mating because one generation of random mating is required to produce Hardy–Weinberg proportions of the genotypes. Afterward, the genotypic frequencies, like allelic frequencies, do not change as long as the population continues to meet the assumptions of the Hardy–Weinberg law. When genotypes are in the expected proportions of p^2, $2pq$, and q^2, the population is said to be in **Hardy–Weinberg equilibrium**.

CONCEPTS

The Hardy–Weinberg law describes how reproduction and Mendelian principles affect the allelic and genotypic frequencies of a population.

✓ **CONCEPT CHECK 2**

Which statement is not an assumption of the Hardy–Weinberg law?
a. The allelic frequencies (p and q) are equal.
b. The population is randomly mating.
c. The population is large.
d. Natural selection has no effect.

Genotypic Frequencies at Hardy–Weinberg Equilibrium

How do the conditions of the Hardy–Weinberg law lead to genotypic proportions of p^2, $2pq$, and q^2? Mendel's principle of segregation says that each individual organism possesses two alleles at a locus and that each of the two alleles has an equal probability of passing into a gamete. Thus, the frequencies of alleles in gametes will be the same as the frequencies of alleles in the parents. Suppose that we have a Mendelian population in which the frequencies of alleles A and a are p and q, respectively. These frequencies will also be those in the gametes. If mating is random (one of the assumptions of the

Conclusion: Random mating will produce genotypes of the next generation in proportions $p^2(AA)$, $2pq(Aa)$, and $q^2(aa)$.

25.2 Random mating produces genotypes in the proportions p^2, $2pq$, and q^2.

Hardy–Weinberg law), the gametes will come together in random combinations, which can be represented by a Punnett square (**Figure 25.2**).

The multiplication rule of probability (Chapter 3) can be used to determine the probability of various gametes pairing. For example, the probability of a sperm containing allele A is p and the probability of an egg containing allele A is p. Applying the multiplication rule, we find that the probability that these two gametes will combine to produce an AA homozygote is $p \times p = p^2$. Similarly, the probability of a sperm containing allele a combining with an egg containing allele a to produce an aa homozygote is $q \times q = q^2$. An Aa heterozygote can be produced in one of two ways: (1) a sperm containing allele A may combine with an egg containing allele a ($p \times q$) or (2) an egg containing allele A may combine with a sperm containing allele a ($p \times q$). Thus, the probability of alleles A and a combining to produce an Aa heterozygote is $2pq$. In summary, whenever the frequencies of alleles in a random mating population are p and q, the frequencies of the genotypes in the next generation will be p^2, $2pq$, and q^2. Figure 25.2 demonstrates that only a single generation of random mating is required to produce the Hardy-Weinberg genotypic proportions.

Closer Examination of the Hardy–Weinberg Law

Before we consider the implications of the Hardy–Weinberg law, we need to take a closer look at the three assumptions that it makes about a population. First, it assumes that the population is large. How big is "large"? Theoretically, the Hardy–Weinberg law requires that a population be infinitely large in size, but this requirement is obviously unrealistic. In practice, many large populations are in the predicted Hardy–Weinberg proportions, and significant deviations arise only when population size is rather small. Later in the chapter, we will examine the effects of small population size on allelic frequencies.

The second assumption of the Hardy–Weinberg law is that members of the population mate randomly, which means that each genotype mates relative to its frequency. For example, suppose that three genotypes are present in a population in the following proportions: $f(AA) = 0.6$, $f(Aa) = 0.3$, and $f(aa) = 0.1$. With random mating, the frequency of mating between two AA homozygotes ($AA \times AA$) will be equal to the multiplication of their frequencies: $0.6 \times 0.6 = 0.36$, whereas the frequency of mating between two aa homozygotes ($aa \times aa$) will be only $0.1 \times 0.1 = 0.01$.

The third assumption of the Hardy–Weinberg law is that the allelic frequencies of the population are not affected by natural selection, migration, and mutation. Although mutation occurs in every population, its rate is so low that it has little short-term effect on the predictions of the Hardy–Weinberg law (although it may largely shape allelic frequencies over long periods of time when no other forces are acting). Although natural selection and migration are significant factors in real populations, we must remember that the purpose of the Hardy–Weinberg law is to examine only the effect of reproduction on the gene pool. When this effect is known, the effects of other factors (such as migration and natural selection) can be examined.

A final point is that the assumptions of the Hardy–Weinberg law apply to a single locus. No real population mates randomly for all traits, and a population is not completely free of natural selection for all traits. The Hardy–Weinberg law, however, does not require random mating and the absence of selection, migration, and mutation for all traits; it requires these conditions only for the locus under consideration. A population may be in Hardy–Weinberg equilibrium for one locus but not for others.

Implications of the Hardy–Weinberg Law

The Hardy–Weinberg law has several important implications for the genetic structure of a population. One implication is that a population cannot evolve if it meets the Hardy–Weinberg assumptions, because evolution consists of change in the allelic frequencies of a population. Therefore the Hardy–Weinberg law tells us that reproduction alone will not bring about evolution. Other processes such as natural selection, mutation, migration, or chance are required for populations to evolve.

A second important implication is that, when a population is in Hardy–Weinberg equilibrium, the genotypic frequencies are determined by the allelic frequencies. When a population is not in Hardy–Weinberg equilibrium, we have no basis for predicting the genotypic frequencies. Although we can always determine the allelic frequencies from the genotypic frequencies (see Equation 25.3), the reverse (determining the genotypic frequencies from the allelic frequencies) is possible only when the population is in Hardy–Weinberg equilibrium.

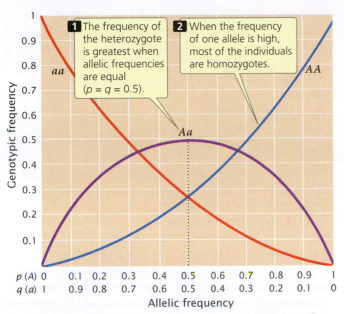

1 The frequency of the heterozygote is greatest when allelic frequencies are equal ($p = q = 0.5$).

2 When the frequency of one allele is high, most of the individuals are homozygotes.

25.3 When a population is in Hardy–Weinberg equilibrium, the proportions of genotypes are determined by the frequencies of alleles.

Table 25.1	Extensions of the Hardy–Weinberg law	
Situation	**Allelic Frequencies**	**Genotypic Frequencies**
Three alleles	$f(A^1) = p$	$f(A^1A^1) = p^2$
	$f(A^2) = q$	$f(A^1A^2) = 2pq$
	$f(A^3) = r$	$f(A^2A^2) = q^2$
		$f(A^1A^3) = 2pr$
		$f(A^2A^3) = 2qr$
		$f(A^3A^3) = r^2$
X-linked alleles	$f(X^1) = p$	$f(X^1X^1 \text{ female}) = p^2$
	$f(X^2) = q$	$f(X^1X^2 \text{ female}) = 2pq$
		$f(X^2X^2 \text{ female}) = q^2$
		$f(X^1Y \text{ male}) = p$
		$f(X^2Y \text{ male}) = q$

Note: For X-linked female genotypes, the frequencies are the proportions among all females; for X-linked male genotypes, the frequencies are the proportions among all males.

For a locus with two alleles, the frequency of the heterozygote is greatest when allelic frequencies are between 0.33 and 0.66 and is at a maximum when allelic frequencies are each 0.5 (**Figure 25.3**). The heterozygote frequency also never exceeds 0.5 when the population is in Hardy–Weinberg equilibrium. Furthermore, when the frequency of one allele is low, homozygotes for that allele will be rare, and most of the copies of a rare allele will be present in heterozygotes. As you can see from Figure 25.3, when the frequency of allele *a* is 0.2, the frequency of the *aa* homozygote is only 0.04 (q^2), but the frequency of *Aa* heterozygotes is 0.32 (*2pq*); 80% of the *a* alleles are in heterozygotes. Use **Animation 25.1** to examine the effect of allelic frequencies on genotypic frequencies when a population is in Hardy–Weinberg equilibrium.

A third implication of the Hardy–Weinberg law is that a single generation of random mating produces the equilibrium frequencies of p^2, *2pq*, and q^2. The fact that genotypes are in Hardy–Weinberg proportions does not prove that the population is free from natural selection, mutation, and migration. It means only that these forces have not acted since the last time random mating took place.

Extensions of the Hardy–Weinberg Law

The Hardy–Weinberg expected proportions can also be applied to multiple alleles and X-linked alleles (**Table 25.1**). With multiple alleles, the genotypic frequencies expected at equilibrium are the square of the allelic frequencies. For an autosomal locus with three alleles, the equilibrium genotypic frequencies will $(p + q + r)^2 = p^2 + 2pq + q^2 + 2pr + 2qr + r^2$. For an X-linked locus with two alleles, X^A and X^a, the equilibrium frequencies of the female genotypes are

$(p + q)^2 = p^2 + 2pq + q^2$, where p^2 is the frequency of X^AX^A, *2pq* is the frequency of X^AX^a, and q^2 is the frequency of X^aX^a. Males have only a single X-linked allele, and so the frequencies of the male genotypes are p (frequency of X^AY) and q (frequency of X^aY). These proportions are those of the genotypes among males and females rather than the proportions among the entire population. Thus, p^2 is the expected proportion of females with the genotype X^AX^A; if females make up 50% of the population, then the expected proportion of this genotype in the entire population is $0.5 \times p^2$.

The frequency of an X-linked recessive trait among males is q, whereas the frequency among females is q^2. When an X-linked allele is uncommon, the trait will therefore be much more frequent in males than in females. Consider hemophilia A, a clotting disorder caused by an X-linked recessive allele with a frequency (q) of approximately 1 in 10,000, or 0.0001. At Hardy–Weinberg equilibrium, this frequency will also be the frequency of the disease among males. The frequency of the disease among females, however, will be $q^2 = (0.0001)^2 = 0.00000001$, which is only 1 in 100 million. Hemophilia is 10,000 times as frequent in males as in females.

Testing for Hardy–Weinberg Proportions

To determine if a population's genotypes are in Hardy–Weinberg equilibrium, the genotypic proportions expected under the Hardy–Weinberg law must be compared with the observed genotypic frequencies. To do so, we first calculate the allelic frequencies, then find the expected genotypic frequencies by using the square of the allelic frequencies, and, finally, compare the observed and expected genotypic frequencies by using a chi-square test.

WORKED PROBLEM

Jeffrey Mitton and his colleagues found three genotypes (R^2R^2, R^2R^3, and R^3R^3) at a locus encoding the enzyme peroxidase in ponderosa pine trees growing at Glacier Lake, Colorado. The observed numbers of these genotypes were:

Genotypes	Number observed
R^2R^2	135
R^2R^3	44
R^3R^3	11

Are the ponderosa pine trees at Glacier Lake in Hardy–Weinberg equilibrium at the peroxidase locus?

Solution Strategy

What information is required in your answer to the problem?

The results of a chi-square test to determine whether the population is in Hardy-Weinberg equilibrium.

What information is provided to solve the problem?

The numbers of the different genotypes in a sample of the population.

Solution Steps

If the frequency of the R^2 allele equals p and the frequency of the R^3 allele equals q, the frequency of the R^2 allele is:

$$p = f(R^2) = \frac{(2n_{R^2R^2}) + (n_{R^2R^3})}{2N} = \frac{2(135) + 44}{2(190)} = 0.826$$

The frequency of the R^3 allele is obtained by subtraction:

$$q = f(R^3) = 1 - p = 0.174$$

The frequencies of the genotypes expected under Hardy–Weinberg equilibrium are then calculated by using p^2, $2pq$, and q^2:

$$R^2R^2 = p^2 = (0.826)^2 = 0.683$$

$$R^2R^3 = 2pq = 2(0.826)(0.174) = 0.287$$

$$R^3R^3 = q^2 = (0.174)^2 = 0.03$$

Multiplying each of these expected genotypic frequencies by the total number of observed genotypes in the sample (190), we obtain the numbers expected for each genotype:

$$R^2R^2 = 0.683 \times 190 = 129.8$$

$$R^2R^3 = 0.287 \times 190 = 54.5$$

$$R^3R^3 = 0.03 \times 190 = 5.7$$

By comparing these expected numbers with the observed numbers of each genotype, we see that there are more R^2R^2 and R^3R^3 homozygotes and fewer R^2R^3 heterozygotes in the population than we expect at equilibrium.

A chi-square goodness-of-fit test is used to determine whether the differences between the observed and the expected numbers of each genotype are due to chance:

$$X^2 = \sum \frac{(\text{observed} - \text{expected})^2}{\text{expected}}$$

$$= \frac{(135 - 129.8)^2}{129.8} + \frac{(44 - 54.5)^2}{54.5} + \frac{(11 - 5.7)^2}{5.7}$$

$$= 0.21 + 2.02 + 4.93 = 7.16$$

The calculated chi-square value is 7.16; to obtain the probability associated with this chi-square value, we determine the appropriate degrees of freedom.

Up to this point, the chi-square test for assessing Hardy–Weinberg equilibrium has been identical with the chi-square tests that we used in Chapter 3 to assess progeny ratios in a genetic cross, where the degrees of freedom were $n - 1$ and n equaled the number of expected genotypes. For the Hardy–Weinberg test, however, we must subtract an additional degree of freedom, because the expected numbers are based on the observed allelic frequencies; therefore, the observed numbers are not completely free to vary. In general, the degrees of freedom for a chi-square test of Hardy–Weinberg equilibrium equal the number of expected genotypic classes minus the number of associated alleles. For this particular Hardy–Weinberg test, the degree of freedom is $3 - 2 = 1$.

After we have calculated both the chi-square value and the degrees of freedom, the probability associated with this value can be sought in a chi-square table (see Table 3.7). With one degree of freedom, a chi-square value of 7.16 has a probability between 0.01 and 0.001. It is very unlikely that the peroxidase genotypes observed at Glacier Lake are in Hardy–Weinberg proportions.

▶ For additional practice, determine whether the genotypic frequencies in Problem 22 at the end of the chapter are in Hardy–Weinberg equilibrium.

CONCEPTS

The observed number of genotypes in a population can be compared with the Hardy–Weinberg expected proportions by using a chi-square goodness-of-fit test.

✓ CONCEPT CHECK 3

What is the expected frequency of heterozygotes in a population with allelic frequencies x and y that is in Hardy–Weinberg equilibrium?

a. $x + y$ c. $2xy$
b. xy d. $(x - y)^2$

Estimating Allelic Frequencies with the Hardy–Weinberg Law

A practical use of the Hardy–Weinberg law is that it allows us to calculate allelic frequencies when dominance is present. For example, cystic fibrosis is an autosomal recessive disorder characterized by respiratory infections, incomplete digestion, and abnormal sweating (see p. 106 in Chapter 5). Among North American Caucasians, the incidence of the disease is approximately 1 person in 2000. The formula for calculating allelic frequency (see Equation 25.3) requires that we know the numbers of homozygotes and heterozygotes, but cystic fibrosis is a recessive disease and so we cannot easily distinguish between homozygous unaffected persons and heterozygous carriers. Although molecular tests are available for identifying heterozygous carriers of the cystic fibrosis gene, the low frequency of the disease makes widespread screening impractical. In such situations, the Hardy–Weinberg law can be used to estimate the allelic frequencies.

If we assume that a population is in Hardy–Weinberg equilibrium with regard to this locus, then the frequency of the recessive genotype (aa) will be q^2, and the allelic frequency is the square root of the genotypic frequency:

$$q = \sqrt{f(aa)} \qquad (25.9)$$

If the frequency of cystic fibrosis in North American Caucasians is approximately 1 in 2000, or 0.0005, then $q = \sqrt{0.0005} = 0.02$. Thus, about 2% of the alleles in the Caucasian population encode cystic fibrosis. We can calculate the frequency of the normal allele by subtracting: $p = 1 - q = 1 - 0.02 = 0.98$. After we have calculated p and q, we can use the Hardy–Weinberg law to determine the frequencies of homozygous unaffected people and heterozygous carriers of the gene:

$$f(AA) = p^2 = (0.98)^2 = 0.960$$
$$f(Aa) = 2pq = 2(0.02)(0.98) = 0.0392$$

Thus, about 4% (1 of 25) of Caucasians are heterozygous carriers of the allele that causes cystic fibrosis. ▶ TRY PROBLEM 25

CONCEPTS

Although allelic frequencies cannot be calculated directly for traits that exhibit dominance, the Hardy–Weinberg law can be used to estimate the allelic frequencies if the population is in Hardy–Weinberg equilibrium for that locus. The frequency of the recessive allele will be equal to the square root of the frequency of the recessive trait.

✓ CONCEPT CHECK 4

In cats, all-white color is dominant over not all-white. In a population of 100 cats, 19 are all-white. Assuming that the population is in Hardy–Weinberg equilibrium, what is the frequency of the all-white allele in this population?

25.3 Nonrandom Mating Affects the Genotypic Frequencies of a Population

An assumption of the Hardy–Weinberg law is that mating is random with respect to genotype. Nonrandom mating affects the way in which alleles combine to form genotypes and alters the genotypic frequencies of a population.

We can distinguish between two types of nonrandom mating. **Positive assortative mating** refers to a tendency for like individuals to mate. For example, humans exhibit positive assortative mating for height: tall people mate preferentially with other tall people; short people mate preferentially with other short people. **Negative assortative mating** refers to a tendency for unlike individuals to mate. If people engaged in negative assortative mating for height, tall and short people would preferentially mate. Assortative mating is usually for a particular trait and will affect only those genes that encode the trait (and genes closely linked to them).

One form of nonrandom mating is **inbreeding**, which is preferential mating between related individuals. Inbreeding is actually positive assortative mating for relatedness, but it differs from other types of assortative mating because it affects all genes, not just those that determine the trait for which the mating preference exists. Inbreeding causes a departure from the Hardy–Weinberg equilibrium frequencies of p^2, $2pq$, and q^2. More specifically, it leads to an increase in the proportion of homozygotes and a decrease in the proportion of heterozygotes in a population. **Outcrossing** is the avoidance of mating between related individuals.

In a diploid organism, a homozygous individual has two copies of the same allele. These two copies may be the same in *state*, which means that the two alleles are alike in structure and function but do not have a common origin. Alternatively, the two alleles in a homozygous individual may be the same because they are identical by *descent*; that is, the copies are descended from a single allele that was present in an ancestor (**Figure 25.4**). If we go back far enough in time, many alleles are likely to be identical by descent, but for calculating the effects of inbreeding we consider identity by descent by going back only a few generations.

Inbreeding is usually measured by the **inbreeding coefficient**, designated F, which is a measure of the probability that two alleles are identical by descent. Inbreeding coefficients can range from 0 to 1. A value of 0 indicates that mating in a large population is random; a value of 1 indicates that all alleles are identical by descent. Inbreeding coefficients can be calculated from analyses of pedigrees or they can be determined from the reduction in the heterozygosity of a population. Although we will not go into the details of how F is calculated, an understanding of how inbreeding affects genotypic frequencies is important.

When inbreeding takes place, the proportion of heterozygotes decreases by $2Fpq$, and half of this value (Fpq) is *added* to the proportion of each homozygote each generation. The frequencies of the genotypes will then be

$$f(AA) = p^2 + Fpq$$
$$f(Aa) = 2pq - 2Fpq \qquad (25.10)$$
$$f(aa) = q^2 + Fpq$$

Consider a population that reproduces by self-fertilization ($F = 1$). We will assume that this population begins with

(a) Alleles identical by descent

25.4 Individuals may be homozygous by descent or by state. Inbreeding is a measure of the probability that two alleles are identical by descent.

These two copies of the A^1 allele are descended from the same copy in a common ancestor; so they are identical by descent.

(b) Alleles identical by state

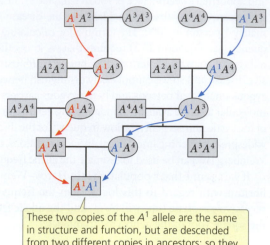

These two copies of the A^1 allele are the same in structure and function, but are descended from two different copies in ancestors; so they are identical in state.

genotypic frequencies in Hardy–Weinberg proportions (p^2, $2pq$, and q^2). With selfing, each homozygote produces progeny only of the same homozygous genotype ($AA \times AA$ produces all AA; and $aa \times aa$ produces all aa), whereas only half the progeny of a heterozygote will be like the parent ($Aa \times Aa$ produces $\frac{1}{4}$ AA, $\frac{1}{2}$ Aa, and $\frac{1}{4}$ aa). Selfing therefore reduces the proportion of heterozygotes in the population by half with each generation, until all genotypes in the population are homozygous (**Table 25.2** and **Figure 25.5**). Although all individuals in the population are homozygous at this point, both types of homozygotes (AA and aa) are still present, provided neither homozygote has an advantage over the other (there are no selective differences). ▶ TRY PROBLEM 28

For most outcrossing species, close inbreeding is harmful because it increases the proportion of homozygotes and thereby boosts the probability that deleterious and lethal

recessive alleles will combine to produce homozygotes with a harmful trait. Assume that a recessive allele (a) that causes a genetic disease has a frequency (q) of 0.01. If the population mates randomly ($F = 0$), the frequency of individuals affected with the disease (aa) will be $q^2 = 0.01^2 = 0.0001$; so only 1 in 10,000 individuals will have the disease. However, if $F = 0.25$ (the equivalent of brother–sister mating), then the expected frequency of the homozygote genotype is $q^2 + Fpq = (0.01)^2 + (0.25)(0.99)(0.01) = 0.0026$; thus, the genetic disease is 26 times as frequent at this level of inbreeding. This increased appearance of lethal and deleterious traits with inbreeding is termed **inbreeding depression**; the more intense

Table 25.2	Generational increase in frequency of homozygotes in a self-fertilizing population starting with $p = q = 0.5$		
	Genotypic Frequencies		
Generation	**AA**	**Aa**	**aa**
1	$\frac{1}{4}$	$\frac{1}{2}$	$\frac{1}{4}$
2	$\frac{1}{4} + \frac{1}{8} = \frac{3}{8}$	$\frac{1}{4}$	$\frac{1}{4} + \frac{1}{8} = \frac{3}{8}$
3	$\frac{3}{8} + \frac{1}{16} = \frac{7}{16}$	$\frac{1}{8}$	$\frac{3}{8} + \frac{1}{16} = \frac{7}{16}$
4	$\frac{7}{16} + \frac{1}{32} = \frac{15}{32}$	$\frac{1}{16}$	$\frac{7}{16} + \frac{1}{32} = \frac{15}{32}$
n	$\dfrac{1 - (\frac{1}{2})^n}{2}$	$(\frac{1}{2})^n$	$\dfrac{1 - (\frac{1}{2})^n}{2}$
	$\frac{1}{2}$	0	$\frac{1}{2}$

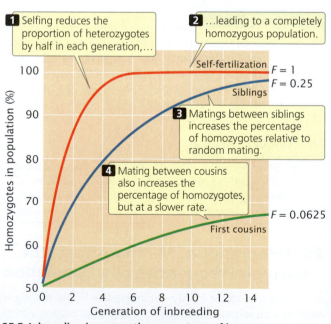

1 Selfing reduces the proportion of heterozygotes by half in each generation,…

2 …leading to a completely homozygous population.

3 Matings between siblings increases the percentage of homozygotes relative to random mating.

4 Mating between cousins also increases the percentage of homozygotes, but at a slower rate.

25.5 Inbreeding increases the percentage of homozygous individuals in a population.

Table 25.3	Effects of inbreeding on Japanese children		
Genetic Relationship of Parents		*F*	Mortality of Children (through 12 years of age)
Unrelated		0	0.082
Second cousins		0.016 ($^1/_{64}$)	0.108
First cousins		0.0625 ($^1/_{16}$)	0.114

Source: After D. L. Hartl and A. G. Clark, *Principles of Population Genetics*, 2d ed. (Sunderland, Mass.: Sinauer, 1989), Table 2. Original data from W. J. Schull and J. V. Neel, *The Effects of Inbreeding on Japanese Children* (New York: Harper & Row, 1965).

25.7 Although inbreeding is generally harmful, a number of inbreeding organisms are successful. Shown here is the terrestrial slug *Arion circumscriptos*, an inbreeding species that causes damage in greenhouses and flower gardens. [J. Meul/ARCO/Age Fotostock America, Inc.]

the inbreeding, the more severe the inbreeding depression. To see the effects of inbreeding on genotypic frequencies, (A) view the Mini-Tutorial in **Animation 25.1**.

The harmful effects of inbreeding have been recognized for thousands of years and may be the basis of cultural taboos against mating between close relatives. William Schull and James Neel found that for each 10% increase in *F* the mean IQ of Japanese children dropped six points. Child mortality also increases with close inbreeding (**Table 25.3**); children of first cousins have a 40% increase in mortality over that seen among the children of unrelated people. Inbreeding also has deleterious effects on crops (**Figure 25.6**) and domestic animals.

Inbreeding depression is most often studied in humans, as well as in plants and animals reared in captivity, but the negative effects of inbreeding may be more severe in natural populations. Julie Jimenez and her colleagues collected wild mice from a natural population in Illinois and bred them in the laboratory for three to four generations. Laboratory matings were chosen so that some mice had no inbreeding,

whereas others had an inbreeding coefficient of 0.25. When both types of mice were released back into the wild, the weekly survival of the inbred mice was only 56% of that of the noninbred mice. Inbred male mice also continously lost body weight after release into the wild, whereas noninbred male mice initially lost weight but then regained it within a few days after release.

In spite of the fact that inbreeding is generally harmful for outcrossing species, a number of plants and animals regularly inbreed and are successful (**Figure 25.7**). As stated earlier, inbreeding increases homozygosity, and eventually all individuals in the population become homozygous. If a species undergoes inbreeding for a number of generations, many deleterious recessive alleles are weeded out by natural or artificial selection so that the population becomes homozygous for beneficial alleles. In this way, the harmful effects of inbreeding may eventually be eliminated, leaving a population that is homozygous for beneficial traits.

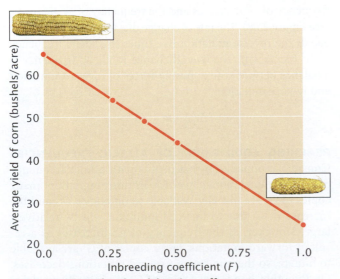

25.6 Inbreeding often has deleterious effects on crops. As inbreeding increases, the average yield of corn, for example, decreases.

CONCEPTS

Nonrandom mating alters the frequencies of the genotypes but not the frequencies of the alleles. Inbreeding is preferential mating between related individuals. With inbreeding, the frequency of homozygotes increases, whereas the frequency of heterozygotes decreases.

✓ CONCEPT CHECK 5

What is the effect of outcrossing on a population?
a. Allelic frequencies change.
b. There will be more heterozygotes than predicted by the Hardy–Weinberg law.
c. There will be fewer heterozygotes than predicted by the Hardy–Weinberg law.
d. Genotypic frequencies will equal those predicted by the Hardy–Weinberg law.

25.4 Several Evolutionary Forces Potentially Cause Changes in Allelic Frequencies

The Hardy–Weinberg law indicates that allelic frequencies do not change as a result of reproduction. Processes that bring about change in allelic frequency include mutation, migration, genetic drift (random effects due to small population size), and natural selection.

Mutation

Before evolution can take place, genetic variation must exist within a population; consequently, all evolution depends on processes that generate genetic variation. Although new *combinations* of existing genes may arise through recombination in meiosis, all genetic variants ultimately arise through mutation.

THE EFFECT OF MUTATION ON ALLELIC FREQUENCIES

Mutation can influence the rate at which one genetic variant increases at the expense of another. Consider a single locus in a population of 25 diploid individuals. Each individual possesses two alleles at the locus under consideration, so the gene pool of the population consists of 50 allele copies. Let us assume that there are two different alleles, designated G^1 and G^2 with frequencies p and q, respectively. If there are 45 copies of G^1 and 5 copies of G^2 in the population, $p = 0.90$ and $q = 0.10$. Now suppose that a mutation changes a G^1 allele into a G^2 allele. After this mutation, there are 44 copies of G^1 and 6 copies of G^2, and the frequency of G^2 has increased from 0.10 to 0.12. Mutation has changed the allelic frequency.

If copies of G^1 continue to mutate to G^2, the frequency of G^2 will increase and the frequency of G^1 will decrease (**Figure 25.8**). The amount that G^2 will change (Δq) as a result of mutation depends on: (1) the rate of G^1-to-G^2 mutation (μ) and (2) p, the frequency of G^1 in the population. When p is large, there are many copies of G^1 available to mutate to G^2, and the amount of change will be relatively large. As more mutations occur and p decreases, there will be fewer copies of G^1 available to mutate to G^2. The change in G^2 as a result of mutation equals the mutation rate times the allelic frequency:

$$\Delta q = \mu p \tag{25.11}$$

As the frequency of p decreases as a result of mutation, the change in frequency due to mutation will be less and less.

So far, we have considered only the effects of $G^1 \rightarrow G^2$ forward mutations. Reverse $G^2 \rightarrow G^1$ mutations also occur at rate v, which will probably be different from the forward mutation rate, μ. Whenever a reverse mutation occurs, the

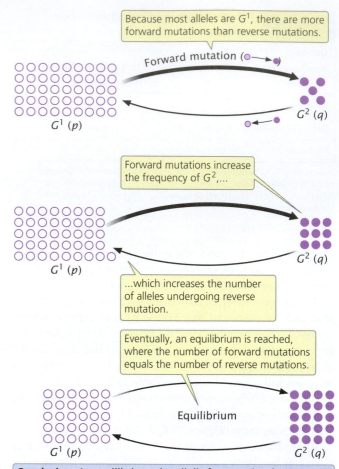

Because most alleles are G^1, there are more forward mutations than reverse mutations.

Forward mutation (o—→•)

G^1 (p) G^2 (q)

Forward mutations increase the frequency of G^2,...

G^1 (p) G^2 (q)

...which increases the number of alleles undergoing reverse mutation.

Eventually, an equilibrium is reached, where the number of forward mutations equals the number of reverse mutations.

Equilibrium

G^1 (p) G^2 (q)

Conclusion: At equilibrium, the allelic frequencies do not change even though mutation in both directions continues.

25.8 Recurrent mutation changes allelic frequencies. Forward and reverse mutations eventually lead to equilibrium.

frequency of G^2 decreases and the frequency of G^1 increases (see Figure 25.8). The rate of change due to reverse mutations equals the reverse mutation rate times the allelic frequency of G^2 ($\Delta q = vq$). The overall change in allelic frequency is a balance between the opposing forces of forward mutation and reverse mutation:

$$\Delta q = \mu p - vq \tag{25.12}$$

REACHING EQUILIBRIUM OF ALLELIC FREQUENCIES

Consider a population that begins with a high frequency of G^1 and a low frequency of G^2. In this population, many copies of G^1 are initially available to mutate to G^2, and the increase in G^2 due to forward mutation will be relatively large. However, as the frequency of G^2 increases as a result of forward mutations, fewer copies of G^1 are available to mutate, so the number of forward mutations decreases. On the other hand, few copies of G^2 are initially available to undergo a reverse mutation to G^1 but as the frequency of G^2 increases, the number of copies of G^2 available to undergo reverse mutation to G^1 increases; so the number of genes undergoing reverse mutation will increase. Eventually, the number of genes undergoing forward mutation will be counterbalanced by the number of genes undergoing reverse mutation. At this point, the increase in q due to forward mutation will be equal to the decrease in q due to reverse mutation, and there will be no net change in allelic frequency ($\Delta q = 0$), in spite of the fact that forward and reverse mutations continue to occur. The point at which there is no change in the allelic frequency of a population is referred to as **equilibrium** (see Figure 25.8). At equilibrium, the frequency of G^2 ($\hat{q}$) will be

$$\hat{q} = \frac{\mu}{\mu + \nu} \qquad (25.13)$$

This final equation tells us that the allelic frequency at equilibrium is determined solely by the forward (μ) and reverse (ν) mutation rates. ▶TRY PROBLEM 30

SUMMARY OF EFFECTS When the only evolutionary force acting on a population is mutation, allelic frequencies change with the passage of time because some alleles mutate into others. Eventually, these allelic frequencies reach equilibrium and are determined only by the forward and reverse mutation rates. When the allelic frequencies reach equilibrium, the Hardy–Weinberg law tells us that genotypic frequencies also will remain the same.

The mutation rates for most genes are low, so change in allelic frequency due to mutation in one generation is very small, and long periods of time are required for a population to reach mutational equilibrium. For example, if the forward and reverse mutation rates for alleles at a locus are 1×10^{-5} and 0.3×10^{-5} per generation, respectively (rates that have actually been measured at several loci in mice), and the allelic frequencies are $p = 0.9$ and $q = 0.1$, then the net change in allelic frequency per generation due to mutation is

$$
\begin{aligned}
\Delta q &= \mu p - \nu q \\
&= (1 \times 10^{-5})(0.9) - (0.3 \times 10^{-5})(0.1) \\
&= 8.7 \times 10^{-6} = 0.0000087
\end{aligned}
$$

Therefore, change due to mutation in a single generation is extremely small and, as the frequency of p drops as a result of mutation, the amount of change will become even smaller (**Figure 25.9**). The effect of typical mutation rates on Hardy–Weinberg equilibrium is negligible, and many generations are required for a population to reach mutational equilibrium. Nevertheless, if mutation is the only force acting on a population for long periods of time, mutation rates will determine allelic frequencies.

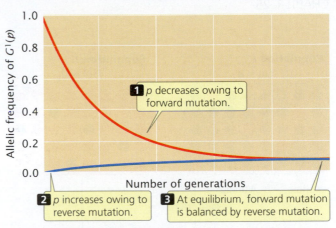

1 p decreases owing to forward mutation.

2 p increases owing to reverse mutation.

3 At equilibrium, forward mutation is balanced by reverse mutation.

25.9 Change due to recurrent mutation slows as the frequency of p drops. Allelic frequencies approach mutational equilibrium at typical low mutation rates. The allelic frequency of G^1 decreases as a result of forward ($G^1 \rightarrow G^2$) mutation at rate of 0.0001 and increases as a result of reverse ($G^2 \rightarrow G^1$) mutation at rate of 0.00001. Owing to the low rate of mutations, eventual equilibrium takes many generations to be reached.

CONCEPTS

Recurrent mutation causes changes in the frequencies of alleles. At equilibrium, the allelic frequencies are determined by the forward and reverse mutation rates. Because mutation rates are low, the effect of mutation on allelic frequencies per generation is very small.

✓ CONCEPT CHECK 6

When a population is in equilibrium for forward and reverse mutation rates, which of the following is true?
a. The number of forward mutations is greater than the number of reverse mutations.
b. No forward or reverse mutations occur.
c. The number of forward mutations is equal to the number of reverse mutations.
d. The population is in Hardy–Weinberg equilibrium.

Migration

Another process that may bring about change in allelic frequencies is the influx of genes from other populations, commonly called **migration** or **gene flow**. One of the assumptions of the Hardy–Weinberg law is that migration does not take place, but many natural populations do experience migration from other populations. The overall effect of migration is twofold: (1) it prevents populations from becoming genetically different from one another and (2) it increases genetic variation within populations.

THE EFFECT OF MIGRATION ON ALLELIC FREQUENCIES Let's consider the effects of migration by looking at a simple, unidirectional model of migration between two populations that differ in the frequency of an allele a. Say that the frequency of this allele in population I is q_I and in population

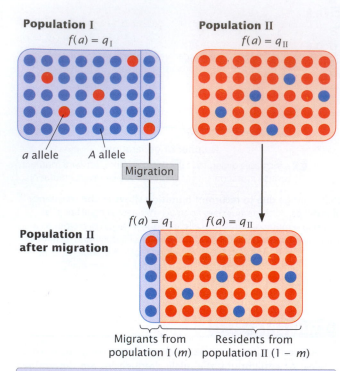

Population I
$f(a) = q_I$

Population II
$f(a) = q_{II}$

a allele   *A* allele

Migration

$f(a) = q_I$ $f(a) = q_{II}$

Population II after migration

Migrants from population I (*m*)   Residents from population II (1 − *m*)

Conclusion: The frequency of allele *a* in population II after migration is $q'_{II} = q_I m + q_{II}(1 − m)$.

25.10 The amount of change in allelic frequency due to migration between populations depends on the difference in allelic frequency and the extent of migration. Shown here is a model of the effect of unidirectional migration on allelic frequencies. The frequency of allele *a* in the source population (population I) is q_I. The frequency of this allele in the recipient population (population II) is q_{II}.

II is q_{II} (**Figure 25.10**). In each generation, a representative sample of the individuals in population I migrates to population II and reproduces, adding its alleles to population II's gene pool. Migration is only from population I to population II (unidirectional), and all the conditions of the Hardy–Weinberg law (large population size, random mating, etc.) apply, except the absence of migration.

After migration, population II consists of two types of individuals. Some are migrants; they make up proportion *m* of population II, and they carry alleles from population I, so the frequency of allele *a* in the migrants is q_I. The other individuals in population II are the original residents. If the migrants make up proportion *m* of population II, then the residents make up 1 − *m*; because the residents originated in population II, the frequency of allele *a* in this group is q_{II}. After migration, the frequency of allele *a* in the merged population II (q'_{II}) is

$$q'_{II} = q_I(m) + q_{II}(1 − m) \qquad (25.14)$$

where $q_I(m)$ is the contribution to *q* made by the copies of allele *a* in the migrants and $q_{II}(1 − m)$ is the contribution to *q* made by copies of allele *a* in the residents. The change in the allelic frequency due to migration (Δq) will be

$$\Delta q = m(q_I − q_{II}) \qquad (25.15)$$

Equation 25.15 summarizes the factors that determine the amount of change in allelic frequency due to migration. The amount of change in *q* is directly proportional to the migration (*m*); as the amount of migration increases, the change in allelic frequency increases. The magnitude of change is also affected by the differences in allelic frequencies of the two populations ($q_I − q_{II}$); when the difference is large, the change in allelic frequency will be large.

With each generation of migration, the frequencies of the two populations become more and more similar until, eventually, the allelic frequency of population II equals that of population I. When $q_I − q_{II} = 0$, there will be no further change in the allelic frequency of population II, in spite of the fact that migration continues. If migration between two populations takes place for a number of generations with no other evolutionary forces present, an equilibrium is reached at which the allelic frequency of the recipient population equals that of the source population.

The simple model of unidirectional migration between two populations just outlined can also be expanded to accommodate multidirectional migration between several populations.

THE OVERALL EFFECT OF MIGRATION   Migration has two major effects. First, it causes the gene pools of populations to become more similar. Later, we will see how genetic drift and natural selection lead to genetic differences between populations; migration counteracts this tendency and tends to keep populations homogeneous in their allelic frequencies. Second, migration adds genetic variation to populations. Different alleles may arise in different populations owing to rare mutational events, and these alleles can be spread to new populations by migration, increasing the genetic variation within the recipient population. ▶ TRY PROBLEM 34

CONCEPTS

Migration causes changes in the allelic frequency of a population by introducing alleles from other populations. The magnitude of change due to migration depends on both the extent of migration and the difference in allelic frequencies between the source and the recipient populations. Migration decreases genetic differences between populations and increases genetic variation within populations.

✓ CONCEPT CHECK 7

Each generation, 10 random individuals migrate from population A to population B. What will happen to allelic frequency *q* as a result of migration when *q* is equal in populations A and B?

a. *q* in A will decrease.   c. *q* will not change in either A or B.
b. *q* in B will increase.   d. *q* in B will become q^2.

Genetic Drift

The Hardy–Weinberg law assumes random mating in an infinitely large population; only when population size is infinite will the gametes carry genes that perfectly represent the parental gene pool. But no real population is infinitely large and when population size is limited, the gametes that unite to form individuals of the next generation carry a sample of alleles present in the parental gene pool. Just by chance, the composition of this sample will often deviate from that of the parental gene pool, and this deviation may cause allelic frequencies to change. The smaller the gametic sample, the greater the chance that its composition will deviate from that of the entire gene pool.

The role of chance in altering allelic frequencies is analogous to the flip of a coin. Each time we flip a coin, we have a 50% chance of getting a head and a 50% chance of getting a tail. If we flip a coin 1000 times, the observed ratio of heads to tails will be very close to the expected 50 : 50 ratio. If, however, we flip a coin only 10 times, there is a good chance that we will obtain not exactly five heads and five tails, but maybe seven heads and three tails or eight tails and two heads. This kind of deviation from an expected ratio due to limited sample size is referred to as **sampling error**.

Sampling error arises when gametes unite to produce progeny. Many organisms produce a large number of gametes but, when population size is small, a limited number of gametes unite to produce the individuals of the next generation. Chance influences which alleles are present in this limited sample and, in this way, sampling error may lead to **genetic drift**, or changes in allelic frequency. Because the deviations from the expected ratios are random, the direction of change is unpredictable. We can nevertheless predict the magnitude of the changes.

THE MAGNITUDE OF GENETIC DRIFT

The effect of genetic drift can be viewed in two ways. First, we can see how it influences the change in allelic frequencies of a single population with the passage of time. Second, we can see how it affects differences that accumulate among series of populations. Imagine that we have 10 small populations, all beginning with the exact same allelic frequencies of $p = 0.5$ and $q = 0.5$. When genetic drift occurs in a population, allelic frequencies within the population will change but, because drift is random, the way in which allelic frequencies change in each population will not be the same. In some populations, p may increase as a result of chance. In other populations, p may decrease as a result of chance. In time, the allelic frequencies in the 10 populations will become different: the populations will genetically diverge. As time passes, the change in allelic frequency within a population and the genetic divergence among populations are due to the same force—the random change in allelic frequencies. The magnitude of genetic drift can be assessed either by examining the change in allelic frequency within a single population or by examining the magnitude of genetic differences that accumulate among populations.

The amount of genetic drift can be estimated from the variance in allelic frequency. Variance, s^2, is a statistical measure that describes the degree of variability in a trait (see p. 691 in Chapter 24). Suppose that we observe a large number of separate populations, each with N individuals and allelic frequencies of p and q. After one generation of random mating, genetic drift expressed in terms of the variance in allelic frequency among the populations (s_p^2) will be

$$s_p^2 = \frac{pq}{2N} \qquad (25.16)$$

The amount of change resulting from genetic drift (the variance in allelic frequency) is determined by two parameters: the allelic frequencies (p and q) and the population size (N). Genetic drift will be maximal when p and q are equal (each 0.5). For example, assume that a population consists of 50 individuals. When the allelic frequencies are equal ($p = q = 0.5$), the variance in allelic frequency (s_p^2) will be $(0.5 \times 0.5)/(2 \times 50) = 0.0025$. In contrast, when $p = 0.9$ and $q = 0.1$, the variance in allelic frequency will be only 0.0009. Genetic drift will also be higher when the population size is small. If $p = q = 0.5$, but the population size is only 10 instead of 50, then the variance in allelic frequency becomes $(0.5 \times 0.5)/(2 \times 10) = 0.0125$, which is five times as great as when population size is 50.

This divergence of populations through genetic drift is strikingly illustrated in the results of an experiment carried out by Peter Buri on fruit flies (**Figure 25.11**). Buri examined the frequencies of two alleles (bw^{75} and bw) that affect eye color in fruit flies. He set up 107 populations, each consisting of eight males and eight females. He began each population with a frequency of bw^{75} equal to 0.5. He allowed the flies within each replicate to mate randomly and, each generation, he randomly selected eight male and eight female flies to be the parents of the next generation. He followed the changes in the frequencies of the two alleles over 19 generations. In one population, the average frequency of bw^{75} (p) over the 19 generations was 0.5312. We can use Equation 25.16 to calculate the expected variance in allelic frequency due to genetic drift. The frequency of the bw allele (q) will be $1 - p = 1 - 0.53125 = 0.46875$. The population size ($N$) equals 16. The expected variance in allelic frequency will be

$$\frac{pq}{2N} \qquad \frac{0.53125 \times 0.46875}{2 \times 16} = 0.0156$$

which was very close to the actual observed variance of 0.0151.

The effect of population size on genetic drift is illustrated by a study conducted by Luca Cavalli-Sforza and his colleagues. They studied variation in blood types among villagers in the Parma Valley of Italy, where the amount of migration between villages was limited. They found that variation in allelic frequency was greatest between small isolated villages in the upper valley but decreased between larger villages and towns farther down the valley. This result is exactly what we expect with genetic drift: there should be more genetic drift and thus more variation among villages when population size is small.

Experiment

Question: What effect does genetic drift have on the genetic composition of populations?

Methods Buri examined the frequencies of two alleles (bw^{75} and bw) that affect *Drosophila* eye color in 107 replicate small populations over 19 generations.

Results

Conclusion: As a result of genetic drift, allelic frequencies in the different populations diverged and often became fixed for one allele or the other.

25.11 Populations diverge in allelic frequency and become fixed for one allele as a result of genetic drift. In Buri's study of two eye-color alleles (bw^{75} and bw) in *Drosophila*, each population consisted of eight males and eight females and began with the frequency of bw^{75} equal to 0.5.

For ecological and demographic studies, population size is usually defined as the number of individuals in a group. However, the evolution of a gene pool depends only on those individuals who contribute genes to the next generation. Population geneticists usually define population size as the

equivalent number of breeding adults, the **effective population size** (N_e). Several factors determine the equivalent number of breeding adults, including the sex ratio, variation between individuals in reproductive success, fluctuations in population size, the age structure of the population, and whether mating is random.

CONCEPTS

Genetic drift is change in allelic frequency due to chance factors. The amount of change in allelic frequency due to genetic drift is inversely related to the effective population size (the equivalent number of breeding adults in a population).

✓ CONCEPT CHECK 8

Which of the following statements is an example of genetic drift?
a. Allele *g* for fat production increases in a small population because birds with more body fat have higher survivorship in a harsh winter.
b. Random mutation increases the frequency of allele *A* in one population but not in another.
c. Allele *R* reaches a frequency of 1.0 because individuals with genotype *rr* are sterile.
d. Allele *m* is lost when a virus kills all but a few individuals and just by chance none of the survivors possess allele *m*.

CAUSES OF GENETIC DRIFT All genetic drift arises from sampling error, but there are several different ways in which sampling error can arise. First, a population may be reduced in size for a number of generations because of limitations in space, food, or some other critical resource. Genetic drift in a small population for multiple generations can significantly affect the composition of a population's gene pool.

A second way that sampling error can arise is through the **founder effect**, which is due to the establishment of a population by a small number of individuals; the population of bighorn sheep at the National Bison Range, discussed in the introduction to this chapter, underwent a founder effect. Although a population may increase and become quite large, the genes carried by all its members are derived from the few genes originally present in the founders (assuming no migration or mutation). Chance events affecting which genes were present in the founders will have an important influence on the makeup of the entire population.

A third way in which genetic drift arises is through a **genetic bottleneck**, which develops when a population undergoes a drastic reduction in population size. An example is seen in northern elephant seals (**Figure 25.12**). Before 1800, thousands of northern elephant seals were found along the California coast, but hunting between 1820 and 1880 devastated the population. By 1884, as few as 20 seals survived on a remote beach of Isla de Guadalupe west of Baja California, Mexico. Restrictions on hunting enacted by the United States and Mexico allowed the seals to recover, and there are now more than 30,000 seals in

25.12 Northern elephant seals underwent a severe genetic bottleneck between 1820 and 1880. Today, these seals have low levels of genetic variation. [PhotoDisc/Getty Images.]

the population. All seals in the population today are genetically similar, because they have genes that were carried by the few survivors of the population bottleneck.

THE EFFECTS OF GENETIC DRIFT Genetic drift has several important effects on the genetic composition of a population. First, it produces change in allelic frequencies within a population. Because drift is random, allelic frequency is just as likely to increase as it is to decrease and will wander with the passage of time (hence the name genetic drift). **Figure 25.13** illustrates a computer simulation of genetic drift in five populations over 30 generations, starting with $q = 0.5$ and maintaining a constant population size of 10 males and 10 females. These allelic frequencies change randomly from generation to generation.

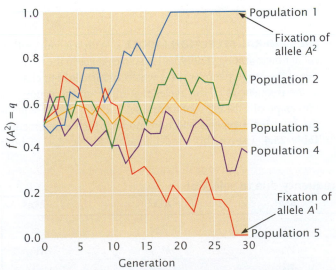

25.13 Genetic drift changes allelic frequencies within populations, leading to a reduction in genetic variation through fixation and genetic divergence among populations. Shown here is a computer simulation of changes in the frequency of allele A^2 (q) in five different populations due to random genetic drift. Each population consists of 10 males and 10 females and begins with $q = 0.5$.

A second effect of genetic drift is to reduce genetic variation within populations. Through random change, an allele may eventually reach a frequency of either 1 or 0, at which point all individuals in the population are homozygous for one allele. When an allele has reached a frequency of 1, we say that it has reached **fixation**. Other alleles are lost (reach a frequency of 0) and can be restored only by migration from another population or by mutation. Fixation, then, leads to a loss of genetic variation within a population. This loss can be seen in the northern elephant seals just described. Today, these seals have low levels of genetic variation; a study of 24 protein-encoding genes found no individual or population differences in these genes. A subsequent study of sequence variation in mitochondrial DNA also revealed low levels of genetic variation. In contrast, southern elephant seals had much higher levels of mitochondrial DNA variation. Southern elephant seals, which are found in Antarctica and sub-Antarctic waters, also were hunted, but their population size never dropped below 1000; therefore, unlike the northern elephant seals, they did not experience a genetic bottleneck.

Given enough time, all small populations will become fixed for one allele or the other. Which allele becomes fixed is random and is determined by the initial frequency of the allele. If a population begins with two alleles, each with a frequency of 0.5, both alleles have an equal probability of fixation. However, if one allele is initially common, it is more likely to become fixed.

A third effect of genetic drift is that different populations diverge genetically with time. In Figure 25.13, all five populations begin with the same allelic frequency ($q = 0.5$) but, because drift is random, the frequencies in different populations do not change in the same way, and so populations gradually acquire genetic differences. Eventually, all the populations will reach fixation; some will become fixed for one allele, and others will become fixed for the alternative allele.

The three results of genetic drift (allelic frequency change, loss of variation within populations, and genetic divergence between populations) take place simultaneously, and all result from sampling error. The first two results take place *within* populations, whereas the third takes place *between* populations. ▶ **TRY PROBLEM 35**

CONCEPTS

Genetic drift results from continuous small population size, the founder effect (establishment of a population by a few founders), and the bottleneck effect (population reduction). Genetic drift causes a change in allelic frequencies within a population, a loss of genetic variation through the fixation of alleles, and genetic divergence between populations.

Natural Selection

A final process that brings about changes in allelic frequencies is natural selection, the differential reproduction of genotypes (see p. 704 in Chapter 24). Natural selection takes place when individuals with adaptive traits produce a greater

25.14 Natural selection leads to adaptations, such as those seen in the polar bears that inhabit the extreme Arctic environment. These bears blend into the snowy background, which helps them in hunting seals. The hairs of their fur stay erect even when wet, and thick layers of blubber provide insulation, which protects against subzero temperatures. Their digestive tracts are adapted to a seal-based carnivorous diet. [Digital Vision.]

number of offspring than that produced by others in the population. If the adaptive traits have a genetic basis, they are inherited by the offspring and appear with greater frequency in the next generation. A trait that provides a reproductive advantage thereby increases with the passage of time, enabling populations to become better suited to their environments—better adapted. Natural selection is unique among evolutionary forces in that it promotes adaptation (**Figure 25.14**).

FITNESS AND THE SELECTION COEFFICIENT   The effect of natural selection on the gene pool of a population depends on the fitness values of the genotypes in the population. **Fitness** is defined as the relative reproductive success of a genotype. Here, the term *relative* is critical: fitness is the reproductive success of one genotype compared with the reproductive successes of other genotypes in the population.

Fitness (W) ranges from 0 to 1. Suppose the average number of viable offspring produced by three genotypes is

Genotypes:	A^1A^1	A^1A^2	A^2A^2
Mean number of offspring produced:	10	5	2

To calculate fitness for each genotype, we take the mean number of offspring produced by a genotype and divide it by the mean number of offspring produced by the most prolific genotype:

$$A^1A^1 \qquad\qquad A^1A^2 \qquad\qquad A^2A^2$$
$$W_{11} = \frac{10}{10} = 1.0 \quad W_{12} = \frac{5}{10} = 0.5 \quad W_{22} = \frac{2}{10} = 0.2 \quad (25.17)$$

The fitness of genotype A^1A^1 is designated W_{11}, that of A^1A^2 is W_{12}, and that of A^2A^2 is W_{22}. A related variable is the **selection coefficient** (s), which is the relative intensity of selection against a genotype. We usually speak of selection for a particular genotype, but keep in mind that, when selection is *for* one genotype, selection is automatically *against* at least one other genotype. The selection coefficient is equal to $1 - W$; so the selection coefficients for the preceding three genotypes are

$$A^1A^1 \qquad A^1A^2 \qquad A^2A^2$$
Selection coefficient ($1 - W$): $s_{11} = 0$ $s_{12} = 0.5$ $s_{22} = 0.8$

CONCEPTS

Natural selection is the differential reproduction of genotypes. It is measured as fitness, which is the reproductive success of a genotype compared with other genotypes in a population.

✓ **CONCEPT CHECK 9**

The average numbers of offspring produced by three genotypes are: $GG = 6$; $Gg = 3$, $gg = 2$. What is the fitness of Gg?

a. 3   b. 0.5   c. 0.3   d. 0.27

THE GENERAL SELECTION MODEL   With selection, differential fitness among genotypes leads to changes in the frequencies of the genotypes over time, which, in turn, lead to changes in the frequencies of the alleles that make up the genotypes. We can predict the effect of natural selection on allelic frequencies by using a general selection model, which is outlined in **Table 25.4**. Use of this model requires knowledge of both

Table 25.4   Method for determining changes in allelic frequency due to selection

	A^1A^1	A^1A^2	A^2A^2
Initial genotypic frequencies	p^2	$2pq$	q^2
Fitnesses	W_{11}	W_{12}	W_{22}
Proportionate contribution of genotypes to population	p^2W_{11}	$2pqW_{12}$	q^2W_{22}
Relative genotypic frequency after selection	$\dfrac{p^2W_{11}}{\overline{w}}$	$\dfrac{2pqW_{12}}{\overline{w}}$	$\dfrac{q^2W_{22}}{\overline{w}}$

Note: $\overline{w} = p^2W_{11} + 2pqW_{12} + q^2W_{22}$

Allelic frequencies after selection: $p' = f(A^1) = f(A^1A^1) + \frac{1}{2}f(A^1A^2)$; $q' = 1 - p'$.

Table 25.5 Formulas for calculating change in allelic frequencies with different types of selection

Type of Selection	Fitness Values			Change in q
	A^1A^1	A^1A^2	A^2A^2	
Selection against a recessive trait	1	1	$1-s$	$\dfrac{-spq^2}{1-sq^2}$
Selection against a dominant trait	1	$1-s$	$1-s$	$\dfrac{-spq^2}{1-s+sq^2}$
Selection against a trait with no dominance	1	$1-\frac{1}{2}s$	$1-s$	$\dfrac{-\frac{1}{2}spq}{1-sq}$
Selection against both homozygotes (overdominance)	$1-s_{11}$	1	$1-s_{22}$	$\dfrac{pq(s_{11}p-s_{22}q)}{1-s_{11}p^2-s_{22}q^2}$

the initial allelic frequencies and the fitness values of the genotypes. It assumes that mating is random and that the only force acting on a population is natural selection. The general selection model can be used to calculate the allelic frequencies after any type of selection. It is also possible to work out formulas for determining the change in allelic frequency when selection is against recessive, dominant, and codominant traits, as well as traits in which the heterozygote has highest fitness (**Table 25.5**).

CONCEPTS

The change in allelic frequency due to selection can be determined for any type of genetic trait by using the general selection model.

WORKED PROBLEM

Let's apply the general selection model in Table 25.4 to a real example. Alcohol is a common substance in rotting fruit, where fruit-fly larvae grow and develop; larvae use the enzyme alcohol dehydrogenase (ADH) to detoxify the effects of the alcohol. In some fruit-fly populations, two alleles are present at the locus that encodes ADH: Adh^F, which encodes a form of the enzyme that migrates rapidly (fast) on an electrophoretic gel; and Adh^S, which encodes a form of the enzyme that migrates slowly on an electrophoretic gel. Female fruit flies with different Adh genotypes produce the following numbers of offspring when alcohol is present:

Genotype	Mean number of offspring
Adh^F/Adh^F	120
Adh^F/Adh^S	60
Adh^S/Adh^S	30

a. Calculate the relative fitnesses of females having these genotypes.

b. If a population of fruit flies has an initial frequency of Adh^F equal to 0.2, what will the frequency be in the next generation when alcohol is present?

Solution Strategy

What information is required in your answer to the problem?

a. The fitnesses of females of each genotype.

b. The frequency of Adh in the next generation when alcohol is present.

What information is provided to solve the problem?

- The mean numbers of offspring produced by each genotype.
- The population has an initial allelic frequency of 0.2.

Solution Steps

a. First, we must calculate the fitnesses of the three genotypes. Fitness is the relative reproductive output of a genotype and is calculated by dividing the mean number of offspring produced by that genotype by the mean number of offspring produced by the most prolific genotype. The fitnesses of the three Adh genotypes therefore are:

Genotype	Mean number of offspring	Fitness
Adh^F/Adh^F	120	$W_{FF} = \dfrac{120}{120} = 1$
Adh^F/Adh^S	60	$W_{FS} = \dfrac{60}{120} = 0.5$
Adh^S/Adh^S	30	$W_{SS} = \dfrac{30}{120} = 0.25$

	$Adh^F Adh^F$	$Adh^F Adh^S$	$Adh^S Adh^S$
Initial genotypic frequencies:	$p^2 = (0.2)^2 = 0.04$	$2pq = 2(0.2)(0.8) = 0.32$	$q^2 = (0.8)^2 = 0.64$
Fitnesses:	$W_{FF} = 1$	$W_{FS} = 0.5$	$W_{SS} = 0.25$
Proportionate contribution of genotypes to population:	$p^2 W_{FF} = 0.04(1) = 0.04$	$2pq W_{FS} = (0.32)(0.5) = 0.16$	$q^2 W_{SS} = (0.64)(0.25) = 0.16$

b. To calculate the frequency of the Adh^F allele after selection, we can apply the table method. In the first row of the table above, we record the initial genotypic frequencies before selection has acted. If mating has been random (an assumption of the model), the genotypes will have the Hardy–Weinberg equilibrium frequencies of p^2, $2pq$, and q^2. In the second row of the table above, we put the fitness values of the corresponding genotypes. The proportion of the population represented by each genotype after selection is obtained by multiplying the initial genotypic frequency times its fitness (third row of Table 25.4). Now the genotypes are no longer in Hardy–Weinberg equilibrium.

The mean fitness ($\bar{w}$) of the population is the sum of the proportionate contributions of the three genotypes: $\bar{w} = p^2 W_{11} + 2pq W_{12} + q^2 W_{22} = 0.04 + 0.16 + 0.16 = 0.36$. The mean fitness $\bar{w}$ is the average fitness of all individuals in the population and allows the frequencies of the genotypes after selection to be obtained.

The frequency of a genotype after selection will be equal to its proportionate contribution divided by the mean fitness of the population ($p^2 W_{11}/\bar{w}$ for genotype $A^1 A^1$, $2pq W_{12}/\bar{w}$ for genotype $A^1 A^2$, and $q^2 W_{22}/\bar{w}$ for genotype $A^2 A^2$) as shown in the fourth line of Table 25.4. We can now add these values to our table as shown below:

	$Adh^F Adh^F$	$Adh^F Adh^S$	$Adh^S Adh^S$
Initial genotypic frequencies:	$p^2 = (0.2)^2 = 0.04$	$2pq = 2(0.2)(0.8) = 0.32$	$q^2 = (0.8)^2 = 0.64$
Fitnesses:	$W_{FF} = 1$	$W_{FS} = 0.5$	$W_{SS} = 0.25$
Proportionate contribution of genotypes to population:	$p^2 W_{FF} = 0.04(1) = 0.04$	$2pq W_{FS} = (0.32)(0.5) = 0.16$	$q^2 W_{SS} = (0.64)(0.25) = 0.16$
Relative genotypic frequency after selection:	$\dfrac{p^2 w_{FF}}{\bar{w}} = \dfrac{0.04}{0.36} = 0.11$	$\dfrac{2pq w_{FS}}{\bar{w}} = \dfrac{0.16}{0.36} = 0.44$	$\dfrac{q^2 w_{SS}}{\bar{w}} = \dfrac{0.16}{0.36} = 0.44$

After the new genotypic frequencies have been calculated, the new allelic frequency of Adh^F (p') can be determined by using the now-familiar formula of Equation 25.4:

$$p' = f(Adh^F) = f(Adh^F/Adh^F) + \tfrac{1}{2}f(Adh^F/Adh^S)$$
$$= 0.11 + \tfrac{1}{2}(0.44) = 0.33$$

and that of q' can be obtained by subtraction:

$$q' = 1 - p'$$
$$= 1 - 0.33 = 0.67$$

We predict that the frequency of Adh^F will increase from 0.2 to 0.33.

▶ For more practice with the selection model, try Problem 37 at the end of this chapter.

THE RESULTS OF SELECTION The results of selection depend on the relative fitnesses of the genotypes. If we have three genotypes ($A^1 A^1$, $A^1 A^2$, and $A^2 A^2$) with fitnesses W_{11}, W_{12}, and W_{22}, we can identify six different types of natural selection (**Table 25.6**). In type 1 selection, a dominant allele

A^1 confers a fitness advantage; in this case, the fitnesses of genotypes $A^1 A^1$ and $A^1 A^2$ are equal and higher than the fitness of $A^2 A^2$ ($W_{11} = W_{12} > W_{22}$). Because both the heterozygote and the $A^1 A^1$ homozygote have copies of the A^1 allele and produce more offspring than the $A^2 A^2$ homozygote does, the frequency of the A^1 allele will increase with time, and the frequency of the A^2 allele will decrease. This form of selection, in which one allele or trait is favored over another, is termed **directional selection**.

Type 2 selection (see Table 25.6) is directional selection against a dominant allele A^1 ($W_{11} = W_{12} < W_{22}$). In this case, the A^2 allele increases and the A^1 allele decreases. Type 3 and type 4 selection also are directional selection but, in these cases, there is incomplete dominance and the heterozygote has a fitness that is intermediate between the two homozygotes ($W_{11} > W_{12} > W_{22}$ for type 3; $W_{11} < W_{12} < W_{22}$ for type 4). When $A^1 A^1$ has the highest fitness (type 3), the A^1 allele increases and the A^2 allele decreases with the passage of time. When $A^2 A^2$ has the highest fitness (type 4), the A^2 allele increases and the A^1 allele decreases with time. Eventually, directional selection leads to fixation of the favored allele and elimination of the other allele, as long as no other evolutionary forces act on the population.

Table 25.6 Types of natural selection

Type	Fitness Relation	Form of Selection	Result
1	$W_{11} = W_{12} > W_{22}$	Directional selection against recessive allele A^2	A^1 increases, A^2 decreases
2	$W_{11} = W_{12} < W_{22}$	Directional selection against dominant allele A^1	A^2 increases, A^1 decreases
3	$W_{11} > W_{12} > W_{22}$	Directional selection against incompletely dominant allele A^2	A^1 increases, A^2 decreases
4	$W_{11} < W_{12} < W_{22}$	Directional selection against incompletely dominant allele A^1	A^2 increases, A^1 decreases
5	$W_{11} < W_{12} > W_{22}$	Overdominance	Stable equilibrium, both alleles maintained
6	$W_{11} > W_{12} < W_{22}$	Underdominance	Unstable equilibrium

Note: W_{11}, W_{12}, and W_{22} represent the fitnesses of genotypes A^1A^1, A^1A^2, and A^2A^2, respectively.

Two types of selection (types 5 and 6) are special situations that lead to equilibrium, where there is no further change in allelic frequency. Type 5 selection is referred to as **overdominance** or heterozygote advantage. Here, the heterozygote has higher fitness than the fitnesses of the two homozygotes ($W_{11} < W_{12} > W_{22}$). With overdominance, both alleles are favored in the heterozygote, and neither allele is eliminated from the population. Initially, the allelic frequencies may change because one homozygote has higher fitness than the other; the direction of change will depend on the relative fitness values of the two homozygotes. The allelic frequencies change with overdominant selection until a stable equilibrium is reached, at which point there is no further change. The allelic frequency at equilibrium ($\hat{q}$) depends on the relative fitnesses (usually expressed as selection coefficients) of the two homozygotes:

$$\hat{q} = f(A^2) = \frac{s_{11}}{s_{11} + s_{22}} \qquad (25.18)$$

where s_{11} represents the selection coefficient of the A^1A^1 homozygote and s_{22} represents the selection coefficient of the A^2A^2 homozygote.

An example of overdominance is the gene for sickle-cell anemia in humans, a disease that results from a mutation in one of the genes that encodes hemoglobin. People who are homozygous for the sickle-cell mutation produce only sickle-cell hemoglobin, have severe anemia, and often have tissue damage. People who are heterozygous—with one normal copy and one mutated copy of the gene—produce both normal and sickle-cell hemoglobin, but their red blood cells contain enough normal hemoglobin to prevent sickle-cell anemia. However, heterozygotes are resistant to malaria and have higher fitness than do homozygotes for normal hemoglobin and homozygotes for sickle-cell anemia.

The last type of selection (type 6) is **underdominance**, in which the heterozygote has lower fitness than both homozygotes ($W_{11} > W_{12} < W_{22}$). Underdominance leads to an unstable equilibrium; here, allelic frequencies will not change as long as they are at equilibrium but, if they are disturbed from the equilibrium point by some other evolutionary force, they will move away from equilibrium until one allele eventually becomes fixed. To see the effects of natural selection on allelic and genotypic frequencies, view the Mini-Tutorial in **Animation 25.1.** ▶ TRY PROBLEM 38

CONCEPTS

Natural selection changes allelic frequencies; the direction and magnitude of change depend on the intensity of selection, the dominance relations of the alleles, and the allelic frequencies. Directional selection favors one allele over another and eventually leads to fixation of the favored allele. Overdominance leads to a stable equilibrium with maintenance of both alleles in the population. Underdominance produces an unstable equilibrium because the heterozygote has lower fitness than those of the two homozygotes.

✔ CONCEPT CHECK 10

How does overdominance differ from directional selection?

CHANGE IN THE ALLELIC FREQUENCY OF A RECESSIVE ALLELE DUE TO NATURAL SELECTION The rate at which selection changes allelic frequencies depends on the allelic frequency itself. If an allele (A^2) is lethal and recessive, $W_{11} = W_{12} = 1$, whereas $W_{22} = 0$. The frequency of the A^2 allele will decrease with time (because the A^2A^2 homozygote produces no offspring), and the rate of decrease will be proportional to the frequency of the recessive allele. When the frequency of the allele is high, the change in each generation is relatively large but, as the frequency of the allele drops, a higher proportion of the alleles are in the heterozygous genotypes, where they are immune to the action of natural selection (the heterozygotes have the same phenotype as the favored homozygote). Thus, selection against a rare recessive allele is very inefficient and its removal from the population is slow.

The relation between the frequency of a recessive allele and its rate of change under natural selection has an important implication. Some people believe that the medical treatment of patients with rare recessive diseases will cause the disease gene to increase, eventually leading to degeneration of the human gene pool. This mistaken belief was the basis of eugenic laws that were passed in the early part of the twentieth century prohibiting the marriage of persons with certain genetic conditions and allowing the involuntary sterilization of others. However, most copies of rare recessive alleles are present in heterozygotes, and selection against the homozygotes will have little effect on the frequency of a recessive allele. Thus, whether the homozygotes for a recessive trait reproduce or not has little effect on the frequency of the disorder.

MUTATION AND NATURAL SELECTION   Recurrent mutation and natural selection act as opposing forces on detrimental alleles; mutation increases their frequency and natural selection decreases their frequency. Eventually, these two forces reach an equilibrium, in which the number of alleles added by mutation is balanced by the number of alleles removed by selection.

The frequency of a deleterious recessive allele at equilibrium ($\hat{q}$) is equal to the square root of the mutation rate divided by the selection coefficient:

$$\hat{q} = \sqrt{\frac{\mu}{s}} \qquad (25.19)$$

For selection acting on a deleterious dominant allele, the frequency of the dominant allele at equilibrium can be shown to be

$$\hat{q} = \frac{\mu}{s} \qquad (25.20)$$

Achondroplasia is a common type of human dwarfism that results from a dominant gene. People with this condition are fertile, although they produce only about 74% as many children as are produced by people without achondroplasia. The fitness of people with achondroplasia therefore averages 0.74, and the selection coefficient (s) is $1 - W$, or 0.26. If we assume that the mutation rate for achondroplasia is about 3×10^{-5} (a typical mutation rate in humans), then we can predict that the equilibrium frequency for the achondroplasia allele will be

$$\hat{q} = \frac{0.00003}{0.26} = 0.0001153$$

This frequency is close to the actual frequency of the condition.

▶ **TRY PROBLEM 41**

CONCEPTS

Mutation and natural selection act as opposing forces on detrimental alleles: mutation tends to increase their frequency and natural selection tends to decrease their frequency, eventually producing an equilibrium.

CONNECTING CONCEPTS

The General Effects of Forces That Change Allelic Frequencies

You now know that four processes bring about change in the allelic frequencies of a population: mutation, migration, genetic drift, and natural selection. Their short- and long-term effects on allelic frequencies are summarized in **Table 25.7**. In some cases, these changes continue until one allele is eliminated and the other becomes fixed in the population. Genetic drift and directional selection will eventually result in fixation, provided these forces are the only ones acting on a population. With the other evolutionary forces, allelic frequencies change until an equilibrium point is reached, and then there is no additional change in allelic frequency. Mutation, migration, and some forms of natural selection can lead to stable equilibria (see Table 25.7).

The different evolutionary forces affect both genetic variation within populations and genetic divergence between populations. Evolutionary forces that maintain or increase genetic variation within populations are listed in the upper-left quadrant of **Figure 25.15**. These forces include some types of natural selection, such as overdominance, in which both alleles are favored. Mutation and migration also increase genetic variation within populations because they introduce new alleles to the population. Evolutionary forces that decrease genetic variation within populations are listed in the lower-left quadrant of Figure 25.15. These forces include genetic drift, which decreases variation through the fixation of alleles, and some forms of natural selection such as directional selection.

The various evolutionary forces also affect the amount of genetic divergence between populations. Natural selection increases divergence between populations if different alleles are favored in the different populations, but it can also decrease divergence between populations by favoring the same allele in the different populations. Mutation almost always increases divergence between populations because different mutations arise in each population. Genetic drift also increases divergence between populations because changes in allelic frequencies due to drift are random and are likely to change in different directions in separate populations. Migration, on the other hand, decreases divergence between populations because it makes populations similar in their genetic composition.

Migration and genetic drift act in opposite directions: migration increases genetic variation within populations and decreases divergence between populations, whereas genetic drift decreases genetic variation within populations and increases divergence among populations. Mutation increases both variation within

Table 25.7	Effects of different evolutionary forces on allelic frequencies within populations	
Force	**Short-Term Effect**	**Long-Term Effect**
Mutation	Change in allelic frequency	Equilibrium reached between forward and reverse mutations
Migration	Change in allelic frequency	Equilibrium reached when allelic frequencies of source and recipient population are equal
Genetic drift	Change in allelic frequency	Fixation of one allele
Natural selection	Change in allelic frequency	Directional selection: fixation of one allele Overdominant selection: equilibrium reached

	Within populations	Between populations
Increase genetic variation	**Mutation Migration Some types of natural selection**	**Mutation Genetic drift Some types of natural selection**
Decrease genetic variation	**Genetic drift Some types of natural selection**	**Migration Some types of natural selection**

25.15 Mutation, migration, genetic drift, and natural selection have different effects on genetic variation within populations and on genetic divergence between populations.

populations and divergence between populations. Natural selection can either increase or decrease variation within populations, and it can increase or decrease divergence between populations.

An important point to keep in mind is that real populations are simultaneously affected by many evolutionary forces. We have examined the effects of mutation, migration, genetic drift, and natural selection in isolation so that the influence of each process would be clear. However, in the real world, populations are commonly affected by several evolutionary forces at the same time, and evolution results from the complex interplay of numerous processes.

CONCEPTS SUMMARY

■ Population genetics examines the genetic composition of groups of individuals and how this composition changes with time.

■ A Mendelian population is a group of interbreeding, sexually reproducing individuals, whose set of genes constitutes the population's gene pool. Evolution takes place through changes in this gene pool.

■ A population's genetic composition can be described by its genotypic and allelic frequencies.

■ The Hardy–Weinberg law describes the effects of reproduction and Mendel's laws on the allelic and genotypic frequencies of a population. It assumes that a population is large, randomly mating, and free from the effects of mutation, migration, and natural selection. When these conditions are met, the allelic frequencies do not change and the genotypic frequencies stabilize after one generation in the Hardy–Weinberg equilibrium proportions p^2, $2pq$, and q^2, where p and q equal the frequencies of the alleles.

■ Nonrandom mating affects the frequencies of genotypes but not those of alleles.

■ Inbreeding, a type of positive assortative mating, increases the frequency of homozygotes while decreasing the frequency of heterozygotes. Inbreeding is frequently detrimental because it increases the appearance of lethal and deleterious recessive traits.

■ Mutation, migration, genetic drift, and natural selection can change allelic frequencies.

■ Recurrent mutation eventually leads to an equilibrium, with the allelic frequencies being determined by the relative rates of forward and reverse mutation. Change due to mutation in a single generation is usually very small because mutation rates are low.

■ Migration, the movement of genes between populations, increases the amount of genetic variation within populations and decreases the number of differences between populations.

■ Genetic drift is change in allelic frequencies due to chance factors. Genetic drift arises when a population consists of a small number of individuals, is established by a small number of founders, or undergoes a major reduction in size. Genetic drift changes allelic frequencies, reduces genetic variation within populations, and causes genetic divergence among populations.

■ Natural selection is the differential reproduction of genotypes; it is measured by the relative reproductive successes (fitnesses) of genotypes. The effects of natural selection on allelic frequency can be determined by applying the general selection model. Directional selection leads to the fixation of one allele. The rate of change in allelic frequency due to selection depends on the intensity of selection, the dominance relations, and the initial frequencies of the alleles.

■ Mutation and natural selection can produce an equilibrium, in which the number of new alleles introduced by mutation is balanced by the elimination of alleles through natural selection.

IMPORTANT TERMS

genetic rescue (p. 716)
Mendelian population (p. 716)
gene pool (p. 716)
genotypic frequency (p. 717)
allelic frequency (p. 717)
Hardy–Weinberg law (p. 719)
Hardy–Weinberg equilibrium (p. 719)

positive assortative mating (p. 723)
negative assortative mating (p. 723)
inbreeding (p. 723)
outcrossing (p. 723)
inbreeding coefficient (p. 723)

inbreeding depression (p. 724)
equilibrium (p. 727)
migration (gene flow) (p. 727)
sampling error (p. 729)
genetic drift (p. 729)
effective population size (p. 730)

founder effect (p. 730)
genetic bottleneck (p. 730)
fixation (p. 731)
fitness (p. 732)
selection coefficient (p. 732)
directional selection (p. 734)
overdominance (p. 735)
underdominance (p. 735)

ANSWERS TO CONCEPT CHECKS

1. There are fewer alleles than genotypes, so the gene pool can be described by fewer parameters when allelic frequencies are used. Additionally, the genotypes are temporary assemblages of alleles that break down each generation; the alleles are passed from generation to generation in sexually reproducing organisms.

2. a

3. c

4. 0.10

5. b

6. c

7. c

8. d

9. b

10. In overdominance, the heterozygote has the highest fitness and both alleles are maintained. In directional selection, selection causes one allele or trait to increase in frequency.

WORKED PROBLEM

Problem

A recessive allele for red hair (*r*) has a frequency of 0.2 in population I and a frequency of 0.01 in population II. A famine in population I causes a number of people in population I to migrate to population II, where they reproduce randomly with the members of population II. Geneticists estimate that, after migration, 15% of the people in population II consist of people who migrated from population I. What will be the frequency of red hair in population II after the migration?

Solution Strategy

What information is required in your answer to the problem?

The frequency of red hair in population II after the migration.

What information is provided to solve the problem?

- The allele for red hair (*r*) is recessive.
- The initial frequency of *r* in population I is 0.2.
- The initial frequency of *r* in population II is 0.01.
- After migration, 15% of the people in population II consist of people who migrated from population I.

For help with this problem, review:

Migration in Section 25.4.

Solution Steps

From Equation 25.14, the allelic frequency in a population after migration (q'_{II}) is

$$q'_{II} = q_I(m) + q_{II}(1 - m)$$

where q_I and q_{II} are the allelic frequencies in population I (migrants) and population II (residents), respectively, and *m* is the proportion of population II that consist of migrants. In this problem, the frequency of red hair is 0.2 in population I and 0.01 in population II. Because 15% of population II consists of migrants, $m = 0.15$. Substituting these values into Equation 25.14, we obtain

$$q'_{II} = 0.2(0.15) + (0.01)(1 - 0.15) = 0.03 + 0.0085 = 0.0385$$

which is the expected frequency of the allele for red hair in population II after migration. Red hair is a recessive trait; if mating is random for hair color, the frequency of red hair in population II after migration will be

$$f(rr) = q^2 = (0.0385)^2 = 0.0015$$

> **Recall:** With random mating the expected frequencies of the genotypes are p^2 (*RR*), $2pq$ (*Rr*) and q^2 (*rr*).

COMPREHENSION QUESTIONS

Section 25.1

1. What is a Mendelian population? How is the gene pool of a Mendelian population usually described?

Section 25.2

2. What are the predictions given by the Hardy–Weinberg law?

3. What assumptions must be met for a population to be in Hardy–Weinberg equilibrium?

4. What is random mating?

5. Give the Hardy–Weinberg expected genotypic frequencies for (a) an autosomal locus with three alleles, and (b) an X-linked locus with two alleles.

Section 25.3

6. Define inbreeding and briefly describe its effects on a population.

Section 25.4

7. What determines the allelic frequencies at mutational equilibrium?

8. What factors affect the magnitude of change in allelic frequencies due to migration?

9. Define genetic drift and give three ways in which it can arise. What effect does genetic drift have on a population?

10. What is effective population size? How does it affect the amount of genetic drift?

11. Define natural selection and fitness.

12. Briefly describe the differences between directional selection, overdominance, and underdominance. Describe the effect of each type of selection on the allelic frequencies of a population.

13. What factors affect the rate of change in allelic frequency due to natural selection?

14. Compare and contrast the effects of mutation, migration, genetic drift, and natural selection on genetic variation within populations and on genetic divergence between populations.

> ▶ For more questions that test your comprehension of the key chapter concepts, go to **LEARNING**Curve for this chapter.

APPLICATION QUESTIONS AND PROBLEMS

Section 25.1

15. How would you respond to someone who said that models are useless in studying population genetics because they represent oversimplifications of the real world?

*16. Voles (*Microtus ochrogaster*) were trapped in fields in southern Indiana and genotyped for a transferrin locus. The following numbers of genotypes were recorded, where T^E and T^F represent different alleles.

[Tom McHugh/Photo Researchers.]

$T^E T^E$	$T^E T^F$	$T^F T^F$
407	170	17

Calculate the genotypic and allelic frequencies of the transferrin locus for this population.

17. Jean Manning, Charles Kerfoot, and Edward Berger studied genotypic frequencies at the phosphoglucose isomerase (GPI) locus in the cladoceran *Bosmina longirostris* (a small crustacean known as a water flea). At one location, they collected 176 of the animals from Union Bay in Seattle, Washington, and determined their GPI genotypes by using electrophoresis (J. Manning, W. C. Kerfoot, and E. M. Berger. 1978. *Evolution* 32:365–374).

Genotype	Number
$S^1 S^1$	4
$S^1 S^2$	38
$S^2 S^2$	134

Determine the genotypic and allelic frequencies for this population.

*18. Orange coat color of cats is due to an X-linked allele (X^O) that is codominant with the allele for black (X^+). Genotypes of the orange locus of cats in Minneapolis and St. Paul, Minnesota, were determined, and the following data were obtained:

$X^O X^O$ females	11
$X^O X^+$ females	70
$X^+ X^+$ females	94
$X^O Y$ males	36
$X^+ Y$ males	112

Calculate the frequencies of the X^O and X^+ alleles for this population.

Section 25.2

19. Use the graph shown in **Figure 25.3** to determine which genotype is most frequent when the frequency of the *A* allele is:

  a. 0.2

  b. 0.5

  c. 0.8

20. A total of 6129 North American Caucasians were blood typed for the MN locus, which is determined by two codominant alleles, L^M and L^N. The following data were obtained:

Blood type	Number
M	1787
MN	3039
N	1303

Carry out a chi-square test to determine whether this population is in Hardy–Weinberg equilibrium at the MN locus.

21. Assume that the phenotypes of lady beetles shown in **Figure 25.1** are encoded by the following genotypes:

Phenotype	Genotype
All black	BB
Some black spots	Bb
No black spots	bb

a. For the lady beetles shown in the figure, calculate the frequencies of the genotypes and frequencies of the alleles.

b. Use a chi-square test to determine if the lady beetles shown are in Hardy Weinberg equilibrium.

*22. Most black bears (*Ursus americanus*) are black or brown in color. However, occasional white bears of this species appear in some populations along the coast of British Columbia. Kermit Ritland and his colleagues determined that white coat color in these bears results from a recessive muta- tion (*G*) caused by a single nucleotide replacement in which guanine substitutes for adenine at the melanocortin-1 receptor locus (*mcr1*), the same locus responsible for red hair in humans (K. Ritland, C. Newton, and H. D. Marshall. 2001. *Current Biology* 11:1468–1472). The wild-type allele at this locus (*A*) encodes black or brown color. Ritland and his colleagues collected samples from bears on three islands and determined their genotypes at the *mcr1* locus.

[Wendy Shattil/Alamy.]

Genotype	Number
AA	42
AG	24
GG	21

a. What are the frequencies of the *A* and *G* alleles in these bears?

b. Give the genotypic frequencies expected if the population is in Hardy–Weinberg equilibrium.

c. Use a chi-square test to compare the number of observed genotypes with the number expected under Hardy–Weinberg equilibrium. Is this population in Hardy–Weinberg equilibrium? Explain your reasoning.

23. Genotypes of leopard frogs from a population in central Kansas were determined for a locus (*M*) that encodes the enzyme malate dehydrogenase. The following numbers of genotypes were observed:

Genotype	Number
M^1M^1	20
M^1M^2	45
M^2M^2	42
M^1M^3	4
M^2M^3	8
M^3M^3	6
Total	125

a. Calculate the genotypic and allelic frequencies for this population.

b. What would the expected numbers of genotypes be if the population were in Hardy–Weinberg equilibrium?

24. Full color (*D*) in domestic cats is dominant over dilute color (*d*). Of 325 cats observed, 194 have full color and 131 have dilute color.

a. If these cats are in Hardy–Weinberg equilibrium for the dilution locus, what is the frequency of the dilute allele?

b. How many of the 194 cats with full color are likely to be heterozygous?

*25. Tay–Sachs disease is an autosomal recessive disorder. Among Ashkenazi Jews, the frequency of Tay–Sachs disease is 1 in 3600. If the Ashkenazi population is mating randomly for the Tay–Sachs gene, what proportion of the population consists of heterozygous carriers of the Tay–Sachs allele?

26. In the plant *Lotus corniculatus*, cyanogenic glycoside protects the plant against insect pests and even grazing by cattle. This glycoside is due to a simple dominant allele. A population of *L. corniculatus* consists of 77 plants that possess cyanogenic glycoside and 56 that lack the compound. What is the frequency of the dominant allele responsible for the presence of cyanogenic glycoside in this population?

27. Color blindness in humans is an X-linked recessive trait. Approximately 10% of the men in a particular population are color blind.

a. If mating is random for the color-blind locus, what is the frequency of the color-blind allele in this population?

b. What proportion of the women in this population are expected to be color blind?

c. What proportion of the women in the population are expected to be heterozygous carriers of the color-blind allele?

Section 25.3

*28. The human MN blood type is determined by two codominant alleles, L^M and L^N. The frequency of L^M in Eskimos on a small Arctic island is 0.80.

a. If random mating takes place in this population, what are the expected frequencies of the M, MN, and N blood types on the island?

b. If the inbreeding coefficient for this population is 0.05, what are the expected frequencies of the M, MN, and N blood types on the island?

29. Demonstrate mathematically that full-sib mating ($F = \frac{1}{4}$) reduces the heterozygosity by $\frac{1}{4}$ with each generation.

Section 25.4

*30. The forward mutation rate for piebald spotting in guinea pigs is 8×10^{-5}; the reverse mutation rate is 2×10^{-6}.

If no other evolutionary forces are assumed to be present, what is the expected frequency of the allele for piebald spotting in a population that is in mutational equilibrium?

31. For three years, Gunther Schlager and Margaret Dickie estimated the forward and reverse mutation rates for five loci in mice that encode various aspects of coat color by examining more than 5 million mice for spontaneous mutations (G. Schlager and M. M. Dickie. 1966. *Science* 151:205–206). The numbers of mutations detected at the dilute locus are as follows:

	Number of gametes examined	Number of mutations detected
Forward mutations	260,675	5
Reverse mutations	583,360	2

Calculate the forward and reverse mutation rates at this locus. If these mutations rates are representative of rates in natural populations of mice, what would the expected equilibrium frequency of dilute mutations be?

32. In **Figure 25.10**, each blue dot represents one copy of the *A* allele and each red dot represents each copy of the *a* allele. Calculate the frequencies of the *A* allele in population II before and after migration. Explain why the frequency of *A* in population II changed after migration.

33. In German cockroaches, curved wing (*cv*) is recessive to normal wing (*cv$^+$*). Bill, who is raising cockroaches in his dorm room, finds that the frequency of the gene for curved wings in his cockroach population is 0.6. In his friend Joe's apartment, the frequency of the gene for curved wings is 0.2. One day Joe visits Bill in his dorm room, and several cockroaches jump out of Joe's hair and join the population in Bill's room. Bill estimates that, now, 10% of the cockroaches in his dorm room are individual roaches that jumped out of Joe's hair. What is the new frequency of curved wings among cockroaches in Bill's room?

*34. A population of water snakes is found on an island in Lake Erie. Some of the snakes are banded and some are unbanded; banding is caused by an autosomal allele that is recessive to an allele for no bands. The frequency of banded snakes on the island is 0.4, whereas the frequency of banded snakes on the mainland is 0.81. One summer, a large number of snakes migrate from the mainland to the island. After this migration, 20% of the island population consists of snakes that came from the mainland.

a. If both the mainland population and the island population are assumed to be in Hardy–Weinberg equilibrium for the alleles that affect banding, what is the frequency of the allele for bands on the island and on the mainland before migration?

b. After migration has taken place, what is the frequency of the banded allele on the island?

*35. Pikas are small mammals that live at high elevation in the talus slopes of mountains. Most populations located on mountain tops in Colorado and Montana in North America are isolated from one another: the pikas don't occupy the low-elevation habitats that separate the mountain tops and don't venture far from the talus slopes. Thus, there is little gene flow between populations. Furthermore, each population is small in size and was founded by a small number of pikas.

A group of population geneticists propose to study the amount of genetic variation in a series of pika populations and to compare the allelic frequencies in different populations. On the basis of the biology and distribution of pikas, predict what the population geneticists will find concerning the within- and between-population genetic variation.

36. What proportion of the populations shown in **Figure 25.13** reached fixation by generations 10, 25, and 30? How does the proportion of populations that reach fixation due to genetic drift change over time?

*37. In a large, randomly mating population, the frequency of the allele (*s*) for sickle-cell hemoglobin is 0.028. The results of studies have shown that people with the following genotypes at the beta-chain locus produce the average numbers of offspring given:

Genotype	Average number of offspring produced
SS	5
Ss	6
ss	0

a. What will the frequency of the sickle-cell allele (*s*) be in the next generation?

b. What will the frequency of the sickle-cell allele be at equilibrium?

*38. Two chromosomal inversions are commonly found in populations of *Drosophila pseudoobscura*: Standard (*ST*) and Arrowhead (*AR*). When treated with the insecticide DDT, the genotypes for these inversions exhibit overdominance, with the following fitnesses:

Genotype	Fitness
ST/ST	0.47
ST/AR	1
AR/AR	0.62

What will the frequencies of *ST* and *AR* be after equilibrium has been reached?

39. In a large, randomly mating population, the frequency of an autosomal recessive lethal allele is 0.20. What will the frequency of this allele be in the next generation if the lethality takes place before reproduction?

40. The fruit fly *Drosophila melanogaster* normally feeds on rotting fruit, which may ferment and contain high levels of alcohol. Douglas Cavener and Michael Clegg studied allelic frequencies at the locus for alcohol dehydrogenase (*Adh*) in experimental populations of *D. melanogaster* (D. R. Cavener and M. T. Clegg. 1981. *Evolution* 35:1–10). The experimental populations were established from wild-caught flies and were raised in cages in the laboratory. Two control populations (C1 and C2) were raised on a standard cornmeal–molasses–agar diet. Two ethanol populations (E1 and E2) were raised on a cornmeal–molasses–agar diet to which was added 10% ethanol. The four populations were periodically sampled to determine the allelic frequencies of two alleles at the alcohol dehydrogenase locus, *Adh*^{S} and *Adh*^{F}. The frequencies of these alleles in the experimental populations are shown in the graph.

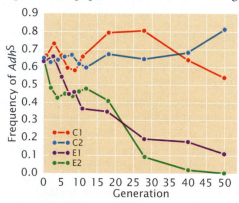

a. On the basis of these data, what conclusion might you draw about the evolutionary forces that are affecting the *Adh* alleles in these populations?

b. Cavener and Clegg measured the viability of the different *Adh* genotypes in the alcohol environment and obtained the following values:

Genotype	Relative viability
Adh^F/*Adh*^{F}	0.932
Adh^F/*Adh*^{S}	1.288
Adh^S/*Adh*^{S}	0.596

Using these relative viabilities, calculate relative fitnesses for the three genotypes. If a population has an initial frequency of $p = f(Adh^F) = 0.5$, what will the expected frequency of *Adh*^{F} be in the next generation on the basis of these fitness values?

*41. A certain form of congenital glaucoma is caused by an autosomal recessive allele. Assume that the mutation rate is 10^{-5} and that persons having this condition produce, on the average, only about 80% of the offspring produced by persons who do not have glaucoma.

a. At equilibrium between mutation and selection, what will the frequency of the gene for congenital glaucoma be?

b. What will the frequency of the disease be in a randomly mating population that is at equilibrium?

42. Examine **Figure 25.15**. Which evolutionary forces:

a. Cause an increase in genetic variation both within and between populations?

b. Cause a decrease in genetic variation both within and between populations?

c. Cause an increase in genetic variation within populations but cause a decrease in genetic variation between populations?

CHALLENGE QUESTIONS

Section 25.2

43. The frequency of allele *A* in a population is 0.8 and the frequency of allele *a* is 0.2. If the population mates randomly for this locus, give all the possible matings among the genotypes at this locus and the expected proportion of each type.

Section 25.4

44. The Barton Springs salamander is an endangered species found only in three adjacent springs in the city of Austin, Texas. There is growing concern that a chemical spill on a nearby freeway could pollute the spring and wipe out the species. To provide a source of salamanders to repopulate the spring in the event of such a catastrophe, a proposal has been made to establish a captive breeding population of the salamander in a local zoo. You are asked to provide a plan for the establishment of this captive breeding population, with the goal of maintaining as much of the genetic variation of the species as possible. What factors might cause loss of genetic variation in the establishment of the captive population? How could loss of such variation be prevented? With the assumption that only a limited number of salamanders can be maintained in captivity, what procedures should be instituted to ensure the long-term maintenance of as much variation as possible?

▶ Go to your **LaunchPad** to find additional learning resources and the Suggested Readings for this chapter.

26

Evolutionary Genetics

Taster Genes in Spitting Apes

Some chimpanzees, like humans, have the ability to taste phenylthiocarbamide (PTC), whereas others do not. Recent research indicates that the PTC taste polymorphism evolved independently in humans and chimpanzees. [FLPA/Alamy.]

Almost every student of biology knows about the taster test. The teacher passes out small pieces of paper impregnated with a compound called phenylthiocarbamide (PTC), and the students, following the teacher's instructions, put the paper in their mouths. The reaction is always the same: a number of the students immediately spit the paper out, repelled by the bitter taste of PTC. A few students, however, can't taste the PTC and continue to suck on the paper, wondering what all the spitting is about. Variation among individuals in a trait such as the ability to taste PTC is termed a polymorphism.

The ability to taste PTC is inherited as an autosomal dominant trait in humans. The frequencies of taster and nontaster alleles have been estimated in hundreds of human populations worldwide. Almost all populations have both tasters and nontasters; the frequency of the two alleles varies widely.

PTC is not found in nature, but the ability to taste it is strongly correlated with the ability to taste other naturally occurring bitter compounds, some of which are toxic. The ability to taste PTC has also been linked to dietary preferences and may be associated with susceptibility to certain diseases, such as thyroid deficiency. These observations suggest that natural selection has played a role in the evolution of the taster trait. Some understanding of the evolution of the taster trait was gained when well-known population geneticist Ronald A. Fisher and his colleagues took a trip to the zoo in 1939.

Fisher wondered whether other primates also might have the ability to taste PTC. To answer this question, he prepared some drinks with different levels of PTC and set off for the zoo with his friends, fellow biologists Edmund (E. B.) Ford and Julian Huxley. At the zoo, the PTC-laced drinks were offered to eight chimpanzees and one orangutan. Fisher and his friends were initially concerned that they might not be able to tell whether the apes could taste the PTC. That concern disappeared, however, when the first one sampled the drink and immediately spat on Fisher. Of the eight chimpanzees tested, six were tasters and two were nontasters.

The observation that chimpanzees and humans both have the PTC taste polymorphism led Fisher and his friends to assume that the polymorphism arose in a common ancestor of humans and chimpanzees, which passed it on to both species. However, they had no way to test their hypothesis. Sixty-five years later, geneticists armed with the latest molecular genetic techniques were able to determine the origin of the PTC taste polymorphism and test the hypothesis of Fisher and his friends.

Molecular studies revealed that our ability to taste PTC is controlled by alleles at the *TAS2R38* locus, a 1000-bp gene found on chromosome 7. This locus encodes receptors for bitter compounds and is expressed in the cells of our taste buds. One common allele encodes a receptor that allows the ability to taste PTC; an alternative allele encodes a receptor that does not respond to PTC.

Recent research has demonstrated that PTC taste sensitivity in chimpanzees also is controlled by alleles at the *TAS2R38* locus. However, much to the investigators' surprise, the taster alleles in humans and chimpanzees are not the same at the molecular level. In the human taster and nontaster alleles, nucleotide differences at three positions affect which amino acids are present in the taste-receptor protein. In chimpanzees, none of these nucleotide differences are present. Instead, a mutation in the initiation codon produces the nontaster allele. This substitution eliminates the normal initiation codon, and the ribosome initiates translation at an alternative downstream initiation codon, resulting in the production of a shortened protein receptor that fails to respond to PTC.

What these findings mean is that Fisher and his friends were correct that humans and chimpanzees both have PTC taste polymorphism but were incorrect in their hypothesis about its origin: humans and chimpanzees independently evolved the PTC taste polymorphism.

This chapter is about the genetic basis of evolution. As illustrated by the PTC taster polymorphism, evolutionary genetics has a long history but has been transformed in recent years by the application of powerful molecular genetic techniques. In Chapter 25, we considered the evolutionary forces that bring about change in the allelic frequencies of a population: mutation, migration, genetic drift, and selection. In this chapter, we examine some specific ways that these forces shape the genetic makeup of populations and bring about long-term evolution. We begin by looking at how the process of evolution depends on genetic variation and how genetic variation is studied in natural populations. We then turn to the evolutionary changes that bring about the appearance of new species and how evolutionary histories (phylogenies) are constructed. We end the chapter by taking a look at patterns of evolutionary change at the molecular level.

26.1 Evolution Occurs Through Genetic Change Within Populations

Evolution is one of the foundational principles of all of biology. Theodosius Dobzhansky, an important early leader in the field of evolutionary genetics, once remarked "Nothing in biology makes sense except in the light of evolution." Indeed, evolution is an all-encompassing theory that helps to make sense of much of the natural world, from the sequences of DNA found in our cells to the types of organisms that surround us. The evidence for evolution is overwhelming. Evolution has been directly observed numerous times; for example, hundreds of different insect species evolved resistance to common pesticides that were introduced after World War II. Evolution is supported by the fossil record, comparative anatomy, embryology, the distribution of plants and animals (biogeography), and molecular genetics.

BIOLOGICAL EVOLUTION In spite of its vast importance to all fields of biology, evolution is often misunderstood and misinterpreted. In our society, the term *evolution* frequently refers to any type of change. However, biological **evolution** refers only to a specific type of change—genetic change taking place in a group of organisms. Two aspects of this definition should be emphasized. First, evolution includes genetic change only. Many nongenetic changes take place in living organisms, such as the development of a complex intelligent person from a single-celled zygote. Although remarkable, this change isn't evolution, because it does not include changes in genes. The second aspect to emphasize is that evolution takes place in *groups* of organisms. An individual organism does not evolve; what evolves is the gene pool common to a group of organisms.

EVOLUTION AS A TWO-STEP PROCESS Evolution can be thought of as a two-step process. First, genetic variation arises. Genetic variation has its origin in the processes of mutation, which produces new alleles, and recombination, which shuffles alleles into new combinations. Both of these processes are random and produce genetic variation continually, regardless of evolution's requirement for it. The second step in the process of evolution is the increase and decrease in the frequencies of genetic variants. Various evolutionary forces discussed in Chapter 25 cause some alleles in the gene pool to increase in frequency and other alleles to decrease in frequency. This shift in the composition of the gene pool common to a group of organisms constitutes evolutionary change.

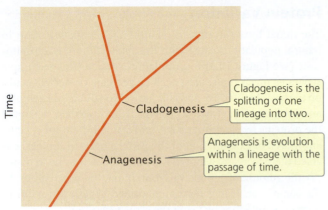

26.1 Anagenesis and cladogenesis are two different types of evolutionary change. Anagenesis is change within an evolutionary lineage; cladogenesis is the splitting of lineages.

TYPES OF EVOLUTION   We can differentiate between two types of evolution that take place within a group of organisms connected by reproduction. **Anagenesis** refers to evolution taking place in a single group (a lineage) with the passage of time (**Figure 26.1**). Another type of evolution is **cladogenesis**, the splitting of one lineage into two. When a lineage splits, the two branches no longer have a common gene pool and evolve independently of one another. New species arise through cladogenesis. ▶ **TRY PROBLEM 21**

CONCEPTS

Biological evolution is genetic change that takes place within a group of organisms. Anagenesis is evolution that takes place within a single lineage; cladogenesis is the splitting of one lineage into two.

✓ CONCEPT CHECK 1

Briefly describe how evolution takes place as a two-step process.

26.2 Many Natural Populations Contain High Levels of Genetic Variation

Because genetic variation must be present for evolution to take place, evolutionary biologists have long been interested in the amounts of genetic variation in natural populations and the forces that control the amount and nature of that variation. For many years, they could not examine genes directly and were limited to studying the phenotypes of organisms. Although genetic variation could not be quantified directly, studies of phenotypes suggested that many populations of organisms harbor considerable genetic variation. Populations of organisms in nature exhibit tremendous phenotypic variation: frogs vary in color patterns, birds differ in size, butterflies vary in spotting patterns, mice differ in coat color, and humans vary in blood types, to mention just a few. Crosses revealed that a few of these traits were inherited as simple genetic traits (**Figure 26.2**) but,

Normal homozygotes

Heterozygotes

Recessive bimacula phenotype

26.2 Early evolutionary geneticists were forced to rely on the phenotypic traits that had a simple genetic basis. Variation in the spotting patterns of the butterfly *Panaxia dominula* is an example.

for most traits, the precise genetic basis was complex and unknown, preventing early evolutionary geneticists from quantifying the amount of genetic variation in natural populations.

As discussed in Chapter 24, a population's response to selection depends on narrow-sense heritability, which is a measure of the additive genetic variation of a trait within a population. Many organisms respond to artificial selection carried out by humans, suggesting that the populations of these organisms contain much additive genetic variation. For example, humans have used artificial selection to produce numerous dog breeds, which vary tremendously in size, shape, color, and behavior (see Figure 24.21). Early studies of chromosome variations in *Drosophila* and plants also suggested that genetic variation in natural populations is plentiful and widespread.

Molecular Variation

Advances in molecular genetics have made it possible to investigate evolutionary change directly by analyzing protein and nucleic acid sequences. These molecular data offer a number of advantages for studying the process and pattern of evolution:

Molecular data are genetic. Evolution results from genetic change with time. Many anatomical, behavioral, and physiological traits have a genetic basis, but the relation between the underlying genes and the trait may be complex. Protein- and nucleic-acid-sequence variation has a clear genetic basis that is often easy to interpret.

Molecular methods can be used with all organisms. Early studies of population genetics relied on simple genetic traits such as human blood types, banding patterns in snails, or spotting patterns in butterflies (see Figure 26.2), which are restricted to a small group of organisms. However, all living organisms have proteins and nucleic acids, so molecular data can be collected from any organism.

Molecular methods can be applied to a huge amount of genetic variation. An enormous amount of data can be accessed by molecular methods. The human genome, for example, contains more than 3 billion base pairs of DNA, which constitutes a large pool of information about our evolution.

All organisms can be compared with the use of molecular data. Trying to assess the evolutionary history of distantly related organisms is often difficult because they have few characteristics in common. The evolutionary relationships between angiosperms were traditionally assessed by comparing floral anatomy, whereas the evolutionary relationships of bacteria were determined by their nutritional and staining properties. Because plants and bacteria have so few structural characteristics in common, evaluating how they are related to one another was difficult in the past. All organisms have certain molecular traits in common, such as ribosomal RNA sequences and some fundamental proteins. These molecules offer a valid basis for comparisons among all organisms.

Molecular data are quantifiable. Protein and nucleic acid sequence data are precise, accurate, and quantifiable, which facilitates the objective assessment of evolutionary relationships.

Molecular data often provide information about the process of evolution. Molecular data can reveal important clues about the process of evolution. For example, the results of a study of DNA sequences have revealed that one type of insecticide resistance in mosquitoes probably arose from a single mutation that subsequently spread throughout the world.

The database of molecular information is large and growing. Today, this database of DNA and protein sequences can be used for making evolutionary comparisons and inferring mechanisms of evolution.

Protein Variation

The initial breakthrough in quantifying genetic variation in natural populations came with the application of electrophoresis (see Figure 19.3) to population studies. This technique separates macromolecules, such as proteins or nucleic acids, on the basis of their size and charge. In 1966, Richard Lewontin and John Hubby extracted proteins from fruit flies, separated the proteins by electrophoresis, and stained them for specific enzymes. An examination of the pattern of bands on gels enabled them to assign genotypes to individual flies and to quantify the amount of genetic variation in natural populations. In the same year, Harry Harris quantified genetic variation in human populations by using the same technique.

Numerous studies have now examined protein variation in a wide variety of organisms and found that most populations possess a large amount of variation in proteins (**Figure 26.3**). However, protein electrophoresis actually underestimates the true amount of genetic variation because it does not detect some amino acid substitutions; nor does it detect genetic variation in DNA that does not alter the amino acids of a protein (synonymous codons and variation in noncoding regions of the DNA).

One hypothesis to account for the extensive molecular variation in proteins is the **neutral-mutation hypothesis**, which proposes that much molecular variation is adaptively neutral; individuals with different molecular variants have equal fitness. This hypothesis, developed by Motoo Kimura, does not propose that the proteins are functionless; rather, it suggests that most variants are functionally equivalent. Because these variants are functionally equivalent, natural selection does not differentiate between them, and their evolution is shaped largely by the random processes of genetic drift and mutation. The neutral-mutation hypothesis accepts that natural selection is an important force in evolution but views selection as a process that favors the "best" allele while eliminating others. It proposes that, when selection is important, there will be *little* genetic variation.

26.3 Molecular variation in proteins is revealed by electrophoresis. Tissue samples from *Drosophila pseudoobscura* were subjected to electrophoresis and stained for esterase. Esterases encoded by different alleles migrate different distances. Shown on the gel are homozygotes for three different alleles. [Courtesy Richard Lewontin.]

Another view, termed the **balance hypothesis,** is that the genetic variation in natural populations is maintained by selection that favors variation (balancing selection). Overdominance, in which the heterozygote has higher fitness than that of either homozygote, is one type of balancing selection. Under this hypothesis, the molecular variants are not physiologically equivalent and do not have the same fitness. Instead, genetic variation within natural populations is shaped largely by selection, and, when selection is important, there will be *much* variation.

Many attempts to disprove one hypothesis or the other failed, because it was unclear precisely how much variation was actually present (remember that protein electrophoresis detects only some genetic variation) and because both hypotheses are capable of explaining many different patterns of genetic variation. The results of recent studies that provide direct information about DNA sequence variation demonstrate that much variation at the level of DNA has little obvious effect on the phenotype and is therefore likely to be neutral.

CONCEPTS

The application of electrophoresis to the study of protein variation in natural populations revealed that most organisms possess large amounts of genetic variation. The neutral-mutation hypothesis proposes that most molecular variation is neutral with regard to natural selection and is shaped largely by mutation and genetic drift. The balance hypothesis proposes that genetic variation is maintained by balancing selection.

✓ CONCEPT CHECK 2

Which statement is true of the neutral-mutation hypothesis?

a. All proteins are functionless.

b. Natural selection plays no role in evolution.

c. Most molecular variants are functionally equivalent.

d. All of the above.

DNA Sequence Variation

The development of techniques for isolating, cutting, and sequencing DNA in the past 30 years has provided powerful tools for detecting, quantifying, and investigating genetic variation. The application of these techniques has provided a detailed view of genetic variation at the molecular level.

RESTRICTION-SITE VARIATION Among the first techniques for detecting and analyzing genetic variation in DNA sequences was the use of restriction enzymes. Each restriction enzyme recognizes and cuts a particular sequence of DNA nucleotides known as that enzyme's restriction site (see Chapter 19). Variation in the presence of a restriction site is called a restriction fragment length polymorphism (RFLP; see Figure 19.22). Each restriction enzyme recognizes a limited number of nucleotide sites in a particular piece of DNA but, if a number of different restriction enzymes are used and the sites recognized by the enzymes are assumed to be random sequences, RFLPs can be used to estimate the amount of variation in the DNA and the proportion of nucleotides that differ between organisms. RFLPs can also be used to analyze the genetic structure of populations and to assess evolutionary relationships among organisms. RFLPs were widely used in evolutionary studies before the development of rapid and inexpensive methods for directly sequencing DNA, and restriction analysis is still employed today in some studies of molecular evolution.

MICROSATELLITE VARIATION Microsatellites are short DNA sequences that exist in multiple copies repeated in tandem (see p. 561 in Chapter 19). Variation in the number of copies of the repeats is common, with individual organisms often differing in the number of repeat copies. Microsatellites can be detected by using the polymerase chain reaction (PCR). Pairs of primers are used that flank a region of repeated copies of the sequence. The DNA fragments that are synthesized in the PCR reaction vary in length, depending on the number of tandem repeats present. DNA from an individual organism with more repeats will produce a longer amplified segment. After PCR has been completed, the amplified fragments are separated with the use of gel electrophoresis and stained, producing a series of bands on a gel (see Figure 19.28). The banding patterns that result represent different alleles and can be used to quantify genetic variation, assess genetic relationships among individual organisms, and quantify genetic differences in a population. An advantage of using microsatellites is that the PCR reaction can be used on very small amounts of DNA and is rapid. The amplified fragments can be fluorescently labeled and detected by a laser, allowing the process to be automated.

David Coltman and his colleagues used microsatellite variation to study paternity in bighorn sheep (**Figure 26.4**; also see the introduction to Chapter 25) and showed that sport hunting of trophy rams has reduced the weight and horn size of the animals. Samples of blood, hair, and ear tissue were collected from bighorn sheep at Ram Mountain in Alberta, Canada—a population that has been monitored since 1971. DNA was extracted from the tissue samples and amplified with PCR, revealing

26.4 Microsatellite variation has been used to study the response of bighorn sheep to selective pressure on horn size due to trophy hunting. [Eyewire.]

variation at 20 microsatellite loci. On the basis of the microsatellite variation, paternity was assigned to 241 rams, and the family relationships of the sheep were worked out. Using these family relationships and the quantitative genetic techniques described in Chapter 24, the geneticists were able to show that ram weight and horn size had high heritability and exhibited a strong positive genetic correlation. Trophy hunters selectively shoot rams with large horns, often before they are able to reproduce. This selective pressure has produced a response to selection: the rams are evolving smaller horns. Between 1971 and 2002, horn size in the population decreased by about one quarter. Because of the positive genetic correlation between horn size and body size, the body size of rams also is decreasing. The killing of trophy rams with large horns has led to a decrease in the very traits that are prized by the hunters. This research illustrates a practical application of the use of microsatellites in evolutionary analysis.

VARIATION DETECTED BY DNA SEQUENCING

Techniques for rapidly and inexpensively sequencing DNA (see Chapter 19) have made this type of data an important tool in population and evolutionary studies. DNA sequence data often reveal processes that influence evolution and are invaluable for determining the evolutionary relationships of different organisms. The use of PCR for producing the DNA used in the sequencing reactions means that data can be obtained from a very small initial sample of DNA.

DNA sequencing was used to reassess the genetic relationships among African elephants. Alfred Roca and his colleagues obtained tissue samples from 195 elephants by shooting them with needlelike darts that fell to the ground after hitting an elephant but retained a small plug of skin. From the skin samples, the scientists sequenced 1732 base pairs of DNA from four nuclear genes. Their analysis revealed large genetic differences between forest elephants and savannah elephants, suggesting that there is limited gene flow (migration) between these two groups of elephants. On the basis of these results, the scientists proposed that two different species of elephants exist in Africa.

Another example of the use of DNA sequence data to decipher evolutionary relationships was the unusual case of HIV infection in a dental practice in Florida. In July 1990, the U.S. Centers for Disease Control and Prevention (CDC) reported that a young woman in Florida had become HIV positive after undergoing an invasive dental procedure performed by a dentist who had AIDS. The woman had no known risk factors for HIV infection and no known contact with other HIV-positive persons. The CDC acknowledged that she might have acquired the infection from her dentist. By 1992, seven of the dentist's patients had tested positive for HIV, and this number eventually increased to ten.

All of the dentist's HIV-positive patients had received invasive dental procedures, such as root canals and tooth extractions, in the period when the dentist was infected. Among the seven patients originally studied by the CDC (patients A–G, **Table 26.1**), two had known risk factors for HIV infection (intravenous drug use, homosexual behavior, or sexual relations with HIV-infected persons), and a third had possible but unconfirmed risk factors.

To determine whether the dentist had infected his patients, the CDC conducted a study of the molecular evolution of HIV isolates from the dentist and from the patients. HIV undergoes rapid evolution, making it possible to trace the path of its transmission.

Blood specimens were collected from the dentist, the patients, and a group of 35 local controls (other HIV-infected people who lived within 90 miles of the dental practice but who had no known contact with the dentist). DNA was extracted from white blood cells, and a 680-bp fragment of the *envelope* gene of the virus was amplified by PCR. The fragments from the dentist, patients, and local controls were then sequenced and compared.

The divergence between the viral sequences taken from the dentist, the seven patients, and the controls is shown in Table 26.1. Viral DNA taken from patients with no confirmed risk factors (patients A, B, C, E, and G) differed from the dentist's viral DNA by 3.4% to 4.9%, whereas the viral

Table 26.1   HIV-positive persons included in study of HIV isolates from a Florida dental practice

Person	Sex	Known Risk Factors	Average Differences in DNA Sequences (%)	
			From HIV from Dentist	From HIV from Controls
Dentist	M	Yes		11.0
Patient A	F	No	3.4	10.9
Patient B	F	No	4.4	11.2
Patient C	M	No	3.4	11.1
Patient E	F	No	3.4	10.8
Patient G	M	No	4.9	11.8
Patient D	M	Yes	13.6	13.1
Patient F	M	Yes	10.7	11.9

Source: After C. Y. Ou et al., *Science* 256:1165–1171, 1992, Table 1.

DNA from the controls differed from the dentist's by an average of 11%. The viral sequences collected from five patients (A, B, C, E, and G) were more closely related to the viral sequences collected from the dentist than to viral sequences from the general population, strongly suggesting that these patients acquired their HIV infection from the dentist. The viral isolates from patients D and F (patients with confirmed risk factors), however, differed from that of the dentist by 10.7% and 13.6%, suggesting that these two patients did not acquire their infection from the dentist.

An analysis of the evolutionary relationships of the viral sequences confirmed that the virus taken from the dentist had a close evolutionary relationship to viruses taken from patients A, B, C, E, and G (**Figure 26.5**). The viruses from patients D and F, with known risk factors, were no more similar to the virus from the dentist than to viruses from local controls. Of three additional HIV-positive patients that have been identified since 1992, only one has viral sequences that are closely related to those from the dentist.

This study provides an excellent example of the relevance of evolutionary studies to real-world problems. How the dentist infected his patients during their visits to his office remains a mystery, but this case is unusual. A study of almost 16,000 patients treated by HIV-positive health-care workers failed to find a single case of confirmed transmission of HIV from the health-care worker to the patient.

CONCEPTS

Variation in DNA nucleotide sequence can be analyzed by using restriction fragment length polymorphisms, microsatellites, and data from direct DNA sequencing.

✔ CONCEPT CHECK 3

What are some of the advantages of using microsatellites for evolutionary studies?

26.3 New Species Arise Through the Evolution of Reproductive Isolation

Earlier we discussed two types of evolution: anagenesis (change within a lineage), and cladogenesis, the splitting of lineages. Cladogenesis arises through speciation, the process by which one population separates into two distinct evolutionary groups.

The term *species* literally means kind or appearance; **species** are different kinds or types of living organisms. In many cases species differences are easy to recognize: a horse is clearly a different species from a chicken. Sometimes, however, species differences are not so clear. Some species of *Plethodon* salamanders are so similar in appearance that they can be distinguished only by looking at their proteins or genes.

The concept of a species has two primary uses in biology. First, species is a name given to a particular type of organism. For effective communication, biologists must use a standard set of names for the organisms that they study, and species names serve that purpose. When a geneticist talks about conducting crosses with *Drosophila melanogaster*, other biologists immediately understand which organism was used. The second use of the term species is in an evolutionary context: a species is considered an evolutionarily independent group of organisms.

The Biological Species Concept

What kinds of differences are required to consider two organisms different species? A widely used definition of species is the **biological species concept**, first fully developed by evolutionary biologist Ernst Mayr in 1942. Mayr was primarily interested in the biological characteristics that are responsible

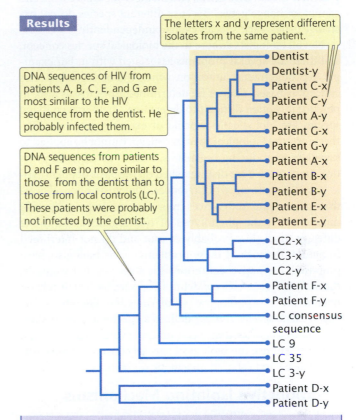

Experiment

Question: Did the dentist transmit HIV to his patients?

Methods Viral samples were collected from the dentist, patients, and local controls. Gene fragments were sequenced and compared.

Results

The letters x and y represent different isolates from the same patient.

DNA sequences of HIV from patients A, B, C, E, and G are most similar to the HIV sequence from the dentist. He probably infected them.

DNA sequences from patients D and F are no more similar to those from the dentist than to those from local controls (LC). These patients were probably not infected by the dentist.

Dentist
Dentist-y
Patient C-x
Patient C-y
Patient A-y
Patient G-x
Patient G-y
Patient A-x
Patient B-x
Patient B-y
Patient E-x
Patient E-y
LC2-x
LC3-x
LC2-y
Patient F-x
Patient F-y
LC consensus sequence
LC 9
LC 35
LC 3-y
Patient D-x
Patient D-y

Conclusion: Viral sequences from patients A, B, C, E, and G cluster with those of the dentist, indicating a close evolutionary relationship. Sequences from patients D and F, along with those of local controls, are more distantly related.

26.5 Evolutionary tree showing the relationships of HIV isolates from a dentist, seven of his patients (A through G), and other HIV-positive persons from the same region (local controls, LC). The phylogeny is based on DNA sequences taken from the *envelope* gene of the virus. [After C. Ou et al., Molecular epidemiology of HIV transmission in a dental practice, *Science* 256:1167, 1992.]

for separating organisms into independently evolving units. He defined a species as a group of organisms whose members are capable of interbreeding with one another but are reproductively isolated from the members of other species. In other words, members of the same species have the biological potential to exchange genes, and members of different species cannot exchange genes. Because different species do not exchange genes, each species evolves independently.

Not all biologists adhere to the biological species concept, and there are several problems associated with it. For example, reproductive isolation, on which the biological species concept is based, cannot be determined from fossils and, in practice it is often difficult to determine whether even living species are biologically capable of exchanging genes. Furthermore, the biological species concept cannot be applied to asexually reproducing organisms, such as bacteria. In practice, most species are distinguished on the basis of phenotypic (usually anatomical) differences. Biologists often assume that phenotypic differences represent underlying genetic differences; if the phenotypes of two organisms are quite different, then they probably cannot and do not interbreed in nature. Because of these problems, some biologists have proposed alternative definitions for a species. For example, the *morphospecies concept* defines a species based entirely on phenotypic (morphological) similarities and differences. The *phylogenetic species concept* defines a species as the smallest recognizable group that has a unique evolutionary history. Here we will use the biological species concept because it is widely used and is based on genetic differences.

Reproductive Isolating Mechanisms

The key to species differences under the biological species concept is reproductive isolation—biological characteristics that prevent genes from being exchanged between different species. Any biological factor or mechanism that prevents gene exchange is termed a **reproductive isolating mechanism**.

PREZYGOTIC REPRODUCTIVE ISOLATING MECHANISMS Some species are separated by **prezygotic reproductive isolating mechanisms,** which prevent gametes from two different species from fusing and forming a hybrid zygote. In **ecological isolation**, members of two species do not encounter one another and therefore do not reproduce with one another: they have different ecological niches, living in different habitats and interacting with the environment in different ways. For example, some species of forest-dwelling birds feed and nest in the forest canopy, whereas others confine their activities to the forest floor. Because they never come into contact, these birds are reproductively isolated from one another. Other species are separated by **behavioral isolation**, differences in behavior that prevent interbreeding. Many male frogs attract females of the same species by using a unique, species-specific call. Two closely related frogs may use the same pond but never interbreed, because females are attracted only to the call of their own species.

Another type of prezygotic reproductive isolation is **temporal isolation**, in which reproduction takes place at different times of the year. Some species of plants do not exchange genes, because they flower at different times of the year. **Mechanical isolation** results from anatomical differences that prevent successful copulation. This type of isolation is seen in many insects, in which closely related species differ in their male and female genitalia, and so copulation is physically impossible. Finally, some species are separated by **gametic isolation**, in which mating between individuals of different species takes place, but the gametes do not form zygotes. Male gametes may not survive in the female reproductive tract or may not be attracted to female gametes. In other cases, male and female gametes meet but are too incompatible to fuse to form a zygote. Gametic isolation is seen in many plants, where pollen from one species cannot fertilize the ovules of another species.

POSTZYGOTIC REPRODUCTIVE ISOLATING MECHANISMS Other species are separated by **postzygotic reproductive isolating mechanisms**, in which gametes of two species fuse and form a zygote, but there is no gene flow between the two species, either because the resulting hybrids are inviable or sterile or because reproduction breaks down in subsequent generations.

If prezygotic reproductive isolating mechanisms fail or have not yet evolved, mating between two organisms of different species may take place, with the formation of a hybrid zygote containing genes from two different species. In many cases, such species are still separated by **hybrid inviability**, in which incompatibility between genomes of the two species prevents the hybrid zygote from developing. Hybrid inviability is seen in some groups of frogs, in which mating between different species and fertilization take place, but the resulting embryos never complete development.

Other species are separated by **hybrid sterility**, in which hybrid embryos complete development but are sterile, so that genes are not passed between species. Donkeys and horses frequently mate and produce a viable offspring—a mule—but most mules are sterile; thus, there is no gene flow between donkeys and horses (but see Problem 47 at the end of Chapter 8).Finally, some closely related species are capable of mating and producing viable and fertile F_1 progeny. However, genes do not flow between the two species because of **hybrid breakdown**, in which further crossing of the hybrids produces inviable or sterile offspring. The different types of reproductive isolating mechanisms are summarized in **Table 26.2**.

CONCEPTS

The biological species concept defines a species as a group of potentially interbreeding organisms that are reproductively isolated from the members of other species. Under this concept, species are separated by prezygotic or postzygotic reproductive isolating mechanisms.

✓ CONCEPT CHECK 4

Which statement is an example of postzygotic reproductive isolation?
a. Sperm of species A dies in the oviduct of species B before fertilization can take place.

b. Hybrid zygotes between species A and B are spontaneously aborted early in development.
c. The mating seasons of species A and B do not overlap.
d. Males of species A are not attracted to the pheromones produced by the females of species B.

Modes of Speciation

Speciation is the process by which new species arise. In regard to the biological species concept, speciation comes about through the evolution of reproductive isolating mechanisms.

There are two principle ways in which new species arise. **Allopatric speciation** arises when a geographic barrier first splits a population into two groups and blocks the exchange of genes between them. The interruption of gene flow then leads to the evolution of genetic differences that result in reproductive isolation. **Sympatric speciation** is speciation that arises in the absence of any external barrier to gene flow; reproductive isolating mechanisms evolve within a single population. We will take a more detailed look at both of these mechanisms next.

ALLOPATRIC SPECIATION Allopatric speciation is initiated when a geographic barrier splits a population into two or more groups and prevents gene flow between the isolated groups (**Figure 26.6a**). Geographic barriers can take a number of forms. Uplifting of a mountain range may split a population of lowland plants into separate groups on each side of the mountains. Oceans serve as effective barriers for many types of terrestrial organisms, separating individuals on different islands from one another and from those on the mainland. Rivers often separate populations of fish located in

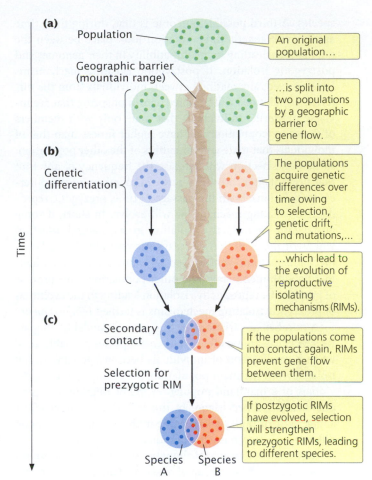

26.6 **Allopatric speciation is initiated by a geographic barrier to gene flow between two populations.**

separate drainages. The erosion of mountains may leave populations of alpine plants isolated on separate mountain peaks.

After two populations have been separated by a geographic barrier that prevents gene flow between them, they evolve independently (**Figure 26.6b**). The genetic isolation allows each population to accumulate genetic differences that are not found in the other population; genetic differences arise through natural selection, unique mutations, and genetic drift (if the populations are small). Genetic differentiation may eventually lead to prezygotic and postzygotic isolation. It is important to note that prezygotic isolation and postzygotic isolation arise simply as a consequence of genetic divergence.

If the geographic barrier that once separated the two populations disappears or individuals are able to disperse over it, the populations come into secondary contact (**Figure 26.6c**). At this point, several outcomes are possible. If limited genetic differentiation has taken place during the separation of the populations, reproductive isolating mechanisms may not have evolved or may be incomplete. Genes will flow between the two populations, eliminating any genetic differences that did arise, and the populations will remain a single species.

A second possible outcome is that genetic differentiation during separation leads to prezygotic reproductive isolating mechanisms; in this case, the two populations are different

Table 26.2	Types of reproductive isolating mechanisms	
Type	**Characteristics**
Prezygotic	**Before a zygote has formed**
Ecological	Differences in habitat; individuals do not meet
Temporal	Reproduction takes place at different times
Mechanical	Anatomical differences prevent copulation
Behavioral	Differences in mating behavior prevent mating
Gametic	Gametes incompatible or not attracted to each other
Postzygotic	**After a zygote has formed**
Hybrid inviability	Hybrid zygote does not survive to reproduction
Hybrid sterility	Hybrid is sterile
Hybrid breakdown	F_1 hybrids are viable and fertile, but F_2 are inviable or sterile

species. A third possible outcome is that, during their time apart, some genetic differentiation took place between the populations, leading to incompatibility in their genomes and postzygotic isolation. If postzygotic isolating mechanisms have evolved, any mating between individuals from the different populations will produce hybrid offspring that are inviable or sterile. Individuals that mate only with members of the same population will have higher fitness than that of individuals that mate with members of the other population; so natural selection will increase the frequency of any trait that prevents interbreeding between members of the different populations. With the passage of time, prezygotic reproductive isolating mechanisms will evolve. In short, if some postzygotic reproductive isolation exists, natural selection will favor the evolution of prezygotic reproductive isolating mechanisms to prevent wasted reproduction by individuals mating with members of the other population. This process of postzygotic reproductive isolation leading to the evolution of prezygotic isolating mechanisms is termed *reinforcement*.

A number of variations in this general model of allopatric speciation are possible. Many new species probably arise when a small group of individuals becomes geographically isolated from the main population; for example, a few individuals of a mainland population might migrate to a geographically isolated island. In this situation, founder effect and genetic drift play a larger role in the evolution of genetic differences between the populations.

An excellent example of allopatric speciation can be found in Darwin's finches, a group of birds that Charles Darwin discovered on the Galápagos Islands during his voyage aboard the *Beagle*. The Galápagos are an archipelago of islands located about 900 km off the coast of South America (**Figure 26.7**). Consisting of more than a dozen large islands and many smaller ones, the Galápagos formed from volcanoes that erupted over a geological hot spot that has remained stationary while the geological plate over it moved eastward in the past 3 million years. The movement of the geological plate pulled newly formed islands eastward, and so the islands to the east (San Cristóbal and Española) are older than those to the west (Isabela and Fernandina). With the passage of time, the number of islands in the archipelago increased as new volcanoes arose.

Darwin's finches consist of 14 species that are found on various islands in the Galápagos archipelago (**Figure 26.8**). The birds vary in the shape and sizes of their beaks, which are adapted for eating different types of food items. Recent studies of the development of finch embryos have helped to reveal some of the molecular details of how differences in beak shapes have evolved (see p. 649 in Chapter 22).

Genetic studies have demonstrated that all the birds are closely related and evolved from a single ancestral species that migrated to the islands from the coast of South America some 2 million to 3 million years ago. The evolutionary relationships among the 14 species, based on studies of microsatellite data, are depicted in Figure 26.8. Most of the species are separated by a behavioral isolating mechanism (song in particular), but some of the species can and occasionally do hybridize in nature.

The first finches to arrive in the Galápagos probably colonized one of the larger eastern islands. A breeding population became established and increased with time. At some point, a few birds dispersed to other islands, where they were effectively isolated from the original population, and established a new population. The new population underwent genetic differentiation owing to genetic drift and adaptation to the local conditions of the island, and it eventually became reproductively isolated from the original population. Individual birds from the new population then dispersed to other islands and gave rise to additional species. This process was repeated many times. Occasionally, newly evolved birds dispersed to

26.7 The Galápagos Islands are young geologically and are volcanic in origin. The oldest islands are to the east. [After *Philosophical Transactions at the Royal Society of London, Series B* 351: 756–772, 1996.]

26.8 Darwin's finches consist of 14 species that evolved from a single ancestral species that migrated to the Galápagos Islands and underwent repeated allopatric speciation. [After B. R. Grant and P. R. Grant, *Bioscience* 53:965–975, 2003.]

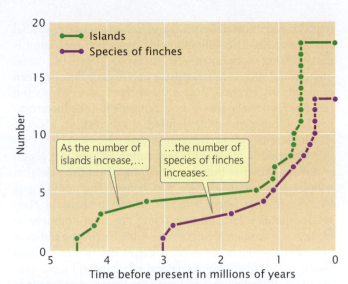

26.9 The number of species of Darwin's finches present at various times in the past corresponds with the number of islands in the Galápagos archipelago. [Data from P. R. Grant, B. R. Grant, and J. C. Deutsch, Speciation and hybridization in island birds, *Philosophical Transactions of the Royal Society of London, Series B* 351: 765–772, 1996.]

an island where another species was already present, giving rise to secondary contact between the species. Today, many of the islands have more than one resident finch.

The age of the 14 species has been estimated with data from mitochondrial DNA. **Figure 26.9** shows that there is a strong correspondence between the number of bird species present at various times in the past and the number of islands in the archipelago. This correspondence is one of the most compelling pieces of evidence for the theory that the different species of finches arose through allopatric speciation.

CONCEPTS

Allopatric speciation is initiated when a geographic barrier to gene flow splits a single population into two or more populations. With the passage of time, the populations evolve genetic differences, which bring about reproductive isolation. After postzygotic reproductive isolating mechanisms have evolved, selection favors the evolution of prezygotic reproductive isolating mechanisms.

✓ CONCEPT CHECK 5

What role does genetic drift play in allopatric speciation?

SYMPATRIC SPECIATION Sympatric speciation arises in the absence of any geographic barrier to gene flow; reproductive isolating mechanisms evolve within a single interbreeding population. Sympatric speciation has long been controversial within evolutionary biology. Ernst Mayr believed that sympatric speciation was impossible, and he demonstrated that many apparent cases of sympatric speciation could be explained by allopatric speciation. More recently, however, evidence has accumulated that sympatric speciation can and has arisen under special circumstances. The difficulty with sympatric speciation is that isolating mechanisms arise as a *consequence* of genetic differentiation, which takes place only if gene flow between groups is interrupted. But, without reproductive isolation (or some external barrier), how can gene flow be interrupted? How can genetic differentiation arise within a single group that is freely exchanging genes?

Most models of sympatric speciation assume that genetic differentiation is initiated by strong disruptive selection taking place within a single population. One homozygote (A^1A^1) is strongly favored on one resource (perhaps the plant species that is host to an insect) and the other homozygote (A^2A^2) is favored on a different resource (perhaps a different host plant). Heterozygotes (A^1A^2) have low fitness on both resources. In this situation, natural selection will favor genotypes at other loci that cause assortative mating (matings between like individuals, see Chapter 25), and so no matings take place between A^1A^1 and A^2A^2, which would produce A^1A^2 offspring with low fitness.

Now imagine that alleles at a second locus affect mating behavior, such that C^1C^1 individuals prefer mating only with other C^1C^1 individuals, and C^2C^2 individuals prefer mating with other C^2C^2 individuals. If alleles at the A locus are nonrandomly associated with alleles at the C locus so that only $A^1A^1\ C^1C^1$ individuals and $A^2A^2\ C^2C^2$ individuals exist, then gene flow will be restricted between individuals using the different resources, allowing the two groups to evolve further genetic differences that might lead to reproductive isolation and sympatric speciation.

The difficulty with this model is that recombination quickly breaks up the nonrandom associations between genotypes at the two loci, producing individuals such as $A^1A^1\ C^2C^2$, which would prefer to mate with $A^2A^2\ C^2C^2$ individuals. This mating would produce all A^1A^2 offspring, which do poorly on both resources. Thus, even limited recombination will prevent the evolution of the mating-preference genes.

Sympatric speciation is more probable if the genes that affect resource utilization also affect mating preferences. Genes that affect both resource utilization and mating preference are indeed present in some host races—populations of specialized insects that feed on different host plants. Guy Bush studied what appeared to be initial stages of speciation in host races of the apple maggot fly (*Rhagoletis pomonella*, **Figure 26.10**). The flies of this species feed on the fruits of a specific host tree. Mating takes place near the fruits, and the flies lay their eggs on the ripened fruits, where their larvae grow and develop. *R. pomonella* originally existed only on the fruits of hawthorn trees, which are native to North America; 150 years ago, *R. pomonella* was first observed on cultivated apples, which are related to hawthorns but a different species. Infestations of apples by this new apple host race of *R. pomonella* quickly spread, and, today, many apple trees throughout North America are infested with the flies.

26.10 Host races of the apple maggot fly, *Rhagoletis pomonella*, have evolved some reproductive isolation without any geographic barrier to gene flow. [Joseph Berger, Bugwood.org.]

The apple host race of *R. pomonella* probably originated when a few flies acquired a mutation that allowed them to feed on apples instead of hawthorn fruits. Because mating takes place on and near the fruits, flies that use apples are more likely to mate with other flies using apples, leading to genetic isolation between flies using hawthorns and those using apples. Indeed, Bush found that some genetic differentiation has already taken place between the two host races. Flies lay their eggs on ripening fruit, and there has been strong selection for the flies to synchronize their reproduction with the period when their host species has ripening fruit. Apples ripen several weeks earlier than hawthorns. Correspondingly, the peak mating period of the apple host races is 3 weeks earlier than that of the hawthorn race. These differences in the timing of reproduction between apple and hawthorn races have further reduced gene flow—to about 2%—between the two host races and have led to significant genetic differentiation between them. These differences have evolved in the past 150 years and evolution appears to be ongoing. Although genetic differentiation has taken place between apple and hawthorn host races of *R. pomonella* and some degree of reproductive isolation has evolved between them, reproductive isolation is not yet complete and speciation has not fully taken place. ▶ TRY PROBLEM 22

SPECIATION THROUGH POLYPLOIDY   A special type of sympatric speciation takes place through polyploidy (see Chapter 8). Polyploid organisms have more than two genomes ($3n$, $4n$, $5n$, etc.). As discussed in Chapter 8, allopolyploidy often arises when two diploid species hybridize, producing $2n$ hybrid offspring. Nondisjunction in one of the hybrid offspring produces a $4n$ tetraploid. Because this tetraploid contains exactly two copies of each chromosome, it is usually fertile and will be reproductively isolated from the two parental species by differences in chromosome number (see Figure 8.28).

Numerous species of flowering plants are allopolyploids. Speciation through polyploidy was observed when it led to a new species of salt-marsh grass that arose along the coast of England around 1870. This polyploid contains genomes of the European salt grass *Spartina maritima* ($2n = 60$) and the American salt grass *S. alterniflora* ($2n = 62$; **Figure 26.11**).

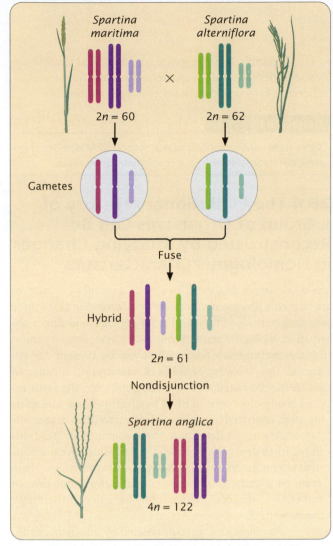

26.11 *Spartina anglica* arose sympatrically through allopolyploidy.

Seeds from the American salt grass were probably transported to England in the ballast of a ship. Regardless of how it got there, *S. alterniflora* grew in an English marsh and eventually crossed with *S. maritima*, producing a hybrid with $2n = 61$. Nondisjunction in the hybrid then led to chromosome doubling, producing a new species *S. anglica* with $4n = 122$ (see Figure 26.11). This new species subsequently spread along the coast of England.

CONCEPTS

Sympatric speciation arises within a single interbreeding population without any geographic barrier to gene flow. Sympatric speciation may arise under special circumstances, such as when resource use is linked to mating preference (in host races) or when species hybridization leads to allopolyploidy.

Genetic Differentiation Associated with Speciation

As we have seen, genetic differentiation leads to the evolution of reproductive isolating mechanisms, which restrict gene flow between populations and lead to speciation. How much genetic differentiation is required for reproductive isolation to take place? This question has received considerable study by evolutionary geneticists, but, unfortunately, there is no universal answer. Some newly formed species differ in many genes, whereas others appear to have undergone divergence in just a few genes.

One group of organisms that has been extensively studied for genetic differences associated with speciation is the genus *Drosophila*. The *Drosophila willistoni* group consists of at least 12 species found in Central and South America in various stages in the process of speciation. Using protein electrophoresis, Francisco Ayala and his colleagues genotyped flies from different geographic populations (populations with limited genetic differences), subspecies (populations with considerable genetic differences), sibling species (newly arisen species), and nonsibling species (older species). For each group, they computed a measure of genetic similarity, which ranges from 1 to 0 and represents the overall level of genetic differentiation (**Table 26.3**). They found that there was a general decrease in genetic similarity as flies evolve from geographic populations to subspecies to sibling species to nonsibling species. These data suggest that considerable genetic differentiation at many loci is required for speciation to arise. A study of *D. simulans* and *D. melanogaster*, two species that produce inviable hybrids when crossed, suggested that at least 200 genes contribute to the inviability of hybrids between the two species.

However, other studies suggest that speciation may have arisen through changes in just a few genes. For example, *D. heteroneura* and *D. silvestris* are two species of Hawaiian fruit flies that exhibit behavioral reproductive isolation. The isolation is determined largely by differences in head shape; *D. heteroneura* has a hammer-shaped head with widely separated eyes that is recognized by females of the same species but rejected by *D. silvestris* females. Genetic studies indicate that only a few loci (about 10) determine the differences in head shape.

In another study, researchers examined genetic differences in two closely related species of birds, the collared flycatcher (*Ficedula albicollis*) and pied flycatcher (*F. hypoleuca*), both of which occur in Europe. These two species are believed to have separated from a common ancestor less than 2 million years ago, probably by allopatric speciation associated with glaciation. In areas where both species

Table 26.3	Genetic similarity in groups of the *Drosophila willistoni* complex
Group	**Mean Genetic Similarity**
Geographic populations	0.970
Subspecies	0.795
Sibling species	0.517
Nonsibling species	0.352

occur together, collared flycatchers and pied flycatcher will sometimes mate, but the two species are separated by both pre- and postzygotic reproductive isolating mechanisms. For example, females prefer males with the plumage and song of their own species, and hybrids do not survive as well as offspring of parents both from the same species.

Researchers at Uppsala University in Sweden completely sequenced the genomes of 10 male birds of each species. They found that large parts of the genomes of the two species were the same: for example, only about 5 out of 1000 nucleotides differed between species, compared with 4 out of 1000 nucleotides within species. The differences that did exist between the species were not uniform across the genome; rather, the differences tended to be concentrated in a few "divergence islands," which had up to 50 times greater sequence differences than the average for the genome. A few divergence islands were found on each chromosome, and they occurred primarily near centromeres and telomeres, suggesting that differences in chromosome structure play an important

role in speciation. The divergence islands tended to show low levels of differences within species, indicating that natural selection had favored different genes in each region.

CONCEPTS

Some newly arising species have a considerable number of genetic differences; others have few genetic differences.

26.4 The Evolutionary History of a Group of Organisms Can Be Reconstructed by Studying Changes in Homologous Characteristics

The evolutionary relationships among a group of organisms are termed a **phylogeny**. Because most evolution takes place over long periods of time and is not amenable to direct observation, biologists must reconstruct phylogenies by inferring the evolutionary relationships among present-day organisms. The discovery of fossils of ancestral organisms can aid in the reconstruction of phylogenies, but the fossil record is often too poor to be of much help. Thus, biologists are often restricted to analyses of characteristics in present-day organisms to determine their evolutionary relationships. In the past, phylogenetic relationships were reconstructed on the basis of phenotypic characteristics—often, anatomical traits. Today, molecular data, including protein and DNA sequences, are frequently used to construct phylogenetic trees.

Phylogenies are reconstructed by inferring changes that have taken place in homologous characteristics. Such characteristics evolved from the same character in a common ancestor. For example, although the front leg of a mouse and the wing of a bat look different and have different functions, close examination of their structure and development reveals that they are indeed homologous; both evolved from the forelimb of an early mammal that was an ancestor to both mouse and bat. And, because mouse and bat have these homologous features and others in common, we know that they are both mammals. Similarly, DNA sequences are homologous if two present-day sequences evolved from a single sequence found in an ancestor. For example, all eukaryotic organisms have a gene for cytochrome *c*, an enzyme that helps carry out oxidative respiration. This gene is assumed to have arisen in a single organism in the distant past and was then passed down to descendants of that early ancestor. Today, all copies of the cytochrome *c* gene are homologous, because they all evolved from the same original copy in the distant ancestor of all organisms that possess this gene.

A graphical representation of a phylogeny is called a **phylogenetic tree**. As shown in **Figure 26.12**, a phylogenetic tree depicts the evolutionary relationships among different organisms, similarly to the way in which a

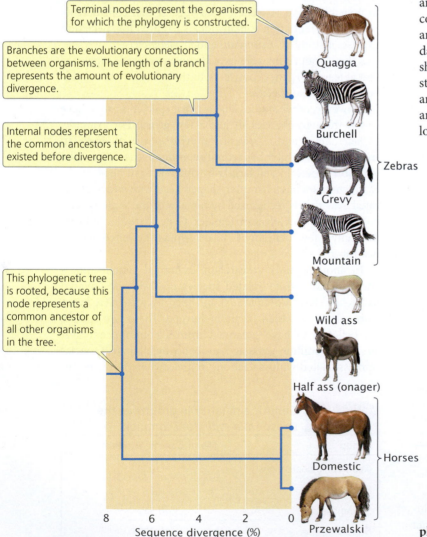

Terminal nodes represent the organisms for which the phylogeny is constructed.

Branches are the evolutionary connections between organisms. The length of a branch represents the amount of evolutionary divergence.

Internal nodes represent the common ancestors that existed before divergence.

This phylogenetic tree is rooted, because this node represents a common ancestor of all other organisms in the tree.

Quagga

Burchell

Grevy ⟩ Zebras

Mountain

Wild ass

Half ass (onager)

Domestic ⟩ Horses

Przewalski

Sequence divergence (%)

8 6 4 2 0

26.12 A phylogenetic tree is a graphical representation of the evolutionary relationships among a group of organisms.

pedigree represents the genealogical relationships among family members. A phylogenetic tree consists of **nodes** that represent the different organisms being compared, which might be different individuals, populations, or species. Terminal nodes (those at the end of the outermost branches of the tree) represent organisms for which data have been obtained, usually present-day organisms. Internal nodes represent common ancestors that existed before divergence between organisms took place. In most cases, the internal nodes represent past ancestors that are inferred from the analysis. The nodes are connected by **branches**, which represent the evolutionary connections between organisms. In some phylogenetic trees, the lengths of the branches represent the amount of evolutionary divergence that has taken place between organisms. When one internal node represents a common ancestor to all other nodes on the tree, the tree is said to be **rooted**. Trees are often rooted by including in the analysis an organism that is distantly related to all the others; this distantly related organism is referred to as an outgroup.

Phylogenetic trees are created to depict the evolutionary relationships among organisms; they are also created to depict the evolutionary relationships among DNA sequences. The latter type of phylogenetic tree is termed a **gene tree** (**Figure 26.13**). ▶ TRY PROBLEM 27

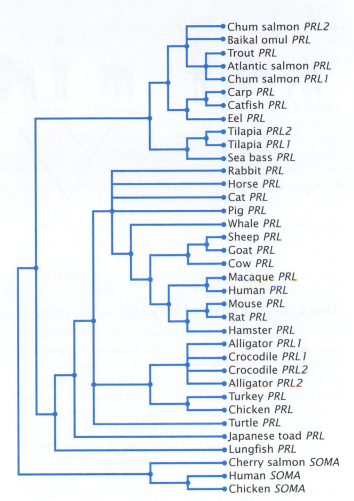

26.13 A gene tree can be used to represent the evolutionary relationships among a group of genes. This gene tree is a rooted tree, in which *PRL* represents a prolactin gene; *PRL1* and *PRL2* are two different prolactin genes found in the same organism; and *SOMA* represents a somatropin gene, which is related to prolactin genes. [After M. P. Simmons and J. V. Freudestein, Uninode coding vs. gene tree parsimony for phylogenetic reconstruction using duplicate genes, *Molecular Phylogenetics and Evolution* 23:488, 2002.]

CONCEPTS

A phylogeny represents the evolutionary relationships among a group of organisms and is often depicted graphically by a phylogenetic tree, which consists of nodes representing the organisms and branches representing their evolutionary connections.

✔ **CONCEPT CHECK 6**

Which feature is found in a rooted tree but not in an unrooted tree?

a. Terminal nodes.

b. Internal nodes.

c. A common ancestor to all other nodes.

d. Branch lengths that represent the amount of evolutionary divergence between nodes.

The Alignment of Homologous Sequences

Today, phylogenetic trees are often constructed from DNA sequence data. This construction requires that homologous sequences be compared. Thus, a first step in constructing phylogenetic trees from DNA sequence data is to identify homologous genes and properly align their nucleotide bases. Consider the following sequences that might be found in two different organisms:

Nucleotide position	1 2 3 4 5 6 7 8
Gene *X* from species A	5′–A T T G C G A A–3′
Gene *X* from species B	5′–A T G C C A A C–3′

These two sequences can be aligned in several possible ways. We might assume that there have been base substitutions at positions 3, 4, 6, and 8:

Nucleotide position	1 2 3 4 5 6 7 8
Gene *X* from species A	5′–A T T G C G A A–3′
Gene *X* from species B	5′–A T G C C A A C–3′

Alternatively, we might assume that a nucleotide at position 3 has been inserted or deleted, generating a gap in the sequence of species B, and that there has been a single nucleotide substitution at position 6:

Nucleotide position	1 2 3 4 5 6 7 8
Gene *X* from species A	5′–A T T G C G A A–3′
Gene *X* from species B	5′–A T – G C C A A C–3′

The second alignment requires fewer evolutionary steps (a deletion or insertion plus one base substitution) than does the first alignment (four base substitutions). Sequence

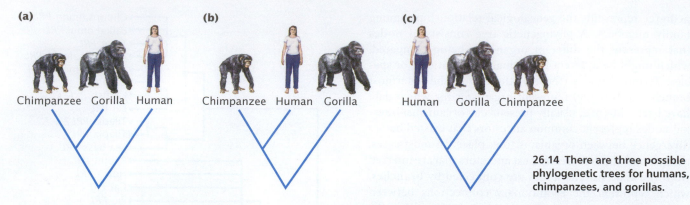

26.14 There are three possible phylogenetic trees for humans, chimpanzees, and gorillas.

alignments are usually made by computer programs that include assumptions about which types of change are more likely to take place. If two sequences have undergone much divergence, then generating alignments can be difficult.

The Construction of Phylogenetic Trees

Consider a simple phylogeny that depicts the evolutionary relationships among three organisms—humans, chimpanzees, and gorillas. Charles Darwin originally proposed that chimpanzees and gorillas were closely related to humans and modern research supports a close relationship between these three species. There are three possible phylogenetic trees for humans, chimpanzees, and gorillas (**Figure 26.14**). The goal of the evolutionary biologist is to determine which of the trees is correct. Molecular data applied to this question strongly suggest a close relationship between humans and chimpanzees. To understand the difficulty in constructing phylogenetic trees, let's consider for a moment the number of all possible trees that might exist for a group of organisms. The number of possible rooted trees for a group of organisms is

$$\text{number of rooted trees} = \frac{(2N - 3)!}{2^{N-2}(N - 2)!}$$

where N equals the number of organisms included in the phylogeny, and the ! symbol stands for factorial, the product of all the integers from N to 1. Substituting values of N into this equation, we find:

Number of organisms included in phylogeny (N)	Number of rooted trees
2	1
3	3
4	15
5	105
10	34,459,425
20	8.2×10^{21}

As the number of organisms in the phylogeny increases beyond just a few, the number of possible rooted trees becomes astronomically large. Clearly, choosing the best tree by directly comparing all the possibilities is impossible.

There are several different approaches to inferring evolutionary relationships and constructing phylogenetic trees.

In one approach, termed the *distance approach*, evolutionary relationships are inferred on the basis of the overall degree of similarity between organisms. Typically, a number of different phenotypic characteristics or gene sequences are examined and the organisms are grouped on the basis of their overall similarity, taking into consideration all the examined characteristics and sequences. A second approach, called the *maximum parsimony approach*, infers phylogenetic relationships on the basis of the fewest number of evolutionary changes that must have taken place since the organisms last had an ancestor in common. A third approach, called *maximum likelihood* and *Bayesian methods*, infers phylogenetic relationships on the basis of which phylogeny maximizes the probability of obtaining the set of characteristics exhibited by the organisms. In this approach, a phylogeny with a higher probability of producing the observed characters in the organisms studied is preferred over a phylogeny with a lower probability.

With all three approaches to constructing phylogenies, several different numerical methods are available for the construction of phylogenetic trees. All include certain assumptions that help limit the number of different trees that must be considered; most rely on computer programs that compare phenotypic characteristics or sequence data to sequentially group organisms in the construction of the tree.

CONCEPTS

Molecular data can be used to infer phylogenies (evolutionary histories) of groups of living organisms. The construction of phylogenies requires the proper alignment of homologous DNA sequences. Several different approaches are used to reconstruct phylogenies, including distance methods, maximum parsimony methods, and maximum likelihood and Bayesian methods.

26.5 Patterns of Evolution Are Revealed by Molecular Changes

The ability to analyze genetic variation at the molecular level has revealed a number of evolutionary processes and features that were formerly unsuspected. This section considers several aspects of evolution at the molecular level.

Rates of Molecular Evolution

Findings from molecular studies of numerous genes have demonstrated that different genes and different parts of the same gene often evolve at different rates.

RATES OF NUCLEOTIDE SUBSTITUTION Rates of evolutionary change in nucleotide sequences are usually measured as the rate of nucleotide substitution, which is the number of substitutions taking place per nucleotide site per year within a population. To calculate the rate of nucleotide substitution, we begin by looking at homologous sequences from different organisms. We first align the homologous sequences and then compare the sequences and determine the number of nucleotides that differ between the two sequences. We might compare the growth-hormone sequences for mice and rats, which diverged from a common ancestor some 15 million years ago. From the number of different nucleotides in their growth-hormone genes, we compute the number of nucleotide substitutions that must have taken place since they diverged. Because the same site may have mutated more than once, the number of nucleotide substitutions is larger than the number of nucleotide differences in two sequences; special mathematical methods have been developed for inferring the actual number of substitutions likely to have taken place.

When we have the number of nucleotide substitutions per nucleotide site, we divide by the amount of evolutionary time that separates the two organisms (usually obtained from the fossil record) to obtain an overall rate of nucleotide substitution. For the mouse and rat growth-hormone gene, the overall rate of nucleotide substitution is approximately 8×10^{-9} substitutions per site per year.

NONSYNONYMOUS AND SYNONYMOUS RATES OF SUBSTITUTION Nucleotide changes in a gene that alter the amino acid sequence of a protein are referred to as nonsynonymous substitutions. Nucleotide changes, particularly those at the third position of a codon, that do not alter the amino acid sequence, are called synonymous substitutions. The rate of nonsynonymous substitution varies widely among mammalian genes. The rate for the α-actin protein is only 0.01×10^{-9} substitutions per site per year, whereas the rate for interferon γ is 2.79×10^{-9}, almost 300 times as high. The rate of synonymous substitution also varies among genes, but not to the extent of variation in the nonsynonymous rate. For most protein-encoding genes, the synonymous rate of change is considerably higher than the nonsynonymous rate because synonymous mutations are tolerated by natural selection (**Table 26.4**). Nonsynonymous mutations, on the other hand, alter the amino acid sequence of the protein and, in many cases, are detrimental to the fitness of the organism; most of these mutations are eliminated by natural selection.

SUBSTITUTION RATES FOR DIFFERENT PARTS OF A GENE Different parts of a gene also evolve at different rates, with the highest rates of substitutions in regions of the gene that have the least effect on function, such as the

Table 26.4 Rates of nonsynonymous and synonymous substitutions in mammalian genes based on human–rodent comparisons

Gene	Nonsynonymous Rate (per site per 10^9 years)	Synonymous Rate (per site per 10^9 years)
α-Actin	0.01	3.68
β-Actin	0.03	3.13
Albumin	0.91	6.63
Aldolase A	0.07	3.59
Apoprotein E	0.98	4.04
Creatine kinase	0.15	3.08
Erythropoietin	0.72	4.34
α-Globin	0.55	5.14
β-Globin	0.80	3.05
Growth hormone	1.23	4.95
Histone 3	0.00	6.38
Immunoglobulin heavy chain (variable region)	1.07	5.66
Insulin	0.13	4.02
Interferon α 1	1.41	3.53
Interferon γ	2.79	8.59
Luteinizing hormone	1.02	3.29
Somatostatin-28	0.00	3.97

Source: After W. Li and D. Graur, *Fundamentals of Molecular Evolution* (Sunderland, Mass.: Sinauer, 1991), p. 69, Table 1.

third position of a codon, flanking regions, and introns (**Figure 26.15** on the next page). The 5′ and 3′ flanking regions of genes are not transcribed into RNA; therefore, substitutions in these regions do not alter the amino acid sequence of the protein, although they may affect gene expression (see Chapters 16 and 17). Rates of substitution in introns are nearly as high. Although these nucleotides do not encode amino acids, introns must be spliced out of the pre-mRNA for a functional protein to be produced, and particular sequences are required at the 5′ splice site, 3′ splice site, and branch point for correct splicing (see Chapter 14).

Substitution rates are somewhat lower in the 5′ and 3′ untranslated regions of a gene. These regions are transcribed into RNA but do not encode amino acids. The 5′ untranslated region contains the ribosome-binding site, which is essential for translation, and the 3′ untranslated region contains sequences that may function in regulating mRNA stability and translation; so substitutions in these regions may have deleterious effects on organismal fitness and may not be tolerated.

The lowest rates of substitution are seen in nonsynonymous changes in the coding region, because these substitutions always alter the amino acid sequence of the protein and are often

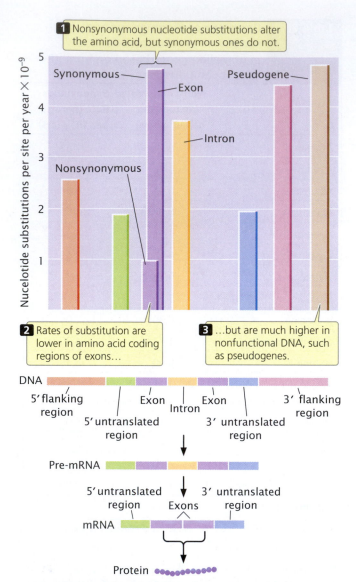

26.15 Different parts of genes evolve at different rates. The highest rates of nucleotide substitution are in sequences that have the least effect on protein function.

deleterious. High rates of substitution occur in pseudogenes, most of which are duplicated nonfunctional copies of genes that have acquired mutations. Such genes usually no longer produce a functional product; so mutations in pseudogenes have little effect on the fitness of the organism.

In summary, there is a relation between the function of a sequence and its rate of evolution; higher rates are found where they have the least effect on function. This observation fits with the neutral-mutation hypothesis, which predicts that molecular variation is not affected by natural selection. ▶ TRY PROBLEM 30

The Molecular Clock

The neutral-mutation hypothesis proposes that evolutionary change at the molecular level takes place primarily through the fixation of neutral mutations by genetic drift. The rate at which one neutral mutation replaces another depends only on the mutation rate, which should be fairly constant for any particular gene. If the rate at which a protein evolves is roughly constant over time, the amount of molecular change that a protein has undergone can be used as a **molecular clock** to date evolutionary events.

For example, the enzyme cytochrome *c* could be examined in two organisms known from fossil evidence to have had a common ancestor 400 million years ago. By determining the number of differences in the cytochrome *c* amino acid sequences in each organism, we could calculate the number of substitutions that have occurred per amino acid site. The occurrence of 20 amino acid substitutions since the two organisms diverged indicates an average rate of 5 substitutions per 100 million years. Knowing how fast the molecular clock ticks allows us to use molecular changes in cytochrome *c* to date other evolutionary events: if we found that cytochrome *c* in two organisms differed by 15 amino acid substitutions, our molecular clock would suggest that they diverged some 300 million years ago. If we assumed some error in our estimate of the rate of amino acid substitution, statistical analysis would show that the true divergence time might range from 160 million to 440 million years. The molecular clock was proposed by Emile Zuckerkandl and Linus Pauling in 1965 as a possible means of dating evolutionary events on the basis of molecules in present-day organisms. A number of studies have examined the rate of evolutionary change in proteins (**Figure 26.16**) and genes, and the molecular clock has been widely used to date evolutionary events when the fossil record is absent or ambiguous. For example, researchers used a molecular clock to estimate when Darwin's finches diverged from a common ancestor that originally colonized the Galapagos Islands. This clock was based on DNA sequence differences in the cytochrome b gene. They concluded that the ancestor of Darwin's finches arrived in the Galapagos and began diverging some 2-3 million years ago. The results of several studies have shown that the molecular clock does not always tick at a constant rate, particularly over shorter time periods, and this method remains controversial.

CONCEPTS

Different genes and different parts of the same gene evolve at different rates. Those parts of genes that have the least effect on function tend to evolve at the highest rates. The idea of the molecular clock is that individual proteins and genes evolve at a constant rate and that the differences in the sequences of present-day organisms can be used to date past evolutionary events.

✓ CONCEPT CHECK 7

In general, which types of sequences are expected to exhibit the slowest evolutionary change?

a. Synonymous changes in amino acid coding regions of exons.
b. Nonsynonymous changes in amino acid coding regions of exons.
c. Introns.
d. Pseudogenes.

(a)

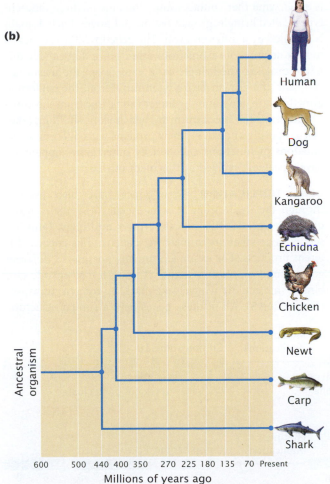

(b)

26.16 The molecular clock is based on the assumption of a constant rate of change in protein or DNA sequence. (a) Relation between the rate of amino acid substitution and time since divergence, based in part on amino acid sequences of alpha hemoglobin from the eight species shown in part b. The constant rate of evolution in protein and DNA sequences has been used as a molecular clock to date past evolutionary events. (b) Phylogeny of eight of the species that are plotted in part a and their approximate times of divergence based on the fossil record.

Evolution Through Changes in Gene Regulation

One of the challenges of evolutionary biology is understanding the genetic basis of adaptation. Many evolutionary changes occur with relatively few genetic differences. For example, humans and chimpanzees differ greatly in anatomy, physiology, and behavior and yet differ at only about 4% of their DNA sequences (see introduction to Chapter 17). Evolutionary biologists have long assumed that many anatomical differences result, not from the evolution of new genes, but rather from relatively small DNA differences that alter the expression of existing genes. Recent research in evolutionary genetics has focused on how evolution occurs through alteration of gene expression.

An example of adaptation that has occurred through changes to regulatory sequences is seen in the evolution of pigmentation in *Drosophila melanogaster* fruit flies in Africa. Most fruit flies are light tan in color, but flies in some populations of Africa exhibit much darker abdomens. These darker flies usually occur in mountainous regions at higher elevations. Indeed, 59% of pigmentation variation among populations within sub-Saharan Africa can be explained by differences in elevation (**Figure 26.17**). Researchers have demonstrated that these differences are genetically determined and that natural selection has favored darker pigmentation at high elevation. High-elevation populations are exposed to lower temperatures, and the darker pigmentation is assumed to help flies absorb more solar radiation and better regulate their body temperature in these environments.

How did flies at high elevation evolve darker color? Genetic studies indicate that the dark abdominal pigmentation seen in flies from these populations results from variation at or near a locus called *ebony*. The *ebony* locus encodes a multifunctional enzyme that produces an exoskeleton which is yellow in color; its absence produces a dark phenotype. Sequencing of the *ebony* locus of flies from light and dark

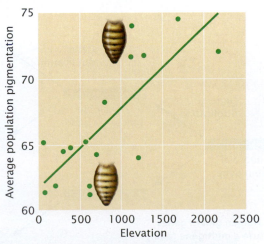

26.17 Sub-Saharan African populations of the fruit fly *Drosophila melanogaster* exhibit a positive association between pigmentation and elevation.

populations found no differences in the coding region of the *ebony* gene. However, molecular analysis revealed a marked reduction in the amount of ebony mRNA in darker flies, suggesting that the differences in pigmentation are not due to mutations at the *ebony* gene itself, but rather in its expression. Further investigation detected genetic differences within an enhancer that is about 3,600 bp upstream of the *ebony* gene. Dark and light flies differed in over 120 nucleotides scattered over 2,400 bp of the enhancer. However, by experimentally creating enhancers with different combinations of these mutations, researchers determined that five of the mutations are responsible for the majority of the differences in pigmentation.

These studies suggested that over time high-elevation populations accumulated multiple mutations in the enhancer, which reduced the expression of the *ebony* locus and caused darker pigmentation. Further analysis suggested that these mutations were added sequentially. Some of the mutations are widespread throughout Africa; it is assumed that these existing mutations were favored by natural selection in high-elevation populations and increased in frequency because they helped the flies thermoregulate in colder environments. Other mutations are only seen in the high-elevation populations, suggesting that they probably arose as new mutations within these populations and were quickly favored by natural selection.

Genome Evolution

The rapid growth of sequence data available in DNA databases has been a source of insight into evolutionary processes. Whole-genome sequences also are providing new information about how genomes evolve and the processes that shape the size, complexity, and organization of genomes.

EXON SHUFFLING Many proteins are composed of groups of amino acids, called domains, that specify discrete functions or contribute to the molecular structure of a protein. For example, in Chapter 16, we considered the DNA-binding domains of proteins that regulate gene expression. Analyses of gene sequences from eukaryotic organisms indicate that exons often encode discrete functional domains of proteins.

Some genes elongated and evolved new functions when one or more exons duplicated and underwent divergence. For example, the human serum-albumin gene is made up of three copies of a sequence that encodes a protein domain consisting of 195 amino acids. Additionally, the genes that encode human immunoglobulins have undergone repeated tandem duplications, creating many similar *V*, *J*, *D*, and *C* segments (see pp. 652–653 in Chapter 22) that enable the immune system to respond to almost any foreign substance that enters the body.

A comparison of DNA sequences from different genes reveals that new genes have repeatedly evolved through a process called **exon shuffling**, in which exons of different genes are exchanged, creating genes that are mosaics of other genes. For example, tissue plasminogen activator (TPA) is an enzyme that contains four domains of three different types, called kringle, growth factor, and finger. Each domain is encoded by a different exon. The gene for TPA is believed to have acquired its exons from other genes that encode different proteins: the kringle exon came from the plasminogen gene, the growth-factor exon came from the epidermal growth-factor gene; and the finger exon came from the fibronectin gene. The mechanism by which exon shuffling takes place is poorly known, but new proteins with different combinations of functions encoded by other genes apparently have repeatedly evolved by this mechanism.

GENE DUPLICATION New genes have also evolved through the duplication of whole genes and their subsequent divergence. This process creates **multigene families**, sets of genes that are similar in sequence but encode different products. For example, humans possess 13 different genes found on chromosomes 11 and 16 that encode globinlike molecules, which take part in oxygen transport (**Figure 26.18**). All of these genes have a similar structure,

26.18 Human globin genes constitute a multigene family that has evolved through successive gene duplications.

with three exons separated by two introns, and are assumed to have evolved through repeated duplication and divergence from a single globin gene in a distant ancestor. This ancestral gene is thought to have been most similar to the present-day myoglobin gene and first duplicated to produce an α/β-globin precursor gene and the myoglobin gene. The α/β-globin gene then underwent another duplication to give rise to a primordial α-globin gene and a primordial β-globin gene. Subsequent duplications led to multiple α-globin and β-globin genes. Similarly, vertebrates contain four clusters of *Hox* genes, each cluster comprising from 9 to 11 genes. *Hox* genes play an important role in development (see pp. 642–643 in Chapter 22).

Some gene families include genes that are arrayed in tandem on the same chromosome; others are dispersed among different chromosomes. Gene duplication is a common occurrence in eukaryotic genomes; for example, about 5% of the human genome consists of duplicated segments.

Gene duplication provides a mechanism for the addition of new genes with novel functions; after a gene duplicates, there are two copies of the sequence, one of which is free to change and potentially take on a new function. The extra copy of the gene may, for example, become active at a different time in development or be expressed in a different tissue or even diverge and encode a protein having different amino acids. However, the most common fate of gene duplication is that one copy acquires a mutation that renders it nonfunctional, giving rise to a pseudogene. Pseudogenes are common in the genomes of complex eukaryotes; the human genome is estimated to contain as many as 20,000 pseudogenes.

WHOLE-GENOME DUPLICATION In addition to the duplication of individual genes, whole genomes of some organisms have duplicated in the past. For example, a comparison of the genome of the yeast *Saccharomyces cerevisiae* with the genomes of other fungi reveals that *S. cerevisiae* or one of its immediate ancestors underwent a whole-genome duplication, generating two copies of every gene. Many of the copies subsequently acquired new functions; others acquired mutations that destroyed the original function and then diverged into random DNA sequences. Whole-genome duplication can take place through polyploidy.

During their evolution, plants have undergone a number of whole-genome duplications through polyploidy. While polyploidy is less common in animals, genetic evidence suggests that several whole-genome duplication events have occurred during animal evolution. In 1970, Susumu Ohno proposed that early vertebrates underwent two rounds of genome duplication. Called the 2R hypothesis, this idea has been controversial, but recent data from genome sequencing has provided support for it.

HORIZONTAL GENE TRANSFER Traditionally, scientists assumed that organisms acquire their genomes through vertical transmission—transfer through the reproduction of genetic information from parents to offspring, and most phylogenetic trees assume vertical transmission of genetic information. Findings from DNA sequence studies reveal that DNA sequences are sometimes exchanged by a horizontal gene transfer, in which DNA is transferred between individuals of different species (see Chapter 9). This process is especially common among bacteria, and there are a number of documented cases in which genes are transferred from bacteria to eukaryotes. The extent of horizontal gene transfer among eukaryotic organisms is controversial, with few well-documented cases. Horizontal gene transfer can obscure phylogenetic relationships and make the reconstruction of phylogenetic trees difficult.

One apparent case of horizontal gene transfer among eukaryotes is the presence in some aphids of genes for enzymes that synthesize carotenoids. Carotenoids are colored compounds produced by bacteria, archaea, fungi, and plants. Many animals also have carotenoids, but they lack the enzymes necessary of make the compounds themselves; in almost all cases, animals obtain carotenoids from their food.

Aphids—small insects that feed on plants—have carotenoids, which are responsible for color differences between and within species. Some aphids are green and contain α-, β-, and γ-carotene, which are all yellow carotenoids. Other aphids are red or brown and contain lycopene or torulene, carotenoids that are red. One species, the pea aphid (*Acyrthosiphon pisum*) has both green and red individuals, and these differences are genetically inherited. Many researchers previously assumed that the color differences were due to carotenoids that were acquired in the aphids' food.

Researchers have recently sequenced the entire genome of *A. pisum*, which provided the opportunity to determine whether the aphids possessed their own genes for carotenoid synthesis. Examination of genomic sequences revealed that pea aphids have several genes that code for carotenoid-synthesizing enzymes. Interestingly, these genes are closely related to carotenoid-synthesizing genes found in some fungi. The evidence suggests that in the distant past, an aphid acquired its carotenoid genes from a fungus through horizontal gene transfer and then passed the genes on to other aphids through vertical transmission.

CONCEPTS

New genes may evolve through the duplication of exons, shuffling of exons, duplication of genes, and duplication of whole genomes. Genes can be passed among distantly related organisms through horizontal gene transfer.

- Evolution is genetic change taking place within a group of organisms. It is a two-step process: (1) genetic variation arises, and (2) genetic variants change in frequency.

- Anagenesis refers to change within a single lineage; cladogenesis is the splitting of one lineage into two.

- Molecular methods offer a number of advantages for the study of evolution.

- The use of protein electrophoresis to study genetic variation in natural populations showed that most natural populations have large amounts of genetic variation in their proteins. The neutral-mutation hypothesis proposes that molecular variation is selectively neutral and is shaped largely by mutation and genetic drift. The balance hypothesis proposes that molecular variation is maintained largely by balancing selection.

- Variation in DNA sequences can be assessed by analyzing restriction fragment length polymorphisms, microsatellites, and data from direct sequencing.

- A species can be defined as a group of organisms that are capable of interbreeding with one another and are reproductively isolated from the members of other species.

- Species are prevented from exchanging genes by prezygotic or postzygotic reproductive isolation.

- Allopatric speciation arises when a geographic barrier prevents gene flow between two populations. With the passage of time, the two populations acquire genetic differences that may lead to reproductive isolating mechanisms.

- Sympatric speciation arises when reproductive isolation exists in the absence of any geographic barrier. It may arise under special circumstances.

- Some species arise only after populations have undergone considerable genetic differences; others arise after changes have taken place in only a few genes.

- Evolutionary relationships (a phylogeny) can be represented by a phylogenetic tree, consisting of nodes that represent organisms and branches that represent their evolutionary connections.

- Approaches to constructing phylogenetic trees include the distance approach, the maximum parsimony approach, and the maximum likelihood and Bayesian methods approach.

- Different parts of the genome show different amounts of genetic variation. In general, those parts that have the least effect on function evolve at the highest rates.

- The molecular-clock hypothesis proposes a constant rate of nucleotide substitution, providing a means of dating evolutionary events by looking at nucleotide differences between organisms.

- Genome evolution takes place through the duplication and shuffling of exons, the duplication of genes to form gene families, whole-genome duplication, and the horizontal transfer of genes between organisms.

IMPORTANT TERMS

evolution (p. 744)
anagenesis (p. 745)
cladogenesis (p. 745)
neutral-mutation hypothesis (p. 746)
balance hypothesis (p. 747)
species (p. 749)
biological species concept (p. 749)
reproductive isolating mechanism (p. 750)
prezygotic reproductive isolating mechanism (p. 750)
ecological isolation (p. 750)
behavioral isolation (p. 750)
temporal isolation (p. 750)
mechanical isolation (p. 750)
gametic isolation (p. 750)
postzygotic reproductive isolating mechanism (p. 750)
hybrid inviability (p. 750)
hybrid sterility (p. 750)
hybrid breakdown (p. 750)
speciation (p. 751)
allopatric speciation (p. 751)
sympatric speciation (p. 751)
phylogeny (p. 756)
phylogenetic tree (p. 756)
node (p. 757)
branch (p. 757)
rooted tree (p. 757)
gene tree (p. 757)
molecular clock (p. 760)
exon shuffling (p. 762)
multigene family (p. 762)

ANSWERS TO CONCEPT CHECKS

1. First genetic variation arises. Then various evolutionary forces cause changes in the frequency of genetic variants.

2. c

3. Microsatellites are often highly variable among individuals. They can be amplified with the use of PCR, and so they can be detected with a small amount of starting DNA. Finally, the detection and analysis of microsatellites can be automated.

4. b

5. Genetic drift can bring about changes in the allelic frequencies of populations and lead to genetic differences among populations. Genetic differentiation is the cause of postzygotic and prezygotic reproductive isolation between populations that leads to speciation.

6. c

7. b

COMPREHENSION QUESTIONS

Section 26.1

1. How is biological evolution defined?

2. What are the two steps in the process of evolution?

3. How is anagenesis different from cladogenesis?

Section 26.2

4. Why does protein variation, as revealed by electrophoresis, underestimate the amount of true genetic variation?

5. What are some of the advantages of using molecular data in evolutionary studies?

6. What is the key difference between the neutral-mutation hypothesis and the balance hypothesis?

7. Describe some of the methods that have been used to study variation in DNA.

Section 26.3

8. What is the biological species concept?

9. What is the difference between prezygotic and postzygotic reproductive isolating mechanisms? List different types of each.

10. What is the basic difference between allopatric and sympatric modes of speciation?

11. Briefly outline the process of allopatric speciation.

12. What are some of the difficulties with sympatric speciation?

13. Briefly explain how switching from hawthorn fruits to apples has led to genetic differentiation and partial reproductive isolation in *Rhagoletis pomonella*.

Section 26.4

14. Draw a simple phylogenetic tree and identify a node, a branch, and an outgroup.

15. Briefly describe differences among the distance approach, the maximum parsimony approach, and the maximum likelihood approach to the reconstruction of phylogenetic trees.

Section 26.5

16. Outline the different rates of evolution that are typically seen in different parts of a protein-encoding gene. What might account for these differences?

17. What is the molecular clock?

18. What is exon shuffling? How can it lead to the evolution of new genes?

19. What is a multigene family? What processes produce multigene families?

20. Define horizontal gene transfer. What problems does it cause for evolutionary biologists?

> ▶ For more questions that test your comprehension of the key chapter concepts, go to **LEARNING***Curve* for this chapter.

APPLICATION QUESTIONS AND PROBLEMS

Section 26.1

*21. The following illustrations represent two different patterns of evolution. Briefly discuss the differences in these two patterns, particularly in regard to the role of cladogenesis in evolutionary change.

Evolutionary change

Evolutionary change

Section 26.3

*22. Which of the isolating mechanisms listed in **Table 26.2** have partly evolved between apple and hawthorn host races of *Rhagoletis pomonella*, the apple maggot fly?

23. We considered the sympatric evolution of reproductive isolating mechanisms in host races of *Rhagoletis pomonella*, the apple maggot fly. The wasp *Diachasma alloeum* parasitizes apple maggot flies, laying its eggs on the larvae of the flies. Immature wasps hatch from the eggs and feed on the fly larvae. Research by Andrew Forbes and his colleagues (Forbes et al. 2009. *Science* 323:776–779) demonstrated that wasps that parasitize apple races of *R. pomonella* are genetically differentiated from those that parasitize hawthorn races of *R. pomonella*. They also found that wasps that prey on the apple race of the flies are attracted to odors from apples, whereas wasps that prey on the hawthorn race are attracted to odors from hawthorn fruit. Propose an explanation for how genetic differences might have evolved between the wasps that parasitize the two races of *R. pomonella*. How might these differences lead to speciation in the wasps?

24. Which of the following in **Figure 26.11** is a polyploid?

   a. *Spartina maritime.*

   b. *Spartina alterniflora.*

   c. Hybrid between *Spartina maritima* and *Spartina alterniflora.*

   d. *Spartina anglica.*

   e. Both c and d above.

Section 26.4

25. How many rooted trees are theoretically possible for a group of seven organisms? How many for 12 organisms?

26. Align the sequences below so as to maximize their similarity. What is the minimum number of evolutionary steps that separate these two sequences?

   TTGCAAAC
   TGAAACTG

*27. Michael Bunce and his colleagues in England, Canada, and the United States extracted and sequenced mitochondrial DNA from fossils of Haast's eagle, a gigantic eagle that was driven to extinction 700 years ago when humans first arrived in New Zealand (M. Bunce et al. 2005. *PLOS Biology* 3:44–46). Using mitochondrial DNA sequences from living eagles and those from Haast eagle fossils, they created the phylogenetic tree at the right. On this phylogenetic tree, identify (a) all terminal nodes; (b) all internal nodes; (c) one example of a branch; and (d) the outgroup.

28. On the basis of the phylogeny of Darwin's finches shown in **Figure 26.8**, predict which two species in each of the following groups will be the most similar genetically.

   a. *Camarhynchus parvulus, Camarhynchus psittacula, Camarhynchus pallida*

   b. *Camarhynchus parvulus, Camarhynchus pallida, Platyspiza crassirostris.*

   c. *Geospiza difficilis, Geospiza conirostris, Geospiza scandens.*

   d. *Camarhynchus parvulus, Certhidea fusca, Pinaroloxias inornata.*

Section 26.5

29. Assume that one of the genes shown in **Table 26.4** had similar nonsynonymous and synonymous rates of substitutions. What might this suggest about the evolution of this gene?

*30. Based on the information provided in **Figure 26.15**, do introns or 3′ untranslated regions of a gene have higher rates of nucleotide substitution? Explain why.

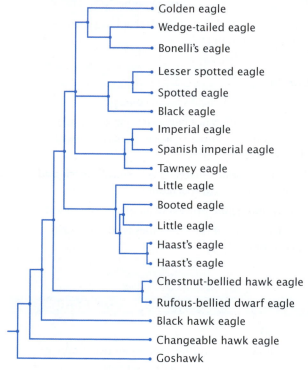

Golden eagle
Wedge-tailed eagle
Bonelli's eagle
Lesser spotted eagle
Spotted eagle
Black eagle
Imperial eagle
Spanish imperial eagle
Tawney eagle
Little eagle
Booted eagle
Little eagle
Haast's eagle
Haast's eagle
Chestnut-bellied hawk eagle
Rufous-bellied dwarf eagle
Black hawk eagle
Changeable hawk eagle
Goshawk

[After M. Bunce et al., *PLOS Biology* 3:44–46, 2005.]

CHALLENGE QUESTIONS

Section 26.3

31. Explain why natural selection may cause prezygotic reproductive isolating mechanisms to evolve if postzygotic reproductive isolating mechanisms are already present but natural selection can never cause the evolution of postzygotic reproductive isolating mechanisms.

32. Polyploidy is very common in flowering plants: approximately 40% of all flowering plant species are polyploids. Although polyploidy exists in many different animal groups, it is much less common. Why is polyploidy more common in plants than in animals? Give one or more possible reasons.

Go to your **LaunchPad** to find additional learning resources and the Suggested Readings for this chapter.

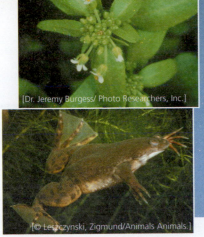
[Dr. Jeremy Burgess/ Photo Researchers, Inc.]

[Eye of Science/Science Source.]

[Steve Gschmeissner/Science Photo Library/Alamy.]

[Biophoto Associates/Photo Researchers.]

[© Gelia I Dreamstime.com]

[© Leszczynski, Zigmund/Animals Animals.]

[Sinclair Stammers/Science Source.]

Model genetic organisms possess characteristics that make them useful for genetic studies. Shown are several organisms commonly used in genetic studies.

Reference Guide to Model Genetic Organisms

What do Lou Gehrig, the finest first baseman in major league history, and Stephen Hawking, the world's most famous theoretical physicist, have in common? They both suffered or suffer from amyotrophic lateral sclerosis (ALS, also known as Lou Gehrig disease), a degenerative neurological disease that leads to progressive weakness and wasting of skeletal muscles. Interestingly, findings from recent research reveal that some athletes with ALS symptoms in fact suffer from brain trauma due to sports injuries; so Lou Gehrig might not have had the disease named for him.

Most cases of ALS are sporadic, appearing in people who have no family history of the disease. However, about 10% of the cases run in families and are inherited. In 2004, geneticists discovered a large Brazilian family with multiple cases of ALS. Genetic analysis revealed that ALS in this family is due to a mutation in a gene called *VABP*, which encodes a vesicle-associated membrane protein.

To better understand how mutations in *VABP* lead to the symptoms of ALS, geneticists turned to an unlikely subject—the fruit fly *Drosophila melanogaster*. Fruit flies don't have ALS, but they do possess a gene very similar to *VABP*. Using a wide array of techniques that have been developed for genetically manipulating fruit flies, geneticists created transgenic flies with the mutant sequence of the *VABP* gene that causes ALS in humans. These flies are a disease model for ALS and are being used to better understand what the gene *VABP* does normally and how its disruption can lead to ALS.

The field of genetics has been greatly influenced and shaped by a few key organisms—called model genetic organisms—whose characteristics make them particularly amenable to genetic studies. The use of *Drosophila* for studying ALS in humans illustrates the power of this approach. Because features of genetic systems are common to many organisms, research conducted on one species can often be a source of insight into the genetic systems of other species. This commonality of genetic function means that geneticists can focus their efforts on model organisms that are easy to work with and likely to yield results.

Model genetic organisms possess life cycles and genomic features that make them well suited to genetic study and analysis. Some key features possessed by many model genetic organisms include:

- a short generation time, and so several generations of genetic crosses can be examined in reasonable time;
- the production of numerous progeny, which allows genetic ratios to be easily observed;
- the ability to carry out and control genetic crosses in the organism;
- the ability to be reared in a laboratory environment, requiring little space and few resources to maintain;
- the availability of numerous genetic variants; and
- an accumulated body of knowledge about their genetic systems.

In recent years, the genomes of many model genetic organisms have been completely sequenced, greatly facilitating their use in genetic research.

Not all model organisms possess all of these characteristics. However, each model genetic organism has one or more features that make it useful for genetic analysis. For example, corn cannot be easily grown in the laboratory (and usually isn't) and it has a relatively long generation time, but it produces numerous progeny and there are many genetic variants of corn available for study.

This reference guide highlights six model genetic organisms with important roles in the development of genetics: the fruit fly (*Drosophila melanogaster*), bacterium (*Escherichia coli*), roundworm (*Caenorhabditis elegans*), thale cress plant (*Arabidopsis thaliana*), house mouse (*Mus musculus*), and yeast (*Saccharomyces cerevisiae*). These six organisms have been widely used in genetic research and instruction. A number of other organisms also are used as model systems in genetics, including corn (*Zea mays*), zebrafish (*Danio rerio*), clawed frog (*Xenopus lavis*), bread mold (*Neurospora crassa*), rat (*Rattus norvegicus*), and Rhesus macaque (*Macaca mulatta*), just to mention a few.

The Fruit Fly *Drosophila melanogaster*

Drosophila melanogaster, a fruit fly, was among the first organisms used for genetic analysis and, today, it is one of the most widely used and best known genetically of all organisms. It has played an important role in studies of linkage, epistasis, chromosome genetics, development, behavior, and evolution. Because all organisms use a common genetic system, understanding a process such as replication or transcription in fruit flies helps us to understand these same processes in humans and other eukaryotes.

Drosophila is a genus of more than 1000 described species of small flies (about 1 to 2 mm in length) that frequently feed and reproduce on fruit, although they rarely cause damage and are not considered economic pests. The best known and most widely studied of the fruit flies is *D. melanogaster,* but genetic studies have been extended to many other species of the genus as well. *D. melanogaster* first began to appear in biological laboratories about 1900. After first taking up breeding experiments with mice and rats, Thomas Hunt Morgan began using fruit flies in experimental studies of heredity at Columbia University. Morgan's laboratory, located on the top floor of Schermerhorn Hall, became known as the Fly Room (see Figure 4.11b). To say that the Fly Room was unimpressive is an understatement. The cramped room, only about 16 by 23 feet, was filled with eight desks, each occupied by a student and his experiments. The primitive laboratory equipment consisted of little more than milk bottles for rearing the flies and hand-held lenses for observing their traits. Later, microscopes replaced the hand-held lenses, and crude incubators were added to maintain the fly cultures, but even these additions did little to increase the physical sophistication of the laboratory. Morgan and his students were not tidy: cockroaches were abundant (living off spilled *Drosophila* food), dirty milk bottles filled the sink, ripe bananas—food for the flies—hung from the ceiling, and escaped fruit flies hovered everywhere. In spite of its physical limitations, the Fly Room was the source of some of the most important research in the history of biology. There was daily excitement among the students, some of whom initially came to the laboratory as undergraduates. The close quarters facilitated informality and the free flow of ideas. Morgan and the Fly Room illustrate the tremendous importance of "atmosphere" in producing good science. Morgan and his students eventually used *Drosophila* to elucidate many basic principles of heredity, including sex-linked inheritance, epistasis, multiple alleles, and gene mapping.

Advantages of *D. melanogaster* as a model genetic organism *Drosophila*'s widespread use in genetic studies is no accident. The fruit fly has a number of characteristics that make it an ideal subject for genetic investigations. Compared with other organisms, it has a relatively short generation time; fruit flies will complete an entire generation in about 10 days at room temperature, and so several generations can be studied within a few weeks. Although *D. melanogaster* has a short generation time, it possesses a complex life cycle, passing through several different developmental stages, including egg, larva, pupa, and adult. A female fruit fly is capable of mating within 8 hours of emergence and typically begins to lay eggs after about 2 days. Fruit flies also produce a large number of offspring, laying as many as 400 to 500 eggs in a 10-day period. Thus, large numbers of progeny can be obtained from a single genetic cross.

Another advantage is that fruit flies are easy to culture in the laboratory. They are usually raised in small glass vials or bottles and are fed easily prepared, pastelike food consisting of bananas or corn meal and molasses. Males and females are readily distinguished and virgin females are easily isolated, facilitating genetic crosses. The flies are small, requiring little space—several hundred can be raised in a half-pint bottle—but they are large enough for many mutations to be easily observed with a hand lens or a dissecting microscope.

Finally, *D. melanogaster* is the organism of choice for many geneticists because it has a relatively small genome consisting of 175 million base pairs of DNA, which is only about 5% of the size of the human genome. It has four pairs of chromosomes: three pairs of autosomes and one pair of sex chromosomes. The X chromosome (designated chromosome 1) is large and acrocentric, whereas the Y chromosome is large and submetacentric, although it contains very little genetic information. Chromosomes 2 and 3 are large and metacentric; chromosome 4 is a very small acrocentric chromosome. In the salivary glands, the chromosomes are very large (see p. 304 in Chapter 11), making *Drosophila* an excellent subject for chromosome studies. In 2000, the complete genome of *D. melanogaster* was sequenced, followed by the sequencing of the genome of *D. pseudoobscura* in 2005 and the genomes of 10 additional *Drosophila* genomes in 2007. *Drosophila* continues today to be one of the most versatile and powerful of all genetic model organisms. ∎

The Fruit Fly
Drosophila melanogaster

ADVANTAGES

- Small size
- Short generation time of 10 days at room temperature
- Each female lays 400–500 eggs
- Easy to culture in laboratory
- Small genome
- Large chromosomes
- Many mutations available

STATS

Taxonomy:	Insect
Size:	2–3 mm in length
Anatomy:	3 body segments, 6 legs, 1 pair of wings
Habitat:	Feeds and reproduces on fruit

Life Cycle

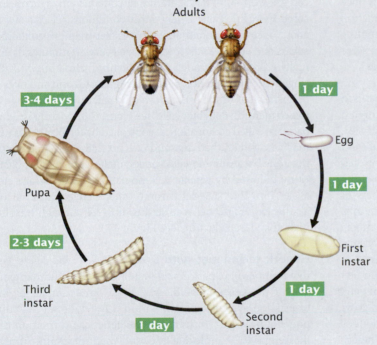

Adults · 1 day · Egg · 1 day · First instar · 1 day · Second instar · 1 day · Third instar · 2-3 days · Pupa · 3-4 days

Autosomes · Sex chromosomes

GENOME

Chromosomes:	3 pairs of autosomes and X and Y ($2n = 8$)
Amount of DNA:	175 million base pairs
Number of genes:	14,000
Percentage of genes in common with humans:	~50%
Average gene size:	3000 base pairs
Genome sequenced in year:	2000

CONTRIBUTIONS TO GENETICS

- Basic principles of heredity including sex-linked inheritance, multiple alleles, epistatsis, gene mapping, etc.
- Mutation research
- Chromosome variation and behavior
- Population genetics
- Genetic control of pattern formation
- Behavioral genetics

The Bacterium *Escherichia coli*

The most widely studied prokaryotic organism and one of the best genetically characterized of all species is the bacterium *Escherichia coli*. Although some strains of *E. coli* are toxic and cause disease, most are benign and reside naturally in the intestinal tracts of humans and other warm-blooded animals. *E. coli* was first described by Theodore Escherich in 1885 but, for many years, the assumption was that all bacteria reproduced only asexually and that genetic crosses were impossible. In 1946, Joshua Lederberg and Edward Tatum demonstrated that *E. coli* undergoes a type of sexual reproduction; their finding initiated the use of *E. coli* as a model genetic organism. A year later, Lederberg published the first genetic map of *E. coli* based on recombination frequencies and, in 1952, William Hays showed that mating between bacteria is asymmetrical, with one bacterium serving as genetic donor and the other as genetic recipient.

Advantages of *E. coli* as a model genetic organism

Escherichia coli is one of the true workhorses of genetics; its twofold advantage is rapid reproduction and small size. Under optimal conditions, this organism can reproduce every 20 minutes and, in a mere 7 hours, a single bacterial cell can give rise to more than 2 million descendants. One of the values of rapid reproduction is that enormous numbers of cells can be grown quickly, and so even very rare mutations will appear in a short period. Consequently, numerous mutations in *E. coli,* affecting everything from colony appearance to drug resistance, have been isolated and characterized.

Escherichia coli is easy to culture in the laboratory in liquid medium (see Figure 9.1a) or on solid medium within petri plates (see Figure 9.1b). In liquid culture, *E. coli* cells will grow to a concentration of a billion cells per milliliter, and trillions of bacterial cells can be easily grown in a single test tube. When *E. coli* cells are diluted and spread onto the solid medium of a petri dish, individual bacteria reproduce asexually, giving rise to a concentrated clump of 10 million to 100 million genetically identical cells, called a colony. This colony formation makes it easy to isolate genetically pure strains of the bacteria.

The *E. coli* genome

The *E. coli* genome is on a single chromosome and—compared with those of humans, mice, plants, and other multicellular organisms—is relatively small, consisting of 4,638,858 base pairs. If stretched out straight, the DNA molecule in the single *E. coli* chromosome would be 1.6 mm long, almost a thousand times as long as the *E. coli* cell within which it resides (see Figure 11.1). To accommodate this huge amount of DNA within the confines of a single cell, the *E. coli* chromosome is highly coiled and condensed. The information within the *E. coli* chromosome also is compact, having little noncoding DNA between and within the genes

and having few sequences for which there is more than one copy. The *E. coli* genome contains an estimated 4300 genes, many of which have no known function. These "orphan genes" may play important roles in adapting to unusual environments, coordinating metabolic pathways, organizing the chromosome, or communicating with other bacterial cells. The haploid genome of *E. coli* makes it easy to isolate mutations because there are no dominant genes at the same loci to suppress and mask recessive mutations.

Life cycle of *E. coli*

Wild-type *E. coli* is prototrophic and can grow on minimal medium that contains only glucose and some inorganic salts. Under most conditions, *E. coli* divides about once an hour, although, in a richer medium containing sugars and amino acids, it will divide every 20 minutes. It normally reproduces through simple binary fission, in which the single chromosome of a bacterium replicates and migrates to opposite sides of the cell, followed by cell division, giving rise to two identical daughter cells (see Figure 2.5). Mating between bacteria, called conjugation, is controlled by fertility genes normally located on the F plasmid (see pp. 248–249). In conjugation, one bacterium donates genetic material to another bacterium, followed by genetic recombination that integrates new alleles into the bacterial chromosome. Genetic material can also be exchanged between strains of *E. coli* through transformation and transduction (see Figure 9.7).

Genetic techniques with *E. coli*

Escherichia coli is used in a number of experimental systems in which fundamental genetic processes are studied in detail. For example, in vitro translation systems contain within a test tube all the components necessary to translate the genetic information of a messenger RNA molecule into a polypeptide chain. Similarly, in vitro systems containing components from *E. coli* cells allow transcription, replication, gene expression, and many other important genetic functions to be studied and analyzed under controlled laboratory conditions.

Escherichia coli is also widely used in genetic engineering (recombinant DNA; see Chapter 19). Plasmids have been isolated from *E. coli* and genetically modified to create effective vectors for transferring genes into bacteria and eukaryotic cells. Often, new genetic constructs (DNA sequences created in the laboratory) are assembled and cloned in *E. coli* before transfer to other organisms. Methods have been developed to introduce specific mutations within *E. coli* genes, and so genetic analysis no longer depends on the isolation of randomly occurring mutations. New DNA sequences produced by recombinant DNA can be introduced by transformation into special strains of *E. coli* that are particularly efficient (competent) at taking up DNA.

Bacterium
Escherichia coli

ADVANTAGES

- Small size
- Rapid reproduction, dividing every 20 minutes under optimal conditions
- Easy to culture in liquid medium or on petri plates
- Small genome
- Many mutants available
- Numerous methods available for genetic engineering

STATS

Taxonomy:	Eubacteria
Size:	1–2 μm in length
Anatomy:	Single cell surrounded by cell wall with nucleoid region
Habitat:	Intestinal tract of warm-blooded animals

Life Cycle

Chromosome

Bacterial chromosome replicates

Asexual reproduction

Genetic exchange

Conjugation

F factor

F⁺

F⁻

Chromosomes separate

Transfer of genetic information

Cell division

F⁺

F⁺

Chromosome

GENOME

Chromosomes:	1 circular chromosome
Amount of DNA:	4.64 million base pairs
Number of genes:	4300
Percentage of genes in common with humans:	8%
Average gene size:	1000 base pairs
Genome sequenced in:	1997

CONTRIBUTIONS TO GENETICS

- Gene regulation
- Molecular biology and biochemistry of genetic processes, such as replication, transcription, translation, recombination
- Gene structure and organization in bacteria
- Workhorse of recombinant DNA
- Gene mutations

Because of its powerful advantages as a model genetic organism, *E. coli* has played a leading role in many fundamental discoveries in genetics, including elucidation of the genetic code, probing the nature of replication, and working out the basic mechanisms of gene regulation. ∎

The Nematode Worm *Caenorhabditis elegans*

You may be asking, What is a nematode, and why is it a model genetic organism? Although rarely seen, nematodes are among the most abundant organisms on Earth, inhabiting soils throughout the world. Most are free living and cause no harm, but a few are important parasites of plants and animals, including humans. Although *Caenorhabditis elegans* has no economic or medical importance, it has become widely used in genetic studies because of its simple body plan, ease of culture, and high reproductive capacity. First introduced to the study of genetics by Sydney Brenner, who formulated plans in 1962 to use *C. elegans* for the genetic dissection of behavior, this species has made important contributions to the study of development, cell death, aging, and behavior.

Advantages of *C. elegans* as a model genetic organism

An ideal genetic organism, *C. elegans* is small, easy to culture, and produces large numbers of offspring. The adult *C. elegans* is about 1 mm in length. Most investigators grow *C. elegans* on agar-filled petri plates that are covered with a lawn of bacteria, which the nematodes devour. Thousands of worms can be easily cultured in a single laboratory. Compared with most multicellular animals, they have a very short generation time, about 3 days at room temperature. And they are prolific reproducers, with a single female producing from 250 to 1000 fertilized eggs in 3 to 4 days.

Another advantage of *C. elegans,* particularly for developmental studies, is that the worm is transparent, allowing easy observation of internal development at all stages. It has a simple body structure, with a small, invariant number of somatic cells: 959 cells in a mature hermaphroditic female and 1031 cells in a mature male.

Life cycle of *C. elegans*

Most mature adults are hermaphrodites, with the ability to produce both eggs and sperm and undergo self-fertilization. A few are male, which produce only sperm and mate with hermaphrodites. The hermaphrodites have two sex chromosomes (XX); the males possess a single sex chromosome (XO). Thus, hermaphrodites that self-fertilize produce only hermaphrodites (with the exception of a few males that result from nondisjunction of the X chromosomes). When hermaphrodites mate with males, half of the progeny are XX hermaphrodites and half are XO males.

Eggs are fertilized internally, either from sperm produced by the hermaphrodite or from sperm contributed by a male. The eggs are then laid, and development is completed externally. Approximately 14 hours after fertilization, a larva hatches from the egg and goes through four larval stages—termed L1, L2, L3, and L4—that are separated by molts. The L4 larva undergoes a final molt to produce the adult worm. Under normal laboratory conditions, worms will live for 2 to 3 weeks.

The *C. elegans* genome

Geneticists began developing plans in 1989 to sequence the genome of *C. elegans,* and the complete genome sequence was obtained in 1998. Compared with the genomes of most multicellular animals, that of *C. elegans,* at 103 million base pairs of DNA, is small, which facilitates genomic analysis. The availability of the complete genome sequence provides a great deal of information about gene structure, function, and organization in this species. For example, the process of programmed cell death (apoptosis, see Chapter 22) plays an important role in development and in the suppression of cancer. Apoptosis in *C. elegans* is remarkably similar to that in humans. Having the complete genome sequence of *C. elegans,* and given its ease of genetic manipulation, geneticists have identified genes that participate in apoptosis, which has increased our understanding of apoptosis in humans and its role in cancer.

Genetic techniques with *C. elegans*

Chemical mutagens are routinely used to generate mutations in *C. elegans*—mutations that are easy to identify and isolate. The ability of hermaphrodites to self-fertilize means that progeny homozygous for recessive mutations can be obtained in a single generation; the existence of males means that genetic crosses can be carried out.

Developmental studies are facilitated by the transparent body of the worms. As stated earlier, *C. elegans* has a small and exact number of somatic cells. Researchers studying the development of *C. elegans* have meticulously mapped the entire cell lineage of the species, and so the developmental fate of every cell in the adult body can be traced to the original single-celled fertilized egg. Developmental biologists often use lasers to destroy (ablate) specific cells in a developing worm and then study the effects on physiology, development, and behavior.

Figure 1 A sequence for the green fluorescent protein (GFP) has been used to visually determine the expression of genes inserted into *C. elegans* (lower photograph). The gene for GFP is injected into the ovary of a worm and becomes incorporated into the worm genome. The expression of this transgene produces GFP, which fluoresces green (upper photograph). [Huaqi Jiang, Rong Guo, and Jo Anne Powell-Coffman. The *Caenorhabditis elegans* hif-1 gene encodes a bHLH-PAS protein that is required for adaptation to hypoxia. *PNAS* 98: 7916-7921, 2001. ©2001 National Academy of Sciences, U.S.A.]

The Nematode Worm
Caenorhabditis elegans

ADVANTAGES

- Small size
- Short generation time of 3 days
- Each female can produce 200–1000 eggs
- Easy to culture in laboratory
- Simple body plan
- Transparent
- Capable of self-fertilization or crossing

STATS

Taxonomy:	Nematode
Size:	1 mm
Anatomy:	Unsegmented, elongated body
Habitat:	Lives and reproduces in soil

Life Cycle

Hermaphrodite

Self-fertilization

Male

Mating with male

Adult

Fertilized egg

Larva

12 hours

26 hours

L1

3 hours

L4

10 hours

L3

11 hours

L2

Chromosomes

GENOME

Chromosomes:	5 pairs of autosomes plus 2 X chromosomes in females (hermaphrodites) or 1 X chromosome in males
Amount of DNA:	103 million base pairs
Number of genes:	20,500
Percentage of genes in common with humans:	25%
Average gene size:	5000 base pairs
Genome sequenced in year:	1998

CONTRIBUTIONS TO GENETICS

- Genetics of development
- Apoptosis (programmed cell death)
- Genetic control of behavior
- Aging

[Photograph courtesy of William Goodyer and Monique Zetka.]

RNA interference has proved to be an effective tool for turning off genes in *C. elegans*. Geneticists inject double-stranded copies of RNA that is complementary to specific genes; the double-stranded RNA then silences the expression of these genes through the RNAi process. The worms can even be fed bacteria that have been genetically engineered to express the double-stranded RNA, thus avoiding the difficulties of microinjection.

Transgenic worms can be produced by injecting DNA into the ovary, where the DNA becomes incorporated into the oocytes. Geneticists have created a special reporter gene that produces the jellyfish green fluorescent protein (GFP). When this reporter gene is injected into the ovary and becomes inserted into the worm genome, its expression produces GFP, which fluoresces green, allowing the expression of the gene to be easily observed (**Figure 1**). ■

The Plant *Arabidopsis thaliana*

Much of the early work in genetics was carried out on plants, including Mendel's seminal discoveries in pea plants as well as important aspects of heredity, gene mapping, chromosome genetics, and quantitative inheritance in corn, wheat, beans, and other plants. However, by the mid-twentieth century, many geneticists had turned to bacteria, viruses, yeast, *Drosophila,* and mouse genetic models. Because a good genetic plant model did not exist, plants were relatively neglected, particularly for the study of molecular genetic processes.

This changed in the last part of the twentieth century with the widespread introduction of a new genetic model organism, the plant *Arabidopsis thaliana. A. thaliana* was identified in the sixteenth century, and the first mutant was reported in 1873; but this species was not commonly studied until the first detailed genetic maps appeared in the early 1980s. Today, *Arabidopsis* figures prominently in the study of genome structure, gene regulation, development, and evolution in plants, and it provides important basic information about plant genetics that is applied to economically important plant species.

Advantages of *Arabidopsis* as a model genetic organism

The thale cress *Arabidopsis thaliana* is a member of the Brassicaceae family and grows as a weed in many parts of the world. Except in its role as a model genetic organism, *Arabidopsis* has no economic importance, but it has a number of characteristics that make it well suited to the study of genetics. As an angiosperm, it has features in common with other flowering plants, some of which play critical roles in the ecosystem or are important sources of food, fiber, building materials, and pharmaceutical agents. *Arabidopsis*'s chief advantages are its small size (maximum height of 10–20 cm), prolific reproduction, and small genome.

Arabidopsis thaliana completes development—from seed germination to seed production—in about 6 weeks. Its small size and ability to grow under low illumination make it ideal for laboratory culture. Each plant is capable of producing from 10,000 to 40,000 seeds, and the seeds typically have a high rate of germination; so large numbers of progeny can be obtained from single genetic crosses.

The *Arabidopsis* genome

A key advantage for molecular studies is *Arabidopsis*'s small genome, which consists of only 125 million base pairs of DNA on five pairs of chromosomes, compared with 2.5 billion base pairs of DNA in the maize genome and 16 billion base pairs in the wheat genome. The genome of *A. thaliana* was completely sequenced in 2000, providing detailed information about gene structure and organization in this species. A number of variants of *A. thaliana*—called ecotypes—that vary in shape, size, physiological characteristics, and DNA sequence are available for study.

Life cycle of *Arabidopsis*

The *Arabidopsis* life cycle is fairly typical of most flowering plants (see Figure 2.22). The main, vegetative part of the plant is diploid; haploid gametes are produced in the pollen and ovaries. When a pollen grain lands on the stigma of a flower, a pollen tube grows into the pistil and ovary. Two haploid sperm nuclei contained in each pollen grain travel down the pollen tube and enter the embryo sac. There, one of the haploid sperm cells fertilizes the haploid egg cell to produce a diploid zygote. The other haploid sperm cell fuses with two haploid nuclei to form the $3n$ endosperm, which provides tissue that will nourish the growing embryonic plant. The zygotes develop within the seeds, which are produced in a long pod.

Under appropriate conditions, the embryo germinates and begins to grow into a plant. The shoot grows upward and the roots downward, a compact rosette of leaves is produced and, under the right conditions, the shoot enlarges and differentiates into flower structures. At maturity, *A. thaliana* is a low-growing plant with roots, a main shoot with branches that bear mature leaves, and small white flowers at the tips of the branches.

Genetic techniques with *Arabidopsis*

A number of traditional and modern molecular techniques are commonly used with *Arabidopsis* and provide it with special advantages for genetic studies. *Arabidopsis* can self-fertilize, which means that any recessive mutation appearing in the germ line can be recovered in the immediate progeny. Cross-fertilization also is possible by removing the anther from one plant and dusting pollen on the stigma of another plant—essentially the same technique used by Gregor Mendel with pea plants (see Figure 3.4).

As already mentioned, many naturally occurring variants of *Arabidopsis* are available for study, and new mutations can be produced by exposing its seeds to chemical mutagens, radiation, or transposable elements that randomly insert into genes. The large number of offspring produced by *Arabidopsis* facilitates screening for rare mutations.

Genes from other organisms can be transferred to *Arabidopsis* by the Ti plasmid from the bacterium *Agrobacterium tumefaciens,* which naturally infects plants and transfers the Ti plasmid to plant cells (see Chapter 19). After transfer to a plant cell, the Ti plasmid randomly inserts into the DNA of the plant that it infects, thereby generating mutations in the plant DNA in a process called *insertional mutagenesis.* Geneticists have modified the Ti plasmid to carry a *GUS* gene, which has no promoter of its own. The *GUS* gene encodes

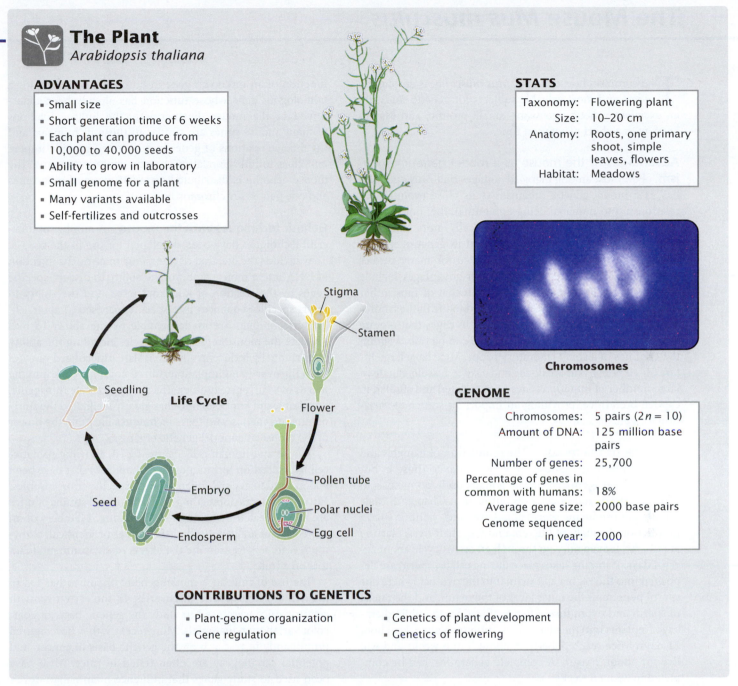

The Plant
Arabidopsis thaliana

ADVANTAGES

- Small size
- Short generation time of 6 weeks
- Each plant can produce from 10,000 to 40,000 seeds
- Ability to grow in laboratory
- Small genome for a plant
- Many variants available
- Self-fertilizes and outcrosses

STATS

Taxonomy:	Flowering plant
Size:	10–20 cm
Anatomy:	Roots, one primary shoot, simple leaves, flowers
Habitat:	Meadows

Chromosomes

GENOME

Chromosomes:	5 pairs ($2n = 10$)
Amount of DNA:	125 million base pairs
Number of genes:	25,700
Percentage of genes in common with humans:	18%
Average gene size:	2000 base pairs
Genome sequenced in year:	2000

Life Cycle

Seedling

Seed

Embryo

Endosperm

Egg cell

Polar nuclei

Pollen tube

Flower

Stigma

Stamen

CONTRIBUTIONS TO GENETICS

- Plant-genome organization
- Gene regulation
- Genetics of plant development
- Genetics of flowering

[Photograph courtesy of Anand P Tyagi and Luca Comai.]

an enzyme that converts a colorless compound (X-Glu) into a blue dye. Because the *GUS* gene has no promoter, it is expressed only when inserted into the coding sequence of a plant gene. When that happens, the enzyme encoded by *GUS* is synthesized and converts X-Glu into a blue dye that stains the cell. This dye provides a means to visually determine the expression pattern of a gene that has been interrupted by Ti DNA, producing information about the expression of genes that are mutated by insertional mutagenesis. ■

The Mouse *Mus musculus*

The common house mouse, *Mus musculus*, is among the oldest and most valuable subjects for genetic study. It's an excellent genetic organism—small, prolific, and easy to keep, with a short generation time.

Advantages of the mouse as a model genetic organism Foremost among many advantages that *Mus musculus* has as a model genetic organism is its close evolutionary relationship to humans. Being a mammal, the mouse is genetically, behaviorally, and physiologically more similar to humans than are other organisms used in genetics studies, making the mouse the model of choice for many studies of human and medical genetics. Other advantages include a short generation time compared with that of most other mammals. *Mus musculus* is well adapted to life in the laboratory and can be easily raised and bred in cages that require little space; thus several thousand mice can be raised within the confines of a small laboratory room. Mice have large litters (8–10 pups), and are docile and easy to handle. Finally, a large number of mutations have been isolated and studied in captive-bred mice, providing an important source of variation for genetic analysis.

Life cycle of the mouse The production of gametes and reproduction in the mouse are very similar to those in humans. Diploid germ cells in the gonads undergo meiosis to produce sperm and oocytes, as outlined in Chapter 2. Male mice begin producing sperm at puberty and continue sperm production throughout the remainder of their lives. Starting at puberty, female mice go through an estrus cycle about every 4 days. If mating takes place during estrus, sperm are deposited into the vagina and swim into the oviduct, where one sperm penetrates the outer layer of the ovum, and the nuclei of sperm and ovum fuse. After fertilization, the diploid embryo implants into the uterus. Gestation typically takes about 21 days. Mice reach puberty in about 5 to 6 weeks and will live for about 2 years. A complete generation can be completed in about 8 weeks.

The mouse genome The mouse genome contains about 2.6 billion base pairs of DNA, which is similar in size to the human genome. Mice and humans also have similar numbers of genes. For most human genes, there are homologous genes in the mouse. An important tool for determining the function of an unknown gene in humans is to search for a homologous gene whose function has already been determined in the mouse. Furthermore, the linkage relations of many mouse genes are similar to those in humans, and the linkage relations of genes in mice often provide important clues to linkage relations among genes in humans. The mouse genome is distributed across 19 pairs of autosomes and one pair of sex chromosomes.

Genetic techniques with the mouse A number of powerful techniques have been developed for use in the mouse. They include the creation of transgenic mice by the injection of DNA into a mouse embryo, the ability to disrupt specific genes by the creation of knockout mice, and the ability to insert specific sequences into specific loci (see Chapter 19). These techniques are made possible by the ability to manipulate the mouse reproductive cycle, including the ability to hormonally induce ovulation, isolate unfertilized oocytes from the ovary, and implant fertilized embryos back into the uterus of a surrogate mother. The ability to create transgenic, knockout, and knock-in mice has greatly facilitated the study of human genetics, and these techniques illustrate the power of the mouse as a model genetic organism.

Mouse and human cells can be fused, allowing somatic-cell hybridization techniques (see Chapter 7) that have been widely used to assign human genes to specific chromosomes. Mice also tolerate inbreeding well, and inbred strains of mice are easily created by brother–sister mating. Members of an inbred strain are genetically very similar or identical, allowing researchers to examine the effects of environmental factors on a trait.

The use of mice as a model genetic organism has led to many important genetic discoveries. In the early twentieth century, mice were used to study the genetic basis of coat-color variation in mammals. More recently, they have figured prominantly in research on the genetic basis of cancer, and potential carcinogens are often tested in mice. Mice have been used to study genes that influence mammalian development, including mutations that produce birth defects in humans. A large number of mouse models of specific human diseases have been created—in some cases, by isolating and inbreeding mice with naturally occurring mutations and, in other cases, by using knockout and knock-in techniques to disable and modify specific genes. ∎

The Mouse
Mus musculus

ADVANTAGES

- Closely related to humans
- Small size
- Rapid reproduction
- Easy to rear in the laboratory
- Tolerates inbreeding

STATS

Taxonomy:	Mammal
Size:	2–3 inches 20 grams
Anatomy:	Typical rodent body plan
Habitat:	Fields, houses, and other human structures

Life Cycle

Maturity — ♀ ♂ — Sperm — Oocyte — Fertilization — Embryo — Birth

5–6 weeks

21 days

Chromosomes

GENOME

Chromosomes:	19 pairs of autosomes and 1 pair of sex chromosomes ($2n = 40$)
Amount of DNA:	2.7 billion base pairs
Number of genes:	26,762
Percentage of genes in common with humans:	99%
Average gene size:	40,000 base pairs
Genome sequenced in year:	2002

CONTRIBUTIONS TO GENETICS

- Model for human diseases
- Cancer genetics
- Immunogenetics

[Photograph courtesy of Ellen C. Akeson and Muriel T. Davisson, The Jackson Laboratory, Bar Harton, Maine.]

The Yeast *Saccharomyces cerevisiae*

Common baker's yeast (*Saccharomyces cerevisiae*) has been widely adopted as a simple model system for the study of eukaryotic genetics. Long used for baking bread and making beer, yeast has more recently been utilized for the production of biofuels. Louis Pasteur identified *S. cerevisiae* as the microorganism responsible for fermentation in 1857, and its use in genetic analysis began about 1935. Having been the subject of extensive studies in classical genetics for many years, yeast genes are well known and characterized. At the same time, yeast's unicellular nature makes it amenable to molecular techniques developed for bacteria. Thus, yeast combines both classical genetics and molecular biology to provide a powerful model for the study of eukaryotic genetic systems.

Advantages of yeast as a model genetic organism

The great advantage of yeast is that it not only is a eukaryotic organism, with genetic and cellular systems similar to those of other, more-complex eukaryotes such as humans, but also is unicellular, with many of the advantages of manipulation found with bacterial systems. Like bacteria, yeast cells require little space, and large numbers of cells can be grown easily and inexpensively in the laboratory.

Yeast exists in both diploid and haploid forms. When haploid, the cells possess only a single allele at each locus, which means that the allele will be expressed in the phenotype; unlike the situation in diploids, there is no dominance through which some alleles mask the expression of others. Therefore, recessive alleles can be easily identified in haploid cells, and then the interactions between alleles can be examined in the diploid cells.

Another feature that makes yeast a powerful genetic model system is that, subsequent to meiosis, all of the products of a meiotic division are present in a single structure called an ascus (see the next subsection) and remain separate from the products of other meiotic divisions. The four cells produced by a single meiotic division are termed a tetrad. In most organisms, the products of different meiotic divisions mix, and so identification of the results of a single meiotic division is impossible. For example, if we were to isolate four sperm cells from the testes of a mouse, it is extremely unlikely that all four would have been produced by the same meiotic division. Having tetrads separate in yeast allows us to directly observe the effects of individual meiotic divisions on the types of gametes produced and to more easily identify crossover events. The genetic analysis of a tetrad is termed tetrad analysis.

Yeast has been subjected to extensive genetic analysis, and thousands of mutants have been identified. In addition, many powerful molecular techniques developed for manipulating genetic sequences in bacteria have been adapted for use in yeast. Yet, in spite of a unicellular structure and ease of manipulation, yeast cells possess many of the same genes found in humans and other complex multicellular eukaryotes, and many of these genes have identical or similar functions in these eukaryotes. Thus, the genetic study of yeast cells often contributes to our understanding of other, more-complex eukaryotic organisms, including humans.

Life cycle of yeast

As stated earlier, *Saccharomyces cerevisiae* can exist as either haploid or diploid cells. Haploid cells usually exist when yeast is starved for nutrients and reproduce mitotically, producing identical, haploid daughter cells through budding. Yeast cells can also undergo sexual reproduction. There are two mating types, a and α; haploid cells of different mating types fuse and then undergo nuclear fusion to create a diploid cell. The diploid cell is capable of budding mitotically to produce genetically identical diploid cells. Starvation induces the diploid cells to undergo meiosis, resulting in four haploid nuclei, which become separated into different cells, producing haploid spores. Because the four products of meiosis (a tetrad) are enclosed in a common structure, the ascus, all the products of a single meiosis can be isolated (tetrad analysis).

The yeast genome

Saccharomyces cerevisiae has 16 pairs of typical eukaryotic chromosomes. The rate of recombination is high, giving yeast a relatively long genetic map compared with those of other organisms. The genome of *S. cerevisiae* contains 12 million base pairs, plus the 2 million to 3 million base pairs of rRNA genes. In 1996, *S. cerevisiae* was the first eukaryotic organism whose genome was completely sequenced.

Genetic techniques with yeast

One advantage of yeast to researchers is the use of plasmids to transfer genes or DNA sequences of interest into cells. Yeast cells naturally possesses a circular plasmid, named 2μ, that is 6300 bp long and is transmitted to daughter cells in mitosis and meiosis. This plasmid has an origin of replication recognized by the yeast replication system, and so it replicates autonomously in the cell. The 2μ plasmid has been engineered to provide an efficient vector for transferring genes into yeast. In other cases, bacterial plasmids have been adapted for use in yeast. Some of them undergo homologous recombination with the yeast chromosome, transferring their sequences to the yeast chromosome. Shuttle vectors, which can be propagated in both bacteria and yeast, are particularly effective. Such vectors make it possible to

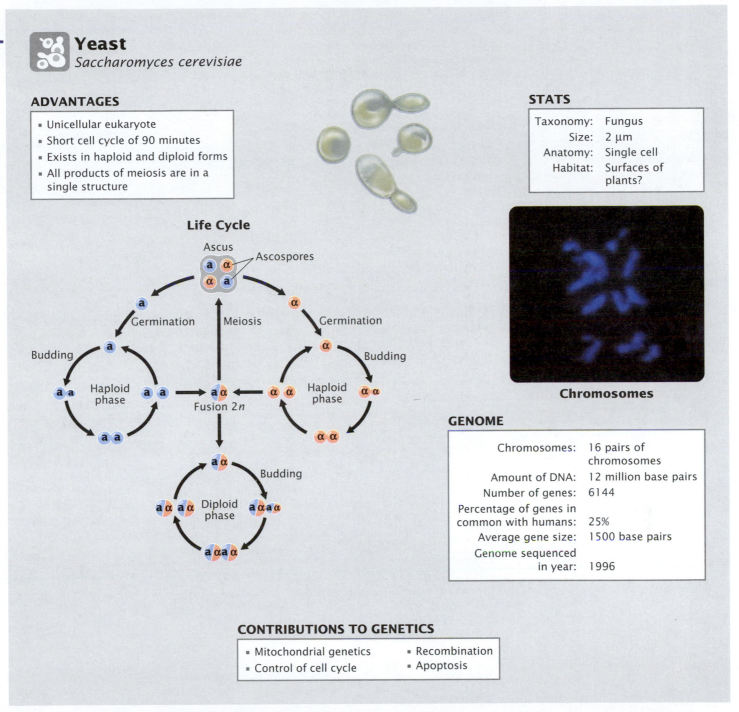

Yeast
Saccharomyces cerevisiae

ADVANTAGES

- Unicellular eukaryote
- Short cell cycle of 90 minutes
- Exists in haploid and diploid forms
- All products of meiosis are in a single structure

STATS

Taxonomy:	Fungus
Size:	2 μm
Anatomy:	Single cell
Habitat:	Surfaces of plants?

Life Cycle

Ascus
Ascospores
Germination · Meiosis · Germination
Budding
Haploid phase
Fusion 2*n*
Budding
Haploid phase
Diploid phase
Budding

Chromosomes

GENOME

Chromosomes:	16 pairs of chromosomes
Amount of DNA:	12 million base pairs
Number of genes:	6144
Percentage of genes in common with humans:	25%
Average gene size:	1500 base pairs
Genome sequenced in year:	1996

CONTRIBUTIONS TO GENETICS

- Mitochondrial genetics
- Control of cell cycle
- Recombination
- Apoptosis

[Photograph courtesy of Mara Stewart and Dean Dawson, Department of Microbiology and Molecular Biology, Sackler School of Biomedical Sciences, Tufts University.]

construct and manipulate gene sequences in bacteria, where often more-powerful techniques are available for genetic manipulation and selection, and then transfer the gene sequences into yeast cells, where their function can be tested.

Plasmids are limited in the size of DNA fragments that they can carry (see p. 542 in Chapter 19). Yeast artificial chromosomes (YACs)—engineered DNA fragments that contain centromeric and telomeric sequences and segregate like chromosomes in meiosis and mitosis—overcome this limitation; YACs can hold DNA fragments as large as several hundred thousand base pairs. ■

Glossary

abortive initiation Process during initiation of transcription in which RNA polymerase repeatedly generates and releases short transcripts, from 2 to 6 nucleotides in length, while still bound to the promoter. Occurs in both prokaryotes and eukaryotes.

acceptor arm The arm in tRNA to which an amino acid attaches.

acentric chromatid Lacks a centromere; produced when crossing over takes place within a paracentric inversion. The acentric chromatid does not attach to a spindle fiber and does not segregate in meiosis or mitosis; so it is usually lost after one or more rounds of cell division.

acidic activation domain Commonly found in some transcriptional activator proteins, a domain that contains multiple amino acids with negative charges and stimulates the transcription of certain genes.

acrocentric chromosome Chromosome in which the centromere is near one end, producing a long arm at one end, and a knob, or satellite, at the other end.

activator *See* transcriptional activator protein.

adaptive mutation Process by which a specific environment induces mutations that enable organisms to adapt to the environment.

addition rule States that the probability of any of two or more mutually exclusive events occurring is calculated by adding the probabilities of the individual events.

additive genetic variance Component of the genetic variance that can be attributed to the additive effect of different genotypes.

adenine (A) Purine base in DNA and RNA.

adenosine-3′,5′-cyclic monophosphate (cAMP) Modified nucleotide that functions in catabolite repression. Low levels of glucose stimulate high levels of cAMP; cAMP then attaches to CAP, which binds to the promoter of certain operons and stimulates transcription.

adjacent-1 segregation Type of segregation that takes place in a heterozygote for a translocation. If the original, nontranslocated chromosomes are N_1 and N_2 and the chromosomes containing the translocated segments are T_1 and T_2, then adjacent-1 segregation takes place when N_1 and T_2 move toward one pole and T_1 and N_2 move toward the opposite pole.

adjacent-2 segregation Type of segregation that takes place in a heterozygote for a translocation. If the original, nontranslocated chromosomes are N_1 and N_2 and the chromosomes containing the translocated segments are T_1 and T_2, then adjacent-2 segregation takes place when N_1 and T_1 move toward one pole and T_2 and N_2 move toward the opposite pole.

A-DNA Right-handed helical structure of DNA that exists when little water is present.

affinity capture Use of an antibody to capture one protein from a complex mixture of proteins. The captured protein will "pull down" with it any proteins with which it interacts, which can then be analyzed by mass spectrometry to identify these proteins.

allele One of two or more alternate forms of a gene.

allelic frequency Proportion of a particular allele in a population.

allopatric speciation Speciation that arises when a geographic barrier first splits a population into two groups and blocks the exchange of genes between them.

allopolyploidy Condition in which the sets of chromosomes of a polyploid individual possessing more than two haploid sets are derived from two or more species.

allosteric protein Protein that changes its conformation on binding with another molecule.

alternate segregation Type of segregation that takes place in a heterozygote for a translocation. If the original, nontranslocated chromosomes are N_1 and N_2 and the chromosomes containing the translocated segments are T_1 and T_2, then alternate segregation takes place when N_1 and N_2 move toward one pole and T_1 and T_2 move toward the opposite pole.

alternation of generations Complex life cycle in plants that alternates between the diploid sporophyte stage and the haploid gametophyte stage.

alternative processing pathway One of several pathways by which a single pre-mRNA can be processed in different ways to produce alternative types of mRNA.

alternative splicing Process by which a single pre-mRNA can be spliced in more than one way to produce different types of mRNA.

Ames test Test in which special strains of bacteria are used to evaluate the potential of chemicals to cause cancer.

amino acid Repeating unit of proteins; consists of an amino group, a carboxyl group, a hydrogen atom, and a variable R group.

aminoacyl (A) site One of three sites in a ribosome occupied by a tRNA in translation. All charged tRNAs (with the exception of the initiator tRNA) first enter the A site in translation.

aminoacyl-tRNA synthetase Enzyme that attaches an amino acid to a tRNA. Each aminoacyl-tRNA synthetase is specific for a particular amino acid.

amniocentesis Procedure used for prenatal genetic testing to obtain a sample of amniotic fluid from a pregnant woman. A long sterile needle is inserted through the abdominal wall into the amniotic sac to obtain the fluid.

amphidiploidy Type of allopolyploidy in which two different diploid genomes are combined such that every chromosome has one and only one homologous partner and the genome is functionally diploid.

anagenesis Evolutionary change within a single lineage.

anaphase Stage of mitosis in which chromatids separate and move toward the spindle poles.

anaphase I Stage of meiosis I. In anaphase I, homologous chromosomes separate and move toward the spindle poles.

anaphase II Stage of meiosis II. In anaphase II, chromatids separate and move toward the spindle poles.

aneuploidy Change from the wild type in the number of chromosomes; most often an increase or decrease of one or two chromosomes.

annotation Linking the sequence information of a gene that has been identified to other information about the gene's function and expression, about the protein encoded by the gene, and about similar genes in other species.

Antennapedia **complex** Cluster of five homeotic genes in fruit flies that affects the development of the adult fly's head and anterior thoracic segments.

antibody Produced by a B cell, a protein that circulates in the blood and other body fluids. An antibody binds to a specific antigen and marks the antigen for destruction by a phagocytic cell.

anticipation Increasing severity or earlier age of onset of a genetic trait in succeeding generations. For example, symptoms of a genetic disease may become more severe as the trait is passed from generation to generation.

anticodon Sequence of three nucleotides in transfer RNA that pairs with the corresponding codon in messenger RNA in translation.

antigen Substance that is recognized by the immune system and elicits an immune response.

antigenic drift Used in reference to a rapidly evolving virus, with new strains appearing frequently because of mutations.

antigenic shift Used in reference to a virus that has undergone major changes in its genome through the reassortment of genetic material from two different strains of the virus.

antiparallel Refers to a characteristic of the DNA double helix in which the two polynucleotide strands run in opposite directions.

antisense RNA Small RNA molecule that base pairs with a complementary DNA or RNA sequence and affects its functioning.

antiterminator Protein or DNA sequence that inhibits the termination of transcription.

apoptosis Programmed cell death, in which a cell degrades its own DNA, the nucleus and cytoplasm shrink, and the cell undergoes phagocytosis by other cells without leakage of its contents.

archaea One of the three primary divisions of life. Archaea consist of unicellular organisms with prokaryotic cells.

artificial selection Selection practiced by humans.

attenuation Type of gene regulation in some bacterial operons, in which transcription is initiated but terminates prematurely before the transcription of the structural genes.

attenuator Secondary structure that forms in the 5′ untranslated region of some operons and causes the premature termination of transcription.

autoimmune disease Characterized by an abnormal immune response to a person's own (self) antigen.

autonomous element Transposable element that is fully functional and able to transpose on its own.

autonomously replicating sequence (ARS) DNA sequence that confers the ability to replicate; contains an origin of replication.

autopolyploidy Condition in which all the sets of chromsomes of a polyploid individual possessing more than two haploid sets are derived from a single species.

autosome Chromosome that is the same in males and females; nonsex chromosome.

auxotrophic bacterium A bacterium or fungus that possesses a nutritional mutation that disrupts its ability to synthesize an essential biological molecule; cannot grow on minimal medium but can grow on minimal medium to which has been added the biological molecule that it cannot synthesize.

backcross Cross between an F_1 individual and one of the parental (P) genotypes.

bacterial artificial chromosome (BAC) Cloning vector used in bacteria that is capable of carrying DNA fragments as large as 500 kb.

bacterial colony Clump of genetically identical bacteria derived from a single bacterial cell that undergoes repeated rounds of division.

bacteriophage Virus that infects bacterial cells.

balance hypothesis Proposes that much of the molecular variation seen in natural populations is maintained by balancing selection that favors genetic variation.

Barr body Condensed, darkly staining structure that is found in most cells of female placental mammals and is an inactivated X chromosome.

basal transcription apparatus Complex of transcription factors, RNA polymerase, and other proteins that assemble on the promoter and are capable of initiating minimal levels of transcription.

base *See* **nitrogenous base.**

base analog Chemical substance that has a structure similar to that of one of the four standard bases of DNA and may be incorporated into newly synthesized DNA molecules in replication.

base-excision repair DNA repair that first excises modified bases and then replaces the entire nucleotide.

base substitution Mutation in which a single pair of bases in DNA is altered.

B cell Particular type of lymphocyte that produces humoral immunity; matures in the bone marrow and produces antibodies.

B-DNA Right-handed helical structure of DNA that exists when water is abundant; the secondary structure described by Watson and Crick and probably the most common DNA structure in cells.

behavioral isolation Reproductive isolation due to differences in behavior that prevent interbreeding.

bidirectional replication Replication at both ends of a replication bubble.

bioinformatics Synthesis of molecular biology and computer science that develops databases and computational tools to store, retrieve, and analyze nucleic-acid- and protein-sequence data.

biological species concept Defines a species as a group of organisms whose members are capable of interbreeding with one another but are reproductively isolated from the members of other species. Because different species do not exchange genes, each species evolves independently. Not all biologists adhere to this concept.

biotechnology Use of biological processes, particularly molecular genetics and recombinant DNA technology, to produce products of commercial value.

bithorax **complex** Cluster of three homeotic genes in fruit flies that influences the adult fly's posterior thoracic and abdominal segments.

bivalent Refers to a synapsed pair of homologous chromosomes.

blending inheritance Early concept of heredity proposing that offspring possess a mixture of the traits from both parents.

branch Evolutionary connections between organisms in a phylogenetic tree.

branch migration Movement of a cross bridge along two DNA molecules.

branch point Adenine nucleotide in nuclear pre-mRNA introns that lies from 18 to 40 nucleotides upstream of the 3′ splice site.

broad-sense heritability Proportion of the phenotypic variance that can be attributed to genetic variance.

cap-binding complex A group of proteins in eukaryotes that binds to the 5′ cap and initiates translation. Aids in exporting mRNA from the nucleus to the cytoplasm and promotes the initial (pioneer) round of translation.

caspase Enzyme that cleaves other proteins and regulates apoptosis. Each caspase is synthesized as a large, inactive precursor (a procaspase) that is activated by cleavage, often by another caspase.

catabolite activator protein (CAP) Protein that functions in catabolite repression. When bound with cAMP, CAP binds to the promoter of certain operons and stimulates transcription.

catabolite repression System of gene control in some bacterial operons in which glucose is used preferentially and the metabolism of other sugars is repressed in the presence of glucose.

cDNA (complementary DNA) library Collection of bacterial colonies or phage colonies containing DNA fragments that have been produced by reverse transcription of cellular mRNA.

cell cycle Stages through which a cell passes from one cell division to the next.

cell line Genetically identical cells that divide indefinitely and can be cultured in the laboratory.

cell theory States that all life is composed of cells, that cells arise only from other cells, and that the cell is the fundamental unit of structure and function in living organisms.

cellular immunity Type of immunity resulting from T cells, which recognize antigens found on the surfaces of self cells.

centiMorgan Another name for map unit.

central dogma Concept that genetic information passes from DNA to RNA to protein in a one-way information pathway.

centriole Cytoplasmic organelle consisting of microtubules; present at each pole of the spindle apparatus in animal cells.

centromere Constricted region on a chromosome that stains less strongly than the rest of the chromosome; region where spindle microtubules attach to a chromosome.

centromeric sequence DNA sequence found in functional centromeres.

centrosome Structure from which the spindle apparatus develops: contains the centriole.

Chargaff's rules Rules developed by Erwin Chargaff and his colleagues concerning the ratios of bases in DNA.

checkpoint A key transition point at which progression to the next stage in the cell cycle is regulated.

chiasma (pl., chiasmata) Point of attachment between homologous chromosomes at which crossing over took place.

chi-square goodness of fit test Statistical test used to evaluate how well a set of observed values fit the expected values. The probability associated with a calculated chi-square value is the probability that the differences between the observed and the expected values may be due to chance.

chloroplast DNA (cpDNA) DNA in chloroplasts; has many characteristics in common with eubacterial DNA and typically consists of a circular molecule that lacks histone proteins and encodes some of the rRNAs, tRNAs, and proteins found in chloroplasts.

chorionic villus sampling (CVS) Procedure used for prenatal genetic testing in which a small piece of the chorion (the outer layer of the placenta) is removed from a pregnant woman. A catheter is inserted through the vagina and cervix into the uterus. Suction is then applied to remove the sample.

chromatin Material found in the eukaryotic nucleus; consists of DNA and proteins.

chromatin-remodeling complex Complex of proteins that alters chromatin structure without acetylating histone proteins.

chromatin-remodeling protein Binds to a DNA sequence and disrupts chromatin structure, causing the DNA to become more accessible to RNA polymerase and other proteins.

chromosomal puff Localized swelling of a polytene chromosome; a region of chromatin in which DNA has unwound and is undergoing transcription.

chromosome Structure consisting of DNA and associated proteins that carries and transmits genetic information.

chromosome deletion Loss of a chromosome segment.

chromosome duplication Mutation that doubles a segment of a chromosome.

chromosome inversion Rearrangement in which a segment of a chromosome has been inverted 180 degrees.

chromosome jumping Method of moving from a gene on a cloned fragment to sequences on distantly linked fragments.

chromosome mutation Difference from the wild type in the number or structure of one or more chromosomes; often affects many genes and has large phenotypic effects.

chromosome rearrangement Change from the wild type in the structure of one or more chromosomes.

chromosome theory of heredity States that genes are located on chromosomes.

chromosome walking Method of locating a gene by using partly overlapping genomic clones to move in steps from a previously cloned, linked gene to the gene of interest.

cis configuration Arrangement in which two or more wild-type genes are on one chromosome and their mutant alleles are on the homologous chromosome; also called coupling configuration.

cladogenesis Evolution in which one lineage is split into two.

clonal evolution Process by which mutations that enhance the ability of cells to proliferate predominate in a clone of cells, allowing the clone to become increasingly rapid in growth and increasingly aggressive in proliferation properties.

cloning vector Stable, replicating DNA molecule to which a foreign DNA fragment can be attached and transferred to a host cell.

cloverleaf structure Secondary structure common to all tRNAs.

coactivator Protein that cooperates with an activator of transcription. In eukaryotic transcriptional control, coactivators often physically interact with transcriptional activators and the basal transcription apparatus.

codominance Type of allelic interaction in which the heterozygote simultaneously expresses traits of both homozygotes.

codon Sequence of three nucleotides that encodes one amino acid in a protein.

coefficient of coincidence Ratio of observed double crossovers to expected double crossovers.

cohesin Molecule that holds the two sister chromatids of a chromosome together. The breakdown of cohesin at the centromeres enables the chromatids to separate in anaphase of mitosis and anaphase II of meiosis.

cohesive end Short, single-stranded overhanging end on a DNA molecule produced when the DNA is cut by certain restriction enzymes. Cohesive ends are complementary and can spontaneously pair to rejoin DNA fragments that have been cut with the same restriction enzyme.

colinearity Concept that there is a direct correspondence between the nucleotide sequence of a gene and the continuous sequence of amino acids in a protein.

colony *See* **bacterial colony.**

comparative genomics Comparative studies of the genomes of different organisms.

competent cell Capable of taking up DNA from its environment (capable of being transformed).

complementary DNA strands The relation between the two nucleotide strands of DNA in which each purine on one strand pairs with a specific pyrimidine on the opposite strand (A pairs with T, and G pairs with C).

complementation Two different mutations in the heterozygous condition are exhibited as the wild-type phenotype; indicates that the mutations are at different loci.

complementation test Test designed to determine whether two different mutations are at the same locus (are allelic) or at different loci (are nonallelic). Two individuals that are homozygous for two independently derived mutations are crossed, producing F_1 progeny that are heterozygous for the mutations. If the mutations are at the same locus, the F_1 will have a mutant phenotype. If the mutations are at different loci, the F_1 will have a wild-type phenotype.

complete dominance Refers to an allele or a phenotype that is expressed in homozygotes (*AA*) and in heterozygotes (*Aa*); only the dominant allele is expressed in a heterozygote phenotype.

complete linkage Linkage between genes that are located close together on the same chromosome with no crossing over between them.

complete medium Used to culture bacteria or some other microorganism; contains all the nutrients required for growth and synthesis, including those normally synthesized by the organism. Nutritional mutants can grow on complete medium.

composite transposon Type of transposable element in bacteria that consists of two insertion sequences flanking a segment of DNA.

compound heterozygote An individual with two different recessive alleles at a locus that results in a recessive phenotype.

concept of dominance Principle of heredity discovered by Mendel stating that, when two different alleles are present in a genotype, only one allele may be expressed in the phenotype. The dominant allele is the allele that is expressed, and the recessive allele is the allele that is not expressed.

concordance Percentage of twin pairs in which both twins have a particular trait.

concordant Refers to a pair of twins both of whom have the trait under consideration.

conditional mutation Expressed only under certain conditions.

conditional probability Probability that is modified by additional information that another event has occurred.

conjugation Mechanism by which genetic material can be exchanged between bacterial cells. In conjugation, two bacteria lie close together and a cytoplasmic connection forms between them. A plasmid or sometimes a part of the bacterial chromosome passes through this connection from one cell to the other.

consanguinity Mating between related individuals.

consensus sequence Comprises the most commonly encountered nucleotides found at a specific location in DNA or RNA.

−10 consensus sequence (Pribnow box) Consensus sequence (TATAAT) found in most bacterial promoters approximately 10 bp upstream of the transcription start site.

−35 consensus sequence Consensus sequence (TTGACA) found in many bacterial promoters approximately 35 bp upstream of the transcription start site.

constitutive gene A gene that is not regulated and is expressed continually.

constitutive mutation Causes the continuous transcription of one or more structural genes.

contig Set of overlapping DNA fragments that have been assembled in the correct order to form a continuous stretch of DNA sequence.

continuous characteristic Displays a large number of possible phenotypes that are not easily distinguished, such as human height.

continuous replication Replication of the leading strand in the same direction as that of unwinding, allowing new nucleotides to be added continuously to the 3′ end of the new strand as the template is exposed.

coordinate induction Simultaneous synthesis of several enzymes that is stimulated by a single environmental factor.

copy-number variation (CNV) Difference among individual organisms in the number of copies of any large DNA sequence (larger than 1000 bp).

core enzyme Part of bacterial RNA polymerase that, during transcription, catalyzes the elongation of the RNA molecule by the addition of RNA nucleotides; consists of four subunits: two copies of alpha (α), a single copy of beta (β), and a single copy of beta prime (β').

corepressor Substance that inhibits transcription in a repressible system of gene regulation; usually a small molecule that binds to a repressor protein and alters it so that the repressor is able to bind to DNA and inhibit transcription.

core promoter Located immediately upstream of the eukaryotic promoter, DNA sequences to which the basal transcription apparatus binds.

correlation Degree of association between two or more variables.

correlation coefficient Statistic that measures the degree of association between two or more variables. A correlation coefficient can range from −1 to +1. A positive value indicates a direct relation between the variables; a negative correlation indicates an inverse relation. The absolute value of the correlation coefficient provides information about the strength of association between the variables.

cosmid Cloning vector that combines the properties of plasmids and phage vectors and is used to clone large pieces of DNA in bacteria. A cosmid is a small plasmid that carries a λ *cos* site, allowing the plasmid to be packaged into a viral coat.

cotransduction Process in which two or more genes are transferred together from one bacterial cell to another. Only genes located close together on a bacterial chromosome will be cotransduced.

cotransformation Process in which two or more genes are transferred together during cell transformation.

coupling configuration *See* **cis configuration.**

CpG island DNA region that contains many copies of a cytosine base followed by a guanine base; often found near transcription start sites in eukaryotic DNA. The cytosine bases in CpG islands are commonly methylated when genes are inactive but are demethylated before the initiation of transcription.

CRISPR RNAs (crRNAs) Small RNA molecules found in prokaryotes that assist in the destruction of foreign DNA.

cross bridge In a heteroduplex DNA molecule, the point at which each nucleotide strand passes from one DNA molecule to the other.

crossing over Exchange of genetic material between homologous but nonsister chromatids.

cruciform Structure formed by the pairing of inverted repeats on both strands of double-stranded DNA.

C value Haploid amount of DNA found in a cell of an organism.

C-value paradox The absence of a relationship between genome size and complexity in eukaryotic organisms and the question of what is the function of extra DNA in species with large genome sizes.

cyclin A key protein in the control of the cell cycle; combines with a cyclin-dependent kinase (CDK). The levels of cyclin rise and fall in the course of the cell cycle.

cyclin-dependent kinase (CDK) A key protein in the control of the cell cycle; combines with cyclin.

cytokinesis Process by which the cytoplasm of a cell divides.

cytoplasmic inheritance Inheritance of characteristics encoded by genes located in the cytoplasm. Because the cytoplasm is usually contributed entirely by only one parent, most cytoplasmically inherited characteristics are inherited from a single parent.

cytosine (C) Pyrimidine base in DNA and RNA.

deamination Loss of an amino group (NH_2) from a base.

degenerate genetic code Refers to the fact that the genetic code contains more information than is needed to specify all 20 common amino acids.

deletion Mutation in which one or more nucleotides are deleted from a DNA sequence.

deletion mapping Technique for determining the chromsomal location of a gene by studying the association of its phenotype or product with particular chromosome deletions.

denaturation (melting) Process that separates the strands of double-stranded DNA when DNA is heated.

deoxyribonucleotide Basic building block of DNA, consisting of deoxyribose, a phosphate, and a nitrogenous base.

deoxyribose Five-carbon sugar in DNA; lacks a hydroxyl group on the 2′-carbon atom.

depurination Break in the covalent bond connecting a purine base to the 1′-carbon atom of deoxyribose, resulting in the loss of the purine base.

determination Process by which a cell becomes committed to developing into a particular cell type.

diakinesis Fifth substage of prophase I in meiosis. In diakinesis, chromosomes contract, the nuclear membrane breaks down, and the spindle forms.

dicentric bridge Structure produced when the two centromeres of a dicentric chromatid are pulled toward opposite poles, stretching the dicentric chromosome across the center of the nucleus. Eventually, the dicentric bridge breaks as the two centromeres are pulled apart.

dicentric chromatid Chromatid that has two centromeres; produced when crossing over takes place within a paracentric inversion. The two centromeres of the dicentric chromatid are frequently pulled toward opposite poles in mitosis or meiosis, breaking the chromosome.

dideoxyribonucleoside triphosphate (ddNTP) Special substrate for DNA synthesis used in the Sanger dideoxy sequencing method; identical with dNTP (the usual substrate for DNA synthesis) except that it lacks a 3′-OH group. The incorporation of a ddNTP into DNA terminates DNA synthesis.

dihybrid cross A cross between two individuals that differ in two characteristics—more specifically, a cross between individuals that are homozygous for different alleles at the two loci (*AA BB* × *aa bb*); also refers to a cross between two individuals that are both heterozygous at two loci (*Aa Bb* × *Aa Bb*).

dioecious organism Belongs to a species whose members have either male or female reproductive structures.

diploid Possessing two sets of chromosomes (two genomes).

diplotene Fourth substage of prophase I in meiosis. In diplotene, centromeres of homologous chromosomes move apart, but the homologs remain attached at chiasmata.

directional selection Selection in which one trait or allele is favored over another.

direct repair DNA repair in which modified bases are changed back into their original structures.

direct-to-consumer genetic test Test for a genetic condition; the test can be purchased directly by a consumer, without the involvement of a physician or other health care provider.

discontinuous characteristic Exhibits only a few, easily distinguished phenotypes. An example is seed shape in which seeds are either round or wrinkled.

discontinuous replication Replication of the lagging strand in the direction opposite that of unwinding, which means that DNA must be synthesized in short stretches (Okazaki fragments).

discordant Refers to a pair of twins of whom one twin has the trait under consideration and the other does not.

displaced duplication Chromosome rearrangement in which the duplicated segment is some distance from the original segment, either on the same chromosome or on a different one.

dizygotic twins Nonidentical twins that arise when two different eggs are fertilized by two different sperm; also called fraternal twins.

D loop Region of mitochondrial DNA that contains an origin of replication and promoters; it is displaced during the initiation of replication, leading to the name displacement, or D, loop.

DNA fingerprinting Technique used to identify individuals by examining their DNA sequences.

DNA gyrase *E. coli* topoisomerase enzyme that relieves the torsional strain that builds up ahead of the replication fork.

DNA helicase Enzyme that unwinds double-stranded DNA by breaking hydrogen bonds.

DNA library Collection of bacterial colonies containing all the DNA fragments from one source.

DNA ligase Enzyme that catalyzes the formation of a phosphodiester bond between adjacent 3′-OH and 5′-phosphate groups in a DNA molecule.

DNA methylation Modification of DNA by the addition of methyl groups to certain positions on the bases.

DNA polymerase Enzyme that synthesizes DNA.

DNA polymerase I Bacterial DNA polymerase that removes and replaces RNA primers with DNA nucleotides.

DNA polymerase II Bacterial DNA polymerase that takes part in DNA repair; restarts replication after synthesis has halted because of DNA damage.

DNA polymerase III Bacterial DNA polymerase that synthesizes new nucleotide strands by using the 3′-OH group provided by the primer.

DNA polymerase IV Bacterial DNA polymerase; probably takes part in DNA repair.

DNA polymerase V Bacterial DNA polymerase; probably takes part in DNA repair.

DNA polymerase α Eukaryotic DNA polymerase that initiates replication.

DNA polymerase β Eukaryotic DNA polymerase that participates in DNA repair.

DNA polymerase δ Eukaryotic DNA polymerase that replicates the lagging strand during DNA synthesis; also carries out DNA repair and translesion DNA synthesis.

DNA polymerase ε Eukaryotic DNA polymerase that replicates the leading strand during DNA synthesis.

DNA polymerase γ Eukaryotic DNA polymerase that replicates mitochondrial DNA. A γ-like DNA polymerase replicates chloroplast DNA.

DNase I hypersensitive site Chromatin region that becomes sensitive to digestion by the enzyme DNase I.

DNA sequencing Process of determining the sequence of bases along a DNA molecule.

DNA transposon *See* **transposable element.**

domain Functional part of a protein.

dominance genetic variance Component of the genetic variance that can be attributed to dominance (interaction between genes at the same locus).

dominant Refers to an allele or a phenotype that is expressed in homozygotes (*AA*) and in heterozygotes (*Aa*); only the dominant allele is expressed in a heterozygote phenotype.

dosage compensation Equalization in males and females of the amount of protein produced by X-linked genes. In placental mammals, dosage compensation is accomplished by the random inactivation of one X chromosome in the cells of females.

double fertilization Fertilization in plants; includes the fusion of a sperm cell with an egg cell to form a zygote and the fusion of a second sperm cell with the polar nuclei to form an endosperm.

double-strand-break model Model of homologous recombination in which a DNA molecule undergoes double-strand breaks.

down mutation Decreases the rate of transcription.

downstream core promoter element Consensus sequence [RG(A or T)CGTG] found in some eukaryotic RNA polymerase II core promoters; usually located approximately 30 bp downstream of the transcription start site.

Down syndrome (trisomy 21) Characterized by variable degrees of intellectual disability, characteristic facial features, some retardation of growth and development, and an increased incidence of heart defects, leukemia, and other abnormalities; caused by the duplication of all or part of chromosome 21.

ecological isolation Reproductive isolation in which different species live in different habitats and interact with the environment in different ways. Thus, their members do not encounter one another and do not reproduce with one another.

Edward syndrome (trisomy 18) Characterized by severe intellectual disability, low-set ears, a short neck, deformed feet, clenched fingers, heart problems, and other disabilities; results from the presence of three copies of chromosome 18.

effective population size Effective number of breeding adults in a population; influenced by the number of individuals contributing genes to the next generation, their sex ratio, variation between individuals in reproductive success, fluctuations in population size, the age structure of the population, and whether mating is random.

egg Female gamete.

egg-polarity gene Determines the major axes of development in an early fruit-fly embryo. One set of egg-polarity genes determines the anterior–posterior axis and another determines the dorsal–ventral axis.

elongation factor G (EF-G) Protein that combines with GTP and is required for movement of the ribosome along the mRNA during translation.

elongation factor Ts (EF-Ts) Protein that regenerates elongation factor Tu in the elongation stage of protein synthesis.

elongation factor Tu (EF-Tu) Protein taking part in the elongation stage of protein synthesis; forms a complex with GTP and a charged amino acid and then delivers the charged tRNA to the ribosome.

endosymbiotic theory States that some membrane-bounded organelles, such as mitochondria and chloroplasts, in eukaryotic cells originated as free-living eubacterial cells that entered into an endosymbiotic relation with a eukaryotic host cell and evolved into the present-day organelles; supported by a number of similarities in structure and sequence between organelle and eubacterial DNAs.

enhancer Sequence that stimulates maximal transcription of distant genes; affects only genes on the same DNA molecule (is cis acting), contains short consensus sequences, is not fixed in relation to the transcription start site, can stimulate almost any promoter in its vicinity, and may be upstream or downstream of the gene. The function of an enhancer is independent of sequence orientation.

environmental variance Component of the phenotypic variance that is due to environmental differences among individual members of a population.

epialleles Alleles that do not differ in their base sequence but have epigenetic differences that produce heritable variations in phenotypes.

epigenetic change Stable alteration of chromatin structure that may be passed on to other cells or to an individual organism. *See also* **epigenetics.**

epigenetic marks Heritable changes to DNA or chromatin structure that do not involve alteration of the base sequence and that bring about changes in gene expression.

epigenetics Phenonmena due to alterations to DNA that do not include changes in the base sequence; often affect the way in which the DNA sequences are expressed. Such alterations are often stable and heritable in the sense that they are passed from one cell to another.

epigenome All epigenetic modifications within the genome of an individual organism.

engineered nuclease Protein consisting of part of a restriction enzyme that cleaves DNA combined with another protein that recognizes and binds to a specific DNA sequence. Capable of making double-stranded cuts to the DNA at a predetermined DNA sequence. Engineered nucleases can be custom designed to bind to and cut any particular DNA sequence.

episome Plasmid capable of integrating into a bacterial chromosome.

epistasis Type of gene interaction in which a gene at one locus masks or suppresses the effects of a gene at a different locus.

epistatic gene Masks or suppresses the effect of a gene at a different locus.

equilibrium Situation in which no further change takes place; in population genetics, refers to a population in which allelic frequencies do not change.

equilibrium density gradient centrifugation Method used to separate molecules or organelles of different density by centrifugation.

eubacteria One of the three primary divisions of life. Eubacteria consist of unicellular organisms with prokaryotic cells and include most of the common bacteria.

euchromatin Chromatin that undergoes condensation and decondensation in the course of the cell cycle.

eukaryote Organism with a complex cell structure including a nuclear envelope and membrane-bounded organelles. One of the three primary divisions of life, eukaryotes include unicellular and multicellular forms.

evolution Genetic change taking place in a group of organisms.

exit (E) site One of three sites in a ribosome occupied by a tRNA. In the elongation stage of translation, the tRNA moves from the peptidyl (P) site to the E site from which it then exits the ribosome.

exon Coding region of a split gene (a gene that is interrupted by introns). After processing, the exons remain in messenger RNA.

exon shuffling Process, important in the evolution of eukaryotic genes, by which exons of different genes are exchanged and mixed into new combinations, creating new genes that are mosaics of other preexisting genes.

expanding nucleotide repeat Mutation in which the number of copies of a set of nucleotides (most often three nucleotides) increases in succeeding generations.

expressed-sequence tag (EST) Unique fragment of DNA from the coding region of a gene, produced by the reverse transcription of cellular RNA. Parts of the fragments are sequenced so that they can be identified.

expression vector Cloning vector containing DNA sequences such as a promoter, a ribosome-binding site, and transcription initiation and termination sites that allow DNA fragments inserted into the vector to be transcribed and translated.

expressivity Degree to which a trait is expressed.

familial Down syndrome Caused by a Robertsonian translocation in which the long arm of chromosome 21 is translocated to another chromosome; tends to run in families.

fertilization Fusion of gametes (sex cells) to form a zygote.

fetal cell sorting Separation of fetal cells from maternal blood. Genetic testing on the fetal cells can provide information about genetic diseases and disorders in the fetus.

F (fertility) factor Episome of *E. coli* that controls conjugation and gene exchange between *E. coli* cells. The F factor contains an origin of replication and genes that enable the bacterium to undergo conjugation.

F_1 (first filial) generation Offspring of the initial parents (P) in a genetic cross.

F_2 (second filial) generation Offspring of the F_1 generation in a genetic cross; the third generation of a genetic cross.

first polar body One of the products of meiosis I in oogenesis; contains half the chromosomes but little of the cytoplasm.

fitness Reproductive success of a genotype compared with that of other genotypes in a population.

5-methylcytosine Modified form of the base cytosine, containing a methyl group (CH_3) on the 5′ carbon.

5′ cap Modified 5′ end of eukaryotic mRNA, consisting of an extra nucleotide (methylated) and methylation of the 2′ position of the ribose sugar in one or more subsequent nucleotides; plays a role in the binding of the ribosome to mRNA and affects mRNA stability and the removal of introns.

5′ end End of the polynucleotide chain where a phosphate is attached to the 5′-carbon atom of the nucleotide.

5′ splice site The 5′ end of an intron where cleavage takes place in RNA splicing.

5′ untranslated (5′ UTR) region Sequence of nucleotides at the 5′ end of mRNA; does not encode the amino acids of a protein.

fixation Point at which one allele reaches a frequency of 1. At this point, all members of the population are homozygous for the same allele.

flanking direct repeat Short, directly repeated sequence produced on either side of a transposable element when the element inserts into DNA.

forward genetics Traditional approach to the study of gene function that begins with a phenotype (a mutant organism) and proceeds to a gene that encodes the phenotype.

forward mutation Alters a wild-type phenotype.

founder effect Sampling error that arises when a population is established by a small number of individuals; leads to genetic drift.

fragile site Constriction or gap that appears at a particular location on a chromosome when cells are cultured under special conditions. One fragile site on the human X chromosome is associated with intellectual disability (fragile-X syndrome) and results from an expanding trinucleotide repeat.

fragile-X syndrome A form of X-linked intellectual disability that appears primarily in males; results from an increase in the number of repeats of a CGG trinucleotide.

frameshift mutation Alters the reading frame of a gene.

fraternal twins Nonidentical twins that arise when two different eggs are fertilized by two different sperm; also called dizygotic twins.

frequency distribution Graphical way of representing values. In genetics, usually the phenotypes found in a group of individuals are displayed as a frequency distribution. Typically, the phenotypes are plotted on the horizontal (x) axis and the numbers (or proportions) of individuals with each phenotype are plotted on the vertical (y) axis.

functional genomics Area of genomics that studies the functions of genetic information contained within genomes.

G_0 (gap 0) Nondividing stage of the cell cycle.

G_1 (gap 1) Stage in interphase of the cell cycle in which the cell grows and develops.

G_2 (gap 2) Stage of interphase in the cell cycle that follows DNA replication. In G_2, the cell prepares for division.

gain-of-function mutation Produces a new trait or causes a trait to appear in inappropriate tissues or at inappropriate times in development.

gametic isolation Reproductive isolation due to the incompatibility of gametes. Mating between members of different species takes place, but the gametes do not form zygotes. Seen in many plants, where pollen from one species cannot fertilize the ovules of another species.

gametophyte Haploid phase of the life cycle in plants.

gap genes In fruit flies, set of segmentation genes that define large sections of the embryo. Mutations in these genes usually eliminate whole groups of adjacent segments.

gel electrophoresis Technique for separating charged molecules (such as proteins or nucleic acids) on the basis of molecular size or charge or both.

gene Genetic factor that helps determine a trait; often defined at the molecular level as a DNA sequence that is transcribed into an RNA molecule.

gene cloning Insertion of DNA fragments into bacteria in such a way that the fragments will be stable and copied by the bacteria.

gene conversion Process of nonreciprocal genetic exchange that can produce abnormal ratios of gametes following meiosis.

gene desert In reference to the density of genes in the genome, a region that is gene poor—that is, a long stretch of DNA possibly consisting of hundreds of thousands to millions of base pairs completely devoid of any known genes or other functional sequences.

gene family *See* **multigene family.**

gene flow Movement of genes from one population to another; also called migration.

gene interaction Interaction between genes at different loci that affect the same characteristic.

gene mutation Affects a single gene or locus.

gene pool Total of all genes in a population.

generalized transduction Transduction in which any gene can be transferred from one bacterial cell to another by a virus.

general transcription factor Protein that binds to eukaryotic promoters near the start site and is a part of the basal transcription apparatus that initiates transcription.

gene regulation Mechanisms and processes that control the phenotypic expression of genes.

gene therapy Use of recombinant DNA to treat a disease or disorder by altering the genetic makeup of the patient's cells.

genetic bottleneck Sampling error that arises when a population undergoes a drastic reduction in population size; leads to genetic drift.

genetic-conflict hypothesis Suggests that genomic imprinting evolved because different and conflicting pressures act on maternal and paternal alleles for genes that affect fetal growth. For example, paternally derived alleles often favor maximum fetal growth, whereas maternally derived alleles favor less than maximum fetal growth because of the high cost of fetal growth to the mother.

genetic correlation Phenotypic correlation due to the same genes affecting two or more characteristics.

genetic counseling Educational process that attempts to help patients and family members deal with all aspects of a genetic condition.

genetic drift Change in allelic frequency due to sampling error.

genetic engineering Common term for recombinant DNA technology.

genetic–environmental interaction variance Component of the phenotypic variance that results from an interaction between genotype and environment. Genotypes are expressed differently in different environments.

Genetic Information Nondiscrimination Act (GINA) U.S. law prohibiting health insurers from using genetic information to make decisions about health-insurance coverage and rates; prevents employers from using genetic information in employment decisions; also prevents health insurers and employers from asking or requiring a person to take a genetic test.

genetic (linkage) map Map of the relative distances between genetic loci, markers, or other chromosome regions determined by rates of recombination; measured in percent recombination or map units.

genetic marker Any gene or DNA sequence used to identify a location on a genetic or physical map.

genetic maternal effect Determines the phenotype of an offspring. With genetic maternal effect, an offspring inherits genes for the characteristics from both parents, but the offspring's phenotype is determined not by its own genotype but by the nuclear genotype of its mother.

genetic rescue Introduction of new genetic variation into an inbred population that often dramatically improves the health of the population in an effort to increase its chances of long-term survival.

genetic variance Component of the phenotypic variance that is due to genetic differences among individual members of a population.

gene tree Phylogenetic tree representing the evolutionary relationships among a set of genes.

genic interaction variance Component of the genetic variance that can be attributed to genic interaction (interaction between genes at different loci).

genic sex determination Sex determination in which the sexual phenotype is specified by genes at one or more loci, but there are no obvious differences in the chromosomes of males and females.

genome Complete set of genetic instructions for an organism.

genome-wide association studies Studies that look for nonrandom associations between the presence of a trait and alleles at many different loci scattered across a genome—that is, for associations between traits and particular suites of alleles in a population.

genomic imprinting Differential expression of a gene that depends on the sex of the parent that transmitted the gene. If the gene is inherited from the father, its expression differs from that if the gene is inherited from the mother.

genomic library Collection of bacterial or phage colonies containing DNA fragments that consist of the entire genome of an organism.

genomics Study of the content, organization, and function of genetic information in whole genomes.

genotype The set of genes possessed by an individual organism.

genotypic frequency Proportion of a particular genotype within a population.

germ-line mutation Mutation in a germ-line cell (one that gives rise to gametes).

germ-plasm theory States that cells in the reproductive organs carry a complete set of genetic information.

G_2/M checkpoint Important point in the cell cycle near the end of G_2. After this checkpoint has been passed, the cell undergoes mitosis.

G overhang A guanine-rich sequence of nucleotides that protrudes beyond the complementary C-rich strand at the end of a chromosome.

group I intron Belongs to a class of introns in some ribosomal RNA genes that are capable of self-splicing.

group II intron Belongs to a class of introns in some protein-encoding genes that are capable of self-splicing and are found in mitochondria, chloroplasts, and a few eubacteria.

G_1/S checkpoint Important point in the cell cycle. After the G_1/S checkpoint has been passed, DNA replicates and the cell is committed to dividing.

guanine (G) Purine base in DNA and RNA.

guide RNA (gRNA) RNA molecule that serves as a template for an alteration made in mRNA during RNA editing.

gynandromorph Individual organism that is a mosaic for the sex chromosomes, possessing tissues with different sex-chromosome constitutions.

gyrase *See* **DNA gyrase.**

hairpin Secondary structure formed when sequences of nucleotides on the same strand are complementary and pair with each other.

haploid Possessing a single set of chromosomes (one genome).

haploinsufficiency The appearance of a mutant phenotype in an individual cell or organism that is heterozygous for a normally recessive trait.

haploinsufficient gene Must be present in two copies for normal function. If one copy of the gene is missing, a mutant phenotype is produced.

haplotype A specific set of linked genetic variants or alleles on a single chromosome or on part of a chromosome.

Hardy–Weinberg equilibrium Frequencies of genotypes when the conditions of the Hardy–Weinberg law are met.

Hardy–Weinberg law Important principle of population genetics stating that, in a large, randomly mating population not affected by mutation, migration, or natural selection, allelic frequencies will not change and genotypic frequencies stabilize after one generation in the proportions p^2 (the frequency of AA), $2pq$ (the frequency of Aa), and q^2 (the frequency of aa), where p equals the frequency of allele A and q equals the frequency of allele a.

H-DNA DNA structure consisting of three nucleotide strands (triplex DNA). Can occur when a single nucleotide strand from one part of the molecule pairs with double-stranded DNA from another part of the molecule.

heat-shock protein Produced by many cells in response to extreme heat and other stresses; helps cells prevent damage from such stressing agents.

helicase *See* **DNA helicase.**

hemizygosity Possession of a single allele at a locus. Males of organisms with XX-XY sex determination are hemizygous for X-linked loci because their cells possess a single X chromosome.

heritability Proportion of phenotypic variation due to genetic differences. *See also* **broad-sense heritability** and **narrow-sense heritability.**

hermaphroditism Condition in which an individual organism possesses both male and female reproductive structures. True hermaphrodites produce both male and female gametes.

heterochromatin Chromatin that remains in a highly condensed state throughout the cell cycle; found at the centromeres and telomeres of most chromosomes.

heteroduplex DNA DNA consisting of two strands, each of which is from a different chromosome.

heterogametic sex The sex (male or female) that produces two types of gametes with respect to sex chromosomes. For example, in the XX-XY sex-determining system, the male produces both X-bearing and Y-bearing gametes.

heterokaryon Cell possessing two nuclei derived from different cells through cell fusion.

heteroplasmy Presence of two or more distinct variants of DNA within the cytoplasm of a single cell.

heterozygote screening Tests members of a population to identify heterozygous carriers of a disease-causing allele who are healthy but have the potential to produce children who have the disease.

heterozygous Refers to an individual organism that possesses two different alleles at a locus.

highly repetitive DNA DNA that consists of short sequences that are present in hundreds of thousands to millions of copies; clustered in certain regions of chromosomes.

histone Low-molecular-weight protein found in eukaryotes that complexes with DNA to form chromosomes.

histone code Modification of histone proteins, such as the addition or removal of phosphate groups, methyl groups, or acetyl groups, that encode information affecting how genes are expressed.

Holliday intermediate Structure that forms in homologous recombination; consists of two duplex molecules connected by a cross bridge.

Holliday junction Model of homologous recombination that is initiated by single-strand breaks in a DNA molecule.

holoenzyme Complex of an enzyme and other protein factors necessary for complete function.

homeobox Conserved subset of nucleotides in homeotic genes. In *Drosophila*, it consists of 180 nucleotides that encode 60 amino acids of a DNA-binding domain related to the helix-turn-helix motif.

homeotic complex (HOM-C) Major cluster of homeotic genes in fruit flies; consists of the *Antennapedia* complex, which affects the development of the adult fly's head and anterior segments, and the *bithorax* complex, which affects the adult fly's posterior thoracic and abdominal segments.

homeotic gene Determines the identity of individual segments or parts in an early embryo. Mutations in such genes cause body parts to appear in the wrong places.

homogametic sex The sex (male or female) that produces gametes that are all alike with regard to sex chromosomes. For example, in the XX-XY sex-determining system, the female produces only X-bearing gametes.

homologous genes Evolutionarily related genes, having descended from a gene in a common ancestor.

homologous pair of chromosomes Two chromosomes that are alike in structure and size and that carry genetic information for the same set of hereditary characteristics. One chromosome of a homologous pair is inherited from the male parent and the other is inherited from the female parent.

homologous recombination Exchange of genetic information between homologous DNA molecules.

homoplasmy Presence of only one version of DNA within the cytoplasm of a single cell.

homozygous Refers to an individual organism that possesses two identical alleles at a locus.

horizontal gene transfer Transfer of genes from one organism to another by a mechanism other than reproduction.

Hox **gene** Gene that contains a homeobox.

human papilloma virus (HPV) Virus associated with cervical cancer.

Human Proteome Project Project with the goal of identifying and characterizing all proteins in the human body.

humoral immunity Type of immunity resulting from antibodies produced by B cells.

hybrid breakdown Reproductive isolating mechanism in which closely related species are capable of mating and producing viable and fertile F_1 progeny, but genes do not flow between the two species, because further crossing of the hybrids produces inviable or sterile offspring.

hybrid dysgenesis Sudden appearance of numerous mutations, chromosome aberrations, and sterility in the offspring of a cross between a male fly that possesses *P* elements and a female fly that lacks them.

hybrid inviability Reproductive isolating mechanism in which mating between two organisms of different species takes place and hybrid offspring are produced but are not viable.

hybridization Pairing of two partly or fully complementary single-stranded nucleotide chains.

hybrid sterility Hybrid embryos complete development but are sterile; exemplified by mating between donkeys and horses to produce a mule, a viable but usually sterile offspring.

hypostatic gene Gene that is masked or suppressed by the action of a gene at a different locus.

identical twins Twins that arise when a single egg fertilized by a single sperm splits into two separate embryos; also called monozygotic twins.

inbreeding Mating between related individuals that takes place more frequently than expected on the basis of chance.

inbreeding coefficient Measure of inbreeding; the probability (ranging from 0 to 1) that two alleles are identical by descent.

inbreeding depression Decreased fitness arising from inbreeding; often due to the increased expression of lethal or deleterious recessive traits.

incomplete dominance Refers to the phenotype of a heterozygote that is intermediate between the phenotypes of the two homozygotes.

incomplete linkage Linkage between genes that exhibit some crossing over; intermediate in its effects between independent assortment and complete linkage.

incomplete penetrance Refers to a genotype that does not always express the expected phenotype. Some individuals possess the genotype for a trait but do not express the phenotype.

incorporated error Incorporation of a damaged nucleotide or mismatched base pair into a DNA molecule.

independent assortment Independent separation of chromosome pairs in anaphase I of meiosis; contributes to genetic variation.

induced mutation Mutation that results from environmental agents, such as chemicals or radiation.

induced pluripotent stem cells (iPSCs) Adult cells that have been artificially induced to dedifferentiate and revert to pluripotent stem cells capable of becoming many types of cells.

inducer Substance that stimulates transcription in an inducible system of gene regulation; usually a small molecule that binds to a repressor protein and alters that repressor so that it can no longer bind to DNA and inhibit transcription.

inducible operon Operon or other system of gene regulation in which transcription is normally off. Something must take place for transcription to be induced, or turned on.

induction Stimulation of the synthesis of an enzyme by an environmental factor, often the presence of a particular substrate.

in-frame deletion Deletion of some multiple of three nucleotides, which does not alter the reading frame of the gene.

in-frame insertion Insertion of some multiple of three nucleotides, which does not alter the reading frame of the gene.

inheritance of acquired characteristics Early notion of inheritance proposing that acquired traits are passed to descendants.

initiation codon The codon in mRNA that specifies the first amino acid (fMet in bacterial cells; Met in eukaryotic cells) of a protein; most commonly AUG.

initiation factor 1 (IF-1) Protein required for the initiation of translation in bacterial cells; enhances the dissociation of the large and small subunits of the ribosome.

initiation factor 2 (IF-2) Protein required for the initiation of translation in bacterial cells; forms a complex with GTP and the charged initiator tRNA and then delivers the charged tRNA to the initiation complex.

initiation factor 3 (IF-3) Protein required for the initiation of translation in bacterial cells; binds to the small subunit of the ribosome and prevents the large subunit from binding during initiation.

initiator protein Binds to an origin of replication and unwinds a short stretch of DNA, allowing helicase and other single-strand-binding proteins to bind and initiate replication.

insertion Mutation in which nucleotides are added to a DNA sequence.

insertion sequence Simple type of transposable element found in bacteria and their plasmids that contains only the information necessary for its own movement.

in situ hybridization Method used to determine the chromosomal location of a gene or other specific DNA fragment or the tissue distribution of an mRNA by using a labeled probe that is complementary to the sequence of interest.

insulator DNA sequence that blocks or insulates the effect of an enhancer; must be located between the enhancer and the promoter to have blocking activity; also may limit the spread of changes in chromatin structure.

integrase Enzyme that inserts prophage, or proviral, DNA into a chromosome.

interactome Complete set of protein interactions in a cell.

intercalating agent Chemical substance that is about the same size as a nucleotide and may become sandwiched between adjacent bases in DNA, distorting the three-dimensional structure of the helix and causing single-nucleotide insertions and deletions in replication.

interference Degree to which one crossover interferes with additional crossovers.

intergenic suppressor mutation Suppressor mutation that occurs in a gene (locus) that is different from the gene containing the original mutation.

interkinesis Period between meiosis I and meiosis II.

internal promoter Promoter located within the sequences of DNA that are transcribed into RNA.

interphase Period in the cell cycle between the cell divisions. In interphase, the cell grows, develops, and prepares for cell division.

interspersed repeat sequence Repeated sequence at multiple locations throughout the genome.

intrachromosomal recombination Recombination among genes located on the same chromosome.

intragenic mapping Maps the locations of mutations within a single locus.

intragenic suppressor mutation Suppressor mutation that occurs in the same gene (locus) as the mutation that it suppresses.

intron Intervening sequence in a split gene; removed from the RNA after transcription.

inverted repeats Sequences on the same strand that are inverted and complementary.

isoaccepting tRNAs Different tRNAs with different anticodons that specify the same amino acid.

isotopes Different forms of an element that have the same number of protons and electrons but differ in the number of neutrons in the nucleus.

junctional diversity Addition or deletion of nucleotides at the junctions of gene segments brought together in the somatic recombination of genes that encode antibodies and T-cell receptors.

karyotype Picture of an individual organism's complete set of metaphase chromosomes.

kinetochore Set of proteins that assemble on the centromere, providing the point of attachment for spindle microtubules.

Klinefelter syndrome Human condition in which cells contain one or more Y chromosomes along with multiple X chromosomes (most commonly XXY but may also be XXXY, XXXXY, or XXYY). Persons with Klinefelter syndrome are male in appearance but frequently possess small testes, some breast enlargement, and reduced facial and pubic hair; often taller than normal and sterile, most have normal intelligence.

knock-in mouse Mouse that carries a foreign sequence inserted at a specific chromosome location.

knockout mouse Mouse in which a normal gene has been disabled ("knocked out").

labeling Method for adding a radioactive or chemical label to a molecule such as DNA or RNA.

lagging strand DNA strand that is replicated discontinuously.

large ribosomal subunit The larger of the two subunits of a functional ribosome.

lariat Looplike structure created in the splicing of nuclear pre-mRNA in which the 5′ end of an intron is attached to a branch point in pre-mRNA.

leading strand DNA strand that is replicated continuously.

leptotene First substage of prophase I in meiosis. In leptotene, chromosomes contract and become visible.

lethal allele Causes the death of an individual organism, often early in development, and so the organism does not appear in the progeny of a genetic cross. Recessive lethal alleles kill individual organisms that are homozygous for the allele; dominant lethals kill both heterozygotes and homozygotes.

lethal mutation Causes premature death.

LINE *See* **long interspersed element.**

linkage analysis Gene mapping based on the detection of physical linkage between genes, as measured by the rate of recombination, in progeny from a cross.

linkage disequilibrium Nonrandom association between genetic variants within a haplotype.

linkage group Genes located together on the same chromosome.

linked genes Genes located on the same chromosome.

linker Small, synthetic DNA fragment that contains one or more restriction sites. Can be attached to the ends of any piece of DNA and used to insert a gene into a plasmid vector when restriction sites are not available.

linker DNA Stretch of DNA separating two nucleosomes.

locus Position on a chromosome where a specific gene is located.

lod (logarithm of odds) score Logarithm of the ratio of the probability of obtaining a set of observations, assuming a specified degree of linkage, to the probability of obtaining the same set of observations with independent assortment; used to assess the likelihood of linkage between genes from pedigree data.

long interspersed element (LINE) Long DNA sequence repeated many times and interspersed throughout the genome.

loss-of-function mutation Causes the complete or partial absence of normal function.

loss of heterozygosity At a locus having a normal allele and a mutant allele, inactivation or loss of the normal allele.

Lyon hypothesis Proposed by Mary Lyon in 1961, this hypothesis proposes that one X chromosome in each female cell becomes inactivated (a Barr body) and suggests that which of the X chromosomes becomes inactivated is random and varies from cell to cell.

lysogenic cycle Life cycle of a bacteriophage in which phage genes first integrate into the bacterial chromosome and are not immediately transcribed and translated.

lytic cycle Life cycle of a bacteriophage in which phage genes are transcribed and translated, new phage particles are produced, and the host cell is lysed.

major histocompatibility complex (MHC) antigen Belongs to a large and diverse group of antigens found on the surfaces of cells that mark those cells as self; encoded by a large cluster of genes known as the major histocompatibility complex. T cells simultaneously bind to foreign and MHC antigens.

malignant tumor Consists of cells that are capable of invading other tissues.

map-based sequencing Method of sequencing a genome in which sequenced fragments are ordered into contigs with the use of genetic or physical maps.

mapping function Relates recombination frequencies to actual physical distances between genes.

map unit (m.u.) Unit of measure for distances on a genetic map; 1 map unit equals 1% recombination.

mass spectrometry Method for precisely determining the molecular mass of a molecule by using the migration rate of an ionized molecule in an electrical field.

maternal blood screening test Tests for genetic conditions in a fetus by analyzing the blood of the mother. For example, the level of α-fetoprotein in maternal blood provides information about the probability that a fetus has a neural-tube defect.

mean Statistic that describes the center of a distribution of measurements; calculated by dividing the sum of all measurements by the number of measurements; also called the average.

mechanical isolation Reproductive isolation resulting from anatomical differences that prevent successful copulation.

mediator Complex of proteins that is one of the components of the basal transcription apparatus.

megaspore One of the four products of meiosis in plants.

megasporocyte In the ovary of a plant, a diploid reproductive cell that undergoes meiosis to produce haploid macrospores.

meiosis Process in which chromosomes of a eukaryotic cell divide to give rise to haploid reproductive cells. Consists of two divisions: meiosis I and meiosis II.

meiosis I First phase of meiosis. In meiosis I, chromosome number is reduced by half.

meiosis II Second phase of meiosis. Events in meiosis II are similar to those in mitosis.

melting *See* **denaturation.**

melting temperature Midpoint of the melting range of DNA.

memory cell Long-lived lymphocyte among the clone of cells generated when a foreign antigen is encountered. If the same antigen is encountered again, the memory cells quickly divide and give rise to another clone of cells specific for that particular antigen.

Mendelian population Group of interbreeding, sexually reproducing individuals.

meristic characteristic Characteristic whose phenotype varies in whole numbers, such as number of vertebrae.

merozygote Bacterial cell that has two copies of some genes—one copy on the bacterial chromosome and a second copy on an introduced F plasmid; also called partial diploid.

messenger RNA (mRNA) RNA molecule that carries genetic information for the amino acid sequence of a protein.

metacentric chromosome Chromosome in which the two chromosome arms are approximately the same length.

metagenomics An emerging field of sequencing technology in which the genome sequences of a group of organisms inhabiting a common environment are sampled and determined.

metaphase Stage of mitosis. In metaphase, chromosomes align in the center of the cell.

metaphase I Stage of meiosis I. In metaphase I, homologous pairs of chromosomes align in the center of the cell.

metaphase II Stage of meiosis II. In metaphase II, individual chromosomes align on the metaphase plate.

metaphase plate Plane in a cell between two spindle poles. In metaphase, chromosomes align on the metaphase plate.

metastasis Refers to cells that separate from malignant tumors and travel to other sites, where they establish secondary tumors.

5′-methylcytosine Modified nucleotide, consisting of cytosine to which a methyl group has been added; predominate form of methylation in eukaryotic DNA.

microarray Ordered array of DNA fragments fixed to a solid support, which serve as probes to detect the presence of complementary sequences; often used to assess the expression of genes in various tissues and under different conditions.

microbiome Complete set of all bacteria found in a particular environment. For example, the human microbiome consists of all bacteria found in and on the human body.

microRNA (miRNA) Small RNA molecule, typically 21 or 22 bp in length, produced by cleavage of double-stranded RNA arising from small hairpins within RNA that is mostly single stranded. The miRNAs combine with proteins to form a complex that binds (imperfectly) to mRNA molecules and inhibits their translation.

microsatellite *See* **variable number of tandem repeats.**

microspore Haploid product of meiosis in plants.

microsporocyte Diploid reproductive cell in the stamen of a plant; undergoes meiosis to produce four haploid microspores.

microtubule Long fiber composed of the protein tubulin; plays an important role in the movement of chromosomes in mitosis and meiosis.

migration Movement of genes from one population to another; also called gene flow.

minimal medium Used to culture bacteria or some other microorganism; contains only the nutrients required by prototrophic (wild-type) cells—typically, a carbon source, essential elements such as nitrogen and phosphorus, certain vitamins, and other required ions and nutrients.

mismatch repair Process that corrects mismatched nucleotides in DNA after replication has been completed. Enzymes excise incorrectly paired nucleotides from the newly synthesized strand and use the original nucleotide strand as a template when replacing them.

missense mutation Alters a codon in mRNA, resulting in a different amino acid in the protein encoded.

mitochondrial DNA (mtDNA) DNA in mitochondria; has some characteristics in common with eubacterial DNA and typically consists of a circular molecule that lacks histone proteins and encodes some of the rRNAs, tRNAs, and proteins found in mitochondria.

mitosis Process by which the nucleus of a eukaryotic cell divides.

mitosis-promoting factor (MPF) Protein functioning in the control of the cell cycle; consists of a cyclin combined with cyclin-dependent kinase (CDK). Active MPF stimulates mitosis.

mitotic spindle Array of microtubules that radiate from two poles; moves chromosomes in mitosis and meiosis.

model genetic organism An organism that is widely used in genetic studies because it has characteristics, such as short generation time and large numbers of progeny, that make it particularly useful for genetic analysis.

moderately repetitive DNA DNA consisting of sequences that are from 150 to 300 bp in length and are repeated thousands of times.

modified base Rare base found in some RNA molecules. Such bases are modified forms of the standard bases (adenine, guanine, cytosine, and uracil).

molecular chaperone Molecule that assists in the proper folding of another molecule.

molecular clock Refers to the use of molecular differences to estimate the time of divergence between organisms; assumes a roughly constant rate at which one neutral mutation replaces another.

molecular genetics Study of the chemical nature of genetic information and how it is encoded, replicated, and expressed.

molecular motor Specialized protein that moves cellular components.

monoecious organism Individual organism that has both male and female reproductive structures.

monohybrid cross A cross between two individuals that differ in a single characteristic—more specifically, a cross between individuals that are homozygous for different alleles at the same locus ($AA \times aa$); also refers to a cross between two individuals that are heterozygous for two alleles at a single locus ($Aa \times Aa$).

monosomy Absence of one of the chromosomes of a homologous pair.

monozygotic twins Identical twins that arise when a single egg fertilized by a single sperm splits into two separate embryos.

Morgan 100 map units.

morphogen Molecule whose concentration gradient affects the developmental fate of surrounding cells.

mosaicism Condition in which regions of tissue within a single individual have different chromosome constitutions.

M (mitotic) phase Period of active cell division; includes mitosis (nuclear division) and cytokinesis (cytoplasmic division).

mRNA surveillance Mechanisms for the detection and elimination of mRNAs that contain errors that may create problems in the course of translation.

multifactorial characteristic Determined by multiple genes and environmental factors.

multigene family Set of genes similar in sequence that arose through repeated duplication events; often encode different proteins.

multiple alleles Presence in a group of individuals of more than two alleles at a locus. However, each member of the group has only two of the possible alleles.

multiple 3′ cleavage sites Refers to the presence of more than one 3′ cleavage site on a single pre-mRNA, which allows cleavage and polyadenylation to take place at different sites, producing mRNAs of different lengths.

multiplication rule States that the probability of two or more independent events occurring together is calculated by multiplying the probabilities of each of the individual events.

mutagen Any environmental agent that significantly increases the rate of mutation above the spontaneous rate.

mutagenesis screen Method for identifying genes that influence a specific phenotype. Random mutations are induced in a population of organisms, and individual organisms with mutant phenotypes are identified. These individual organisms are crossed to determine the genetic basis of the phenotype and to map the location of mutations that cause the phenotype.

mutation Heritable change in genetic information.

mutation driver A mutation found in a cancer cell that contributes to the process of cancer development.

mutation passenger A mutation found in a cancer cell that does not contribute to the development of cancer.

mutation rate Frequency with which a gene changes from the wild type to a specific mutant; generally expressed as the number of mutations per biological unit (that is, mutations per cell division, per gamete, or per round of replication).

narrow-sense heritability Proportion of the phenotypic variance that can be attributed to additive genetic variance.

natural selection Differential reproduction of genotypes.

negative assortative mating Mating between unlike individuals that is more frequent than would be expected on the basis of chance.

negative control Gene regulation in which the binding of a regulatory protein to DNA inhibits transcription (the regulatory protein is a repressor).

negative supercoiling *See* **supercoiling.**

neutral mutation Changes the amino acid sequence of a protein but does not alter the function of the protein.

neutral-mutation hypothesis Proposes that much of the molecular variation seen in natural populations is adaptively neutral and unaffected by natural selection. Under this hypothesis, individuals with different molecular variants have equal fitnesses.

newborn screening Tests newborn infants for certain genetic disorders.

next-generation sequencing technologies Sequencing methods, such as pyrosequencing, that are capable of simultaneously determining the sequences of many DNA fragments; the technologies are much faster and less expensive than the Sanger method.

nitrogenous base Nitrogen-containing base that is one of the three parts of a nucleotide.

node Point in a phylogenetic tree that represents an organism. Terminal nodes are those that are at the outermost branches of the tree and represent organisms for which data have been obtained. Internal nodes represent ancestors common to organisms on different branches of the tree.

no-go decay A mRNA surveillance system in eukaryotes that helps remove stalled ribosomes resulting from secondary structures in mRNA, chemical damage to mRNA, premature stop codons, and ribosomal defects.

nonautonomous element Transposable element that cannot transpose on its own but can transpose in the presence of an autonomous element of the same family.

nondisjunction Failure of homologous chromosomes or sister chromatids to separate in meiosis or mitosis.

nonhistone chromosomal protein One of a heterogeneous assortment of nonhistone proteins in chromatin.

nonidentical twins Twins that arise when two different eggs are fertilized by two different sperm; also called dizygotic twins or fraternal twins.

noninvasive prenatal genetic diagnosis A genetic test performed on an unborn fetus without taking a tissue sample from the fetus. Usually performed by testing fetal DNA that occurs within the maternal blood.

nonoverlapping genetic code Refers to the fact that, generally, each nucleotide is a part of only one codon and encodes only one amino acid in a protein.

nonreciprocal translocation Movement of a chromosome segment to a nonhomologous chromosome or region without any (or with unequal) reciprocal exchange of segments.

nonrecombinant (parental) gamete Contains only the original combinations of genes present in the parents.

nonrecombinant (parental) progeny Possesses the original combinations of traits possessed by the parents.

nonreplicative transposition Type of transposition in which a transposable element excises from an old site and moves to a new site, resulting in no net increase in the number of copies of the transposable element.

nonsense codon Codon in mRNA that signals the end of translation; also called a stop codon or termination codon. The three common nonsense codons are UAA, UAG, and **UGA.**

nonsense-mediated mRNA decay (NMD) Process that brings about the rapid elimination of mRNA that has a premature stop codon.

nonsense mutation Changes a sense codon (one that specifies an amino acid) into a stop codon.

nonstop RNA decay Mechanism in eukaryotic cells for dealing with ribosomes stalled at the 3′ end of an mRNA that lacks a termination codon. A protein binds to the A site of the stalled ribosome and recruits other proteins that degrade the mRNA from the 3′ end.

nontemplate strand The DNA strand that is complementary to the template strand; not ordinarily used as a template during transcription.

normal distribution Common type of frequency distribution that exhibits a symmetrical, bell-shaped curve; usually arises when a large number of independent factors contribute to the measurement.

Northern blotting Process by which RNA is transferred from a gel to a solid support such as a nitrocellulose or nylon filter.

nuclear envelope Membrane that surrounds the genetic material in eukaryotic cells to form a nucleus; segregates the DNA from other cellular contents.

nuclear matrix Network of protein fibers in the nucleus; holds the nuclear contents in place.

nuclear pre-mRNA intron Belongs to a class of introns in protein-encoding genes that reside in the nuclei of eukaryotic cells; removed by spliceosomal-mediated splicing.

nucleoid Bacterial DNA confined to a definite region of the cytoplasm.

nucleoside Ribose or deoxyribose bonded to a base.

nucleosome Basic repeating unit of chromatin, consisting of a core of eight histone proteins (two each of H2A, H2B, H3, and H4) and about 146 bp of DNA that wraps around the core about two times.

nucleotide Repeating unit of DNA or RNA made up of a sugar, a phosphate, and a base.

nucleotide-excision repair DNA repair that removes bulky DNA lesions and other types of DNA damage.

nucleus Space in eukaryotic cells that is enclosed by the nuclear envelope and contains the chromosomes.

nullisomy Absence of both chromosomes of a homologous pair $(2n - 2)$.

Okazaki fragment Short stretch of newly synthesized DNA. Produced by discontinuous replication on the lagging strand, these fragments are eventually joined together.

oligonucleotide-directed mutagenesis Method of site-directed mutagenesis that uses an oligonucleotide to introduce a mutant sequence into a DNA molecule.

oncogene Dominant-acting gene that stimulates cell division, leading to the formation of tumors and contributing to cancer; arises from mutated copies of a normal cellular gene (proto-oncogene).

one gene, one enzyme hypothesis Idea proposed by Beadle and Tatum that each gene encodes a separate enzyme.

one gene, one polypeptide hypothesis Modification of the one gene, one enzyme hypothesis; proposes that each gene encodes a separate polypeptide chain.

oogenesis Egg production in animals.

oogonium Diploid cell in the ovary; capable of undergoing meiosis to produce an egg cell.

open reading frame (ORF) Continuous sequence of DNA nucleotides that contains a start codon and a stop codon in the same reading frame; is assumed to be a gene that encodes a protein but, in many cases, the protein has not yet been identified.

operator DNA sequence in the operon of a bacterial cell. A regulator protein binds to the operator and affects the rate of transcription of structural genes.

operon Set of structural genes in a bacterial cell along with a common promoter and other sequences (such as an operator) that control the transcription of the structural genes.

origin of replication Site where DNA synthesis is initiated.

orthologous genes Homologous genes found in different species because the two species have a common ancestor that also possessed the gene.

outcrossing Mating between unrelated individuals that is more frequent than would be expected on the basis of chance.

overdominance Selection in which the heterozygote has higher fitness than that of either homozygote; also called heterozygote advantage.

ovum Final product of oogenesis.

pachytene Third substage of prophase I in meiosis. The synaptonemal complex forms during pachytene.

pair-rule genes Set of segmentation genes in fruit flies that define regional sections of the embryo and affect alternate segments. Mutations in these genes often cause the deletion of every other segment.

palindrome Sequence of nucleotides that reads the same on complementary strands; inverted repeats.

pangenesis Early concept of heredity proposing that particles carry genetic information from different parts of the body to the reproductive organs.

paracentric inversion Chromosome inversion that does not include the centromere in the inverted region.

paralogous genes Homologous genes in the same species that arose through the duplication of a single ancestral gene.

paramutation Epigenetic change in which one allele of a genotype alters the expression of another allele; the altered expression persists for several generations, even after the altering allele is no longer present.

parental gamete *See* **nonrecombinant (parental) gamete.**

parental progeny *See* **nonrecombinant (parental) progeny.**

partial diploid Bacterial cell that possesses two copies of genes, including one copy on the bacterial chromosome and the other on an extra piece of DNA (usually a plasmid); also called merozygote.

Patau syndrome (trisomy 13) Characterized by severe intellectual disability, a small head, sloping forehead, small eyes, cleft lip and palate, extra fingers and toes, and other disabilities; results from the presence of three copies of chromosome 13.

pedigree Pictorial representation of a family history outlining the inheritance of one or more traits or diseases.

penetrance Percentage of individuals with a particular genotype that express the phenotype expected of that genotype.

pentaploidy Possession of five haploid sets of chromosomes ($5n$).

peptide bond Chemical bond that connects amino acids in a protein.

peptidyl (P) site One of three sites in a ribosome occupied by a tRNA in translation. In the elongation stage of protein synthesis, tRNAs move from the aminoacyl (A) site into the P site.

peptidyl transferase Activity in the ribosome that creates a peptide bond between two amino acids. Evidence suggests that this activity is carried out by one of the RNA components of the ribosome.

pericentric inversion Chromosome inversion that includes the centromere in the inverted region.

P (parental) generation First set of parents in a genetic cross.

phage *See* **bacteriophage.**

phenocopy Phenotype that is produced by environmental effects and is the same as the phenotype produced by a genotype.

phenotype Appearance or manifestation of a characteristic.

phenotypic correlation Correlation between two or more phenotypes in the same individual.

phenotypic variance Measures the degree of phenotypic differences among a group of individuals; composed of genetic, environmental, and genetic–environmental interaction variances.

phenylketonuria (PKU) Genetic disease characterized by intellectual disability, light skin, and eczema; caused by mutations in the gene that encodes phenylalanine hydroxylase (PAH), a liver enzyme that normally metabolizes the amino acid phenylalanine. When the enzyme is defective, phenylalanine is not metabolized and builds up to high levels in the body, eventually causing intellectual disability and other characteristics of the disease. The disease is inherited as an autosomal recessive disorder and can be effectively treated by limiting phenylalanine in the diet.

phosphate group A phosphorus atom attached to four oxygen atoms; one of the three components of a nucleotide.

phosphodiester Molecule containing R–O–P–O–R, where R is a carbon-containing group, O is oxygen, and P is phosphorus.

phosphodiester linkage Phosphodiester bond connecting two nucleotides in a polynucleotide strand.

phylogenetic tree Graphical represenation of the evolutionary connections between organisms or genes.

phylogeny Evolutionary relationships among a group of organisms or genes, usually depicted as a family tree or branching diagram.

physical map Map of physical distances between loci, genetic markers, or other chromosome segments; measured in base pairs.

pilus (pl., **pili**) Extension of the surface of some bacteria that allows conjugation to take place. When a pilus on one cell makes contact with a receptor on another cell, the pilus contracts and pulls the two cells together.

Piwi-interacting RNA (piRNA) Small RNA molecule belonging to a class named after Piwi proteins with which these molecules interact; similar to microRNA and small interfering RNA and thought to have a role in the regulation of sperm development.

plaque Clear patch of lysed cells on a continuous layer of bacteria on the agar surface of a petri plate. Each plaque represents a single original phage that multiplied and lysed many cells.

plasmid Small, circular DNA molecule found in bacterial cells that is capable of replicating independently from the bacterial chromosome.

pleiotropy A single genotype influences multiple phenotypes.

pluripotency In embryonic stem cells, the property of being undifferentiated, with the capacity to form every type of cell in an organism.

poly(A)-binding protein (PABP) Binds to the poly(A) tail of eukaryotic mRNA and makes the mRNA more stable. There are several types of PABP.

poly(A) tail String of adenine nucleotides added to the 3′ end of a eukaryotic mRNA after transcription.

polycistronic mRNA Single bacterial RNA molecule that encodes more than one polypeptide chain; uncommon in eukaryotes.

polygenic characteristic Encoded by genes at many loci.

polymerase chain reaction (PCR) Method of enzymatically amplifying DNA fragments.

polynucleotide strand Series of nucleotides linked together by phosphodiester bonds.

polypeptide Chain of amino acids linked by peptide bonds; also called a protein.

polyploidy Possession of more than two haploid sets of chromosomes.

polyribosome Messenger RNA molecule with several ribosomes attached to it.

polytene chromosome Giant chromosome in the salivary glands of *Drosophila melanogaster*. Each polytene chromosome consists of a number of DNA molecules lying side by side.

population The group of interest; often represented by a subset called a sample. Also, a group of members of the same species.

population genetics Study of the genetic composition of populations (groups of members of the same species) and how a population's collective group of genes changes with the passage of time.

positional cloning Method that allows for the isolation and identification of a gene by examining the cosegregation of a phenotype with previously mapped genetic markers.

position effect Dependence of the expression of a gene on the gene's location in the genome.

positive assortative mating Mating between like individuals that is more frequent than would be expected on the basis of chance.

positive control Gene regulation in which the binding of a regulatory protein to DNA stimulates transcription (the regulatory protein is an activator).

positive supercoiling *See* **supercoiling.**

posttranslational modification Alteration of a protein after translation; may include cleavage from a larger precursor protein, the removal of amino acids, and the attachment of other molecules to the protein.

postzygotic reproductive isolating mechanism Reproductive isolation that arises after a zygote is formed, either because the resulting hybrids are inviable or sterile or because reproduction breaks down in subsequent generations.

preformationism Early concept of inheritance proposing that a miniature adult (homunculus) resides in either the egg or the sperm and increases in size in development, with all traits being inherited from the parent that contributes the homunculus.

preimplantation genetic diagnosis (PGD) Genetic testing on an embryo produced by in vitro fertilization before implantation of the embryo in the uterus.

pre-messenger RNA (pre-mRNA) Eukaryotic RNA molecule that is modified after transcription to become mRNA.

presymptomatic genetic testing Tests people to determine whether they have inherited a disease-causing gene before the symptoms of the disease have appeared.

prezygotic reproductive isolating mechanism Reproductive isolation in which gametes from two different species are prevented from fusing and forming a hybrid zygote.

primary Down syndrome Caused by the presence of three copies of chromosome 21.

primary immune response Initial clone of cells specific for a particular antigen and generated when the antigen is first encountered by the immune system.

primary oocyte Oogonium that has entered prophase I.

primary spermatocyte Spermatogonium that has entered prophase I.

primary structure of a protein The amino acid sequence of a protein.

primase Enzyme that synthesizes a short stretch of RNA on a DNA template; functions in replication to provide a 3′-OH group for the attachment of a DNA nucleotide.

primer Short stretch of RNA on a DNA template; provides a 3′-OH group for the attachment of a DNA nucleotide at the initiation of replication.

principle of independent assortment (Mendel's second law) Important principle of heredity discovered by Mendel that states that genes encoding different characteristics (genes at different loci) separate independently; applies only to genes located on different chromosomes or to genes far apart on the same chromosome.

principle of segregation (Mendel's first law) Important principle of heredity discovered by Mendel that states that each diploid individual possesses two alleles at a locus and that these two alleles separate when gametes are formed, one allele going into each gamete.

probability Likelihood of the occurrence of a particular event; more formally, the number of times that a particular event occurs divided by the number of all possible outcomes. Probability values range from 0 to 1.

proband A person having a trait or disease for whom a pedigree is constructed.

probe Known sequence of DNA or RNA that is complementary to a sequence of interest and will pair with it; used to find specific DNA sequences.

prokaryote Unicellular organism with a simple cell structure. Prokaryotes include eubacteria and archaea.

prometaphase Stage of mitosis. In prometaphase, the nuclear membrane breaks down and the spindle microtubules attach to the chromosomes.

promoter DNA sequence to which the transcription apparatus binds so as to initiate transcription; indicates the direction of transcription, which of the two DNA strands is to be read as the template, and the starting point of transcription.

proofreading Ability of DNA polymerases to remove and replace incorrectly paired nucleotides in the course of replication.

prophage Phage genome that is integrated into a bacterial chromosome.

prophase Stage of mitosis. In prophase, the chromosomes contract and become visible, the cytoskeleton breaks down, and the mitotic spindle begins to form.

prophase I Stage of meiosis I. In prophase I, chromosomes condense and pair, crossing over takes place, the nuclear membrane breaks down, and the spindle forms.

prophase II Stage of meiosis after interkinesis. In prophase II, chromosomes condense, the nuclear membrane breaks down, and the spindle forms. Some cells skip this stage.

protein-coding region The part of mRNA consisting of the nucleotides that specify the amino acid sequence of a protein.

protein domain Region of a protein that has a specific shape or function.

protein kinase Enzyme that adds phosphate groups to other proteins.

protein microarray Large number of different proteins applied to a glass slide as a series of spots, each spot containing a different protein; used to analyze protein–protein interactions.

proteome Set of all proteins encoded by a genome.

proteomics Study of the proteome, the complete set of proteins found in a given cell.

proto-oncogene Normal cellular gene that controls cell division. When mutated, it may become an oncogene and contribute to cancer progression.

prototrophic bacterium A wild-type bacterium that can use a carbon source, essential elements such as nitrogen and phosphorus, certain vitamins, and other required ions and nutrients to synthesize all the compounds that they need for growth and reproduction.

provirus DNA copy of viral DNA or viral RNA; integrated into the host chromosome and replicated along with the host chromosome.

pseudoautosomal region Small region of the X and Y chromosomes that contains homologous gene sequences.

pseudodominance Expression of a normally recessive allele owing to a deletion on the homologous chromosome.

Punnett square Shorthand method of determining the outcome of a genetic cross. On a grid, the gametes of one parent are written along the upper edge and the gametes of the other parent are written along the left-hand edge. Within the cells of the grid, the alleles in the gametes are combined to form the genotypes of the offspring.

purine Type of nitrogenous base in DNA and RNA. Adenine and guanine are purines.

pyrimidine Type of nitrogenous base in DNA and RNA. Cytosine, thymine, and uracil are pyrimidines.

pyrimidine dimer Structure in which a bond forms between two adjacent pyrimidine molecules on the same strand of DNA; disrupts normal hydrogen bonding between complementary bases and distorts the normal configuration of the DNA molecule.

quantitative characteristic Continuous characteristic; displays a large number of possible phenotypes, which must be described by a quantitative measurement.

quantitative genetics Genetic analysis of complex characteristics or characteristics influenced by multiple genetic factors.

quantitative trait locus (QTL) A gene or chromosomal region that contributes to the expression of quantitative characteristics.

quaternary structure of a protein Interaction of two or more polypeptides to form a functional protein.

reading frame Particular way in which a nucleotide sequence is read in groups of three nucleotides (codons) in translation. Each reading frame begins with a start codon and ends with a stop codon.

realized heritability Narrow-sense heritability measured from a response-to-selection experiment.

real-time PCR Modification of the polymerase chain reaction that quantitatively determines the amount of starting nucleic acid; the amount of DNA amplified is measured as the reaction proceeds.

reannealing *See* **renaturation.**

recessive Refers to an allele or phenotype that is expressed only when the recessive allele is present in two copies (homozygous). The recessive allele is not expressed in the heterozygote phenotype.

reciprocal crosses Crosses in which the phenotypes of the male and female parents are reversed. For example, in one cross, a tall male is crossed with a short female and, in the other cross, a short male is crossed with a tall female.

reciprocal translocation Reciprocal exchange of segments between two nonhomologous chromosomes.

recombinant DNA technology Set of molecular techniques for locating, isolating, altering, combining, and studying DNA segments.

recombinant gamete Possesses new combinations of genes.

recombinant progeny Possesses new combinations of traits formed from recombinant gametes.

recombination Process that produces new combinations of alleles.

recombination frequency Proportion of recombinant progeny produced in a cross.

regression Analysis of how one variable changes in response to another variable.

regression coefficient Statistic that measures how much one variable changes, on average, with a unit change in another variable.

regulator gene Gene associated with an operon in bacterial cells that encodes a protein or RNA molecule that functions in controlling the transcription of one or more structural genes.

regulator protein Produced by a regulator gene, a protein that binds to another DNA sequence and controls the transcription of one or more structural genes.

regulatory element DNA sequence that affects the transcription of other DNA sequences to which it is physically linked.

regulatory gene DNA sequence that encodes a protein or RNA molecule that interacts with DNA sequences and affects their transcription or translation or both.

regulatory promoter DNA sequence located immediately upstream of the core promoter that affects transcription; contains consensus sequences to which transcriptional regulator proteins bind.

relaxed state of DNA Energy state of a DNA molecule when there is no structural strain on the molecule.

release factor Protein required for the termination of translation; binds to a ribosome when a stop codon is reached and stimulates the release of the polypeptide chain, the tRNA, and the mRNA from the ribosome. Eukaryotic cells require two release factors (eRF-1 and eRF-2), whereas *E. coli* requires three (RF-1, RF-2, and RF-3).

renaturation The process by which two complementary single-stranded DNA molecules pair; also called reannealing.

repetitive DNA Sequences that exist in multiple copies in a genome.

replicated error Replication of an incorporated error in which a change in the DNA sequence has been replicated and all base pairings in the new DNA molecule are correct.

replication Process by which DNA is synthesized from a single-stranded nucleotide template.

replication bubble Segment of a DNA molecule that is unwinding and undergoing replication.

replication fork Point at which a double-stranded DNA molecule separates into two single strands that serve as templates for replication.

replication licensing factor Protein that ensures that replication takes place only once at each origin; required at the origin before replication can be initiated and removed after the DNA has been replicated.

replication origin Sequence of nucleotides where replication is initiated.

replication terminus Point at which replication stops.

replicative segregation Random segregation of organelles into progeny cells in cell division. If two or more versions of an organelle are present in the original cell, chance determines the proportion of each type that will segregate into each progeny cell.

replicative transposition Type of transposition in which a copy of the transposable element moves to a new site while the original copy remains at the old site; increases the number of copies of the transposable element.

replicon Unit of replication, consisting of DNA from the origin of replication to the point at which replication on either side of the origin ends.

repressible operon Operon or other system of gene regulation in which transcription is normally on. Something must take place for transcription to be repressed, or turned off.

repressor Regulatory protein that binds to a DNA sequence and inhibits transcription.

reproductive isolating mechanism Any biological factor or mechanism that prevents gene exchange.

repulsion *See* **trans configuration.**

resolvase Enzyme required for some types of transposition; brings about resolution—that is, crossing over between sites located within the transposable element. Resolvase may be encoded by the transposable element or by a cellular enzyme that normally functions in homologous recombination.

response element Common DNA sequence found upstream of some groups of eukaryotic genes. A regulatory protein binds to a response element and stimulates the transcription of a gene. The presence of the same response element in several promoters or enhancers allows a single factor to simultaneously stimulate the transcription of several genes.

response to selection The amount that a characteristic changes in one generation owing to selection; equals the selection differential times the narrow-sense heritability.

restriction endonuclease Technical term for a restriction enzyme, which recognizes particular base sequences in DNA and makes double-stranded cuts nearby.

restriction enzyme Recognizes particular base sequences in DNA and makes double-stranded cuts nearby; also called restriction endonuclease.

restriction fragment length polymorphism (RFLP) Variation in the pattern of fragments produced when DNA molecules are cut with the same restriction enzyme; represents a heritable difference in DNA sequences and can be used in genetic analyses.

restriction mapping Determines the locations of sites cut by restriction enzymes in a piece of DNA.

retrotransposon Type of transposable element in eukaryotic cells that possesses some characteristics of retroviruses and transposes through an RNA intermediate.

retrovirus RNA virus capable of integrating its genetic material into the genome of its host. The virus injects its RNA genome into the host cell, where reverse transcription produces a complementary, double-stranded DNA molecule from the RNA template. The DNA copy then integrates into the host chromosome to form a provirus.

reverse duplication Duplication of a chromosome segment in which the sequence of the duplicated segment is inverted relative to the sequence of the original segment.

reverse genetics A molecular approach that begins with a genotype (a DNA sequence) and proceeds to the phenotype by altering the sequence or by inhibiting its expression.

reverse mutation (reversion) Mutation that changes a mutant phenotype back into the wild type.

reverse transcriptase Enzyme capable of synthesizing complementary DNA from an RNA template.

reverse transcription Synthesis of DNA from an RNA template.

reverse-transcription PCR Amplifies sequences corresponding to RNA. Reverse transcriptase is used to convert RNA into complementary DNA, which can then be amplified by the usual polymerase chain reaction.

rho-dependent terminator Sequence in bacterial DNA that requires the presence of the rho subunit of RNA polymerase to terminate transcription.

rho factor Subunit of bacterial RNA polymerase that facilitates the termination of transcription of some genes.

rho-independent terminator Sequence in bacterial DNA that does not require the presence of the rho subunit of RNA polymerase to terminate transcription.

ribonucleoside triphosphate (rNTP) Substrate of RNA synthesis; consists of ribose, a nitrogenous base, and three phosphates linked to the $5'$-carbon atom of the ribose. In transcription, two of the phosphates are cleaved, producing an RNA nucleotide.

ribonucleotide Nucleotide containing ribose; present in RNA.

ribose Five-carbon sugar in RNA.

ribosomal RNA (rRNA) RNA molecule that is a structural component of the ribosome.

riboswitch Regulatory sequences in an RNA molecule. When an inducer molecule binds to the riboswitch, the binding changes the configuration of the RNA molecule and alters the expression of the RNA, usually by affecting the termination of transcription or affecting translation.

ribozyme RNA molecule that can act as a biological catalyst.

RNA-coding region Sequence of DNA nucleotides that encodes an RNA molecule.

RNA editing Process in which the protein-coding sequence of an mRNA is altered after transcription. The amino acids specified by the altered mRNA are different from those predicted from the nucleotide sequence of the gene encoding the protein.

RNA-induced silencing complex (RISC) Combination of a small interfering RNA (siRNA) molecule or a microRNA (miRNA) molecule and proteins that can cleave mRNA, leading to the degradation of the mRNA, or affecting transcription, or repressing translation of the mRNA.

RNA interference (RNAi) Process in which cleavage of double-stranded RNA produces small interfering RNAs (siRNAs) that bind to mRNAs containing complementary sequences and bring about their cleavage and degradation.

RNA polymerase Enzyme that synthesizes RNA from a DNA template during transcription.

RNA polymerase I Eukaryotic RNA polymerase that transcribes large ribosomal RNA molecules (18 S rRNA and 28 S rRNA).

RNA polymerase II Eukaryotic RNA polymerase that transcribes pre-messenger RNA, some small nuclear RNAs, and some microRNAs.

RNA polymerase III Eukaryotic RNA polymerase that transcribes transfer RNA, small ribosomal RNAs (5 S rRNA), some small nuclear RNAs, and some microRNAs.

RNA polymerase IV Transcribes small interfering RNAs in plants.

RNA polymerase V Transcribes RNA that has a role in heterochromatin formation in plants.

RNA replication Process in some viruses by which RNA is synthesized from an RNA template.

RNA silencing Mechanism by which double-stranded RNA is cleaved and processed to yield small single-stranded interfering RNAs (siRNAs), which bind to complementary sequences in mRNA and bring about the cleavage and degradation of mRNA; also known as RNA interference and posttranscriptional RNA gene silencing. Some siRNAs also bind to complementary sequences in DNA and guide enzymes to methylate the DNA.

RNA splicing Process by which introns are removed and exons are joined together.

Robertsonian translocation Translocation in which the long arms of two acrocentric chromosomes become joined to a common centromere, resulting in a chromosome with two long arms and usually another chromosome with two short arms.

rolling-circle replication Replication of circular DNA that is initiated by a break in one of the nucleotide strands, producing a double-stranded circular DNA molecule and a single-stranded linear DNA molecule, the latter of which may circularize and serve as a template for the synthesis of a complementary strand.

rooted tree Phylogenetic tree in which one internal node represents the common ancestor of all other organisms (nodes) on the tree. In a rooted tree, all the organisms depicted have a common ancestor.

R plasmid (R factor) Plasmid having genes that confer antibiotic resistance to any cell that contains the plasmid.

sample Subset used to describe a population.

sampling error Deviations from expected ratios due to chance occurrences when the number of events is small.

secondary immune response Clone of cells generated when a memory cell encounters an antigen; provides long-lasting immunity.

secondary oocyte One of the products of meiosis I in female animals; receives most of the cytoplasm.

secondary spermatocyte Product of meiosis I in male animals.

secondary structure of a protein Regular folding arrangement of amino acids in a protein. Common secondary structures found in proteins include the alpha helix and the beta pleated sheet.

second polar body One of the products of meiosis II in oogenesis; contains a set of chromosomes but little of the cytoplasm.

segmental duplications Regions larger than 1000 bp that are almost identical in sequence in eukaryotic genomes.

segmentation genes Set of about 25 genes in fruit flies that control the differentiation of the embryo into individual segments, affecting the number and organization of the segments. Mutations in these genes usually disrupt whole sets of segments.

segment-polarity genes Set of segmentation genes in fruit flies that affect the organization of segments. Mutations in these genes cause part of each segment to be deleted and replaced by a mirror image of part or all of an adjacent segment.

selection coefficient Measure of the relative intensity of selection against a genotype; equals 1 minus fitness.

selection differential Difference in phenotype between the selected individuals and the average of the entire population.

semiconservative replication Replication in which the two nucleotide strands of DNA separate, each serving as a template for the synthesis of a new strand. All DNA replication is semiconservative.

sense codon Codon that specifies an amino acid in a protein.

separase Molecule that cleaves cohesin molecules, which hold the sister chromatids together.

sequence tagged site (STS) Short (200-500 bp) sequence of DNA, present only once in the genome, whose chromosomal location has been determined. Often used to determine the genomic location of a DNA clone.

sequential hermaphroditism Phenomenon in which the sex of an individual organism changes in the course of its lifetime; the organism is male at one age or developmental stage and female at a different age or stage.

70S initiation complex Final complex formed in the initiation of translation in bacterial cells; consists of the small and large subunits of the ribosome, mRNA, and initiator tRNA charged with fMet.

sex Male or female.

sex chromosomes Chromosomes that differ morphologically or in number in males and females.

sex determination Specification of sex (male or female). Sex-determining mechanisms include chromosomal, genic, and environmental sex-determining systems.

sex-determining region Y (*SRY*) gene On the Y chromosome, a gene that triggers male development.

sex-influenced characteristic Characteristic encoded by autosomal genes that are more readily expressed in one sex. For example, an autosomal dominant gene may have higher penetrance in males than in females or an autosomal gene may be dominant in males but recessive in females.

sex-limited characteristic Characteristic encoded by autosomal genes and expressed in only one sex. Both males and females carry genes for sex-limited characteristics, but the characteristics appear in only one of the sexes.

sex-linked characteristic Characteristic determined by a gene or genes on sex chromosomes.

shelterin A multiprotein complex that binds to mammalian telomeres and protects the ends of the DNA from being inadvertently repaired as a double-stranded break in the DNA.

Shine–Dalgarno sequence Consensus sequence found in the bacterial 5′ untranslated region of mRNA; contains the ribosome-binding site.

short interspersed element (SINE) Short DNA sequence repeated many times and interspersed throughout the genome.

short tandem repeat (STR) Very short DNA sequence repeated in tandem and found widely in the human genome.

shuttle vector Cloning vector that allows DNA to be transferred to more than one type of host cell.

sigma factor Subunit of bacterial RNA polymerase that allows the RNA polymerase to recognize a promoter and initiate transcription.

signal sequence From 15 to 30 amino acids that are found at the amino end of some eukaryotic proteins and direct the protein to specific locations in the cell; usually cleaved from the protein.

signal-transduction pathway System in which an external signal (initiated by a hormone or growth factor) triggers a cascade of intracellular reactions that ultimately produce a specific response.

silencer Sequence that has many of the properties possessed by an enhancer but represses transcription.

silent mutation Change in the nucleotide sequence of DNA that does not alter the amino acid sequence of a protein.

single-nucleotide polymorphism (SNP) Single-base-pair differences in DNA sequence between individual members of a species.

single-strand-binding (SSB) protein Binds to single-stranded DNA in replication and prevents it from annealing with a complementary strand and forming secondary structures.

sister chromatids Two copies of a chromosome that are held together at the centromere. Each chromatid consists of a single DNA molecule.

site-directed mutagenesis Produces specific nucleotide changes at selected sites in a DNA molecule.

small cytoplasmic RNA (scRNA) Small RNA molecule found in the cytoplasm of eukaryotic cells.

small interfering RNA (siRNA) Single-stranded RNA molecule (usually from 21 to 25 nucleotides in length) produced by the cleavage and processing of double-stranded RNA; binds to complementary sequences in mRNA and brings about the cleavage and degradation of the mRNA. Some siRNAs bind to complementary sequences in DNA and bring about their methylation.

small nuclear ribonucleoprotein (snRNP) Structure found in the nuclei of eukaryotic cells that consists of small nuclear RNA (snRNA) and protein; functions in the processing of pre-mRNA.

small nuclear RNA (snRNA) Small RNA molecule found in the nuclei of eukaryotic cells; functions in the processing of pre-mRNA.

small nucleolar RNA (snoRNA) Small RNA molecule found in the nuclei of eukaryotic cells; functions in the processing of rRNA and in the assembly of ribosomes.

small ribosomal subunit The smaller of the two subunits of a functional ribosome.

somatic-cell hybridization Fusion of different cell types.

somatic hypermutation High rate of somatic mutation such as that in genes encoding antibodies.

somatic mutation Mutation in a cell that does not give rise to a gamete.

somatic recombination Recombination in somatic cells, such as maturing lymphocytes, among segments of genes that encode antibodies and T-cell receptors.

SOS system System of proteins and enzymes that allow a cell to replicate its DNA in the presence of a distortion in DNA structure; makes numerous mistakes in replication and increases the rate of mutation.

Southern blotting Process by which DNA is transferred from a gel to a solid support such as a nitrocellulose or nylon filter.

specialized transduction Transduction in which genes near special sites on the bacterial chromosome are transferred from one bacterium to another; requires lysogenic bacteriophages.

speciation Process by which new species arise. *See also* **biological species concept, allopatric speciation,** and **sympatric speciation.**

species Term applied to different kinds or types of living organisms.

spermatid Immediate product of meiosis II in spermatogenesis; matures to sperm.

spermatogenesis Sperm production in animals.

spermatogonium Diploid cell in the testis; capable of undergoing meiosis to produce a sperm.

S (synthesis) phase Stage of interphase in the cell cycle. In S phase, DNA replicates.

spindle microtubule Microtubule that moves chromosomes in mitosis and meiosis.

spindle pole Point from which spindle microtubules radiate.

spliceosome Large complex consisting of several RNAs and many proteins that splices protein-encoding pre-mRNA; contains five small ribonucleoprotein particles (U1, U2, U4, U5, and U6).

spontaneous mutation Arises from natural changes in DNA structure or from errors in replication.

sporophyte Diploid phase of the life cycle in plants.

SR proteins Group of serine- and arginine-rich proteins that regulate alternative splicing of pre-mRNA.

standard deviation Statistic that describes the variability of a group of measurements; the square root of the variance.

stem cells Undifferentiated cells that are capable of forming every type of cell in an organism.

stop (termination or nonsense) codon Codon in mRNA that signals the end of translation. The three common stop codons are UAA, UAG, and UGA.

strand slippage Slipping of the template and newly synthesized strands in replication in which one of the strands loops out from the other and nucleotides are inserted or deleted on the newly synthesized strand.

structural gene DNA sequence that encodes a protein that functions in metabolism or biosynthesis or that has a structural role in the cell.

structural genomics Area of genomics that studies the organization and sequence of information contained within genomes; sometimes used by protein chemists to refer to the determination of the three-dimensional structure of proteins.

submetacentric chromosome Chromosome in which the centromere is displaced toward one end, producing a short arm and a long arm.

supercoiling Coiled tertiary structure that forms when strain is placed on a DNA helix by overwinding or underwinding of the helix. An overwound DNA exhibits positive supercoiling; an underwound DNA exhibits negative supercoiling.

suppressor mutation Mutation that hides or suppresses the effect of another mutation at a site that is distinct from the site of the original mutation.

sympatric speciation Speciation arising in the absence of any geographic barrier to gene flow; reproductive isolating mechanisms evolve within a single interbreeding population.

synapsis Close pairing of homologous chromosomes.

synaptonemal complex Three-part structure that develops between synapsed homologous chromosomes.

synonymous codons Different codons that specify the same amino acid.

tag single-nucleotide polymorphism (tag-SNP) Used to identify a haplotype.

tandem duplication Chromosome rearrangement in which a duplicated chromosome segment is adjacent to the original segment.

tandem repeat sequences DNA sequences repeated one after another; tend to be clustered at specific locations on a chromosome.

***Taq* polymerase** DNA polymerase commonly used in PCR reactions. Isolated from the bacterium *Thermus aquaticus*, the enzyme is stable at high temperatures, and so it is not denatured during the strand-separation step of the cycle.

TATA-binding protein (TBP) Polypeptide chain found in several different transcription factors that recognizes and binds to sequences in eukaryotic promoters.

TATA box Consensus sequence (TATAAAA) commonly found in eukaryotic RNA polymerase II promoters; usually located from 25 to 30 bp upstream of the transcription start site. The TATA box determines the start point for transcription.

T cell Particular type of lymphocyte that produces cellular immunity; originates in the bone marrow and matures in the thymus.

T-cell receptor Found on the surface of a T cell, a receptor that simultaneously binds a foreign and a self-antigen on the surface of a cell.

telocentric chromosome Chromosome in which the centromere is at or very near one end.

telomerase Enzyme that is made up of both protein and RNA and replicates the ends (telomeres) of eukaryotic chromosomes. The RNA part of the enzyme has a template that is complementary to repeated sequences in the telomere and pairs with them, providing a template for the synthesis of additional copies of the repeats.

telomere Stable end of a chromosome.

telomere-associated sequence Sequence found at the ends of a chromosome next to the telomeric sequence; consists of long, complex repeated sequences.

telomeric sequence Sequence found at the ends of a chromosome; consists of many copies of short, simple sequences repeated one after the other.

telophase Stage of mitosis. In telophase, the chromosomes arrive at the spindle poles, the nuclear membrane re-forms, and the chromosomes relax and lengthen.

telophase I Stage of meiosis I. In telophase I, chromosomes arrive at the spindle poles.

telophase II Stage of meiosis II. In telophase II, chromosomes arrive at the spindle poles.

temperate phage Bacteriophage that utilizes the lysogenic cycle, in which the phage DNA integrates into the bacterial chromosome and remains in an inactive state.

temperature-sensitive allele Expressed only at certain temperatures.

template strand The strand of DNA that is used as a template during transcription. The RNA synthesized during transcription is complementary and antiparallel to the template strand.

temporal isolation Reproductive isolation in which the reproduction of different groups takes place at different times of the year, and so there is no gene flow between groups; exemplified by species of plants that flower at different times of the year and thus do not exchange genes.

terminal inverted repeats Sequences found at both ends of a transposable element that are inverted complements of one another.

termination codon Codon in mRNA that signals the end of translation; also called nonsense codon or stop codon. The three common termination codons are UAA, UAG, and UGA.

terminator Sequence of DNA nucleotides that causes the termination of transcription.

tertiary structure of a protein Higher-order folding of amino acids in a protein to form the overall three-dimensional shape of the molecule.

testcross A cross between an individual with an unknown genotype and an individual with the homozygous recessive genotype.

tetrad The four products of meiosis; all four chromatids of a homologous pair of chromosomes.

tetrad analysis Genetic analysis of a tetrad, the products of a single meiosis.

tetraploidy Possession of four haploid sets of chromosomes ($4n$).

tetrasomy Presence of two extra copies of a chromosome ($2n + 2$).

TFIIB recognition element (BRE) Consensus sequence [(G or C)(G or C)(G or C)CGCC] found in some RNA polymerase II core promoters; usually located from 32 to 38 bp upstream of the transcription start site.

theory of clonal selection Explains the generation of primary and secondary immune responses. The binding of a B cell to an antigen stimulates the cell to divide, giving rise to a clone of genetically identical cells, all of which are specific for the antigen.

theta replication Replication of circular DNA that is initiated by the unwinding of the two nucleotide strands, producing a replication bubble. Unwinding continues at one or both ends of the bubble, making it progressively larger. DNA replication on both of the template strands is simultaneous with unwinding until the two replication forks meet.

30S initiation complex Initial complex formed in the initiation of translation in bacterial cells; consists of the small subunit of the ribosome, mRNA, initiator tRNA charged with fMet, GTP, and initiation factors 1, 2, and 3.

three-point testcross Cross between an individual heterozygous at three loci and an individual homozygous for recessive alleles at those loci.

3′ end End of a polynucleotide chain where an OH group is attached to the 3′-carbon atom of the nucleotide.

3′ splice site The 3′ end of an intron where cleavage takes place in RNA splicing.

3′ untranslated (3′ UTR) region Sequence of nucleotides at the 3′ end of mRNA; does not encode the amino acids of a protein but affects both the stability of the mRNA and its translation.

threshold characteristic Discontinuous characteristic whose expression depends on an underlying susceptibility that varies continuously.

thymine (T) Pyrimidine base in DNA but not in RNA.

Ti plasmid Large plasmid isolated from the bacterium *Agrobacterium tumefaciens* and used to transfer genes to plant cells.

topoisomerase Enzyme that adds or removes rotations in a DNA helix by temporarily breaking nucleotide strands; controls the degree of DNA supercoiling.

totipotency The potential of a cell to develop into any other cell type.

trans configuration Arrangement in which each chromosome contains one wild-type (dominant) gene and one mutant (recessive) gene; also called repulsion.

transcription Process by which RNA is synthesized from a DNA template.

transcriptional activator protein Protein in eukaryotic cells that binds to consensus sequences in regulatory promoters or enhancers and affects transcription initiation by stimulating the assembly of the basal transcription apparatus.

transcription bubble Region of a DNA molecule that has unwound to expose a single-stranded template, which is being transcribed into RNA.

transcription factor Protein that binds to DNA sequences in eukaryotic cells and affects transcription.

transcription start site The first DNA nucleotide that is transcribed into an RNA molecule.

transcription unit Sequence of nucleotides in DNA that encodes a single RNA molecule, along with the sequences necessary for its transcription; normally contains a promoter, an RNA-coding sequence, and a terminator.

transcriptome Set of all RNA molecules transcribed from a genome.

transducing phage Contains a piece of the bacterial chromosome inside the phage coat. *See also* **generalized transduction.**

transductant Bacterial cell that has received genes from another bacterium through transduction.

transduction Type of gene exchange that takes place when a virus carries genes from one bacterium to another. After it is inside the cell, the newly introduced DNA may undergo recombination with the bacterial chromosome.

transesterification Chemical reaction in some RNA-splicing reactions.

transfer–messenger RNA (tmRNA) An RNA molecule that has properties of both mRNA and tRNA; functions in rescuing ribosomes that are stalled at the end of mRNA.

transfer RNA (tRNA) RNA molecule that carries an amino acid to the ribosome and transfers it to a growing polypeptide chain in translation.

transfer RNA intron Belongs to a class of introns in tRNA genes. The splicing of these genes relies on enzymes.

transformant Cell that has received genetic material through transformation.

transformation Mechanism by which DNA found in the medium is taken up by the cell. After transformation, recombination may take place between the introduced genes and the cellular chromosome.

transforming principle Substance responsible for transformation. DNA is the transforming principle.

transgene Foreign gene or other DNA fragment carried in germline DNA.

transgenic mouse Mouse whose genome contains a foreign gene or genes added by employing recombinant DNA methods.

transition Base substitution in which a purine is replaced by a different purine or a pyrimidine is replaced by a different pyrimidine.

translation Process by which a protein is assembled from information contained in messenger RNA.

translesion DNA polymerase Specialized DNA polymerases that are able to replicate through distorted structures and bulky lesions that halt other DNA polymerases. They often make more errors during DNA synthesis than other DNA polymerases.

translocation Movement of a chromosome segment to a nonhomologous chromosome or to a region within the same chromosome. Also, movement of a ribosome along mRNA in the course of translation.

translocation carrier Individual organism heterozygous for a chromosome translocation.

transmission genetics Field of genetics that encompasses the basic principles of genetics and how traits are inherited.

transposable element DNA sequence capable of moving from one site to another within the genome through a mechanism that differs from that of homologous recombination.

transposase Enzyme encoded by many types of transposable elements that is required for their transposition. The enzyme makes single-strand breaks at each end of the transposable element and on either side of the target sequence where the element inserts.

transposition Movement of a transposable genetic element from one site to another. Replicative transposition increases the number of copies of the transposable element; nonreplicative transposition does not increase the number of copies.

trans-splicing The process of splicing together exons from two or more pre-mRNAs.

transversion Base substitution in which a purine is replaced by a pyrimidine or a pyrimidine is replaced by a purine.

trihybrid cross A cross between two individuals that differ in three characteristics ($AA\ BB\ CC \times aa\ bb\ cc$); also refers to a cross between two individuals that are both heterozygous at three loci ($Aa\ Bb\ Cc \times Aa\ Bb\ Cc$).

triplet code Refers to the fact that three nucleotides encode each amino acid in a protein.

triploidy Possession of three haploid sets of chromosomes ($3n$).

triplo-X syndrome Human condition in which cells contain three X chromosomes. A person with triplo-X syndrome has a female phenotype without distinctive features other than a tendency to be tall and thin; a few such women are sterile, but many menstruate regularly and are fertile.

trisomy Presence of an additional copy of a chromosome ($2n + 1$).

trisomy 8 Presence of three copies of chromosome 8; in humans, results in intellectual disability, contracted fingers and toes, low-set malformed ears, and a prominent forehead.

trisomy 13 Presence of three copies of chromosome 13; in humans, results in Patau syndrome.

trisomy 18 Presence of three copies of chromosome 18; in humans, results in Edward syndrome.

trisomy 21 Presence of three copies of chromosome 21; in humans, results in Down syndrome.

tRNA charging Chemical reaction in which an aminoacyl-tRNA synthetase attaches an amino acid to its corresponding tRNA.

tRNA-modifying enzyme Creates a modified base in RNA by catalyzing a chemical change in the standard base.

tubulin Protein found in microtubules.

tumor-suppressor gene Gene that normally inhibits cell division. Recessive mutations in such genes often contribute to cancer.

Turner syndrome Human condition in which cells contain a single X chromosome and no Y chromosome (XO). Persons with Turner syndrome are female in appearance but do not undergo puberty and have poorly developed female secondary sex characteristics; most are sterile but have normal intelligence.

two-dimensional polyacrylamide gel electrophorsis (2D-PAGE) Method for separating proteins into spots in which the proteins are separated in one dimension by charge, separated in a second dimension by mass, and then stained. Each spot is proportional to the amount of protein present.

two-point testcross Cross between an individual heterozygous at two loci and an individual homozygous for recessive alleles at those loci.

ultrasonography Procedure for visualizing the fetus. High-frequency sound is beamed into the uterus. Sound waves that encounter dense tissue bounce back and are transformed into a picture of the fetus.

unbalanced gamete Gamete that has a variable number of chromosomes; some chromosomes may be missing and others may be present in more than one copy.

underdominance Selection in which the heterozygote has lower fitness than that of either homozygote.

unequal crossing over Misalignment of the two DNA molecules during crossing over, resulting in one DNA molecule with an insertion and the other with a deletion.

uniparental disomy Inheritance of both chromosomes of a homologous pair from a single parent.

unique-sequence DNA Sequence present only once or a few times in a genome.

universal genetic code Refers to the fact that particular codons specify the same amino acids in almost all organisms.

up mutation Mutation that increases the rate of transcription.

upstream element Consensus sequence found in some bacterial promoters that contains a number of A–T pairs and is found about 40 to 60 bp upstream of the transcription start site.

uracil (U) Pyrimidine base in RNA but not normally in DNA.

variable number of tandem repeats (VNTRs) Short sequences repeated in tandem that vary greatly in number among individuals; also called microsatellites. Because they are quite variable, VNTRs are commonly used in DNA fingerprinting.

variance Statistic that describes the variability of a group of measurements.

virulent phage Bacteriophage that reproduces only through the lytic cycle and kills its host cell.

virus Noncellular replicating agent consisting of nucleic acid surrounded by a protein coat; can replicate only within its host cell.

Western blotting Process by which protein is transferred from a gel to a solid support such as a nitrocellulose or nylon filter.

whole-genome shotgun sequencing Method of sequencing a genome in which sequenced fragments are assembled into the correct sequence in contigs by using only the overlaps in sequence.

wild type The trait or allele that is most commonly found in natural (wild) populations.

wobble Base pairing between codon and anticodon in which there is nonstandard pairing, usually at the third (3′) position of the codon; allows more than one codon to pair with the same anticodon.

X-linked characteristic Characteristic determined by a gene or genes on the X chromosome.

X-ray diffraction Method for analyzing the three-dimensional shape and structure of chemical substances. Crystals of a substance are bombarded with X-rays, which hit the crystals, bounce off, and produce a diffraction pattern on a detector. The pattern of the spots produced on the detector provides information about the molecular structure.

yeast artificial chromosome (YAC) Cloning vector consisting of a DNA molecule with a yeast origin of replication, a pair of telomeres, and a centromere. YACs can carry very large pieces of DNA (as large as several hundred thousand base pairs) and replicate and segregate as yeast chromosomes do.

Y-linked characteristic Characteristic determined by a gene or genes on the Y chromosome.

Z-DNA Secondary structure of DNA characterized by 12 bases per turn, a left-handed helix, and a sugar–phosphate backbone that zigzags back and forth.

zygotene Second substage of prophase I in meiosis. In zygotene, chromosomes enter into synapsis.

Index